Introduction to Control Systems

SECOND EDITION

ARUN K GHOSH

Visiting Professor
Sir JC Bose School of Engineering, Hooghly

Formerly
Head, Instrumentation Centre, University of Kalyani
Principal, Murshidabad College of Engineering and Technology, Berhampore
Principal, Bengal College of Engineering and Technology, Durgapur

PHI Learning Private Limited
Delhi-110092
2020

₹ 1295.00

INTRODUCTION TO CONTROL SYSTEMS, Second Edition
Arun K Ghosh

ISBN-978-81-203-4820-2

The export rights of this book are vested solely with the publisher.

Fourth Printing (Second Edition) ... ... April, 2020

Published by Asoke K. Ghosh, PHI Learning Private Limited, Rimjhim House, 111, Patparganj Industrial Estate, Delhi-110092 and Printed by Syndicate Binders, A-20, Hosiery Complex, Noida, Phase-II Extension, Noida-201305 (N.C.R. Delhi).

To

the fond memory of my mother

Manorama

Contents

Preface

It's a pleasure for me to present the second edition of this book to the students and teachers of undergraduate engineering colleges.

In this edition, I have changed the name of the book, introduced a new chapter on Preliminary Mathematics and incorporated many more worked-out examples, mostly selected from the GATE question papers in Instrumentation Engineering of the last 18 years. The presentation of a few chapters has also been changed wherever it was felt necessary.

My fervent hope is that the book in its present form will be able to cater to the requirements of the student community. I shall appreciate receiving any comments or suggestions for improvement of the book.

Arun K Ghosh
arunkghosh_bect@yahoo.com

Preface to the First Edition

A course on linear and digital control systems is part and parcel of almost all disciplines of undergraduate engineering curricula. This text purports to provide students with an easy access to the subject.

While writing it, I was guided by about half a dozen syllabi of different universities, mostly belonging to West Bengal. They compare well among themselves, owing presumably to the reason that they were framed after some model syllabus. I will, therefore, not be mistaken if I presume that my coverage will satisfy many a syllabus of universities of other states of this country.

I have tried to make the presentation student-friendly by showing step-by-step procedures wherever necessary and incorporating many worked-out examples and review questions culled from question papers of university examinations rather than generating them myself. My experience of teaching the subject has also stood me good stead.

I have assumed a knowledge of higher secondary mathematics on the part of the student and, therefore, included extra relevant mathematics in appendices. Also a primer of the very useful software MATLAB has been included with an eye to helping students to explore the subject themselves as well as assisting them in their lab work. I shall be happy if the book is found useful by those for whom it is written.

I would like to thank Professor BN Biswas, Director, Academy of Technology, Adisaptagram, West Bengal for encouraging me to write this book. My thanks are due to Professor Arup Bhowmick, Mr Adhish Chakraborty, Mr Indrayudh Mandal, Mr Supratik Paul of GNIT, Dr Partha Pratim Sarkar of University of Kalyani, and Mrs A Indira and Mr Jayanta Ghosh of BCET for helping me collect question papers and syllabi. I thank my daughter Rumi, son-in-law Abhijit, wife Giti and my elder brother Atin who helped me in one way or another during the preparation of the manuscript. Last, but not the least, I must thank Ms Pushpita Ghosh (Managing Editor and Marketing Director) of PHI Learning for her kind help and cooperation.

Arun K Ghosh

Acronyms

ADC	Analogue to Digital Converter
BIBO	Bounded Input Bounded Output
CCF	Controllable Canonical Form
CCW	Counter-ClockWise
CLTF	Closed-Loop Transfer Function
CW	ClockWise
DAC	Digital to Analogue Converter
DTFT	Discrete Time Fourier Transform
emf	electro-motive force
EMI	Equated Monthly Instalments
Gm	Gain margin
GUI	Graphical User Interface
IAE	Integral Absolute Error
ISE	Integral Square Error
KCL	Kirchhoff's Current Law
KVL	Kirchhoff's Voltage Law
LTI	Linear Time-Invariant
MIMO	Multiple Input Multiple Output
OCF	Observable Canonical Form
OLTF	Open-Loop Transfer Function
Pm	Phase margin
RHP	Right-Hand Plane
RHS	Right-Hand Side
RL	Root Locus
ROC	Region Of Convergence
SISO	Single Input Single Output
STM	State Transition Matrix
ZOH	Zero-Order Hold

CHAPTER 1

Preliminary Mathematics

The study of control theory and its implementation needs some background in the following areas of Mathematics:

1. Complex variables
2. Laplace transform
3. Matrices
4. Difference equations
5. z-transform

Hence, before we embark on the study of control systems, we will refresh ourselves with the use and applications of these mathematical tools and without proofs of theorems we will be using.

In the following sections, we will discuss these tools to the extent we need them for our study.

1.1 COMPLEX VARIABLES

If a and b are two real quantities and z is defined as

$$z = a + \jmath b$$

where $\jmath = \sqrt{-1}$, then z is called a complex quantity. Either or both of a and b may be a variable to make z a *complex variable*.

Geometrical Representation

The complex plane is called an Argand[1] diagram which is used to plot a complex number. The real numbers are plotted on the horizontal axis while the imaginary numbers on the vertical axis. Figure 1.1 shows Argand diagrams.

As in polar coordinates, a complex number can be defined by its distance from the origin and the angle (in radians) it subtends to the positive real axis. The distance is called *modulus* (M) and the angle is called *argument* (ϕ). The argument is not unique, but is determined up

[1]Jean-Robert Argand (1768–1822) was a gifted amateur mathematician. He worked as a manager in a bookstore in Paris when he published the idea of geometrical interpretation of complex numbers known as the *Argand diagram*.

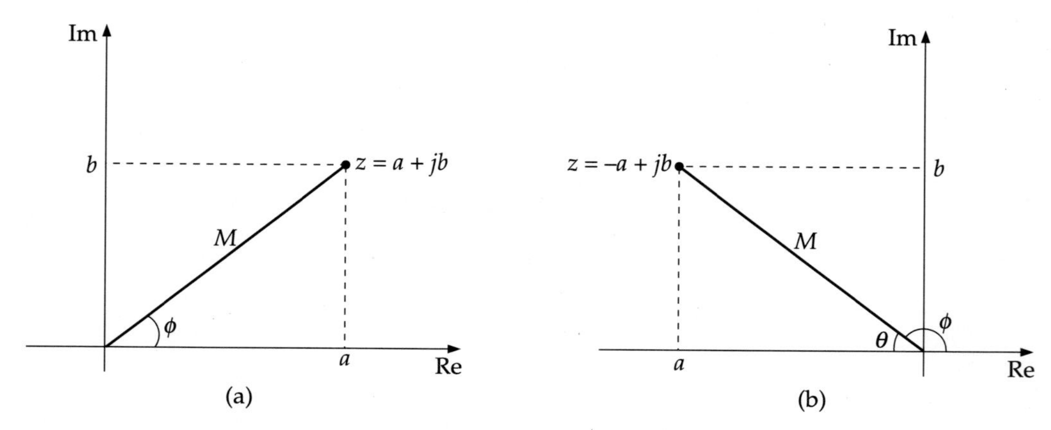

Figure 1.1 Argand diagram for: (a) $z = a + \jmath b$; (b) $z = -a + \jmath b$.

to a multiple of 2π. The complex quantity can be written as $M\angle\phi$. Thus, for $z = a + \jmath b$ [see Figure 1.1(a)],

$$M = \sqrt{a^2 + b^2}, \qquad \phi = \tan^{-1}\frac{b}{a}$$

However, for $z = -a + \jmath b$ [see Figure 1.1(b)],

$$\phi = \pi - \theta = \pi - \tan^{-1}\frac{b}{a}$$

The real part of z is written as $\text{Re}\{z\}$ or $\Re\{z\}$ and the imaginary part as $\text{Im}\{z\}$ or $\Im\{z\}$.

With the above defnition of M and ϕ, we can write

$$z = M(\cos\phi + \jmath\sin\phi)$$

It follows from trigonometric identities that

$$|z_1 z_2| = |z_1||z_2|$$

$$\arg(z_1 z_2) = \arg(z_1) + \arg(z_2)$$

$$\arg\left(\frac{z_1}{z_2}\right) = \arg(z_1) - \arg(z_2), \qquad z_2 \neq 0$$

If $z = a + \jmath b$, then the conjugate of z is defined as z^* (or $\overline{z}$) $= a - \jmath b$. The following properties hold true for z and z^*:

$$|z^*| = |z|, \quad \arg(z^*) = -\arg z, \quad (z_1 \pm z_2)^* = z_1^* \pm z_2^*$$

$$(z_1 z_2)^* = z_1^* z_2^*, \quad \text{Re}\{z\} = \frac{z + z^*}{2}, \quad \text{Im}\{z\} = \frac{z - z^*}{2\jmath}, \quad zz^* = z^2$$

Complex Algebra

Complex numbers can undergo addition, subtraction, multiplication and division. The following are the basic rules:

Addition: $(a + jb) + (c + jd) = (a + c) + j(b + d)$

Subtraction: $(a + jb) - (c + jd) = (a - c) + j(b - d)$

Multiplication: $(a + jb) \cdot (c + jd) = (ac - bd) + j(ad + bc)$

Division: $(a + jb) \div (c + jd) = \dfrac{(ac + bd) + j(bc - ad)}{c^2 - b^2}$

Euler's Theorem

Euler's[2] theorem states that

$$\cos\theta \pm j\sin\theta = e^{\pm j\theta} \tag{1.1}$$

From Eq. (1.1) we can get the following equations:

$$\cos\theta = \frac{e^{j\theta} + e^{-j\theta}}{2}$$

$$\sin\theta = \frac{e^{j\theta} - e^{-j\theta}}{2j}$$

De Moivre's Theorem

De Moivre's[3] theorem states that

$$(\cos\theta + j\sin\theta)^n = \cos n\theta + j\sin n\theta$$

for any integral value of n.

1.2 LAPLACE TRANSFORM

The Laplace[4] transform is a very powerful tool for studying differential equations. The control problems concern dynamic analogue systems. Such situations are aptly described by differential equations. Normally, differential equations are solved by individually handling the corresponding homogeneous equation and the particular integral. And it is not easy to handle a number of simultaneous differential equations.

The Laplace transform helps us

1. to solve the homogeneous equation and the particular integral in one operation
2. to convert differential equations to algebraic equations, solve such simultaneous equations with ease and then get the final solution through inverse Laplace transform.

[2]Leonhard Euler (pronounced as *oiler*, 1707–1783) was a pioneering Swiss mathematician and physicist.

[3]Abraham de Moivre (1667–1754) was a French mathematician.

[4]Pierre-Simon, Marquis de Laplace (1749–1827) was a French mathematician and astronomer who contributed much to the development of mathematical astronomy and statistics.

Definition. Let $f(t)$ be a function of t specified for $t > 0$. Then, the Laplace transform of $f(t)$, denoted by $\mathscr{L}\{f(t)\}$, is defined as

$$\mathscr{L}\{f(t)\} \equiv F(s) = \int_0^\infty e^{-st} f(t)\, dt \tag{1.2}$$

The quantity s, which is often referred to as the Laplace operator, may be real or complex. It is more useful to consider s complex as $s = \sigma + \jmath\omega$, where ω is the frequency.

Actually, in mathematics, the Laplace transform limits are from $-\infty$ to ∞. But here in Eq. (1.2), we have chosen the limits from 0 to ∞. This may be called the *one-sided Laplace transform*. The one-sided Laplace transform implies that the initial conditions of the system at $t = 0$ were zero. Physically that is a perfectly acceptable situation. Because, a system is not expected to produce a response before an input is applied at $t = 0$.

Important Properties of Laplace Transform

We now consider a few important properties of the Laplace transform. These will be of help to us while working out Laplace transforms of functions.

Multiplication by a constant. If $f(t)$ is a real function in the t-domain and $\mathscr{L}\{f(t)\} \equiv F(s)$, where $\mathscr{L}$ is the Laplace transform operator, then

$$\mathscr{L}\{kf(t)\} = kF(s) \tag{1.3}$$

Linearity. If c_1 and c_2 are constants and $\mathscr{L}\{f_1(t)\} \equiv F_1(s)$, $\mathscr{L}\{f_2(t)\} \equiv F_2(s)$, then

$$\begin{aligned}\mathscr{L}\{c_1 f_1(t) \pm c_2 f_2(t)\} &= c_1 \mathscr{L}\{f_1(t)\} \pm c_2 \mathscr{L}\{f_2(t)\} \\ &\equiv c_1 F_1(s) \pm c_2 F_2(s)\end{aligned}$$

Because of this property, $\mathscr{L}$ is called a *linear operator.*

EXAMPLE 1.1. Find the Laplace transform of $3t^2 - 2\cos 2t + 4e^{-t}$.

Solution

$$\begin{aligned}\mathscr{L}\{3t^2 - 2\cos 2t + 4e^{-t}\} &= 3\mathscr{L}\{t^2\} - 2\mathscr{L}\{\cos 2t\} + 4\mathscr{L}\{e^{-t}\} \\ &= 3\left(\frac{2!}{s^3}\right) - 2\left(\frac{s}{s^2+4}\right) + 4\left(\frac{1}{s+1}\right) \\ &= \frac{6}{s^3} - \frac{2s}{s^2+4} + \frac{4}{s+1}\end{aligned}$$

For the Laplace transforms of different functions used here, see Appendix A.

Exponential shifting. If $\mathscr{L}\{f(t)\} \equiv F(s)$, then

$$\mathscr{L}\{e^{\mp at} f(t)\} = F(s \pm a) \tag{1.4}$$

EXAMPLE 1.2. Find the Laplace transform of $e^{-t}\cos 2t$.

Solution Since $\mathscr{L}\{\cos 2t\} = \dfrac{s}{s^2+4}$, we have

$$\mathscr{L}\{e^{-t}\cos 2t\} = \frac{s+1}{(s+1)^2+4} = \frac{s+1}{s^2+2s+5}$$

Time shifting. If $\mathscr{L}\{f(t)\} \equiv F(s)$, and

$$g(t) = f(t-a)u(t-a) \quad a \geq 0$$

then

$$\mathscr{L}\{g(t)\} = e^{-as}F(s) \tag{1.5}$$

EXAMPLE 1.3. Find the Laplace transform of

$$g(t) = \begin{cases} (t-2)^3 & t > 2 \\ 0 & t < 2 \end{cases}$$

Solution Since $\mathscr{L}\{t^3\} = \dfrac{3!}{s^4} = \dfrac{6}{s^4}$, it follows that

$$\mathscr{L}\{g(t)\} = \frac{6e^{-2s}}{s^4}$$

Change of scale. If $\mathscr{L}\{f(t)\} \equiv F(s)$, then

$$\mathscr{L}\{f(at)\} = \frac{1}{a}F\left(\frac{s}{a}\right) \tag{1.6}$$

EXAMPLE 1.4. Find the Laplace transform of $\sin 3t$.

Solution Since $\mathscr{L}\{\sin t\} = \dfrac{1}{s^2+1}$, we have

$$\mathscr{L}\{\sin 3t\} = \frac{1}{3}\frac{1}{\left(\dfrac{s}{3}\right)^2 + 1} = \frac{3}{s^2+9}$$

Real differentiation. If $\mathscr{L}\{f(t)\} \equiv F(s)$, then

(a) $\mathscr{L}\{f'(t)\} = sF(s) - f(0)$ provided, $f(t)$ is continuous in $0 \leq t \leq N$ and of exponential order for $t > N$, while $f'(t)$ is sectionally continuous for $0 \leq t \leq N$.

(b) $\mathscr{L}\{f''(t) = s^2F(s) - sf(0) - f''(0)$

(c) $\mathscr{L}\{f^{(n)}(t)\} = s^nF(s) - s^{n-1}f(0) - s^{n-2}f'(0) - \cdots sf^{(n-2)}(0) - f^{(n-1)}(0)$

EXAMPLE 1.5. If $f(t) = \cos 3t$, then find the Laplace transform of $f'(t)$.

Solution $\mathscr{L}\{f(t)\} \equiv F(s) = \dfrac{s}{s^2+9}$. Therefore,

$$\mathscr{L}\{f'(t)\} = sF(s) - f(0) = \frac{s^2}{s^2+9} - 1 = -\frac{9}{s^2+9}$$

Laplace transform of integrals. If $\mathscr{L}\{f(t)\} \equiv F(s)$, then

$$\mathscr{L}\left\{\int_0^t f(u)\,du\right\} = \frac{F(s)}{s} \tag{1.7}$$

EXAMPLE 1.6. Find the Laplace transform of $\int_0^t \sin 2u\,du$.

Solution Since we have $\mathscr{L}\{\sin 2t\} = \dfrac{2}{s^2+4}$

$$\mathscr{L}\left\{\int_0^t \sin 2u\,du\right\} = \frac{\frac{2}{s^2+4}}{s} = \frac{2}{s(s^2+4)}$$

Initial value theorem. If $\mathscr{L}\{f(t)\} \equiv F(s)$, then

$$\lim_{t\to 0} f(t) = \lim_{s\to\infty} sF(s) \tag{1.8}$$

EXAMPLE 1.7. If

$$F(s) = \frac{1}{s\left(s^2+3s+2\right)} \tag{i}$$

then find the initial value of the corresponding $f(t)$.

Solution The required initial value is

$$\begin{aligned}\lim_{t\to 0} f(t) &= \lim_{s\to\infty} s\frac{1}{s\left(s^2+3s+2\right)}\\ &= \lim_{s\to\infty}\frac{1}{s^2+3s+2}\\ &= 0\end{aligned}$$

Final value theorem. If $\mathscr{L}\{f(t)\} \equiv F(s)$ and if $sF(s)$ is analytic on the imaginary axis and in the right half of the s-plane, then

$$\lim_{t\to\infty} f(t) = \lim_{s\to 0} sF(s) \tag{1.9}$$

EXAMPLE 1.8. Find the final value of the corresponding $f(t)$ of Eq. (i) of Example 1.7.

Solution The required final value is

$$\begin{aligned}\lim_{t\to\infty} f(t) &= \lim_{s\to 0} s\frac{1}{s\left(s^2+3s+2\right)}\\ &= \lim_{s\to 0}\frac{1}{s^2+3s+2}\\ &= \frac{1}{2}\end{aligned}$$

Note: (a) By virtue of the initial and final value theorems, we need not find out the t-space function $f(t)$ to figure out its initial and final values, if its Laplace transformed form is known.

(b) The final value theorem is very helpful in evaluating the steady-state response of systems represented by their transfer functions. But the condition stated above must be kept in mind before applying it. Let us see what happens if the condition is not satisfied.

EXAMPLE 1.9. If $F(s) = \dfrac{\alpha}{s^2 + \alpha^2}$ find the steady-state value of $f(t)$ with the help of the final value theorem.

Solution If we apply the final value theorem to find the steady state value of the function, we get

$$\lim_{t \to \infty} f(t) = \lim_{s \to 0} \left(s \cdot \frac{\alpha}{s^2 + \alpha^2} \right) = 0$$

Note: We know that $f(t) = \sin t$, which has no final value. The result obtained by applying the final value theorem is erroneous because the function $sF(s)$ has two poles on the imaginary axis and, therefore, the final value theorem cannot be applied here.

Complex differentiation. If $F(s) \equiv \mathscr{L}\{f(t)\}$, then

$$\mathscr{L}\{tf(t)\} = -\frac{d}{ds} F(s)$$

at all values of s except the poles of $F(s)$. The general form of this result, known as the *complex differentiation*, is given by

$$\mathscr{L}\{t^n f(t)\} = (-1)^n \frac{d^n}{ds^n} F(s), \quad n = 1,\ 2,\ 3, \ldots$$

EXAMPLE 1.10. Find the Laplace transform of $t \sin at$.

Solution The required Laplace transform is as follows:

$$\begin{aligned}
\mathscr{L}\{t \sin at\} &= -\frac{d}{ds} \mathscr{L}\{\sin at\} \\
&= -\frac{d}{ds} \left[\frac{a}{s^2 + a^2} \right] \\
&= -a \left[-\frac{2s}{\left(s^2 + a^2\right)^2} \right] \\
&= \frac{2as}{\left(s^2 + a^2\right)^2}
\end{aligned}$$

Real convolution (or Complex multiplication). If $F_1(s) \equiv \mathscr{L}\{f_1(t)\}$ and $F_2(s) \equiv \mathscr{L}\{f_2(t)\}$, and $f_1(t) = f_2(t) = 0$ for $t < 0$, then

$$\begin{aligned} F_1(s)F_2(s) &= \mathscr{L}\left\{\int_0^t f_1(\tau)f_2(t-\tau)d\tau\right\} \\ &= \mathscr{L}\left\{\int_0^t f_2(\tau)f_1(t-\tau)d\tau\right\} \\ &= \mathscr{L}\left\{f_1(t) \star f_2(t)\right\} \end{aligned} \tag{1.10}$$

where the asterisk ($\star$) indicates convolution[5] in the t-domain. Thus, we observe from Eq. (1.10) that the convolution of two real functions in the t-domain is equivalent to multiplication of their Laplace-transformed counterparts.

Note: The inverse Laplace transform of the product of two s-domain functions is not equal to the product of corresponding t-domain functions. Stated mathematically,

$$\mathscr{L}^{-1}\{F_1(s)F_2(s)\} \neq f_1(t)f_2(t) \tag{1.11}$$

Complex convolution (or Real multiplication). This is a dual relation to the real convolution. It states that the Laplace transformed product of two real functions in the t-domain is equal to their convolution in the s-domain, i.e.

$$\mathscr{L}\{f_1(t)f_2(t)\} = F_1(s) \star F_2(s) \tag{1.12}$$

The properties of Laplace transform are summarized in Table 1.1.

Inverse Laplace Transform

We have already seen that, to get the real t-space solution from the Laplace transformed functions, we have to perform inverse Laplace transform. Stated mathematically,

$$f(t) = \mathscr{L}^{-1}\{F(s)\} \tag{1.13}$$

In its explicit form,

$$f(t) = \frac{1}{2\pi\jmath}\int_{k-\jmath\infty}^{k+\jmath\infty} F(s)e^{st}ds \tag{1.14}$$

where k is a real constant which is greater than all the singularities of $F(s)$. Equation (1.14) is a line integral to be solved in the s-plane. This is not an easy task. However, the silver lining under the cloud is, for most of our tasks, the inversion can be performed with the help of the Laplace transform table[6].

While discussing control theory, we will deal with transfer functions[7] of systems. Let a transfer function be given by

$$G(s) = \frac{N(s)}{D(s)} = \frac{N(s)}{(s+a_1)(s+a_2)\cdots(s+a_n)}$$

where $D(s)$ is a higher order polynomial than $N(s)$. To carry out inverse Laplace transform, it is often necessary and convenient to break the transfer function into partial fractions. This can be done by the following methods.

[5]See Appendix C at page 665.

[6]See Appendix A at page 661.

[7]See Section 3.2 at page 96.

Table 1.1 Properties of Laplace Transform

S.No.	*Property*	*Mathematical equation/relation*
1.	Multiplication by a constant	$\mathscr{L}\{af(t)\} = aF(s)$
2.	Linearity	$\mathscr{L}\{c_1 f_1(t) \pm c_2 f_2(t)\} = c_1 F_1(s) \pm c_2 F_2(s)$
3.	Exponential shifting	$\mathscr{L}\{e^{\mp at} f(t)\} = F(s \pm a)$
4.	Time shifting	If $g(t) = f(t-a)u(t-a) \quad a \geq 0$, then $\mathscr{L}\{g(t)\} = e^{-as}F(s)$
5.	Change of scale	$\mathscr{L}\{f(at)\} = \frac{1}{a} F\left(\frac{s}{a}\right)$
6.	Real differentiation	$\mathscr{L}\{f^{(n)}(t)\} = s^n F(s) - s^{n-1} f(0) - s^{n-2} f'(0) - \cdots$ $sf^{(n-2)}(0) - f^{(n-1)}(0), \quad n = 1, 2, 3, \ldots$
7.	Integration	$\mathscr{L}\left\{\int_0^t f(u)\, du\right\} = \frac{F(s)}{s}$
8.	Division by t	$\mathscr{L}\left\{\frac{1}{t} f(t)\right\} = \int_s^{\infty} F(s)\, ds$ if $\lim_{t \to 0} f(t)$ exists
9.	Initial value theorem	$\lim_{t \to 0} f(t) = \lim_{s \to \infty} sF(s)$
10.	Final value theorem	$\lim_{t \to \infty} f(t) = \lim_{s \to 0} sF(s)$
11.	Complex differentiation	$\mathscr{L}\{t^n f(t)) = (-1)^n \frac{d^n}{ds^n} F(s), \quad n = 1, 2, 3, \ldots$
12.	Real convolution	$F_1(s)F_2(s) = \mathscr{L}\{f_1(t) \star f_2(t)\}$
13.	Complex convolution	$\mathscr{L}\{f_1(t)f_2(t)\} = F_1(s) \star F_2(s)$

Partial fraction expansion for transfer functions. While expanding transfer functions by the partial fraction method, we may encounter any of the following two situations:

- When all poles of $G(s)$ are simple.
- When some poles are of multiple order.

(i) *When all poles are simple.* The situation can be divided into two categories:

(a) When the denominator polynomial is of higher order.

(b) When the numerator polynomial is of higher order.

We consider two examples to show the procedures.

(a) *When the denominator polynomial is of higher order.*

EXAMPLE 1.11. Express the following function into partial fractions:

$$G(s) = \frac{4s+2}{s(s+1)(s+2)} \tag{i}$$

Solution We write Eq. (i), splitting the denominator fractions, as

$$G(s) \equiv \frac{A}{s} + \frac{B}{s+1} + \frac{C}{s+2} \tag{ii}$$

where A, B and C are unknown constants. These constants can be evaluated as follows:

$$A = s \cdot G(s)|_{s=0} = s \cdot \left.\frac{4s+2}{s(s+1)(s+2)}\right|_{s=0} = \left.\frac{4s+2}{(s+1)(s+2)}\right|_{s=0} = 1$$

$$B = (s+1) \cdot G(s)|_{s=-1} = (s+1) \cdot \left.\frac{4s+2}{s(s+1)(s+2)}\right|_{s=-1} = \left.\frac{4s+2}{s(s+2)}\right|_{s=-1} = 2$$

$$C = (s+2) \cdot G(s)|_{s=-2} = (s+2) \cdot \left.\frac{4s+2}{s(s+1)(s+2)}\right|_{s=-2} = \left.\frac{4s+2}{s(s+1)}\right|_{s=-2} = -3$$

Substituting the values of A, B and C in identity (ii), we have

$$G(s) = \frac{1}{s} + \frac{2}{s+1} - \frac{3}{s+2}$$

(b) *When the numerator polynomial is of higher order.*

EXAMPLE 1.12. Express the following function into appropriate partial fractions:

$$x(s) = \frac{4s^3 + s^2 + 8s + 4}{s^2 + 4}$$

Solution We note that the function cannot be solved using partial fractions as such because the numerator is of 3rd order and the denominator is only of 2nd order. Therefore, long division can be used to reduce the order of the equation as follows:

$$\begin{array}{r|llll}
 & 4s & +1 & & \\ \hline
s^2+4) & 4s^3 & + s^2 & + 8s & + 4 \\
 & 4s^3 & & + 16s & \\ \hline
 & & s^2 & - 8s & + 4 \\
 & & s^2 & & + 4 \\ \hline
 & & & - 8s &
\end{array}$$

The result helps us write a new function that has a reduced fraction. Thus,

$$x(s) = 4s + 1 - \frac{8s}{s^2 + 4}$$

This fraction can be solved with partial fractions as

$$\frac{8s}{s^2+4} = \frac{j\,2s}{s + j\,2} - \frac{j\,2s}{s - j\,2}$$

(ii) *When some poles are of multiple order.* In this case, the constants of the multiple order pole C_n are obtained from the formula

$$C_n = \frac{1}{(n-1)\,!}\frac{d^{n-1}}{ds^{n-1}}\left[(s+a)^n G(s)\right] \tag{1.15}$$

The following example will demonstrate how to handle this situation.

EXAMPLE 1.13. Find partial fractions for

$$G(s) = \frac{1}{s(s+1)^3(s+2)} \tag{i}$$

Solution We have to write Eq. (i) as

$$G(s) \equiv \frac{A}{s} + \frac{B}{s+2} + \frac{C_1}{(s+1)^3} + \frac{C_2}{(s+1)^2} + \frac{C_3}{s+1} \tag{ii}$$

A, B, C_1, C_2 and C_3 in identify (ii) are found out in the following way:

$$A = s \cdot G(s)\big|_{s=0} = \left.\frac{1}{(s+1)^3(s+2)}\right|_{s=0} = \frac{1}{2}$$

$$B = (s+2)\cdot G(s)\big|_{s=-2} = \left.\frac{1}{s(s+1)^3}\right|_{s=-2} = \frac{1}{2}$$

$$C_1 = (s+1)^3 \cdot G(s)\big|_{s=-1} = \left.\frac{1}{s(s+2)}\right|_{s=-1} = -1$$

$$C_2 = \left.\frac{d}{ds}\left[(s+1)^3 \cdot G(s)\right]\right|_{s=-1} = \left.\frac{d}{ds}\left[\frac{1}{s(s+2)}\right]\right|_{s=-1} = \left.\frac{-(2s+2)}{s^2(s+2)^2}\right|_{s=-1} = 0$$

$$C_3 = \left.\frac{1}{2!}\frac{d^2}{ds^2}\left[(s+1)^3 \cdot G(s)\right]\right|_{s=-1} = \left.\frac{1}{2!}\frac{d}{ds}\left[\frac{-(2s+2)}{s^2(s+2)^2}\right]\right|_{s=-1}$$

$$= \frac{1}{2}\left[\frac{-2}{s^2(s+2)^2} + \frac{2(2s+2)}{s^3(s+2)^2} + \frac{2(2s+2)}{s^2(s+2)^3}\right]_{s=-1} = -1$$

Thus,

$$G(s) = \frac{1}{2s} + \frac{1}{2(s+2)} - \frac{1}{(s+1)^3} - \frac{1}{s+1}$$

General method for all situations. We consider the foregoing examples to demonstrate the procedure.

EXAMPLE 1.14. Find the partial fractions of

$$G(s) = \frac{4s+2}{s(s+1)(s+2)} \tag{i}$$

by the general method.

Solution We write it as

$$G(s) = \frac{A}{s} + \frac{B}{s+1} + \frac{C}{s+2}$$

$$= \frac{A(s+1)(s+2) + Bs(s+2) + Cs(s+1)}{s(s+1)(s+2)} \tag{ii}$$

Comparing the numerators of Eqs. (i) and (ii), we get

$$A(s+1)(s+2) + Bs(s+2) + Cs(s+1) = 4s + 2$$

or

$$(A+B+C)s^2 + (3A+2B+C)s + 2A = 4s + 2 \tag{iii}$$

Comparing coefficients of equal powers of s on both sides of Eq. (iii), we have

$$A + B + C = 0 \tag{iv}$$

$$3A + 2B + C = 4 \tag{v}$$

$$2A = 2 \tag{vi}$$

Solving Eqs. (iv) to (vi), we obtain

$$A = 1, \quad B = 2, \quad C = -3$$

EXAMPLE 1.15. Find the partial fractions of

$$G(s) = \frac{1}{s(s+1)^3(s+2)} \tag{i}$$

by the general method.

Solution Equation (i) above can be rewriten as

$$G(s) = \frac{A}{s} + \frac{B}{s+2} + \frac{C_1}{(s+1)^3} + \frac{C_2}{(s+1)^2} + \frac{C_3}{s+1}$$

$$= \frac{A(s+1)^3(s+2) + Bs(s+1)^3 + C_1 s(s+2) + C_2 s(s+1)(s+2) + C_3 s(s+1)^2(s+2)}{s(s+1)^3(s+2)} \tag{ii}$$

Comparing the numerators of Eqs. (i) and (ii), we get

$$(A+B+C_3)s^4 + (5A+3B+C_2+5C_3)s^3 + (9A+3B+C_1+3C_2+5C_3)s^2$$
$$+ (7A+B+2C_1+2C_2+2C_3)s + 2A = 1 \tag{iii}$$

Comparing the coefficients of equal powers of s on both sides of Eq. (iii), we obtain

$$A + B + C_3 = 0$$

$$5A + 3B + C_2 + 5C_3 = 0$$

$$9A + 3B + C_1 + 3C_2 + 5C_3 = 0$$

$$7A + B + 2C_1 + 2C_2 + 2C_3 = 0$$

$$2A = 1$$

On solving the above simultaneous equations, we get the following five values for the constants:

$$A = \frac{1}{2}, \quad B = \frac{1}{2}, \quad C_1 = -1, \quad C_2 = 0, \quad C_3 = -1$$

The transfer function of overdamped second order system[8] can also be handled similarly as the following example shows.

EXAMPLE 1.16. Find the partial fractions of

$$G(s) = \frac{K}{\dfrac{s^2}{\omega_n^2} + \dfrac{2\zeta s}{\omega_n} + 1} = \frac{K\omega_n^2}{s^2 + 2\zeta\omega_n s + \omega_n^2} \tag{i}$$

Solution For $\zeta > 1$, the denominator of Eq. (i) can be factorised as

$$G(s) = \frac{K\omega_n^2}{s\left(s + \zeta\omega_n - \omega_n\sqrt{\zeta^2 - 1}\right)\left(s + \zeta\omega_n + \omega_n\sqrt{\zeta^2 - 1}\right)}$$

$$\equiv K\omega_n^2\left(\frac{A}{s} + \frac{B}{s + \zeta\omega_n - \omega_n\sqrt{\zeta^2 - 1}} + \frac{C}{s + \zeta\omega_n + \omega_n\sqrt{\zeta^2 - 1}}\right) \tag{ii}$$

where A, B and C are arbitrary coefficients to be determined.

We get the following condition to determine A, B and C:

$$A(s^2 + 2\zeta\omega_n s + \omega_n^2) + Bs(s + \zeta\omega_n + \omega_n\sqrt{\zeta^2 - 1}) + Cs(s + \zeta\omega_n - \omega_n\sqrt{\zeta^2 - 1}) = 1$$

or

$$(A + B + C)s^2 + \left[(2A + B + C)\zeta\omega_n + (B - c)\omega_n\sqrt{\zeta^2 - 1}\right]s + A\omega_n^2 = 1 \tag{iii}$$

Comparing coefficients of different powers of s on both sides of Eq. (iii), we obtain

$$A\omega_n^2 = 1$$

or

$$A = \frac{1}{\omega_n^2}$$

$$A + B + C = 0$$

or

$$B + C = -A = -\frac{1}{\omega_n^2} \tag{iv}$$

$$(2A + B + C)\zeta\omega_n + (B - C)\omega_n\sqrt{\zeta^2 - 1} = 0 \tag{v}$$

Substituting $A + B + C = 0$ in Eq. (v), we have

$$A\zeta + (B - C)\sqrt{\zeta^2 - 1} = 0$$

or

$$B - C = -\frac{\zeta}{\omega_n^2\sqrt{\zeta^2 - 1}} \tag{vii}$$

[8]See Section 7.3 at page 236.

From Eqs. (iv) and (vii),

$$B = \frac{\zeta + \sqrt{\zeta^2 - 1}}{2\omega_n^2 \sqrt{\zeta^2 - 1}}, \quad C = \frac{\zeta - \sqrt{\zeta^2 - 1}}{2\omega_n^2 \sqrt{\zeta^2 - 1}}$$

Alternatively, the partial fraction coefficients A, B and C can be found out by the matrix inversion method (see Example 1.34).

Let us now work out a few problems concerning Laplace and inverse Laplace transforms before considering other mathematical methods.

EXAMPLE 1.17.

(a) The Laplace transform of a function $f(t)$ is defined by

$$F(s) \equiv \mathscr{L}\{f(t)\} = \int_0^\infty e^{-st} f(t)\, dt$$

Find the inverse Laplace transform of $F(s - a)$.

(b) Using the result obtained in part (a), find

$$\mathscr{L}\{e^{at} \cos \omega t\} \quad \text{when} \quad \mathscr{L}\{\cos \omega t\} = \frac{s}{s^2 + \omega^2}$$

Solution (a) We observe that

$$\begin{aligned} \mathscr{L}\{e^{at} f(t)\} &= \int_0^\infty e^{at} e^{-st} f(t)\, dt \\ &= \int_0^\infty e^{-(s-a)t} f(t)\, dt \\ &\equiv F(s-a) \end{aligned}$$

Therefore, the inverse Laplace transform of $F(s - a)$ is $e^{at} f(t)$.

Note: This is the time shifting property of the Laplace transform.

(b) Given $\mathscr{L}\{\cos \omega t\} = \dfrac{s}{s^2 + \omega^2}$. Therefore, using the above derivation, we get

$$\mathscr{L}\{e^{at} \cos \omega t\} = \frac{s - a}{(s - a)^2 + \omega^2}$$

EXAMPLE 1.18. Given $g(t)$, as shown in Figure 1.2. Find the Laplace transform of $g(t)$.

Solution $g(t)$ can be written as

$$g(t) = \begin{cases} 0, & t < 3 \\ 1, & 3 \le t \le 5 \\ 0, & t > 5 \end{cases}$$

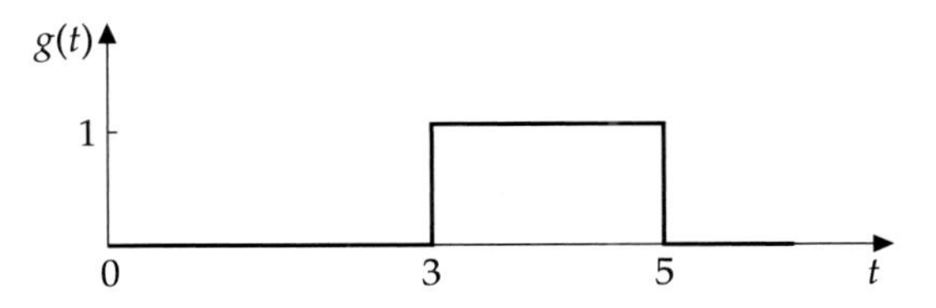

Figure 1.2 $g(t)$ (Example 1.18).

So, $g(t)$ can be rewritten as

$$g(t) = g_1(t) - g_2(t) = u(t-3) - u(t-5)$$

Therefore,

$$\mathscr{L}\{g(t)\} = \mathscr{L}\{g_1(t)\} - \mathscr{L}\{g_2(t)\} = \frac{e^{-3s} - e^{-5s}}{s}$$

EXAMPLE 1.19. $F(s) = \left(1 - e^{-sT}\right)/s$ is the Laplace transform of

(a) a pulse of width T
(b) a square wave of period T
(c) a unit step delayed by T
(d) a ramp delayed by T

Solution Given

$$F(s) = \frac{1 - e^{-sT}}{s} = \frac{1}{s} - \frac{e^{-sT}}{s}$$

Its inverse Laplace transform is

$$f(t) = 1 - u(t - T) \tag{i}$$

The plot of Eq. (i) is given in Figure 1.3. Obviously, it is a pulse of width T.

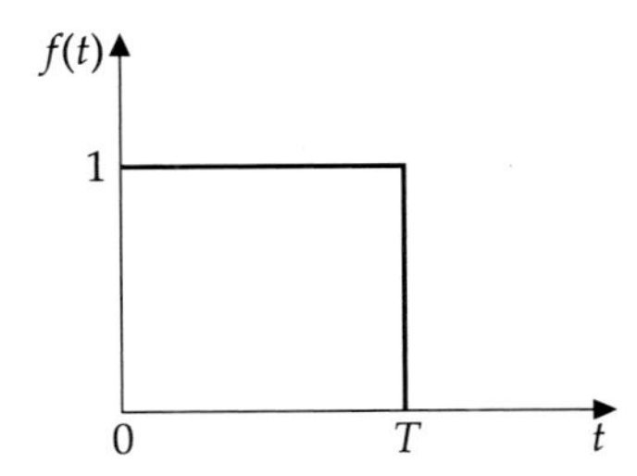

Figure 1.3 Plot of Eq. (i) (Example 1.19).

Therefore, the answer is (a).

EXAMPLE 1.20. Find the Laplace transform of the function $f(t)$ given by Figure 1.4.

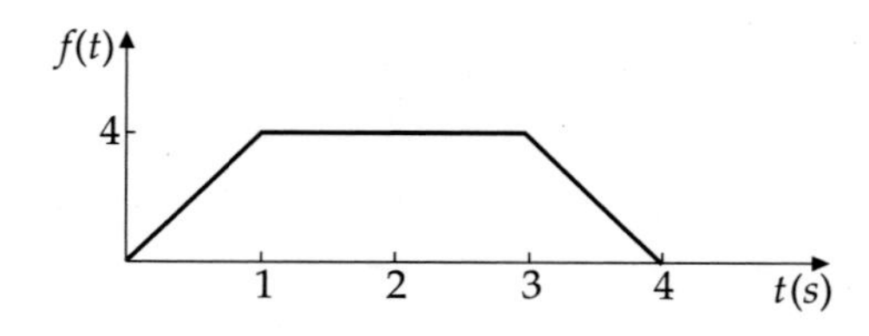

Figure 1.4 Function $f(t)$ (Example 1.20).

Solution The function consists of the following four segments:

1. From $t = 0$ to $t = 1$, $f(t) = 4tu(t)$.
2. From $t = 1$ to $t = 3$, $f(t) = 4u(t)$.
3. From $t = 3$ to $t = 4$, $f(t) = -4tu(t)$.
4. From $t = 4$, $f(t) = 0$.

Therefore, we can write it as follows:

$$\begin{aligned} f(t) =& [4tu(t) - 4(t-1)u(t-1)] + [4u(t-1) - 4u(t-3)] \\ &- [4(t-3)u(t-3) - 4(t-4)u(t-4)] \end{aligned}$$

Its Laplace transform yields

$$\begin{aligned} F(s) &= \frac{4}{s^2} - \frac{4e^{-s}}{s^2} + \frac{4e^{-s}}{s} - \frac{4e^{-3s}}{s} - \frac{4e^{-3s}}{s^2} + \frac{4e^{-4s}}{s^2} \\ &= \frac{4}{s}\left(e^{-s} - e^{-3s}\right) + \frac{4}{s^2}\left(1 - e^{-s} - e^{-3s} + e^{-4s}\right) \end{aligned}$$

EXAMPLE 1.21. For the differential equation

$$\frac{d^2x}{dt^2} + 6\frac{dx}{dt} + 8x = 0 \tag{i}$$

with the initial conditions

$$x(0) = 1, \qquad \left.\frac{dx}{dt}\right|_{t=0} = 0 \tag{ii}$$

the solution is

(a) $x(t) = 2e^{-6t} - e^{-2t}$.
(b) $x(t) = 2e^{-2t} - e^{-4t}$.
(c) $x(t) = -e^{-6t} + 2e^{-4t}$.
(d) $x(t) = e^{-2t} + 2e^{-4t}$.

Solution Taking the Laplace transform of Eq. (i), we get

$$s^2X(s) - sx(0) - x'(0) + 6[sX(s) - x(0)] + 8X(s) = 0$$

Using the initial conditions given in Eq. (ii), we obtain

$$(s^2 + 6s + 8)X(s) = s + 6$$

or

$$X(s) = \frac{s+6}{(s+2)(s+4)} \tag{iii}$$

Splitting Eq. (iii) into partial fractions, we have

$$\frac{s+6}{(s+2)(s+4)} \equiv \frac{A}{s+2} + \frac{B}{s+4}$$

$$A = \left.\frac{s+6}{s+4}\right|_{s=-2} = 2$$

$$B = \left.\frac{s+6}{s+2}\right|_{s=-4} = -1$$

Therefore,

$$X(s) = \frac{2}{s+2} - \frac{1}{s+4} \tag{iv}$$

Taking the inverse Laplace transform of Eq. (iv), we finally get

$$x(t) = 2e^{-2t} - e^{-4t}$$

Therefore, the answer is (b).

Note: If the initial conditions are zero, simply substitute s for each differentiation and s^{-1} for each integration, to Laplace transform a differential equation.

EXAMPLE 1.22. Find the Laplace transform of $5e^{-3t} + u(t-1)u(t-2)$.

Solution

$$\mathscr{L}\{5e^{-3t} + u(t-1)u(t-2)\} = \mathscr{L}\{5e^{-3t}\} + \mathscr{L}\{u(t-1)\} - \mathscr{L}\{u(t-2)\}$$

$$= \frac{5}{s+3} + \frac{e^{-s}}{s} - \frac{e^{-2s}}{s}$$

EXAMPLE 1.23. Find the Laplace transform of $e^{-2t} + 5t[u(t-2) - u(t)]$.

Solution Let $f(t)$ be the given function. Then,

$$\mathscr{L}\{f(t)\} = \mathscr{L}\{e^{-2t}\} + \mathscr{L}\{5tu(t-2)\} - \mathscr{L}\{5tu(t)\}$$

$$= \frac{1}{s+2} + 5\mathscr{L}\{(t-2)u(t-2) + 2u(t-2)\} - \frac{5}{s^2}$$

$$= \frac{1}{s+2} + 5e^{-2s}\mathscr{L}\{tu(t)\} + 10\mathscr{L}\{u(t-2)\} - \frac{5}{s^2}$$

$$= \frac{1}{s+2} + \frac{5e^{-2s}}{s^2} + 10e^{-2s}\mathscr{L}\{u(t)\} - \frac{5}{s^2}$$

$$= \frac{1}{s+2} + \frac{5e^{-2s}}{s^2} + \frac{10e^{-2s}}{s} - \frac{5}{s^2}$$

EXAMPLE 1.24. Determine the Laplace transform of

$$u_2(t) = \lim_{\tau \to 0} \frac{u(t) - 2u(t-\tau) + u(t-2\tau)}{\tau^2}$$

Solution The Laplace transform of the given function is

$$\begin{aligned}
\mathscr{L}\{u_2(t)\} &= \lim_{\tau \to 0} \frac{1}{\tau^2}\left(\frac{1}{s} - \frac{2}{s}e^{-s\tau} + \frac{1}{s}e^{-2s\tau}\right) \\
&= \lim_{\tau \to 0} \frac{1}{s\tau^2}\left[1 - 2\left\{1 - s\tau + \frac{(s\tau)^2}{2\,!} - \cdots\right\} + \left\{1 - 2s\tau + \frac{(2s\tau)^2}{2\,!} - \cdots\right\}\right] \\
&= \lim_{\tau \to 0} \frac{\left[(s\tau)^2 + (\text{terms of higher order in } s\tau)\right]}{s\tau^2} \\
&= s
\end{aligned}$$

EXAMPLE 1.25. Convert the following function in the s-space to a function in the t-space:

$$\frac{s^3 + 4s^2 + 4s + 4}{s^3 + 4s}$$

Solution The function can be written as

$$\begin{aligned}
1 + \frac{4s^2 + 4}{s^3 + 4s} &\equiv 1 + \frac{A}{s} + \frac{Bs + C}{s^2 + 4} \\
&= 1 + \frac{(A + B)s^2 + Cs + 4A}{s^3 + 4s} \qquad \text{(i)}
\end{aligned}$$

Comparing $(A + B)s^2 + Cs + 4A$ of Eq. (i) with $4s^2 + 4$ of the left hand side, we get

$$4A = 4, \quad A + B = 4, \quad C = 0$$

or

$$A = 1, \quad B = 3, \quad C = 0$$

So,

$$F(s) = 1 + \frac{1}{s} + \frac{3}{s^2 + 4}$$

Its inverse Laplace transform yields

$$f(t) = \delta(t) + 1 + 3\cos(2t)$$

EXAMPLE 1.26. Find the inverse Laplace transform of

$$\frac{s^2 + 4}{s^4 + 10s^3 + 35s^2 + 50s + 24}$$

Solution To factorize the denominator of the given expression, we observe that it is 0 for $s = -1, -2, -3, -4$. Therefore, if

$$\frac{s^2 + 4}{s^4 + 10s^3 + 35s^2 + 50s + 24} = \frac{A}{s + 1} + \frac{B}{s + 2} + \frac{C}{s + 3} + \frac{D}{s + 4}$$

We have

$$A = \left.\frac{s^2+4}{(s+2)(s+3)(s+4)}\right|_{s=-1} = \frac{5}{6}$$

$$B = \left.\frac{s^2+4}{(s+1)(s+3)(s+4)}\right|_{s=-2} = \frac{8}{-2} = -4$$

$$C = \left.\frac{s^2+4}{(s+1)(s+2)(s+4)}\right|_{s=-3} = \frac{13}{2}$$

$$D = \left.\frac{s^2+4}{(s+1)(s+2)(s+3)}\right|_{s=-4} = \frac{20}{-6} = -\frac{10}{3}$$

Therefore, the required inverse Laplace transform is

$$\frac{5}{6}e^{-t} - 4e^{-2t} + \frac{13}{2}e^{-3t} - \frac{10}{3}e^{-4t}$$

EXAMPLE 1.27. Find the Laplace transform of $\dfrac{\sin t}{t}$.

Solution Let $f(t) = \sin t$. Then,

$$F(s) \equiv \mathscr{L}\{f(t)\} = \frac{1}{s^2+1} \tag{i}$$

By using the property # 8 of Table 1.1, we get

$$\begin{aligned}\mathscr{L}\left\{\frac{\sin t}{t}\right\} &= \int_s^{\infty} \frac{1}{\sigma^2+1} d\sigma \qquad \text{[Using Eq. (i)]}\\ &= \tan^{-1}\sigma\Big|_s^{\infty}\\ &= \frac{\pi}{2} - \tan^{-1}(s)\end{aligned}$$

EXAMPLE 1.28. Using the convolution theorem, find the inverse Laplace transform of the following function:

$$\frac{1}{s(s+2)}$$

Solution Let

$$F_1(s) \equiv \frac{1}{s}, \quad F_2(s) \equiv \frac{1}{s+2}$$

Then, $$\mathscr{L}^{-1}\{F_1(s)\} \equiv f_1(t) = 1, \quad \mathscr{L}^{-1}\{F_2(s)\} \equiv f_2(t) = e^{-2t}$$

Using the real convolution theorem, we get

$$\begin{aligned}\mathscr{L}^{-1}\{F_1(s)\cdot F_2(s)\} = f_1(t)\star f_2(t) &\equiv \int_0^t f_1(\tau)\cdot f_2(t-\tau)d\tau \\ &= \int_0^t 1\cdot e^{-2(t-\tau)}d\tau = e^{-2t}\int_0^t e^{2\tau}d\tau \\ &= e^{-2t}\left[\frac{1}{2}e^{2\tau}\right]_0^t = \frac{1}{2}e^{-2t}(e^{2t}-1) \\ &= \frac{1}{2}(1-e^{-2t})\end{aligned}$$

EXAMPLE 1.29. Find the inverse Laplace transform of the following:

$$\frac{(s+4)e^{-s}}{s^2+16}$$

Solution The given function can be written as:

$$F(s) \equiv \frac{se^{-s}}{s^2+16} + \frac{4e^{-s}}{s^2+16}$$

We know that

$$\mathscr{L}^{-1}\left\{\frac{s}{s^2+16}\right\} = \cos 4t, \quad \mathscr{L}^{-1}\left\{\frac{4}{s^2+16}\right\} = \sin 4t$$

Now, using the time shifting property, we get

$$\mathscr{L}^{-1}\left\{\frac{se^{-s}}{s^2+16}\right\} = \cos 4(t-1)$$

$$\mathscr{L}^{-1}\left\{\frac{4e^{-s}}{s^2+16}\right\} = \sin 4(t-1)$$

Therefore, the inverse Laplace transform of the given function is

$$\cos 4(t-1) + \sin 4(t-1)$$

1.3 MATRICES

Definitions

A collection of elements—be they real or complex numbers, functions or operators—arranged in a rectangular or square array, is called a *matrix*.

The elements of a matrix are written with two subscripts, e.g. a_{ij}, where the first subscript i indicates the row number and the second subscript j denotes the column number. The established norm is to write the row number first and then the column number. Thus, the matrix $\mathbf{A}$ can be written as

$$\mathbf{A} = \begin{bmatrix} a_{11} & a_{12} & a_{13} \\ a_{21} & a_{22} & a_{23} \\ a_{31} & a_{32} & a_{33} \end{bmatrix}$$

Column matrix. A matrix having only one column but more than one row, is called a *column matrix* or *column vector.* For example,

$$\text{Column matrix } \mathbf{A} = \begin{bmatrix} a_{11} \\ a_{21} \\ \vdots \\ a_{n1} \end{bmatrix}$$

Row matrix. A matrix having only one row but more than one column is called a *row matrix* or *row vector.* For example,

$$\text{Row matrix } \mathbf{B} = \begin{bmatrix} b_{11} & b_{12} & \ldots & b_{1m} \end{bmatrix}$$

Diagonal matrix. A square matrix having only non-zero diagonal and zero elements everywhere else is called a *diagonal matrix.* For example,

$$\text{Diagonal matrix } \mathbf{A} = \begin{bmatrix} a_{11} & 0 & 0 & \ldots & 0 \\ 0 & a_{22} & 0 & \ldots & 0 \\ 0 & 0 & a_{33} & \ldots & 0 \\ \vdots & \vdots & \vdots & \ddots & \vdots \\ 0 & 0 & 0 & \ldots & a_{nm} \end{bmatrix}$$

Thus, for a diagonal matrix,

$$a_{ij} = 0 \quad \text{for all } i \neq j$$

Identity (or unity) matrix. A diagonal matrix whose elements are 1 is an *identity* or *unity matrix.* It is written as **I** or **U**. For example,

$$\text{Identity matrix } \mathbf{I} = \begin{bmatrix} 1 & 0 & 0 \\ 0 & 1 & 0 \\ 0 & 0 & 1 \end{bmatrix}$$

Null matrix. If all the elements of a matrix are zeros, it is called a *null matrix.* For example,

$$\text{Null or zero matrix } = \begin{bmatrix} 0 & 0 \\ 0 & 0 \end{bmatrix}$$

Order of a Matrix

If a matrix has n rows and m columns, its order is $n \times m$ or n by m. For example, the order of

$$\begin{bmatrix} a_{11} & a_{12} \\ a_{21} & a_{22} \\ a_{31} & a_{32} \end{bmatrix}$$

is 3×2

Determinant of a Matrix

A determinant may be defined for each square matrix. It is written as det **A**, or $\Delta\mathbf{A}$ or $|\mathbf{A}|$.

For example, for the matrix

$$\mathbf{A} = \begin{bmatrix} 1 & 0 & 2 \\ 3 & 1 & 4 \\ 0 & -1 & 5 \end{bmatrix}$$

the determinant is

$$\begin{aligned} \Delta\mathbf{A} &= \begin{vmatrix} 1 & 0 & 2 \\ 3 & 1 & 4 \\ 0 & -1 & 5 \end{vmatrix} \\ &= (1)[(1)(5) - (4)(-1)] + (0)[(4)(0) - (3)(5)] + (2)[(3)(-1) - (1)(0)] \\ &= 3 \end{aligned}$$

Trace of a Matrix

The sum of all the diagonal elements of a square matrix is called the *trace* of the matrix. Symbolically, it is written as Tr $\mathbf{A} = \sum_1^n A_{ii}$. It can be shown[9] that if $\lambda_1,\ \lambda_2, \cdots \lambda_n$ are the eigenvalues[10] of $\mathbf{A}$, then

$$\text{Tr}\ \mathbf{A} = \sum_1^n \lambda_i$$

Also,

$$\det\ \mathbf{A} = \prod_1^n \lambda_i$$

Singular Matrix

If the determinant of a square matrix is zero, it is called a *singular matrix*. Singularity of a matrix indicates that not all its rows or columns are independent of each other. Suppose, we have a set of simultaneous equations given by

$$\begin{aligned} x_1 + 2x_2 + x_3 &= 0 \\ -2x_1 - x_2 + 2x_3 &= 0 \\ -x_1 + x_2 + 3x_3 &= 0 \end{aligned} \tag{i}$$

It can be written in matrix multiplication form as[11]

$$\begin{bmatrix} 1 & 2 & 1 \\ -2 & -1 & 2 \\ -1 & 1 & 3 \end{bmatrix} \begin{bmatrix} x_1 \\ x_2 \\ x_3 \end{bmatrix} = 0$$

We observe that

$$\begin{vmatrix} 1 & 2 & 1 \\ -2 & -1 & 2 \\ -1 & 1 & 3 \end{vmatrix} = 0$$

[9]See, for example, LA Pipes and LR Harvill, *Applied Mathematics for Engineers and Physicists*, McGraw-Hill International Edition (1971), Singapore, p. 106.

[10]See page 29.

[11]See section on *Matrix Algebra* at page 25 for matrix multiplication.

which means that the matrix representing the coefficients of the simultaneous equations is singular and, therefore, the rows are not independent of each other. It is also vindicated by the fact that the third equation of relations (i) is not independent—it is merely the sum of the first two equations.

Transpose of a Matrix

If the elements of rows and columns of a matrix **A** are interchanged, the resulting matrix is called the transpose of **A** and written as $\mathbf{A}'$ or $\mathbf{A}^{\mathrm{T}}$. For example, let

$$\mathbf{A} = \begin{bmatrix} 1 & 4 & 7 \\ 2 & 5 & 8 \\ 3 & 6 & 9 \end{bmatrix}$$

Then,

$$\mathbf{A}' \text{ or } \mathbf{A}^{\mathrm{T}} = \begin{bmatrix} 1 & 2 & 3 \\ 4 & 5 & 6 \\ 7 & 8 & 9 \end{bmatrix}$$

Minor of a Matrix

Minor of a matrix, M_{ij}, is obtained by finding the determinant of the matrix after removing the ith row and jth column from it. Let

$$\mathbf{A} = \begin{bmatrix} a_{11} & a_{12} & a_{13} \\ a_{21} & a_{22} & a_{23} \\ a_{31} & a_{32} & a_{33} \end{bmatrix}$$

Then,

$$M_{22} = \begin{vmatrix} a_{11} & \not{a}_{12} & a_{13} \\ \not{a}_{21} & \not{a}_{22} & \not{a}_{23} \\ a_{31} & \not{a}_{32} & a_{33} \end{vmatrix} = \begin{vmatrix} a_{11} & a_{13} \\ a_{31} & a_{33} \end{vmatrix} = a_{11}a_{33} - a_{13}a_{31}$$

which is obtained by removing the 2nd row and 2nd column from **A** and defining a determinant with the remaining elements, as shown.

Cofactor

The cofactor of the element a_{ij} of a matrix is defined as

$$\alpha_{ij} = (-1)^{i+j} M_{ij}$$

where M_{ij} is the minor of the matrix. The cofactor of a_{22} of matrix **A** above is

$$\alpha_{22} = (-1)^{2+2}(a_{11}a_{33} - a_{13}a_{31}) = a_{11}a_{33} - a_{13}a_{31}$$

EXAMPLE 1.30. Find the cofactor of b_{32} of matrix **B** defined as:

$$\mathbf{B} = \begin{bmatrix} 1 & 0 & 2 \\ 3 & 1 & 4 \\ 0 & 1 & 5 \end{bmatrix}$$

Solution The minor is

$$M_{32} = \begin{vmatrix} 1 & 2 \\ 3 & 4 \end{vmatrix}$$

Therefore, the cofactor is

$$\alpha_{32} = (-1)^{3+2}(-2) = 2$$

Adjoint of a Matrix

The adjoint of a matrix is obtained after replacing the elements of the matrix by their cofactors and then taking the transpose of the resulting cofactor matrix.

EXAMPLE 1.31. Find the adjoint of the matrix $\mathbf{B}$ of Example 1.30.

Solution The minors of the matrix $\mathbf{B}$ are

$$M_{11} = \begin{vmatrix} 1 & 4 \\ 1 & 5 \end{vmatrix} = 5 - 4 = 1, \quad M_{12} = \begin{vmatrix} 3 & 4 \\ 0 & 5 \end{vmatrix} = 15 - 0 = 15, \quad M_{13} = \begin{vmatrix} 3 & 1 \\ 0 & 1 \end{vmatrix} = 3 - 0 = 3$$

$$M_{21} = \begin{vmatrix} 0 & 2 \\ 1 & 5 \end{vmatrix} = 0 - 2 = -2, \quad M_{22} = \begin{vmatrix} 1 & 2 \\ 0 & 5 \end{vmatrix} = 5 - 0 = 5, \quad M_{23} = \begin{vmatrix} 1 & 0 \\ 0 & 1 \end{vmatrix} = 1 - 0 = 1$$

$$M_{31} = \begin{vmatrix} 0 & 2 \\ 1 & 4 \end{vmatrix} = 0 - 2 = -2, \quad M_{32} = \begin{vmatrix} 1 & 2 \\ 3 & 4 \end{vmatrix} = 4 - 6 = -2, \quad M_{33} = \begin{vmatrix} 1 & 0 \\ 3 & 1 \end{vmatrix} = 1 - 0 = 1$$

The cofactors are as follows:

$$\alpha_{11} = (-1)^{1+1}M_{11} = 1, \quad \alpha_{12} = (-1)^{1+2}M_{12} = -15, \quad \alpha_{13} = (-1)^{1+3}M_{13} = 3$$

$$\alpha_{21} = (-1)^{2+1}M_{21} = 2, \quad \alpha_{22} = (-1)^{2+2}M_{22} = 5, \quad \alpha_{23} = (-1)^{2+3}M_{23} = -1$$

$$\alpha_{31} = (-1)^{3+1}M_{31} = -2, \quad \alpha_{32} = (-1)^{3+2}M_{32} = 2, \quad \alpha_{33} = (-1)^{3+3}M_{33} = 1$$

Therefore,

$$\text{Adj}\,(\mathbf{B}) = \begin{bmatrix} 1 & -15 & 3 \\ 2 & 5 & -1 \\ -2 & 2 & 1 \end{bmatrix}^{\mathrm{T}} = \begin{bmatrix} 1 & 2 & -2 \\ -15 & 5 & 2 \\ 3 & -1 & 1 \end{bmatrix}$$

Rank of a Matrix

A matrix $\mathbf{A}$ is said to be of rank r if there exists an $r \times r$ submatrix of $\mathbf{A}$ which is non-singular and all other $q \times q$ (where $q \geq r + 1$) submatrices are singular. Let us consider the 4×4 matrix

$$\mathbf{A} = \begin{bmatrix} 1 & 2 & 3 & 4 \\ 0 & 1 & -1 & 0 \\ 1 & 0 & 1 & 2 \\ 1 & 1 & 0 & 2 \end{bmatrix}$$

Here, det $\mathbf{A} = 0$, but the determinant of the submatrix consisting of first three columns by first three rows is not zero. Therefore, the rank of $\mathbf{A}$ is 3.

Matrix Algebra

Equality of matrices. Two matrices **A** and **B** are said to be equal if they are of the same order and each element of **A** equals the corresponding element of **B**.

Matrix addition and subtraction. Two matrices **A** and **B**, both of the same order, can be added or subtracted from one another by adding or subtracting their respective elements. If

$$\mathbf{A} = \begin{bmatrix} a_{11} & a_{12} & a_{13} \\ a_{21} & a_{22} & a_{23} \\ a_{31} & a_{32} & a_{33} \end{bmatrix}, \quad \mathbf{B} = \begin{bmatrix} b_{11} & b_{12} & b_{13} \\ b_{21} & b_{22} & b_{23} \\ b_{31} & b_{32} & b_{33} \end{bmatrix}$$

Then,

$$\mathbf{C} = \mathbf{A} \pm \mathbf{B} = \begin{bmatrix} a_{11} \pm b_{11} & a_{12} \pm b_{12} & a_{13} \pm b_{13} \\ a_{21} \pm b_{21} & a_{22} \pm b_{22} & a_{23} \pm b_{23} \\ a_{31} \pm b_{31} & a_{32} \pm b_{32} & a_{33} \pm b_{33} \end{bmatrix}$$

Matrix addition and subtraction obey the associative and cumulative laws:

$$\text{Associative law} : (\mathbf{A} \pm \mathbf{B}) \pm \mathbf{C} = \mathbf{A} \pm (\mathbf{B} \pm \mathbf{C})$$

$$\text{Cumulative law} : \pm\mathbf{A} \pm \mathbf{B} \pm \mathbf{C} = \pm\mathbf{B} \pm \mathbf{C} \pm \mathbf{A} = \pm\mathbf{C} \pm \mathbf{A} \pm \mathbf{B}$$

Matrix multiplication. Two matrices **A** and **B** can be multiplied if they are conformable, i.e. if the number of columns of **A** equals the number of rows of **B**. Each element of the product matrix is obtained through the sum of multiplication of the row elements of **A** by corresponding column elements of **B**. Stated mathematically,

$$\mathbf{C} = \mathbf{AB} \quad \text{when} \quad c_{ij} = \sum_k a_{ik} b_{kj}$$

Let

$$\mathbf{A} = \begin{bmatrix} a_{11} & a_{12} & a_{13} \\ a_{21} & a_{22} & a_{23} \end{bmatrix}, \quad \mathbf{B} = \begin{bmatrix} b_{11} \\ b_{21} \\ b_{31} \end{bmatrix}$$

Then,

$$\mathbf{AB} = \begin{bmatrix} a_{11}b_{11} + a_{12}b_{21} + a_{13}b_{31} \\ a_{21}b_{11} + a_{22}b_{21} + a_{23}b_{31} \end{bmatrix}$$

In general, matrix multiplication is

- Not commutative, i.e., $\mathbf{AB} \neq \mathbf{BA}$.
- Distributive, i.e., $\mathbf{A}(\mathbf{B} + \mathbf{C}) = \mathbf{AB} + \mathbf{AC}$.
- Associative, i.e., $\mathbf{A}(\mathbf{BC}) = (\mathbf{AB})\mathbf{C}$ if the multiplicand matrices are conformable.

EXAMPLE 1.32. If

$$\mathbf{A} = \begin{bmatrix} 1 & 0 \\ 5 & -1 \\ 2 & 3 \end{bmatrix} \qquad \mathbf{B} = \begin{bmatrix} 2 & 1 & 4 \\ 1 & 0 & 3 \end{bmatrix}$$

then show that $\mathbf{AB} \neq \mathbf{BA}$.

Solution The required products are found as follows:

$$\mathbf{AB} = \begin{bmatrix} 1 & 0 \\ 5 & -1 \\ 2 & 3 \end{bmatrix} \begin{bmatrix} 2 & 1 & 4 \\ 1 & 0 & 3 \end{bmatrix}$$

$$= \begin{bmatrix} (1)(2)+(0)(1) & (1)(1)+(0)(0) & (1)(4)+(0)(3) \\ (5)(2)+(-1)(1) & (5)(1)+(-1)(0) & (5)(4)+(-1)(3) \\ (2)(2)+(3)(1) & (2)(1)+(3)(0) & (2)(4)+(3)(3) \end{bmatrix}$$

$$= \begin{bmatrix} 2 & 1 & 4 \\ 9 & 5 & 17 \\ 7 & 2 & 17 \end{bmatrix}$$

$$\mathbf{BA} = \begin{bmatrix} 2 & 1 & 4 \\ 1 & 0 & 3 \end{bmatrix} \begin{bmatrix} 1 & 0 \\ 5 & -1 \\ 2 & 3 \end{bmatrix}$$

$$= \begin{bmatrix} (2)(1)+(1)(5)+(4)(2) & (2)(0)+(1)(-1)+(4)(3) \\ (1)(1)+(0)(5)+(3)(2) & (1)(0)+(0)(-1)+(3)(3) \end{bmatrix}$$

$$= \begin{bmatrix} 15 & 11 \\ 7 & 9 \end{bmatrix}$$

Obviously, $\mathbf{AB} \neq \mathbf{BA}$.

Multiplication by a scalar. Multiplication of a matrix $\mathbf{A}$ by a scalar k means multiplying each element of $\mathbf{A}$ by k. Therefore, if

$$\mathbf{A} = \begin{bmatrix} a_{11} & a_{12} & a_{13} \\ a_{21} & a_{22} & a_{23} \\ a_{31} & a_{32} & a_{33} \end{bmatrix}$$

then

$$k\mathbf{A} = \begin{bmatrix} ka_{11} & ka_{12} & ka_{13} \\ ka_{21} & ka_{22} & ka_{23} \\ ka_{31} & ka_{32} & ka_{33} \end{bmatrix}$$

Inverse of a matrix. In matrix algebra, if

$$\mathbf{Ax} = \mathbf{y}, \quad \text{then} \quad \mathbf{x} = \mathbf{A}^{-1}\mathbf{y}$$

where $\mathbf{A}^{-1}$ is called the inverse of $\mathbf{A}$. The inverse of a matrix $\mathbf{A}$ exists if (i) $\mathbf{A}$ is a square matrix, and (ii) $\mathbf{A}$ is non-singular which implies that none of its eigenvalues[12] is zero. $\mathbf{A}^{-1}$ can be obtained from the relation

$$\mathbf{A}^{-1} = \frac{\text{Adj}\,(\mathbf{A})}{|\mathbf{A}|} \tag{1.16}$$

[12]See page 29.

EXAMPLE 1.33. Find out the inverse of the matrix

$$\mathbf{B} = \begin{bmatrix} 1 & 0 & 2 \\ 3 & 1 & 4 \\ 0 & 1 & 5 \end{bmatrix}$$

Solution We have already worked out Adj ($\mathbf{B}$) in Example 1.31 at page 24 as follows:

$$\text{Adj}\,(\mathbf{B}) = \begin{bmatrix} 1 & 2 & -2 \\ -15 & 5 & 2 \\ 3 & -1 & 1 \end{bmatrix}$$

Now, $|\mathbf{B}| = 7$. Therefore,

$$\mathbf{B}^{-1} = \frac{\text{Adj}\,(\mathbf{B})}{|\mathbf{B}|} = \frac{\begin{bmatrix} 1 & 2 & -2 \\ -15 & 5 & 2 \\ 3 & -1 & 1 \end{bmatrix}}{7} = \begin{bmatrix} \frac{1}{7} & \frac{2}{7} & -\frac{2}{7} \\ -\frac{15}{7} & \frac{5}{7} & \frac{2}{7} \\ \frac{3}{7} & -\frac{1}{7} & \frac{1}{7} \end{bmatrix}$$

While discussing partial fraction expansions, we mentioned that the coefficients of partial fractions could be found by the matrix inversion method. Here is an example.

EXAMPLE 1.34. Find the partial fractions of

$$G(s) = \frac{K}{\frac{s^2}{\omega_n^2} + \frac{2\zeta s}{\omega_n} + 1} = \frac{K\omega_n^2}{s^2 + 2\zeta\omega_n s + \omega_n^2} \tag{i}$$

by the matrix inversion method.

Solution Let $\zeta\omega_n = x$ and $\omega_n\sqrt{1-\zeta^2} = y$. Then

$$G(s) \equiv K\omega_n^2\left[\frac{A}{s} + \frac{B}{s+x-y} + \frac{C}{s+x+y}\right]$$

$$= K\omega_n^2\left[\frac{A\big[(s+x)^2 - y^2\big] + Bs(s+x+y) + Cs(s+x-y)}{s(s+x-y)(s+x+y)}\right] \tag{ii}$$

Comparing the numerator on the RHS of Eq. (ii) with that of Eq. (i), we get

$$A\big[(s+x)^2 - y^2\big] + Bs(s+x+y) + Cs(s+x-y) = 1$$

$$\Rightarrow \quad A\big(s^2 + 2xs + x^2 - y^2\big) + B\big[s^2 + s(x+y)\big] + C\big[s^2 + s(x-y)\big] = 1$$

$$\Rightarrow \quad (A+B+C)s^2 + [2xA + (x+y)B + (x-y)C]s + \big(x^2 - y^2\big)A = 1 \tag{iii}$$

Equation (iii) implies that

$$A + B + C = 0$$
$$2xA + (x+y)B + (x-y)C = 0$$
$$(x^2 - y^2)A = 1$$

which can be written in the matrix form as

$$\begin{bmatrix} 1 & 1 & 1 \\ 2x & x+y & x-y \\ x^2-y^2 & 0 & 0 \end{bmatrix} \begin{bmatrix} A \\ B \\ C \end{bmatrix} = \begin{bmatrix} 0 \\ 0 \\ 1 \end{bmatrix}$$

Therefore,

$$\begin{bmatrix} A \\ B \\ C \end{bmatrix} = \begin{bmatrix} 1 & 1 & 1 \\ 2x & x+y & x-y \\ x^2-y^2 & 0 & 0 \end{bmatrix}^{-1} \begin{bmatrix} 0 \\ 0 \\ 1 \end{bmatrix} \tag{iii}$$

Let

$$\mathbf{P} = \begin{bmatrix} 1 & 1 & 1 \\ 2x & x+y & x-y \\ x^2-y^2 & 0 & 0 \end{bmatrix}$$

Then,

$$\det \mathbf{P} = (x-y)(x^2-y^2) - (x+y)(x^2-y^2) = -2y(x^2-y^2)$$

The cofactors of $\mathbf{P}$ are as follows:

$$\alpha_{11} = (-1)^{1+1}\begin{vmatrix} x+y & x-y \\ 0 & 0 \end{vmatrix} = 0$$

$$\alpha_{12} = (-1)^{1+2}\begin{vmatrix} 2x & x-y \\ x^2-y^2 & 0 \end{vmatrix} = (x-y)(x^2-y^2)$$

$$\alpha_{13} = (-1)^{1+3}\begin{vmatrix} 2x & x+y \\ x^2-y^2 & 0 \end{vmatrix} = -(x+y)(x^2-y^2)$$

$$\alpha_{21} = (-1)^{2+1}\begin{vmatrix} 1 & 1 \\ 0 & 0 \end{vmatrix} = 0$$

$$\alpha_{22} = (-1)^{2+2}\begin{vmatrix} 1 & 1 \\ x^2-y^2 & 0 \end{vmatrix} = -(x^2-y^2)$$

$$\alpha_{23} = (-1)^{2+3}\begin{vmatrix} 1 & 1 \\ x^2-y^2 & 0 \end{vmatrix} = (x^2-y^2)$$

$$\alpha_{31} = (-1)^{3+1}\begin{vmatrix} 1 & 1 \\ x+y & x-y \end{vmatrix} = -2y$$

$$\alpha_{32} = (-1)^{3+2}\begin{vmatrix} 1 & 1 \\ 2x & x-y \end{vmatrix} = (x+y)$$

$$\alpha_{33} = (-1)^{3+3}\begin{vmatrix} 1 & 1 \\ 2x & x+y \end{vmatrix} = -(x-y)$$

Thus,

$$\mathbf{P}^{-1} = \frac{1}{\det \mathbf{P}}\begin{bmatrix} \alpha_{11} & \alpha_{12} & \alpha_{13} \\ \alpha_{21} & \alpha_{22} & \alpha_{23} \\ \alpha_{31} & \alpha_{32} & \alpha_{33} \end{bmatrix}^{\mathrm{T}}$$

$$= -\frac{1}{2y\left(x^2-y^2\right)}\begin{bmatrix} 0 & (x-y)\left(x^2-y^2\right) & -(x+y)\left(x^2-y^2\right) \\ 0 & -\left(x^2-y^2\right) & \left(x^2-y^2\right) \\ -2y & x+y & -(x-y) \end{bmatrix}^{\mathrm{T}}$$

$$= \begin{bmatrix} 0 & 0 & \dfrac{1}{x^2-y^2} \\ -\dfrac{x-y}{2y} & \dfrac{1}{2y} & -\dfrac{x+y}{2y(x^2-y^2)} \\ -\dfrac{x+y}{2y} & -\dfrac{1}{2y} & \dfrac{x-y}{2y(x^2-y^2)} \end{bmatrix} \tag{iv}$$

Substituting the value of $\mathbf{P}^{-1}$ from Eq. (iv) in Eq. (iii), we get

$$A = \frac{1}{x^2-y^2} = \frac{1}{\omega_n^2}$$

$$B = -\frac{x+y}{2y\left(x^2-y^2\right)} = -\frac{\zeta+\sqrt{\zeta^2-1}}{2\omega_n^2\sqrt{\zeta^2-1}}$$

$$C = \frac{x-y}{2y\left(x^2-y^2\right)} = \frac{\zeta-\sqrt{\zeta^2-1}}{2\omega_n^2\sqrt{\zeta^2-1}}$$

Thus,

$$G(s) = K\left[\frac{1}{s} - \frac{\dfrac{1}{2}\left(1+\dfrac{\zeta}{\sqrt{\zeta^2-1}}\right)}{s+\zeta\omega_n-\omega_n\sqrt{\zeta^2-1}} - \frac{\dfrac{1}{2}\left(1-\dfrac{\zeta}{\sqrt{\zeta^2-1}}\right)}{s+\zeta\omega_n+\omega_n\sqrt{\zeta^2-1}}\right]$$

Eigenvalues and Eigenvectors of Matrices

If $\mathbf{A}$ is an $n \times n$ matrix, $\mathbf{x}$ is a nonzero vector and λ is any scalar so that

$$\mathbf{Ax} = \lambda\mathbf{x} \tag{1.17}$$

then $\mathbf{x}$ is called an eigenvector and λ, an eigenvalue of $\mathbf{A}$. Eigenvalues and eigenvectors are not independent of each other. Hence, it is better said that $\mathbf{x}$ is the eigenvector associated with or corresponding to λ and vice versa.

Suppose

$$\mathbf{A} = \begin{bmatrix} -4 & 2 \\ 3 & -5 \end{bmatrix}$$

then

$$\mathbf{x} = \begin{bmatrix} 1 \\ 1 \end{bmatrix}$$

is an eigenvector associated with the eigenvalue -2 because

$$\begin{bmatrix} -4 & 2 \\ 3 & -5 \end{bmatrix} \begin{bmatrix} 1 \\ 1 \end{bmatrix} = -2 \begin{bmatrix} 1 \\ 1 \end{bmatrix}$$

conforming to the definition given by Eq. (1.17).

Eigenvalues. To find the eigenvalues of a matrix $\mathbf{A}$, we have to solve the characteristic equation given by

$$\Delta\,(\lambda\mathbf{I} - \mathbf{A}) = 0 \tag{1.18}$$

EXAMPLE 1.35. Find the eigenvalues of the matrix

$$\mathbf{A} = \begin{bmatrix} -4 & 2 \\ 3 & -5 \end{bmatrix}$$

Solution For the matrix $\mathbf{A}$ defined above, the characteristic function is

$$\Delta(\lambda) = \begin{vmatrix} \lambda + 4 & -2 \\ -3 & \lambda + 5 \end{vmatrix}$$

The characteristic equation $\Delta(\lambda) = 0$ gives

$$(\lambda + 4)(\lambda + 5) - 6 = 0$$

or

$$\lambda^2 + 9\lambda + 14 = 0$$

yielding

$$\lambda_1 = -2, \quad \lambda_2 = -7$$

Thus, the eigenvalues of $\mathbf{A}$ are -2 and -7.

Eigenvectors. To find eigenvectors, we have to proceed as shown in the following example.

EXAMPLE 1.36. Find the eigenvectors of the matrix

$$\mathbf{A} = \begin{bmatrix} -4 & 2 \\ 3 & -5 \end{bmatrix}$$

Solution We know from Example 1.35 that the eigenvalues of the matrix are -7 and -2. Suppose $\mathbf{c}_1$ is the *column eigenvector* associated with the eigenvalue $\lambda_1 = -7$. Then, $\mathbf{c}_1$ must satisfy the equation

$$(\lambda_1 \mathbf{I} - \mathbf{A}) = 0 \qquad \text{(i)}$$

or
$$\begin{vmatrix} \lambda + 4 & -2 \\ -3 & \lambda + 5 \end{vmatrix} \begin{bmatrix} c_1 \\ c_2 \end{bmatrix} = 0 \qquad \text{(ii)}$$

Substituting $\lambda_1 = -7$, we get, from Eq. (ii), the relations

$$\begin{aligned} -3c_1 - 2c_2 &= 0 \\ -3c_1 - 2c_2 &= 0 \end{aligned} \qquad \text{(iii)}$$

Both of the equations in Eq. (iii) give the relation $3c_1 = -2c_2$. Let us put $c_1 = 2$ which gives $c_2 = -3$. Then the eigenvector corresponding to the eigenvalue -7 is

$$\mathbf{c}_1 = \begin{bmatrix} 2 \\ -3 \end{bmatrix} \qquad \text{(iv)}$$

In a similar way, for the eigenvalue -2, we get the equations

$$\begin{aligned} 2c_1 - 2c_2 &= 0 \\ -3c_1 + 3c_2 &= 0 \end{aligned}$$

leading to the relation $c_1 = c_2$. Let us put $c_1 = c_2 = 1$. Then the column eigenvector of $\mathbf{A}$ corresponding to the eigenvalue -2 is

$$\mathbf{c}_2 = \begin{bmatrix} 1 \\ 1 \end{bmatrix} \qquad \text{(v)}$$

Hence, the eigenvector matrix is given by

$$\mathbf{P} = \begin{bmatrix} \mathbf{c}_1 & \mathbf{c}_2 \end{bmatrix} = \begin{bmatrix} 2 & 1 \\ -3 & 1 \end{bmatrix}$$

The eigenvectors that we presented in Eqs. (iv) and (v) are often called *right vectors.*

The matrix $\mathbf{A}$ also possesses *row* or *left* eigenvectors which differ from the column vectors. These are found out as follows: For $\lambda = -7$,

$$\begin{bmatrix} r_1 & r_2 \end{bmatrix} \begin{bmatrix} -3 & -2 \\ -3 & -2 \end{bmatrix} = 0$$

This yields the relation

$$-3r_1 - 3r_2 = 0$$

$\Rightarrow$
$$r_1 = -r_2$$

Hence,
$$\mathbf{r_1} = \begin{bmatrix} 1 & -1 \end{bmatrix} \qquad \text{(vi)}$$

Similarly, for $\lambda = -2$,

$$\begin{bmatrix} r_1 & r_2 \end{bmatrix} \begin{bmatrix} 2 & -2 \\ 3 & 3 \end{bmatrix} = 0$$

$$\Rightarrow \quad 2r_1 - 3r_2 = 0$$

$$\Rightarrow \quad 2r_1 = 3r_2$$

Therefore,

$$\mathbf{r_2} = \begin{bmatrix} 3 & 2 \end{bmatrix} \tag{vii}$$

Useful properties. We list a few useful properties of the eigenvalues and eigenvectors in Table 1.2.

Table 1.2 Useful Properties of Eigenvalues and Eigenvectors

Eigenvalue	*Eigenvector*
1. The product of the eigenvalues is equal to the determinant of the matrix. $\det \mathbf{A} = \prod_i \lambda_i$	1. Only for distinct eigenvalues, the eigenvectors of a real *symmetric matrix* are mutually orthogonal.
2. The sum of the eigenvalues is equal to the trace of the matrix. $\text{Tr}\ \mathbf{A} = \sum_i \lambda_i$	2. The eigenvectors of $\mathbf{A}^{-1}$ are the same as the eigenvectors of $\mathbf{A}$.
3. If the eigenvalues of $\mathbf{A}$ are λ_i, and $\mathbf{A}$ is invertible (i.e. not singular), then the eigenvalues of $\mathbf{A}^{-1}$ are simply λ_i^{-1}.	3. The eigenvectors of $f(\mathbf{A})$ are the same as the eigenvectors of $\mathbf{A}$.
4. If $\mathbf{A}$ is an $n \times n$ *triangular matrix* (upper /lower triangular) or *diagonal matrix*, the eigenvalues of $\mathbf{A}$ are the diagonal entries of $\mathbf{A}$.	
5. $\mathbf{A}$ and its *transpose matrix* have the same eigenvalues.	
6. Eigenvalues of a *symmetric matrix* are all real.	

Normalization of eigenvectors. We may express Eqs. (iv) to (vii) by the following equations:

$$\begin{aligned} \mathbf{Ac_1} &= \lambda_1 \mathbf{c_1} \quad \mathbf{Ac_2} = \lambda_2 \mathbf{c_2} \\ \mathbf{r_1 A} &= \lambda_1 \mathbf{r_1}, \quad \mathbf{r_2 A} = \lambda_2 \mathbf{r_2} \end{aligned} \tag{viii}$$

We may observe that $\mathbf{r}$ or $\mathbf{c}$ can be multiplied by an arbitrary scale factor and Eq. (viii) will still be valid. The eigenvectors are said to be normalized if they are scaled so that $\mathbf{r_1} \cdot \mathbf{c_1} = 1$ and $\mathbf{r_2} \cdot \mathbf{c_2} = 1$. Thus, our normalized eigenvector sets are as follows:

$$\mathbf{r_1} \cdot \mathbf{c_1} = \begin{bmatrix} 1 & -1 \end{bmatrix} \begin{bmatrix} 2 \\ -3 \end{bmatrix} = 2 + 3 = 5$$

Hence,
$$\mathbf{r_1} = \begin{bmatrix} \frac{1}{5} & -\frac{1}{5} \end{bmatrix}, \quad \mathbf{c_1} = \begin{bmatrix} 2 \\ -3 \end{bmatrix} \tag{ix}$$

$$\mathbf{r_2} \cdot \mathbf{c_2} = \begin{bmatrix} 3 & 2 \end{bmatrix} \begin{bmatrix} 1 \\ 1 \end{bmatrix} = 3 + 2 = 5$$

Therefore,
$$\mathbf{r_2} = \begin{bmatrix} 3 & 2 \end{bmatrix}, \quad \mathbf{c_2} = \begin{bmatrix} 1/5 \\ 1/5 \end{bmatrix}$$

Note: 1. The matrix $\mathbf{c_1 r_1}$ is *idempotent*[13], which means any power of this matrix is the matrix itself. Let us check its square.

$$\mathbf{c_1 r_1} = \begin{bmatrix} 2 \\ -3 \end{bmatrix} \begin{bmatrix} \frac{1}{5} & -\frac{1}{5} \end{bmatrix} = \frac{1}{5} \begin{bmatrix} 2 & -2 \\ -3 & 3 \end{bmatrix}$$

$$(\mathbf{c_1 r_1})^2 = \frac{1}{25} \begin{bmatrix} 2 & -2 \\ -3 & 3 \end{bmatrix} \begin{bmatrix} 2 & -2 \\ -3 & 3 \end{bmatrix} = \frac{1}{25} \begin{bmatrix} 4+6 & -4-6 \\ -6-9 & 6+9 \end{bmatrix} = \frac{1}{5} \begin{bmatrix} 2 & -2 \\ -3 & 3 \end{bmatrix}$$

This happens because

$$(\mathbf{c_1 r_1})(\mathbf{c_1 r_1}) = \mathbf{c_1}(\mathbf{r_1} \cdot \mathbf{c_1})\mathbf{r_1} = \mathbf{c_1 r_1}$$

Of course, $\mathbf{c_2 r_2}$ possesses the same property.

2. The matrix $\mathbf{c_1 r_1}$ is orthogonal to the matrix $\mathbf{c_2 r_2}$, i.e.

$$\mathbf{c_2 r_2} = \begin{bmatrix} 1/5 \\ 1/5 \end{bmatrix} \begin{bmatrix} 3 & 2 \end{bmatrix} = \frac{1}{5} \begin{bmatrix} 3 & 2 \\ 3 & 2 \end{bmatrix}$$

$$(\mathbf{c_1 r_1})(\mathbf{c_2 r_2}) = \frac{1}{25} \begin{bmatrix} 2 & -2 \\ -3 & 3 \end{bmatrix} \begin{bmatrix} 3 & 2 \\ 3 & 2 \end{bmatrix} = \frac{1}{25} \begin{bmatrix} 6-6 & 4-4 \\ -9+9 & -6+6 \end{bmatrix} = \frac{1}{25} \begin{bmatrix} 0 & 0 \\ 0 & 0 \end{bmatrix}$$

Now, we will discuss three very useful tools related to matrices:

1. Sylvester's theorem
2. The Cayley-Hamilton theorem
3. Eigendecomposition of matrices

Sylvester's Theorem

Sylvester's[14] theorem states that if k eigenvalues of the square matrix $\mathbf{A}$ are all distinct, and $p(\mathbf{A})$ is a polynomial in $\mathbf{A}$ of the following form:

$$p(\mathbf{A}) = c_0 \mathbf{A}^n + c_1 \mathbf{A}^{n-1} + \cdots + c_{n-1}\mathbf{A} + c_n \mathbf{I}$$

[13]Latin word, meaning *self-power*.

[14]James Joseph Sylvester (1814–1897) was an English mathematician, who made fundamental contributions to matrix theory, invariant theory, number theory, partition theory, and combinatorics.

where $\mathbf{I}$ is the kth order unit matrix and c_n's are constants, then the polynomial can be expressed in the following form:

$$p(\mathbf{A}) = \sum_{i}^{k} p(\lambda_i) Z_0(\lambda_i), \text{ where } Z_0(\lambda_i) = \frac{\mathbf{M}(\lambda_i)}{\Delta'(\lambda_i)} \tag{1.19}$$

In Eq. (1.19), $\Delta(\lambda_i)$ is the characteristic function of $\mathbf{A}$ as defined in Eq. (1.18),

$$\Delta'(\lambda_i) = \left.\frac{d\Delta}{d\lambda}\right|_{\lambda=\lambda_i}$$

and $\mathbf{M}(\lambda)$ is the adjoint matrix of the *characteristic matrix* of $\mathbf{A}$ defined as

$$\mathbf{M}(\lambda) = \Delta(\lambda)[\lambda\mathbf{I} - \mathbf{A}]^{-1}$$

We now demonstrate the usefulness of Sylvester's theorem as follows. Consider the following matrix which we have already defined:

$$\mathbf{A} = \begin{bmatrix} -4 & 2 \\ 3 & -5 \end{bmatrix}$$

As already seen (see Example 1.35 at page 30),

$$\Delta(\lambda) = \lambda^2 + 9\lambda + 14 \tag{i}$$

and the eigenvalues are -7 and -2. Equation (i) yields

$$\Delta'(\lambda) = 2\lambda + 9$$

To apply Sylvester's theorem, let us consider the special polynomial

$$p(\mathbf{A}) = \mathbf{A}^n$$

Then, according to Sylvester's theorem, we have

$$\mathbf{A}^n = \lambda_1^n Z_0(\lambda_1) + \lambda_2^n Z_0(\lambda_2) \tag{ii}$$

The parameters in Eq. (ii) are calculated as follows:

$$[\lambda\mathbf{I} - \mathbf{A}] = \begin{bmatrix} \lambda+4 & -2 \\ -3 & \lambda+5 \end{bmatrix}$$

Its cofactors are $\alpha_{11} = \lambda + 5, \alpha_{12} = 3, \alpha_{21} = 2, \alpha_{22} = \lambda + 4$, and

$$\mathbf{M}(\lambda) = \begin{bmatrix} \lambda+5 & 2 \\ 3 & \lambda+4 \end{bmatrix}$$

$$\Delta'(\lambda_1) = 2(-7) + 9 = -5, \quad \Delta'(\lambda_2) = 2(-2) + 9 = 5$$

$$Z_0(-7) = -\frac{1}{5}\begin{bmatrix} -2 & 2 \\ 3 & -3 \end{bmatrix}, \quad Z_0(-2) = \frac{1}{5}\begin{bmatrix} 3 & 2 \\ 3 & 2 \end{bmatrix}$$

The simplest demonstration is to show that $p(\mathbf{A}) = \mathbf{A}$, and hence

$$\begin{aligned}
\mathbf{A} &= \lambda_1 Z_0(\lambda_1) + \lambda_2 Z_0(\lambda_2) \\
&= \frac{7}{5}\begin{bmatrix} -2 & 2 \\ 3 & -3 \end{bmatrix} - \frac{2}{5}\begin{bmatrix} 3 & 2 \\ 3 & 2 \end{bmatrix} \\
&= \frac{1}{5}\begin{bmatrix} -14 & 14 \\ 21 & -21 \end{bmatrix} + \frac{1}{5}\begin{bmatrix} -6 & -4 \\ -6 & -4 \end{bmatrix} \\
&= \frac{1}{5}\begin{bmatrix} -20 & 10 \\ 15 & -25 \end{bmatrix} = \begin{bmatrix} -4 & 2 \\ 3 & -5 \end{bmatrix}
\end{aligned}$$

Consider $p(\mathbf{A}) = \mathbf{A}^2$. Then,

$$\begin{aligned}
\mathbf{A}^2 &= \lambda_1^2 Z_0(\lambda_1) + \lambda_2^2 Z_0(\lambda_2) \\
&= -\frac{7^2}{5}\begin{bmatrix} -2 & 2 \\ 3 & -3 \end{bmatrix} + \frac{2^2}{5}\begin{bmatrix} 3 & 2 \\ 3 & 2 \end{bmatrix} \\
&= \frac{1}{5}\begin{bmatrix} 98 & -98 \\ -147 & 147 \end{bmatrix} + \frac{1}{5}\begin{bmatrix} 12 & 8 \\ 12 & 8 \end{bmatrix} \\
&= \frac{1}{5}\begin{bmatrix} 110 & -90 \\ -135 & 155 \end{bmatrix} \\
&= \begin{bmatrix} 22 & -18 \\ -27 & 31 \end{bmatrix}
\end{aligned}$$

The result can be verified by simple multiplication:

$$\begin{aligned}
\begin{bmatrix} -4 & 2 \\ 3 & -5 \end{bmatrix}\begin{bmatrix} -4 & 2 \\ 3 & -5 \end{bmatrix} &= \begin{bmatrix} 16+6 & -8-10 \\ -12-15 & 6+25 \end{bmatrix} \\
&= \begin{bmatrix} 22 & -18 \\ -27 & 31 \end{bmatrix}
\end{aligned}$$

Suppose $p(\mathbf{A}) = \mathbf{A}^{-1}$. Then,

$$\begin{aligned}
\mathbf{A}^{-1} &= \lambda_1^{-1} Z_0(\lambda_1) + \lambda_2^{-1} Z_0(\lambda_2) \\
&= \frac{1}{(7)(5)}\begin{bmatrix} -2 & 2 \\ 3 & -3 \end{bmatrix} - \frac{1}{(2)(5)}\begin{bmatrix} 3 & 2 \\ 3 & 2 \end{bmatrix} \\
&= \begin{bmatrix} -\dfrac{2}{35} - \dfrac{2}{10} & \dfrac{2}{35} - \dfrac{2}{10} \\ \dfrac{3}{35} - \dfrac{3}{10} & -\dfrac{3}{35} - \dfrac{2}{10} \end{bmatrix} \\
&= -\frac{1}{14}\begin{bmatrix} 5 & 2 \\ 3 & 4 \end{bmatrix}
\end{aligned}$$

Let us verify it by the direct method as follows:

$$\det \mathbf{A} = 20 - 6 = 14$$

The cofactors: $\alpha_{11} = -5, \alpha_{12} = -3, \alpha_{21} = -2, \alpha_{22} = -4$, and

$$\text{Adj}\,(\mathbf{A}) = \begin{bmatrix} -5 & -2 \\ -3 & -4 \end{bmatrix}$$

Therefore

$$\mathbf{A}^{-1} = -\frac{1}{14}\begin{bmatrix} 5 & 2 \\ 3 & 4 \end{bmatrix}$$

The Cayley–Hamilton Theorem

The Cayley[15]–Hamilton[16] theorem states that any square matrix satisfies its own characteristic equation. This means that if $\mathbf{A}$ has the characteristic polynomial $p(\lambda)$ then $p(\mathbf{A}) = 0$.

We can verify the theorem from an example. Consider the matrix

$$\mathbf{A} = \begin{bmatrix} -4 & 2 \\ 3 & -5 \end{bmatrix}$$

Its characteristic polynomial and square are given by

$$\lambda^2 + 9\lambda + 14 = 0$$

$$\mathbf{A}^2 = \begin{bmatrix} 22 & -18 \\ -27 & 31 \end{bmatrix}$$

Therefore,

$$\begin{aligned} p(\mathbf{A}) &= \mathbf{A}^2 + 9\mathbf{A} + 14\mathbf{I} \\ &= \begin{bmatrix} 22 & -18 \\ -27 & 31 \end{bmatrix} + 9\begin{bmatrix} -4 & 2 \\ 3 & -5 \end{bmatrix} + 14\begin{bmatrix} 1 & 0 \\ 0 & 1 \end{bmatrix} \\ &= \begin{bmatrix} 22 - 36 + 14 & -18 + 18 + 0 \\ -27 + 27 + 0 & 31 - 45 + 14 \end{bmatrix} \\ &= \begin{bmatrix} 0 & 0 \\ 0 & 0 \end{bmatrix} \qquad \text{Verified.} \end{aligned}$$

The characteristic polynomial of an $n \times n$ polynomial is of degree n. The Cayley–Hamilton theorem provides a way to express $\mathbf{A}^k$, where k is an integer, in terms of a polynomial in $\mathbf{A}$ of degree less than n. The method will be clear from the following example.

EXAMPLE 1.37. Find $\mathbf{A}^5$ using the Cayley–Hamilton theorem and verify the result from Sylvester's theorem. The matrix $\mathbf{A}$ is the same as defined in Example 1.35.

[15]Arthur Cayley (1821–1895) was a British mathematician.

[16]William Rowan Hamilton (1805–1865) was an Irish physicist, astronomer, and mathematician, who made important contributions to classical mechanics, optics, and algebra.

Solution From the equation

$$\mathbf{A}^2 + 9\mathbf{A} + 14\mathbf{I} = 0$$

we have

$$\mathbf{A}^2 = -9\mathbf{A} - 14\mathbf{I}$$

Now,

$$\begin{aligned}
\mathbf{A}^5 &= \mathbf{A}^2 \cdot \mathbf{A}^2 \cdot \mathbf{A} \\
&= (-9\mathbf{A} - 14\mathbf{I}) \cdot (-9\mathbf{A} - 14\mathbf{I}) \cdot \mathbf{A} \\
&= \left(81\mathbf{A}^2 + 252\mathbf{A} + 196\mathbf{I}\right) \cdot \mathbf{A} \\
&= \left[81(-9\mathbf{A} - 14\mathbf{I}) + 252\mathbf{A} + 196\mathbf{I}\right] \cdot \mathbf{A} \\
&= (-477\mathbf{A} - 938\mathbf{I}) \cdot \mathbf{A} \\
&= -477\mathbf{A}^2 - 938\mathbf{A} \\
&= -477(-9\mathbf{A} - 14\mathbf{I}) - 938\mathbf{A} \\
&= (4293\mathbf{A} - 938\mathbf{A}) + 6678\mathbf{I} \\
&= 3355\mathbf{A} + 6678\mathbf{I}
\end{aligned}$$

Therefore,

$$\begin{aligned}
\mathbf{A}^5 &= 3355\begin{bmatrix} -4 & 2 \\ 3 & -5 \end{bmatrix} + 6678\begin{bmatrix} 1 & 0 \\ 0 & 1 \end{bmatrix} \\
&= \begin{bmatrix} -13420 & 6710 \\ 10065 & -16775 \end{bmatrix} + \begin{bmatrix} 6678 & 0 \\ 0 & 6678 \end{bmatrix} \\
&= \begin{bmatrix} -6742 & 6710 \\ 10065 & -10097 \end{bmatrix}
\end{aligned}$$

Let us check the result with the help of Sylvester's theorem as follows:

$$\begin{aligned}
\mathbf{A}^5 &= \frac{(-7)^5}{5}\begin{bmatrix} 2 & -2 \\ -3 & 3 \end{bmatrix} + \frac{(-2)^5}{5}\begin{bmatrix} 3 & 2 \\ 3 & 2 \end{bmatrix} \\
&= -\frac{16807}{5}\begin{bmatrix} 2 & -2 \\ -3 & 3 \end{bmatrix} - \frac{32}{5}\begin{bmatrix} 3 & 2 \\ 3 & 2 \end{bmatrix} \\
&= \frac{1}{5}\begin{bmatrix} -33710 & 33550 \\ 50325 & -50485 \end{bmatrix} \\
&= \begin{bmatrix} -6742 & 6710 \\ 10065 & -10097 \end{bmatrix}
\end{aligned}$$

EXAMPLE 1.38. Find the inverse of the matrix $\mathbf{A}$ through the Cayley–Hamilton theorem.

Solution From the equation for $\mathbf{A}^2$ of the previous example, we have

$$\mathbf{A}^2 + 9\mathbf{A} = -14\mathbf{I}$$

$$\Rightarrow \qquad \mathbf{A}(\mathbf{A} + 9\mathbf{I}) = -14\mathbf{I}$$

$$\Rightarrow \qquad \mathbf{A} \cdot \left[-\frac{1}{14}(\mathbf{A} + 9\mathbf{I}) \right] = \mathbf{I}$$

$$\Rightarrow \qquad \mathbf{A}^{-1} = -\frac{1}{14}(\mathbf{A} + 9\mathbf{I})$$

Now,
$$\mathbf{A} + 9\mathbf{I} = \begin{bmatrix} -4 & 2 \\ 3 & -5 \end{bmatrix} + 9\begin{bmatrix} 1 & 0 \\ 0 & 1 \end{bmatrix} = \begin{bmatrix} 5 & 2 \\ 3 & 4 \end{bmatrix}$$

Therefore,
$$\mathbf{A}^{-1} = -\frac{1}{14}\begin{bmatrix} 5 & 2 \\ 3 & 4 \end{bmatrix}$$

Eigendecomposition of a Matrix

If $\mathbf{A}$ is an $n \times n$ matrix with n linearly independent column vectors c_i $(i = 1 \ldots n)$, then $\mathbf{A}$ can be factorized as

$$\mathbf{A} = \mathbf{P}\boldsymbol{\Lambda}\mathbf{P}^{-1}$$

where

$\mathbf{P}$ is an $n \times n$ matrix whose ith column is the column vector c_i of $\mathbf{A}$, and
$\boldsymbol{\Lambda}$ is the diagonal matrix whose diagonal elements are the corresponding eigenvalues, i.e.,

$$\boldsymbol{\Lambda}_{ii} = \lambda_i.$$

Note: (a) Only diagonalizable matrices can be factorized in this way.

(b) If n_v is the total number of linearly independent eigenvectors, $\mathbf{A}$ can be eigendecomposed *iff*[17] $n_v = n$. This implies that the characteristic polynomial $p(\lambda)$ has no repeated roots.

(c) If it is said that "**A** has an inverse", it does *not* imply that **A** can be eigendecomposed.

(d) If it is said that "**A** can be eigendecomposed", it does *not* imply that **A** has an inverse.

Practical implications of eigendecomposition. 1. If a matrix $\mathbf{A}$ can be eigendecomposed and its inverse exists, then the inverse is given by

$$\mathbf{A}^{-1} = \mathbf{P}\boldsymbol{\Lambda}^{-1}\mathbf{P}^{-1} \tag{1.20}$$

Since $\boldsymbol{\Lambda}$ is a diagonal matrix, its inverse is given by

$$\left[\boldsymbol{\Lambda}^{-1}\right]_{ii} = \frac{1}{\lambda_i}$$

which is easy to calculate.

2. From Eq. (1.20), we observe that

$$f(\mathbf{A}) = \mathbf{P}f(\boldsymbol{\Lambda})\mathbf{P}^{-1} \tag{1.21}$$

[17] if and only if.

The validity of Eq. (1.21) can be verified as follows:

$$\begin{aligned}\mathbf{A}^2 &= (\mathbf{P}\boldsymbol{\Lambda}\mathbf{P}^{-1})(\mathbf{P}\boldsymbol{\Lambda}\mathbf{P}^{-1})\\ &= \mathbf{P}\boldsymbol{\Lambda}(\mathbf{P}^{-1}\mathbf{P})\boldsymbol{\Lambda}\mathbf{P}^{-1}\\ &= \mathbf{P}\boldsymbol{\Lambda}^2\mathbf{P}^{-1}\end{aligned} \tag{1.22}$$

Equation (1.22) can be generalized to give

$$\mathbf{A}^n = \mathbf{P}\boldsymbol{\Lambda}^n\mathbf{P}^{-1}$$

EXAMPLE 1.39. Find $\mathbf{A}^{-1}$ and $\mathbf{A}^5$ by the method of eigendecomposition when

$$\mathbf{A} = \begin{bmatrix} -4 & 2 \\ 3 & -5 \end{bmatrix}$$

Solution We have seen in Example 1.36 at page 30 that for the matrix $\mathbf{A}$ the linearly independent column vectors $\mathbf{c}_i$ $(i = 1,\ 2)$ are

$$\mathbf{c}_1 = \begin{bmatrix} 2 \\ -3 \end{bmatrix}, \quad \mathbf{c}_2 = \begin{bmatrix} 1 \\ 1 \end{bmatrix}$$

Therefore, the matrix $\mathbf{P}$ is given by

$$\mathbf{P} = \begin{bmatrix} 2 & 1 \\ -3 & 1 \end{bmatrix}$$

Its inverse is calculated as follows:

$$\det \mathbf{P} = 2 + 3 = 5$$

$$\text{Cofactors: } \alpha_{11} = 1,\ \alpha_{12} = 3,\ \alpha_{21} = -1,\ \alpha_{22} = 2$$

$$\text{Adj}\,(\mathbf{P}) = \begin{bmatrix} 1 & -1 \\ 3 & 2 \end{bmatrix}$$

$$\mathbf{P}^{-1} = \frac{1}{5}\begin{bmatrix} 1 & -1 \\ 3 & 2 \end{bmatrix}$$

The eigenvalues of $\mathbf{A}$ are -7 and -2. Hence,

$$\boldsymbol{\Lambda} = \begin{bmatrix} -7 & 0 \\ 0 & -2 \end{bmatrix}$$

$$\mathbf{A} = \mathbf{P}\begin{bmatrix} -7 & 0 \\ 0 & -2 \end{bmatrix}\mathbf{P}^{-1}$$

Thus,

$$\mathbf{A}^n = \mathbf{P}\begin{bmatrix} (-7)^n & 0 \\ 0 & (-2)^n \end{bmatrix}\mathbf{P}^{-1} \tag{i}$$

Let us calculate the values of $\mathbf{A}^{-1}$ and $\mathbf{A}^5$ with the help of Eq. (i) as follows:

$$\mathbf{A}^{-1} = \frac{1}{5}\begin{bmatrix} 2 & 1 \\ -3 & 1 \end{bmatrix}\begin{bmatrix} -\frac{1}{7} & 0 \\ 0 & -\frac{1}{2} \end{bmatrix}\begin{bmatrix} 1 & -1 \\ 3 & 2 \end{bmatrix}$$

$$= \frac{1}{5}\begin{bmatrix} 2 & 1 \\ -3 & 1 \end{bmatrix}\begin{bmatrix} -\frac{1}{7} & \frac{1}{7} \\ -\frac{3}{2} & -1 \end{bmatrix}$$

$$= \frac{1}{5}\begin{bmatrix} -\frac{2}{7} - \frac{3}{2} & \frac{2}{7} - 1 \\ \frac{3}{7} - \frac{3}{2} & -\frac{3}{7} - 1 \end{bmatrix}$$

$$= -\frac{1}{14}\begin{bmatrix} 5 & 2 \\ 3 & 4 \end{bmatrix}$$

$$\mathbf{A}^5 = \frac{1}{5}\begin{bmatrix} 2 & 1 \\ -3 & 1 \end{bmatrix}\begin{bmatrix} (-7)^5 & 0 \\ 0 & (-2)^5 \end{bmatrix}\begin{bmatrix} 1 & -1 \\ 3 & 2 \end{bmatrix}$$

$$= \frac{1}{5}\begin{bmatrix} 2 & 1 \\ -3 & 1 \end{bmatrix}\begin{bmatrix} (-7)^5 & -1(-7)^5 \\ 3(-2)^5 & 2(-2)^5 \end{bmatrix}$$

$$= \frac{1}{5}\begin{bmatrix} 2(-7)^5 + 3(-2)^5 & -2(-7)^5 + 2(-2)^5 \\ 3(-7)^5 + 3(-2)^5 & 3(-7)^5 + 2(-2)^5 \end{bmatrix}$$

$$= \begin{bmatrix} -6742 & 6710 \\ 10065 & -10097 \end{bmatrix}$$

EXAMPLE 1.40. (a) Find the eigenvalues and eigenvectors of the matrix

$$\mathbf{A} = \begin{bmatrix} 3 & -1 \\ -1 & 3 \end{bmatrix}$$

(b) Find $\mathbf{A}^{100}$ for the value given in part (a).

Solution (a) The characteristic equation corresponding to $\mathbf{A}$ is

$$\det(\lambda\mathbf{I} - \mathbf{A}) = 0$$

$$\Rightarrow \qquad \begin{vmatrix} \lambda - 3 & 1 \\ 1 & \lambda - 3 \end{vmatrix} = 0$$

$$\Rightarrow \qquad (\lambda - 3)^2 - 1 = 0 \qquad \text{(i)}$$

Equation (i) yields

$$\lambda - 3 = \pm 1$$

$$\Rightarrow \qquad \lambda = 2,\ 4$$

So, the eigenvalues are 2 and 4. The column vector associated with eigenvalue 2 is found from the equation

$$\begin{bmatrix} \lambda - 3 & 1 \\ 1 & \lambda - 3 \end{bmatrix} \begin{bmatrix} x_{11} \\ x_{21} \end{bmatrix} = 0$$

When $\lambda = 2$, the above equation yields

$$\begin{bmatrix} -1 & 1 \\ 1 & -1 \end{bmatrix} \begin{bmatrix} x_{11} \\ x_{21} \end{bmatrix} = 0$$

$\Rightarrow$

$$x_{11} = x_{21}$$

So, let us put $x_{11} = 1$ and $x_{21} = 1$ to get the eigenvector as

$$\mathbf{x}_1 = \begin{bmatrix} 1 \\ 1 \end{bmatrix}$$

For the column vector associated with the eigenvalue 4, we have

$$\begin{bmatrix} 1 & 1 \\ 1 & 1 \end{bmatrix} \begin{bmatrix} x_{21} \\ x_{22} \end{bmatrix} = 0$$

$\Rightarrow$

$$x_{21} = -x_{22}$$

So, the eigenvector may be written as

$$\mathbf{x}_2 = \begin{bmatrix} 1 \\ -1 \end{bmatrix}$$

(b) Here,

$$\mathbf{P} = \begin{bmatrix} 1 & 1 \\ 1 & -1 \end{bmatrix}$$

$$\det \mathbf{P} = -1 - 1 = -2$$

$$\text{Cofactors: } \alpha_{11} = -1,\ \alpha_{12} = -1,\ \alpha_{21} = -1,\ \alpha_{22} = 1$$

$$\text{Adj}\,(\mathbf{P}) = \begin{bmatrix} -1 & -1 \\ -1 & 1 \end{bmatrix}$$

$$\mathbf{P}^{-1} = \frac{1}{2} \begin{bmatrix} 1 & 1 \\ 1 & -1 \end{bmatrix}$$

From eigendecomposition,

$$\begin{aligned}
\mathbf{A}^{100} &= \frac{1}{2} \begin{bmatrix} 1 & 1 \\ 1 & -1 \end{bmatrix} \begin{bmatrix} 2^{100} & 0 \\ 0 & 4^{100} \end{bmatrix} \begin{bmatrix} 1 & 1 \\ 1 & -1 \end{bmatrix} \\
&= \frac{1}{2} \begin{bmatrix} 1 & 1 \\ -1 & 1 \end{bmatrix} \begin{bmatrix} 2^{100} & 2^{100} \\ 4^{100} & -4^{100} \end{bmatrix} \\
&= \frac{1}{2} \begin{bmatrix} 2^{100} + 4^{100} & 2^{100} - 4^{100} \\ 2^{100} - 4^{100} & 2^{100} + 4^{100} \end{bmatrix} \\
&= 4^{100} \begin{bmatrix} 1 & -1 \\ -1 & 1 \end{bmatrix} \quad \text{since other terms} \ll 4^{100}
\end{aligned}$$

Alternative method. From Sylvester's theorem,

$$\lambda \mathbf{I} - \mathbf{A} = \begin{bmatrix} \lambda - 3 & 1 \\ 1 & \lambda - 3 \end{bmatrix}$$

$$\text{Cofactors: } \alpha_{11} = \lambda - 3,\ \alpha_{12} = -1,\ \alpha_{21} = -1,\ \alpha_{22} = \lambda - 3$$

$$\mathbf{M} = \begin{bmatrix} \lambda - 3 & -1 \\ -1 & \lambda - 3 \end{bmatrix}$$

$$\Delta(\lambda) = \lambda^2 - 6\lambda + 8$$

$$\Delta' = 2\lambda - 6$$

$$Z_0(2) = \frac{1}{2}\begin{bmatrix} 1 & 1 \\ 1 & 1 \end{bmatrix}, \quad Z_0(4) = \frac{1}{2}\begin{bmatrix} 1 & -1 \\ -1 & 1 \end{bmatrix}$$

$$\mathbf{A}^{100} = \frac{(2)^{100}}{2}\begin{bmatrix} 1 & 1 \\ 1 & 1 \end{bmatrix} + \frac{(4)^{100}}{2}\begin{bmatrix} 1 & -1 \\ -1 & 1 \end{bmatrix} \approx 4^{100}\begin{bmatrix} 1 & -1 \\ -1 & 1 \end{bmatrix}$$

EXAMPLE 1.41. Find the rank of the matrix

$$\mathbf{A} = \begin{bmatrix} 1 & 2 & 3 \\ 3 & 4 & 5 \\ 4 & 6 & 8 \end{bmatrix}$$

Solution Here,

$$\det \mathbf{A} = 1(32 - 30) + 2(20 - 24) + 3(18 - 16) = 0$$

So, $\mathbf{A}$ is singular. But the determinant of the submatrix

$$\begin{bmatrix} 1 & 2 \\ 3 & 4 \end{bmatrix} = 4 - 6 = -2$$

Therefore, the submatrix of order 2×2 is not singular. Hence, the rank of the matrix $\mathbf{A}$ is 2.

EXAMPLE 1.42. Let $\mathbf{A} = [a_{ij}]$, $1 \le i,\ j \le n$, with $n \le 3$ and $a_{ij} = i \cdot j$. The rank of $\mathbf{A}$ is

(a) 0 (b) 1 (c) $n - 1$ (d) n

Solution Evidently, the matrix is given by

$$\mathbf{A} = \begin{bmatrix} a_{11} & a_{12} & a_{13} \\ a_{21} & a_{22} & a_{23} \\ a_{31} & a_{32} & a_{33} \end{bmatrix} = \begin{bmatrix} 1 \times 1 & 1 \times 2 & 1 \times 3 \\ 2 \times 1 & 2 \times 2 & 2 \times 3 \\ 3 \times 1 & 3 \times 2 & 3 \times 3 \end{bmatrix} = \begin{bmatrix} 1 & 2 & 3 \\ 2 & 4 & 6 \\ 3 & 6 & 9 \end{bmatrix}$$

It is easy to find that $\det \mathbf{A} = 0$. Also, $\begin{vmatrix} 1 & 2 \\ 2 & 4 \end{vmatrix} = 0$.

Ans: (a).

EXAMPLE 1.43. For a given 2×2 matrix $\mathbf{A}$, it is observed that

$$\mathbf{A}\begin{bmatrix} 1 \\ -1 \end{bmatrix} = -\begin{bmatrix} 1 \\ -1 \end{bmatrix}, \quad \mathbf{A}\begin{bmatrix} 1 \\ -2 \end{bmatrix} = -2\begin{bmatrix} 1 \\ -2 \end{bmatrix}$$

Then the matrix $\mathbf{A}$ is

(a) $\mathbf{A} = \begin{bmatrix} 2 & 1 \\ -1 & -1 \end{bmatrix}\begin{bmatrix} -1 & 0 \\ 0 & -2 \end{bmatrix}\begin{bmatrix} 1 & 1 \\ -1 & -2 \end{bmatrix}$

(b) $\mathbf{A} = \begin{bmatrix} 1 & 1 \\ -1 & -2 \end{bmatrix}\begin{bmatrix} 1 & 0 \\ 0 & 2 \end{bmatrix}\begin{bmatrix} 2 & 1 \\ -1 & -1 \end{bmatrix}$

(c) $\mathbf{A} = \begin{bmatrix} 1 & 1 \\ -1 & -2 \end{bmatrix}\begin{bmatrix} -1 & 0 \\ 0 & -2 \end{bmatrix}\begin{bmatrix} 2 & 1 \\ -1 & -1 \end{bmatrix}$

(d) $\mathbf{A} = \begin{bmatrix} 0 & -2 \\ 1 & -3 \end{bmatrix}$

Solution From the given conditions, we find that the eigenvalues of $\mathbf{A}$ are -1 and -2 for the eigenvectors $\left[\begin{smallmatrix} 1 \\ -1 \end{smallmatrix}\right]$ and $\left[\begin{smallmatrix} 1 \\ -2 \end{smallmatrix}\right]$, respectively.

Now,

$$\mathbf{\Lambda} = \begin{bmatrix} -1 & 0 \\ 0 & -2 \end{bmatrix}$$

$$\mathbf{P} = \begin{bmatrix} 1 & 1 \\ -1 & -2 \end{bmatrix}, \qquad \det \mathbf{P} = -2 + 1 = -1$$

The cofactors: $\alpha_{11} = -2, \alpha_{12} = 1, \alpha_{21} = -1$, and $\alpha_{22} = 1$, and

$$\text{Adj}\,(\mathbf{P}) = \begin{bmatrix} -2 & -1 \\ 1 & 1 \end{bmatrix}, \quad \mathbf{P}^{-1} = \begin{bmatrix} 2 & 1 \\ -1 & -1 \end{bmatrix}$$

$$\mathbf{A} = \mathbf{P\Lambda P}^{-1} = \begin{bmatrix} 1 & 1 \\ -1 & -2 \end{bmatrix}\begin{bmatrix} -1 & 0 \\ 0 & -2 \end{bmatrix}\begin{bmatrix} 2 & 1 \\ -1 & -1 \end{bmatrix}$$

Ans: (c).

EXAMPLE 1.44. An eigenvector of

$$\begin{bmatrix} 1 & 1 & 0 \\ 0 & 2 & 2 \\ 0 & 0 & 3 \end{bmatrix}$$

is

(a) $[-1 \quad 1 \quad 1]^T$ (b) $[1 \quad 2 \quad 1]^T$ (c) $[1 \quad -1 \quad 2]^T$ (d) $[2 \quad 1 \quad -1]^T$

Solution Since the given matrix is upper triangular, the eigenvalues of the matrix are 1, 2 and 3. Now,

(a) $\begin{bmatrix} 1 & 1 & 0 \\ 0 & 2 & 2 \\ 0 & 0 & 3 \end{bmatrix}\begin{bmatrix} -1 \\ 1 \\ 1 \end{bmatrix} = \begin{bmatrix} 0 \\ 4 \\ 6 \end{bmatrix}$

(b) $\begin{bmatrix} 1 & 1 & 0 \\ 0 & 2 & 2 \\ 0 & 0 & 3 \end{bmatrix} \begin{bmatrix} 1 \\ 2 \\ 1 \end{bmatrix} = 3 \begin{bmatrix} 1 \\ 2 \\ 1 \end{bmatrix}$

(c) $\begin{bmatrix} 1 & 1 & 0 \\ 0 & 2 & 2 \\ 0 & 0 & 3 \end{bmatrix} \begin{bmatrix} 1 \\ -1 \\ 2 \end{bmatrix} = \begin{bmatrix} 0 \\ 2 \\ 6 \end{bmatrix}$

(d) $\begin{bmatrix} 1 & 1 & 0 \\ 0 & 2 & 2 \\ 0 & 0 & 3 \end{bmatrix} \begin{bmatrix} 2 \\ 1 \\ -1 \end{bmatrix} = \begin{bmatrix} 3 \\ 0 \\ 6 \end{bmatrix}$

Of the four choices, (b) satisfies the condition $\mathbf{Ax} = \lambda\mathbf{x}$. Therefore, the correct answer is (b).

EXAMPLE 1.45. The trace and determinant of a 2 × 2 matrix are known to be −2 and −35, respectively. Its eigenvalues are

(a) −30 and −5 (b) −37 and −1 (c) −7 and 5 (d) 17.5 and −2

Solution If λ_1 and λ_2 are the eigenvalues of the matrix, we have

$$\text{Tr}\,\mathbf{M} \equiv \lambda_1 + \lambda_2 = -2 \tag{i}$$

$$\det\,\mathbf{M} \equiv \lambda_1 \times \lambda_2 = -35 \tag{ii}$$

From Eqs. (i) and (ii),

$$\lambda_1 - \lambda_2 = \sqrt{(\lambda_1 + \lambda_2)^2 - 4\lambda_1\lambda_2}$$
$$= \sqrt{4 + 4 \times 35} = 12 \tag{iii}$$

From Eqs. (i) and (iii),

$$\lambda_1 + \lambda_2 = -2, \quad \lambda_1 - \lambda_2 = 12$$

It is easy to find that $\lambda_1 = 5$ and $\lambda_2 = -7$.
Ans: (c).

EXAMPLE 1.46. A real $n \times n$ matrix $\mathbf{A} = [a_{ij}]$ is defined as

$$a_{ij} = \begin{cases} i & \text{for } i = j \\ 0 & \text{otherwise} \end{cases}$$

The summation of all n eigenvalues of $\mathbf{A}$ is

(a) $\dfrac{n(n+1)}{2}$ (b) $\dfrac{n(n-1)}{2}$

(c) $\dfrac{n(n+1)(2n+1)}{6}$ (d) n^2

Solution Evidently, $\mathbf{A}$ is a diagonal matrix with elements 1, 2, …, n. Therefore, its eigenvalues are also 1, 2, …, n, their sum being $n(n+1)/2$.
Ans: (a).

EXAMPLE 1.47. A system of linear simultaneous equations is given as $\mathbf{Ax} = \mathbf{B}$, where

$$\mathbf{A} = \begin{bmatrix} 1 & 0 & 1 & 0 \\ 0 & 1 & 0 & 1 \\ 1 & 1 & 0 & 0 \\ 0 & 0 & 0 & 1 \end{bmatrix}, \quad \mathbf{B} = \begin{bmatrix} 0 \\ 0 \\ 0 \\ 1 \end{bmatrix}$$

(i) The rank of the matrix **A** is

(a) 1 (b) 2 (c) 3 (d) 4

(ii) Which of the following statements is true?

(a) **x** is a null vector (b) **x** is unique

(c) **x** does not exist (d) **x** has infinitely many values

Solution (i)
$$\det \mathbf{A} = 1 \times \begin{bmatrix} 1 & 0 & 1 \\ 1 & 0 & 0 \\ 0 & 0 & 1 \end{bmatrix} + 0 \times \begin{bmatrix} 0 & 1 & 0 \\ 0 & 0 & 1 \\ 0 & 1 & 0 \end{bmatrix} + 1 \times \begin{bmatrix} 1 & 0 & 1 \\ 0 & 1 & 1 \\ 1 & 0 & 0 \end{bmatrix} + 0 \times \begin{bmatrix} 0 & 1 & 0 \\ 1 & 1 & 0 \\ 0 & 0 & 0 \end{bmatrix}$$
$$= 0 - 1 = -1$$

Therefore, the rank of the matrix is 4.
Ans: (c).

(ii) $\mathbf{x} = \mathbf{A}^{-1}\mathbf{B}$. The cofactors of **A** are:

$$\alpha_{11} = \begin{vmatrix} 1 & 0 & 1 \\ 1 & 0 & 0 \\ 0 & 0 & 1 \end{vmatrix} = 0, \quad \alpha_{12} = -\begin{vmatrix} 0 & 0 & 1 \\ 1 & 0 & 0 \\ 0 & 0 & 1 \end{vmatrix} = 0, \quad \alpha_{13} = \begin{vmatrix} 0 & 1 & 1 \\ 1 & 1 & 0 \\ 0 & 0 & 1 \end{vmatrix} = -1, \quad \alpha_{14} = -\begin{vmatrix} 0 & 1 & 0 \\ 1 & 1 & 0 \\ 0 & 0 & 0 \end{vmatrix} = 0$$

$$\alpha_{21} = -\begin{vmatrix} 0 & 1 & 0 \\ 1 & 0 & 0 \\ 0 & 0 & 1 \end{vmatrix} = 1, \quad \alpha_{22} = \begin{vmatrix} 1 & 1 & 0 \\ 1 & 0 & 0 \\ 0 & 0 & 1 \end{vmatrix} = -1, \quad \alpha_{23} = -\begin{vmatrix} 1 & 0 & 0 \\ 1 & 0 & 0 \\ 0 & 0 & 1 \end{vmatrix} = 0, \quad \alpha_{24} = \begin{vmatrix} 1 & 0 & 1 \\ 1 & 1 & 0 \\ 0 & 0 & 0 \end{vmatrix} = 0$$

$$\alpha_{31} = \begin{vmatrix} 0 & 1 & 0 \\ 1 & 0 & 1 \\ 0 & 0 & 1 \end{vmatrix} = -1, \quad \alpha_{32} = -\begin{vmatrix} 1 & 1 & 0 \\ 0 & 0 & 1 \\ 0 & 0 & 1 \end{vmatrix} = 0, \quad \alpha_{33} = \begin{vmatrix} 1 & 0 & 0 \\ 0 & 1 & 1 \\ 0 & 0 & 1 \end{vmatrix} = 1, \quad \alpha_{34} = -\begin{vmatrix} 1 & 0 & 1 \\ 0 & 1 & 0 \\ 0 & 0 & 0 \end{vmatrix} = 0$$

$$\alpha_{41} = -\begin{vmatrix} 0 & 1 & 0 \\ 1 & 0 & 1 \\ 1 & 0 & 0 \end{vmatrix} = -1, \quad \alpha_{42} = \begin{vmatrix} 1 & 1 & 0 \\ 0 & 0 & 1 \\ 1 & 0 & 0 \end{vmatrix} = 1, \quad \alpha_{43} = -\begin{vmatrix} 1 & 0 & 0 \\ 0 & 1 & 1 \\ 1 & 1 & 0 \end{vmatrix} = 1, \quad \alpha_{44} = \begin{vmatrix} 1 & 0 & 1 \\ 0 & 1 & 0 \\ 1 & 1 & 0 \end{vmatrix} = -1$$

Therefore,

$$\text{Adj}\,(\mathbf{A}) = \begin{bmatrix} 0 & 1 & -1 & -1 \\ 0 & -1 & 0 & 1 \\ -1 & 0 & 1 & 1 \\ 0 & 0 & 0 & -1 \end{bmatrix}, \quad \mathbf{A}^{-1} = \begin{bmatrix} 0 & -1 & 1 & 1 \\ 0 & 1 & 0 & -1 \\ 1 & 0 & -1 & -1 \\ 0 & 0 & 0 & 1 \end{bmatrix}$$

Thus,

$$\mathbf{x} = \mathbf{A}^{-1}\mathbf{B} = \begin{bmatrix} 0 & -1 & 1 & 1 \\ 0 & 1 & 0 & -1 \\ 1 & 0 & -1 & -1 \\ 0 & 0 & 0 & 1 \end{bmatrix} \begin{bmatrix} 0 \\ 0 \\ 0 \\ 1 \end{bmatrix} = \begin{bmatrix} 1 \\ -1 \\ -1 \\ -1 \end{bmatrix}$$

which is unique.
Ans: (b).

1.4 DIFFERENCE EQUATIONS

We know that a continuous-time dynamic system is represented by a differential equation. But, if it is a discrete-time dynamic system, then a *difference equation* is necessary to represent it.

A difference equation may be constructed as an approximation to samples drawn at regular intervals from an analogue system described by a differential equation. Let us consider an example.

EXAMPLE 1.48. Construct a difference equation for the circuit given in Figure 1.5.

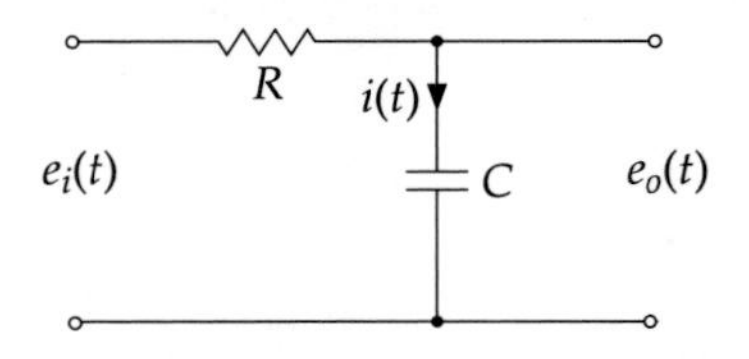

Figure 1.5 Circuit (Example 1.48).

Solution From the KVL,

$$e_i(t) = Ri(t) + \frac{\int i(t)\,dt}{C} \tag{i}$$

$$e_o(t) = \frac{\int i(t)\,dt}{C} \tag{ii}$$

From Eq. (ii),

$$\frac{de_o(t)}{dt} = \frac{i(t)}{C} \tag{iii}$$

Substituting the value of $i(t)$ from Eq. (iii) in Eq. (i), we get

$$e_i(t) = RC\frac{de_o(t)}{dt} + e_o(t) \tag{iv}$$

We can write Eq. (iv) in the difference equation form as follows:

$$e_i(nT) = RC\frac{e_o[(n+1)T] - e_o(nT)}{T} + e_o(nT)$$

where T is the sampling interval and $n = 0,\ 1,\ 2, \cdots$.

Substituting $nT \equiv n$ and $(n+1)T \equiv (n+1)$, since this is a sequence, we have

$$\frac{T}{RC}e_i(n) = e_o(n+1) - e_o(n) + \frac{T}{RC}e_o(n)$$

$$\Rightarrow \qquad e_o(n+1) = \left(1 - \frac{T}{RC}\right)e_o(n) + \frac{T}{RC}e_i(n)$$

$$\Rightarrow \qquad e_o(n+1) = ae_o(n) + (1-a)e_i(n) \tag{v}$$

where $a = 1 - (T/RC)$.

Equation (v) is of the general form

$$y(n+1) = ay(n) + (1-a)u(n) \tag{1.23}$$

where y is the output and u, the input.

Note: (a) Where differential equations have derivatives, difference equations have delays. If the derivative is of higher order, the delay is also of higher order.

(b) Figure 1.5 actually is a simple circuit for a low-pass filter. Therefore, Eq. (1.23) represents a difference equation for a low-pass filter.

From the above example it is clear that, in general, a linear constant-coefficient differential equation of order n can be written as

$$a_0 y(n) + a_1 y(n-1) + \cdots + a_N y(n-N) = b_0 u(n) + b_1 u(n-1) + \cdots + b_L u(n-L)$$

$$\Rightarrow \qquad \sum_{k=0}^{N} a_k(n-k) = \sum_{k=0}^{L} b_k u(n-k)$$

where we have L possible delays of input and N possible delays of output.

We can think that it is a process by which a sequence $u(n)$ is transformed into another sequence $y(n)$. If $u(n) = 0$ for all n, then it is called a *homogeneous difference equation.* A linear difference equation of order n is also called a *linear recurrence relation* of order n because it can be used to compute recursively each $y(n+1)$ from the preceding value $y(n)$. Let us consider the following example which is the well known *loan repayment problem.*

EXAMPLE 1.49. A new flat is purchased with a bank loan of ₹ 1,00,000 over a period of 72 months at a monthly interest of 1 per cent. The principal and interest are to be repaid in 72 equal monthly instalments (EMI). Determine the EMI amount.

Solution Let

$$\begin{aligned}
y(0) &= \text{loan amount} = ₹\ 1{,}00{,}000 \\
N &= \text{number of EMIs} = 72 \\
\alpha &= \text{rate of interest per month} = 1\% = 0.01 \\
y(n) &= \text{amount due after the } n\text{-th month} \\
u &= \text{EMI amount}
\end{aligned}$$

The difference equation for this problem is given by

$$\begin{aligned}
y(n+1) &= y(n) + \alpha y(n) - u \\
&= (1+\alpha)y(n) - u
\end{aligned} \tag{i}$$

Substituting $1+\alpha=\beta$, we observe from Eq. (i) that

$$
\begin{aligned}
y(n+1) &= \beta y(n) - u \\
&= \beta\{\beta y(n-1) - u\} = \beta^2 y(n-1) - u(1+\beta) = \beta^2 y(n-1) - u\sum_{i=0}^{1} \beta^i \\
&= \beta[\beta\{\beta y(n-2) - u\} - u] - u = \beta^3 y(n-2) - u(1+\beta+\beta^2) = \beta^3 y(n-2) - u\sum_{i=0}^{2} \beta^i \\
&\ \ \vdots \\
&= \beta^{n+1} y(0) - u\sum_{i=0}^{n} \beta^i \qquad \text{(ii)}
\end{aligned}
$$

Now, if $S=\sum_{i=0}^{n} \beta^i$, then

$$
\begin{aligned}
S &= 1 + \beta + \beta^2 + \cdots + \beta^n \\
\beta S &= \phantom{1 +{}} \beta + \beta^2 + \cdots + \beta^n + \beta^{n+1} \\
\hline
(1-\beta)S &= 1 \qquad\qquad\qquad\qquad - \beta^{n+1}
\end{aligned}
$$

Thus,

$$S = \sum_{i=0}^{n} \beta^i = \frac{1-\beta^{n+1}}{1-\beta}$$

Therefore, from Eq. (ii) we have

$$
\begin{aligned}
y(n+1) &= \beta^{n+1} y(0) - u \cdot \frac{\beta^{n+1}-1}{\beta - 1} \\
\Rightarrow \qquad &= (1+a)^{n+1} y(0) - u\frac{(1+a)^{n+1}}{a} + \frac{u}{a} \\
&= \frac{u}{a} + (1+a)^{n+1}\left[y(0) - \frac{u}{a}\right] \qquad \text{(iii)}
\end{aligned}
$$

Since the amount due after N months is zero, $y(N)=0$. Therefore, putting $(n+1)=N$ in Eq. (iii), we get

$$y(N) = \frac{u}{a} + (1+a)^{N}\left[y(0) - \frac{u}{a}\right] \qquad \text{(iv)}$$

or

$$
\begin{aligned}
u &= \frac{(1+a)^N y(0) a}{(1+a)^N - 1} \\
&= \frac{(1.01)^{72}(10^5)(0.01)}{(1.01)^{72} - 1} \\
&= ₹\ 1955.02
\end{aligned}
$$

Note: Equation (i) can be represented by a block diagram as shown in Figure 1.6.

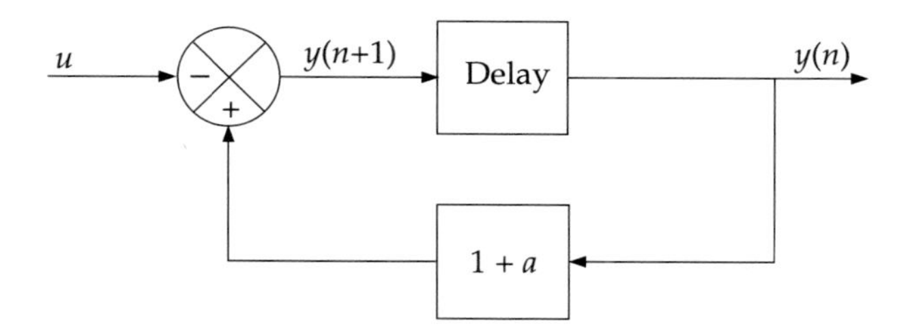

Figure 1.6 Block diagram illustrating Eq. (i) of Example 1.49.

Calculating Responses from Difference Equations

As we know, responses can be static or dynamic.

Static responses. A static response means the steady-state value of the output when the input is constant. It can be calculated from the static version of the difference equation. The static version, in turn, is obtained by neglecting all time dependencies in the difference equation. For example, the output $y(n)$ is replaced with y_s, where the subscript s indicates a static condition.

EXAMPLE 1.50. Calculate the static response for the low-pass filter of Example 1.48.

Solution The difference equation for the low-pass filter is given by

$$y(n+1) = ay(n) + (1-a)u(n)$$

In the static condition, it assumes the following form:

$$y_s = (1-a)y_s + au$$

$\Rightarrow$ $$y_s(1-1+a) = au$$

Therefore, $$y_s = u$$

The above equation indicates that the output is equal to the input in the static condition which is true for a low-pass filter.

Dynamic response. A difference equation is, by itself, an algorithm for calculating responses in the form of time function. The following example illustrates this.

EXAMPLE 1.51. Calculate the dynamic response for a low-pass RC filter when a step input of 1 V is applied to it. Assume $R = 1$ MΩ and $C = 1$ μF.

Solution Let the sampling interval be 1 s. $RC = (1 \times 10^6)(1 \times 10^{-6}) = 1$ s. Then, from Eq. (v) of Example 1.48, we get

$$e_o(1) = 0.5e_o(0) + 0.5 \times 1 = 0.5 \text{ V} \qquad \text{since } e_o(0) = 0$$

$$e_o(2) = 0.5e_o(1) + 0.5 = 0.75 \text{ V}$$

$$e_o(3) = 0.5e_o(2) + 0.5 = 0.875 \text{ V}$$

$$\vdots$$

The response of first 10 s is plotted in Figure 1.7.

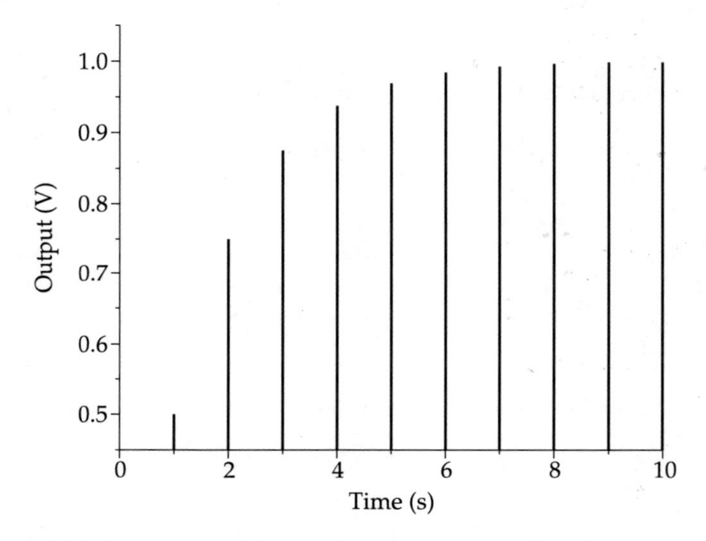

Figure 1.7 Illustration of the dynamic response of a filter (Example 1.51).

1.5 THE *z*-TRANSFORM

The Laplace transform converts linear differential equations to algebraic equations and thus makes their manipulation easy. The z-transform is a similar mathematical tool to deal with linear difference equations or digital data represented by sequences.

Let a discrete set of data be represented by a series of impulses of amplitudes $f(kT)$, where T is the time interval of sampling of data and $k = 0, 1, 2, \ldots \infty$. Then the equation for the entire series of impulses can be written as

$$f^*(t) = f(0)\delta(t) + f(T)\delta(t - T) + f(2T)\delta(t - 2T) + \cdots$$

$$= \sum_{k=0}^{\infty} f(kT)\delta(t - kT) \quad \text{for } t > 0 \tag{1.24}$$

The Laplace transform of Eq. (1.24) yields

$$\mathscr{L}\{f^*(t)\} \equiv F^*(s) = \sum_{k=0}^{\infty} f(kT)e^{-ksT} \tag{1.25}$$

Equation (1.25) represents an infinite series involving the factor e^{-sT} and its powers. The transfer functions of continuous-data systems contain rational functions of s and are thus easy to manipulate. Here, the appearance of e^{-sT} shows the difficulty of using Laplace transform for the general treatment of discrete-data systems because the transfer function relations will no longer be algebraic as they are in continuous-data systems. When the terms involving e^{-sT} appear in a transfer function, other than as a multiplying factor, they make the inverse Laplace transform difficult.

Incidently, we can make operations on infinite sequences a lot easier by transforming the irrational function $F^*(s)$ to a rational function, say $F(z)$, through a transformation from the complex s-plane to another complex z-plane by defining

$$z = e^{sT} \tag{1.26}$$

which leads to

$$s = \frac{1}{T} \ln z \tag{1.27}$$

From Eq. (1.26), we observe that the real and imaginary parts of the complex variable z are related to those of s by the following equations:

$$\begin{aligned} \text{Re}\{z\} &= e^{\sigma T} \cos \omega T \\ \text{Im}\{z\} &= e^{\sigma T} \sin \omega T \end{aligned}$$

since $s = \sigma + \jmath\omega$.

The relationship between z and s, enunciated by Eq. (1.26), is called the *z-transformation.* Substituting Eq. (1.26) in Eq. (1.25), we get

$$F^*\left(s = \frac{1}{T} \ln z\right) \equiv F(z) = \sum_{k=0}^{\infty} f(kT)z^{-k} \tag{1.28}$$

Equation (1.28) will be a rational function of z when expressed in a closed form. Thus, the z-transform of $f(t)$ is written as

$$\mathcal{Z}\{f(t)\} \equiv F(z) = \mathscr{L}\{f^*(t)\}_{s=\ln z/T} = F^*(s)_{s=\ln z/T}$$

The procedure for obtaining $F(z)$ from the discrete function $f^*(t)$ ensures that, in general, any function $f(t)$ that has a Laplace transform also has a z-transform.

To summarize, the steps to obtain the z-transform of a continuous-data function $f(t)$ are as follows:

Step 1. Sample $f(t)$ through an ideal sampler to yield $f^*(t)$. This implies that we need to generate the sequence $f(kT)$, where $k = 0,\ 1,\ 2\cdots$, indicate the sampled values of $f(t)$ at time $t = kT$, T being the sampling interval.

Step 2. Laplace transform $f^*(t)$ to yield $F^*(s)$ as follows:

$$F^*(s) = \mathscr{L}\{f^*(t)\} = \sum_{k=0}^{\infty} f(kT)e^{-skT} \tag{1.29}$$

Step 3. Replace e^{sT} with z to obtain $F(z)$:

$$F(z) = \sum_{k=0}^{\infty} f(kT)z^{-k} \tag{1.30}$$

It is obvious that Eq. (1.30) expresses an infinite series of z^{-k} which needs to be summed up to arrive at a close form for $F(z)$. The following is an example of z-transform with step-by-step procedure.

EXAMPLE 1.52. Determine the z-transform of $x(t) = e^{-at}$.

Solution *Step 1.* We sample the signal $x(t) = e^{at}$ to get

$$x^*(t) = \sum_{k=0}^{\infty} e^{-akT}\delta(t - kT)$$

Step 2. Taking its Laplace transform, we obtain

$$X^*(s) = \sum_{k=0}^{\infty} e^{-akT}e^{-skT} = \sum_{k=0}^{\infty} e^{-(s+a)kT} \tag{i}$$

We can write this equation as

$$X^*(s) = 1 + \sum_{k=1}^{\infty} e^{-(s+a)kT} \tag{ii}$$

Multiplying both sides of Eq. (i) by $e^{-(s+a)T}$, we get

$$\left[e^{-(s+a)T}\right]X^*(s) = \left[e^{-(s+a)T}\right]\sum_{k=0}^{\infty} e^{-(s+a)kT} = \sum_{k=0}^{\infty} e^{-(s+a)(k+1)T}$$

$$= \sum_{k=1}^{\infty} e^{-(s+a)kT} \tag{iii}$$

Subtracting Eq. (iii) from Eq. (ii), we obtain

$$\left[1 - e^{-(s+a)T}\right] X^*(s) = 1$$

$$\Rightarrow \qquad X^*(s) = \frac{1}{1 - e^{-(s+a)T}} \tag{iv}$$

Step 3. Substituting $z = e^{sT}$ in Eq. (iv), we get

$$X(z) = \frac{1}{1 - \dfrac{1}{ze^{aT}}} = \frac{z}{z - e^{aT}}$$

EXAMPLE 1.53. Find the z-transform of

$$f(t) = t^2$$

Solution

$$\mathcal{Z}\{t^2\} = \mathcal{Z}\left\{(kT)^2\right\} = \sum_{k=0}^{\infty}(kT)^2 z^{-k}$$

$$= T^2 \sum_{k=0}^{\infty} k^2 z^{-k}$$

$$= T^2\left(z^{-1} + 4z^{-2} + 9z^{-3} + \cdots\right)$$

Now, we observe

$$(z-1)\mathcal{Z}\left\{(kT)^2\right\} = z\left[T^2\left(z^{-1} + 4z^{-2} + 9z^{-3} + \cdots\right)\right] - 1\left[T^2\left(z^{-1} + 4z^{-2} + 9z^{-3} + \cdots\right)\right]$$

$$= T^2\left(1 + 3z^{-1} + 5z^{-1} + 7z^{-3} + \cdots\right)$$

$$= T^2\left(1 + z^{-1} + z^{-2} + z^{-3} + \cdots\right) + T^2 \cdot 2z\left(1 + 2z^{-1} + 3z^{-2} + 4z^{-3} + \cdots\right)$$

$$= T^2\left[\frac{z}{z-1} + \frac{2}{z} \cdot \frac{z^2}{(z-1)^2}\right]$$

$$= T^2 \cdot \frac{z(z+1)}{(z-1)^2}$$

Therefore,

$$\mathcal{Z}\left\{(kT)^2\right\} = \frac{T^2 z(z+1)}{(z-1)^3}$$

EXAMPLE 1.54. Find the z-transform of

$$f(t) = u(t) + t$$

Solution The given function can be written as

$$f(kT) = u(kT) + kT, \quad k = 0,\ 1,\ 2\ldots$$

Therefore,

$$\mathcal{Z}\{f(kT)\} = \sum_{k=0}^{\infty} z^{-k} + \sum_{k=0}^{\infty} z^{-k}(kT)z^{-k}$$

$$= \frac{1}{1 - z^{-1}} + T\left(z^{-1} + 2z^{-2} + 3z^{-3} + \cdots\right)$$

$$= \frac{z}{z-1} + Tz^{-1}\left(1 + 2z^{-1} + 3z^{-2} + \cdots\right)$$

$$= \frac{z}{z-1} + \frac{Tz^{-1}}{\left(1 - z^{-1}\right)^2}$$

$$= \frac{z^2 + z(T-1)}{\left(1 - z^{-1}\right)^2}$$

The z-transform is usually defined in either of the following two ways:

1. If values of $f(t)$ are defined as the sequence $f(kT)$, where $k = 0,\ 1,\ 2\cdots$, indicate the sampled values of $f(t)$ at time $t = kT$, then $F^*(s)$ and $F(z)$ are obtained through Eqs. (1.29) and (1.30), respectively.
2. If the sequence of numbers of events of $f[k]$ for $k = 0,\ 1,\ 2\ldots$ is given, the z-transform of $f[k]$ is defined as

$$\mathcal{Z}\{f[k]\} = \sum_{k=0}^{\infty} f[k]z^{-k} \tag{1.31}$$

We note that in this case the sampling process and the sampling interval are not explicit.

The z-transform, along with the discrete time Fourier transform (DTFT), is widely used in digital signal processing as well as probability theory. There it is used for two kinds of sequences—*unilateral* and *bilateral* (or *two-sided*). Unilateral sequences are divided into two categories—*causal* and *anti-causal*.

Causal sequence. A sequence is said to be causal when its output does not depend on any 'future' inputs. In more colourful terms, a sequence is causal if it does not 'laugh' before it is 'tickled'. In practical terms, a causal sequence is right-sided [see Figure 1.8(a)].

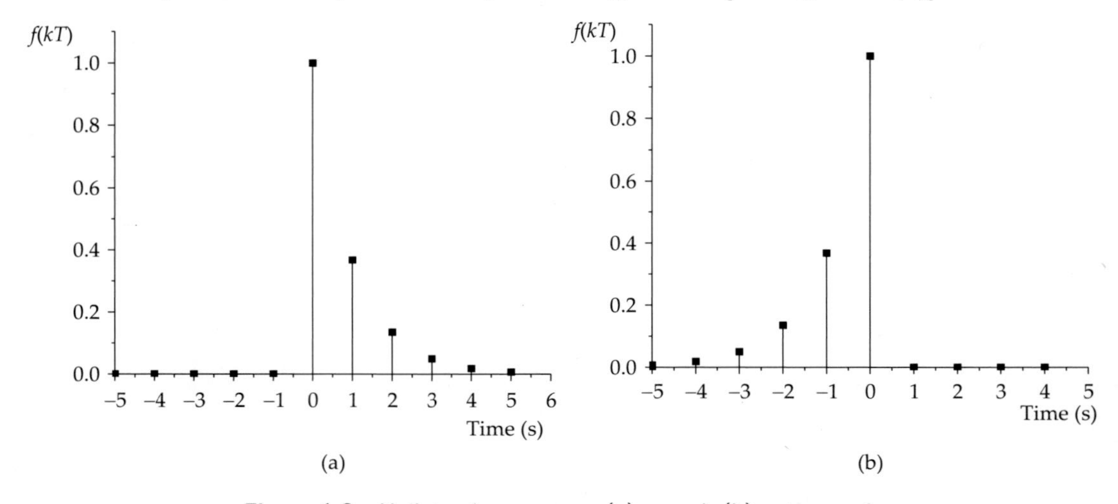

Figure 1.8 Unilateral sequences: (a) causal; (b) anti-causal.

Anti-causal sequence. In an anti-causal sequence, on the other hand, the output anticipates the input one sample into the future. For example,

$$y(n) = x(n+1)$$

where y is the output and x, the input, is an anti-causal sequence. An anti-causal sequence is, therefore, left-sided [Figure 1.8(b)].

All the sequences, in turn, can be finite or infinite. What we have so far discussed deals with infinite causal unilateral sequences. They are called *unilateral transforms.*

The bilateral transform exists for infinite two-sided sequences (Figure 1.9) extending in the range $-\infty < k < +\infty$.

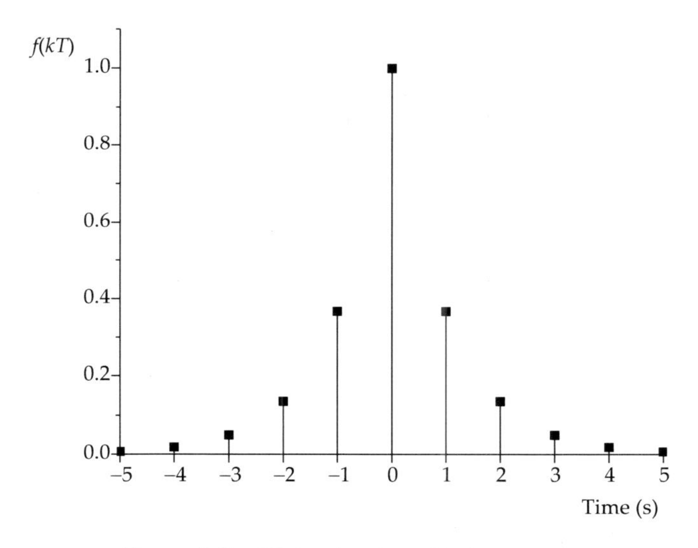

Figure 1.9 Bilateral or two-sided sequence.

Region of Convergence

The region of convergence (ROC) is where the z-transform of a sequence has a finite sum for a region in the complex z-plane. Stated mathematically,

$$\text{ROC} = \left\{ z : \sum_{n=-\infty}^{\infty} f[k]z^{-k} < \infty \right\} \tag{1.32}$$

The z-transform $F(z)$ of $f[k]$ is unique when and only when the ROC is specified.

Properties of the ROC

1. The ROC of $F(z)$ consists of a ring in the z-plane centred around the origin (equivalent to a vertical strip in the s-plane)

2. The ROC does not contain any poles.
3. If $f[k]$ is finite, then the ROC is the entire z-plane, except possibly at $z = 0$ and/or $z = \infty$.
   Examples include:

$$\begin{aligned} \mathcal{Z}\{\delta[k]\} &= 1, && \text{ROC: Entire } z\text{-plane} \\ \mathcal{Z}\{\delta[k-1]\} &= z^{-1}, && \text{ROC: } z \neq 0 \\ \mathcal{Z}\{\delta[k+1]\} &= z, && \text{ROC: } z \neq \infty \end{aligned}$$

4. If $f[k]$ is a right-sided sequence, and if $|z| = r_o$ is in the ROC, then all *finite* values of z for which $|z| > r_o$ are also in the ROC [see Figure 1.10(a)].
5. If $f[k]$ is a left-sided sequence, and if $|z| = r_o$ is in the ROC, then all *finite* values of z for which $0 < |z| < r_o$ are also in the ROC. However, if $f[k] = 0$ for all $k > 0$, then the ROC will include $z = 0$ [see Figure 1.10(b)].
6. If $f[k]$ is two-sided, and if $|z| = r_o$ is in the ROC, then the ROC consists of a ring in the z-plane including the circle $|z| = r_o$ [see Figure 1.10(c)].

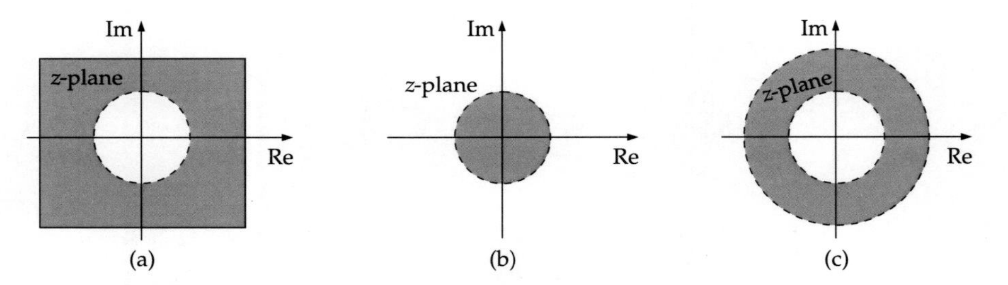

Figure 1.10 ROCs of $F(z)$ for (a) right-sided; (b) left-sided; (c) two-sided infinite duration sequences.

We now work out z-transforms of a few typical sequences.

EXAMPLE 1.55. Find the z-transform of the function defined as

$$\delta(kT) = \begin{cases} 1 & \text{for } k = 0 \\ 0 & \text{for } k > 0 \end{cases}$$

Solution The z-transform of the function is

$$\mathcal{Z}\{\delta(kT)\} = \sum_{k=0}^{\infty} \delta(kT) z^{-k} = (1)z^{-0} + \sum_{k=1}^{\infty} (0) z^{-k} = 1$$

Clearly, this transform exists only when $|z| > 0$ which is the ROC for this transform.

EXAMPLE 1.56. Determine the z-transform of the function given by

$$u(kT) = \begin{cases} 1 & \text{for } k \geq 0 \\ 0 & \text{for } k < 0 \end{cases}$$

Solution The z-transform can be obtained as follows:

$$\mathcal{Z}\{u(kT)\} = \sum_{k=0}^{\infty} u(kT)z^{-k} = \sum_{k=0}^{\infty} z^{-k}$$
$$= 1 + \frac{1}{z} + \frac{1}{z^2} + \cdots$$
$$= \frac{1}{1 - \dfrac{1}{z}} = \frac{z}{z-1} \qquad |z| > 1$$

The condition for the valid region of transform, i.e. $|z| > 1$, is the ROC here.

Note: The function is the unit step sequence.

EXAMPLE 1.57. Determine the z-transform for $a^{kT}, k \geq 0$.

Solution The z-transform is found as follows:

$$\mathcal{Z}\{a^{kT}\} = \sum_{k=0}^{\infty} a^{kT} z^{-k} = \sum_{k=0}^{\infty} \left(a^T z^{-1}\right)^k$$
$$= \frac{1}{1 - a^T z^{-1}}, \qquad \left|a^T z^{-1}\right| < 1$$
$$= \frac{z}{z - a^T}, \qquad |z| > a^T$$

Causal geometric sequence. The sampled sequence for $k \geq 0$ is known as the *causal geometric sequence*. It can be written in the notation specified in Eq. (1.31) as

$$f[k] = a^k u[k] \tag{1.33}$$

Since $u[k] = 0$ for $k < 0$, the sequence can be written as

$$f[k] = \{\ldots 0,\ 0,\ 1,\ a,\ a^2,\ a^3,\ \ldots\}$$

The z-transform of Eq. (1.33) is

$$F(z) = \sum_{k=0}^{\infty} \left(\frac{a}{z}\right)^k = \frac{1}{1 - az^{-1}}$$

The equality holds only if $|az^{-1}| < 1$, which can be rewritten in terms of z as $|z| > a$. Thus, the ROC is $|z| > a$. In this case, the ROC is the complex plane with a disc of radius a at the origin "punched out" as shown in Figure 1.11(a).

Anti-causal geometric sequence. The sampled geometric sequence for $-\infty < k < 0$ is called the *anti-causal geometric sequence*. It can be written as

$$f[k] = -a^k u[-k-1] \tag{1.34}$$

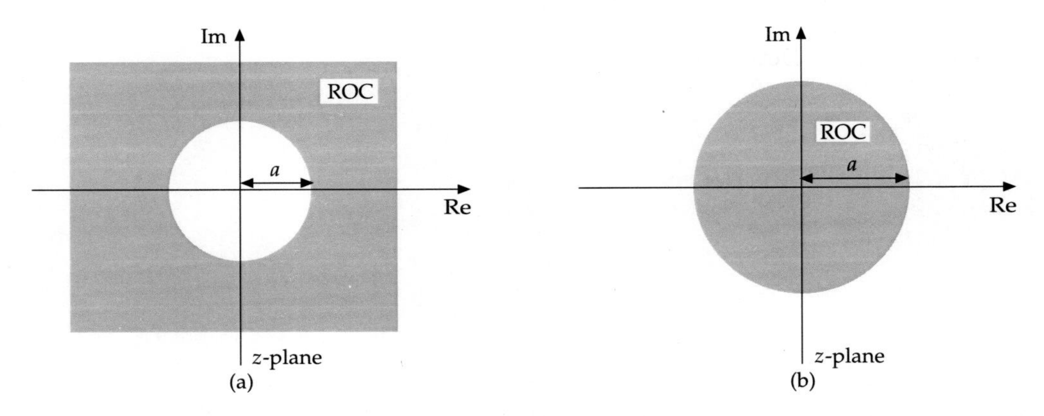

Figure 1.11 ROC for: (a) the casual geometric sequence; (b) the anti-causal geometric sequence.

where u is a unit step function. Here, the sequence is

$$f[k] = \{\ldots,\ -a^{-3},\ -a^{-2},\ -a^{-1},\ 0,\ 0,\ 0,\ \ldots\}$$

The z-transform of Eq. (1.34) is

$$\begin{aligned}F(z) &= \sum_{n=-\infty}^{\infty} f[k]z^{-k} = -\sum_{k=-\infty}^{-1} a^k z^{-k} \\ &= -\sum_{k=-\infty}^{-1} \left(\frac{z}{a}\right)^{-k} \\ &= -\sum_{m=1}^{\infty} \left(\frac{z}{a}\right)^{m} = -\frac{\dfrac{z}{a}}{1-\dfrac{z}{a}}, \quad \left|\frac{z}{a}\right| < 1 \\ &= \frac{z}{z-a}\end{aligned}$$

Clearly, the transform is valid for $|z| < a$ which is the ROC. In this case, it is a disc centred at the origin of the z-plane having a radius of a [see Figure 1.11(b)].

EXAMPLE 1.58. Determine the z-transform of the sampled exponential sequence given by $e^{akT},\ k \geq 0$.

Solution The z-transform is obtained as follows:

$$\begin{aligned}\mathcal{Z}\{e^{-akT}\} &= \sum_{k=0}^{\infty} e^{-akT} z^{-k} = \sum_{k=0}^{\infty} \left(e^{-aT} z^{-1}\right)^k \\ &= \frac{1}{1-(ze^{aT})^{-1}}, \quad \left|\frac{1}{ze^{aT}}\right| < 1 \\ &= \frac{z}{z-e^{-aT}}, \quad \left|e^{aT}\right| > \frac{1}{|z|}\end{aligned}$$

EXAMPLE 1.59. Determine the z-transform of the sampled sinusoidal sequence given by sin (ωkT), where $k \geq 0$.

Solution We know that

$$\sin\,\omega kT = \frac{e^{j\omega kT} - e^{-j\omega kT}}{2j}$$

Hence,

$$\mathcal{Z}\{\sin\,\omega kT\} = \frac{1}{2j}\mathcal{Z}\left\{e^{j\omega kT}\right\} - \frac{1}{2j}\mathcal{Z}\left\{e^{-j\omega kT}\right\} = \frac{1}{2j}\frac{z}{z - e^{j\omega T}} - \frac{1}{2j}\frac{z}{z - e^{-j\omega T}}$$

$$= \frac{1}{2j}\frac{z^2 - ze^{-j\omega T} - z^2 + ze^{j\omega T}}{\left(z - e^{j\omega T}\right)\left(z - e^{-j\omega T}\right)}$$

$$= \frac{z\sin\omega T}{z^2 - 2z\cos\omega T + 1}$$

Residue Method of z-Transform

A powerful tool for obtaining z-transform of the sampled function $f^*(t)$, when its Laplace transform is possible, is the *residue method.* The method is expressed as

$$F(z) \equiv \mathcal{Z}\{f^*(t)\} = \text{sum of residues of } \left.\frac{F(s)z}{z - e^{sT}}\right|_{\text{poles of } F(s)} \tag{1.35}$$

If the denominator of $F(s)$ contains a factor of the form $(s - p)$ so as to indicate that $F(s)$ has a pole at $s = p$, the corresponding residue ρ is obtained from the relation

$$\rho = \left[(s - p)F(s)\frac{z}{z - e^{sT}}\right]_{s=p} \tag{1.36}$$

But, if there are repeated poles of multiplicity n in $F(s)$, the residue is found from the equation:

$$\rho = \frac{1}{(n-1)!}\frac{\partial^{n-1}}{\partial s^{n-1}}\left[(s - p)^n F(s)\frac{z}{z - e^{sT}}\right]_{s=p} \tag{1.37}$$

The following worked out examples will make the method clear.

EXAMPLE 1.60. Find the z-transform of a unit step function by the residue method.

Solution We know that

$$F(s) = \mathscr{L}\{u(t)\} = \frac{1}{s}$$

So, $F(s)$ has only one pole at $s = 0$, which means $p = 0$ in Eq. (1.36). Therefore,

$$\rho = \left[s \cdot \frac{1}{s} \cdot \frac{z}{z - e^{sT}}\right]_{s=0} = \frac{z}{z - 1}$$

Since there is only one residue, from Eq. (1.35), we have the relation

$$F(z) = \mathcal{Z}\{u^*(t)\} = \frac{z}{z - 1}$$

EXAMPLE 1.61. Find the z-transform of the sampled exponential sequence $f(t) = e^{-at}$ by the residue method.

Solution We know that

$$F(s) = \mathscr{L}\{e^{-at}\} = \frac{1}{s+a}$$

So, $F(s)$ has one pole corresponding to $p = -a$ in Eq. (1.36). Therefore,

$$\rho = \left[(s+a)\frac{1}{s+a}\frac{z}{z-e^{sT}}\right]_{s=-a} = \frac{z}{z-e^{-aT}} = F(z)$$

EXAMPLE 1.62. If the Laplace transform of a function is given by

$$F(s) = \frac{1}{s(s+2)}$$

find its z-transform by the residue method.

Solution Here, the poles are at $s = 0,\ -2$ corresponding to $p = 0$ and $p = 2$ of Eq. (1.36). The corresponding residues are

$$\rho\big|_{s=0} = \left[s\cdot\frac{1}{s(s+2)}\frac{z}{z-e^{sT}}\right]_{s=0} = \frac{1}{2}\cdot\frac{z}{z-1}$$

$$\rho\big|_{s=-2} = \left[(s+2)\cdot\frac{1}{s(s+2)}\frac{z}{z-e^{sT}}\right]_{s=-2} = -\frac{1}{2}\cdot\frac{z}{z-e^{-2T}}$$

Therefore,

$$F(z) = \rho_{s=0} + \rho_{s=-2} = \frac{1}{2}\cdot z\left(\frac{1}{z-1} - \frac{1}{z-e^{-2T}}\right) = \frac{1}{2}\cdot\frac{z(1-e^{-2T})}{(z-1)(z-e^{-2T})}$$

EXAMPLE 1.63. Find the z-transform of the Laplace transformed function

$$F(s) = \frac{\omega}{s^2+\omega^2}$$

Solution We note that

$$F(s) = \frac{\omega}{(s+\jmath\omega)(s-\jmath\omega)}$$

So, $F(s)$ has two poles at $s = -\jmath\omega$ and $s = \jmath\omega$. The corresponding residues are

$$\rho\big|_{s=-\jmath\omega} = \left[(s+\jmath\omega)\frac{\omega}{(s+\jmath\omega)(s-\jmath\omega)}\frac{z}{z-e^{sT}}\right]_{s=-\jmath\omega}$$

$$= -\frac{\omega}{2\jmath\omega}\cdot\frac{z}{z-e^{-\jmath\omega T}}$$

$$\rho\big|_{s=\jmath\omega} = \left[(s-\jmath\omega)\cdot\frac{\omega}{(s+\jmath\omega)(s-\jmath\omega)}\cdot\frac{z}{z-e^{sT}}\right]_{s=\jmath\omega}$$

$$= \frac{\omega}{2\jmath\omega}\cdot\frac{z}{z-e^{\jmath\omega T}}$$

Therefore,

$$F(z) = \frac{z}{2\jmath}\left[\frac{1}{z - e^{\jmath\omega T}} - \frac{1}{z - e^{-\jmath\omega T}}\right]$$

$$= \frac{\dfrac{z\left(e^{\jmath\omega T} - e^{-\jmath\omega T}\right)}{2\jmath}}{z^2 - 2z\dfrac{e^{\jmath\omega T} + e^{-\jmath\omega T}}{2} + 1}$$

$$= \frac{z \sin \omega T}{z^2 - 2z\cos\omega T + 1}$$

Note: $F(s) = \dfrac{\omega}{(s + \jmath\omega)(s - \jmath\omega)}$ indicates that $f(t) = \sin\omega t$. In Example 1.59 at page 59 we worked out the z-transform by the general method.

EXAMPLE 1.64. Find the z-transform of the Laplace transform function

$$F(s) = \frac{1}{s^2}$$

Solution The function indicates that it has two poles, i.e. $n = 2$, at $s = 0$. So, we need to apply Eq. (1.37) to find the residue as follows:

$$\rho = \frac{1}{(2-1)!}\frac{\partial^{2-1}}{\partial s^{2-1}}\left[s^2 \cdot \frac{1}{s^2}\frac{z}{z - e^{sT}}\right]_{s=0}$$

$$= \frac{\partial}{\partial s}\left[\frac{z}{z - e^{sT}}\right]_{s=0} = \left.\frac{zT}{\left(z - e^{sT}\right)^2}\right|_{s=0}$$

$$= \frac{Tz}{(z-1)^2}$$

Note: $F(s) = 1/s^2$ implies that $f(t) = t$.

Useful Theorems of *z*-Transform

In what follows, we will use $f(t)$ and $f(kT)$ interchangeably assuming that the continuous-time function $f(t)$ has been sampled to get $f(kT)$.

Linearity. The z-transform of the linear combination of two signals is the linear combination of the individual z-transforms, i.e.

$$\mathcal{Z}\{a_1 f_1(t) \pm a_2 f_2(t)\} = a_1\mathcal{Z}\{f_1(t)\} \pm a_2\mathcal{Z}\{f_2(t)\}$$
$$\equiv a_1 F_1(z) \pm a_2 F_2(z)$$

EXAMPLE 1.65. Determine the z-transform of

$$x(t) = 2\delta(t) + \delta(t+T) + 4\delta(t-3T) - \delta(t-5T) \qquad \text{(i)}$$

Solution From the linearity theorem, we have

$$X(z) = \mathcal{Z}\{2\delta(t)\} + \mathcal{Z}\{\delta(t+T)\} + \mathcal{Z}\{4\delta(t-3T)\} - \mathcal{Z}\{\delta(t-5T)\}$$
$$= 2 + z + 4z^{-3} - z^{-5}$$
$$= z + 2 + 4z^{-3} - z^{-5} \tag{ii}$$

Note: In the sequence representation, Eq. (ii) is written as

$$x[n] = 1,\ \underset{\uparrow}{2},\ 0,\ 0,\ 4,\ 0,\ -1$$

where the symbol $\uparrow$ indicates the time of origin, i.e. $n = 0$, of the sequence. The sequence can be obtained by substituting $t = -T,\ 0,\ T,\ 2T,\ 3T,\ 4T,$ and $5T$ in Eq. (i).

Multiplication by a constant. If $F(z) = \mathcal{Z}\{f(kT)\}$, then

$$\mathcal{Z}\{cf(t)\} = c\mathcal{Z}\{f(t)\} = cF(z)$$

Multiplication by *t*.

$$\mathcal{Z}\{tf(t)\} = -zT\frac{d}{dz}\mathcal{Z}\{f(t)\} = -zT\frac{dF(z)}{dz} \tag{1.38}$$

The proof of Eq. (1.38) now follows.

Proof. We know that

$$F(z) = \sum_{k=0}^{\infty} f(kT)z^{-k}$$

Therefore,

$$\frac{dF(z)}{dz} = \sum_{k=0}^{\infty} f(kT)\left(-kz^{-k-1}\right)$$
$$= -z^{-1}\sum_{k=0}^{\infty} f(kT)\left(kz^{-k}\right)$$
$$= -\frac{1}{zT}\sum_{k=0}^{\infty} kT\ f(kT)z^{-k}$$
$$= -\frac{1}{zT}\mathcal{Z}\{tf(t)\} \quad [\text{putting } kT = t]$$

$$\Rightarrow \qquad \mathcal{Z}\{tf(t)\} = -zT\frac{dF(z)}{dz}$$

Time shifting (or Real translation). A signal can be shifted to the right or left (see Figure 1.12). The z-transform for the left-shifting is just not putting $-z$ for z in the z-transform of the right-shifting.

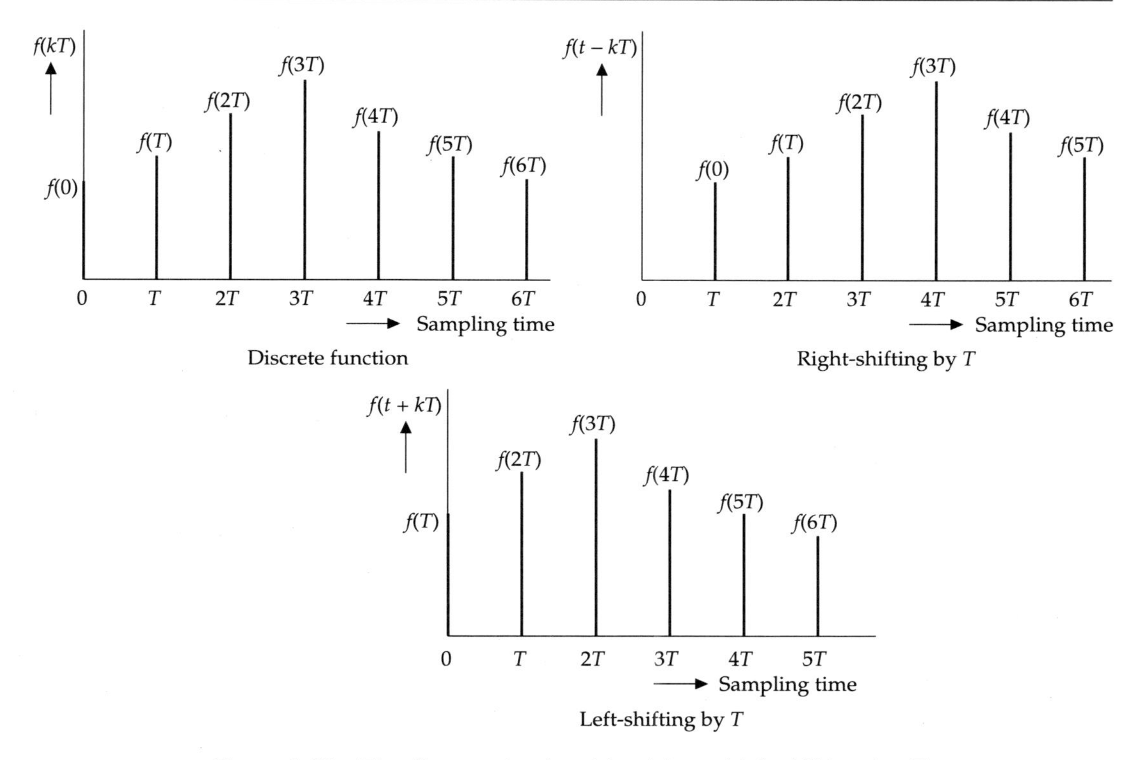

Figure 1.12 The discrete signal and its right and left-shiftings by T.

Right shift. If $F(z) \equiv \mathcal{Z}\{f(t)\}$, then time-shifting the signal by a distance of kT to the right results in multiplying the z-transform by z^{-k}. Consequently,

$$\mathcal{Z}\{f(t-kT)\} = z^{-k}\mathcal{Z}\{f(t)\} \equiv z^{-k}F(z)$$

Note: The transform variable z^{-1} corresponds to a delay of one sampling period in the time domain. This is why z^{-1} is usually referred to as a *delay operator* in digital control systems.

EXAMPLE 1.66. Determine the z-transform of $f(t) = u(t-2T)$, where u indicates a unit step function.

Solution We know that

$$\mathcal{Z}\{u(t)\} = \frac{z}{z-1}$$

Therefore, from the right-shifting property, we get

$$\mathcal{Z}\{u(t-2T)\} = z^{-2}\,\frac{z}{z-1} = \frac{1}{z(z-1)}$$

Left shift. If $F(z) \equiv \mathcal{Z}\{f(t)\}$, then the z-transform of $f(t)$ shifted to the left by kT, where k is a positive integer, results in the following relation:

$$\mathcal{Z}\{f(t+kT)\} \equiv z^k\left[F(z) - \sum_{n=0}^{k-1} f(nT)z^{-n}\right] \tag{1.39}$$

EXAMPLE 1.67. Determine the z-transform of $f(t) = u(t+2T)$.

Solution Here, $k = 2$. Therefore, from Eq. (1.39),

$$\begin{aligned}\mathcal{Z}\{u(t+2T)\} &= z^2\left[F(z) - \sum_{n=0}^{1} f(nT)z^{-n}\right] \\ &= z^2\left[F(z) - \{f(0)z^{-0} + f(T)z^{-1}\}\right]\end{aligned}$$

Now, $F(z) \equiv \mathcal{Z}u(t) = \dfrac{z}{z-1}$, $f(0) = u(2T) = 1$, $f(T) = u(3T) = 1$. Therefore,

$$\begin{aligned}\mathcal{Z}\{u(t+2T)\} &= z^2\left[\frac{z}{z-1} - \left(1 + z^{-1}\right)\right] \\ &= \frac{z}{z-1}\end{aligned}$$

Note: The z-transform is the same as that of a unit step function. This is what is expected. Because an infinite unit step function, when shifted to the left, remains an infinite unit step function starting from $t = 0$.

Exponential shifting. Multiplication by $e^{\pm at}$ is called the *exponential shifting* or *complex translation.* The relevant z-transform is given by

$$\mathcal{Z}\left\{e^{\pm at}f(t)\right\} = F\left(ze^{\mp aT}\right) \tag{1.40}$$

EXAMPLE 1.68. Find the z-transform for

$$x(t) = te^{-at} \quad \text{for } t \geq 0$$

using the complex translation theorem.

Solution We know that

$$X(z) = \frac{Tz}{(z-1)^2} \quad \text{for } x(t) = t$$

Therefore,

$$\begin{aligned}\mathcal{Z}\left\{e^{-at}t\right\} &= \frac{T\left(ze^{aT}\right)}{\left[\left(ze^{aT}\right) - 1\right]^2} \quad \text{[applying Eq. (1.40)]} \\ &= \frac{Tze^{-aT}}{\left(z - e^{-aT}\right)^2}\end{aligned}$$

Partial differentiation. If $F(z,a)$ is the z-transform of $f(t,a)$, where a is an independent variable, the z-transform of the partial derivative of $f(t,a)$ with respect to a is given by

$$\mathcal{Z}\left\{\frac{\partial}{\partial a}[f(t,a)]\right\} = \frac{\partial}{\partial a}F(z,a) \tag{1.41}$$

EXAMPLE 1.69. Determine the z-transform of $f(t) = te^{-at}$ by using the partial differentiation theorem.

Solution From the table in Appendix B at page 663, we observe that

$$\mathcal{Z}\{e^{-at}\} = \frac{z}{z - e^{-aT}}$$

Now,

$$\begin{aligned}\mathcal{Z}\{te^{-at}\} &= \mathcal{Z}\left\{-\frac{\partial}{\partial a}e^{-at}\right\} \\ &= -\frac{\partial}{\partial a}\mathcal{Z}\{e^{-at}\} \quad \text{[using Eq. (1.41)]} \\ &= -\frac{\partial}{\partial a}\left[\frac{z}{z - e^{-aT}}\right] \\ &= \frac{Tze^{-aT}}{\left(z - e^{-aT}\right)^2}\end{aligned}$$

Scaling. If a is an arbitrary constant, real or complex, then the z-transform of $a^t f(t)$ is given by

$$\mathcal{Z}\{a^t f(t)\} = F\left(\frac{z}{a}\right)$$

EXAMPLE 1.70. Using the scaling theorem, determine the z-transform of $a^t \cos \omega t$.

Solution From Appendix B, we see that

$$\begin{aligned}\mathcal{Z}\{\cos \omega t\} &= \frac{z^2 - z\cos \omega T}{z^2 - 2z\cos \omega T + 1} \\ &= \frac{1 - z^{-1}\cos \omega T}{1 - 2z^{-1}\cos \omega T + z^{-2}}\end{aligned}$$

According to the scaling theorem, substituting $z = za^{-1}$ or $z^{-1} = z^{-1}a$, we get

$$\begin{aligned}\mathcal{Z}\{a^t \cos \omega t\} &= \frac{1 - az^{-1}\cos \omega T}{1 - 2az^{-1}\cos \omega T + a^2 z^{-2}} \\ &= \frac{z^2 - az\cos \omega T}{z^2 - 2az\cos \omega T + a^2}\end{aligned}$$

EXAMPLE 1.71. Determine the z-transform of $f(t) = 2^t u(t-2)$.

Solution We know that

$$\mathcal{Z}\{u(t)\} = \frac{z}{z-1}$$

By the time-shifting theorem,

$$\mathcal{Z}\{u(t-2)\} = z^{-2} \cdot \frac{z}{z-1} = \frac{z^{-1}}{z-1}$$

Now using the scaling theorem, we get

$$\mathcal{Z}\{2^t u(t-2)\} = \frac{2z^{-1}}{\frac{z}{2}-1} = \frac{4}{z(z-2)}$$

Real convolution of two sequences. The convolution of two sequences $f_1(kT)$ and $f_2(kT)$, generating a sequence $f(kT)$ is defined as

$$f(kT) \equiv f_1(kT) \star f_2(kT) = \sum_{n=0}^{k} f_1[(k-n)T]f_2(nT) \tag{1.42}$$

The z-transform of the convolution of two sequences is the product of the individual z-transforms.

$$\begin{aligned}\mathcal{Z}\{f_1(kT) \star f_2(kT)\} &= \mathcal{Z}\{f_1(kT)\}\mathcal{Z}\{f_2(kT)\}\\ &\equiv F_1(z)F_2(z)\end{aligned}$$

EXAMPLE 1.72. If

$$\begin{aligned}f_1(t) &= \{2,\ 2,\ 0,\ 1\}\\ f_2(t) &= \{3,\ 3,\ 1,\ 1\}\end{aligned}$$

find $f_1(t) \star f_2(t)$.

Solution Taking z-transform of $f_1(t)$ and $f_2(t)$, we get

$$\begin{aligned}F_1(z) &= 2 + 2z^{-1} + z^{-3}\\ F_2(z) &= z + 3z^{-1} + z^{-2} + z^{-3}\end{aligned}$$

Therefore,

$$F_1(z)F_2(z) = 6 + 12z^{-1} + 8z^{-2} + 7z^{-3} + 5z^{-4} + z^{-5} + z^{-6} \tag{i}$$

Clearly, Eq. (i) is the z-transform of

$$\underset{\uparrow}{\{6},\ 12,\ 8,\ 7,\ 5,\ 1,\ 1\} \equiv f_1(t) \star f_2(t)$$

Alternative method We construct the convolution table as follows:

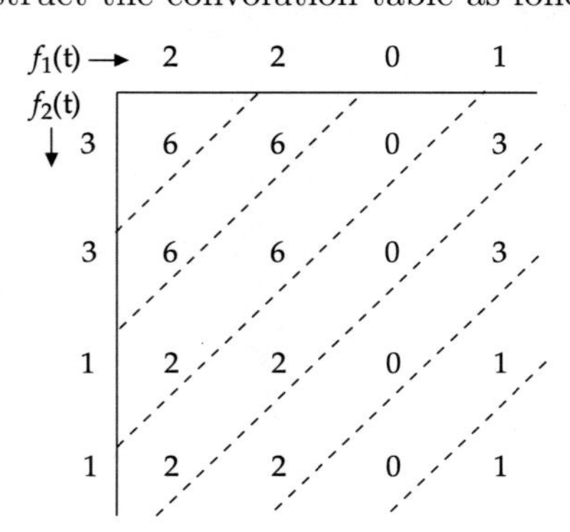

We arrange the $f_1(t)$ sequence horizontally and the $f_2(t)$ sequence vertically. Then we multiply each element of the $f_1(t)$ sequence by an element of the $f_2(t)$ sequence and place the resulting sequence horizontally. The convolution $f_1(t)\star f_2(t)$ is obtained by adding elements of sequences along the slanted lines. Thus,

$$\begin{aligned} f_1(t) \star f_2(t) &= \{6,\ (6+6),\ (2+6+0),\ (2+2+0+3),\ (2+0+3),\ (0+1),\ 1\} \\ &= \{6,\ 12,\ 8,\ 7,\ 5,\ 1,\ 1\} \\ &\quad\ \uparrow \end{aligned}$$

Note: Obviously, this method can be useful only for finite sequences.

Initial value theorem. In case the z-transformed function $f(z)$ is known, the initial value theorem helps us to find the initial value $f(0)$ of the continuous-time function $f(t)$ without inverting $F(z)$. The theorem is stated as follows:

$$\lim_{t\to 0} f(t) = \lim_{z\to\infty} F(z) \tag{1.43}$$

if the limit exists.

EXAMPLE 1.73. Find the initial value of the signal $f(t)$ when its z-transform is given by

$$f(z) = \frac{2z^2+3z+4}{z^2}$$

Solution Using Eq. (1.43), the required initial value is given by

$$\begin{aligned} f(0) &= \lim_{z\to\infty} \frac{2z^2+3z+4}{z^2} \\ &= \lim_{z\to\infty} \left(2+\frac{3}{z}+\frac{4}{z^2}\right) \\ &= 2 \end{aligned}$$

Final value theorem. The final value theorem helps us compute the final value of the continuous-time function $f(t)$ at $t\to\infty$ without inverting the z-transformed function $F(z)$ to the t-space. The theorem states that

$$\lim_{t\to\infty} f(t) = \lim_{z\to 1} \left(1-z^{-1}\right) F(z) \tag{1.44}$$

if the limit exists and the system is stable, i.e. all poles of $\left(1-z^{-1}\right) F(z)$ are inside the unit circle $|z|=1$ on the z-plane.

Note: The theorem is useful to obtain the steady-state value of a discrete control system.

EXAMPLE 1.74. Find the value of $f(kT)$ as $k\to\infty$ when

$$\mathcal{Z}\{f(kT\} \equiv F(z) = \frac{0.7z}{(z-1)(z^2-0.8z+0.15)}$$

Solution We observe that

$$F(z) = \frac{0.7z}{(z-1)(z-0.3)(z-0.5)}$$

or,
$$\left(1 - z^{-1}\right)F(z) = \frac{0.7}{(z-0.3)(z-0.5)}$$

Since $\left(1 - z^{-1}\right)F(z)$ does not have any pole on or outside the unit circle defined by $|z| = 1$ on the z-plane, we can apply the final value theorem defined by Eq. (1.44). Thus,

$$\lim_{k\to\infty} f(kT) = \lim_{z\to 1} \frac{0.7}{(z-0.3)(z-0.5)} = 2$$

The properties of z-transform are summarized in Table 1.3.

Table 1.3 Properties of z-transform

Serial No.	*Property*	*Mathematical expression*
1	Linearity	$\mathcal{Z}\{f_1(t) \pm f_2(t)\} = F_1(z) \pm F_2(z)$
2	Multiplication by a constant	$\mathcal{Z}\{af(t)\} = aF(z)$
3	Multiplication by t	$\mathcal{Z}\{tf(t)\} = -zT\dfrac{dF(z)}{dz}$
4	Division by t	$\mathcal{Z}\left\{\dfrac{f(t)}{t}\right\} = -\dfrac{1}{T}\displaystyle\int_0^z \dfrac{F(z)}{z}dz$
5	Time shifting	Right: $\mathcal{Z}\{f(t-nT)\} = z^{-n}F(z)$ Left: $\mathcal{Z}\{f(t+nT)\} = z^n\left[F(z) - \displaystyle\sum_{k=0}^{n-1} f(nT)z^{-k}\right]$
6	Exponential shifting	$\mathcal{Z}\{e^{\mp at}f(t)\} = F\left(ze^{\pm aT}\right)$
7	Partial differentiation	$\mathcal{Z}\left\{\dfrac{\partial}{\partial a}[f(t,a)]\right\} = \dfrac{\partial}{\partial a}F(z,a)$
8	Scaling	$\mathcal{Z}\{a^t f(t)\} = F\left(\dfrac{z}{a}\right)$
9	Real convolution	$\mathcal{Z}\{f_1(t) \star f_2(t)\} = F_1(z)F_2(z)$
10	Initial value theorem	$\lim_{t\to 0} f(t) = \lim_{z\to\infty} F(z)$
11	Final value theorem	$\lim_{t\to\infty} f(t) = \lim_{z\to 1} \left(1 - z^{-1}\right) F(z)$

We now work out a few examples.

EXAMPLE 1.75. Find the z-transform of the following sequences

(a) $x[n] = \left(\frac{1}{2}\right)^n u[n]$

(b) $x[n] = [(0.4)^n - 3(0.2)^n]u[n]$

(c) $x[n] = \begin{cases} (-1)^n & n = 2,\ 3,\ 4 \cdots \\ 0 & n < 2 \end{cases}$

Solution

(a)
$$X(z) = \sum_{n=0}^{\infty} \left(\frac{1}{2}\right) u[n] z^{-n} = \sum_{n=0}^{\infty} \left(\frac{1}{2z}\right)^{-n} = \frac{1}{1-(1/2z)} = \frac{2z}{2z-1}$$

(b)
$$X(z) = \sum_{n=0}^{\infty} [(0.4)^n - 3(0.2)^n]\, u[n] z^{-n} = \sum_{n=0}^{\infty} \left[\left(\frac{0.4}{z}\right)^n - 3\left(\frac{0.2}{z}\right)^n\right]$$
$$= \frac{1}{1-(0.4/z)} - 3 \cdot \frac{1}{1-(0.2/z)} = \frac{z}{z-0.4} - \frac{3z}{z-0.2}$$

(c)
$$X(z) = \sum_{n=2}^{\infty} (-1)^n z^{-n} = \sum_{n=0}^{\infty} \left(-\frac{1}{z}\right)^n - \left[\left(-\frac{1}{z}\right)^0 + \left(-\frac{1}{z}\right)^1\right]$$
$$= \sum_{n=0}^{\infty} \left(-\frac{1}{z}\right)^n + \left(1 + \frac{1}{z}\right) = \frac{1}{1+(1/z)} + \frac{z+1}{z}$$
$$= \frac{z}{z+1} + \frac{z+1}{z} = \frac{2z^2+2z+1}{z(z+1)}$$

EXAMPLE 1.76. Find the z-transform of the sequence

$$x[n] = e^{-an} u[n]$$

where a is a constant.

Solution

$$X(z) = \sum_{n=0}^{\infty} e^{-an} z^{-n} = \sum_{n=0}^{\infty} \left(\frac{1}{ze^a}\right)^n$$
$$= \frac{1}{1 - \dfrac{1}{ze^a}}, \quad \left|\frac{1}{ze^a}\right| < 1$$
$$= \frac{z}{z - e^a}, \quad |z| > e^a$$

EXAMPLE 1.77. Find the right-handed (i.e. k ranging from 0 to ∞) z-transform $F(z)$ of the sequence $f(k)$, where

$$f(k) = \begin{cases} 1 & \text{for } k = 0 \text{ and even number} \\ -1 & \text{for } k = \text{odd number} \end{cases}$$

Solution By adding the two sequences, we get the sequence

$$f(k) = \sum_{k=0}^{\infty}(-1)^k$$

Therefore, the z-transform is given by

$$\begin{aligned}
F(z) &= \sum_{k=0}^{\infty}(-1)^k z^{-k} \\
&= \sum_{k=0}^{\infty}\left(-\frac{1}{z}\right)^k = \frac{1}{1+(1/z)} \\
&= \frac{z}{z+1}
\end{aligned}$$

EXAMPLE 1.78. Find the z-transform of the following sequence:

$$f(k) = \begin{cases} 2^k & k < 0 \\ \left(\dfrac{1}{2}\right)^k & k = 0 \text{ and even} \\ \left(\dfrac{1}{3}\right)^k & k = \text{ odd} \end{cases}$$

Solution The z-transformed function can be written as

$$\begin{aligned}
F(z) &= \sum_{k=-\infty}^{-1} 2^k z^{-k} + \sum_{k=0,\ k \text{ even}}^{\infty}\left(\frac{1}{2}\right)^k z^{-k} + \sum_{k=\text{ odd}}^{\infty}\left(\frac{1}{3}\right)^k z^{-k} \\
&= \sum_{l=1}^{\infty} 2^{-l} z^l + \sum_{m=0}^{\infty}\left(\frac{1}{2}\right)^{2m} z^{-2m} + \sum_{n=0}^{\infty}\left(\frac{1}{3}\right)^n z^{-(2n+1)} \\
&\equiv F_1(z) + F_2(z) + F_3(z)
\end{aligned}$$

Now,

$$F_1(z) = \sum_{l=1}^{\infty}\left(\frac{z}{2}\right)^l = \frac{z/2}{1-(z/2)} = \frac{z}{2-z}$$

$$F_2(z) = \sum_{m=0}^{\infty}\left(\frac{1}{4z^2}\right)^m = \frac{1}{1-(1/4z^2)} = \frac{4z^2}{4z^2-1}$$

$$\begin{aligned}
F_3(z) &= \sum_{n=0}^{\infty}\left(\frac{1}{3z}\right)^{2n+1} = \frac{1}{3z} + \left(\frac{1}{3z}\right)^3 + \left(\frac{1}{3z}\right)^5 + \cdots = \frac{1/3z}{1-(1/3z)^2} \\
&= \frac{3z}{9z^2-1}
\end{aligned}$$

Therefore,

$$F(z) = F_1(z) + F_2(z) + F_3(z) = \frac{z}{2-z} + \frac{4z^2}{4z^2-1} + \frac{3z}{9z^2-1}$$

EXAMPLE 1.79. Find $F(z)$ if

$$f(s) = \frac{1}{s(s+a)}$$

Solution Let

$$F(s) \equiv \frac{A}{s} + \frac{B}{s+a}$$

Then,

$$A = sF(s)\big|_{s=0} = \left.\frac{1}{s+a}\right|_{s=0} = \frac{1}{a}$$

$$B = (s+a)F(s)\big|_{s=-a} = \left.\frac{1}{s}\right|_{s=-a} = -\frac{1}{a}$$

Therefore,

$$\begin{aligned} F(z) &= \frac{1}{a}\cdot \mathcal{Z}\left\{\frac{1}{s}\right\} - \frac{1}{a}\cdot \mathcal{Z}\left\{\frac{1}{s+a}\right\} \\ &= \frac{1}{a}\left[\frac{z}{z-1} - \frac{z}{z-e^{-aT}}\right] \\ &= \frac{1-e^{-aT}}{a}\cdot\frac{z}{(z-1)\left(z-e^{-aT}\right)} \end{aligned}$$

EXAMPLE 1.80. Find the z-transform of the following function:

$$X(s) = \frac{1}{s^2+2s+2}$$

Solution Factorizing the given function, we get

$$X(s) = \frac{1}{(s+1-\jmath)(s+1+\jmath)} \equiv \frac{A}{s+1-\jmath} + \frac{B}{s+1+\jmath}$$

Then,

$$A = (s+1-\jmath)X(s)\big|_{s=-1+\jmath} = \left.\frac{1}{s+1+\jmath}\right|_{s=-1+\jmath} = \frac{1}{2\jmath}$$

$$B = (s+1+\jmath)X(s)\big|_{s=-1-\jmath} = \left.\frac{1}{s+1-\jmath}\right|_{s=-1-\jmath} = -\frac{1}{2\jmath}$$

Thus,

$$X(s) = \frac{1}{2\jmath}\left[\frac{1}{s+1-\jmath} - \frac{1}{s+1+\jmath}\right]$$

Therefore,

$$\begin{aligned} X(z) &= \frac{1}{2\jmath}\left[\frac{z}{z-e^{-(1-\jmath)T}} - \frac{z}{z-e^{-(1+\jmath)T}}\right] \\ &= \frac{z}{2\jmath}\left[\frac{z-e^{-(1+\jmath)T}-z+e^{-(1-\jmath)T}}{\left\{z-e^{-(1-\jmath)T}\right\}\left\{z-e^{-(1+\jmath)T}\right\}}\right] \end{aligned}$$

Now,

$$\begin{aligned} e^{-(1-\jmath)T} - e^{-(1+\jmath)T} &= e^{-T}\left[e^{\jmath T} - e^{-\jmath T}\right] \\ &= e^{-T}[\cos T + \jmath\sin T - \cos T + \jmath\sin T] \\ &= 2\jmath e^{-T}\sin T \end{aligned}$$

$$e^{-(1-\jmath)T} + e^{-(1+\jmath)T} = 2e^{-T}\cos T$$

Therefore,

$$X(z) = \frac{z}{e^T}\cdot\frac{\sin T}{z^2 - 2ze^{-T}\cos T + e^{-2T}}$$

Inverse *z*-transform

In Laplace transform, we can manipulate equations in the s-space and then revert to the t-space by the inverse Laplace transform. Similarly, in the z-transform we can revert to the t-space through the inverse z-transform though with the difference that the reverted t-space is discrete here. This means that we can revert to the kT-space rather than continuous t-space through the inverse z-transform.

The inverse z-transform can be performed by three methods:

1. Partial fraction expansion method
2. Power series method
3. Residue method

Let us consider the three methods with examples.

Partial fraction expansion method. The procedure is almost the same as we do in the case of inverse Laplace transform except that here we break $F(z)/z$, rather than $F(z)$ itself, into partial fractions. The reason is that almost all z-transformed functions contain z in the numerator as is evident from Appendix B at page 663. So, we actually need to break $F(z)$ into partial fractions of the form

$$F(z) \equiv \frac{Az}{(z\text{-function})_1} + \frac{Bz}{(z\text{-function})_2} + \cdots \tag{1.45}$$

To do that, it is helpful to follow our standard procedure of finding coefficients A, B, ... if we write Eq. (1.45) as

$$\frac{F(z)}{z} \equiv \frac{A}{(z\text{-function})_1} + \frac{B}{(z\text{-function})_2} + \cdots$$

Let us work out a couple of examples.

EXAMPLE 1.81. Find the sequence corresponding to the z-transformed function

$$X(z) = \frac{z^3 - 4.8}{z(z-0.2)(z-0.4)}$$

Solution We rewrite the function as

$$\frac{X(z)}{z} = \frac{z^3 - 4.8}{z^2(z-0.2)(z-0.4)}$$

Let

$$\frac{X(z)}{z} \equiv \frac{A}{z^2} + \frac{B}{z} + \frac{C}{z-0.2} + \frac{D}{z-0.4} \tag{i}$$

Then,

$$A = \left.\frac{z^3 - 4.8}{(z-0.2)(z-0.4)}\right|_{z=0} = \frac{-4.8}{0.08} = -60$$

$$B = \frac{d}{dz}\left[\frac{z^3 - 4.8}{(z-0.2)(z-0.4)}\right]_{z=0}$$

$$= \left[\frac{3z^2}{(z-0.2)(z-0.4)} - \frac{(z^3-4.8)(2z-0.6)}{(z^2-0.6z+0.08)^2}\right]_{z=0} = 450$$

$$C = \left.\frac{z^3 - 4.8}{z^2(z-0.4)}\right|_{z=0.2} = 599$$

$$D = \left.\frac{z^3 - 4.8}{z^2(z-0.2)}\right|_{z=0.4} = -148$$

Therefore, from Eq. (i),

$$\frac{X(z)}{z} = -\frac{60}{z^2} + \frac{450}{z} + \frac{599}{z-0.2} - \frac{148}{z-0.4}$$

or

$$X(z) = -\frac{60}{z} + 450 + \frac{599z}{z-0.2} - \frac{148z}{z-0.4} \tag{ii}$$

On inversion, Eq. (ii) yields

$$x[n] = -60\delta[n-1] + 450\delta[n] + [599(0.2)^n - 148(0.4)^n]u[n]$$

EXAMPLE 1.82. Given the z-transformed function

$$X(z) = \frac{\left(1 - e^{-aT}\right)z}{(z-1)\left(z - e^{-aT}\right)}$$

find out the sampled sequence.

Solution Dividing the function by z and writing it in the form of partial fractions, we have

$$\frac{X(z)}{z} = \frac{\left(1 - e^{-aT}\right)}{(z-1)\left(z - e^{-aT}\right)} \equiv \frac{A}{z-1} + \frac{B}{z - e^{-aT}}$$

A and B are found out in the usual way as

$$A = \left.\frac{1 - e^{-aT}}{z - e^{-aT}}\right|_{z=1} = 1$$

$$B = \left.\frac{1 - e^{-aT}}{z - 1}\right|_{z=e^{-aT}} = -1$$

Therefore,

$$\frac{X(z)}{z} = \frac{1}{z-1} - \frac{1}{z - e^{-aT}}$$

or

$$X(z) = \frac{z}{z-1} - \frac{z}{z - e^{-aT}}$$

or

$$x(kT) = u(kT) - e^{-akT} = 1 - e^{-akT}$$

EXAMPLE 1.83. Find $y(k)$ when

$$Y(z) = \frac{2z^2 - 1.5z}{z^2 - 1.5z + 0.5}$$

Solution Let

$$Y(z) = \frac{z(2z - 1.5)}{(z-1)(z-0.5)} \equiv z\left[\frac{A}{z-1} + \frac{B}{z-0.5}\right]$$

Then,

$$A = (z-1)\left.\frac{Y(z)}{z}\right|_{z=1} = \left.\frac{2z - 1.5}{z - 0.5}\right|_{z=1} = 1$$

$$B = (z-0.5)\left.\frac{Y(z)}{z}\right|_{z=0.5} = \left.\frac{2z - 1.5}{z - 1}\right|_{z=0.5} = 1$$

Therefore,

$$Y(z) = \frac{z}{z-1} + \frac{z}{z-0.5}$$

Its inverse z-transform yields

$$y(kT) = u(kT) + 0.5^{kT}$$

or

$$y[k] = 1 + 0.5^k \qquad k \geq 0$$

Power series method. In the power series method, the z-transformed function $F(z)$ is expanded into a power series of z^{-1}. We have seen in Eq. (1.28) that the coefficients of z^{-k} of such a series are the values of $f(kT)$.

EXAMPLE 1.84. Given

$$X(z) = 3z^3 - z + 2z^{-4}$$

Obtain $x[k]$.

Solution The z-transformed function is already in a power series of

$$X(z) = \sum x[k]z^{-k}$$

i.e.

$$X(z) = 3z^{-(-3)} - 1z^{-(-1)} + 2z^{-(4)}$$

Therefore,

$$x[-3] = 3, \qquad x[-1] = -1, \qquad x[4] = 2, \qquad x[k] = 0 \quad \text{for all other } k\text{'s}$$

Thus,

$$x[k] = \{3,\ 0,\ -1,\ 0,\ 0,\ 0,\ 0,\ 2\}$$
$$\uparrow$$

EXAMPLE 1.85. Find the inverse z-transform of

$$F(z) = \frac{z - ze^{-aT}}{z^2 - ze^{-aT} - z + e^{aT}}$$

by the power series method.

Solution We have to perform a long division to find out the power series as follows. Writing the numerator and denominator in polynomial form, we get

$$\begin{array}{r|l}
 & +\frac{1}{z} - \frac{e^{-aT}}{z} + \frac{1}{z^2} - \frac{e^{-2aT}}{z^2} + \frac{1}{z^3} - \frac{e^{-3aT}}{z^3} + \cdots \\
\hline
z^2 - ze^{-aT} - z + e^{-aT} & +z - ze^{-aT} \\
 & +z \quad - e^{-aT} - 1 \quad + \frac{e^{-aT}}{z} \\
\cline{2-2}
 & - ze^{-aT} + e^{-aT} + 1 \quad - \frac{e^{-aT}}{z} \\
 & - ze^{-aT} + e^{-aT} \quad + e^{-2aT} - \frac{e^{-2aT}}{z} \\
\cline{2-2}
 & +1 \quad - \frac{e^{-aT}}{z} - e^{2aT} + \frac{e^{-2aT}}{z} \\
 & +1 \quad - \frac{e^{-aT}}{z} - \frac{1}{z} + \frac{e^{-aT}}{z^2} \\
\cline{2-2}
 & \text{and so on} \quad \ldots
\end{array}$$

The result of the long division shows that

$$
\begin{aligned}
F(z) &= \left(1 - e^{-aT}\right) z^{-1} + \left(1 - e^{-2aT}\right) z^{-2} + \left(1 - e^{-3aT}\right) z^{-3} + \cdots \\
&= \sum_{k=0}^{\infty} \left(1 - e^{-akT}\right) z^{-k}
\end{aligned}
$$

Therefore, the inverse z-transform is given by

$$f(kT) = 1 - e^{-akT}$$

EXAMPLE 1.86. Find inverse z-transform of

$$F(z) = \frac{Tze^{-aT}}{(z - e^{-aT})^2}$$

by the power series method.

Solution As in the previous example, performing the long division

$$
\begin{array}{r|llll}
 & +\frac{T}{z}e^{-aT} & +\frac{2T}{z^2}e^{-2aT} & +\frac{3T}{z^3}e^{-3aT} & + \cdots \\
\hline
z^2 - 2ze^{-aT} + e^{-2aT} & +Tze^{-aT} & & & \\
 & +Tze^{-aT} & -2Te^{-2aT} & +\frac{T}{z}e^{-3aT} & \\
\cline{2-4}
 & & +2Te^{-2aT} & -\frac{T}{z}e^{-3aT} & \\
 & & +2Te^{-2aT} & -\frac{4T}{z}e^{-3aT} & +\frac{2T}{z^2}e^{-4aT} \\
\cline{3-5}
 & & & +\frac{3T}{z}e^{-3aT} & -\frac{2T}{z^2}e^{-4aT} \\
 & & & +\frac{3T}{z}e^{-3aT} & -\frac{6T}{z^2}e^{-4aT} + \frac{3T}{z^2}e^{-5aT} \\
\cline{4-5}
\end{array}
$$

we find

$$
\begin{aligned}
F(z) &= \frac{T}{z}e^{-aT} + \frac{2T}{z^2}e^{-2aT} + \frac{3T}{z^3}e^{-3aT} + \cdots \\
&= \sum_{k=0}^{\infty} \left(kTe^{-akT}\right) z^{-k}
\end{aligned}
$$

Therefore,

$$f(kT) = kTe^{-akT}$$

The residue method. The time sequence $x(kT)$ and the z-transformed function $X(z)$ are related by the formula

$$f(kT) = \frac{1}{2\pi \jmath} \oint_{\Gamma} F(z)\, z^{k-1} dz$$

It involves a contour integration along the circular path Γ of radius $|z| = e^{cT}$ having its centre at the origin of the z-plane and the value of c is so chosen that all poles of $F(z)$ lie within the circle. It can be shown that

$$f(kT) = \sum \text{ residues of } F(z)\, z^{k-1} \text{ at the poles of } F(z)\, z^{k-1} \tag{1.46}$$

Equation (1.46) is used to find the inverse z-transform.

The residue to a repeated pole at $z = p$ of multiplicity n is given by

$$\rho = \frac{1}{(n-1)!} \lim_{z \to p} \left\{ \frac{\partial^{n-1}}{\partial z^{n-1}} \left[(z-p)^n F(z) z^{k-1} \right] \right\} \tag{1.47}$$

EXAMPLE 1.87. Find the inverse z-transform of

$$X(z) = \frac{z}{z - a^T}$$

by the residue method.

Solution The function has only one pole at $z = a^T$. Therefore,

$$x(kT) = z \cdot z^{k-1}\Big|_{z=a^T} = a^{kT}$$

EXAMPLE 1.88. Find the inverse z-transform of

$$F(z) = \frac{\left(1 - e^{-aT}\right) z}{(z-1)\left(z - e^{-aT}\right)}$$

by the residue method.

Solution The function has two poles at $z = 1$ and $z = e^{-aT}$. Therefore,

$$\begin{aligned} f(kT) &= \left.\frac{\left(1 - e^{-aT}\right) z \cdot z^{k-1}}{z - e^{-aT}}\right|_{z=1} + \left.\frac{\left(1 - e^{-aT}\right) z \cdot z^{k-1}}{z - 1}\right|_{z=e^{-aT}} \\ &= \frac{\left(1 - e^{-aT}\right) \cdot 1^{k-1}}{1 - e^{-aT}} + \frac{\left(1 - e^{-aT}\right) e^{-akT}}{e^{-aT} - 1} \\ &= 1 - e^{-akT} \end{aligned}$$

EXAMPLE 1.89. Using the residue method, find the continuous-time function corresponding to

$$F(z) = \frac{Tz}{(z-1)^2}$$

Solution We see that $F(z)$ has a pole at $z = 1$ with a multiplicity of 2. Therefore, from Eq. (1.47),

$$\begin{aligned}\rho &= \frac{1}{(2-1)!}\lim_{z\to 1}\left\{\frac{\partial^{2-1}}{\partial z^{2-1}}\left[(z-1)^2\cdot\frac{Tz}{(z-1)^2}z^{k-1}\right]\right\}\\ &= \lim_{z\to 1}\frac{\partial}{\partial z}\left(Tz^k\right)\\ &= kTz^{k-1}\Big|_{z=1}\\ &= kT\end{aligned}$$

Therefore, $f(kT) = kT$, or the continuous-time function is $f(t) = t$.

EXAMPLE 1.90. Using convolution, determine the inverse of the function given by

$$F(z) = \frac{z^2}{\left(z-\frac{1}{2}\right)\left(z+\frac{1}{4}\right)}$$

Solution The given function can be written as

$$\begin{aligned}F(z) &= \frac{1}{1-\frac{1}{2}z^{-1}}\cdot\frac{1}{1+\frac{1}{4}z^{-1}}\\ &\equiv F_1(z)\cdot F_2(z)\end{aligned}$$

Performing inverse z-transform of $F_1(z)$ and $F_2(z)$, we get

$$\mathcal{Z}^{-1}\{F_1(z)\} = \mathcal{Z}^{-1}\left\{\frac{1}{1-\frac{1}{2}z^{-1}}\right\} = \left(\frac{1}{2}\right)^{kT}u(kT) \equiv f_1(kT)$$

$$\mathcal{Z}^{-1}\{F_2(z)\} = \mathcal{Z}^{-1}\left\{\frac{1}{1+\frac{1}{4}z^{-1}}\right\} = \left(-\frac{1}{4}\right)^{kT}u(kT) \equiv f_2(kT)$$

Now, we know from Eq. (1.42) that

$$f(kT) = \sum_{n=0}^{k} f_1[(k-n)T]f_2(nT)$$

Therefore,

$$f(kT) = \sum_{n=0}^{k} \left(\frac{1}{2}\right)^{(k-n)T} \left(-\frac{1}{4}\right)^{nT} u(kT)$$

$$= \left(\frac{1}{2}\right)^{kT} \sum_{n=0}^{k} \left[\frac{-1/4}{1/2}\right]^{nT} u(kT)$$

$$= \left(\frac{1}{2}\right)^{kT} \sum_{n=0}^{k} \left(-\frac{1}{2}\right)^{nT} u(kT)$$

$$= \left(\frac{1}{2}\right)^{kT} \cdot \frac{1-\left(-\frac{1}{2}\right)^{(k+1)T}}{1-\left(-\frac{1}{2}\right)} \cdot u(kT) \qquad \left[\because \sum_{n=0}^{k} a^n = \frac{1-a^{k+1}}{1-a}\right]$$

$$= \left(\frac{1}{2}\right)^{kT} \cdot \frac{2}{3}\left[1-\left(-\frac{1}{2}\right)\left(-\frac{1}{2}\right)^{kT}\right] u(kT)$$

$$= \left[\frac{2}{3}\left(\frac{1}{2}\right)^{kT} + \frac{1}{3}\left(-\frac{1}{4}\right)^{kT}\right] u(kT)$$

A few z-transforms are summarized in Appendix B at page 663. The table helps perform both the z-transform and its inverse. We now give a typical example.

EXAMPLE 1.91. Determine the inverse z-transform of $F(z)$ given by

$$F(z) = \frac{z^2+2z}{z^2-2z+4}$$

Solution From the table given in Appendix B at page 663, we find

$$\mathcal{Z}\{e^{-at}\sin \omega t\} = \frac{ze^{-aT}\sin \omega T}{z^2 - 2ze^{-aT}\cos \omega T + e^{-2aT}}$$

$$\mathcal{Z}\{e^{-at}\cos \omega t\} = \frac{z(z - e^{-aT}\cos \omega T)}{z^2 - 2ze^{-aT}\cos \omega T + e^{-2aT}}$$

If we put $e^{-at} = a^t$ and $e^{-aT} = a^T$, we get

$$\mathcal{Z}\{a^t\sin \omega t\} = \frac{za^T\sin \omega T}{z^2 - 2za^T\cos \omega T + a^{2T}} \tag{i}$$

$$\mathcal{Z}\{a^t\cos \omega t\} = \frac{z(z - a^T\cos \omega T)}{z^2 - 2za^T\cos \omega T + a^{2T}} \tag{ii}$$

The denominators of Eqs. (i) and (ii) are the same and they compare with the denominator of $F(z)$. Comparing the denominators, we obtain

$$a^{2T} = 4 \tag{iii}$$

$$a^T \cos \omega T = 1 \tag{iv}$$

Using Eqs. (iii) and (iv), we have

$$\begin{aligned} a^T &= 2 & \Rightarrow \quad & a = 2^{\frac{1}{T}} \\ \cos \omega T &= \frac{1}{2} & \Rightarrow \quad & \omega T = 60^\circ = \frac{\pi}{3} \quad \Rightarrow \quad \omega = \frac{\pi}{3T} \\ \sin \omega T &= \sin \frac{\pi}{3} = \frac{\sqrt{3}}{2} \end{aligned} \tag{v}$$

Now, let us write the numerator of $F(z)$ as the sum of the numerators at the RHS of both Eqs. (i) and (ii) as follows:

$$z^2 + 2z \equiv z^2 - za^T \cos \omega T + A\left(za^T \sin \omega T\right) \tag{vi}$$

$$= z^2 - z(2)\left(\frac{1}{2}\right) + A\left(za^T \sin \omega T\right)$$

$$\Rightarrow \qquad 3z = A\left(za^T \sin \omega T\right)$$

$$\Rightarrow \qquad A = \frac{3}{a^T \sin \omega T} = \frac{3}{(2)\left(\frac{\sqrt{3}}{2}\right)} = \sqrt{3}$$

Substituting the values of a^T, ω obtained in Eq. (v), we get from Eq. (vi)

$$z^2 + 2z = \left(z^2 - 2z \cos \frac{\pi}{3}\right) + \sqrt{3}\left(2z \sin \frac{\pi}{3}\right)$$

$$z^2 - 2z + 4 = z^2 - 2za^T \cos \omega T + a^{2T} = z^2 - 2z\left(2 \cos \frac{\pi}{3}\right) + 4$$

Thus,

$$F(z) = \frac{\left(z^2 - 2z \cos \frac{\pi}{3}\right) + \sqrt{3}\left(2z \sin \frac{\pi}{3}\right)}{z^2 - 2z\left(2 \cos \frac{\pi}{3}\right) + 4} \tag{vii}$$

According to the table given in Appendix B, Eq (vii) gives, on inversion, the relation

$$f(t) = 2^{(t/T)}\left(\cos \frac{\pi t}{3T} + \sqrt{3} \sin \frac{\pi t}{T}\right) \tag{1.48}$$

Solution of difference equations through *z*-transform. We have seen that Laplace transform helps us to solve *differential* equations. We will now see that the z-transform plays a similar role in solving *difference* equations. Let us consider the following examples.

EXAMPLE 1.92. Solve the Example 1.49 of page 47 by the z-transform method.

Solution The corresponding difference equation was

$$y(n+1) = (1+a)y(n) - u \tag{i}$$

Taking z-transform of both sides of Eq. (i), we get

$$zY(z) - zy(0) = (1+a)Y(z) - u \cdot \frac{z}{z-1}$$

$$\Rightarrow \qquad [z-(1+a)]Y(z) = zy(0) - u \cdot \frac{z}{z-1}$$

$$\Rightarrow \qquad Y(z) = \frac{zy(0)}{z-(1+a)} - \frac{uz}{[z-(1+a)](z-1)}$$

To facilitate the inverse z-transform, let us break the last quantity on the RHS into partial fractions. Let

$$\frac{1}{[z-(1+a)](z-1)} \equiv \frac{A}{[z-(1+a)]} + \frac{B}{z-1}$$

Then,

$$A = \left.\frac{1}{z-1}\right|_{z=1+a} = \frac{1}{a}, \qquad B = \left.\frac{1}{z-(1+a)}\right|_{z=1} = -\frac{1}{a}$$

Therefore,

$$\begin{aligned} Y(z) &= \frac{zy(0)}{z-(1+a)} - \frac{uz}{a[z-(1+a)]} + \frac{uz}{a(z-1)} \\ &= \frac{[y(0)-u/a]z}{z-(1+a)} + \frac{uz}{z-1} \\ &= \frac{uz}{z-1} + \frac{[y(0)-u/a]z}{z-(1+a)} \qquad \text{(ii)} \end{aligned}$$

Taking the inverse z-transform on both sides of Eq. (ii), we get

$$y(k) = \frac{u}{a} + \left[y(0) - \frac{u}{a}\right](1+a)^k$$

$$\Rightarrow \qquad y(N) = \frac{u}{a} + \left[y(0) - \frac{u}{a}\right](1+a)^N \quad [\text{putting } k = N] \qquad \text{(iii)}$$

We observe that Eq. (iii) is identical to Eq. (iv) of Example 1.49.

EXAMPLE 1.93. The relation between the input $x[n]$ and the output $y[n]$ of a system is given by

$$y(n) - \frac{1}{2}y(n-1) = x(n)$$

Evaluate the output for the input

$$x[n] = \begin{cases} 1 & n \geq -2 \\ 0 & n < -2 \end{cases}$$

Solution Taking z-transform on both sides of the given difference equation, we get

$$Y(z) - \frac{1}{2}z^{-1}Y(z) = X(z)$$

$$\Rightarrow \qquad \left(1 - \frac{1}{2z}\right)Y(z) = X(z)$$

$$\Rightarrow \qquad \frac{Y(z)}{X(z)} = \frac{z}{z - (1/2)}$$

Now, the input is given by the sequence

$$x[n] = \{1, \ 1, \ \underset{\uparrow}{1}, \ 1, \ \cdots\} = u(t + 2T)$$

Its z-transform is (see Example 1.67 at page 64)

$$X(z) = \frac{z}{z-1}$$

Therefore, the output corresponding to this input is

$$\begin{aligned} Y(z) &= \frac{z}{z - \frac{1}{2}}\left[\frac{z}{z-1}\right] = \frac{z^2}{(z - \frac{1}{2})(z-1)} \\ &= \frac{2z^2}{z-1} - \frac{2z^2}{z - \frac{1}{2}} \\ &= 2z \cdot \frac{z}{z-1} - 2z \cdot \frac{z}{z - \frac{1}{2}} \end{aligned}$$

From its inverse z-transform , we get

$$\begin{aligned} y(k) &= 2u(k+1) - 2\left(\frac{1}{2}\right)^k u(k+1) \\ &= 2\left[1 - \left(\frac{1}{2}\right)^k\right]u(k+1) \end{aligned}$$

EXAMPLE 1.94. Consider the difference equation

$$x(k+2) = x(k+1) + x(k)$$

with $x(0) = 0, \ x(1) = 1$.

(a) Derive an expression for the z-transformed $X(z)$ of $[x(k)]$.

(b) Obtain an expression for $x(k)$ by inverting $X(z)$.

(c) Determine $\lim\limits_{k \to \infty} \dfrac{x(k+1)}{x(k)}$.

Solution (a) Considering the z-transform of the three terms of the given difference equation, we get

$$\sum_{k=0}^{\infty} x(k+2)z^{-k} = z^2\left[X(z) - x(0) - x(1)z^{-1}\right]$$
$$= z^2X(z) - z \qquad [\because x(0) = 0,\ x(1) = 1]$$
$$\sum_{k=0}^{\infty} x(k+1)z^{-k} = z[X(z) - x(0)] = zX(z) \tag{i}$$
$$\sum_{k=0}^{\infty} x(k)z^{-k} = X(z) \tag{ii}$$

Therefore,
$$z^2X(z) - z = zX(z) + X(z)$$
$$\Rightarrow \qquad (z^2 - z - 1)X(z) = z$$
$$\Rightarrow \qquad X(z) = \frac{z}{z^2 - z - 1} \tag{iii}$$

(b) From Eq. (iii),

$$X(z) = \frac{z}{z^2 - z - 1} = \frac{z}{\left(z - \frac{1+\sqrt{5}}{2}\right)\left(z - \frac{1-\sqrt{5}}{2}\right)} \tag{iv}$$

From Eq. (iv), we observe that the poles of multiplicity 1 are at $z = \dfrac{1+\sqrt{5}}{2}$ and $z = \dfrac{1-\sqrt{5}}{2}$. The corresponding residues are

$$\rho_1 = \left.\frac{z^n}{z - \frac{1-\sqrt{5}}{2}}\right|_{z=\frac{1+\sqrt{5}}{2}} = \frac{\left(\frac{1+\sqrt{5}}{2}\right)^n}{\sqrt{5}}$$
$$\rho_2 = \left.\frac{z^n}{z - \frac{1+\sqrt{5}}{2}}\right|_{z=\frac{1-\sqrt{5}}{2}} = \frac{\left(\frac{1-\sqrt{5}}{2}\right)^n}{-\sqrt{5}}$$

Therefore,
$$x(k) = \rho_1 + \rho_2 = \frac{1}{\sqrt{5}}\left[\left(\frac{1+\sqrt{5}}{2}\right)^k - \left(\frac{1-\sqrt{5}}{2}\right)^k\right]u(k)$$

(c) We know from Eqs. (i) and (ii) that

$$\mathcal{Z}\{x(k+1)\} = zX(z)$$
$$\mathcal{Z}\{x(k)\} = X(z)$$

Thus,

$$\lim_{x\to\infty} \frac{x(k+1)}{x(k)} = \lim_{z\to 1} \frac{zX(z)}{X(z)} = 1$$

REVIEW QUESTIONS

1.1 Find the Laplace transform of the following continuous-time functions:

(a) $2(t-1)+e^{-(t+1)}$

(b) $t^2 \sin 2t$

(c) $\cos(5t+1)$

(d) $\frac{d}{dt}\left[t^2 e^{-3t}\right]$

(e) $\frac{d}{dt}[\sin 6t]$

(f) $\frac{d^3 t^2}{dt^3}$

(g) $3t^3(t-1)+e^{-5t}$

(h) $u(t-1)-u(t-2)$

(i) $e^{-2t}u(t-2)$

(j) $e^{-(t-3)}u(t-1)$

(k) $5e^{-3t}+u(t-1)-u(t-2)$

(l) $\ddot{x}+5\dot{x}+3x, \quad \dot{x}(0)=8,\ x(0)=7$

1.2 Find the Laplace transform of the function shown in the following figure.

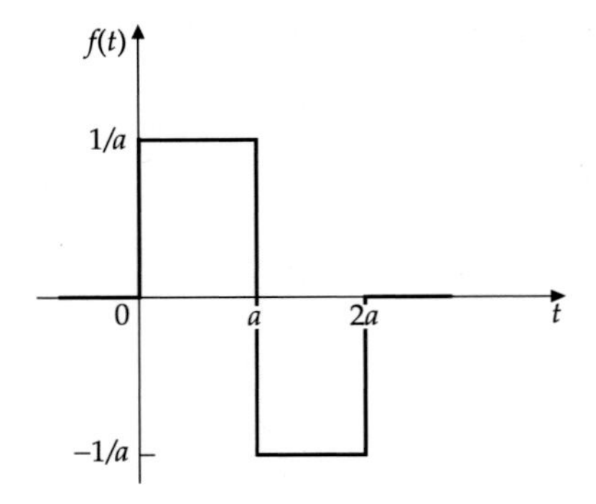

1.3 Find the inverse Laplace transform of the following functions:

(a) $\frac{s+2}{s^2(s+1)(s+3)}$

(b) $\frac{5}{s^2+5}$

(c) $\frac{1}{s(s^2+25)}$

(d) $\frac{6}{s}\left(1-e^{-4.5s}\right)$

(e) $\frac{4+j3}{s+1-j2}+\frac{4-j3}{s+1+j2}$

(f) $\frac{5}{s^4}+\frac{9}{s^2+9}$

(g) $\frac{10}{4s^2+20s+24}$

(h) $\frac{s^4+2s^3+3s^2+4s+5}{s(s+1)}$

1.4 Express the following simultaneous equations in matrix form and then find the values of a, b and c by the matrix inversion method.

$$3a+4b-c=8$$
$$a-2b+3c=6$$
$$a+b-c=0$$

1.5 (a) Invert the matrix

$$\mathbf{A} = \begin{bmatrix} 1 & 1 \\ 1 & 2 \end{bmatrix}$$

(b) Find the eigenvalues of $\mathbf{A}$.

(c) Find the normalized column eigenvectors of $\mathbf{A}$.

1.6 Determine the z-transform of the following functions/sequences:

(a) $f(t) = \delta(t+T) + 3\delta(t) + 6\delta(t-3T) - \delta(t-4T)$

(b) $f(t) = \left(\frac{1}{2}\right)^t [\delta(t) + 2u(t-T)]$

(c) $x(t) = (3)^{t-T}u(t-T)$

(d) $x(t) = t^2 u(t)$

(e) $x[n] = -a^n u[n-1]$

(f) $x[n] = \cos \omega n$ [use Euler's theorem]

1.7 Determine the inverse z-transform of the following functions:

(a) $F(z) = \dfrac{z^2}{z^2 - 1.5z + 0.5}$ $|z| > 1$

(b) $X(z) = \dfrac{-4 + 8z^{-1}}{1 + 6z^{-1} + 8z^{-2}}$

(c) $X(z) = \dfrac{z}{(z-1)(z-2)}$

(d) $F(z) = \dfrac{z^2}{(z-1)(z-0.2)}$

(e) $F(z) = \dfrac{z(1-e^{-T})}{(z-1)(z-e^{-T})}$

(f) $X(z) = \dfrac{2 + 3z^{-1}}{(1+z^{-1})\left(1 + \frac{1}{2}z^{-1}\right)\left(1 - \frac{1}{4}z^{-1}\right)}$

1.8 Consider a discrete-time system for which the input $x[n]$ and the output $y[n]$ are related as

$$y[n] = x[n] - \frac{1}{3}y[n-1]$$

If $y[n] = 0$ for $n < 0$ and $x[n] = \delta[n]$, then express $y[n]$ in terms of the unit sequence $u[n]$.

Answers to Selected Questions

1.1 (a) $2\left(\dfrac{1}{s^2} - \dfrac{1}{s}\right) + \dfrac{e^{-1}}{s+1}$

(b) $-\dfrac{4}{(s^2+4)^2} + \dfrac{16s^2}{(s^2+4)^3} = \dfrac{12s^2 - 16}{(s^2+4)^3}$

(c) $\dfrac{s\cos 1 - 5\sin 1}{s^2 + 25}$

(d) $\dfrac{2s}{(s+3)^3}$

(e) $\dfrac{6s}{s^2 + 36}$

(f) 2

(g) $\dfrac{72}{s^5} - \dfrac{18}{s^4} + \dfrac{1}{s+5}$

(h) $\dfrac{e^{-s}}{s} - \dfrac{e^{-2s}}{s}$

(i) $\dfrac{e^{-(s+2)}}{s+2}$

(j) $\dfrac{e^{2-s}}{s+1}$

(k) $\dfrac{5}{s+3} + \dfrac{e^{-s}}{s} - \dfrac{e^{-2s}}{s}$

(l) $\left[s^2X(s) - 7s - 8\right] + [5sX(s) - 35] + 3X(s)$

1.2 $\dfrac{1}{as}\left(1 - 2e^{-as} + e^{-2as}\right)$

1.3 (a) $\dfrac{2}{3}t - \dfrac{5}{9}u(t) + \dfrac{e^{-t}}{2} + \dfrac{e^{-3t}}{18}$

(b) $\sqrt{5}\sin\sqrt{5}t$

(c) $\dfrac{1}{25}[u(t) - \cos 5t]$

(d) $6\,u(t) - 6\,u(t-4.5)$

(e) $10e^{-t}\cos\left(2t - \tan^{-1}\dfrac{3}{4}\right)$

(f) $t^3 + 2\sin 3t$

(g) $1.5[e^{-2t} - e^{-3t}]$

(h) $\dfrac{d^2}{dt^2}\delta(t) + \dfrac{d}{dt}\delta(t) + 2\delta(t) + 5\,u(t) - 3e^{-t}$

1.4 $\begin{bmatrix} 3 & 4 & -1 \\ 1 & -2 & 3 \\ 1 & 1 & -1 \end{bmatrix}\begin{bmatrix} a \\ b \\ c \end{bmatrix} = \begin{bmatrix} 8 \\ 6 \\ 0 \end{bmatrix}$ $\quad a = 1,\ b = 2,\ c = 3$

1.5 (a) $\mathbf{A}^{-1} = \begin{bmatrix} 2 & -1 \\ -1 & 1 \end{bmatrix}$

(b) $\lambda_1 = 0.382,\ \lambda_2 = 2.618$

(c) $\mathbf{c}_1 = \begin{bmatrix} -0.8507 \\ 0.5257 \end{bmatrix}$ $\quad \mathbf{c}_2 = \begin{bmatrix} 0.5257 \\ 0.8507 \end{bmatrix}$

1.6 (a) $F(z) = z + 3 + 6z^{-3} - z^{-4}$

(b) $\dfrac{1 + (1/2z)}{1 - (1/2z)}$

(c) $X(z)\dfrac{1}{z-3}$

(d) $X(z) = z^{-1}\dfrac{1 + z^{-1}}{\left(1 - z^{-1}\right)^2}$

(e) $X(z) = \dfrac{z}{z-a}$

(f) $X(z) = \dfrac{z^2 - z\cos\omega}{z^2 - 2z\cos\omega + 1}$

1.7 (a) $\left\{\underset{\uparrow}{1}, \dfrac{3}{2}, \dfrac{7}{4}, \dfrac{15}{8}, \dfrac{31}{16}, \ldots\right\}$

(b) $x(kT) = \left[-12(-4)^{kT} + 8(-2)^{kT}\right]u(kT)$

(c) $x(t) = \left(-1 + 2^t\right)u(t)$

(d) $x(t) = \left[1.25 - 0.25(0.2)^t\right]u(t)$

(e) $x(kT) = \left(1 - e^{-kT}\right)u(kT)$

(f) $x(t) = \left[-\dfrac{8}{5}(-1)^t + \dfrac{8}{3}\left(-\dfrac{1}{2}\right)^t + \dfrac{14}{15}\left(\dfrac{1}{4}\right)^t\right]u(t)$

1.8 $y[n] = \left(-\dfrac{1}{3}\right)^n u[n]$

CHAPTER 2

Elementary Concepts

2.1 PREAMBLE

The concept of control engineering was initially introduced for electrical and mechanical applications. But today it has developed to such an extent that its field of application extends not only to every branch of engineering but also to such diverse fields as Economics or Biology. In our everyday life, we encounter so many applications of controls, for example, control of temperature in a refrigerator, maintenance of water levels in toilet flush cisterns or steering a motor car's course. Some of these controls are automatic, though many are manual. The biological world presents a huge variety of automatic controls. Our body is an excellent example. To mention only a few such elementary controls, we can say that the body normally maintains a steady temperature throughout the summer and the winter when external temperature varies widely; our eyes automatically limit light to the retina by squinting in high sun or fully opening in darkness. There are many, many more highly complicated control processes running in our body and we have so far been able to unravel only a few them. The primary objective of this study, of course, is engineering applications.

2.2 DEFINITIONS

Before we embark on the study of the subject, we will consider a few definitions. These are given in Table 2.1.

Figure 2.1 gives a diagrammatic representation of the last six definitions. With this basic background, we will now look into two important concepts in control systems: *open-loop* and *closed-loop systems*.

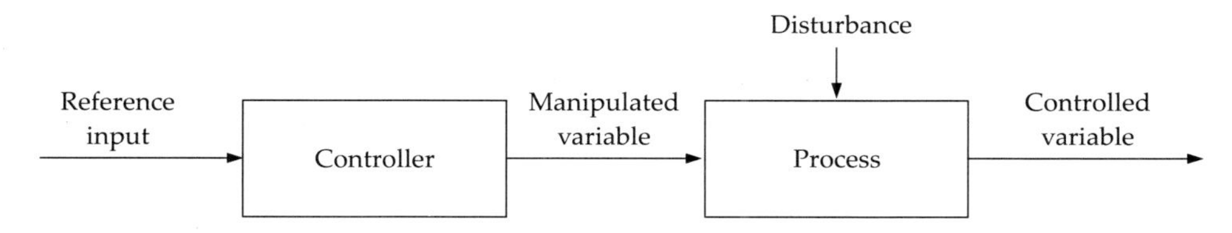

Figure 2.1 Process input-output diagram.

Table 2.1 A Few Definitions Related to The Study of Control Systems

Name	*Meaning*
1. Physical systems	Systems that are governed by the laws of Physics.
2. Plant	A set of machine parts that function together to perform a particular operation. It may be one piece of equipment or a combination of many units. Examples: heating furnace, chemical reactor, power plant, spacecraft, etc.
3. System	A collection of interacting objects—physical or non-physical—that act together to perform a certain objective. Thus, any dynamic phenomenon, be it physical or biological or economic, can be called a system.
4. Process	A progressively continuing operation—natural, artificial or voluntary—that consists of a series of controlled actions or movements systematically directed to achieve a particular result or end.
5. Manipulated variable	The quantity or condition that the controller introduces in order to manipulate the value of the controlled variable so that it reaches a desired value.
6. Controlled variable	The quantity or condition which is generally measured and controlled through some manoeuvre. Normally, it is the output or response of the process.
7. Control	Limiting the deviations of the output of the process within a specified range through a measurement of the output, its comparison with a desired value and then by application of a manipulated variable.
8. Reference (or command) input	The desired value with which the output is compared.
9. Disturbance	The unwanted input to the system that adversely affects the value of the output or response.

2.3 OPEN-LOOP SYSTEM

An open-loop system uses an actuating device to control the system directly without measuring the output and giving it as a feedback to the system. Obviously, in such systems the output has no direct effect on the control action. Figure 2.2 is an example of an open-loop system.

Suppose we want to control the temperature of water in a bucket. We may dip an electric heater into water and control the current through it in order that the heating effect and the radiation of heat from the bath balance in such a way that the water maintains a steady temperature. This is an example of an open-loop control.

Consider, for example, the traffic lights on the road. In the intersection of two roads, three lights—green, amber, and red—allow the traffic to flow, or warn that the signal will change soon, or stop the flow in that direction altogether. These controls are done with the

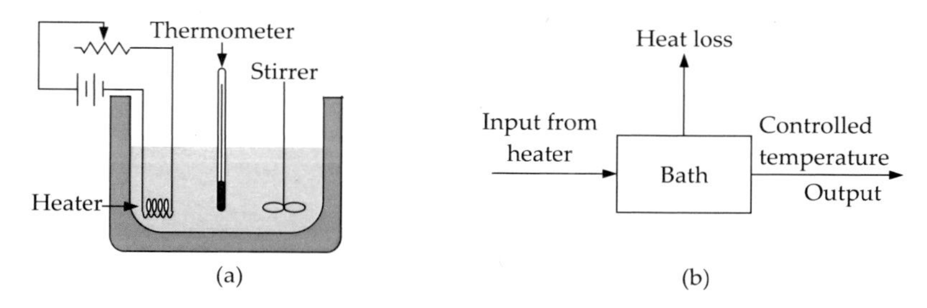

Figure 2.2 Open-loop control system: (a) actual arrangement; (b) block diagram.

help of preset timers which operate irrespective of the traffic situation at a particular instant. That means there is no feedback to the lights from the traffic situation on the road. So, this is another example of an open-loop control. The same open-loop control works in the washing machine where the machine has no feedback from its own system about how dirty the washables are and therefore how long to rinse them.

2.4 CLOSED-LOOP SYSTEM

The temperature control in a bucket of water, where the heater is switched 'ON' or 'OFF' with the help of a contact thermometer and relay combination [Figure 2.3(a)], is an example of a closed-loop control. Here, the output is continuously being measured and compared with the desired value (reference input) through a feedback of the output. The difference between the actual output and the reference input, called the *error signal*, is used to actuate a control element such that the output maintains the desired value. Figure 2.3(b) gives a block diagrammatic representation of the closed-loop control. The *summing point* indicated in the diagram is basically the *error detector*.

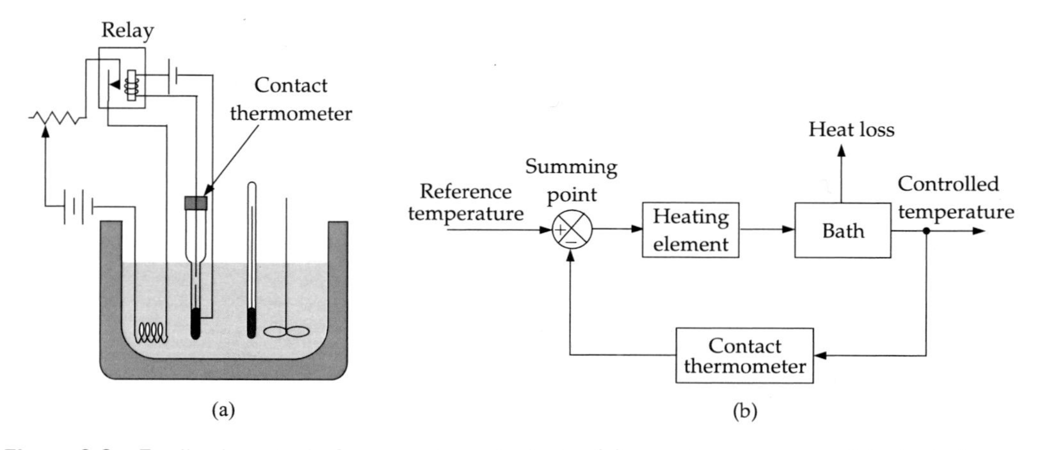

Figure 2.3 Feedback control of temperature of a bath: (a) actual arrangement; (b) block diagram.

Similar automatic temperature controls are maintained in electric irons, geysers or air-conditioners. So, they are also examples of closed-loop controls. The keyword for diagnosing closed-loop control is 'feedback'. If the physical quantity that has to be controlled is sensed by the system itself and the feedback generates a control, it is a closed-loop one (Figure 2.4).

Some electronic calculators, for example, automatically switch off if they are not in use for some fixed time. Is it an example of open-loop control? The answer is no because the calculator has a sensing and feedback mechanism to tell the circuitry whether or not it is in use.

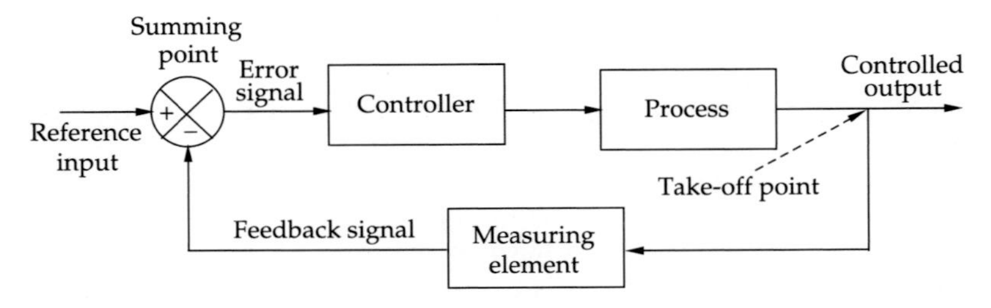

Figure 2.4 Closed-loop control.

In this context, we want to mention that a closed-loop system with one block in the forward path, one block in the feedback path, one take-off point and one summing-point is called a *canonical*[1] closed-loop system.

Table 2.2 lists the differences between open-loop and closed-loop control systems.

Table 2.2 Distinction between Open-loop and Closed-loop Control Systems

Open-Loop	*Closed-Loop*
1. Simplest and economical	Somewhat complicated and therefore costly
2. Less stability problem	May become unstable and generate oscillations
3. Less accurate control	More accurate control
4. Affected by disturbances	Almost independent of disturbances
5. Convenient when the output is difficult to measure	Cannot be installed if the output is not amenable to measurement

A more involved comparison is provided in Table 6.1 at page 219 in Chapter 6.

REVIEW QUESTIONS

2.1 (i) What do you mean by feedback control system?

(ii) Distinguish between an open-loop and a closed-loop system.

2.2 Choose the correct answer:

(i) In an open-loop control system,

(a) the output is measured and compared with a reference input

(b) the output has no control on the input

[1]Means *simple or standard form* (Webster's Universal Collegiate Dictionary).

(c) the control action is actuated by a feedback signal
(d) none of the above statements is correct

(ii) In a closed-loop control system,
(a) the output is measured and compared with a reference input to produce an error signal
(b) the output has no control over the input
(c) the output signal is added to the input signal
(d) none of the above statements is correct

(iii) Automatic traffic lights on roads are examples of
(a) closed-loop system
(b) open-loop system
(c) both open-loop and closed-loop system
(d) none of the above

2.3 Indicate which category — open-loop or closed-loop — the following systems belong to:
(i) A bicycle dynamo generating electricity for the head lamp
(ii) An automatic electric iron
(iii) A sprinkler watering a lawn
(iv) A microwave oven
(v) A household refrigerator
(vi) A household geyser
(vii) A self-operated household pump lifting water to the tank and maintaining water level automatically
(viii) An electronic calculator which turns off if not in use for six minutes
(ix) The household electric energy meter
(x) The speed regulation in electric fans

Answers to Selected Questions

2.2. (i) (b) (ii) (a) (iii) (b).

2.3. (i) closed-loop (ii) closed-loop (iii) open-loop (iv) open-loop (v) closed-loop (vi) closed-loop (vii) closed-loop (viii) closed-loop (ix) open-loop (x) open-loop.

CHAPTER 3

Control Systems and Transfer Functions

3.1 CONTROL SYSTEM CATEGORIES

On the basis of certain criteria, control systems can be divided into the following categories:

1. Linear and nonlinear systems
2. Time-varying and time-invariant systems
3. Deterministic and stochastic systems
4. SISO and MIMO

We consider them in the above order and see what the criteria are.

Linear and Nonlinear Systems

A system is called linear if it satisfies both (i) additivity and (ii) homogeneity properties. If it does not satisfy any one of them, it is called *nonlinear*. Now, the question is: What are additivity and homogeneity properties?

Additivity property. If x and y belong to the domain of the function f, additivity demands

$$f(x+y) = f(x) + f(y)$$

It may appear that this definition looks more mathematical than real-life control problems. But, we have to remember that all control systems should be amenable to mathematical modelling; else, they cannot be tackled systematically. So, this definition, though mathematical, is applicable to real-life control situations.

Let us now examine a few functions and see if they satisfy the additivity criterion.

EXAMPLE 3.1. Let a function be defined by

$$u(y) = \frac{d^2y(t)}{dt^2} + 8\frac{dy(t)}{dt} + y(t) \tag{i}$$

Does it satisfy the additivity property?

Solution Let us check what happens to the function if we substitute $x + y$ for y in Eq. (i).

$$\begin{aligned} u(x+y) &= \frac{d^2}{dt^2}[x(t)+y(t)] + 8\frac{d}{dt}[x(t)+y(t)] + [x(t)+y(t)] \\ &= \left[\frac{d^2x(t)}{dt^2} + 8\frac{dx(t)}{dt} + x(t)\right] + \left[\frac{d^2y(t)}{dt^2} + 8\frac{dy(t)}{dt} + y(t)\right] \\ &= u(x) + u(y) \end{aligned}$$

Therefore, $u(y)$ is an additive function.

EXAMPLE 3.2. Let a function be defined by

$$f(x) = x^2 \tag{i}$$

Does it satisfy the additivity criterion?

Solution As before, we substitute $x + y$ for x in Eq. (i). Then,

$$f(x+y) = (x+y)^2 \neq x^2 + y^2$$

Thus, $f(x)$ is not an additive function.

EXAMPLE 3.3. Let

$$u(y) = y\frac{d^2y}{dt^2} + 7\frac{dy}{dt} \tag{i}$$

Is it an additive function?

Solution On substituting $x + y$ for y in Eq. (i), we get

$$u(x+y) = (x+y)\frac{d^2}{dt^2}(x+y) + 7\frac{d}{dt}(x+y) \neq \left(x\frac{d^2x}{dt^2} + 7\frac{dx}{dt}\right) + \left(y\frac{d^2y}{dt^2} + 7\frac{dy}{dt}\right)$$

Thus, $u(y)$ is not an additive function.

Homogeneity property. The homogeneity property demands that for any y belonging to the domain of the function f, if α is a scalar constant, then

$$f(\alpha y) = \alpha f(y)$$

Let us now examine whether the functions that we tested for additivity satisfy the homogeneity criterion.

EXAMPLE 3.4. Is the function defined by

$$u(y) = \frac{d^2y(t)}{dt^2} + 8\frac{dy(t)}{dt} + y(t) \tag{i}$$

homogeneous?

Solution We substitute αy for y in Eq. (i) and observe

$$\frac{d^2(\alpha y)}{dt^2} + 8\frac{d(\alpha y)}{dt} + \alpha y = \alpha\left(\frac{d^2 y}{dt^2} + 8\frac{dy}{dt} + y\right)$$

Therefore, $u(y)$ defined by Eq. (i) is a homogeneous function.

EXAMPLE 3.5. Does the function defined by

$$f(x) = x^2$$

satisfy the homogeneity test?

Solution We see the result of substitution of x by αx in the above equation that

$$f(\alpha x) = (\alpha x)^2 \neq \alpha x^2$$

Therefore, $f(x)$ defined in the equation does not stand the homogeneity test.

EXAMPLE 3.6. What about the homogeneity of the function defined in the following equation?

$$u(y) = y\frac{d^2 y}{dt^2} + 7\frac{dy}{dt}$$

Solution The same test of substitution of y by αy yields

$$(\alpha y)\frac{d^2(\alpha y)}{dt^2} + 7\frac{d(\alpha y)}{dt} = \alpha^2 y\frac{d^2 y}{dt^2} + 7\alpha\frac{dy}{dt} \neq \alpha\left(y\frac{d^2 y}{dt^2} + 7\frac{dy}{dt}\right)$$

Hence, the function $u(y)$ defined in the above equation is not homogeneous.

Considering all the three functions cited above, we arrive at the conclusion that only the one given in Example 3.4 is linear because it satisfies both additivity and homogeneity properties, while the functions given in Examples 3.5 and 3.6 do not and hence are nonlinear.

EXAMPLE 3.7. The response of a system to a unit impulse is $y(t) = e^{-(t-1)}$. Which one of the following is the correct statement about the system?

(a) The system is nonlinear

(b) The system is unstable in open-loop

(c) The steady-state gain of the system for a unit step input is 0.2

(d) The steady-state gain of the system for a unit step input is 1.0

Solution We observe that for $t \to \infty$, $y(t) = 0$. Therefore, the options (b), (c) and (d) are not true. To test the linearity, let us check if $y(t)$ is additive.

$$\begin{aligned} y(t+T) &= e^{-5(t+T-1)} = e^{-5(t-1)} \cdot e^{-5T} \\ &\neq y(t) + y(T) \end{aligned}$$

Ans: (a).

Time-varying and Time-invariant Systems

If the parameters of a control system vary with time, it is a *time-varying* system. If they do not vary with time, the system is *time-invariant.*

A simple rocket is a time-varying system because, while it is moving in its trajectory, one parameter of its mathematical modelling—mass—is continuously changing with time because of the burning of its fuel. However, a simple pendulum of mass m, executing a simple harmonic motion about its mean position of rest, is a time-invariant system and, therefore, its trajectory can be described by a simple equation such as

$$m\frac{d^2x}{dt^2} = -\alpha x$$

where its parameters—m and α—are independent of time.

An LR circuit, within its current tolerance, is a time-invariant system. It is described by

$$e = L\frac{di}{dt} + Ri$$

where L and R are constants. However, if the current is high, L and R become $L(t)$ and $R(t)$ because they get hot with time and their values change. Obviously, then, the LR circuit is not time-invariant.

Deterministic and Stochastic Control Systems

If a control system produces a predictable and repetitive output when it is subjected to known input and disturbance, it is called a *deterministic* control system. A swinging simple pendulum, for example, is a deterministic system because, at a given place, it will oscillate with the same time-period for a small displacement of its bob.

When the output of a control system becomes statistically predictable, it is called a *stochastic system.* A cluster of molecules of a gas in an enclosure, for example, forms a stochastic system, because for a given temperature of the enclosure, all the molecules will not acquire the same velocity. Some molecules will have very high velocities, some very low, and most of them will acquire a certain velocity, called the *most probable velocity.* In other words, the velocity of molecules is statistically predictable. There are many more examples of stochastic systems. In fact, most of the natural systems are stochastic systems. Take, for example, the temperature at a certain place on the earth's surface. At a certain date of the year, the temperature has a statistically predictable value, called the *normal temperature*, although energy supply and energy reception systems are the same comprising the sun and the earth.

SISO and MIMO

SISO is the acronym for Single Input Single Output system while MIMO for Multiple Input Multiple Output system. The terms are self-explanatory and, therefore, require no elaborate discussion. We may mention here that we will be considering mostly SISO systems in our studies.

3.2 TRANSFER FUNCTION

The transfer function of a system is defined as the ratio of the Laplace transform of the output and the input. Symbolically, if $r(t)$ and $c(t)$ are the (reference) input and (controlled) output, respectively and if $R(s)$ and $C(s)$ are their Laplace transforms, then the transfer function $G(s)$ is given by

$$G(s) = \left.\frac{C(s)}{R(s)}\right|_{\text{initial conditions}=0} \tag{3.1}$$

where[1], $\mathscr{L}\{c(t)\} = C(s)$ and $\mathscr{L}\{r(t)\} = R(s)$. Diagrammatically, it can be represented by Figure 3.1.

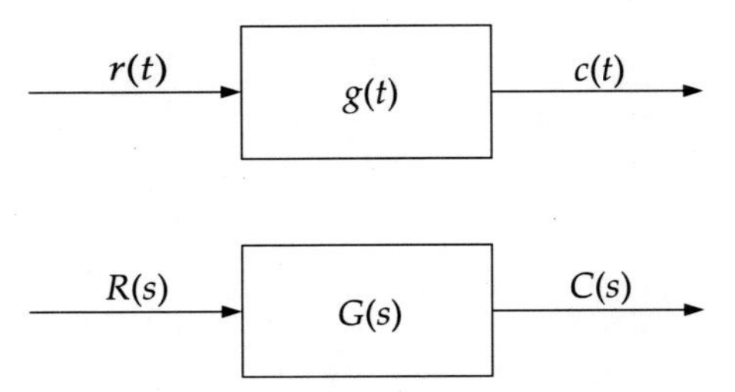

Figure 3.1 Input-output and their Laplace transforms.

The generalized input-output relation of a dynamic system can be written as

$$a_n \frac{d^n c(t)}{dt^n} + a_{n-1}\frac{d^{n-1}c(t)}{dt^{n-1}} + \cdots + a_1 \frac{dc(t)}{dt} + a_0 c(t)$$
$$= b_m \frac{d^m r(t)}{dt^m} + b_{m-1}\frac{d^{m-1}r(t)}{dt^{m-1}} + \cdots + b_1 \frac{dr(t)}{dt} + b_0 r(t)$$

where a's and b's are system parameters which are constants for a linear time-invariant (LTI) system. Taking the Laplace transform of both sides and assuming all initial conditions $= 0$, we get

$$(a_n s^n + a_{n-1}s^{n-1} + \cdots + a_1 s + a_0)C(s) = (b_m s^m + b_{m-1}s^{m-1} + \cdots + b_1 s + b_0)R(s)$$

Therefore,

$$G(s) = \frac{C(s)}{R(s)} = \frac{b_m s^m + b_{m-1}s^{m-1} + \cdots + b_1 s + b_0}{a_n s^n + a_{n-1}s^{n-1} + \cdots + a_1 s + a_0} \tag{3.2}$$

Thus, Eq. (3.2) gives the generalized transfer function for an LTI system. A question may be asked here: Why is the transfer function defined in terms of Laplace transforms of the input and output in the hypothetical s-space rather than the input and output in the real t-space? The answer lies in the very property of the Laplace transform which converts differential and integral equations into simple algebraic equations. All dynamic systems can be described in terms of differential equations. The Laplace transform is very convenient for handling them

[1] $\mathscr{L}\{f(t)\}$ indicates Laplace transform of $f(t)$.

because it transforms them to algebraic equations. Once the manipulations are over, we can go back to the original t-space through the inverse Laplace transform. We will consider a few systems and see how to calculate their transfer functions.

EXAMPLE 3.8. Find the transfer function of the simple LCR circuit given in Fig. 3.2.

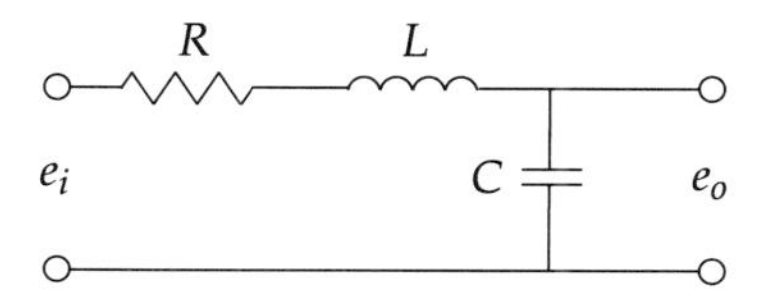

Figure 3.2 Simple LCR circuit (Example 3.8).

Solution KVL gives

$$e_i = L\frac{di}{dt} + Ri + \frac{\int idt}{C} \tag{i}$$

$$e_o = \frac{\int idt}{C} \tag{ii}$$

Taking the Laplace transforms of both Eqs. (i) and (ii) and assuming the system initially relaxed, i.e., all initial conditions = 0, we get

$$E_i(s) = \left(Ls + R + \frac{1}{Cs}\right)I(s) \tag{iii}$$

$$E_o(s) = \frac{1}{Cs}I(s) \tag{iv}$$

Dividing Eq. (iv) by Eq. (iii), after a little manipulation, we get

$$\frac{E_o(s)}{E_i(s)} = G(s) = \frac{1}{s^2LC + sRC + 1} = \frac{1}{(s^2/\omega_n^2) + s\tau + 1}$$

where $\omega_n = \dfrac{1}{\sqrt{LC}}$ and $\tau = RC$.

EXAMPLE 3.9. For the electrical network shown in Fig. 3.3, determine the transfer function $V_o(s)/V_i(s)$:

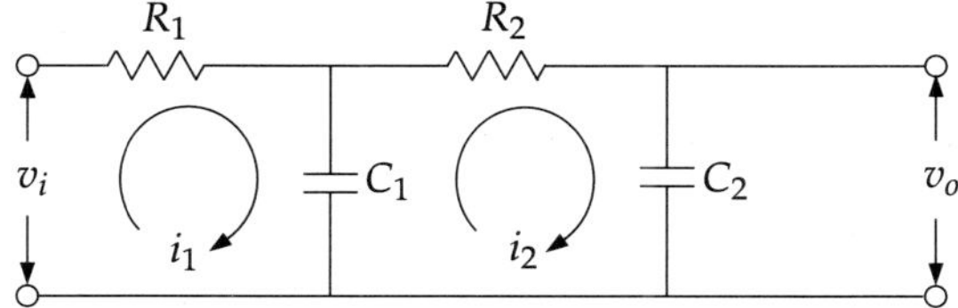

Figure 3.3 Electrical network (Example 3.9).

Solution Applying KVL, we get

$$\begin{aligned} v_i &= i_1 R_1 + \frac{\int (i_1 - i_2)\, dt}{C_1} \\ 0 &= i_2 R_2 + \frac{\int i_2 dt}{C_2} + \frac{\int (i_2 - i_1)\, dt}{C_1} \\ v_o &= \frac{\int i_2\, dt}{C_2} \end{aligned} \tag{i}$$

The Laplace transforms of Eqs. (i) yield

$$\begin{aligned} V_i(s) &= R_1 I_1(s) + \frac{I_1(s) - I_2(s)}{sC_1} \\ 0 &= R_2 I_2(s) + \frac{I_2(s)}{sC_2} + \frac{I_2(s) - I_1(s)}{sC_1} \\ V_o(s) &= \frac{I_2(s)}{sC_2} \end{aligned} \tag{ii}$$

Eliminating $I_1(s)$ and $I_2(s)$ from Eqs. (ii), after a little algebraic manipulation, we obtain

$$\frac{V_o(s)}{V_i(s)} = \frac{1}{s^2 R_1 R_2 C_1 C_2 + s(R_1 C_1 + R_1 C_2 + R_2 C_2) + 1}$$

Note: Instead of writing KVL equations in t-space, we can write them directly in s-space by re-drawing the circuit diagrams with sL for L's and $1/sC$ for C's.

EXAMPLE 3.10. Find the transfer function for the network given in Figure 3.4(a).

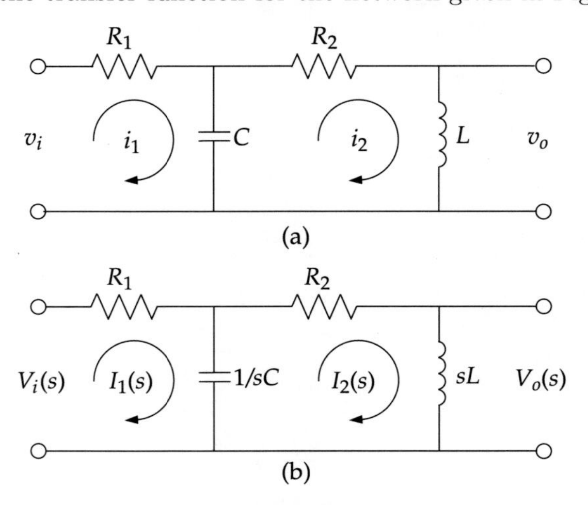

Figure 3.4 Network (Example 3.10): (a) original; (b) Laplace transformed.

Solution Applying KVL to the redrawn network [Figure 3.4(b)], we get

$$\begin{aligned} V_i(s) &= \left(R_1 + \frac{1}{sC}\right) I_1(s) - \frac{1}{sC} I_2(s) \\ 0 &= \left(R_2 + sL + \frac{1}{sC}\right) I_2(s) - \frac{1}{sC} I_1(s) \\ V_o(s) &= sLI_2(s) \end{aligned} \tag{i}$$

Eliminating $I_1(s)$ and $I_2(s)$ from Eqs. (i), after a little algebraic manipulation, we obtain

$$\frac{V_o(s)}{V_i(s)} = \frac{sL}{s^2 R_1 LC + s(R_1 R_2 C + L) + R_1 + R_2}$$

EXAMPLE 3.11. Find the transfer function of a system described by

$$\frac{d^2y}{dt^2} + 3\frac{dy}{dt} + 4y = 2\frac{d^2x}{dt^2} + 6x$$

Solution Taking the Laplace transform of the equation, we get

$$(s^2 + 3s + 4)Y(s) = (2s^2 + 6)X(s)$$

or

$$\frac{Y(s)}{X(s)} = \frac{2s^2 + 6}{s^2 + 3s + 4} \tag{i}$$

Equation (i) is the required transfer function.

EXAMPLE 3.12. Find the transfer function of a system given by

$$3\frac{d^2c(t)}{dt^2} + 5\frac{dc(t)}{dt} + c(t) = r(t) + 3r(t-2) \tag{i}$$

Solution The second function on the RHS of Eq. (i) has a different time base. So, we have to apply the second shifting property when we take the Laplace transform of the equation. Thus, the Laplace transform of the equation is

$$(3s^2 + 5s + 1)C(s) = [1 + 3e^{-2s}]R(s)$$

Therefore,

$$\frac{C(s)}{R(s)} = \frac{1 + 3e^{-2s}}{3s^2 + 5s + 1}$$

EXAMPLE 3.13. Determine V_2/V_1 of the circuit[2] given in Figure 3.5.

[2]This ratio is *not* the transfer function. It is rather the *amplification factor*.

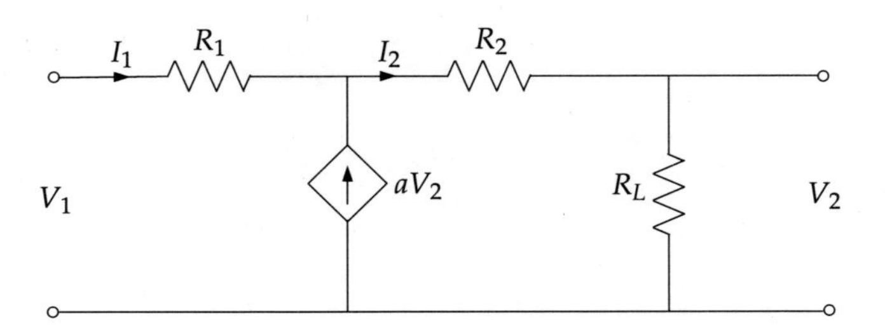

Figure 3.5 Voltage amplifier circuit (Example 3.13).

Solution Using KVL, we get

$$V_1 = R_1 I_1 + (R_2 + R_L) I_2$$

$$I_2 = I_1 + \alpha V_2 \tag{i}$$

$$V_2 = I_2 R_L \tag{ii}$$

$\therefore$

$$V_1 = R_1(I_2 - \alpha V_2) + (R_2 + R_L) I_2 \qquad \text{[using Eq. (i)]}$$

$$= (R_1 + R_2 + R_L) I_2 - \alpha R_1 V_2 \qquad \text{[using Eq. (ii)]}$$

$$= \left[\frac{R_1 + R_2}{R_L} + 1 - \alpha R_1 \right] V_2$$

Thus,

$$\frac{V_2}{V_1} = \frac{R_L}{R_1 + R_2 + R_L - \alpha R_1 R_L} = \frac{R_L}{(1 - \alpha R_L) R_1 + R_2 + R_L}$$

EXAMPLE 3.14. Find $V_2(s)/V_1(s)$ of the circuit given in Figure 3.6, where the op-amp is ideal.

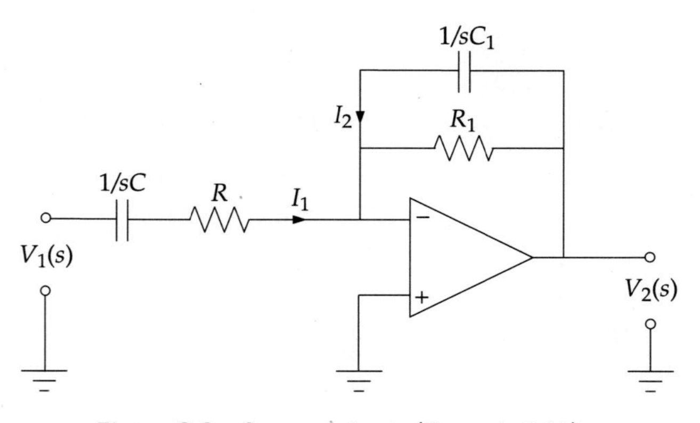

Figure 3.6 Op-amp circuit (Example 3.14).

Solution Since the op-amp is ideal, $I_1 = -I_2$. This gives

$$\frac{V_1(s)}{R + \dfrac{1}{sC}} = -\frac{V_2(s)}{\left(\dfrac{R_1}{sC_1}\right) \div \left(R_1 + \dfrac{1}{sC_1}\right)} = -\frac{V_2(s)}{\dfrac{R_1}{sC_1R_1 + 1}}$$

or

$$\frac{V_2(s)}{V_1(s)} = -\frac{\dfrac{R_1}{sC_1R_1 + 1}}{R + \dfrac{1}{sC}} = -\frac{sCR_1}{(sC_1R_1 + 1)(sCR + 1)}$$

Properties of Transfer Functions

The following are the properties of the transfer function of a control system:

1. Transfer function is the ratio of Laplace transforms of the output and the input. It is not just the ratio of the output and the input. While defining transfer functions, all initial conditions of the system are set to zero values. In other words, the system is considered to be relaxed.
2. It offers a symbolic picture of the dynamic characteristics of a linear time-invariant system.
3. If the transfer functions of individual components of a system are known, the overall transfer function can be determined by taking the product of individual transfer functions. This is, of course, true if there is no loading of the system by the connected devices.
4. Poles and zeros (see below) of a system can be determined from the transfer function. Poles and zeros play a pivotal role in the overall behaviour of a system.
5. The transfer function, with the knowledge of feedback factor, helps us to find out the characteristic equation, which, in turn, gives us an insight into the stability of the system.

Poles and Zeros

Our formulation of transfer function, as given in Eq. (3.2), can be rewritten as

$$G(s) = \frac{C(s)}{R(s)} = \frac{b_m s^m + b_{m-1}s^{m-1} + \cdots + b_1 s + b_0}{a_n s^n + a_{n-1}s^{n-1} + \cdots + a_1 s + a_0}$$

$$\equiv K\frac{(s - b'_1)(s - b'_2)\cdots(s - b'_m)}{(s - a'_1)(s - a'_2)\cdots(s - a'_n)} \tag{3.3}$$

where K is called the *gain factor*. Poles and zeros of the transfer function, as defined from Eq. (3.3), are given in Table 3.1.

Poles and zeros are plotted in a complex plane treating $s = \sigma + \jmath\omega$. It is customary to indicate poles by crosses ($\times$) and zeros by small circles ($\circ$). A typical plot is shown in Figure 3.7.

Table 3.1 Definitions of Pole and Zero

Pole	*Zero*
1. Values of s for which $G(s) \to \infty$ are called poles. From Eq. (3.3) it is clear that a_1', a_2' $\ldots a_n'$ are such values.	1. Values of s for which $G(s) \to 0$ are called zeros. From Eq. (3.3) it is clear that b_1', b_2' $\ldots b_m'$ are such values.
2. If a value is not repeated, it is called a *simple pole.* In case it is repeated, it is called *multiple pole.*	2. If a value is not repeated, it is called a *simple zero.* In case it is repeated, it is called *multiple zero.*

EXAMPLE 3.15. Find the poles and zeros of the transfer function

$$G(s) = \frac{(s^2 + 3s + 2)(s^2 - 2s + 2)}{(s^2 + 7s + 12)(s + 3)(s^2 + 9)}$$

Solution The transfer function can be rewritten as

$$G(s) = \frac{(s + 1)(s + 2)(s - 1 + \jmath 1)(s - 1 - \jmath 1)}{(s + 3)^2(s + 4)(s + \jmath 3)(s - \jmath 3)} \tag{i}$$

From Eq. (i) we find that the poles are at $s = -3,\ -4,\ -\jmath 3$ and $+\jmath 3$, where the pole at $s = -4$ is simple, whereas the pole at $s = -3$ is multiple, having a multiplicity of 2. Zeros are at $s = -1,\ -2,\ 1 + \jmath 1$ and $1 - \jmath 1$. The pole-zero plot is depicted in Figure 3.7.

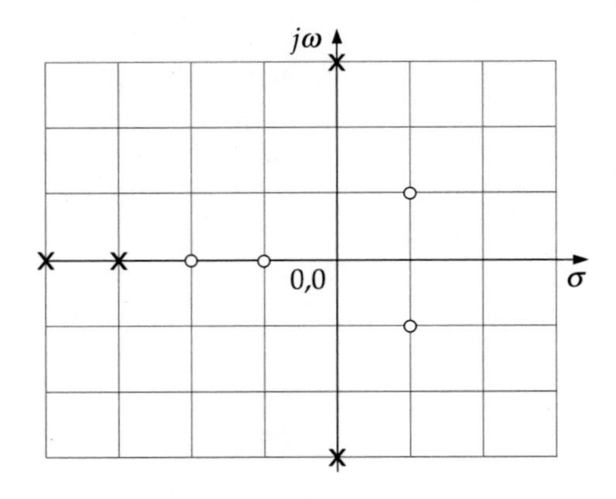

Figure 3.7 Pole-zero plot (Example 3.15).

Note: Complex poles occur in conjugate pairs.

EXAMPLE 3.16. A system has poles at $s = 4,\ -2$ and zero at $s = -3$. Sketch the pole-zero plot and find the transfer function if the gain factor is 4.

Solution The plot is given in Figure 3.8. The transfer function is given by

$$G(s) = \frac{4(s + 3)}{(s + 2)(s - 4)}$$

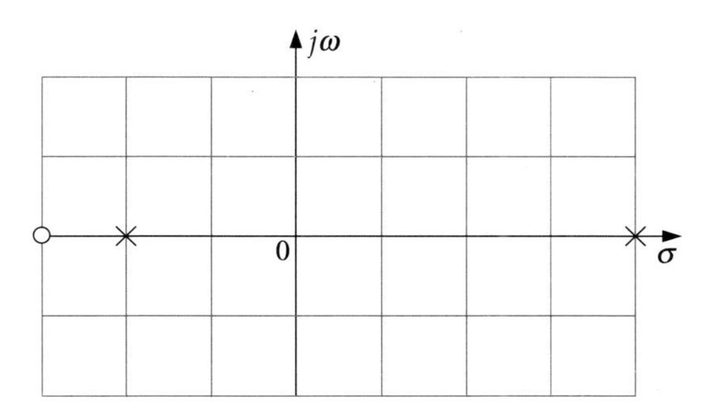

Figure 3.8 Pole-zero plot (Example 3.16).

EXAMPLE 3.17. Draw poles and zeros for $V(s) = \dfrac{s+1}{s^2+7s+12}$ and evaluate $v(t)$.

Solution The numerator and denominator of the function can be factorized to give it the form of

$$V(s) = \frac{(s+1)}{(s+3)(s+4)} \equiv \frac{A}{s+3} + \frac{B}{s+4}$$

Then, by applying rules of partial fractions (see Section 1.2 at page 9), we get

$$A = (s+3)V(s)|_{s=-3} = \left.\frac{s+1}{s+4}\right|_{s=-3} = -2$$

$$B = (s+4)V(s)|_{s=-4} = \left.\frac{s+1}{s+3}\right|_{s=-4} = 3$$

Therefore,
$$V(s) = \frac{3}{s+4} - \frac{2}{s+3} \tag{i}$$

Taking the inverse Laplace transform of Eq. (i), we obtain

$$v(t) = 3e^{-4t} - 2e^{-3t}$$

The pole-zero plot is given in Figure 3.9.

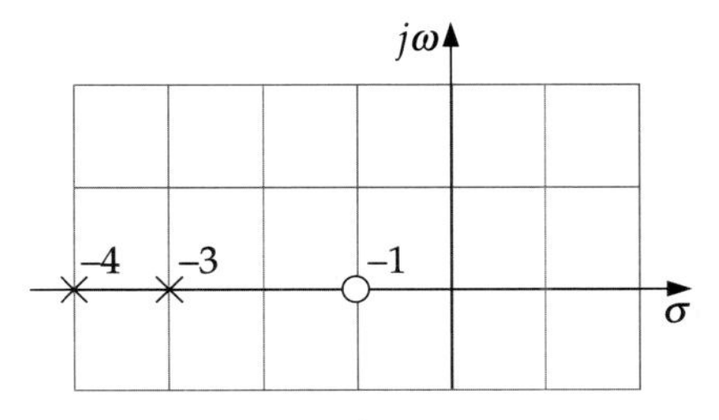

Figure 3.9 Pole-zero plot (Example 3.17).

EXAMPLE 3.18. Find the poles and zeros for the circuit given in Figure 3.10(a).

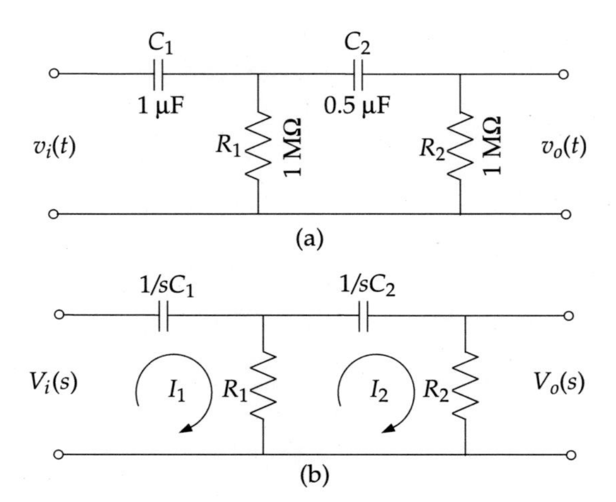

Figure 3.10 Example 3.18: (a) circuit; (b) its Laplace transform equivalent.

Solution Applying KVL to the Laplace transform of the circuit [see Figure 3.10(b)], we get

$$V_i(s) = \left(\frac{1}{sC_1} + R_1\right) I_1(s) - R_1 I_2(s) \tag{i}$$

$$0 = \left(\frac{1}{sC_2} + R_2 + R_1\right) I_2(s) - R_1 I_1(s) \tag{ii}$$

$$V_o(s) = R_2 I_2(s) \tag{iii}$$

Eliminating $I_1(s)$ from Eq. (i) with the help of Eq. (ii), we obtain

$$V_i(s) = \left[\left(\frac{1}{sC_1} + R_1\right)\left(\frac{1}{sR_1C_2} + \frac{R_2}{R_1} + 1\right) - R_1\right] I_2(s)$$

$$= \left(\frac{1}{s^2R_1C_1C_2} + \frac{1}{sC_2} + \frac{R_2}{sC_1R_1} + R_2 + \frac{1}{sC_1}\right) I_2(s) \tag{iv}$$

Putting $R_1 = R_2 = 1\ \text{M}\Omega$, $C_1 = 1\ \mu\text{F}$ and $C_2 = 0.5\ \mu\text{F}$, we have from Eq. (iv)

$$V_i(s) = \left(\frac{10^6}{0.5s^2} + \frac{10^6}{0.5s} + \frac{10^6}{s} + 10^6 + \frac{10^6}{s}\right) I_2(s)$$

$$= 10^6\left(\frac{2}{s^2} + \frac{4}{s} + 1\right) I_2(s) \tag{v}$$

From Eqs. (iii) and (v)

$$\frac{V_o(s)}{V_i(s)} = \frac{s^2}{s^2 + 4s + 2} = \frac{s^2}{(s + 0.586)(s + 3.414)} \tag{vi}$$

From Eq. (vi) we find that there are two zeros at $s = 0$ and poles are at $s = -0.586$ and -3.414.

EXAMPLE 3.19. Find the impulse response of the system, the transfer function of which is given by

$$G(s) = \frac{s}{s^2 + 4s + 2}$$

Solution The transfer function can be written as

$$G(s) = \frac{s}{(s + 0.586)(s + 3.414)} \equiv \frac{A}{s + 0.586} + \frac{B}{s + 3.414}$$

Now,
$$A = (s + 0.586)C(s)|_{s=-0.586} = \left.\frac{s}{s + 3.414}\right|_{s=-0.586} = -0.21$$

$$B = (s + 3.414)C(s)|_{s=-3.414} = \left.\frac{s}{s + 0.586}\right|_{s=-3.414} = 1.21$$

Therefore,
$$C(s) = G(s) = \frac{1.21}{s + 3.414} - \frac{0.21}{s + 0.586}$$

On taking inverse Laplace transform of the above equation, we get

$$c(t) = 1.21e^{-3.414t} - 0.21e^{-0.586t}$$

EXAMPLE 3.20. The transfer function of a system possesses a zero at $s = -4$ and poles at $s = 0$ (twice), $\jmath 2$, $-\jmath 2$, -2 and -3. If the value of the transfer function is 0.1 at $s = 1$, find the gain factor of the system.

Solution The transfer function of the system is given by

$$G(s) = \frac{K(s + 4)}{s^2(s - \jmath 2)(s + \jmath 2)(s + 2)(s + 3)} = \frac{K(s + 4)}{s^2(s^2 + 4)(s^2 + 5s + 6)}$$

Now, from the given condition,

$$G(s) = 0.1 = \left.\frac{K(s + 4)}{s^2(s^2 + 4)(s^2 + 5s + 6)}\right|_{s=1} = \frac{5K}{(5)(12)}$$

Therefore,
$$K = 1.2$$

Type and Order of Transfer Functions

The number of *poles at the origin* denotes the *type* of transfer function while the highest power of s in the *rationalized denominator polynomial* denotes the *order* of the transfer function.

Suppose, the transfer function of a control system is given by

$$G(s) = \frac{s + 9}{s^2 + 2s + 9}$$

Here, the denominator is given by

$$s^2 + 2s + 9$$

which consists of a power of 2 for the Laplace variable s. Therefore, it represents a second order system. Now, the same transfer function can be written as

$$G(s) = \frac{1 + \dfrac{9}{s}}{s + 2 + \dfrac{9}{s}}$$

Does it indicate that it is a first order system? The answer is obviously 'no' because the denominator contains a fraction.

EXAMPLE 3.21. A transfer function is given by

$$G(s) = \frac{100(s+5)(s+50)}{s^4(s+10)(s^2+3s+10)}$$

What are its type and order?

Solution The transfer function indicates four poles at the origin and if the denominator of the transfer function is expressed in a polynomial form, the highest power of s will be 7. Therefore, its type is 4, and the order is 7.

Kinds of Transfer Functions

Transfer functions are considered to be of three kinds:

1. Strictly proper
2. Proper
3. Improper

In our definition of the transfer function in Eq. (3.2), the orders of polynomials of the numerator and denominator are m and n, respectively. This means that there are m zeros and n poles of the transfer function given by Eq. (3.2). Three cases may arise in this respect:

1. $m < n$: This implies that number of poles exceeds that of zeros. Then,

$$\lim_{s \to \infty} G(s) \approx \frac{s^m}{s^n} = 0$$

   This means that, when there is no input to the system, there is no output (see *Initial Value Theorem* at page 6). All physical control systems conform to this property. This is why such a transfer function is called a *strictly proper transfer function.*
2. $m = n$: In this case, for $s \to \infty$, $G(s) \to K$. This situation may arise in some physical systems. Which is why such transfer functions are called *proper transfer functions.*
3. $m > n$: This implies $G(s) \to \infty$ as $s \to \infty$. That does not correspond to physical systems and, therefore, the transfer function is called *improper transfer function.*

Characteristic Equation

The denominator polynomial of the closed-loop transfer function is called the *characteristic polynomial.* The characteristic polynomial equated to zero is called the *characteristic equation.* Thus,

$$a_n s^n + a_{n-1} s^{n-1} + \cdots + a_1 s + a_0 = 0$$

is the characteristic equation of the transfer function represented by Eq. (3.2). Obviously, the roots of the characteristic equation are poles of the closed-loop transfer function.

EXAMPLE 3.22. If the open-loop transfer function of negative feedback control system is

$$G(s)H(s) = \frac{K}{(s+1)^3}$$

then the gain K for a closed-loop pole to be at $-\frac{1}{2} + j\frac{\sqrt{3}}{2}$ is

(a) 1 (b) 2 (c) 3 (d) 4

Solution From the characteristic equation, we get

$$1 + G(s)H(s) = 1 + \frac{K}{(s+1)^3} = 0$$

or

$$K = -(s+1)^3$$

Therefore,

$$K\Big|_{s=-\frac{1}{2}+j\frac{\sqrt{3}}{2}} = \left|\frac{1}{2} + j\frac{\sqrt{3}}{2}\right|^3 = \left|\sqrt{\frac{1}{4}+\frac{3}{4}}\right|^3 = 1 \qquad [\because K \text{ is real}]$$

Ans: (a).

Minimum Phase System

If no pole or zero of a transfer function lies on the right half of the s-plane, the transfer function is called the *minimum phase transfer function* and the system, *minimum phase system.* Conversely the transfer functions having poles or zeros in the right half of the s-plane are called *non-minimum phase transfer functions* and the systems, *non-minimum phase systems.* For example, a system having transfer function as

$$G(s) = \frac{1 + s\tau_1}{1 + s\tau_2}$$

is a minimum phase system, while the one having a transfer function of

$$G(s) = \frac{1 - s\tau_1}{1 + s\tau_2}$$

is a non-minimum phase system. The corresponding pole-zero diagrams are shown in Figure 3.11. We will have occasions to discuss the reasons for such nomenclature later.

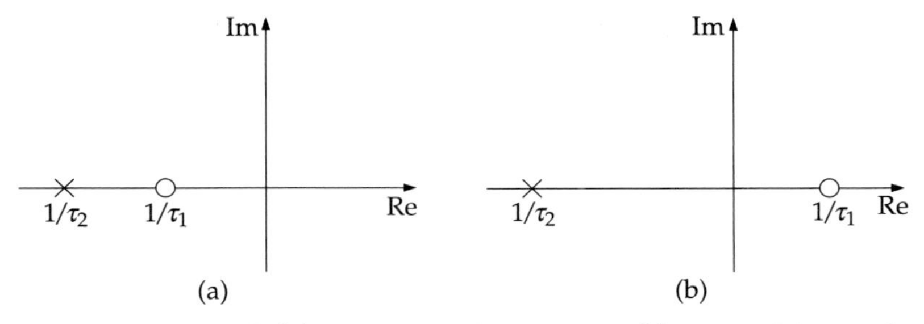

Figure 3.11 Pole-zero plot of: (a) a minimum-phase system; (b) a non-minimum phase system.

EXAMPLE 3.23. Of the following transfer functions, the function that has a minimum phase characteristic is

(a) $\dfrac{1+s}{1-s}$ (b) $\dfrac{1-s}{1+s}$ (c) $\dfrac{e^{-3}}{s}$ (d) $\dfrac{s}{(s+1)(s+2)}$

Solution Of the four choices, the first two are ruled out because they possess pole/zero in the RHP. For the remaining two, the phase factors are calculated as follows:

(c) $$\frac{e^{-3}}{s} \equiv \frac{e^{-3}}{\jmath\omega} = \frac{e^{-3}}{\omega}\angle -90^\circ$$

(d) $$\frac{s}{(s+1)(s+2)} \equiv \frac{\jmath\omega}{(\jmath\omega+1)(\jmath\omega+2)} = \frac{\omega}{\sqrt{(\omega^2+1)(\omega^2+4)}}\angle\left(90^\circ - \tan^{-1}\omega - \tan^{-1}\frac{\omega}{2}\right)$$

Obviously, the answer is (c) which has a fixed phase with no variation.

EXAMPLE 3.24. The relationship between the input $x(t)$ and output $y(t)$ of a system is given by
$$\frac{d^2y}{dt^2} = x(t-2) + \frac{d^2x}{dt^2}$$

The transfer function of the system is

(a) $1+\dfrac{e^{-2s}}{s^3}$ (b) $1+\dfrac{e^{2s}}{s^3}$ (c) $1+s^2e^{-2s}$ (d) $1+s^2e^{2s}$

Solution From the given equation, we get, on Laplace transform, the equation
$$s^2Y(s) = \left(\frac{e^{-2s}}{s} + s^2\right)X(s)$$

$$\Rightarrow \qquad \text{Transfer function} = \frac{Y(s)}{X(s)} = \frac{(e^{-2s}/s)+s^2}{s^2} = 1+\frac{e^{-2s}}{s^3}$$

Ans: (a).

EXAMPLE 3.25. The transfer function of a passive circuit has its poles and zeros on

(a) left and right halves, respectively, of the s-plane
(b) right and left halves, respectively, of the s-plane
(c) right half of the s-plane
(d) left half of the s-plane

Solution Ans: (d) (see Examples 3.8–3.10).

EXAMPLE 3.26. An analogue LTI system has a negative phase-shift which varies linearly with frequency. The magnitude of response is frequency independent. The system transfer function is

(a) $\dfrac{1-s\tau}{1+s\tau}$ (b) $e^{-s\tau}$ (c) $\dfrac{1-s^2\tau^2}{1+s^2\tau^2}$ (d) $\dfrac{1-s\tau}{1+s\tau}e^{-s\tau}$

Solution In frequency domain the transfer functions can be written as

(a) $$\frac{1-\jmath\omega\tau}{1+\jmath\omega\tau} \equiv 1\angle -2\tan^{-1}\omega\tau$$

(b) $$e^{-\jmath\omega\tau} = \cos\omega\tau - \jmath\,\sin\omega\tau \equiv 1\angle -\tan^{-1}\tan\omega\tau = 1\angle -\omega\tau$$

(c) $\dfrac{1+\omega^2\tau^2}{1-\omega^2\tau^2} \equiv \dfrac{1+\omega^2\tau^2}{1-\omega^2\tau^2}\angle 0^\circ$

(d) $\dfrac{1-\jmath\omega\tau}{1+\jmath\omega\tau}e^{-\jmath\omega\tau} \equiv 1\angle -2\tan^{-1}\omega\tau - \omega\tau$

Thus, of the four choices, only (b) satisfies the given conditions.

EXAMPLE 3.27. A transfer function with unity dc gain has three poles at -1, -2 and -3, and no finite zeros. A plant with transfer function is connected with a proportional controller of gain K in the forward path, in a unity feedback configuration. The transfer function is

(a) $\dfrac{s}{(s-1)(s-2)(s-3)}$ (b) $\dfrac{6}{(s+1)(s+2)(s+3)}$

(c) $\dfrac{s}{(s+1)(s+2)(s+3)}$ (d) $\dfrac{6}{(s-1)(s-2)(s-3)}$

Solution We eliminate (a) and (d) which suggest existence of poles in the RHP. We also rule out (c) which consists of a zero at the origin. Obviously, then, (b) is the answer.

EXAMPLE 3.28. As shown in the following figure, a negative feedback system has an amplifier of gain 100 with $\pm 10\%$ tolerance in the forward path, and an attenuator of value $9/100$ in the feedback path.

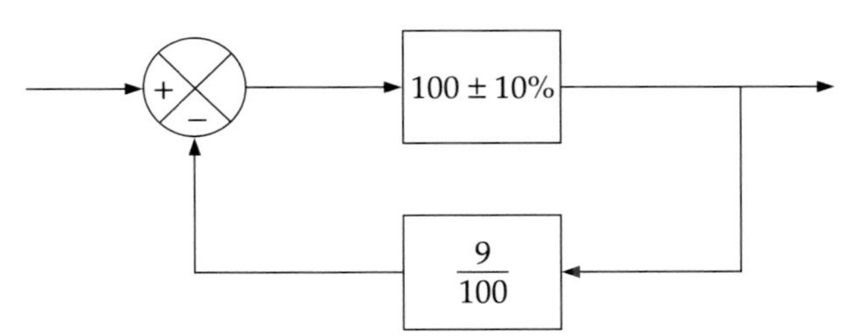

The overall system gain is approximately:

(a) $10 \pm 1\%$ (b) $10 \pm 2\%$ (c) $10 \pm 5\%$ (d) $10 \pm 10\%$

Solution The amplifier gain may vary from 90 to 110. The corresponding variations of the overall gain are as follows:

For 90, $$\text{Gain} = \frac{C(s)}{R(s)} = \frac{90}{1+90\times\dfrac{9}{100}} = 9.89$$

For 110, $$= \frac{110}{1+110\times\dfrac{9}{100}} = 10.09$$

The arithmetic mean of the overall gain is $\dfrac{9.89+10.09}{2} = 9.99$, and the deviation from the mean is ± 0.1. So, the per cent variation is $\dfrac{\pm 0.1}{9.99}\times 100 \approx \pm 1$.

Ans: (a).

3.3 STANDARD INPUTS AND THEIR LAPLACE TRANSFORMS

Unit Step Input

Mathematically, the unit step input is expressed as

$$u(t) = \begin{cases} 0 & \text{if } t \leq 0 \\ 1 & \text{if } t > 0 \end{cases}$$

Graphically, the input looks similar to that in Figure 3.12(a).

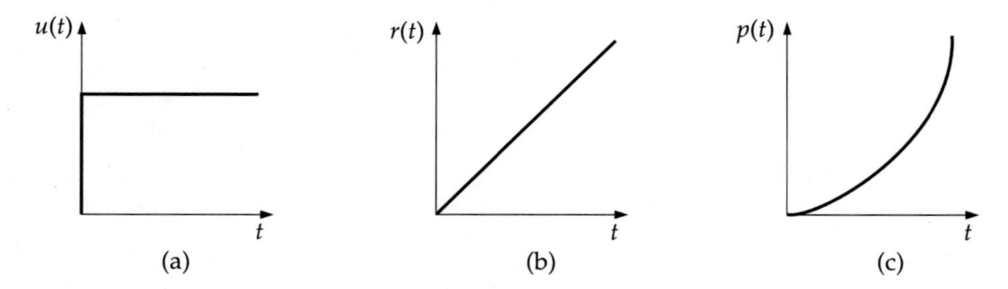

Figure 3.12 Standard inputs: (a) unit step; (b) unit ramp; (c) unit parabolic.

The Laplace transform of the unit step input can be worked out as follows:

$$\mathscr{L}\{u(t)\} = \int_0^\infty u(t)e^{-st}dt$$

$$= \int_0^\infty e^{-st}dt = -\frac{1}{s}\left[e^{-st}\right]_0^\infty$$

$$\therefore \qquad \mathscr{L}\{u(t)\} = \frac{1}{s}$$

Unit Ramp Input

The unit ramp input is represented as

$$r(t) = t$$

Graphically, it is similar to that in Figure 3.12(b). Its Laplace transform can be worked out as follows:

$$\mathscr{L}\{r(t)\} = \int_0^\infty te^{-st}dt = \left[-\frac{te^{-st}}{s} + \int \frac{e^{-st}}{s}dt\right]_0^\infty$$

$$= \left[-\frac{te^{-st}}{s}\right]_0^\infty - \left[\frac{e^{-st}}{s^2}\right]_0^\infty$$

Therefore, $$\mathscr{L}\{r(t)\} = \frac{1}{s^2}$$

Unit Parabolic Input

The mathematical form of the parabolic input is

$$p(t) = \frac{t^2}{2}$$

Its graphical form is shown in Figure 3.12(c). The Laplace transform is

$$\mathscr{L}\{p(t)\} = \frac{1}{2}\int_0^\infty t^2 e^{-st}dt = \frac{1}{2}\left(-\frac{t^2}{s}e^{-st}\right)_0^\infty + \frac{1}{s}\int_0^\infty te^{-st}dt$$

$$= 0 + \frac{1}{s}\left(\frac{1}{s^2}\right)$$

Therefore, $$\mathscr{L}\{p(t)\} = \frac{1}{s^3}$$

Unit Impulse Input

This input is defined in terms of the Dirac delta function which is given by

$$\delta(t) = \begin{cases} \lim\limits_{\varepsilon \to 0} \dfrac{1}{\varepsilon} & \text{for } 0 \le t \le \varepsilon \\ 0 & \text{for } t > \varepsilon \end{cases}$$

The delta function can also be written as

$$\delta(t) = \frac{du(t)}{dt}$$

where $u(t)$ is the unit step function.

Thus, the Laplace transform of the impulse input works out as

$$\mathscr{L}\{\delta(t)\} = \mathscr{L}\left\{\frac{du(t)}{dt}\right\} = s \cdot \frac{1}{s}$$

Therefore, $$\mathscr{L}\{\delta(t)\} = 1 \tag{3.4}$$

Impulse response and transfer function. We have seen in Eq. (3.4) that the Laplace transform of the impulse input is 1. We know that the transfer function is defined in Eq. (3.1) as

$$G(s) = \frac{C(s)}{R(s)}$$

which means that the Laplace transform of the response (or output) of a system is given by

$$C(s) = G(s)R(s)$$

For an impulse input, $R(s) = 1$. So, the response of an impulse input equals the transfer function, i.e., $C(s) = G(s)$. This is an important result. Now, let us work out a couple of examples.

EXAMPLE 3.29. The impulse response of a system is e^{-5t}. Determine its transfer function.

Solution Here, $c(t) = e^{-5t}$. Its Laplace transform gives

$$C(s) = \frac{1}{s+5} = G(s)$$

EXAMPLE 3.30. The unit impulse response of a system is $e^{-t}(1-\cos 3t)$. Determine its transfer function.

Solution Here, $c(t) = e^{-t}(1 - \cos 3t) = e^{-t} - e^{-t}\cos 3t$. Therefore,

$$\mathscr{L}\{c(t)\} = \mathscr{L}\{e^{-t}\} - \mathscr{L}\{e^{-t}\cos 3t\} = \frac{1}{s+1} - \frac{s+1}{(s+1)^2+9}$$

$$= \frac{9}{(s+1)(s^2+2s+10)} = G(s)$$

REVIEW QUESTIONS

3.1 Choose the correct answer:

(i) A system has the transfer function of $\frac{1-s}{1+s}$. Its gain at 1 rad/s is

(a) 1 (b) 0 (c) −1 (d) 0.5

(ii) For the characteristic equation $s(s+4)(s+7) + k(s+1)(s+5) = 0$, the roots of the system at $k \to \infty$ will be at

(a) $-1,\ -5$ and ∞ (b) $-4,\ -7$ and ∞
(c) $-4,\ \infty$ and $-\infty$ (d) $0,\ -1$ and -4

(iii) A system has the transfer function of $(1-s)/(1+s)$. It is known as:

(a) minimum phase system (b) non-minimum phase system
(c) low-pass system (d) high-pass system

(iv) The transfer function of a system is given by $G(s) = \frac{10(s+1)(s+10)}{s^4(s+5)(s^2 3s+10)}$. The type and order of the system are, respectively,

(a) 4 and 7 (b) 4 and 9 (c) 7 and 5 (d) 5 and 7

(v) The type number of a transfer function denotes the number of

(a) poles at the origin (b) zeros at the origin
(c) poles at infinity (d) none of these

(vi) Given

$$G(s) = \frac{K}{s^2(s+2)(s+3)}.$$

The type and order of the system are

(a) 3 and 3 (b) 2 and 4 (c) 3 and 1 (d) 3 and 0

(vii) A system has a transfer function $(1 - s)/(1 + s)$. It is known as

(a) low-pass system (b) high-pass system

(c) all pass system (d) minimum phase system

(viii) The system that can be represented by a transfer function is a

(a) nonlinear time dependent system

(b) nonlinear time varying system

(c) linear time varying system

(d) linear time independent system

(ix) The closed-loop gain of the system in the given figure is

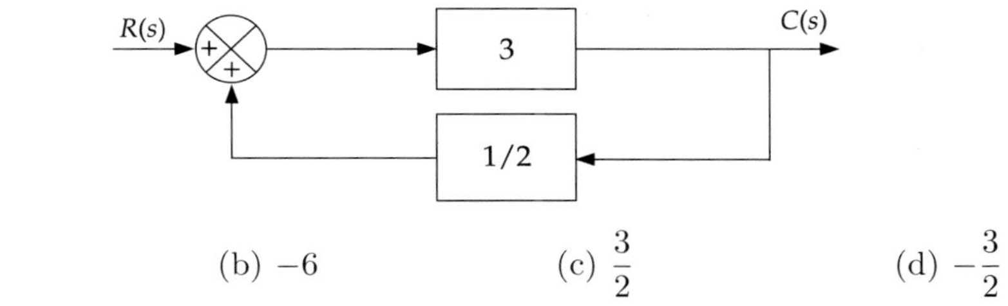

(a) 6 (b) -6 (c) $\frac{3}{2}$ (d) $-\frac{3}{2}$

(x) The transfer function of a simple RC integrator circuit, shown in the figure, is

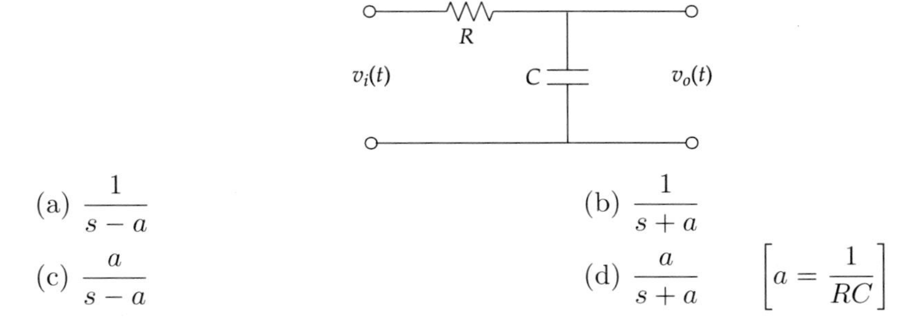

(a) $\frac{1}{s-a}$ (b) $\frac{1}{s+a}$

(c) $\frac{a}{s-a}$ (d) $\frac{a}{s+a}$ $\left[a = \frac{1}{RC}\right]$

3.2 What is a linear time-invariant system?

3.3 Consider the system given in Figure 3.13, where

$$G(s) = \frac{s + 25}{s^4 + 10s^3 + 35s^2 + 50s + 24}$$

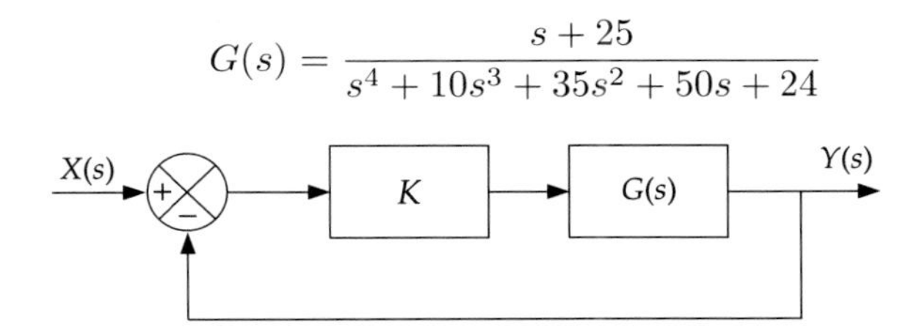

Figure 3.13 System of Question 3.3.

(a) If one pole is at $s = -1$, what are other open-loop poles and zeros?
(b) What is the characteristic equation of the system?

3.4 The dynamical behaviours of a few systems are represented by the following differential equations. Find which of them are linear time-invariant systems.

(a) $M\dfrac{d^2x}{dt^2} + D\dfrac{dx}{dt} + Kx + f(x)$ (b) $L\dfrac{di}{dt} + Ri + \dfrac{\int i\,dt}{C} = e(i)$

(c) $M(t)\dfrac{d^2x}{dt^2} + K(t)x = f(t,x)$ (d) $Li\dfrac{di}{dt} + Ri^2 = P(i)$

3.5 A system transfer function is given by

$$G(s) = \frac{10(s+1)(s+3)}{s(s+2)(s+1-j2)(s+1+j2)}$$

Plot poles and zeros in the s-plane and determine the value of the transfer function at $s = 1$.

3.6 The pole-zero plot of a system is given by Figure 3.14. If $G(s) = 10$ for $s = 1$, find the transfer function of the system.

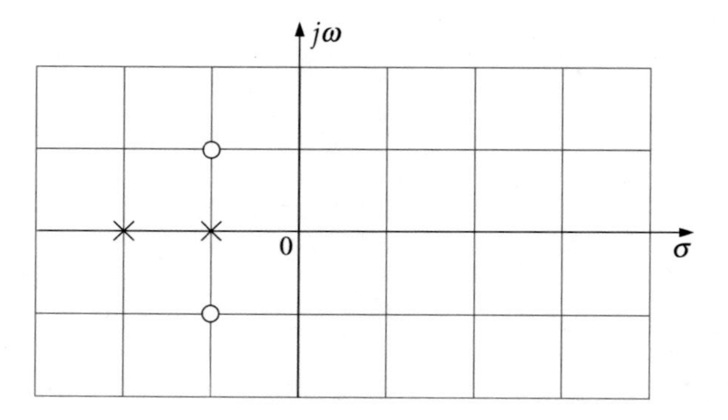

Figure 3.14 Pole-zero plot of Question 3.6.

3.7 The unit impulse response of a system is $e^{-it} \sin \omega t$. Find the transfer function of the system.

3.8 The transfer function of a system is given by

$$G(s) = \frac{Y(s)}{X(s)} = \frac{6}{s^2 + 5s + 2}$$

Find the differential equation that describes the dynamical behaviour of the system.

3.9 What is the difference between order and type of control system? Discuss with the help of examples.

3.10 A system has an open-loop transfer function of $\frac{K}{s^3(Ts+1)}$. What are the type and order of the system?

3.11 Find the transfer function of the following network:

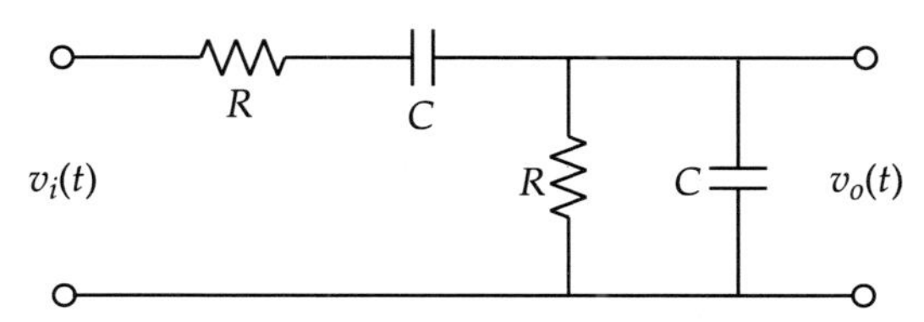

3.12 Obtain the transfer function of the circuit shown:

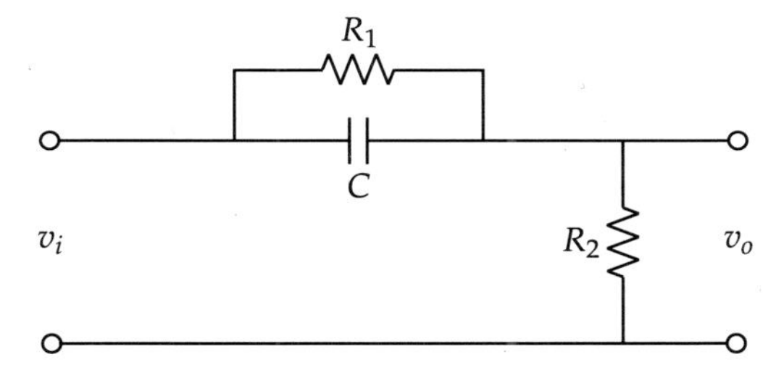

Answers to Selected Questions

3.1 (i) (a), (ii) (a), (iii) (b), (iv) (a), (v) (a), (vi) (b), (vii) (c), (viii) (d), (ix) (b), (x) (d)

3.3 (a) The poles at $s = -2, -3, -4$. Zero at $s = -25$

(b) The characteristic equation is $s^4 + 10s^3 + 35s^2 + (50 + K)s + (24 + 25K) = 0$.

3.4 (a) and (b)

3.5 3.33

3.6 $\frac{12(s^2 + 2s + 2)}{(s+1)(s+2)}$

3.7 $\frac{\omega}{s^2 + 2s + (1 + \omega^2)}$

3.8 $\frac{d^2y}{dt^2} + 5\frac{dy}{dt} + 2y = 6x$

3.10 Type 3, order 4

3.11 $\frac{sR_1R_2C + R_2}{R_1 + R_2 + sR_1R_2C}$

3.12 $\frac{sRC}{s^2R^2C^2 + 2sRC + R + 1}$

CHAPTER 4

Block Diagrams and Signal Flow Graphs

4.1 BLOCK DIAGRAMS

So far we have seen that automatic control systems are concerned with dynamic systems and that such systems can be mathematically represented by a set of simultaneous differential equations. It is convenient to take the Laplace transform of differential equations because this transform converts them into linear algebraic equations which are rather easy to handle. Then transfer functions are defined to establish relations between input and output.

We have also seen that transfer functions can be represented by blocks in a block diagram to give a visual presentation of the relationship between different constituents of the control system. A block diagram can therefore be termed as a shorthand pictorial presentation of the cause and effect relationship between the output and input of a control system (see Figure 4.1).

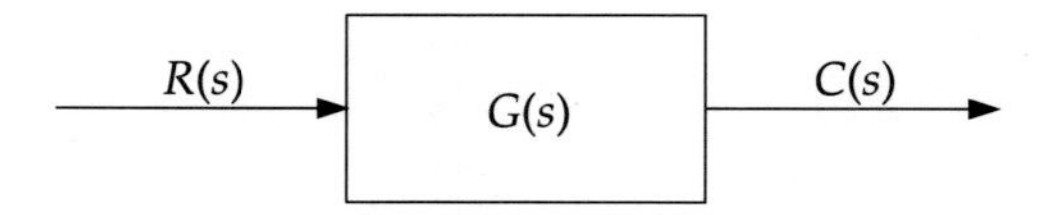

Figure 4.1 An example of a block diagram.

The block diagram technique has been developed not only to show the relationship between constituents of a control system but also to do algebraic manipulations through their reduction procedure. Before we proceed to do that, let us familiarize ourselves with a few definitions on block diagram presentation of control systems.

Definitions

Summing point. Represented by a circle with inner[1] indications of signs (see Figure 4.2), these points show the addition or subtraction of signals from different sources.

Here, the algebraic relation between the different variables at the summing point is

$$C(s) = X(s) - Y(s) + Z(s)$$

[1]Some authors prefer *outer* indication of signs.

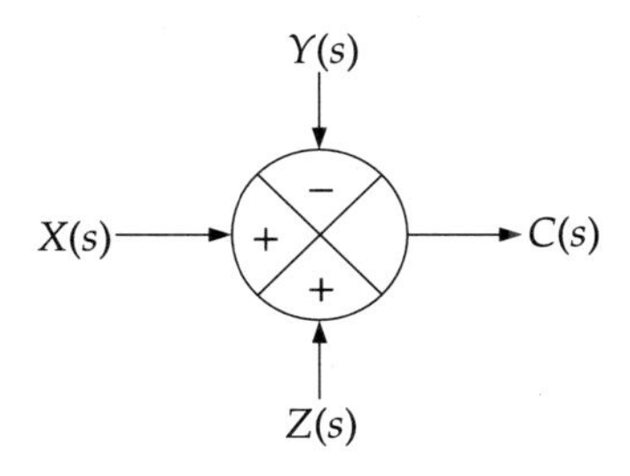

Figure 4.2 Summing point in a block diagram.

Take-off point.[2] If the output of a signal branches off from one point, as in the case of parallel application of voltage to a circuit, the point of branching off is called the *take-off point* (see Figure 4.3).

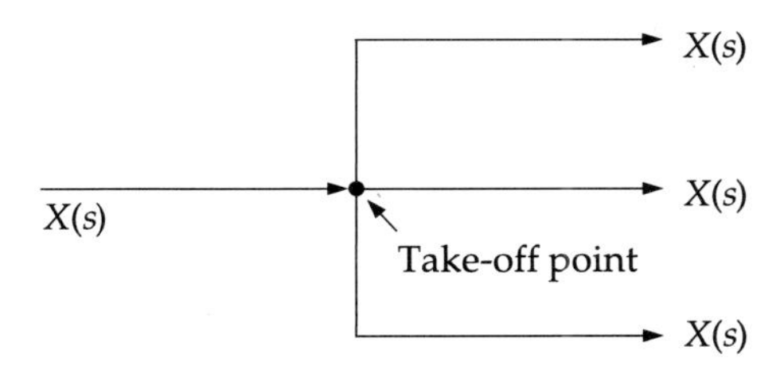

Figure 4.3 Take-off point in a block diagram.

Forward and feedback paths. Forward and feedback paths are indicated in a block diagram (Figure 4.4). A forward path is indicated by an arrow pointing towards right while a left pointing arrow indicates a feedback path.

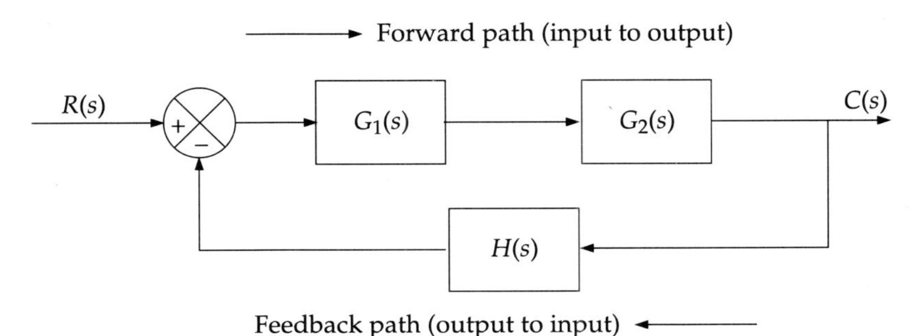

Figure 4.4 Forward and feedback paths in a block diagram.

Table 4.1 lists different symbols that are used in a block diagram.

[2] Also called *pick-off* point.

Table 4.1 List of Symbols used in Block Diagrams

Symbol	*Meaning*
$R(s)$	Laplace transform of the reference signal
$C(s)$	Laplace transform of the controlled output
$G(s)$	Transfer function of the forward path
$H(s)$	Transfer function of the feedback path
$B(s)$	Laplace transform of the feedback signal
$E(s)$	Laplace transform of the error signal
$T(s)$	Closed-loop transfer function $=C(s)/R(s)$

Drawing Block Diagrams

We are familiar with drawing block diagrams of simple control systems—e.g. open-loop and closed-loop systems—right from the beginning. Here we consider how to draw block diagrams if electrical circuits are given or dynamical equations are provided. A few examples will help us understand the procedure.

EXAMPLE 4.1. Draw the block diagram for the electrical circuit given in Figure 4.5.

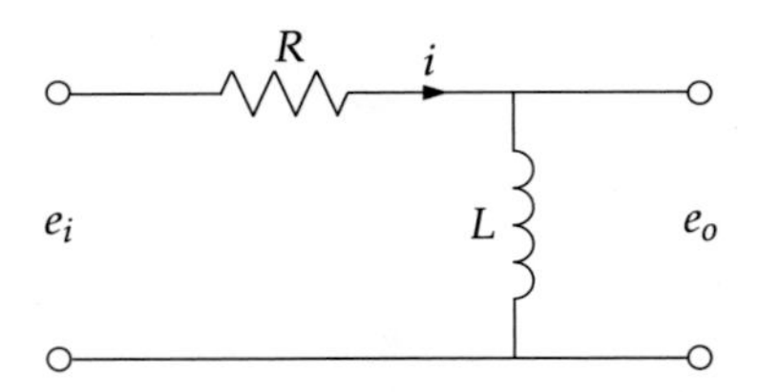

Figure 4.5 Electrical circuit (Example 4.1).

Solution Using KCL and KVL, we can write the following equations:

$$i = \frac{e_i - e_o}{R}$$

$$e_o = L\frac{di}{dt}$$

The Laplace transform of these equations, assuming all initial conditions = 0, are

$$I(s) = \frac{E_i(s) - E_o(s)}{R}$$

$$E_o(s) = sLI(s)$$

The two block diagrams given in Figure 4.6 represent the two equations individually.

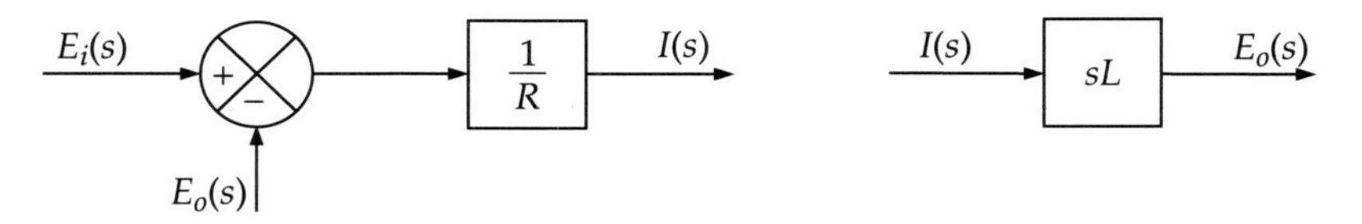

Figure 4.6 Block diagrams of Example 4.1.

These diagrams can be put together as in Figure 4.7 to represent the given circuit.

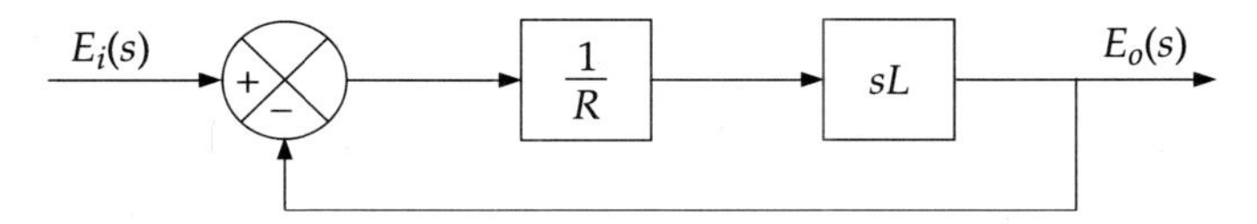

Figure 4.7 Complete block diagram of Example 4.1.

EXAMPLE 4.2. Draw the block diagram for the circuit given in Figure 4.8.

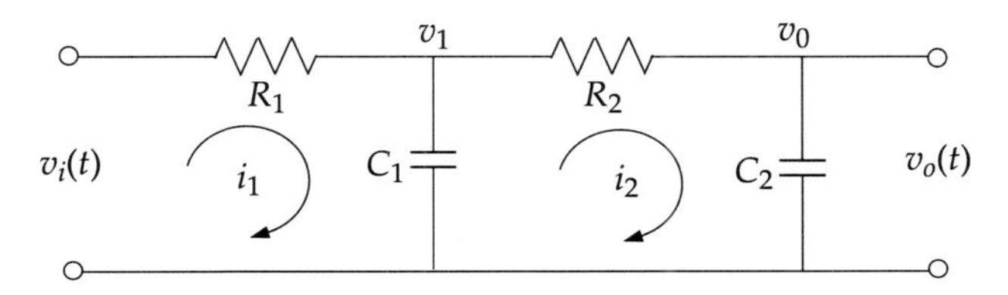

Figure 4.8 Electrical circuit of Example 4.2.

Solution Applying KVL and KCL to the circuit, we get

$$i_1 = \frac{v_i - v_1}{R_1}, \qquad v_1 = \frac{\int (i_1 - i_2)dt}{C_1}$$

$$i_2 = \frac{v_1 - v_o}{R_2}, \qquad v_o = \frac{\int i_2\, dt}{C_2}$$

Their Laplace transforms, under relaxed conditions, yield

$$I_1(s) = \frac{1}{R_1}[V_i(s) - V_1(s)] \tag{i}$$

$$V_1(s) = \frac{1}{sC_1}[I_1(s) - I_2(s)] \tag{ii}$$

$$I_2(s) = \frac{1}{R_2}[V_1(s) - V_o(s)] \tag{iii}$$

$$V_o(s) = \frac{1}{sC_2}I_2(s) \tag{iv}$$

Equation (i) can be converted to a block diagram as follows. We note that $V_i(s)$ is the input corresponding to $v_i(t)$ of the given circuit. So, putting this in the input of a summing point, we draw Figure 4.9 to represent Eq. (i).

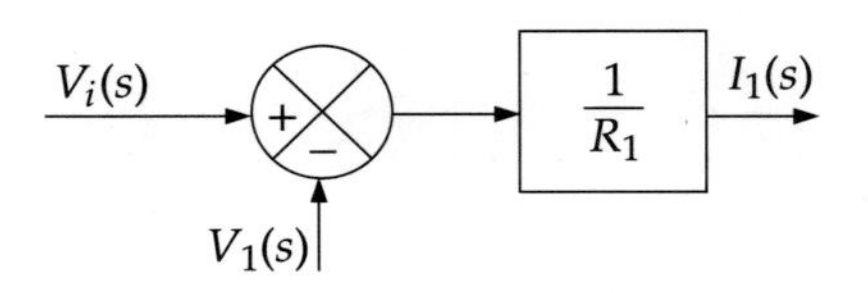

Figure 4.9 Part of block diagram to represent Eq. (i).

Equation (ii) modifies the diagram to its form given in Figure 4.10.

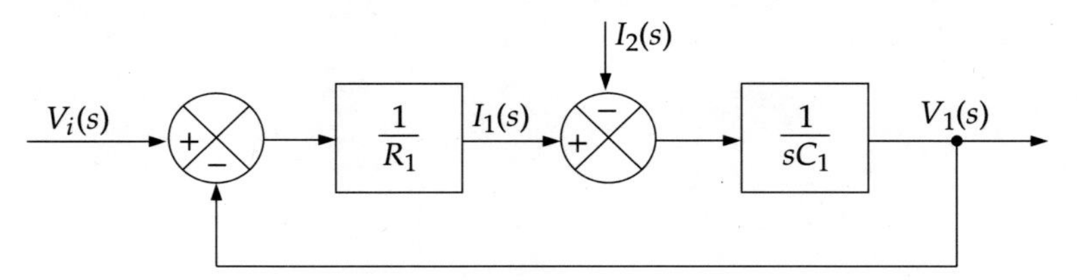

Figure 4.10 Part of block diagram to represent Eqs. (i) and (ii).

Equation (iii) extends the diagram to representation in Figure 4.11.

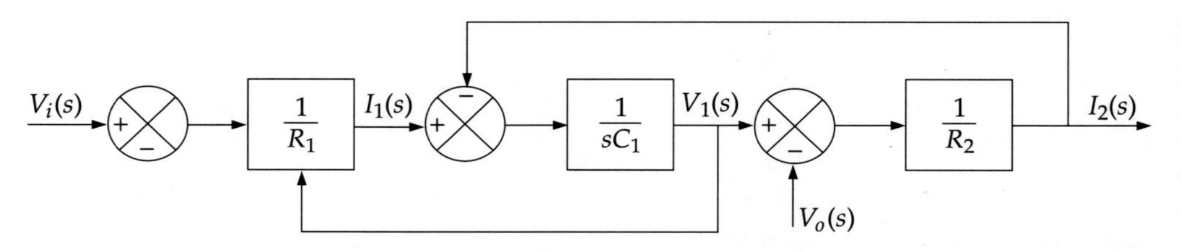

Figure 4.11 Part of block diagram to represent Eqs. (i) to (iii).

The last equation, Eq. (iv), completes the diagram to look like to the one given in Figure 4.12.

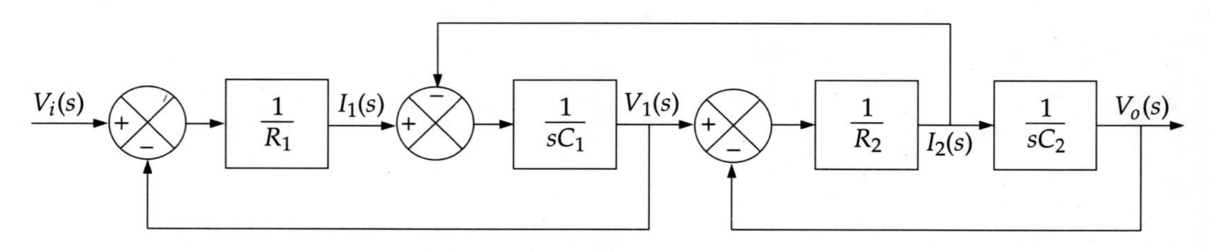

Figure 4.12 Complete block diagram representing Eqs. (i) to (iv).

Reduction of a Closed-Loop Block Diagram

Figure 4.13 represents the block diagram of a closed-loop control system. Different quantities, all Laplace transformed, that we will use are defined in Table 4.1.

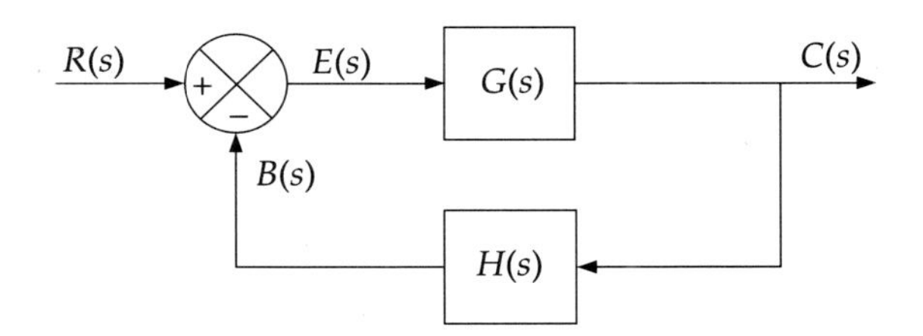

Figure 4.13 Closed-loop control system.

From Figure 4.13,

$$C(s) = E(s)G(s) \tag{4.1}$$

$$B(s) = C(s)H(s) \tag{4.2}$$

$$E(s) = R(s) - B(s) \tag{4.3}$$

Substituting Eq. (4.3) in Eq. (4.1), we get

$$C(s) = [R(s) - B(s)]G(s) \tag{4.4}$$

Plugging the value of $B(s)$ from Eq. (4.2) in Eq. (4.4), we get

$$C(s) = R(s)G(s) - C(s)H(s)G(s) \tag{4.5}$$

On rearrangement, Eq. (4.5) yields the closed-loop transfer function as

$$T(s) = \frac{C(s)}{R(s)} = \frac{G(s)}{1 + G(s)H(s)} \tag{4.6}$$

Equation (4.6) is an important result which indicates that the closed-loop of Figure 4.13 can be reduced to give Figure 4.14.

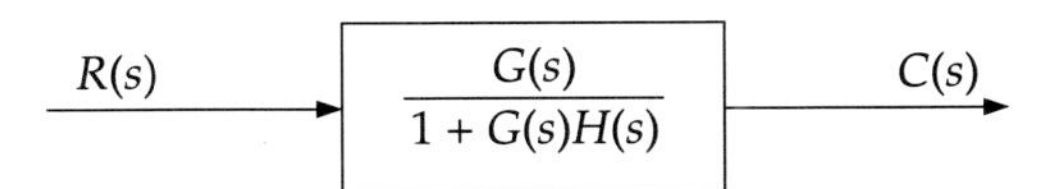

Figure 4.14 Reduced form of closed-loop of Figure 4.13.

Incidentally, the quantity $G(s)H(s)$ appearing in the denominator of Eq. (4.6) is often called the *loop transfer function* or *open-loop transfer function.*

Reduction Rules

The block diagram presentation of a given system may be reduced to a simplified block diagram with fewer blocks than the original diagram. This is done by applying block diagram reduction rules which are presented in Table 4.2.

Table 4.2 Block Diagram Reduction Rules

Transformation	*Original diagram*	*Equivalent diagram*
Combining blocks in cascade	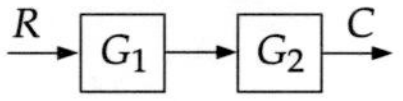 R → G_1 → G_2 → C	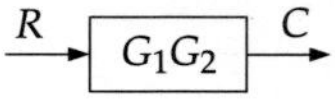 R → G_1G_2 → C
Combining blocks in parallel	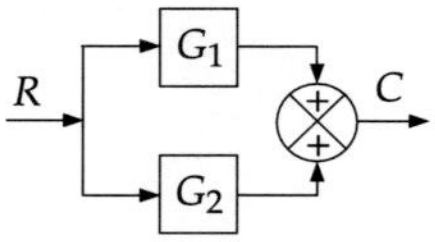 R → G_1, G_2 → (+ +) → C	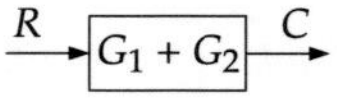 R → $G_1 + G_2$ → C
Moving a summing point beyond a block	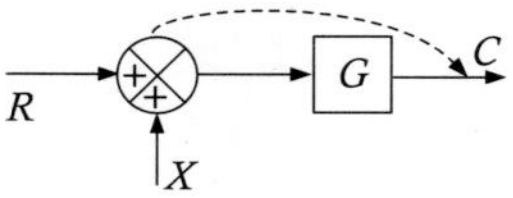 $C = (R + X)G$	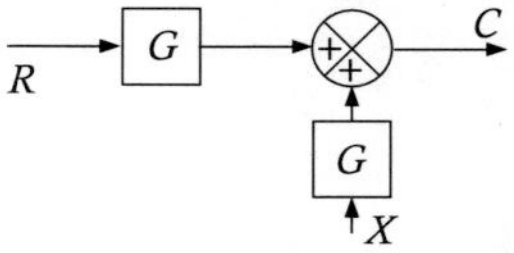 $C = RG + XG$
Moving a summing point behind a block	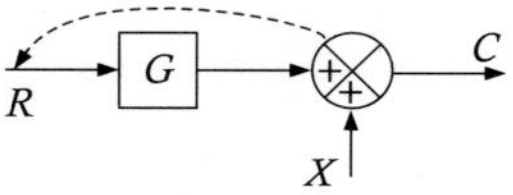 R → G → (+ +) → C; X	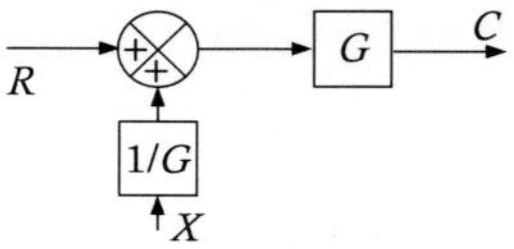 R → (+ +) → G → C; X → $1/G$
Eliminating a feedback loop	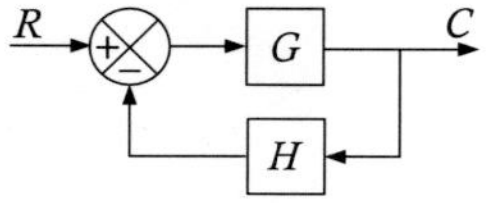 R → (+ −) → G → C; H	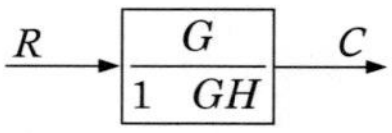 R → $\frac{G}{1\ GH}$ → C
Moving a take-off point beyond a block	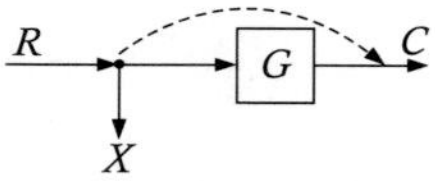 $X = R \quad C = RG$	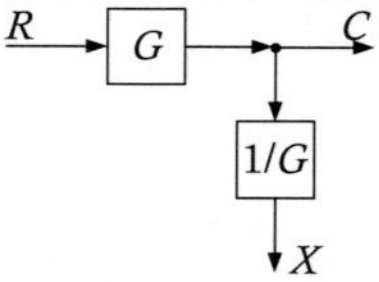 $X = R \quad C = RG$
Moving a take-off point behind a block	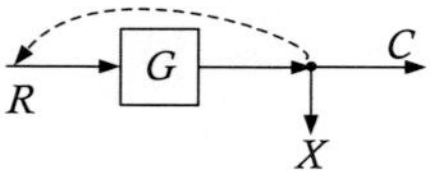 $X = C = RG$	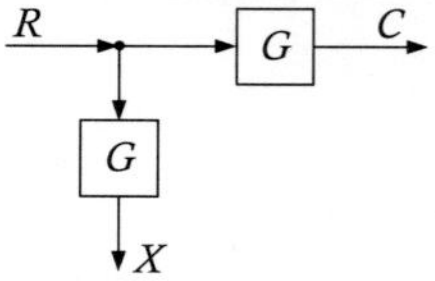 $X = C = RG$

It is better to follow a general procedure for block diagram reduction. The steps needed for the procedure are:

Step 1. Reduce cascade blocks.

Step 2. Reduce parallel blocks.

Step 3. Eliminate feedback loops.

Step 4. Move summing points to right and take-off points to left.

Step 5. Repeat steps 1 to 4 until a simple form is obtained.

However, these steps cannot be followed in all cases in the given order. Reduction of block diagrams requires a little bit of practice when one develops a skill as to what step has to be taken at what time.

Multi-input system. In case a multi-input system has to be reduced, the following steps should be taken (see Example 4.5):

Step 1. Put all but one inputs to zero. Find the resultant output.

Step 2. Repeat Step 1 until all inputs are covered.

Step 3. Find the resultant output by superposition.

With this background, we now consider a few block diagram reductions.

EXAMPLE 4.3. Determine the transfer function of the block diagram given in Figure 4.15.

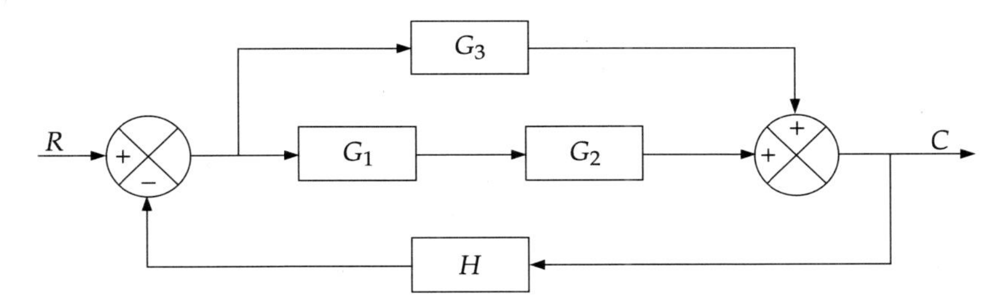

Figure 4.15 Block diagram (Example 4.3).

Solution First we combine cascaded and parallel blocks. As a result, we get the feedback block of Figure 4.15(a). Next, we reduce this feedback loop to arrive at the final reduced form as given in Figure 4.15(b).

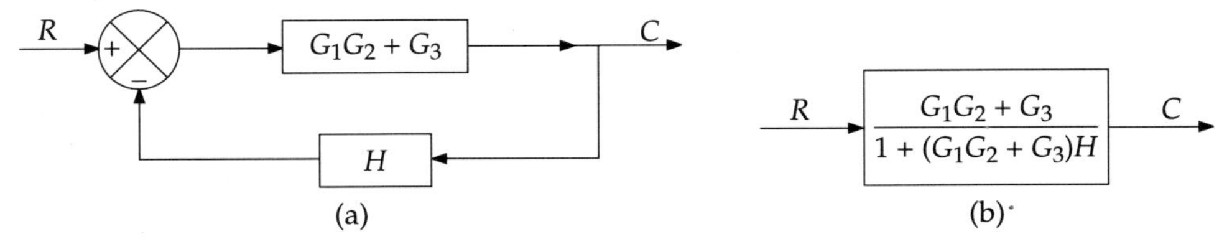

Figure 4.15 (a) Feedback block; (b) final reduced form (Example 4.3).

The transfer function is given by

$$\frac{C(s)}{R(s)} = \frac{G_1G_2 + G_3}{1 + (G_1G_2 + G_3)H}$$

EXAMPLE 4.4. Reduce the block diagram given in Figure 4.16 and obtain the transfer function.

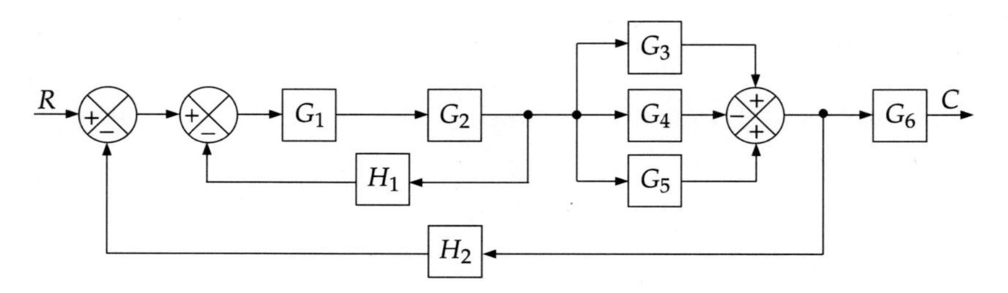

Figure 4.16 Block diagram (Example 4.4).

Solution By inspection, we find that after the first summing point there is a feedback loop and then a parallel block. These are independent. So, we take the following steps: (i) reduce the feedback loop and the parallel block, (ii) reduce the resulting feedback loop on the left, and (iv) multiply the resultant by $G6$. The successive steps are shown in 1, 2 and 3 in Figure 4.16(a).

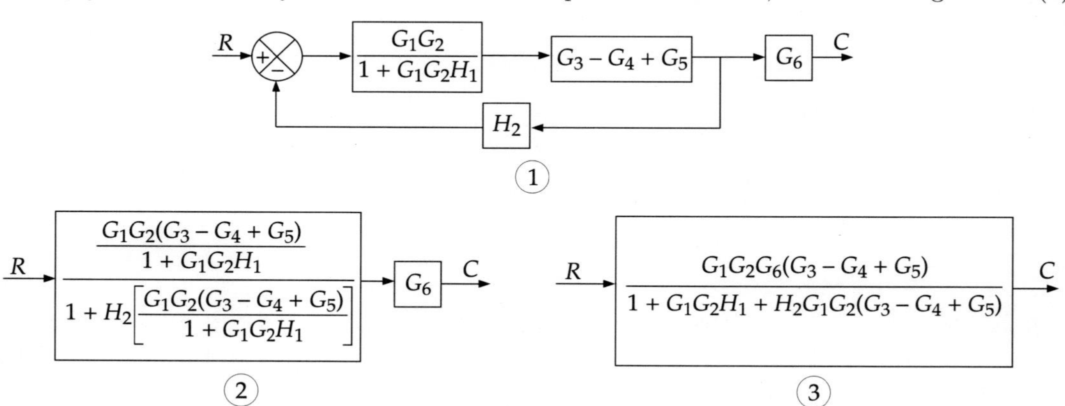

Figure 4.16(a) Successive stages of block diagram reduction.

EXAMPLE 4.5. Find the output of the system shown in Figure 4.17.

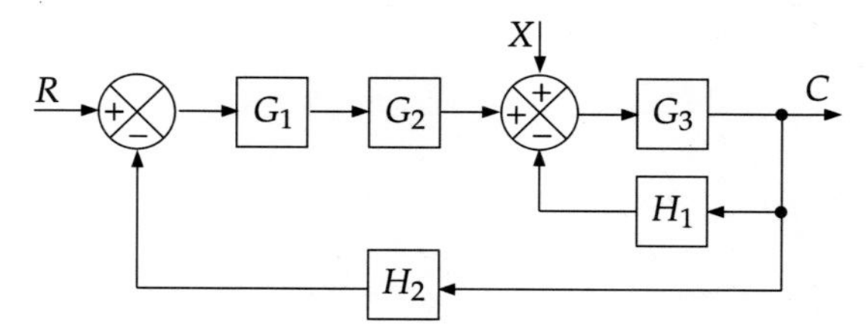

Figure 4.17 System block diagram (Example 4.5).

Solution This is a multi-input system. So, initially, we put $X = 0$ and reduce the resultant diagram. Successive stages of block diagram reduction are shown in Figure 4.17(a), where 1, 2 and 3 indicate successive steps.

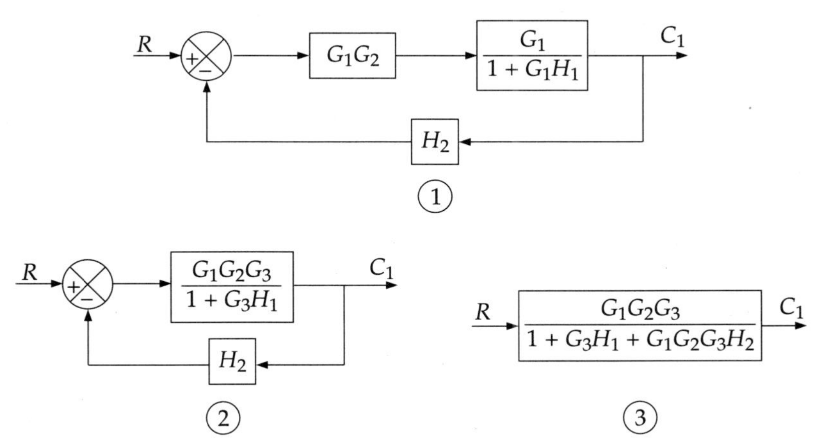

Figure 4.17(a) Successive stages of block diagram reduction with $X = 0$.

Next, we put $R = 0$. Then the block diagram reduction process looks like one shown in Figure 4.17(b).

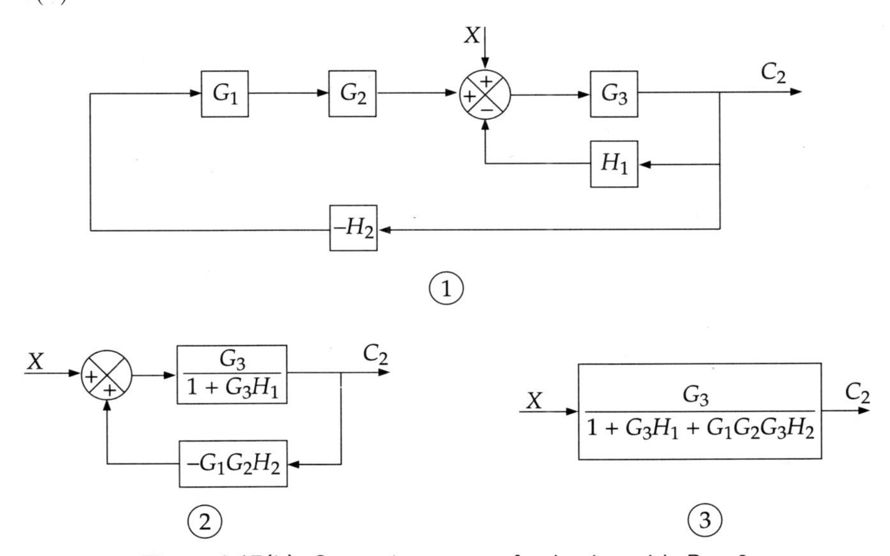

Figure 4.17(b) Successive stages of reduction with $R = 0$.

The final output, as discussed before, is obtained by superposition of the two individual outputs. Thus,

$$C = C_1 + C_2 = \frac{G_1G_2G_3R + G_3X}{1 + G_3H_1 + G_1G_2G_3H_2}$$

EXAMPLE 4.6. Find $C(s)/R(s)$ of the multiple-loop feedback control system shown in Figure 4.18 by the block diagram reduction technique.

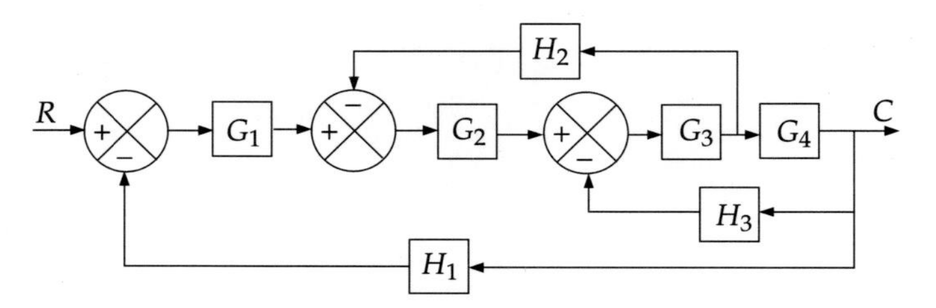

Figure 4.18 Feedback control system (Example 4.6).

Solution By inspection, we observe that if the take-off point after G_3 can be moved to that after G_4, the diagram will result in three feedback loops. Figure 4.18(a) shows how the point is moved to the desired place.

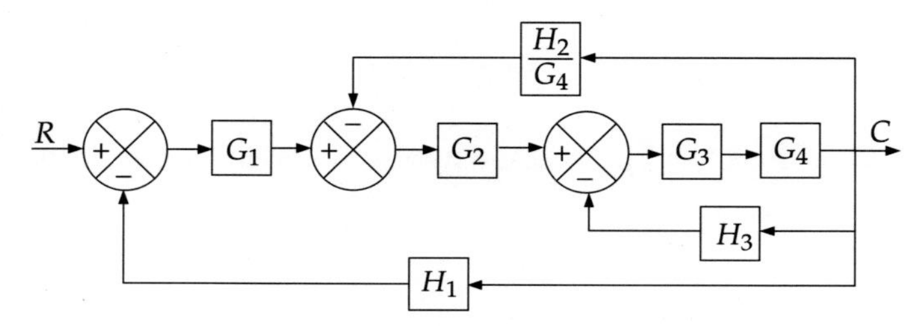

Figure 4.18(a) Movement of the take-off point.

Now we can reduce the feedback loop involving H_3 when the diagram assumes the form of Figure 4.18(b).

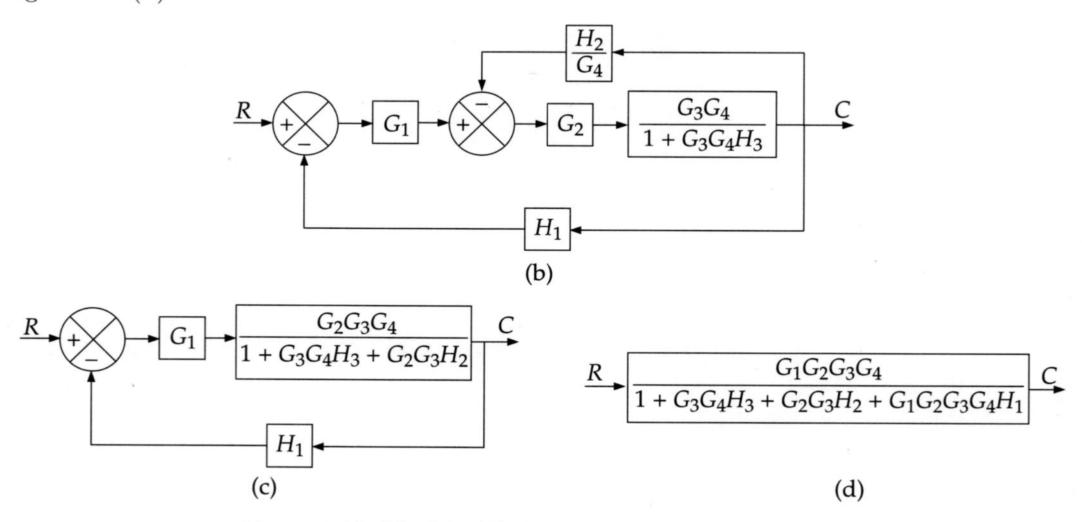

Figure 4.18 (b), (c), (d) Successive stages of reduction.

Next, the feedback loops involving H_2/G_4 and H_1 are reduced in succession. The corresponding diagrams are shown in Figures 4.18(c) and 4.18(d).

Hence,

$$\frac{C(s)}{R(s)} = \frac{G_1G_2G_3G_4}{1 + G_3G_4H_3 + G_2G_3H_2 + G_1G_2G_3G_4H_1}$$

EXAMPLE 4.7. Convert the system given in Figure 4.19 into a unity feedback system and find the characteristic equation.

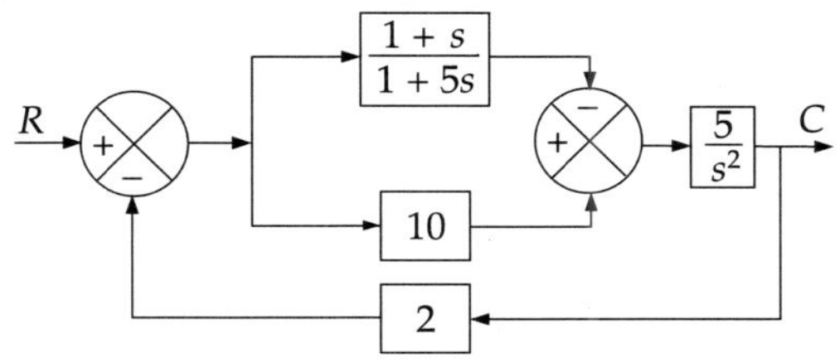

Figure 4.19 Feedback control system (Example 4.7).

Solution First we reduce the parallel blocks to get Figure 4.19(a).

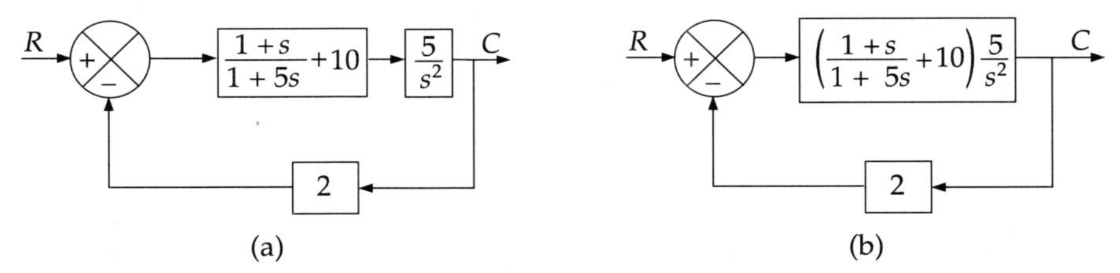

Figure 4.19 (a) and (b) Successive stages of reduction.

The next cascade reduction is shown in Figure 4.19(b). Now, we have to convert it to a unity feedback system. Let the present forward transfer function be G. We need to find G' such that

$$\frac{G}{1+2G} = \frac{G'}{1+G'}$$

$$\Rightarrow \qquad G' = \frac{G}{1+G}$$

Hence,

$$G' = \frac{\left(\dfrac{1+s}{1+5s} + 10\right)\dfrac{5}{s^2}}{1 + \left(\dfrac{1+s}{1+5s} + 10\right)\dfrac{5}{s^2}} = \frac{55 + 255s}{5s^3 + s^2 + 255s + 55}$$

Therefore, the required unity feedback system is as shown in Figure 4.19(c).

Thus characteristic equation is

$$5s^3 + s^2 + 510s + 110 = 0$$

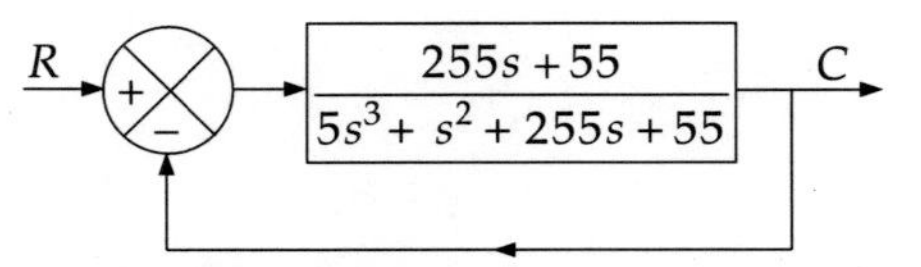

Figure 4.19(c) Unity feedback system.

EXAMPLE 4.8. Reduce the block diagram of Figure 4.20 and find the transfer function.

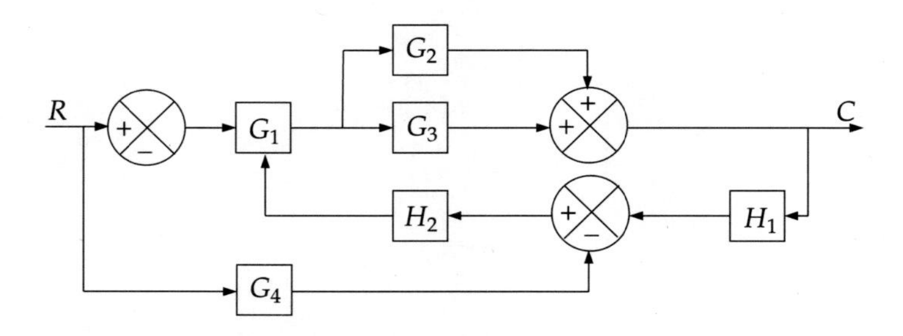

Figure 4.20 Block diagram of Example 4.8.

Solution First we reduce the parallel blocks and combine the resultant with the series block. The resulting diagram is shown in Figure 4.20(a).

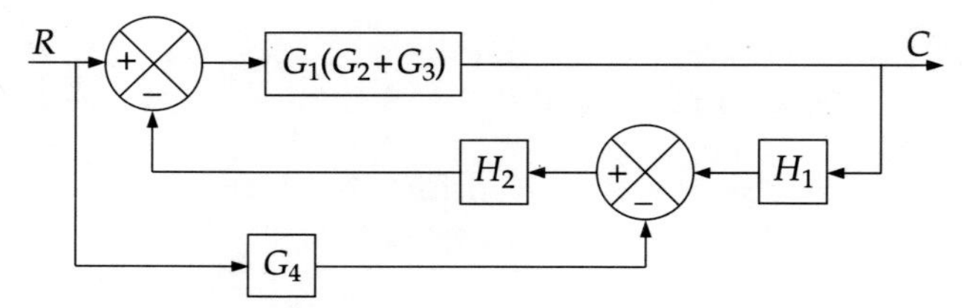

Figure 4.20(a) Reduction of parallel blocks.

Now we observe that one forward path through G_4 and H_2, and a feedback path through H_1 and H_2 merge at the first summing point. Shown explicitly, the situation is as in Figure 4.20(b).

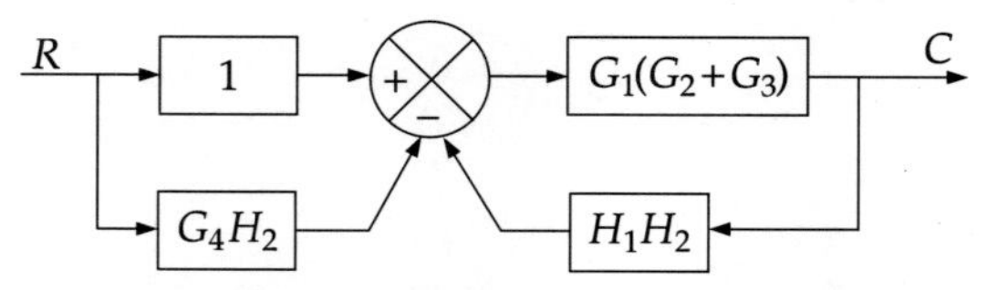

Figure 4.20(b) Reorganization of Figure 4.20(a).

In this diagram, we have two parallel blocks on the left and a feedback loop on the right. On reduction, these yield Figure 4.20(c).

Figure 4.20(c) Final reduced form.

The transfer function is therefore given by

$$\frac{C(s)}{R(s)} = (1 + G_4H_2)\left[\frac{G_1(G_2 + G_3)}{1 + G_1(G_2 + G_3)H_1H_2}\right]$$

$$= \frac{G_1G_2 + G_1G_3 + G_1G_2G_4H_2 + G_1G_3G_4H_2}{1 + G_1G_2H_1H_2 + G_1G_3H_1H_2}$$

EXAMPLE 4.9. Using the block diagram reduction technique, find the transfer function for the block diagram given in Figure 4.21.

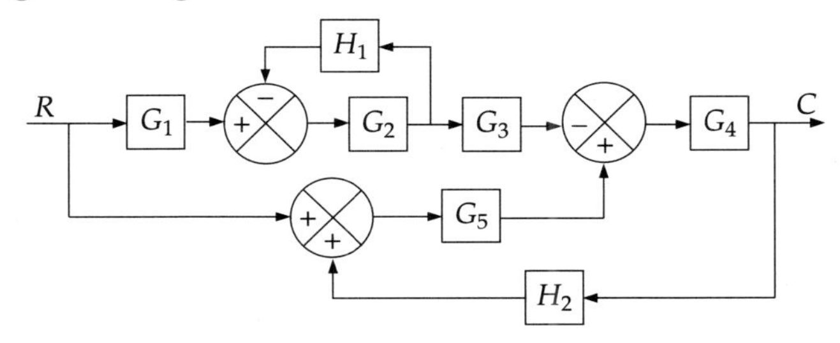

Figure 4.21 Block diagram of Example 4.9.

Solution First we reduce the feedback loop involving H_1. The diagram then assumes the form of Figure 4.21(a).

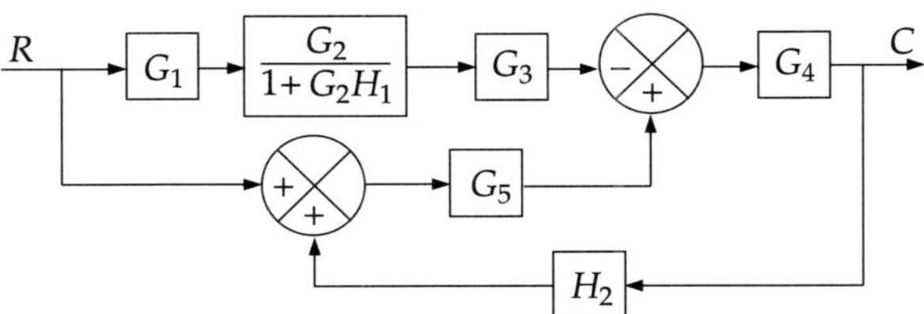

Figure 4.21(a) Reduction of the feedback loop.

We observe that one forward path through G_5 and a feedback path H_2G_5 merge at the first line summing point. The situation is as shown in Figure 4.21(b). The reduction, given in Figure 4.21(c), follows from this diagram

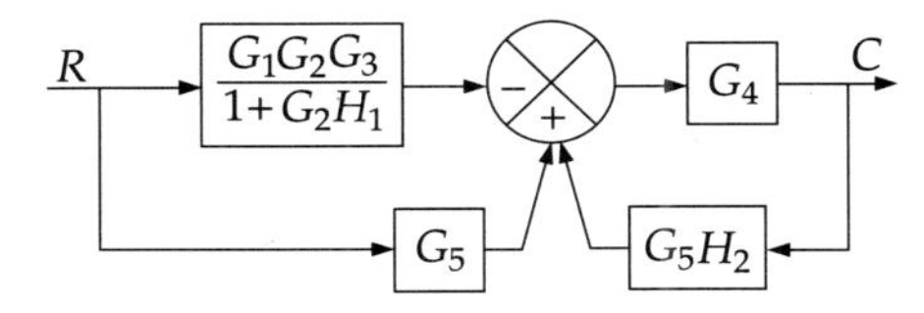

Figure 4.21(b) Reorganization of Figure 4.21(a).

$$R \rightarrow \boxed{G_5 - \frac{G_1G_2G_3}{1+G_2H_1}} \rightarrow \boxed{\frac{G_4}{1 - G_4G_5H_2}} \rightarrow C$$

Figure 4.21(c) Final reduced form.

Therefore, the transfer function is given by

$$\frac{C(s)}{R(s)} = \frac{G_4G_5 + G_2G_4G_5H_1 - G_1G_2G_3G_4}{1 - G_4G_5H_2 + G_2H_1 - G_2G_4G_5H_1H_2}$$

EXAMPLE 4.10. Find out the closed loop transfer function for the system given in Figure 4.22 by the block diagram reduction method.

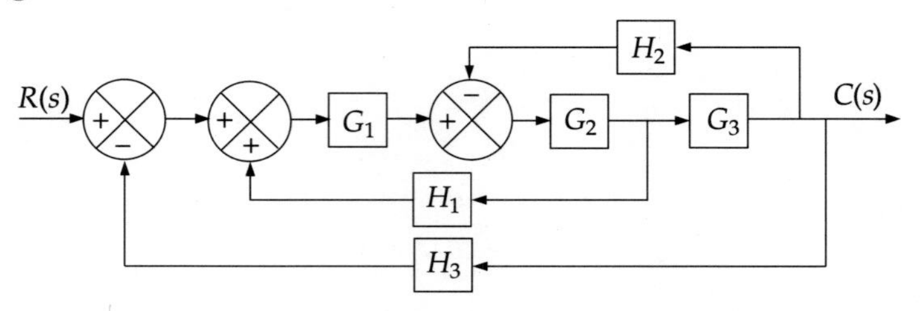

Figure 4.22 Block diagram (Example 4.10).

Solution By inspection we find that if the take-off point of H_1 is moved beyond G_3, only the feedback loops remain. After this operation, the diagram looks similar to the one in Figure 4.22(a).

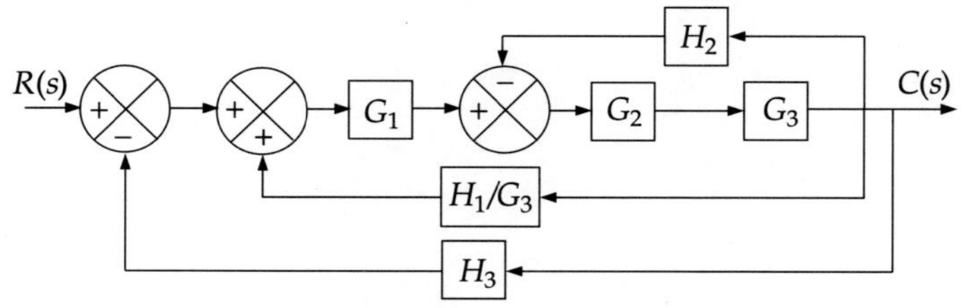

Figure 4.22(a) Take-off point H_1 is moved beyond G_3.

Next, we reduce the feedback loop involving H_2 when the diagram is altered to Figure 4.22(b). Next, we remove the inner feedback loop to arrive at Figure 4.22(c)

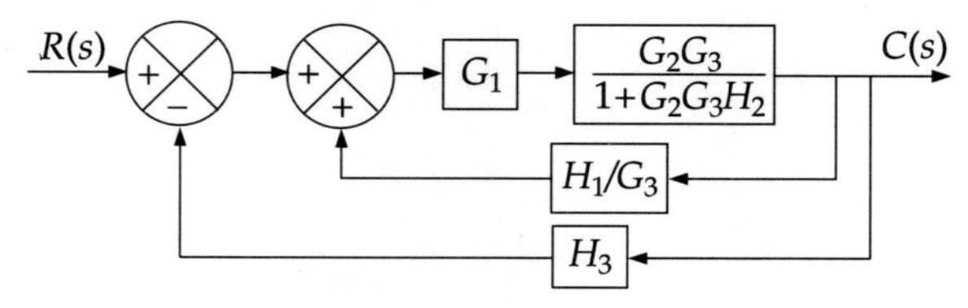

Figure 4.22(b) Feedback loop involving H_2 is reduced.

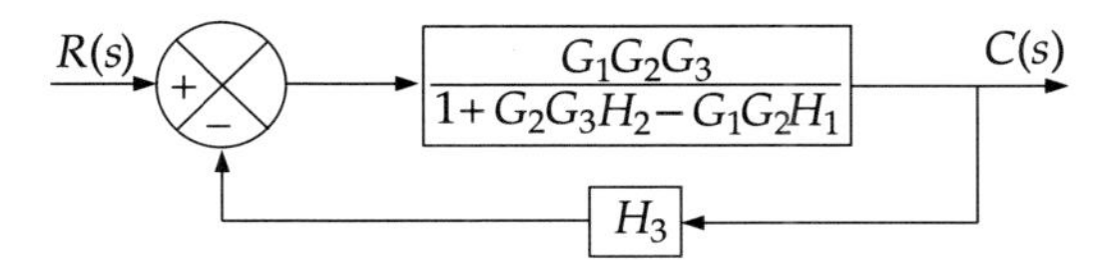

Figure 4.22(c) Inner feedback loop is removed.

Removal of the final feedback loop yields Figure 4.22(d).

R(s)
$$\frac{G_1G_2G_3}{1+G_2G_3H_2-G_1G_2H_1+G_1G_2G_3H_3}$$
C(s)

Figure 4.22(d) Final feedback loop is removed.

Hence, the closed loop transfer function is

$$\frac{C(s)}{R(s)} = \frac{G_1G_2G_3}{1+G_2G_3H_2-G_1G_2H_1+G_1G_2G_3H_3}$$

EXAMPLE 4.11. With the help of the block diagram reduction algebra, find out the transfer function of the system shown in Figure 4.23.

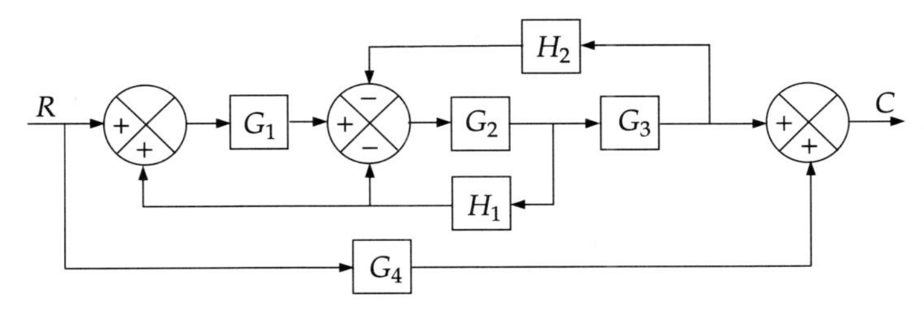

Figure 4.23 Block diagram (Example 4.11).

Solution First we redraw Figure 4.23(a) to separate the feedback H_1 to two summing points.

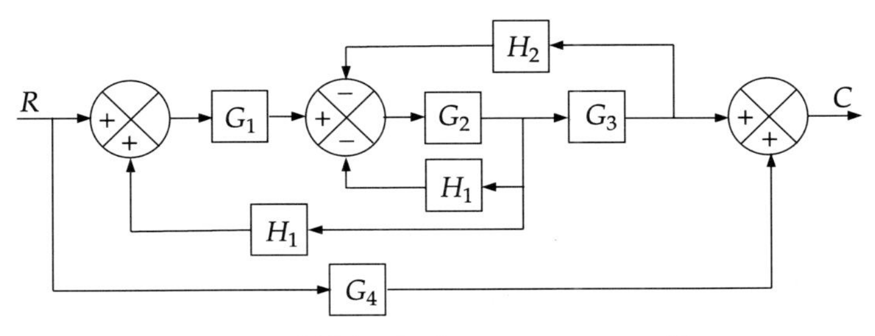

Figure 4.23(a) Redrawn to separate H_1.

Next, we reduce the feedback loop involving H_1 and then remove cascading with G_3. While removing the cascade, we find that the take-off point of H_1 that goes to the first summing point has to be moved to a point beyond G_3. The result is as shown in Figure 4.23(b).

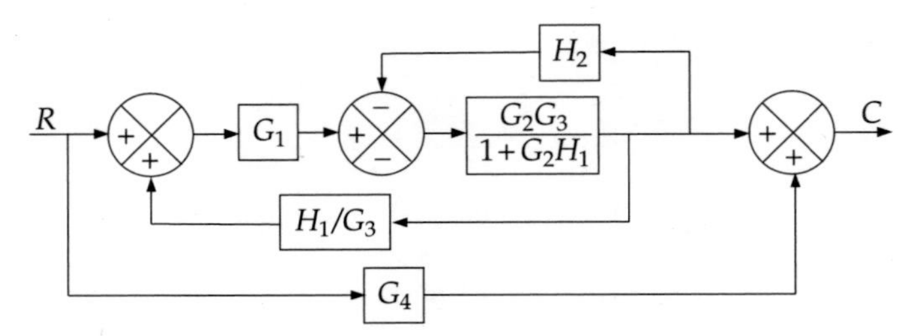

Figure 4.23(b) Feedback loop H_1 reduced and cascading with G_3 removed.

Now, the feedback loop involving H_2 can be reduced and then the cascade with G_1 can be removed to produce Figure 4.23(c).

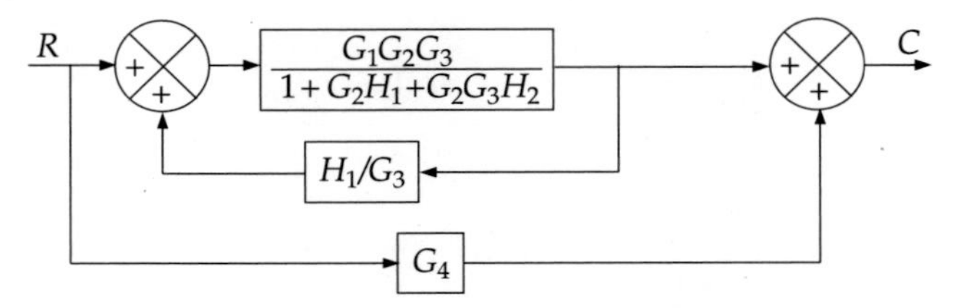

Figure 4.23(c) Feedback loop H_2 reduced and cascading with G_1 removed.

Next, we reduce the feedback loop and add the parallel block G_4 to the resultant to get Figure 4.23(d).

$$R \rightarrow \boxed{\frac{G_1G_2G_3}{1+G_2H_1+G_2G_3H_2-G_1G_2H_1} + G_4} \rightarrow C$$

Figure 4.23(d) Feedback loop reduced and parallel block added.

The transfer function is, therefore, given by

$$\frac{C(s)}{R(s)} = \frac{G_4 + G_2G_4H_1 + G_2G_3G_4H_2 - G_1G_2G_4H_1 + G_1G_2G_3}{1 + G_2H_1 + G_2G_3H_2 - G_1G_2H_1}$$

EXAMPLE 4.12. Using the block diagram reduction techniques, find the closed-loop transfer function $C(s)/R(s)$ whose block diagram is given in Figure 4.24.

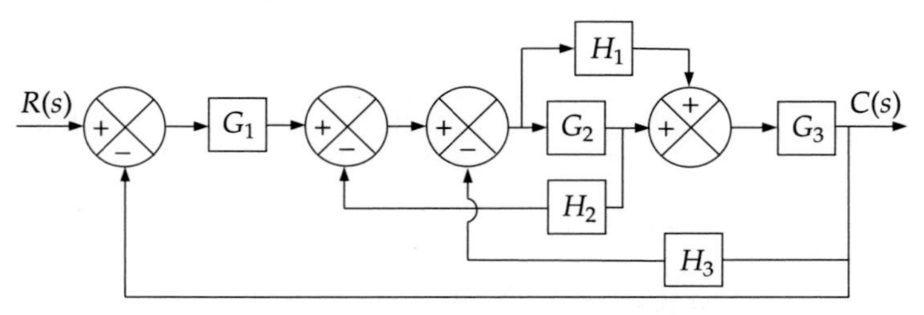

Figure 4.24 Block diagram (Example 4.12).

Solution On inspection we find that (i) the second and third summing points from the left can be merged, (ii) parallel blocks G_2 and H_1 can be reduced; and (iii) the take-off point of H_2 can be moved after the third summing point to generate Figure 4.24(a).

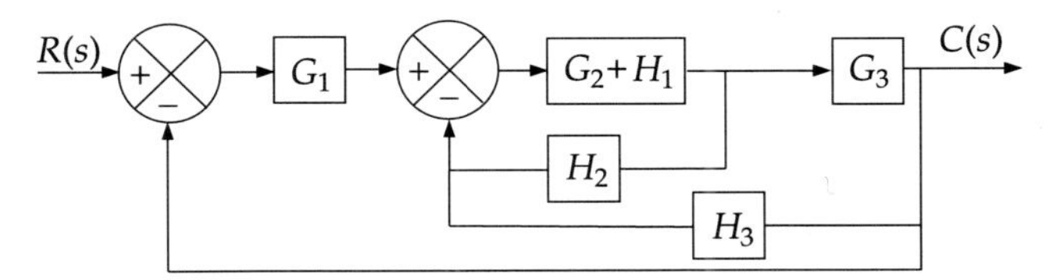

Figure 4.24(a) Summing points merged, parallel blocks reduced, and take-off point moved.

Next, we reduce the inner feedback loop involving H_2 and remove cascading of the resultant with G_3 to arrive at Figure 4.24(b).

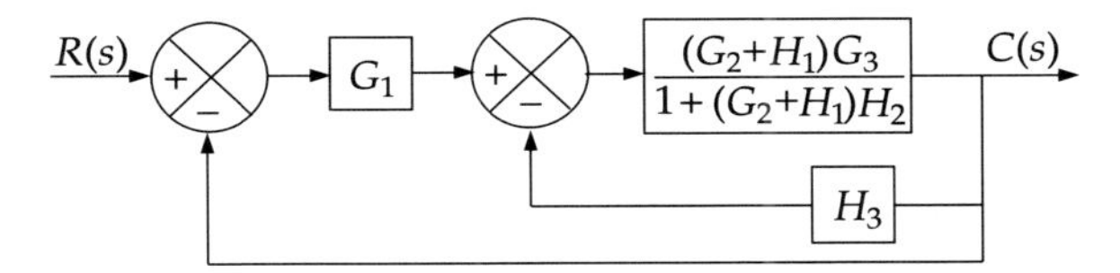

Figure 4.24(b) Inner feedback loop reduced and cascading removed.

Now we eliminate the two feedback loops in succession to generate Figure 4.24(c).

$$R(s) \rightarrow \boxed{\frac{G_1(G_2+H_1)G_3}{1+(G_2+H_1)H_2+(G_2+H_1)G_3H_3+G_1(G_2+H_1)G_3}} \rightarrow C(s)$$

Figure 4.24(c) Final reduced form.

The transfer function is, therefore, given by

$$\frac{C(s)}{R(s)} = \frac{G_1G_2G_3 + G_1G_3H_1}{1 + G_2H_2 + H_1H_2 + G_2G_3H_3 + G_3H_1H_3 + G_1G_2G_3 + G_1G_3H_1}$$

EXAMPLE 4.13. Find the transfer function for the system whose block diagram representation is shown in Figure 4.25.

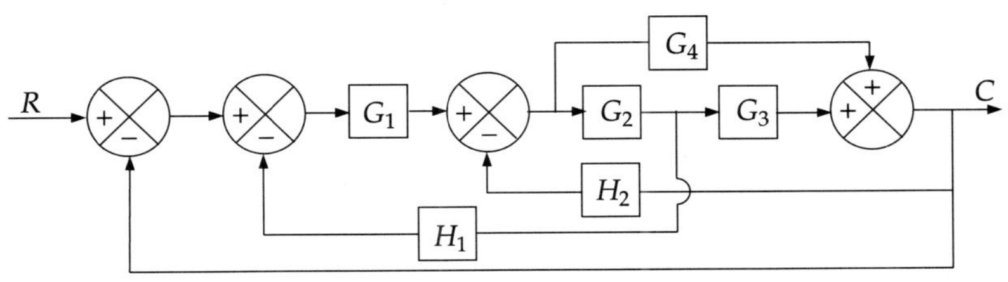

Figure 4.25 Block diagram (Example 4.13).

Solution We observe that (i) the first and second summing points on the left can be merged; (ii) the take-off point of H_1 can be moved before G_2 block; and (iii) cascaded G_2, G_3 and parallel G_4 can be reduced. The diagram then looks like the one in Figure 4.25(a).

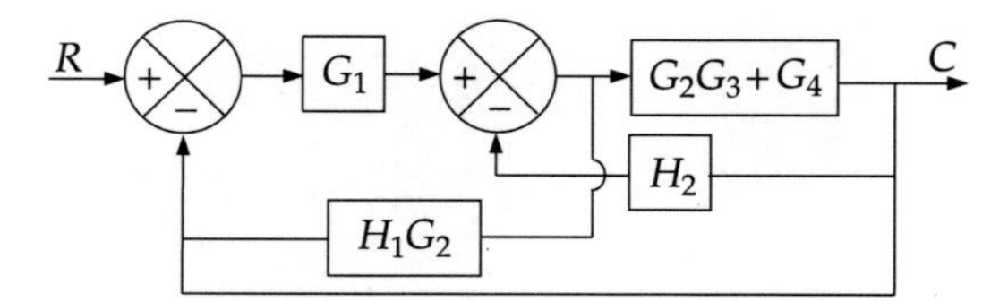

Figure 4.25(a) Summing points merged, take-off point moved, and cascaded blocks reduced.

Next, we move the take-off point of H_1G_2 beyond $G_2G_3 + G_4$. The resulting diagram is as shown in Figure 4.25(b).

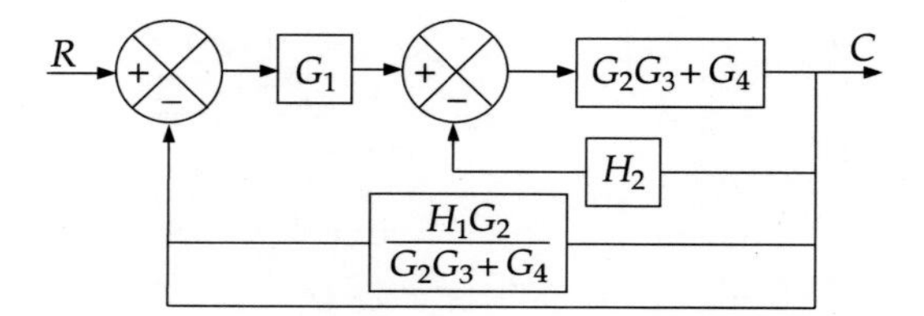

Figure 4.25(b) Take-off point moved.

The next step is to eliminate the feedback loop involving H_2 and then combining the resultant with G_1 so that the diagram looks like that in Figure 4.25(c).

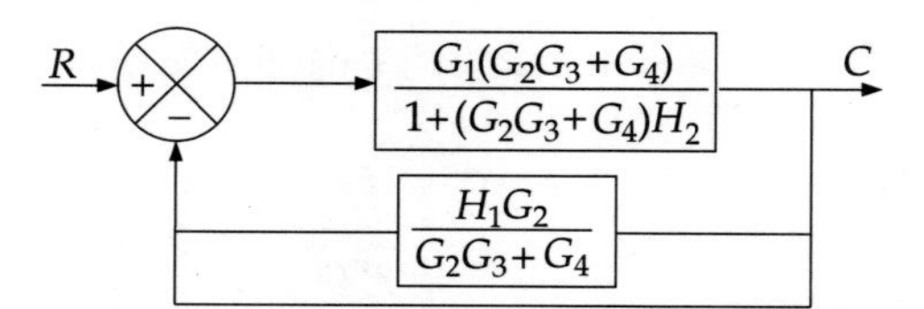

Figure 4.25(c) Feedback eliminated and combined with G_1.

Now we eliminate the two feedback loops to arrive at the final diagram [Figure 4.25(d)].

$$R \rightarrow \boxed{\frac{G_1(G_2G_3 + G_4)}{1 + (G_2G_3 + G_4)H_2 + G_1(G_2G_3 + G_4)\dfrac{H_1G_2}{G_2G_3 + G_4} + G_1(G_2G_3 + G_4)}} \rightarrow C$$

Figure 4.25(d) Final reduced form.

The transfer function is, therefore,

$$\frac{C(s)}{R(s)} = \frac{G_1G_2G_3 + G_1G_4}{1 + G_1G_2G_3 + G_1G_4 + G_2G_3H_2 + G_4H_2 + G_1G_2H_1}$$

EXAMPLE 4.14. Using the block diagram reduction technique, find the transfer function of the block diagram shown in Figure 4.26.

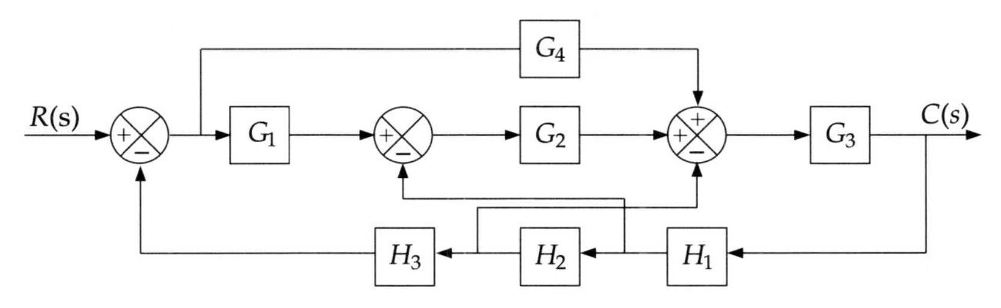

Figure 4.26 Block diagram (Example 4.14).

Solution We note that the given block diagram is equivalent to the one given by Figure 4.27.

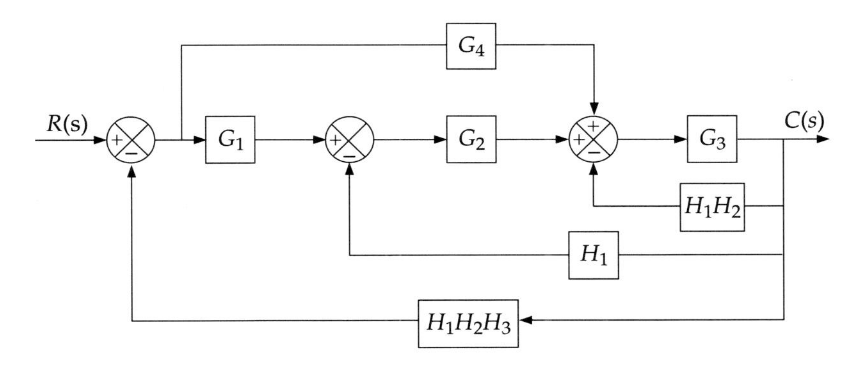

Figure 4.27 Equivalent block diagram (Example 4.14).

Now the closed-loop on the extreme right-hand side can be reduced and the second summing point can be merged with the third one as indicated in Figure 4.28.

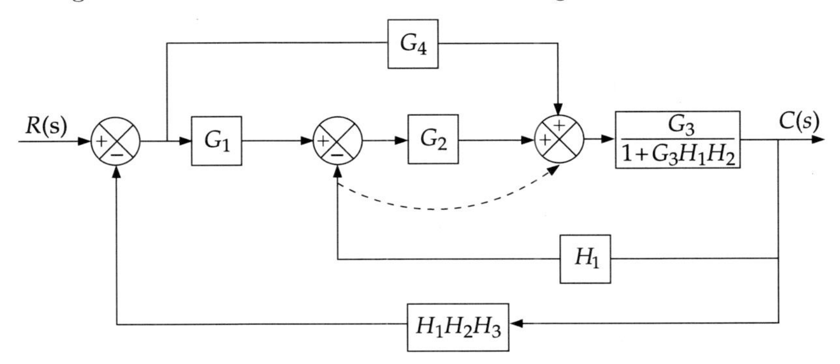

Figure 4.28 Reduction of the extreme RHS closed-loop and indication of movement of summing point.

In Figures 4.29 and 4.30, we see how two successive steps can be taken to reduce the block diagram to a single closed-loop.

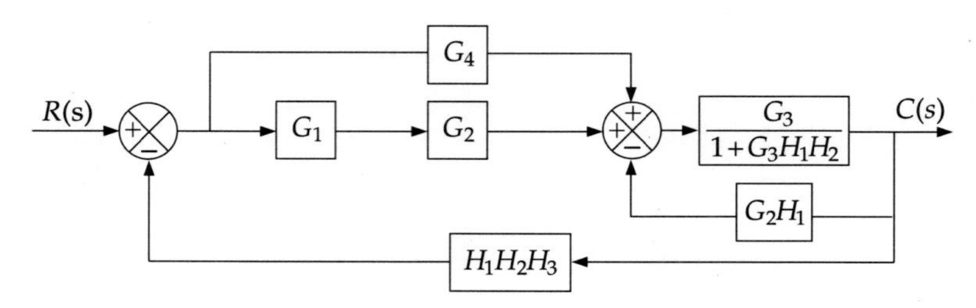

Figure 4.29 Movement of the summing point as indicated in Figure 4.28.

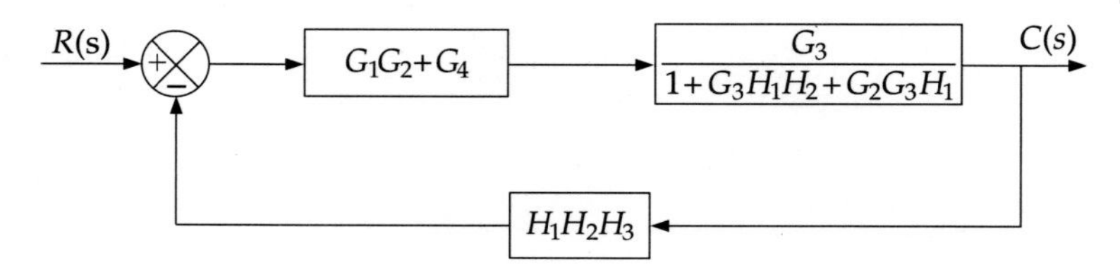

Figure 4.30 Reduction of the closed-loop.

The transfer function is therefore given by

$$\frac{C(s)}{R(s)} = \frac{G_3(G_1G_2 + G_4)}{1 + G_3H_1H_2 + G_2G_3H_1 + (G_1G_2 + G_4)G_3H_1H_2H_3}$$

EXAMPLE 4.15. Reduce the complete block diagram obtained in Example 4.2 at page 120 by the block diagram reduction technique.

Solution Successive steps of the reduction are shown in Figure 4.31. From Figure 4.31(e), we get the transfer function as follows by eliminating the feedback loop:

$$\frac{V_o(s)}{V_i(s)} = \frac{1}{s^2R_1R_2C_1C_2 + s(R_1C_1 + R_2C_2 + R_1C_2) + 1}$$

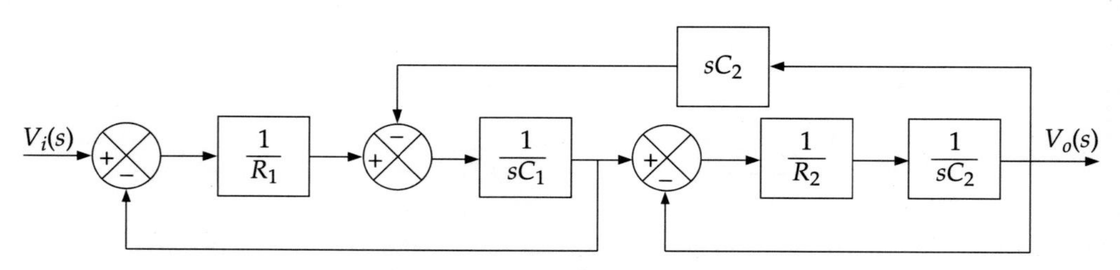

(a) Moving the take-off point beyond a block.

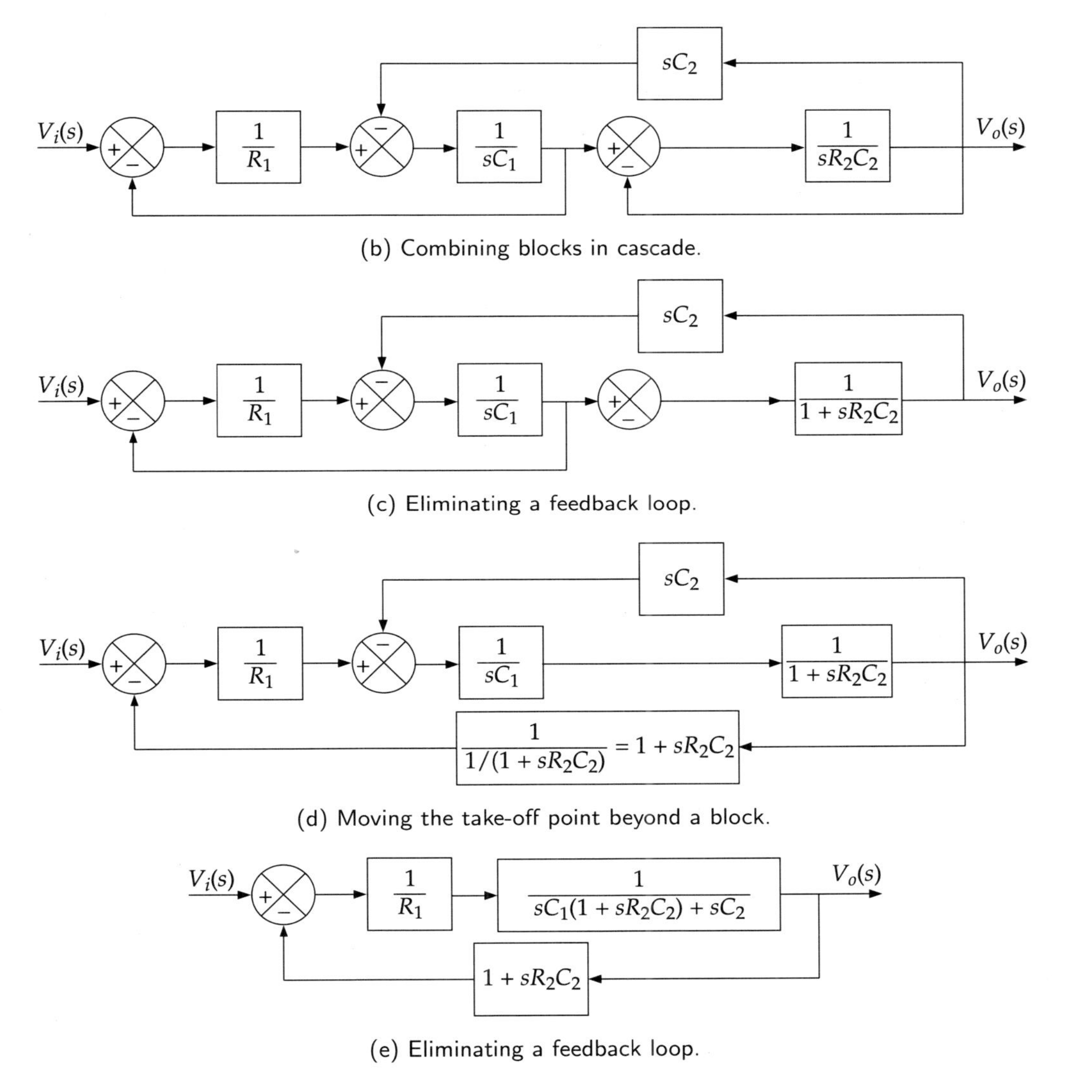

(b) Combining blocks in cascade.

(c) Eliminating a feedback loop.

(d) Moving the take-off point beyond a block.

(e) Eliminating a feedback loop.

Figure 4.31 Different steps of the block diagram reduction (Example 4.15).

4.2 SIGNAL FLOW GRAPHS

Introduced by SJ Mason about half a century ago, signal flow graphs are more elegant and powerful for dealing with more complex interrelationships where the block diagram technique becomes cumbersome. However, it can be utilized for linear systems only.

A signal flow graph (often called *flow graph*) represents a system by directed line segments originating from or terminating at points, called *nodes*. Before we discuss how to construct signal flow graphs, let us consider its constituent elements.

Elements of Signal Flow Graph

The elements of signal flow graph are given in Table 4.3.

Table 4.3 Signal Flow Graph Elements

Element	*What it is*
1. Node	A node represents a dependent or independent variable. x_1, $x_2, \ldots x_9$ in Figure 4.32 are nodes. Nodes can be divided into the following three categories: 1. Source (or Input) node—node having only outgoing branches (x_1). 2. Sink (or Output) node—node having only incoming branches (x_7). 3. Chain node—node having both outgoing and incoming branches (x_3, x_4, x_5, x_6)
2. Branch	Lines joining nodes are branches. An arrow on the branch indicates the direction of flow of the signal and the number written on it indicates the *gain* of the branch. In Figure 4.32, $x_3 = Ax_2$, where A is the gain of the branch.
3. Forward path	All paths leading to output and originating at the input are forward paths. If we denote forward paths in Figure 4.32 by p_1, and p_2, they are: $p_1 \; : \; x_1x_2x_3x_4x_5x_6x_7$ $\qquad$ $p_2 \; : \; x_1x_2x_3x_5x_6x_7$
4. Loop or feedback path	A path originating from and terminating at the same node, without crossing any other node twice is a loop. There are four loops in Figure 4.32, namely, $L_1 \; : \; x_3x_4x_3$ $\qquad$ $L_2 \; : \; x_4x_5x_6x_4$ $L_3 \; : \; x_5x_5$ $\qquad$ $L_4 \; : \; x_3x_5x_6x_4x_3$ Of these, L_3 is a self-loop.
5. Forward path gain	Product of all gains associated with a forward path is the forward path gain. Gains for two paths are: $p_1 \; : \; ACEH$ $\qquad$ $p_2 \; : \; AGH$
6. Loop gain	Product of gains associated with a loop is the loop gain. Gains for four loops are: $L_1 \; : \; BC$ $\qquad$ $L_2 \; : \; -EHD$ $L_3 \; : \; F$ $\qquad$ $L_4 \; : \; -GHDB$
7. Non-touching loops	Loops having no common nodes are non-touching loops. L_1 and L_3 is a pair of non-touching loops.

Figure 4.32 gives a pictorial representation of these elements.

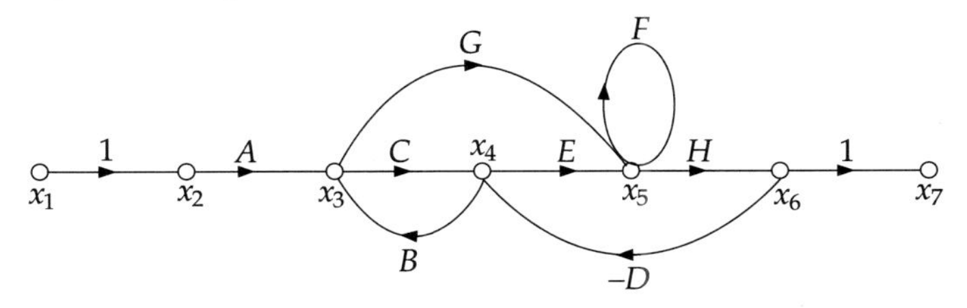

Figure 4.32 A signal flow graph showing different elements.

The transfer function of a system can be determined from signal flow graph by using *Mason's gain formula.*

Mason's Gain Formula

Mason's gain formula is given by

$$\text{Transfer function} = \frac{C(s)}{R(s)} = \frac{\sum_{i=1}^{N} p_i \Delta_i}{\Delta} \tag{4.7}$$

where

N = the total number of forward paths

p_i = the gain of the ith forward path

Δ = the determinant of the graph

Δ_i = the path-factor for the ith path

The determinant of the graph Δ and the path-factor for the ith path Δ_i are defined as follows:

$\Delta = 1 - \Sigma$ (all individual loop gains)

$+ \Sigma$ (gain products of all possible combinations of *two* non-touching loops)

$- \Sigma$ (gain products of all possible combinations of *three* non-touching loops) $+ \cdots$

$\Delta_i = 1 -$ (part of Δ that is non-touching to the path p_i)

We can find out the transfer function of signal flow graph of Figure 4.32 from the following data (see Table 4.3 for clarification):

$$N = 2$$

$$p_1 = ACEH$$

$$\Delta_1 = 1 \text{ (since all loops are touching } p_1)$$

$$p_2 = AGH$$

$$\Delta_2 = 1 \text{ (since all loops are touching } p_2)$$

$$\Delta = 1 - (BC - EHD + F - GHDB) + BCF$$

Now we work out a few examples to familiarize ourselves with the procedure.

EXAMPLE 4.16. Draw a signal flow graph for the following algebraic equations:

$$x_2 = ax_1 + bx_3 \tag{i}$$

$$x_3 = cx_2 + dx_4 \tag{ii}$$

$$x_4 = ex_2 + fx_3 + gx_4 \tag{iii}$$

$$x_5 = hx_2 + kx_4 \tag{iv}$$

Solution We note that there are five variables in total in the four equations. So, first we plot five nodes and then draw the flow graph for Eq. (i):

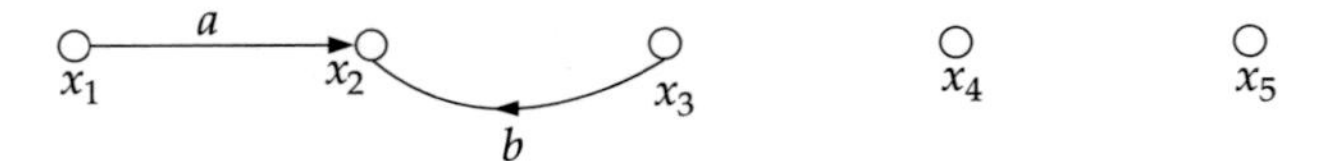

Next, we consider Eq. (ii) and draw additional flow lines:

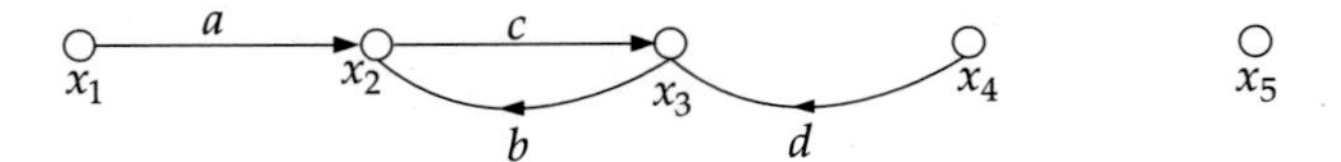

Our next diagram shows how we draw flow lines according to Eq. (iii).

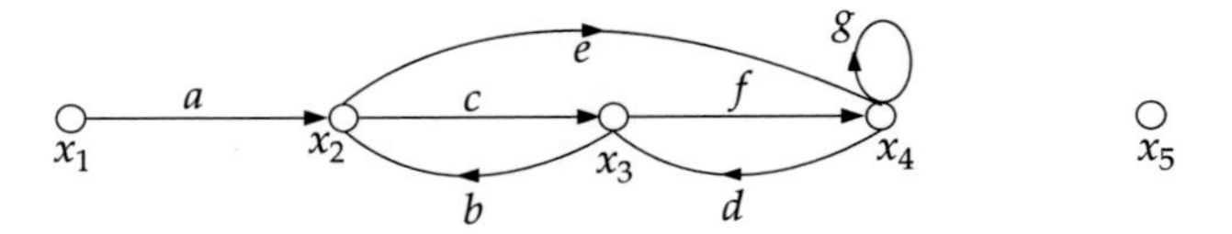

Incorporating flow lines according to Eq. (iv), the complete signal flow graph is as shown in Figure 4.33.

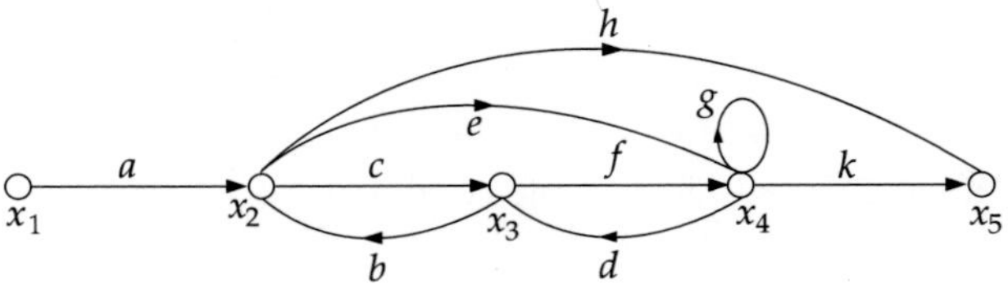

Figure 4.33 Signal flow graph (Example 4.16).

EXAMPLE 4.17. For the signal flow graph shown in the following figure, the transfer function is $\dfrac{C(s)}{R(s)} = \cdots$

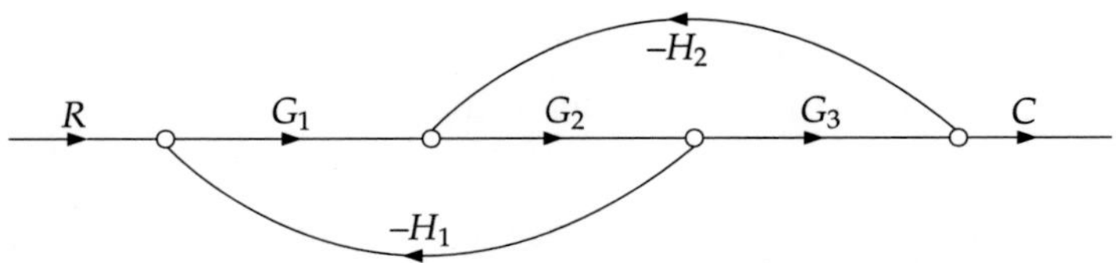

Solution The forward path is only one whereas there are two loops. There is no non-touching loop. So, we have

$$p_1 = G_1G_2G_3$$
$$L_1 = -G_1G_2H_1$$
$$L_2 = -G_2G_3H_2$$
$$\Delta = 1 - (-G_1G_2H_1 - G_2G_3H_2)$$
$$\Delta_1 = 1$$

Therefore, from Mason's gain formula, we have

$$\frac{C(s)}{R(s)} = \frac{G_1G_2G_3}{1 + G_1G_2H_1 + G_2G_3H_2}$$

EXAMPLE 4.18. The transfer function between y_2 and y_1 in the following figure is

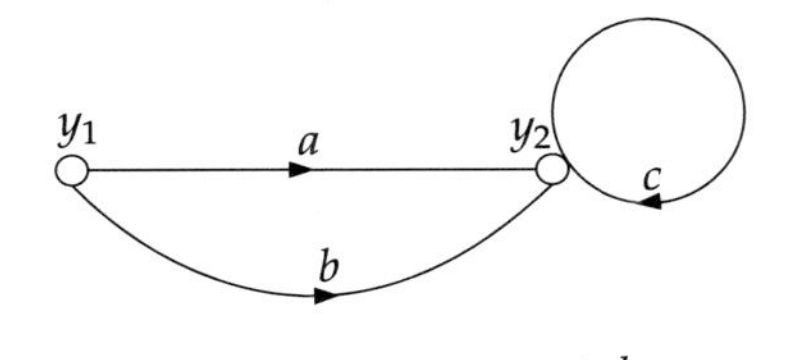

(a) $a + b$ (b) $(a + b)c$ (c) $\dfrac{a+b}{1-c}$ (d) $\dfrac{a+b}{1+c}$

Solution There are two forward paths and one loop. So, we have

$$p_1 = a \qquad p_2 = b$$
$$L_1 = c$$
$$\Delta = 1 - c \qquad \Delta_1 = 1 = \Delta_2$$

Therefore,
$$\frac{C(s)}{R(s)} = \frac{a+b}{1-c}$$

Ans: (c).

EXAMPLE 4.19. The transfer function of the open-loop system $G(s)$ can be diagrammatically represented by the following signal flow graph is:

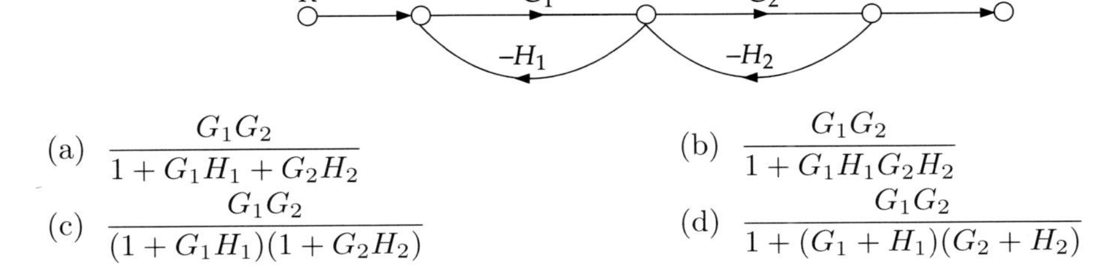

(a) $\dfrac{G_1G_2}{1 + G_1H_1 + G_2H_2}$ (b) $\dfrac{G_1G_2}{1 + G_1H_1G_2H_2}$

(c) $\dfrac{G_1G_2}{(1 + G_1H_1)(1 + G_2H_2)}$ (d) $\dfrac{G_1G_2}{1 + (G_1 + H_1)(G_2 + H_2)}$

Solution There exist one forward path and two loops. So, we have

$$p_1 = G_1G_2$$

$$L_1 = -G_1H_1 \qquad L_2 = -G_2H_2$$

$$\Delta = 1 + G_1H_1 + G_2H_2 \qquad \Delta_1 = 1$$

Therefore,
$$\frac{C}{R} = \frac{G_1G_2}{1 + G_1H_1 + G_2H_2}$$

Ans: (a).

EXAMPLE 4.20. The following figure is the signal flow graph representation of a control system:

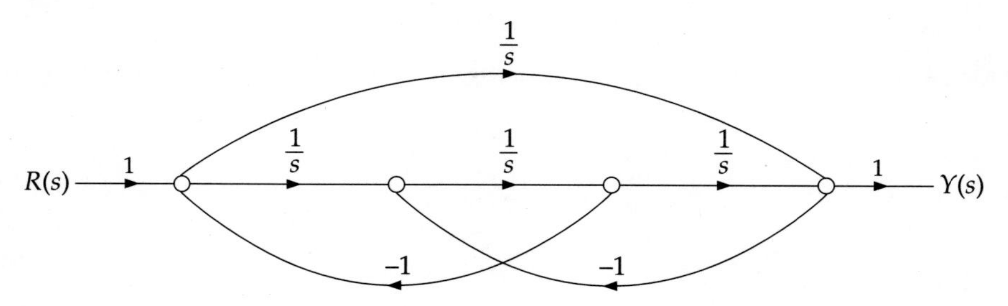

The transfer function $\dfrac{Y(s)}{R(s)}$ is computed as

(a) $\dfrac{1}{s}$ (b) $\dfrac{s^2+1}{s(s^2+2)}$ (c) $\dfrac{s(s^2+1)}{s^2+2}$ (d) $1 - \dfrac{1}{s}$

Solution There are two forward paths and two loops. Therefore, we have

$$p_1 = \frac{1}{s^3} \qquad L_1 = -\frac{1}{s^2}$$

$$p_2 = \frac{1}{s} \qquad L_2 = -\frac{1}{s^2}$$

$$\Delta = 1 + \frac{2}{s^2} \qquad \Delta_1 = 1 = \Delta_2$$

Therefore,

$$\frac{Y(s)}{R(s)} = \frac{\dfrac{1}{s^3} + \dfrac{1}{s}}{1 + \dfrac{2}{s^2}} = \frac{1 + s^2}{s(s^2+2)}$$

Ans: (b).

EXAMPLE 4.21. A filter is represented by the signal flow graph shown in the following figure.

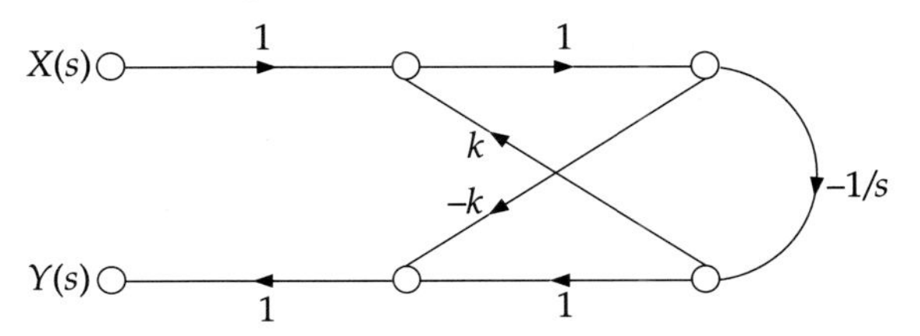

Its input is $x(t)$ and output is $y(t)$. The transfer function of the filter is

(a) $\dfrac{-(1+ks)}{s+k}$ (b) $\dfrac{(1+ks)}{s+k}$ (c) $\dfrac{-(1-ks)}{s+k}$ (d) $\dfrac{(1-ks)}{s+k}$

Solution The signal flow graph contains two forward paths and one loop. Therefore, we have

$$p_1 = -\frac{1}{s} \qquad p_2 = -k$$

$$L_1 = -\frac{k}{s}$$

$$\Delta = 1 + \frac{k}{s} \qquad \Delta_1 = 1 = \Delta_2$$

Therefore,

$$\frac{Y(s)}{X(s)} = \frac{-\dfrac{1}{s} - k}{1 + \dfrac{k}{s}} = \frac{-(1+ks)}{s+k}$$

Ans: (a).

EXAMPLE 4.22. Find the transfer function for the system given by the signal flow graph of Figure 4.34.

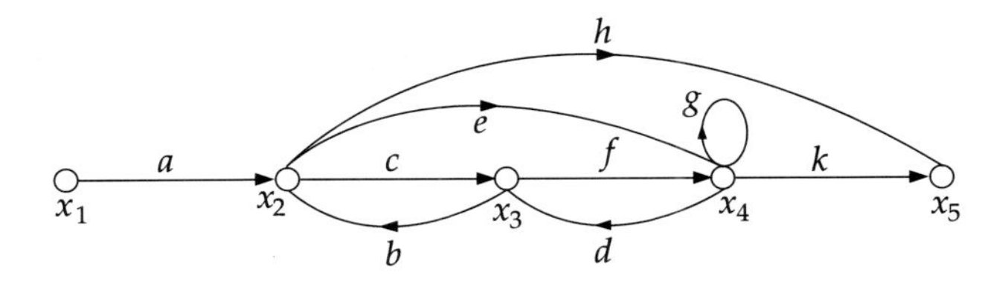

Figure 4.34 Signal flow graph (Example 4.22).

Solution The number of forward paths, $N = 3$. Their gains are

$$p_1 = acfk \qquad p_2 = aek \qquad p_3 = ah$$

The loops and their gains are

$$L_1 = bc \qquad L_2 = df \qquad L_3 = g \qquad L_4 = bed$$

The non-touching loop pair is $(L_1,\ L_3)$. Therefore,

$$\Delta = 1 - (bc + df + g + bed) + bc(g)$$

All loops are touching p_1 and p_2. Hence, $\Delta_1 = \Delta_2 = 1$. L_2 and L_3 do not touch p_3. Therefore, $\Delta_3 = 1 - df - g$. Thus, the transfer function is given by

$$\begin{aligned}\text{Transfer function} &= \frac{p_1\Delta_1 p_2\Delta_2 + p_3\Delta_3}{\Delta} \\ &= \frac{acfk + aek + ah(1 - df - g)}{1 - bc - df - g - bed - bcg}\end{aligned}$$

EXAMPLE 4.23. Find the transfer function for the system given by the signal flow graph of Figure 4.35.

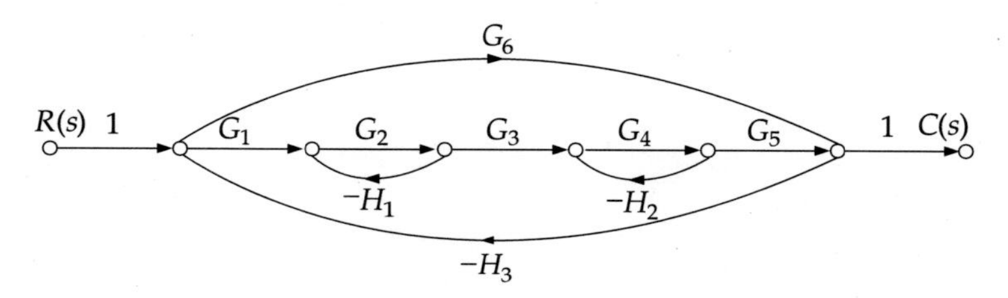

Figure 4.35 Signal flow graph (Example 4.23).

Solution Here, the forward paths, their gains and number are

$$p_1 = G_1G_2G_3G_4G_5 \qquad p_2 = G_6 \qquad N = 2$$

Loops and their gains are

$$L_1 = -G_2H_1 \qquad\qquad L_2 = -G_4H_2$$

$$L_3 = -G_6H_3 \qquad\qquad L_4 = -G_1G_2G_3G_4G_5H_3$$

The first three loops are non-touching. The possible combinations of *two* non-touching loops and the corresponding loop-gains are

$$L_1L_2 = G_2G_4H_1H_2, \qquad L_1L_3 = G_2G_6H_1H_3, \qquad L_2L_3 = G_4G_6H_2H_3$$

The possible combination of *three* non-touching loops and the corresponding loop-gain are

$$L_1L_2L_3 = -G_2G_4G_6H_1H_2H_3$$

Therefore,

$$\begin{aligned}\Delta = 1 &+ (G_2H_1 + G_4H_2 + G_6H_3 + G_1G_2G_3G_4G_5H_3) \\ &+ (G_2G_4H_1H_2 + G_2G_6H_1H_3 + G_4G_6H_2H_3) + (G_2G_4G_6H_1H_2H_3)\end{aligned}$$

$$\Delta_1 = 1 - 0 \qquad (\because \text{ all loops touch } p_1)$$

$$\Delta_2 = 1 + G_2H_1 + G_4H_2 + G_2G_4H_1H_2$$

Thus,

$$\text{Transfer function} = \frac{G_1G_2G_3G_4G_5 + G_6(1 + G_2H_1 + G_4H_2 + G_2G_4H_1H_2)}{\begin{array}{l}1 + (G_2H_1 + G_4H_2 + G_6H_3 + G_1G_2G_3G_4G_5H_3) \\ + (G_2G_4H_1H_2 + G_2G_6H_1H_3 + G_4G_6H_2H_3) \\ + (G_2G_4G_6H_1H_2H_3)\end{array}}$$

EXAMPLE 4.24. Find the transfer function of the network given in Figure 4.36 using signal flow graph and Mason's gain formula.

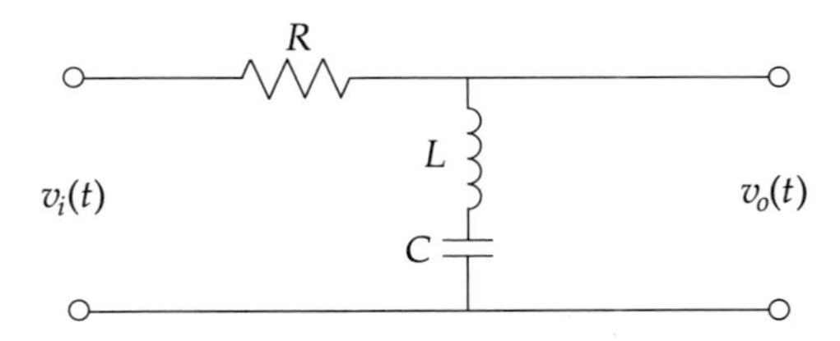

Figure 4.36 Network (Example 4.24).

Solution From the equivalent network in s-space, we get

$$I(s) = \frac{V_i(s) - V_o(s)}{R}$$

$$V_o(s) = \left(sL + \frac{1}{sC}\right) I(s)$$

Taking $V_i(s)$, $I(s)$ and $V_o(s)$ as three nodes, we draw the signal flow graph as shown in Figure 4.37.

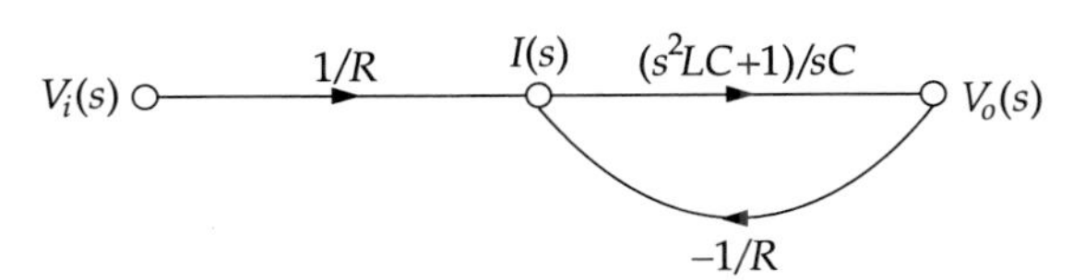

Figure 4.37 Signal flow graph (Example 4.24).

The forward path and gain are

$$p_1 = \frac{s^2LC + 1}{sRC}$$

The loop and gain are

$$L_1 = -\frac{s^2LC + 1}{sRC}$$

Therefore,

$$\Delta = 1 + \frac{s^2LC + 1}{sRC}$$

$$\Delta_1 = 1 \quad (\because \text{ no non-touching loop exists})$$

$$\text{Transfer function} = \frac{V_o(s)}{V_i(s)} = \frac{\dfrac{s^2LC+1}{sRC}}{1 + \dfrac{s^2LC+1}{sRC}}$$

$$= \frac{s^2LC + 1}{s^2LC + 1 + sRC}$$

EXAMPLE 4.25. Find the transfer function of the RC-network given in Figure 4.38 by the signal flow graph method.

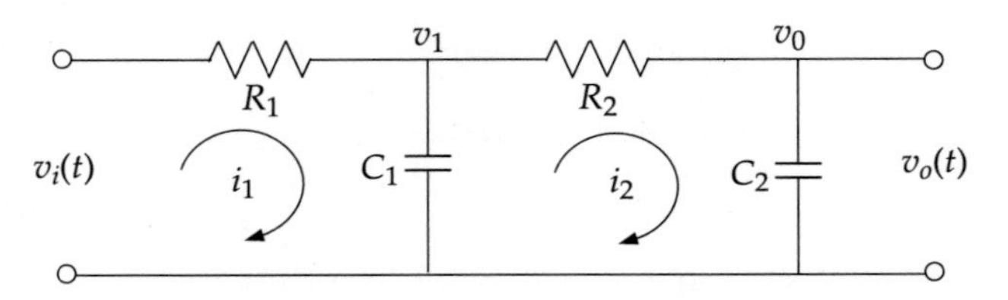

Figure 4.38 RC-network (Example 4.25).

Solution From the s-space conversion of the network, we get

$$I_1(s) = \frac{1}{R_1}[V_i(s) - V_1(s)]$$

$$V_1(s) = \frac{1}{sC_1}[I_1(s) - I_2(s)]$$

$$I_2(s) = \frac{1}{R_2}[V_1(s) - V_o(s)]$$

$$V_o(s) = \frac{1}{sC_2}I_2(s)$$

The corresponding signal flow graph is given in Figure 4.39.

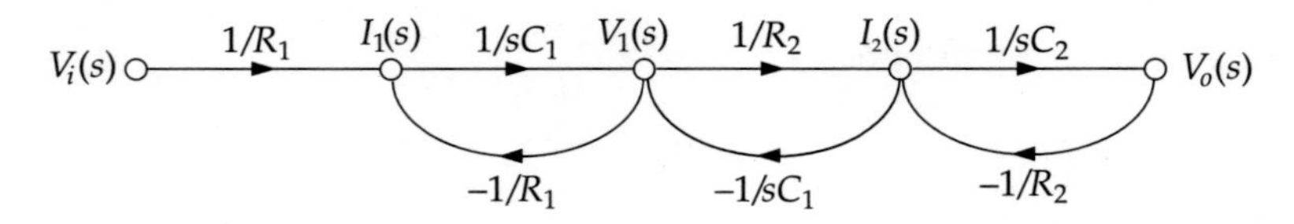

Figure 4.39 Signal flow graph (Example 4.25).

The forward path and gain are

$$p_1 = \frac{1}{s^2 R_1 R_2 C_1 C_2}$$

The loops and gains are

$$L_1 = -\frac{1}{sR_1C_1} \qquad L_2 = -\frac{1}{sR_2C_1} \qquad L_3 = -\frac{1}{sR_2C_2}$$

All loops are touching the forward path. Therefore, $\Delta_1 = 1$, and

$$\begin{aligned}\Delta &= 1 - (L_1 + L_2 + L_3) + L_1L_3 \\ &= 1 + \frac{1}{sR_1C_1} + \frac{1}{sR_2C_1} + \frac{1}{sR_2C_2} + \frac{1}{s^2R_1R_2C_1C_2}\end{aligned}$$

Thus, the transfer function is given by

$$\frac{V_o(s)}{V_i(s)} = \frac{p_1\Delta_1}{\Delta} = \frac{1}{s^2R_1R_2C_1C_2 + sR_2C_2 + sR_1C_2 + sR_1C_1 + 1}$$

EXAMPLE 4.26. Draw the signal flow graph and determine C/R for the block diagram shown in Figure 4.40.

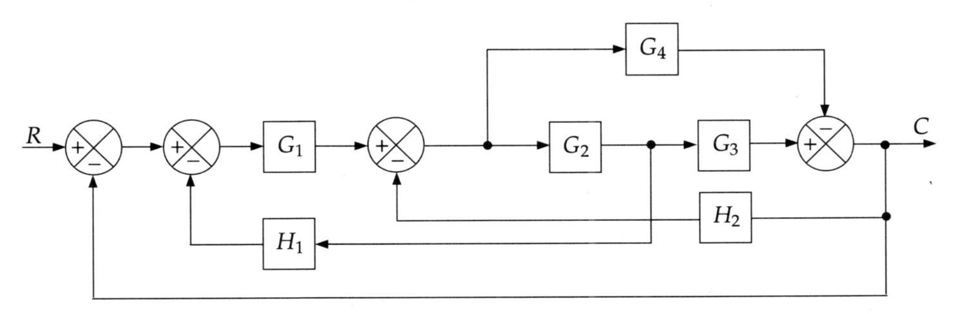

Figure 4.40 Block diagram (Example 4.26).

Solution We observe that the block diagram consists of four summing points and three take-off points. So, to draw the signal flow graph, we place seven nodes, name them $x_1 \cdots x_7$ for our convenience, and draw links between them as shown in Figure 4.41.

The forward paths and loops are

$$p_1 = x_1x_2x_3x_4x_5x_6x_7 \qquad p_2 = x_1x_2x_3x_4x_6x_7$$

$$L_1 = x_2x_3x_4x_5x_2 \qquad L_2 = x_3x_4x_5x_6x_7x_3$$

$$L_3 = x_3x_4x_6x_7x_3 \qquad L_4 = x_1x_2x_3x_4x_5x_6x_7x_1$$

$$L_5 = x_1x_2x_3x_4x_6x_7x_1$$

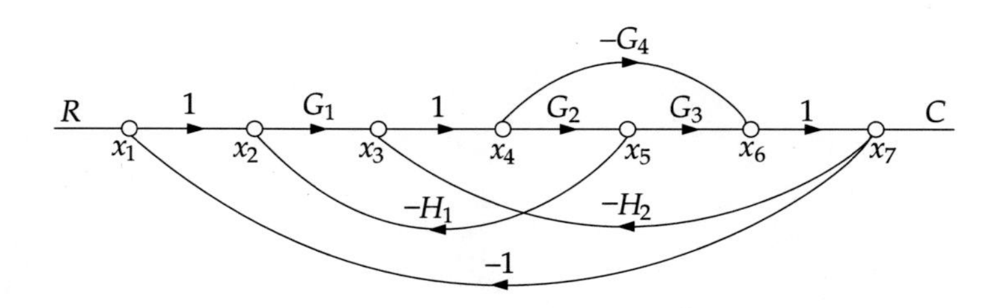

Figure 4.41 Signal flow graph (Example 4.26).

The forward path gains are

$$p_1 = G_1G_2G_3 \qquad p_2 = -G_1G_4$$

The loop gains are

$$L_1 = -G_1G_2H_1 \qquad L_2 = -G_2G_3H_2 \qquad L_3 = G_4H_2 \qquad L_4 = -G_1G_2G_3 \qquad L_5 = G_1G_4$$

There is no non-touching loop and all loops touch the two forward paths. Therefore,

$$\Delta_1 = 1 \qquad \Delta_2 = 1$$

$$\begin{aligned}\Delta &= 1 - (L_1 + L_2 + L_3 + L_4 + L_5)\\ &= 1 + G_1G_2H_1 + G_2G_3H_2 - G_4H_2 + G_1G_2G_3 - G_1G_4\end{aligned}$$

Therefore, $$\text{Transfer function} = \frac{C}{R} = \frac{p_1\Delta_1 + p_2\Delta_2}{\Delta}$$

$$= \frac{G_1G_2G_3 - G_1G_4}{1 + G_1G_2H_1 + G_2G_3H_2 - G_4H_2 + G_1G_2G_3 - G_1G_4}$$

EXAMPLE 4.27. Draw the signal flow graph from the given differential equation

$$3\frac{d^3y}{dx^3} + 11\frac{d^2y}{dx^2} + \frac{5}{2}\frac{dy}{dx} + 16y = x$$

Solution The Laplace transform of the given equation, assuming all initial conditions = 0, is

$$3s^3Y(s) + 11s^2Y(s) + \frac{5}{2}sY(s) + 16Y(s) = X(s)$$

Therefore, $$\text{Transfer function} = \frac{Y(s)}{X(s)} = \frac{1}{3s^3 + 11s^2 + \frac{5}{2}s + 16}$$

$$= \frac{\frac{1}{3s^3}}{1 - \left(-\frac{11}{3s} - \frac{5}{6s^2} - \frac{16}{3s^3}\right)}$$

$$= \frac{p_1\Delta_1}{1 - (L_1 + L_2 + L_3)}$$

where

$$p_1 = \frac{1}{3s^3} \qquad \Delta_1 = 1$$

$$L_1 = \frac{1}{s}\left(-\frac{11}{3}\right) \qquad L_2 = \frac{1}{s^2}\left(-\frac{5}{6}\right) \qquad L_3 = \frac{1}{s^3}\left(-\frac{16}{3}\right)$$

The signal flow graph, drawn accordingly, is presented in Figure 4.42.

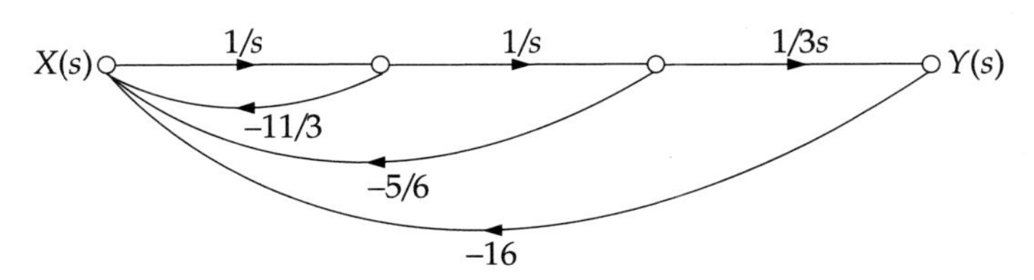

Figure 4.42 Signal flow graph (Example 4.27).

EXAMPLE 4.28. Find the overall transfer function through a signal flow graph and Mason's gain formula for the following equations:

$$x_2 = x_1 - H_3 x_5$$
$$x_3 = G_1 x_2 - H_1 x_4$$
$$x_4 = G_2 x_3 - H_2 x_5$$
$$x_5 = G_3 x_4 - G_5 x_6$$
$$x_6 = -H_4 x_6 + G_4 x_2$$

Solution The signal flow graph is shown in Figure 4.43.

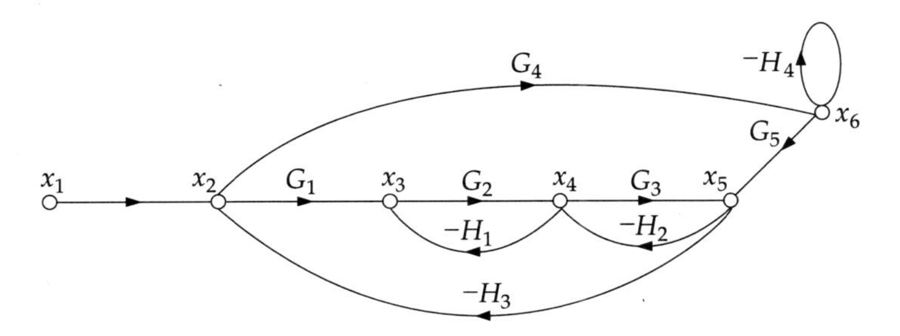

Figure 4.43 Signal flow graph (Example 4.28).

We may note here that though x_1 is the input, the output is x_5. Hence, the forward paths and gains are

$$p_1 = G_1 G_2 G_3 \qquad p_2 = G_4 G_5$$

The loops and gains are

$$L_1 = -G_2H_1, \qquad L_2 = -G_3H_2$$
$$L_3 = -H_4, \qquad L_4 = -G_4G_5H_3$$
$$L_5 = -G_1G_2G_3H_3$$

Now,

$$\begin{aligned}\Delta &= 1 - (L_1 + L_2 + L_3 + L_4 + L_5) + (L_1L_3 + L_1L_4 + L_2L_3 + L_3L_5)\\ &= 1 + (G_2H_1 + G_3H_2 + H_4 + G_4G_5H_3 + G_1G_2G_3H_3)\\ &\quad + (G_2H_1H_4 + G_2G_4G_5H_1H_3 + G_3H_2H_4 + G_1G_2G_3H_3H_4)\\ \Delta_1 &= 1 + H_4, \Delta_2 = 1 + G_2H_1\end{aligned}$$

Therefore,

$$\frac{x_5}{x_1} = \frac{G_1G_2G_3(1 + H_4) + G_4G_5(1 + G_2H_1)}{(1 + H_4)(1 + G_2H_1 + G_3H_2 + G_1G_2G_3H_3) + (1 + G_2H_1)G_4G_5H_3}$$

EXAMPLE 4.29. Find the transfer function of the signal flow graph given in Figure 4.44.

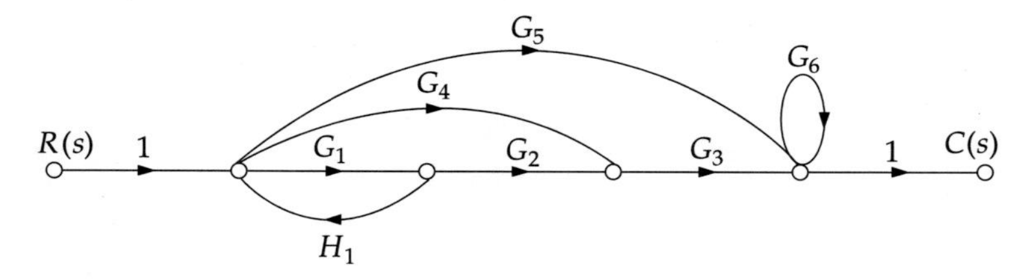

Figure 4.44 Signal flow graph (Example 4.29).

Solution The forward paths and gains are

$$p_1 = G_1G_2G_3 \qquad p_2 = G_3G_4 \qquad p_3 = G_5$$

The loops and gains are $L_1 = G_1H_1$ and $L_2 = G_6$.

The non-touching loop-pair and gain are $L_1L_2 = G_1G_6H_1$

All loops and pairs touch the three paths. Therefore,

$$\Delta_1 = \Delta_2 = \Delta_3 = 1$$
$$\Delta = 1 - (G_1H_1 + G_6) + G_1G_6H_1$$

Thus,

$$\text{Transfer function} = \frac{C(s)}{R(s)} = \frac{G_1G_2G_3 + G_3G_4 + G_5}{1 - G_1H_1 - G_6 + G_1G_6H_1}$$

EXAMPLE 4.30. Find the overall transfer function using Mason's gain formula of the system whose block diagram is shown in Figure 4.45.

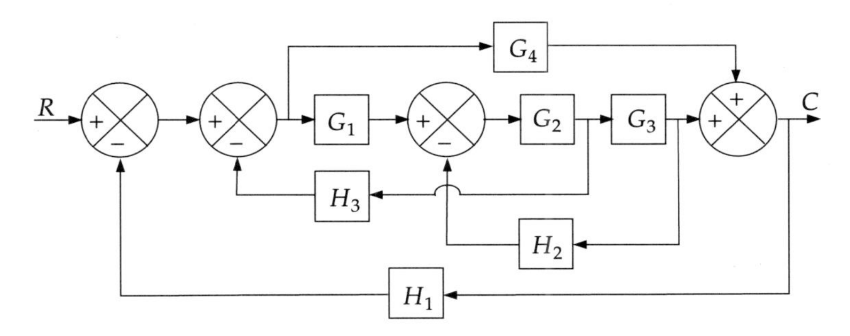

Figure 4.45 Block diagram (Example 4.30).

Solution The signal flow graph of the block diagram is given in Figure 4.46.

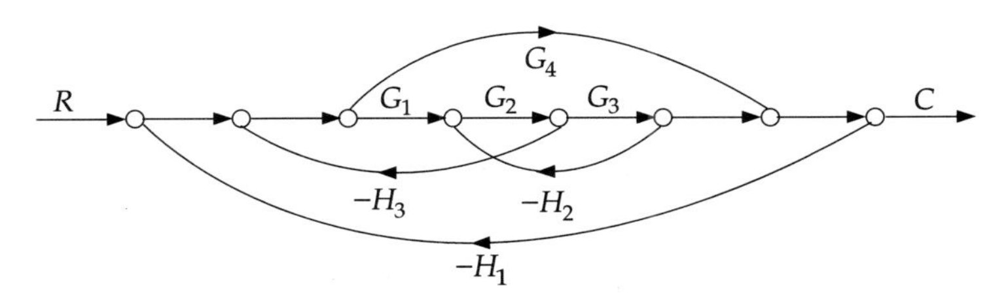

Figure 4.46 Signal flow graph for Figure 4.45.

Here, the forward paths and gains are $p_1 = G_1G_2G_3$ and $p_2 = G_4$.
The loops and gains are

$$L_1 = -G_1G_2H_3 \qquad L_2 = -G_2G_3H_2$$
$$L_3 = -G_1G_2G_3H_1 \qquad L_4 = -G_4H_1$$

The non-touching loop-pair and gain are

$$L_2L_4 = G_2G_3G_4H_1H_2$$

All loops and pairs touch p_1, but L_2 does not touch p_2. Therefore,

$$\Delta_1 = 1 \qquad \Delta_2 = 1 + G_2G_3H_2$$
$$\Delta = 1 + G_1G_2H_3 + G_2G_3H_2 + G_1G_2G_3H_1 + G_4H_1 + G_2G_3G_4H_1H_2$$

Thus, the transfer function is given by

$$\frac{C(s)}{R(s)} = \frac{G_1G_2G_3 + G_2G_3G_4H_2 + G_4}{1 + G_1G_2H_3 + G_2G_3H_2 + G_1G_2G_3H_1 + G_4H_1 + G_2G_3G_4H_1H_2}$$

EXAMPLE 4.31. Draw the signal flow graph of the system of Figure 4.18 (Example 4.6 at page 126) and find $C(s)/R(s)$ using Mason's gain formula.

Solution The signal flow graph is given by Figure 4.47.

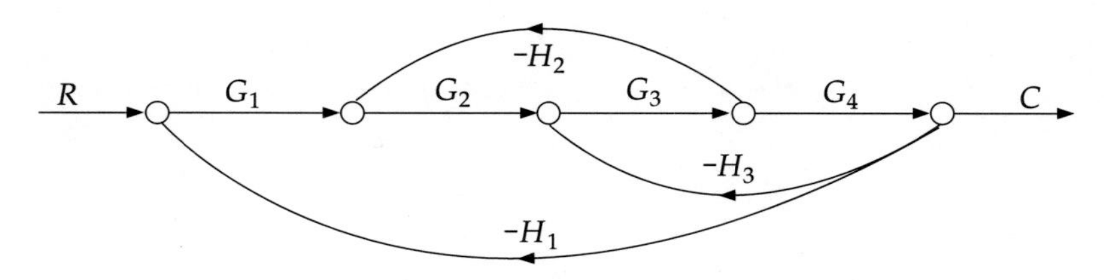

Figure 4.47 Signal flow graph for Figure 4.18.

The forward path, its gain product and path factor are given by $p_1 = G_1G_2G_3G_4$ and $\Delta_1 = 1$.
The loops and their gains are

$$L_1 = -G_2G_3H_2 \qquad L_2 = -G_3G_4H_3 \qquad L_3 = -G_1G_2G_3G_4H_1$$

There exists no non-touching loop pair. Hence, the determinant of the graph is

$$\Delta = 1 + G_2G_3H_2 + G_3G_4H_3 + G_1G_2G_3G_4H_1$$

Therefore,

$$\frac{C(s)}{R(s)} = \frac{G_1G_2G_3G_4}{1 + G_2G_3H_2 + G_3G_4H_3 + G_1G_2G_3G_4H_1}$$

EXAMPLE 4.32. Convert the block diagram of Figure 4.23 (at page 131, Example 4.11) into its signal flow graph and find the transfer function using the gain formula.

Solution The signal flow graph is given by Figure 4.48.

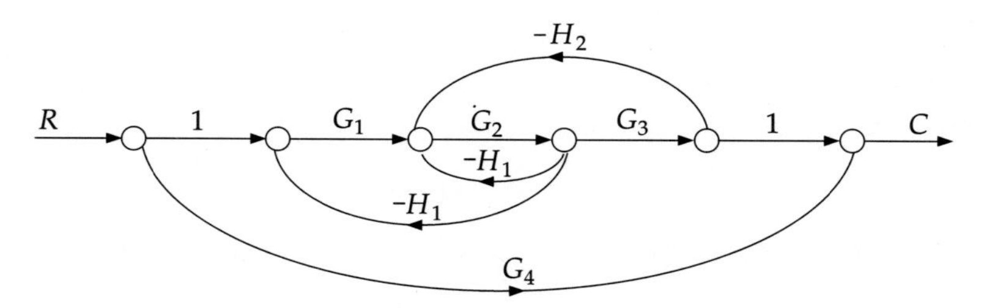

Figure 4.48 Signal flow graph for Figure 4.23.

Forward paths, their gain products and path factors are given by

$$p_1 = G_1G_2G_3 \qquad \Delta_1 = 1$$

$$p_2 = G_4 \qquad \Delta_2 = 1 - G_1G_2H_1 + G_2H_1 + G_2G_3H_2$$

The loops and their gains are

$$L_1 = G_1G_2H_1 \qquad L_2 = -G_2H_1 \qquad L_3 = -G_2G_3H_2$$

There exists no non-touching loop pair. Hence, the determinant of the graph is

$$\Delta = 1 - G_1G_2H_1 + G_2H_1 + G_2G_3H_2$$

Therefore, the transfer function is given by

$$\frac{C(s)}{R(s)} = \frac{G_1G_2G_3 + G_4(1 - G_1G_2H_1 + G_2H_1 + G_2G_3H_2)}{1 - G_1G_2H_1 + G_2H_1 + G_2G_3H_2}$$

EXAMPLE 4.33. Consider the following set of equations:

$$x_2 = t_{12}x_1 + t_{32}x_2$$

$$x_3 = t_{23}x_3 + t_{43}x_4$$

$$x_4 = t_{24}x_2 + t_{34}x_3 + t_{44}x_4$$

$$x_5 = t_{45}x_4$$

(a) First construct the signal flow graphs for individual equations.
(b) Then construct the signal flow graph as a whole.
(c) With the help of Mason's gain formula, determine the overall gain of the system.

Solution

(a) Individual signal flow graphs are given in Figure 4.49 in the given order of equations.

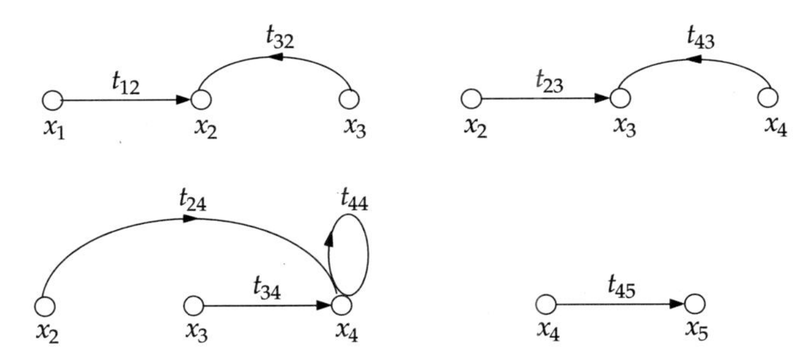

Figure 4.49 Signal flow graphs of individual equations.

(b) The signal flow graph as a whole is given in Figure 4.50.
(c) The forward paths and their gain products are

$$p_1 = t_{12}t_{23}t_{34}t_{45} \qquad p_2 = t_{12}t_{24}t_{45}$$

The loops and their gains are

$$L_1 = t_{23}t_{32} \qquad L_2 = t_{34}t_{43} \qquad L_3 = t_{44}$$

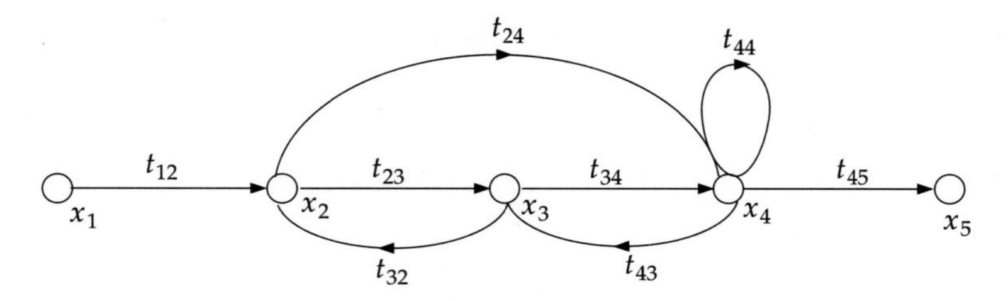

Figure 4.50 Signal flow graph as a whole (Example 4.33).

No non-touching loop-pair exists and all the loops are touched by both the forward paths. Hence, the path factors and the determinant of the graph are

$$\Delta_1 = \Delta_2 = 1 \qquad \Delta = 1 - t_{23}t_{32} - t_{34}t_{43} - t_{44}$$

The overall gain, i.e. the transfer function is, therefore, given by

$$\frac{C(s)}{R(s)} = \frac{t_{12}t_{23}t_{34}t_{45} + t_{12}t_{24}t_{45}}{1 - t_{23}t_{32} - t_{34}t_{43} - t_{44}}$$

EXAMPLE 4.34. The signal flow graph of a system is given by Figure 4.51. Find out the overall gain of the system.

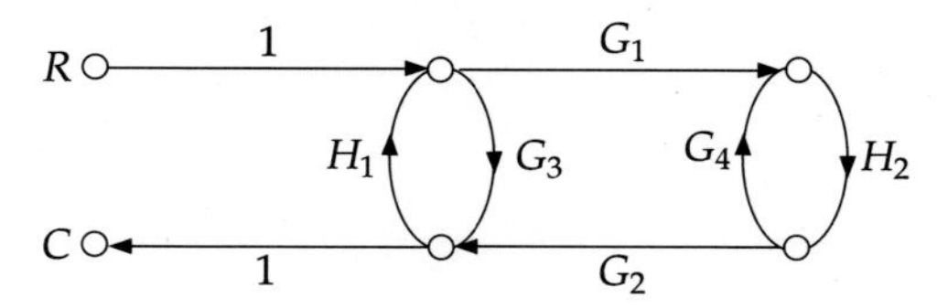

Figure 4.51 Signal flow graph (Example 4.34).

Solution Here, the forward paths, their gain-products, and the path factors are

$$p_1 = G_1H_2G_2 \qquad \Delta_1 = 1$$

$$p_2 = G_3 \qquad \Delta_2 = 1 - G_4H_2$$

The loops and their gains are

$$L_1 = G_3H_1 \qquad L_2 = G_4H_2 \qquad L_3 = G_1H_2G_2H_1$$

There exists one pair of non-touching loops, namely L_1 and L_2. Therefore, the determinant of the graph is

$$\Delta = 1 - G_3H_1 - G_4H_2 - G_1G_2H_1H_2 + G_3G_4H_1H_2$$

Thus, the overall gain, i.e. the transfer function, is

$$\frac{C(s)}{R(s)} = \frac{G_1G_2H_2 + G_3(1 - G_4H_2)}{1 - G_3H_1 - G_4H_2 - G_1G_2H_1H_2 + G_3G_4H_1H_2}$$

EXAMPLE 4.35. Find $C(s)/V(s)$ of the block diagram given in Figure 4.52 by signal flow graph and Mason's gain formula.

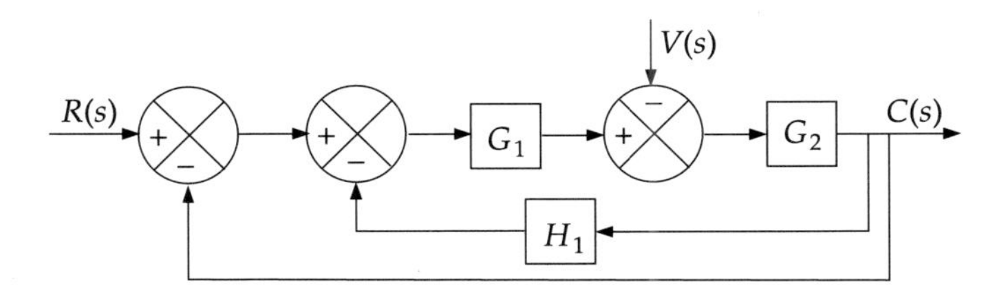

Figure 4.52 Block diagram (Example 4.35).

Solution The signal flow graph is given in Figure 4.53.

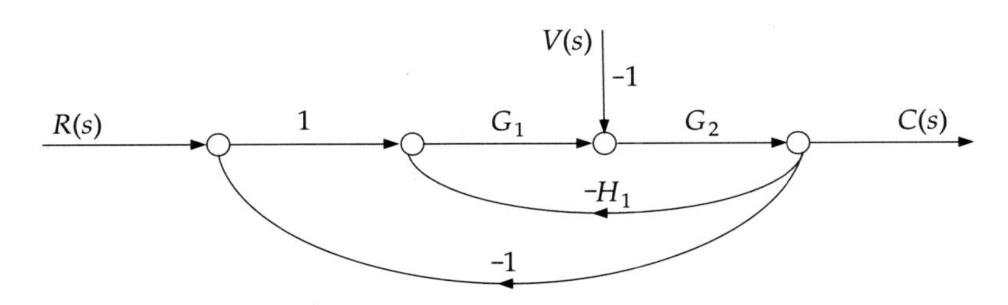

Figure 4.53 Signal flow graph for Figure 4.52.

We note that this is a multi-input system. So, when considering the transfer function between $C(s)$ and $V(s)$, we put $R(s) = 0$. Then, the only forward path, its gain product, and path factor are

$$p_1 = (-1)G_2 = -G_2 \qquad \Delta_1 = 1$$

The loops and their gains are $L_1 = -G_1G_2H_1$ and $L_2 = -G_1G_2$
There is no non-touching loop pair. So, the determinant of the graph is

$$\Delta = 1 + G_1G_2 + G_1G_2H_1$$

The required transfer function is, therefore, given by

$$\frac{C(s)}{V(s)} = \frac{G_2}{1 + G_1G_2 + G_1G_2H_1}$$

EXAMPLE 4.36. Find the transfer function of the system represented by the signal flow graph shown in Figure 4.54.

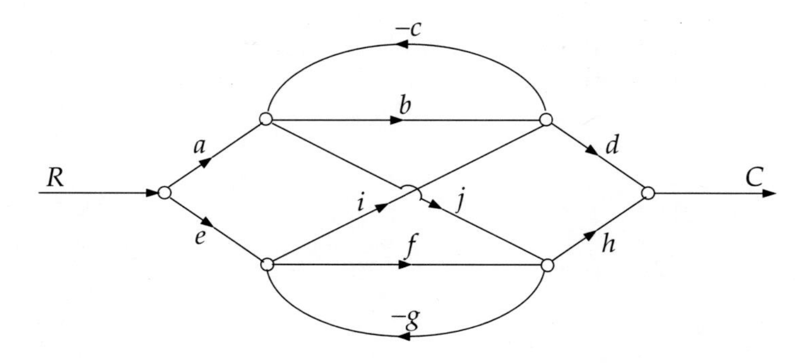

Figure 4.54 Signal flow graph (Example 4.36).

Solution The forward path-gains and their path-factors are

$$\begin{aligned}
p_1 &= abd & \Delta_1 &= 1 + fg \\
p_2 &= ajh & \Delta_2 &= 1 \\
p_3 &= efh & \Delta_3 &= 1 + bc \\
p_4 &= eid & \Delta_4 &= 1 \\
p_5 &= -ajgid & \Delta_5 &= 1 \\
p_6 &= -eicjh & \Delta_6 &= 1
\end{aligned}$$

The loops and their gains are

$$L_1 = -bc$$

$$L_2 = -fg$$

$$L_3 = cjgi$$

The non-touching loops are L_1 and L_2.
Therefore, the determinant of the graph is

$$\Delta = 1 + bc + fg - cgij + bcfg$$

Using Mason's gain formula, the required transfer function is given by

$$\frac{C(s)}{R(s)} = \frac{abd(1 + fg) + ahj + efh(1 + bc) + edi - adgij - cehij}{1 + bc + fg - cgij + bcfg}$$

REVIEW QUESTIONS

4.1 The block diagram of a control system is given in Figure Q4.1.

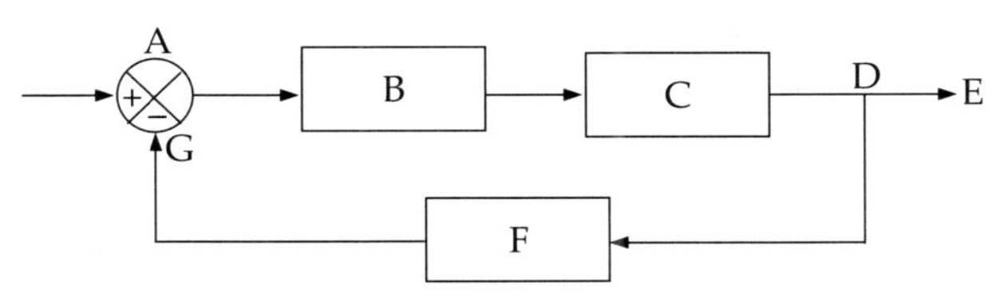

Figure Q4.1

Match the following:

Column I	*Column II*
A	(i) Forward path transfer function
B	(ii) Feedback signal
C	(iii) Summing point
D	(iv) Feedback transfer function
E	(v) Control element
F	(vi) Take-off point
G	(vii) Controlled output

4.2 Choose the correct answer:

(i) The signal flow graph of a system is shown in Figure Q4.2. The number of forward paths and the number of pairs of non-touching loops are, respectively,

(a) 3, 1 (b) 3, 2 (c) 4, 2 (d) 2, 4

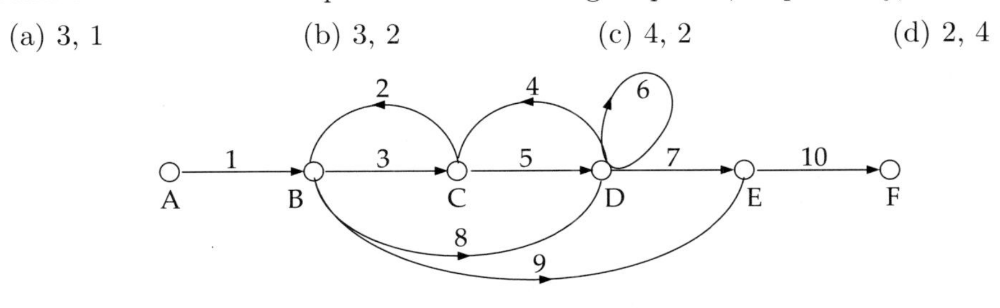

Figure Q4.2

(ii) The signal flow graph is used to obtain the

(a) stability of a system (b) transfer function of a system
(c) controllability of a system (d) observability of a system

(iii) In a signal flow graph, loops are said to be non-touching if they

(a) do not touch alternate nodes (b) do not touch any node
(c) do not possess any common node (d) possess at least one common node

(iv) The signal flow graph approach is applicable to

(a) linear system only (b) nonlinear system only

(c) both linear and nonlinear systems (d) none of these

4.3 (a) What do you mean by the block diagram of a system?

(b) What is Mason's gain formula?

(c) Obtain the overall transfer function of the block diagram given in Figure Q4.3 by block diagram reduction technique.

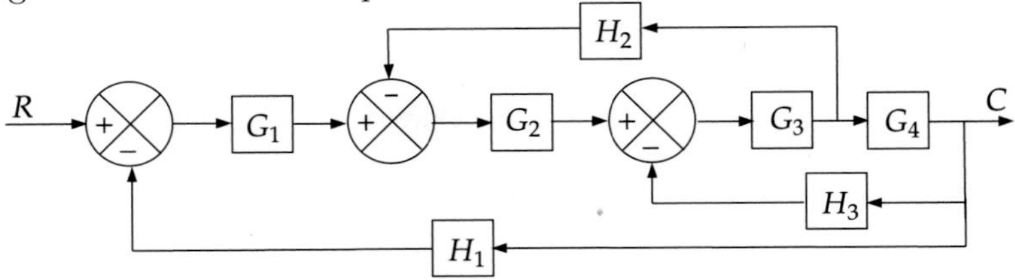

Figure Q4.3

(d) Convert the block diagram into its signal flow graph and find the transfer function using the gain formula.

4.4 Calculate the transfer function of the closed-loop system shown in Figure Q4.4.

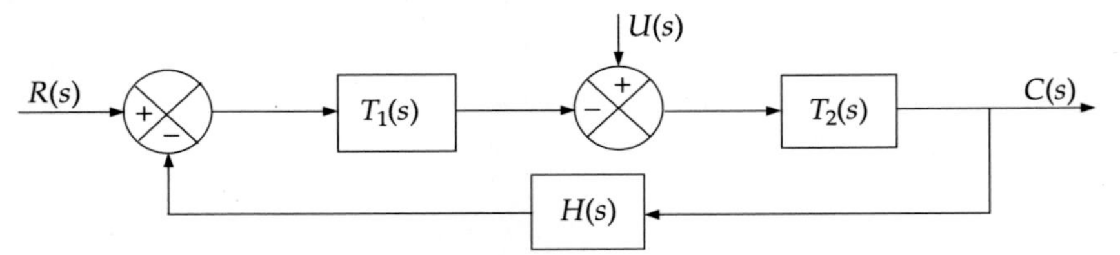

Figure Q4.4

4.5 Find the transfer functions of the systems represented by signal flow graphs shown in Figure Q4.5.

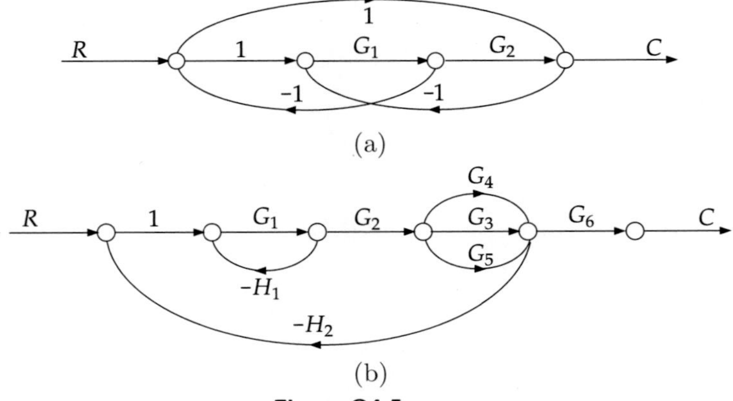

Figure Q4.5

4.6 Simplify the following block diagram (Figure Q4.6) using block diagram reduction techniques.

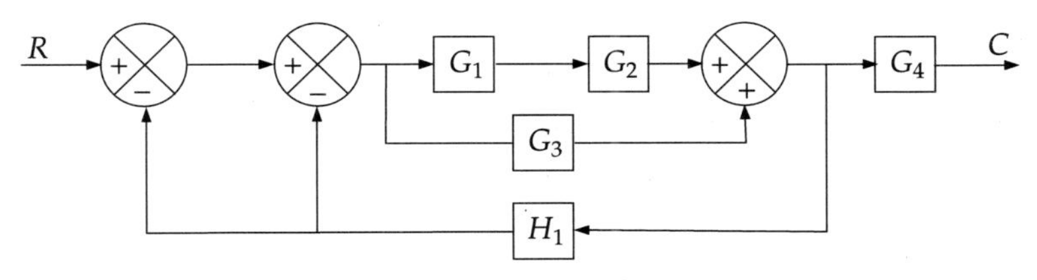

Figure Q4.6

4.7 Find C/R for the signal flow graph in Figure Q4.7:

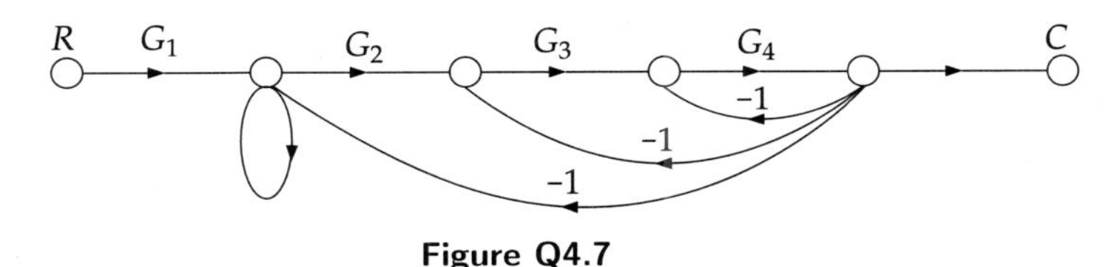

Figure Q4.7

[*Hint:* Assume gain = 1 for the segments where gains are not mentioned.]

Answers to Selected Questions

4.1 A→(iii), B→(v), C→(i), D→(vi), E→(vii), F→(iv), G→(ii)

4.2 (i) (a), (ii) (b), (iii) (c), (iv) (a)

4.3 (d) $\dfrac{C}{R} = \dfrac{G_1G_2G_3G_4}{1 + G_2G_3H_2 + G_3G_4H_3 + G_1G_2G_3G_4H_1}$

4.4 $C = \dfrac{UT_2 - RT_1T_2}{1 + T_1T_2H}$

4.5 (a) $\dfrac{C}{R} = 1$, (b) $\dfrac{C}{R} = \dfrac{G_1G_2G_6(G_3 + G_4 + G_5)}{1 + G_1H_1 + G_1G_2G_3H_2 + G_1G_2G_4H_2 + G_1G_2G_5H_2}$

4.6 $\dfrac{C}{R} = \dfrac{G_1G_2 + G_3}{1 + 2(G_1G_2 + G_3)H_1}$

4.7 $\dfrac{C}{R} = G_1$

CHAPTER 5

Systems and System Components

In Chapter 2, we gave the definition of a *system*. If a system has to be controlled to our desire, we have to build its mathematical model with its controllable parameters. So, in this chapter we will deal with the methods of building mathematical models of systems and then a few components that are of importance for controlling systems.

5.1 MATHEMATICAL MODELLING OF PHYSICAL SYSTEMS

Mathematical modelling of a system generally means describing its dynamical behaviour through differential equations. Such equations are obtained by utilizing the physical laws associated with the process. This approach applies equally to mechanical, electrical, fluid and thermodynamic systems. In mechanical systems, two kinds of motion—translational and rotational—take place. We recapitulate relevant mathematics of these motions from the modelling standpoint.

Translational Systems

If the motion takes place along a straight line, we call it a translational system. When a translation takes place, it may face three kinds of opposing forces originating from (i) inertia, (ii) frictional or damping force, and (iii) elasticity or spring action. Let us consider them in some detail.

Inertial force. Newton's law states that when a force is applied to a body at rest or of uniform motion in a straight line, it generates an acceleration in it. But the mass of the body tends to remain in its original state which is called *inertia*. That means, if the body was at rest before the application of the force, it has a tendency to remain at rest. Thus, inertia is an opposing force which is given by

$$F_M(t) = M\frac{d^2x}{dt^2} = M\frac{dv}{dt}$$

where

M = mass of the body

x = displacement of the body

v = velocity of the body

It can be diagrammatically represented as in Figure 5.1.

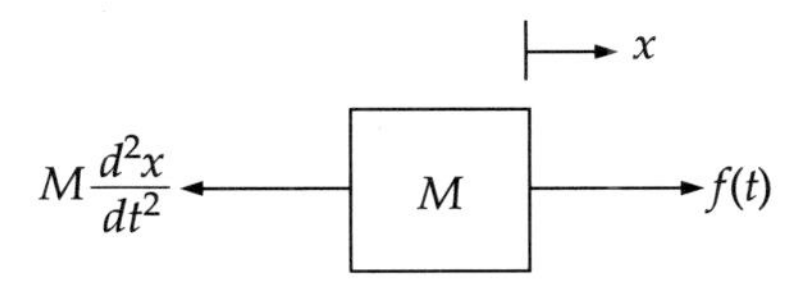

Figure 5.1 Schematic representation of inertial force.

Frictional or damping force. Since a body does not move in absolute vacuum, it experiences a frictional force from its surroundings, be it from the solid on which it moves or from the fluid which it wades through. This force is called *viscous friction* in general because it is proportional to the velocity v with which the body moves at a particular instant. So, the damping force F_D is given by

$$F_D(t) = Dv = D\frac{dx}{dt}$$

where D is the *damping coefficient.*

The damping force is represented by a dashpot as shown in Figure 5.2.

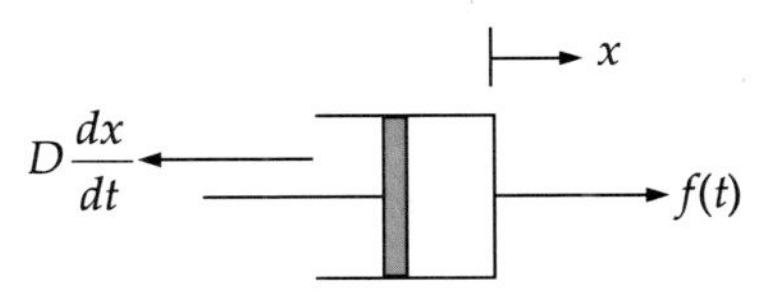

Figure 5.2 Schematic representation of damping force.

Damping forces not necessarily come into play owing to natural causes. Sometimes they are deliberately introduced to prevent or subdue oscillation in some systems, or increase stability in some others. Imagine riding a bike having frictionless tyres on frozen ice!

Elastic or spring force. When a body, such as a spring, suffers elongation, a restoring force, originating from the elasticity of the material, tries to put it back to its original shape. This force of restoration F_k, proportional to the displacement of the body, is given by

$$F_k = kx$$

where k is called the *spring constant* or *stiffness* whose diagrammatic form is given in Figure 5.3.

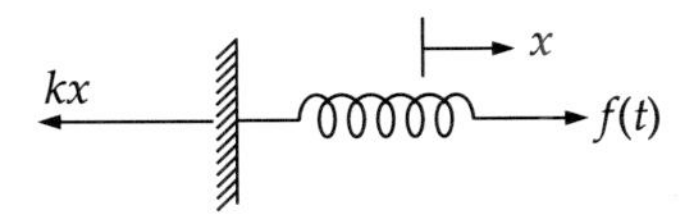

Figure 5.3 Schematic representation of spring force.

Rotational Systems

The angular displacement of a body about a fixed axis generates its rotational motion. Like force in translational motion, a rotation of a body at rest is caused by an externally applied torque which is resisted by three kinds of opposing torques. They are—(i) inertial torque, (ii) damping torque and (iii) spring torque.

Inertial torque. Like mass in translational motion, moment of inertia J stores the inertia of the existing state in rotational motion. Thus, the inertial torque is written as

$$T_I(t) = J\frac{d\omega}{dt} = J\frac{d^2\theta}{dt^2}$$

where θ and ω are angular displacement and angular velocity, respectively.

Damping torque. Damping torque, which is proportional to the angular velocity of the body, is given by

$$T_D(t) = D\omega = D\frac{d\theta}{dt}$$

Spring torque. The spring torque is proportional to the angular displacement of the body and is given by

$$T_k(t) = k\theta$$

Comparable parameters of these two kinds of motion are given in Table 5.1.

Table 5.1 Comparable Parameters of Translational and Rotational Motion

Translation		*Rotation*	
Parameter	*Symbol*	*Parameter*	*Symbol*
Mass	M	Inertia	J
Force	f	Torque	T
Velocity	v	Angular velocity	ω
Displacement	x	Angular displacement	θ

Modelling Mechanical Systems

Differential equations, describing mechanical systems, can be set up in two ways:

- From free body diagrams
- From nodal representations

Each method has its own advantages. While free body diagrams can be set up intuitively, nodal representations are helpful in drawing equivalent electrical networks. We consider both the methods and solve a few examples.

The free body diagram method. In the free body diagram method, the force balance at equilibrium condition of each associated mass (i.e. body) is considered and the equation is set up for each. To find out the transfer function of the whole system, intermediate variables are eliminated.

EXAMPLE 5.1. Deduce the equation governing the motion of the mass M in Figure 5.4. Find out also the transfer function.

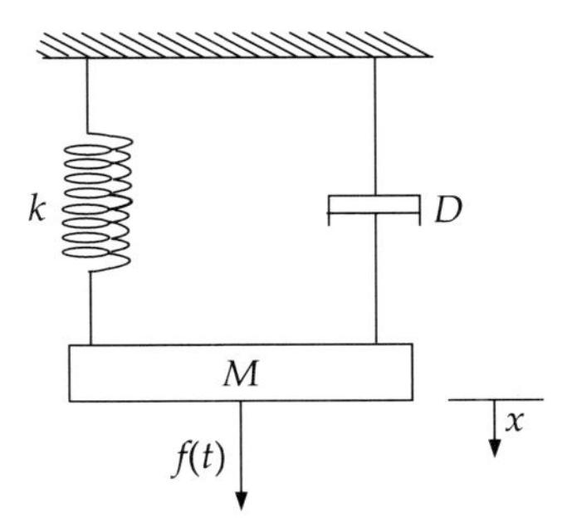

Figure 5.4 Mechanical system (Example 5.1).

Solution When the force f acting on the mass M produces a displacement x, three opposing forces act on it. So, as already discussed, the differential equation governing the motion is given by the force balance condition

$$f = M\frac{d^2x}{dt^2} + D\frac{dx}{dt} + kx \tag{i}$$

Taking the Laplace transform of Eq. (i) with all the initial conditions = 0 (i.e. the system is relaxed)[1] and rearranging the terms, we get the transfer function as

$$F(s) = s^2MX(s) + sDX(s) + kX(s)$$

or

$$\frac{X(s)}{F(s)} = \frac{1}{s^2M + sD + k}$$

EXAMPLE 5.2. Find the transfer function of the system involving two bodies M_1 and M_2 given in Figure 5.5.

[1]Unless stated otherwise, we will take Laplace transforms with all initial conditions = 0.

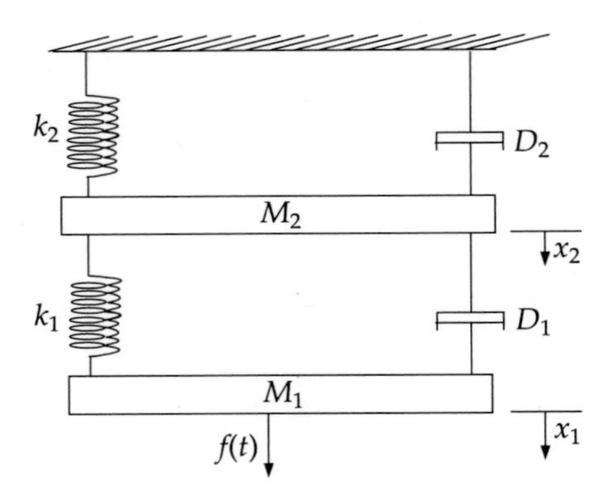

Figure 5.5 Two-body mechanical system (Example 5.2).

Solution We start from the mass M_1 on which the external force $f(t)$ acts. Treating it as a free body, its force balance is shown in Figure 5.6.

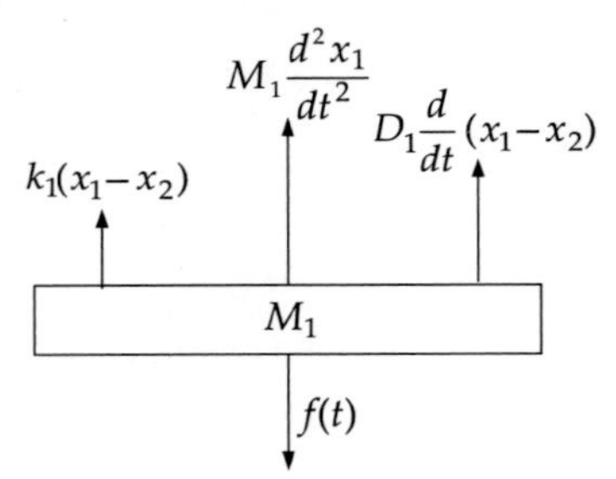

Figure 5.6 Force balance of free body M_1 (Example 5.2).

While drawing the figure, we need to keep it in mind that the inertial force is actuated by the displacement x_1 of the mass M_1, but viscous friction and spring forces are actuated by the differential displacement between M_1 and M_2. The corresponding differential equation and its Laplace transform are

$$M_1\frac{d^2x_1}{dt^2} + D_1\frac{d}{dt}(x_1 - x_2) + k_1(x_1 - x_2) = f(t) \tag{i}$$

$$s^2M_1X_1(s) + sD_1[X_1(s) - X_2(s)] + k[X_1(s) - X_2(s)] = F(s) \tag{ii}$$

On rearranging Eq. (i), we get

$$X_1(s) = \frac{F(s) + sD_1X_2(s) + k_1X_2(s)}{s^2M_1 + sD_1k_1} \tag{iii}$$

Next, taking M_2 as a free body, its force balance diagram is drawn as in Figure 5.7.

While drawing this diagram, we have kept it in mind that the damping and spring force which were reactions (opposing forces) on M_1, work as action (i.e. forward force) on M_2 in keeping with Newton's third law of motion. From Figure 5.7, the following differential equation and its Laplace transform can be written as

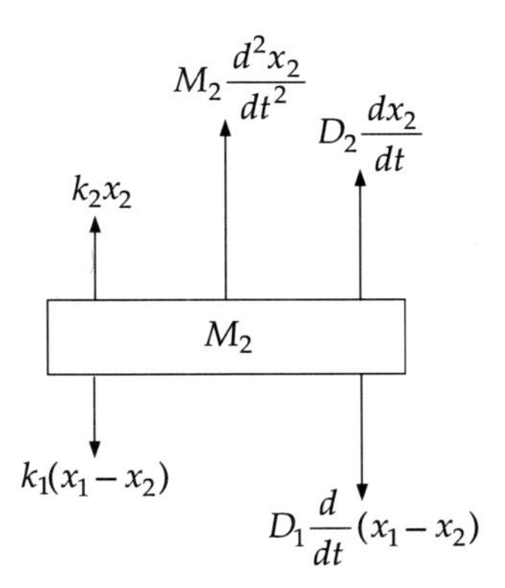

Figure 5.7 Force balance of free body M_2 (Example 5.2).

$$M_2\frac{d^2x_2}{dt^2}+D_2\frac{dx_2}{dt}+k_2x_2=D_1\frac{d}{dt}(x_1-x_2)+k_1(x_1-x_2) \tag{iv}$$

$$(s^2M_2+sD_2+k_2)X_2(s)=(sD_1+k_1)[X_1(s)-X_2(s)] \tag{v}$$

On rearranging Eq. (iv) and eliminating $X_1(s)$ with the help of Eq. (iii), we get

$$\{[s^2M_2+s(D_1+D_2)+(k_1+k_2)](s^2M_1+sD_1+k_1)-(sD_1+k_1)^2\}X_2(s)=F(s)(sD_1+k_1)$$

Therefore, the transfer function is given by

$$\frac{X_2(s)}{F(s)}=\frac{sD_1+k_1}{[s^2M_2+s(D_1+D_2)+(k_1+k_2)](s^2M_1+sD_1+k_1)-(sD_1+k_1)^2}$$

EXAMPLE 5.3. Determine the differential equations describing the dynamics of the mechanical system shown in Figure 5.8.

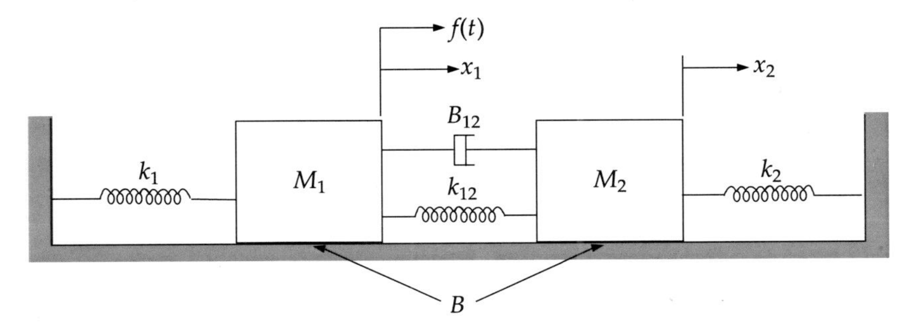

Figure 5.8 Mechanical system (Example 5.3).

Solution The force balance of free body M_1 can be drawn as shown in Figure 5.9.

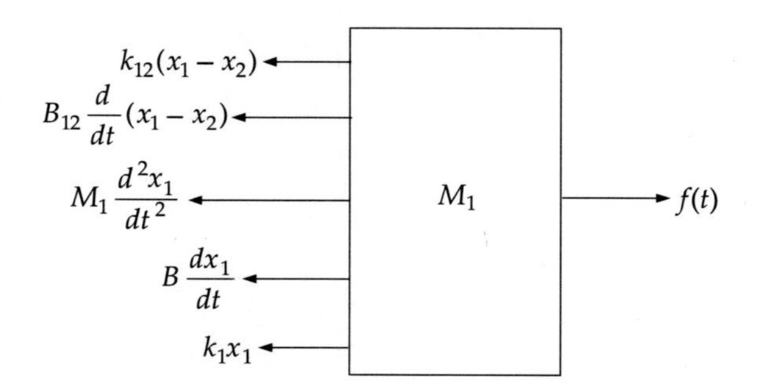

Figure 5.9 Force balance of free body M_1 (Example 5.3).

The corresponding differential equation is given by

$$M_1\frac{d^2x_1}{dt^2} + B_{12}\frac{d}{dt}(x_1 - x_2) + B\frac{dx_1}{dt} + k_{12}(x_1 - x_2) + k_1x_1 = f(t) \qquad \text{(i)}$$

The force balance diagram of free body M_2 is given in Figure 5.10.

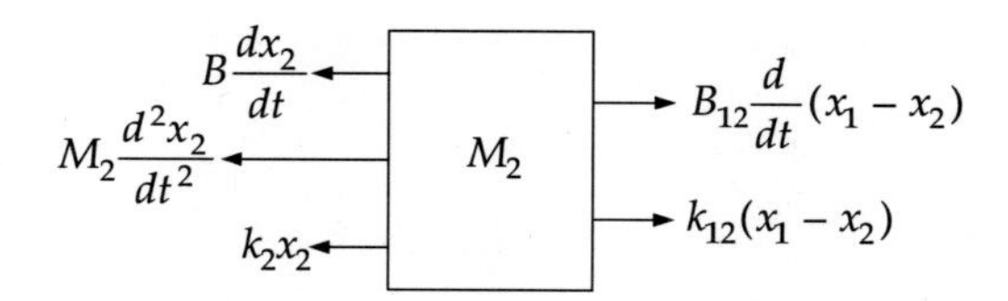

Figure 5.10 Force balance of free body M_2 (Example 5.3).

The corresponding differential equation is

$$M_2\frac{d^2x_2}{dt^2} + B\frac{dx_2}{dt} + k_2x_2 = B_{12}\frac{d}{dt}(x_1 - x_2) + k_{12}(x_1 - x_2) \qquad \text{(ii)}$$

The system dynamics is described by Eqs. (i) and (ii).

Nodal representation method. In the nodal representation method, a network diagram for the mechanical system is first drawn following the steps.

Step 1. A reference line is drawn. It embodies fixed point(s) of the system.

Step 2. Nodes[2] corresponding to displacements are plotted.

Step 3. Masses are connected between nodes and the reference line.

[2]Means "labelled points in a tree diagram at which subordinate points branch off". (Webster's Universal College Dictionary). The literal meaning of the word is "knot".

Step 4. Viscous frictions and springs associated with each mass are connected between nodes and reference line, parallel to the respective mass.

Step 5. Viscous frictions and springs between two displacements are connected between the two respective nodes.

Step 6. External applied force is connected between the relevant node and the reference line.

Step 7. For each node, force balance equations are written.

The procedure will be clear from the following examples, especially from Example 5.4, where we show it step by step explicitly.

EXAMPLE 5.4. Find, by the nodal representation method, the transfer function for the mechanical system shown in Figure 5.5.

Solution The step-by-step procedure of drawing Figure 5.11 is given as follows:

Step 1. Corresponding to the fixed point at the top, from where the system is suspended, we draw a reference line at the bottom.

Step 2. Corresponding to two displacements x_1 and x_2, two nodes are plotted.

x_1 x_2

Step 3. M_1 and M_2 are connected between x_1, x_2 and the reference line.

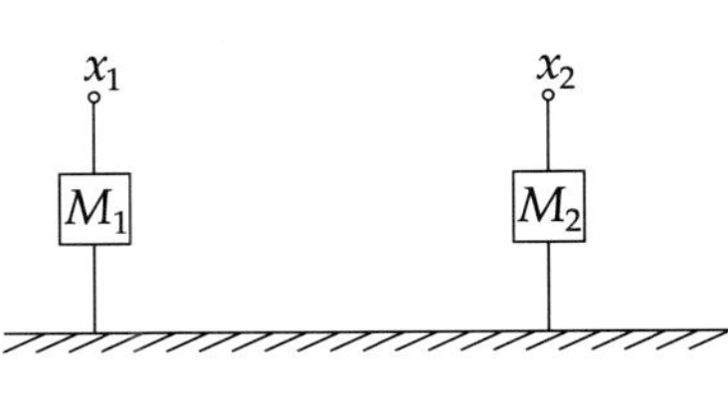

Step 4. D_2 and k_2 are associated with the mass M_2. So, they are connected between x_2 and the reference line parallel to M_2.

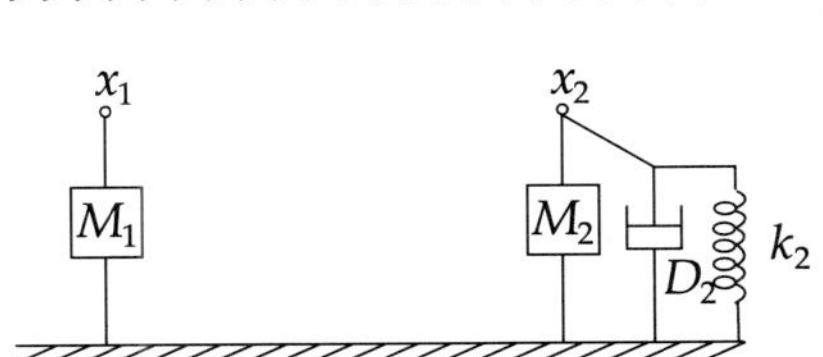

Step 5. D_1 and k_1 act between x_1 and x_2. So, they are connected accordingly.

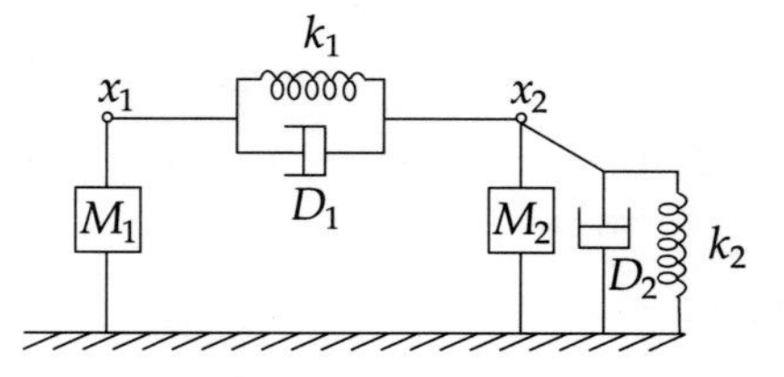

Step 6. $f(t)$ is connected between x_1 and the reference line.

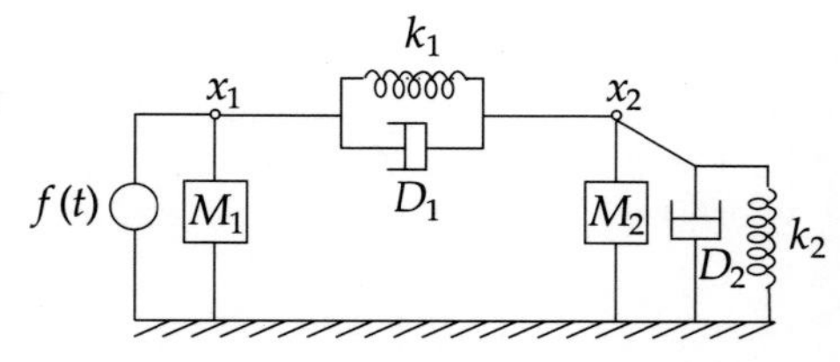

The complete diagram is shown in Figure 5.11.

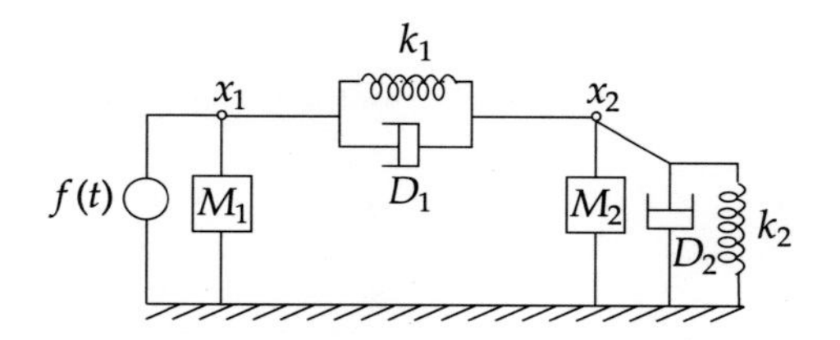

Figure 5.11 Complete nodal diagram (Example 5.4).

Once the diagram is drawn, it is easy to set up differential equations balancing the left and right side of each node separately. For node at x_1, the force balance equation is given by

$$f(t) = M_1\frac{d^2x_1}{dt^2} + D_1\frac{d}{dt}(x_1 - x_2) + k_1(x_1 - x_2) \tag{i}$$

and for the node at x_2, it is given by

$$D_1\frac{d}{dt}(x_1 - x_2) + k_1(x_1 - x_2) = M_2\frac{d^2x_2}{dt^2} + D_2\frac{dx_2}{dt} + k_2x_2 \tag{ii}$$

We observe that Eqs. (i) and (ii) are identical to Eqs. (i) and (iv) of Example 5.2, respectively. The rest of the procedure of finding out the transfer function is the same as given in Example 5.2.

EXAMPLE 5.5. Find the differential equations for Figure 5.8 by the nodal representation method.

Solution The relevant nodal diagram is shown in Figure 5.12. At the force balance equation is node x_1

$$f(t) = M_1\frac{d^2x_1}{dt^2} + B\frac{dx_1}{dt} + k_1x_1 + B_{12}\frac{d}{dt}(x_1 - x_2) + k_{12}(x_1 - x_2) \tag{i}$$

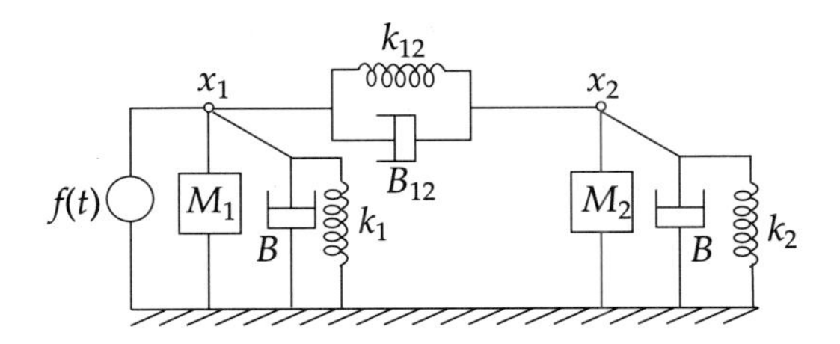

Figure 5.12 Nodal diagram (Example 5.5).

and at node x_2, it is

$$B_{12}\frac{d}{dt}(x_1 - x_2) + k_{12}(x_1 - x_2) = M_2\frac{d^2x_2}{dt^2} + B\frac{dx_2}{dt} + k_2x_2 \tag{ii}$$

We note that Eqs. (i) and (ii) are respectively, identical to Eqs. (i) and (ii) of Example 5.3.

Analogous Systems

In electrical systems, voltage V, current i and magnetic flux linkage, Φ are considered three important parameters. Table 5.2 shows how these parameters can be interrelated through passive components of circuits, namely, R, L and C.

Table 5.2 Interrelationships of Voltage, Current and Magnetic Flux Linkage

Passive component	*Relation between*		
	Voltage-current	*Current-voltage*	*Current-flux linkage*
Resistor	$V = iR$	$i = \frac{V}{R}$	$i = \frac{1}{R}\frac{d\Phi}{dt}$
Inductor	$V = L\frac{di}{dt}$	$i = \frac{1}{L}\int V\ dt$	$i = \frac{\Phi}{L}$
Capacitor	$V = \frac{1}{C}\int i\ dt$	$i = C\frac{dV}{dt}$	$i = C\frac{d^2\Phi}{dt^2}$

We observe that similar differential equations for dynamic systems can be written for mechanical as well and electrical systems as follows:

Mechanical

Translation: $$M\frac{d^2x}{dt^2} + D\frac{dx}{dt} + kx = f$$

Rotation: $$J\frac{d^2\theta}{dt^2} + D\frac{d\theta}{dt} + k\theta = T$$

Electrical

Voltage: $$L\frac{d^2Q}{dt^2} + R\frac{dQ}{dt} + \frac{Q}{C} = V$$

Current: $$C\frac{d^2\Phi}{dt^2} + \frac{1}{R}\frac{d\Phi}{dt} + \frac{\Phi}{L} = i$$

Comparing these equations, we can draw an analogy between different physical quantities as shown in Table 5.3.

Table 5.3 Analogous Quantities

Translation	*Rotation*	*Electrical*	
		Force-voltage analogy	*Force-current analogy*
f	T	V	i
M	J	L	C
D	D	R	$1/R$
k	k	$1/C$	$1/L$
x	θ	Q	Φ

We will now see how these analogies can be exploited to frame differential equations for dynamic systems.

EXAMPLE 5.6. By using force-current $(f - i)$ analogy, draw the electrical analogy of the mechanical system given in Figure 5.13(a).

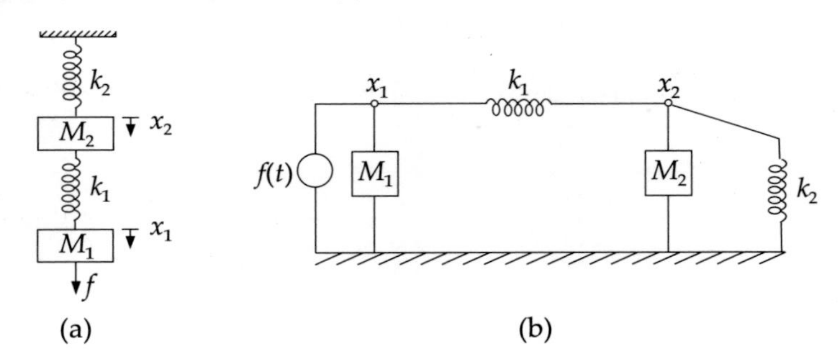

Figure 5.13 (a) Mechanical system; (b) its nodal diagram (Example 5.6).

Solution The nodal diagram for the system is given in Figure 5.13(b). The force-balance equations at x_1 and x_2 are

$$f = M_1 \frac{d^2 x_1}{dt^2} + k_1(x_1 - x_2) \tag{i}$$

$$k_1(x_1 - x_2) = M_2 \frac{d^2 x_2}{dt^2} + k_2 x_2 \tag{ii}$$

To arrive at their electrical analogues, we substitute $f \to i$, $M \to C$, $k \to 1/L$ and $x \to \Phi$ in Eqs. (i) and (ii) to obtain

$$i = C_1 \frac{d^2 \Phi_1}{dt^2} + \frac{1}{L_1}(\Phi_1 - \Phi_2) = C_1 \frac{dV_1}{dt} + \frac{1}{L_1}\left(\int V_1 \, dt - \int V_2 \, dt\right) \tag{iii}$$

$$\frac{1}{L_1}\left(\int V_1\, dt - \int V_2\, dt\right) = C_2\frac{d^2\Phi_2}{dt^2} + \frac{\Phi_2}{L_2} = C_2\frac{dV_2}{dt} + \frac{\int V_2 dt}{L_2} \tag{iv}$$

To draw the analogous electrical circuit,

1. We plot two nodes, V_1 and V_2 and an earth line.
2. Then the left and right loops (see Figure 5.14) are drawn from Eqs. (iii) and (iv), respectively.
3. Observe, as already discussed, that the electrical analogue looks similar to the nodal diagram in the $f - i$ analogy, though L_1 and L_2 and not their inverse, replace k_1 and k_2 in the analogous diagram.

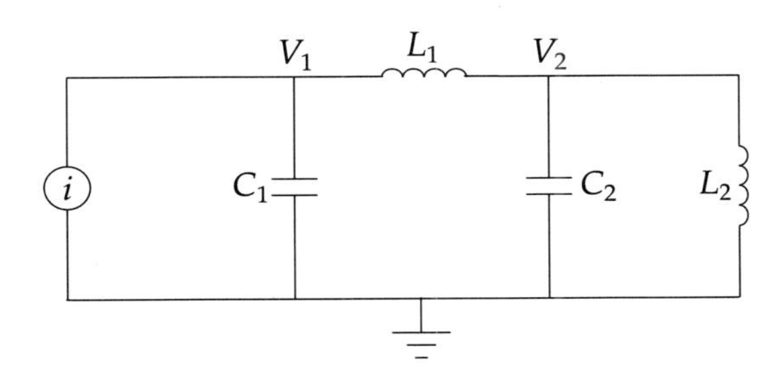

Figure 5.14 Analogous electrical circuit for Figure 5.13.

EXAMPLE 5.7. Find the electrical analogous circuits of the mechanical system shown in Figure 5.15.

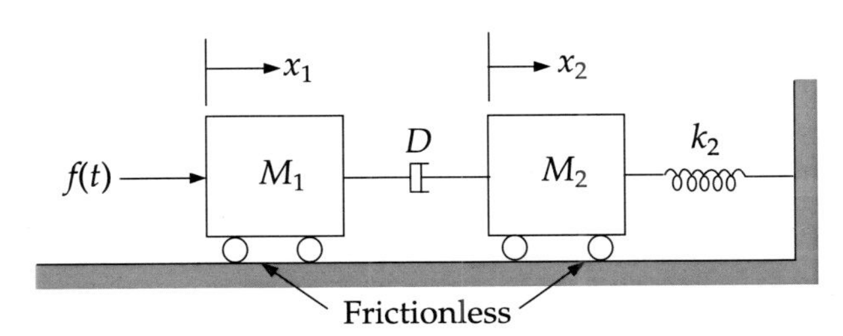

Figure 5.15 Mechanical system (Example 5.7).

Solution The nodal diagram for the system is given in Figure 5.16. The force-balance equations at nodes x_1 and x_2 are

$$f = M_1\frac{d^2x_1}{dt^2} + D\frac{d}{dt}(x_1 - x_2) \tag{i}$$

$$D\frac{d}{dt}(x_1 - x_2) = M_2\frac{d^2x_2}{dt^2} + k_2x_2 \tag{ii}$$

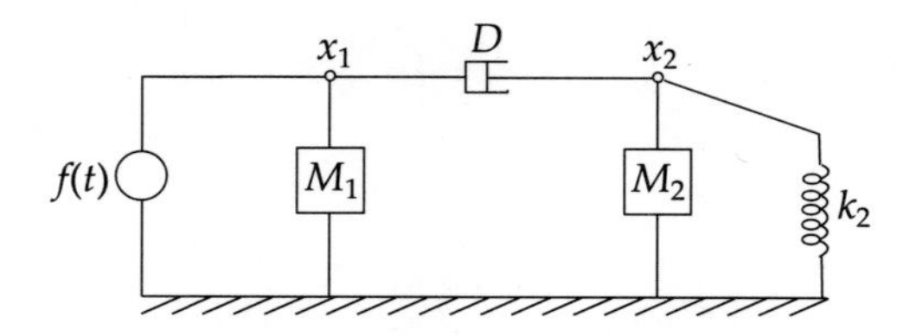

Figure 5.16 Nodal diagram for Figure 5.15.

For the f–V analogy equations, we put $f \rightarrow V$, $M \rightarrow L$, $D \rightarrow R$, $k \rightarrow 1/C$ and $x \rightarrow Q$ to get from Eqs. (i) and (ii).

$$V = L_1 \frac{d^2Q_1}{dt^2} + R\frac{d}{dt}(Q_1 - Q_2)$$

$$= L_1 \frac{di_1}{dt} + R(i_1 - i_2) \tag{iii}$$

$$R(i_1 - i_2) = L_2 \frac{d^2Q_2}{dt^2} + \frac{Q_2}{C_2}$$

$$= L_2 \frac{di_2}{dt} + \frac{\int i_2 \, dt}{C_2} \tag{iv}$$

Equations (iii) and (iv) are represented by Figure 5.17.

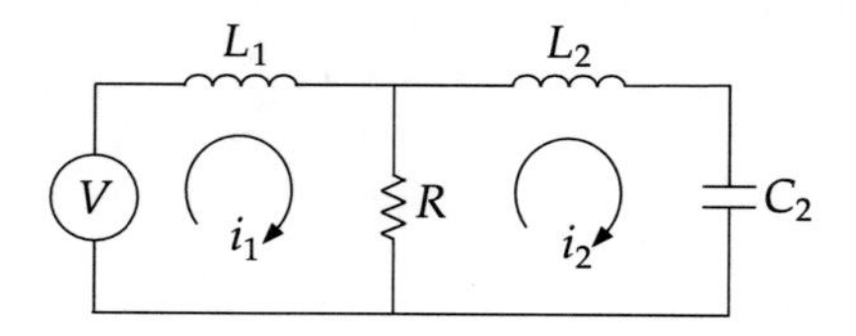

Figure 5.17 Electrical f–V analogue of Figure 5.15.

We note that

1. While M_1 and M_2 are connected *in parallel* in the nodal diagram, their analogous quantities L_1 and L_2 are connected *in series* in the f–V analogue diagram.
2. Similar series-parallel inversion occurs for $D \rightarrow R$ and $k \rightarrow C$.

Next, we consider the f–i analogy. Here, we substitute $f \rightarrow i$, $M \rightarrow C$, $D \rightarrow 1/R$ and $k \rightarrow 1/L$ in Eqs. (i) and (ii) to get the following equations:

$$i = C_1 \frac{d^2\Phi_1}{dt^2} + \frac{1}{R}\frac{d}{dt}(\Phi_1 - \Phi_2)$$

$$= C_1 \frac{dV_1}{dt} + \frac{1}{R}(V_1 - V_2) \tag{v}$$

$$\frac{1}{R}(V_1 - V_2) = C_2\frac{d^2\Phi_2}{dt^2} + \frac{\Phi_2}{L_2}$$

$$= C_2\frac{dV_2}{dt} + \frac{\int V_2\,dt}{L_2} \qquad \text{(vi)}$$

Equations (v) and (vi) are represented by Figure 5.18. As in Example 5.6, we note the similarity between the nodal diagram and the f–i analogous diagram.

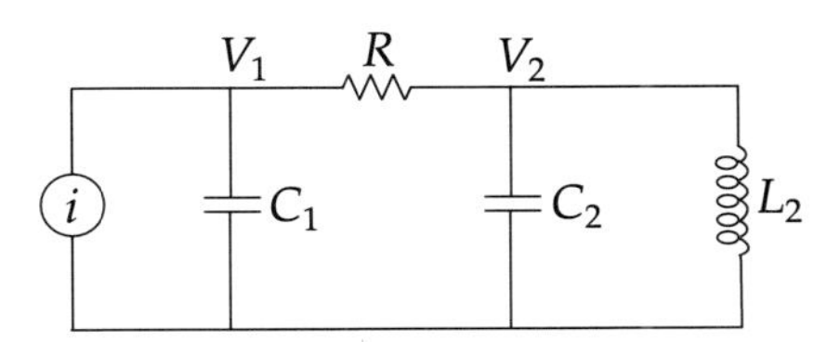

Figure 5.18 Electrical f–i analogue of Figure 5.15.

EXAMPLE 5.8. Set up differential equations for the rotational mechanical system shown in Figure 5.19 and draw the electrical analogue from T–i analogy.

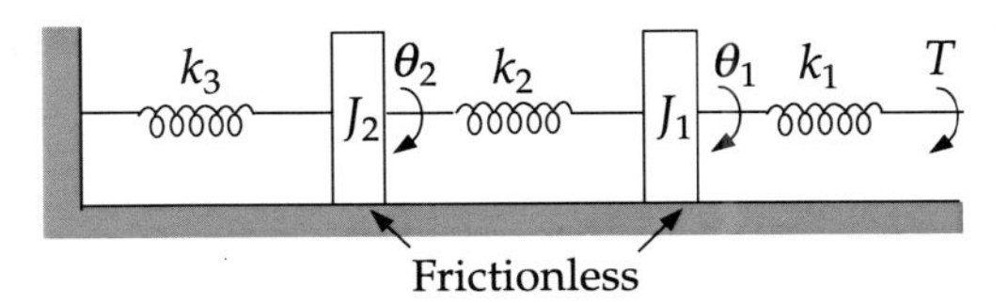

Figure 5.19 Rotational mechanical system (Example 5.8).

Solution The nodal representation diagram of the rotational system is given in Figure 5.20.

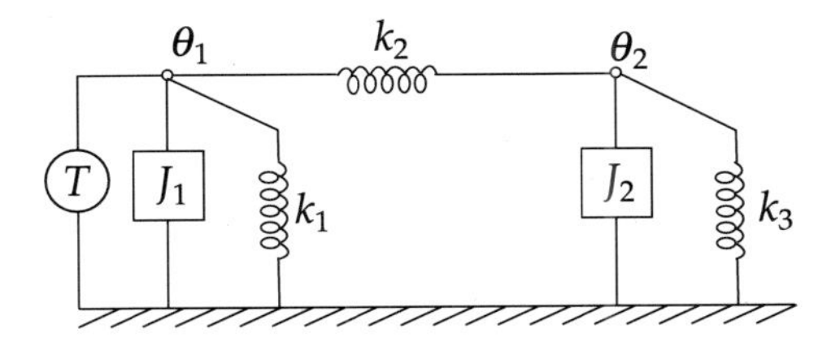

Figure 5.20 Nodal diagram for Figure 5.19.

The torque-balance equations corresponding to nodes θ_1 and θ_2 are as follows:

$$T = J_1\frac{d^2\theta_1}{dt^2} + k_1\theta_1 + k_2(\theta_1 - \theta_2) \qquad \text{(i)}$$

$$k_2(\theta_1 - \theta_2) = J_2\frac{d^2\theta_2}{dt^2} + k_3\theta_2 \qquad \text{(ii)}$$

Substituting $T \to i,\ J \to C,\ k \to 1/L$ and $\theta \to \Phi$ from T–i analogy, from Eqs. (i) and (ii), we get

$$i = C_1\frac{d^2\Phi_1}{dt^2} + \frac{\Phi_1}{L_1} = \frac{\Phi_1 - \Phi_2}{L_2}$$

$$= C_1\frac{dV_1}{dt} + \frac{\int V_1\,dt}{L_1} + \frac{\int (V_1 - V_2)dt}{L_2} \tag{iii}$$

$$\frac{\int (V_1 - V_2)dt}{L_2} = C_2\frac{d^2\Phi_2}{dt^2} + \frac{\Phi_2}{L_3} = C_2\frac{dV_2}{dt} + \frac{\int V_2\,dt}{L_3} \tag{iv}$$

Figure 5.21 represents Eqs. (iii) and (iv).

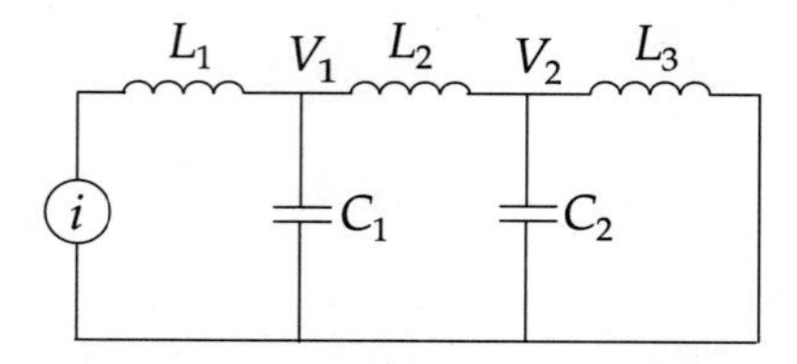

Figure 5.21 T–i analogue of Figure 5.20.

EXAMPLE 5.9. A rotational mechanical system is shown in Figure 5.22. Determine the system equations and find electrically analogous systems from (a) torque-current and (b) torque-voltage analogies.

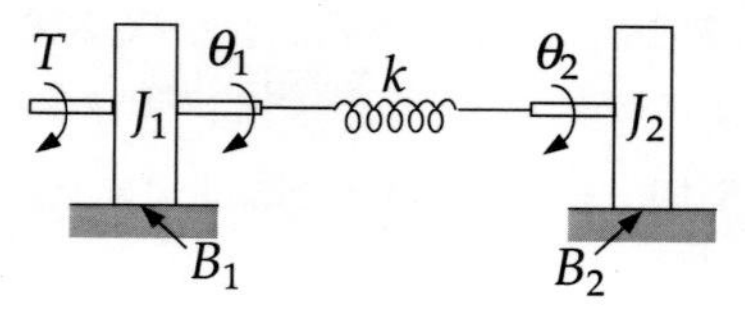

Figure 5.22 Rotational mechanical system (Example 5.9).

Solution The nodal representation diagram of the system is given in Figure 5.23.

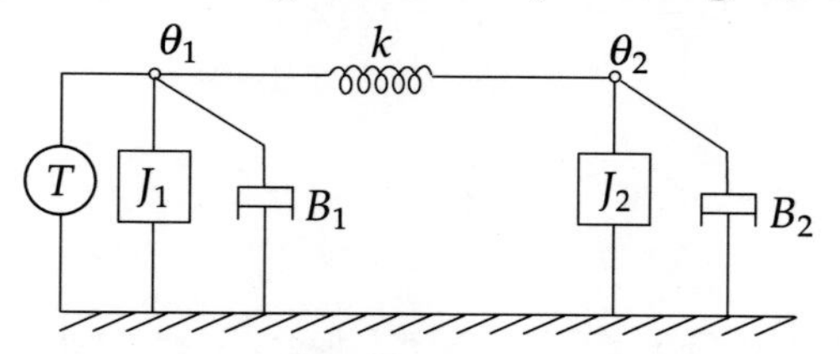

Figure 5.23 Nodal diagram of Figure 5.22.

The torque-balance equations corresponding to nodes q_1 and q_2 are given by

$$T = J_1\frac{d^2\theta_1}{dt^2} + B_1\frac{d\theta_1}{dt} + k(\theta_1 - \theta_2) \tag{i}$$

$$k(\theta_1 - \theta_2) = J_2\frac{d^2\theta_2}{dt^2} + B_2\frac{d\theta_2}{dt} \tag{ii}$$

Equations (i) and (ii) are the required system equations.

(a) To find electrically analogous circuit from the T–i analogy, we substitute $T \to i$, $J \to C$, $B \to 1/R$, $k \to 1/L$ and $q \to F$ in Eqs. (i) and (ii) to get

$$i = C_1 \frac{d^2\Phi_1}{dt^2} + \frac{1}{R_1}\frac{d\Phi_1}{dt} + \frac{\Phi_1 - \Phi_2}{L}$$

$$= C_1 \frac{dV_1}{dt} + \frac{V_1}{R_1} + \frac{\int (V_1 - V_2)\, dt}{L} \quad \text{(iii)}$$

$$\frac{\int (V_1 - V_2)\, dt}{L} = C_2 \frac{dV_2}{dt} + \frac{V_2}{R_2} \quad \text{(iv)}$$

Equations (iii) and (iv) are represented by the circuit given in Figure 5.24.

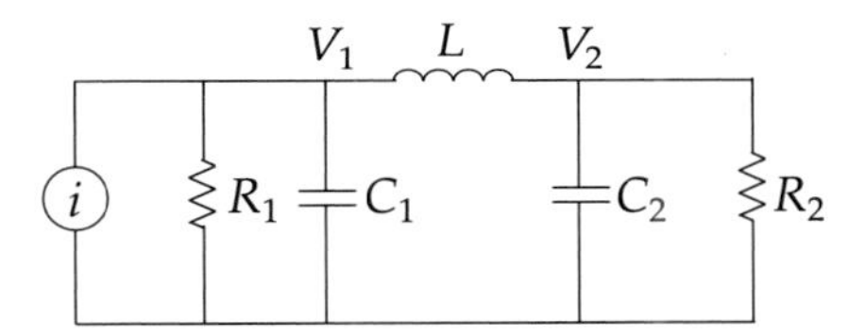

Figure 5.24 *T–i* analogous circuit of Figure 5.22.

(b) To draw the T–V analogous circuit, we first substitute $T \to V$, $J \to L$, $B \to R$, $k \to 1/C$ and $q \to Q$ in Eqs. (i) and (ii) to obtain

$$V = L_1 \frac{d^2Q_1}{dt^2} + R_1 \frac{dQ_1}{dt} + \frac{Q_1 - Q_2}{C}$$

$$= L_1 \frac{di_1}{dt} + R_1 i_1 + \frac{\int (i_1 - i_2)\, dt}{C} \quad \text{(v)}$$

$$\frac{\int (i_1 - i_2)\, dt}{C} = L_2 \frac{di_2}{dt} + R_2 i_2 \quad \text{(vi)}$$

When translated into an electrical network, the yield of Eqs. (v) and (vi) is shown in Figure 5.25.

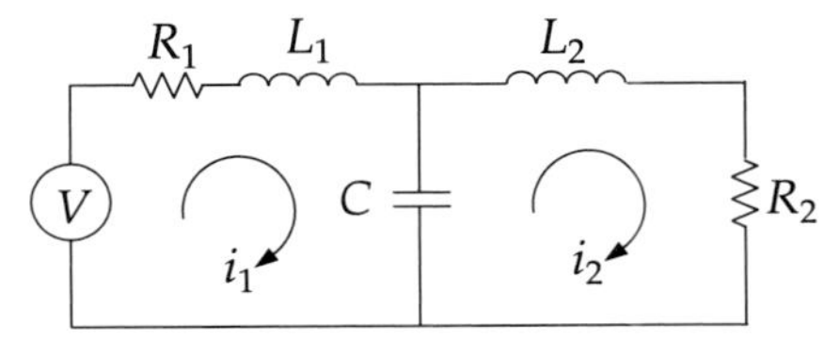

Figure 5.25 *T–V* analogous circuit of Figure 5.22.

Mechanical Coupling

In mechanical systems, parts may be connected through gears. We consider a connection between two toothed wheels A and B of radii r_A and r_B having a number of teeth n_A and n_B, respectively (Figure 5.26).

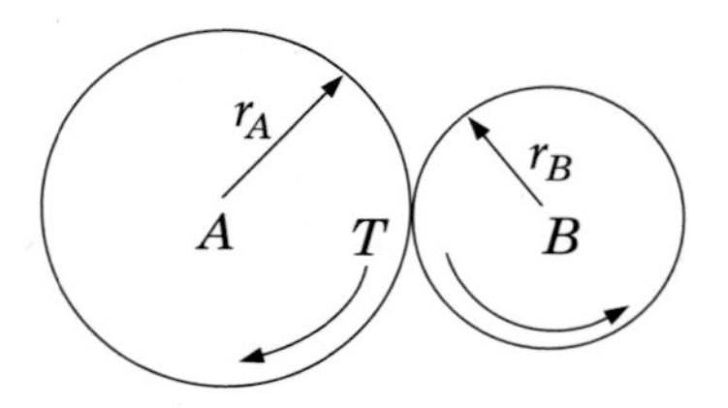

Figure 5.26 Gear train.

The wheels touch at the point T. Let, at any instant, the angular displacements of A and B be θ_A and θ_B. Assuming no friction loss, the work done by A will equal that done by B. So, if T_A and T_B be the torques acting on them,

$$T_A\theta_A = T_B\theta_B$$

or
$$\frac{T_A}{T_B} = \frac{\theta_B}{\theta_A} \tag{5.1}$$

Secondly, the linear distances covered by A and B will be equal. This condition gives us

$$r_A\theta_A = r_B\theta_B$$

or
$$\frac{r_A}{r_B} = \frac{\theta_B}{\theta_A} \tag{5.2}$$

Thirdly, because the size of teeth of both the wheels must match, their numbers must be proportional to their radii. That gives

$$\frac{r_A}{r_B} = \frac{n_A}{n_B} \tag{5.3}$$

Again, from Eq. (5.2), we get

$$\theta_B = \frac{r_A}{r_B}\theta_A \tag{5.4}$$

Therefore,

$$\frac{d\theta_B}{dt} = \frac{r_A}{r_B}\frac{d\theta_A}{dt}$$

$\Rightarrow$
$$\frac{\omega_B}{\omega_A} = \frac{r_A}{r_B} \tag{5.5}$$

where ω_A and ω_B are angular velocities of A and B. Combining Eqs. (5.1) to (5.3) and (5.5), we get

$$\frac{T_A}{T_B} = \frac{r_A}{r_B} = \frac{n_A}{n_B} = \frac{\theta_B}{\theta_A} = \frac{\omega_B}{\omega_A} \tag{5.6}$$

Now, consider a gear system as shown in Figure 5.27, where a load is being driven by a motor through a gear train.

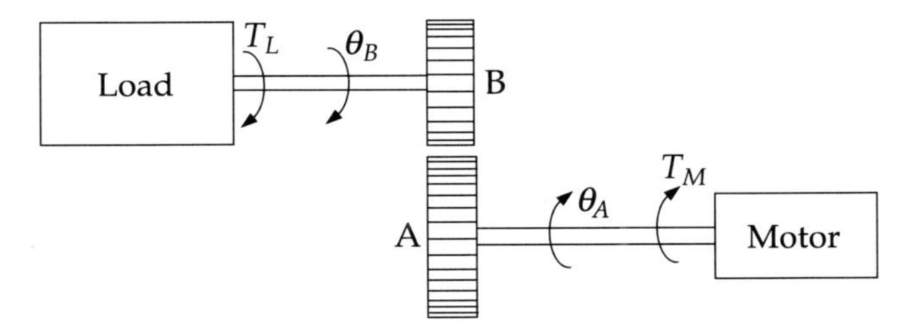

Figure 5.27 Motor driving a load through a gear train.

If T_M = torque generated by the motor
J_A = moment of inertia of the motor shaft
D_A = viscous friction of the motor shaft
T_A = transmitted torque to the load side

then we can set up the torque-balance equation for the motor shaft as

$$T_M = J_A \frac{d^2\theta_A}{dt^2} + D_A \frac{d\theta_A}{dt} + T_A \tag{5.7}$$

A similar equation for the load side can be written as

$$T_B = J_B \frac{d^2\theta_B}{dt^2} + D_B \frac{d\theta_A}{dt} + T_L \tag{5.8}$$

where J_B, D_B correspond to load inertia and viscous friction, and T_L, T_B are torques developed at the load and B. Assuming no friction loss at the gear transmission, from Eq. (5.6), we have

$$T_B = \frac{n_B}{n_A}$$

Substituting this value in Eq. (5.8), on rearrangement, we get

$$T_A = \left(\frac{n_A}{n_B}\right)\left(J_B \frac{d^2\theta_B}{dt^2} + D_B \frac{d\theta_A}{dt} + T_L\right) \tag{5.9}$$

Using $\dfrac{n_A}{n_B} = \dfrac{\theta_B}{\theta_A}$ of Eq. (5.6), we obtain from Eq. (5.9), the relation

$$T_A = \left(\frac{n_A}{n_B}\right)^2 \left(J_B \frac{d^2\theta_B}{dt^2} + D_B \frac{d\theta_A}{dt}\right) + \left(\frac{n_A}{n_B}\right) T_L$$

Substituting the value of T_A in Eq. (5.7), we have

$$T_M = \left[J_A + \left(\frac{n_A}{n_B}\right)^2 J_B\right] \frac{d^2\theta_A}{dt^2} + \left[D_A + \left(\frac{n_A}{n_B}\right)^2 D_B\right] \frac{d\theta_A}{dt} + \left(\frac{n_A}{n_B}\right) T_L$$

$$\equiv J'_A \frac{d^2\theta_A}{dt^2} + D'_A \frac{d\theta_A}{dt} + T'_L$$

where

$$J'_A = J_A + \left(\frac{n_A}{n_B}\right)^2 J_B$$

$$D'_A = D_A + \left(\frac{n_A}{n_B}\right)^2 D_B \tag{5.10}$$

$$T'_L = \left(\frac{n_A}{n_B}\right) T_L$$

The relations given in Eq. (5.10) pertain to the primary of the gears. To arrive at similar relations for the secondary of the gears, we multiply both sides of Eq. (5.7) by (n_B/n_A) to get

$$\left(\frac{n_B}{n_A}\right) T_M = \left(\frac{n_B}{n_A}\right) J_A \frac{d^2\theta_A}{dt^2} + \left(\frac{n_B}{n_A}\right) D_A \frac{d\theta_A}{dt} + \left(\frac{n_B}{n_A}\right) T_A \tag{5.11}$$

Using $\dfrac{\theta_A}{\theta_B} = \dfrac{n_B}{n_A} = \dfrac{T_B}{T_A}$ of Eq. (5.6), from Eq. (5.11), we get

$$\left(\frac{n_B}{n_A}\right) T_M = \left(\frac{n_B}{n_A}\right)^2 J_A \frac{d^2\theta_B}{dt^2} + \left(\frac{n_B}{n_A}\right)^2 D_A \frac{d\theta_B}{dt} + T_B \tag{5.12}$$

Substituting the value of T_B from Eq. (5.8) into Eq. (5.12), on rearranging, we obtain

$$\left(\frac{n_B}{n_A}\right) T_M = \left[J_B + \left(\frac{n_B}{n_A}\right)^2 J_A\right] \frac{d^2\theta_B}{dt^2} + \left[D_B + \left(\frac{n_B}{n_A}\right)^2 D_A\right] \frac{d\theta_B}{dt} + T_L$$

or

$$T'_M = J'_B \frac{d^2\theta_B}{dt^2} + D'_B \frac{d\theta_B}{dt} + T_L$$

where

$$J'_B = J_B + \left(\frac{n_B}{n_A}\right)^2 J_A$$

$$D'_B = D_B + \left(\frac{n_B}{n_A}\right)^2 D_A \tag{5.13}$$

$$T'_M = \left(\frac{n_B}{n_A}\right) T_M$$

The three relations in Eq. (5.13) are the desired relations for the secondary of the gears. We will find application of these relations later.

5.2 CONTROL SYSTEM COMPONENTS

A general feedback control system is shown in Figure 5.28 in the block diagram form.

From the figure, it is clear that a general feedback control system consists of four components:

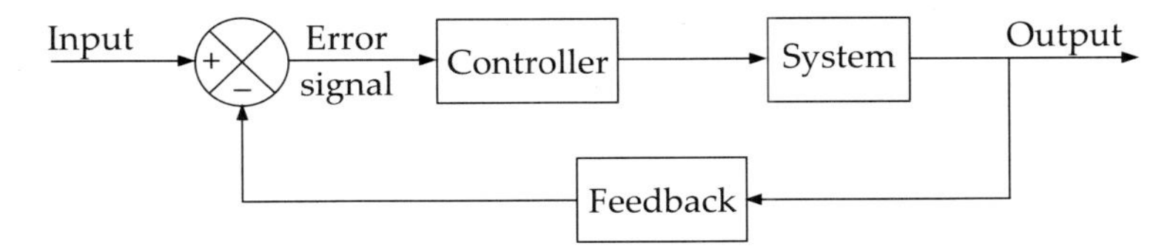

Figure 5.28 General feedback control system.

1. System to be controlled (or Plant)
2. Feedback element (or Transducer)
3. Error signal generator
4. Control signal generator

In the following sections we will consider a few representative devices for the first three of these components. Control signal generators are mostly electronic devices, the discussion of which is beyond the scope of this text.

System to be Controlled

Servomotor. Servomotors are used in control systems to generate a high acceleration from a stationary condition. Depending on their input or control, these motors can be classified as shown in Table 5.4.

Table 5.4 Classification of Servomotors

Current type	*Motor type*
dc	Armature-controlled Field-controlled
ac	

dc servomotors. First we consider dc servomotors as control system components. As already seen, they are of two types.

Armature-controlled dc servomotor. The schematic diagram of an armature-controlled dc motor is shown in Figure 5.29.

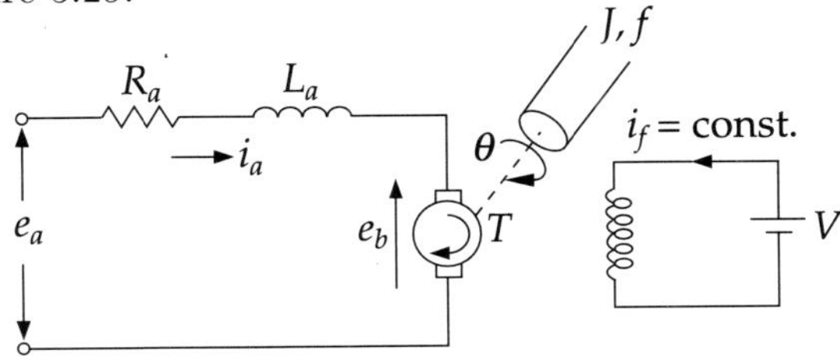

Figure 5.29 Schematic diagram of an armature-controlled dc motor.

Here,

$$\begin{aligned}
R_a &= \text{armature resistance}\\
L_a &= \text{armature inductance}\\
i_a &= \text{armature current}\\
i_f &= \text{field current}\\
e_a &= \text{applied voltage}\\
e_b &= \text{back emf}\\
J &= \text{moment of inertia of the motor shaft and load}\\
f &= \text{coefficient of friction of the motor shaft and load}\\
T &= \text{torque generated by the motor}\\
\theta &= \text{angular deflection of the motor shaft}\\
V &= \text{voltage applied in the field coil}
\end{aligned}$$

Air-gap flux Φ is proportional to the field current. Hence,

$$\Phi = k_f i_f \tag{5.14}$$

where k_f is a constant. The torque is proportional to the flux and armature current. Therefore,

$$\begin{aligned}
T &= k_t \Phi i_a\\
&= k_t k_f i_f i_a \qquad \text{[using Eq. (5.14)]}\\
&= K_T i_a
\end{aligned} \tag{5.15}$$

where $K_T = k_t k_f i_f$, i_f being constant. The back emf is proportional to the rotational speed. Hence,

$$e_b = K_b \frac{d\theta}{dt} \tag{5.16}$$

The Laplace transform of Eq. (5.16) is

$$E_b(s) = K_b s \Theta(s) \tag{5.17}$$

Applying KVL to the armature circuit, we get

$$e_a = i_a R_a + L_a \frac{di_a}{dt} + e_b$$

From the Laplace transform of this equation, we have

$$E_a(s) - E_b(s) = (sL_a + R_a) I_a(s) \tag{5.18}$$

The load-torque equation is

$$J\frac{d^2\theta}{dt^2} + f\frac{d\theta}{dt} = T = K_T i_a \qquad \text{[using Eq. (5.15)]} \tag{5.19}$$

The Laplace transform of Eq. (5.19) is

$$(s^2 J + sf)\Theta(s) = K_T I_a(s) \tag{5.20}$$

Eliminating $I_a(s)$ from Eq. (5.20) with the help of Eq. (5.18), we get

$$(s^2J + sf)\Theta(s) = \left[\frac{E_a(s) - E_b(s)}{sL_a + R_a}\right] K_T$$

$$= \left[\frac{E_a(s) - sK_b\Theta(s)}{sL_a + R_a}\right] K_T \qquad \text{[using Eq. (5.17)]} \tag{5.21}$$

On rearranging Eq. (5.21), we obtain

$$[(s^2J + sf)(sL_a + R_a) + sK_bK_T]\Theta(s) = K_TE_a(s)$$

which, on further rearrangement, yields the transfer function

$$G(s) = \frac{\Theta(s)}{E_a(s)} = \frac{K_T}{s[(sJ + f)(sL_a + R_a) + K_bK_T]} \tag{5.22}$$

To draw a block diagram of the system, we observe that the transfer function can be written as

$$G(s) = \frac{\dfrac{K_T}{(sJ + f)(sL_a + R_a)}}{1 + \dfrac{K_TK_b}{(sJ + f)(sL_a + R_a)}} \cdot \frac{1}{s}$$

It is now clear that the block diagram should look like Figure 5.30. Obviously, it is a *closed-loop system.*

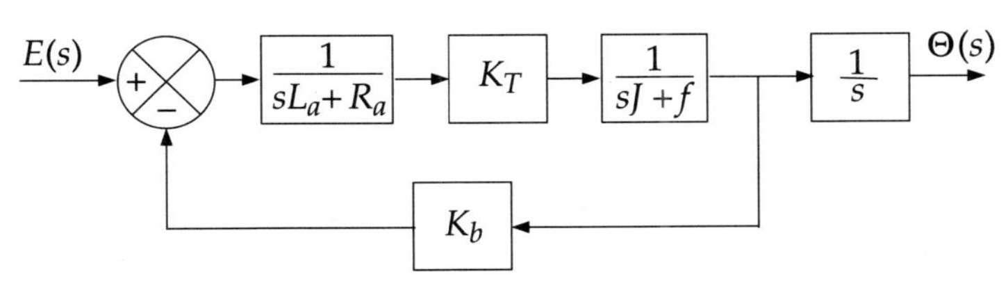

Figure 5.30 Block diagram of armature-controlled dc motor.

If the armature inductance L_a is small, from Eq. (5.22), we get

$$G(s) = \frac{\Theta(s)}{E_a(s)} = \frac{K_T}{s[(sJ + f)R_a + K_bK_T]} = \frac{K_T/R_a}{s^2J + s\left(f + \dfrac{K_bK_t}{R_a}\right)} \equiv \frac{K_T/R_a}{s^2J + sf'} \equiv \frac{K}{s(s\tau + 1)}$$

where

$$f' = f + \frac{K_bK_t}{R_a} = \text{effective coefficient of friction}$$

$$K = \frac{K_T}{R_af'} = \text{motor gain}$$

$$\tau = \frac{1}{f'} = \text{motor time-constant}$$

The equation for f' shows that the back emf effectively increases the friction of the motor.

Field-controlled dc motor. A schematic diagram of the field-controlled dc motor is shown in Figure 5.31.

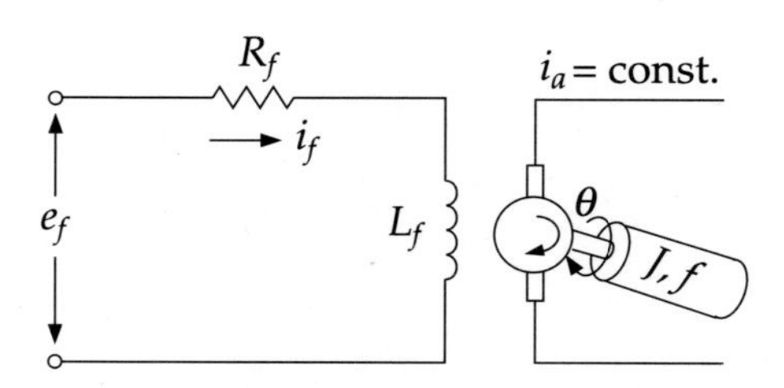

Figure 5.31 Schematic diagram of a field-controlled dc motor.

Here,

R_f = field winding resistance
L_f = field winding inductance
i_a = armature current (constant)
i_f = field current
e_f = field voltage
e_b = back emf
J = moment of inertia of the motor shaft and load
f = coefficient of friction of the motor shaft and load
T = torque generated by the motor
θ = angular displacement of the motor shaft

In contrast to the armature-controlled motor, here the armature current is constant while the field current is variable. Therefore, Eq. (5.15) can be rewritten as

$$T = K_T' i_f$$

Applying KVL to the field circuit, we get

$$L_f \frac{di_f}{dt} + R_f i_f = e_f \tag{5.23}$$

The Laplace transform of Eq. (5.23), with initial conditions = 0, is

$$(sL_f + R_f)I_f(s) = E_f(s) \tag{5.24}$$

The load torque equation and its Laplace transform are

$$J\frac{d^2\theta}{dt^2} + f\frac{d\theta}{dt} = T = K_T' i_f$$

$$(s^2 J + sf)\Theta(s) = K_t' I_f(s) \tag{5.25}$$

Eliminating $I_f(s)$ from Eq. (5.24) with the help of Eq. (5.25) and on rearranging, we obtain

$$\frac{\Theta(s)}{E_f(s)} = \frac{K'_T}{s(sJ+f)(sL_f+R_f)}$$

$$= \frac{K'_T/(R_f f)}{s\,[s(J/f)+1]\,[s(L_f/R_f)+1]}\,.$$

or

$$G(s) = \frac{K'}{s(s\tau_m+1)(s\tau_f+1)}$$

where

$$K' = \frac{K'_T}{R_f f} = \text{motor gain}$$

$$\tau_m = \frac{J}{f} = \text{motor time-constant}$$

$$\tau_f = \frac{L_f}{R_f} = \text{field circuit time-constant}$$

The block diagram for the field-controlled servomotor is given in Figure 5.32. It may be noted from the figure that the field-controlled servomotor is an *open-loop system*.

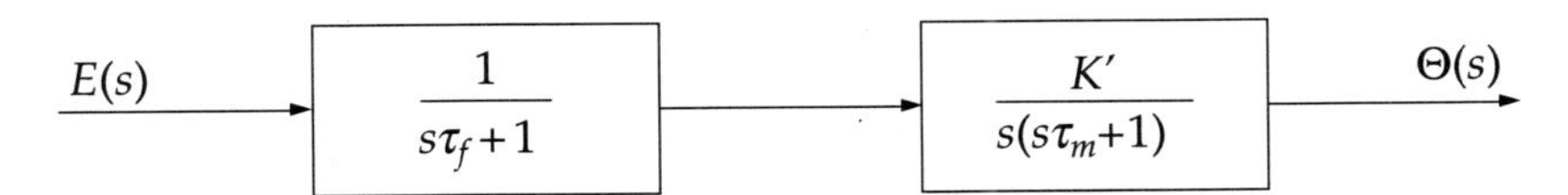

Figure 5.32 Block diagram for the field-controlled servomotor.

ac servomotor. An ac servomotor is basically a two-phase induction motor with low X/R ratio, thus making the torque-speed characteristic nearly linear. The comparison of torque-speed characteristics between an ordinary two-phase induction motor and a servomotor is shown in Figure 5.33.

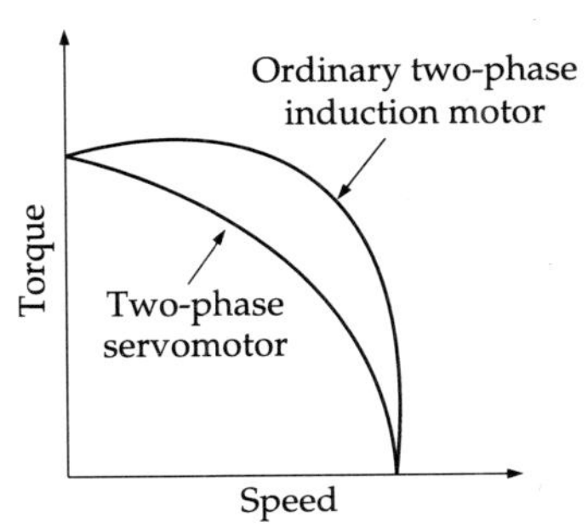

Figure 5.33 Torque-speed characteristics comparison.

An ac servomotor is used for low-power, high acceleration applications. A schematic diagram of an ac servomotor is shown in Figure 5.34.

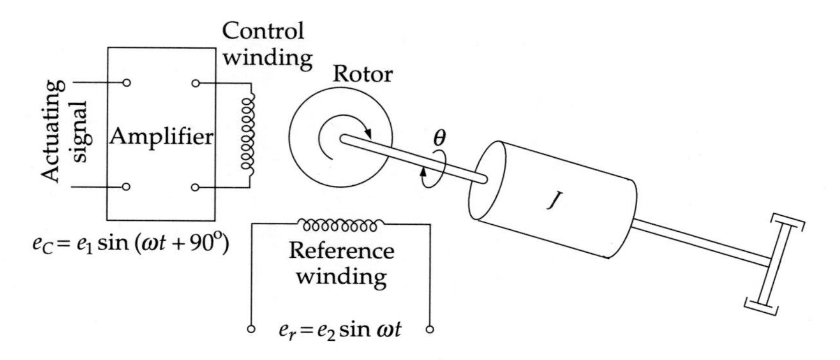

Figure 5.34 Structure of an ac servomotor.

A two-phase motor has two windings at right angles to each other. While the reference winding is given a steady voltage, the amplitude of the control winding voltage varies according to the actuating signal. But the latter voltage always remains 90° out of phase with the former (see Figure 5.35). This can be achieved by feeding voltage to the control winding through a capacitor.

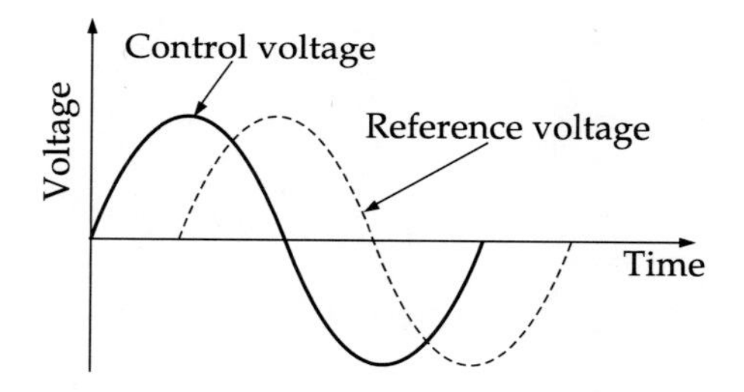

Figure 5.35 Phases of reference and control voltages.

This variation in phase of voltage between the two mutually perpendicular windings generates a rotational motion of the rotor owing to the rotation of the resultant magnetic flux (see Figure 5.36).

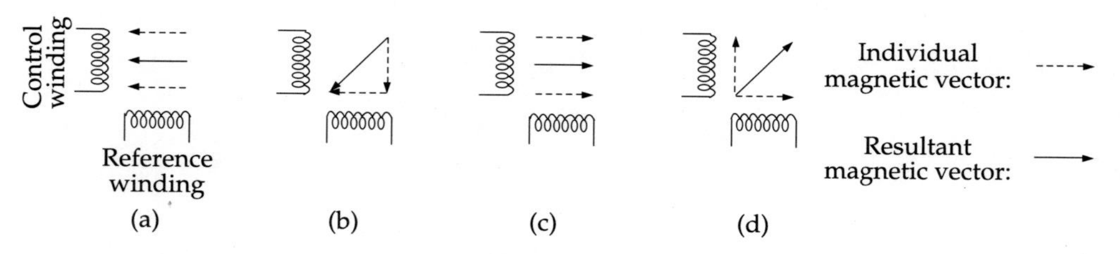

Figure 5.36 Rotation of the resultant magnetic flux—(a), (b), (c) and (d) show the individual magnetic vectors and their resultants which cause the rotation.

The torque-speed characteristics, as a function of the control voltage, are shown in Figure 5.37. The torque corresponding to $\omega = 0$ (i.e. no speed) is called *stall torque*, while the speed corresponding to $\tau = 0$ (i.e. no torque) is called *no-load speed.*

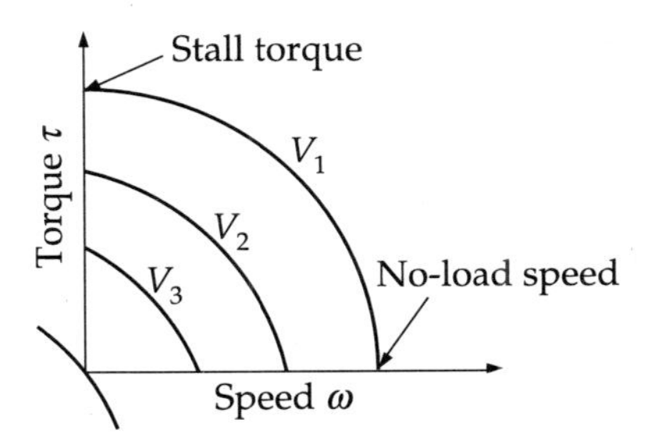

Figure 5.37 Torque-speed characteristics.

Assuming a linear relationship between torque and speed, we can write

$$\tau = m\omega + c \tag{5.26}$$

At $\omega = 0$, let $\tau = \tau_0$. Then,

$$\tau_0 = c = Ke_c \tag{5.27}$$

where e_c is the control voltage and K is a constant. So,

$$K = \frac{\tau_0}{e_c}$$

Again, at $\tau = 0$, let $\omega = \omega_0$. Therefore,

$$m = -\frac{c}{\omega_0} = -\frac{\tau_0}{\omega_0}$$

Thus, Eq. (5.26) can be rewritten as

$$\tau = -\frac{\tau_0}{\omega_0}\omega + Ke_c \tag{5.28}$$

The Laplace transform of Eq. (5.28) yields

$$T(s) = -D\Omega(s) + KE_c(s) \tag{5.29}$$

where $D(= \tau_0/\omega_0)$ is called the *damping factor*[3]. Also, $\omega = (d\theta/dt)$. So, its Laplace transform gives $\Omega(s) = s\Theta(s)$. Substituting this value in Eq. (5.29), we get

$$T(s) = -sD\Theta(s) + KE_c(s) \tag{5.30}$$

From mechanical considerations, we obtain

$$\tau = J\frac{d^2\theta}{dt^2} + D\frac{d\theta}{dt} \tag{5.31}$$

[3]aka *coefficient of viscous friction.*

The Laplace transform of Eq. (5.31) gives

$$T(s) = (s^2 J + sD)\Theta(s) \tag{5.32}$$

From Eqs. (5.30) and (5.32), after a little rearrangement, we have

$$s(sJ + 2D)\Theta(s) = KE_c(s)$$

$$\frac{\Theta(s)}{E_c(s)} = \frac{K}{s(sJ + 2D)} = \frac{K/(2D)}{s\left(s\dfrac{J}{2D} + 1\right)} \tag{5.33}$$

or

$$G(s) = \frac{K_m}{s(s\tau_t + 1)} \tag{5.34}$$

where $K_m = (K/2D)$ = motor gain constant and $\tau_t = (J/2D)$ = motor time-constant.

The block diagram of an ac servomotor is given in Figure 5.38. We note that this is also an *open-loop system* like the dc field-controlled servomotor.

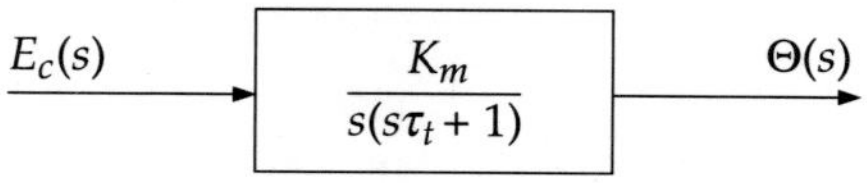

$E_c(s)$ → $\dfrac{K_m}{s(s\tau_t + 1)}$ → $\Theta(s)$

Figure 5.38 Block diagram of an ac servomotor.

EXAMPLE 5.10. An ac servomotor has both its windings excited with 115 V AC. Its stall torque is 2 lb-ft and the coefficient of viscous friction = 0.2 ft-lb-s.

(a) Determine its no load speed.

(b) If it is connected to a load of 0.8 ft-lb (the coefficient of viscous friction = 0.05 ft-lb-s) through a gear pass with a ratio of 4, calculate the speed at which the motor will run.

Solution

(a) From Eq. (5.28) $\tau = -D\omega + Ke_c$. For $\tau = 0$, $\omega = \omega_0$. Therefore, the no-load speed is

$$\omega_0 = \frac{Ke_c}{D} = \frac{\tau_0}{D} = \frac{2}{0.2} = 10 \text{ rad/s}$$

(b) When N = gear ratio and D_L = coefficient of viscous friction due to load, the total viscous friction D_T is [see Eq. (5.10)]

$$D_T = D + N^2 D_L$$

$$= 0.2 + 4^2 \times 0.05$$

$$= 1.0 \text{ ft-lb-s}$$

Therefore, from $\tau_L = Ke_c - D_T\omega = \tau_0 - D_T\omega$,

$$\omega = \frac{\tau_0 - \tau_L}{D_T} = \frac{2 - 0.8}{1} = 1.2 \text{ rad/s}$$

Stepper motor. Stepper motors move in steps in response to input pulses, i.e. they are actuators in incremental motion control systems. They are widely used in digitally controlled machines, such as computer printers and disc- and tape-drives. Generally, stepper motors are of two types:

- Variable reluctance type
- Permanent magnet type

We will consider only the first type here.

In a variable reluctance stepper motor, the stator and rotor are made of one or more stacks. Stators are fixed to a common frame while the rotors have a common shaft [see Figure 5.39(a)]. Both stators and rotors have a toothed structure [see Figure 5.39(b)]. Stators are excited by pulses while rotors are unexcited.

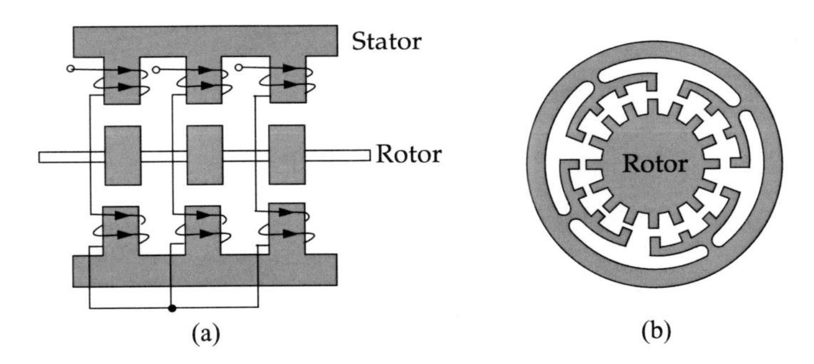

Figure 5.39 Stepper motor: (a) sectional view; (b) top view.

If n_t is the number of teeth of the rotor and θ is the angle of misalignment between the rotor and stator teeth, then there are two positions:

(i) $\theta = 0°$

(ii) $\theta = \pm\dfrac{360°}{2n_t} = \pm\dfrac{180°}{n_t}$

when no torque will be acting on the rotor. Of these two positions, the first constitutes a stable equilibrium, i.e. a small disturbance from here will bring the rotor back to this position. The second position, on the other hand, is unstable, meaning thereby that a small disturbance from here makes the rotor move away from this position (see Figure 5.40 for the torque vs. angle curve). As a consequence, the rotor locks into $\theta = 0°$ position.

The construction of the motor is such that rotor teeth of all stacks are perfectly aligned, but the stator teeth of stacks are displaced by an angle of

$$\phi = \frac{360°}{Nn_t}$$

where N is the number of stacks.

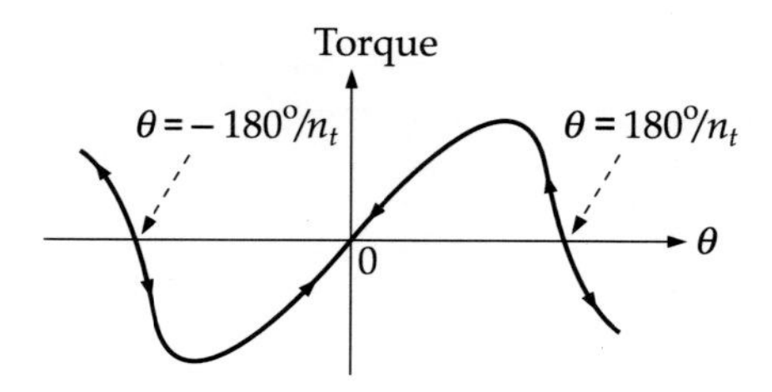

Figure 5.40 Torque-angle curve for stepper motor.

Consider a 3-stack 24-teeth motor where the stack x is aligned with the rotor. Then the angle of displacement of stacks y and z is

$$\phi = \frac{360^\circ}{3 \times 24} = 5^\circ$$

Now, if the y stator is excited by a pulse, the rotor will move 5° in one direction while a pulse excitation of the z stator will make the rotor move by the same angle in the opposite direction. Thus, motions corresponding to pulse trains of $yzxyz$ and $zyxzy$ will be opposite to each other. Care should be taken to time the pulses so that the rotor finds enough time to lock itself to each step rather than skipping any.

Feedback Element: Tachometer

Tachometers[4] are used to measure rotational speed. These basically miniature low voltage generators are often used as feedback elements. They can be of two types—dc and ac.

The dc tachometer. In an ordinary generator, the rotor has inertia. But a tachometer rotor should have minimum inertia. To achieve this end, moving coil tachometers use ironless rotors. The rotor along with the armature winding is shaped as a cup and is hung between the permanent magnet poles and the inner iron structure like a cantilever. Figure 5.41 gives a diagrammatic presentation of a dc tachometer.

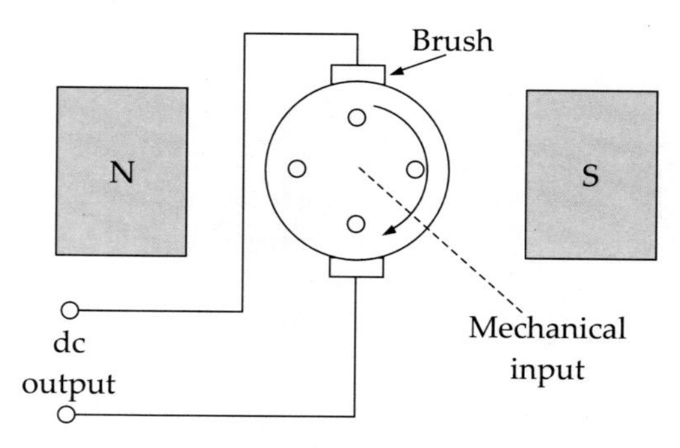

Figure 5.41 Schematic diagram of a dc tachometer.

[4]The Greek word *takhos* means speed.

The generated voltage is given by the familiar Faraday relation

$$e(t) \propto \frac{d\Phi}{dt}$$

where Φ is the flux linkage. Since the change in flux linkage varies with the angular velocity $\omega(t)$ of the rotor, we have

$$e(t) = K\omega(t) \tag{5.35}$$

Taking the Laplace transform of Eq. (5.35), we get the following expression for the transfer function of the dc tachometer

$$\frac{E(s)}{\Omega(s)} = K$$

The unit of K is V/krpm. Typically, K value is $\sim$ 6 V/krpm. As in dc generators, the ac output is converted to dc by a commutator.

The ac tachometer. Structurally, an ac tachometer resembles a two-phase induction motor. It consists of two stator windings arranged in space quadrature along with a rotor which is just an aluminium cup that rotates in an air-gap that contains magnetic flux. Figure 5.42 is a diagrammatic representation of the dc tachometer.

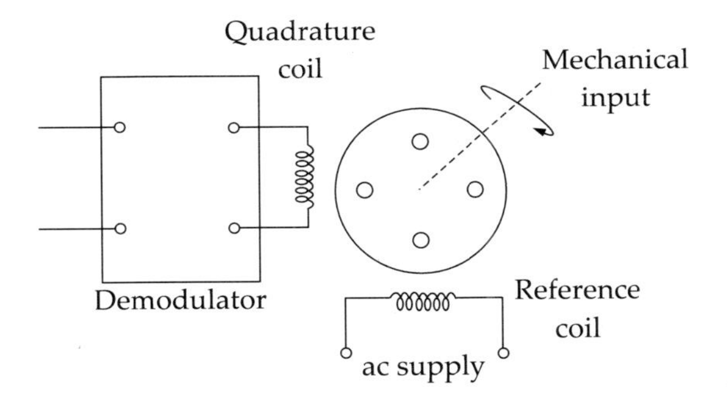

Figure 5.42 Schematic diagram of an ac tachometer.

A sinusoidal voltage given by

$$e_r(t) = E_r \sin \omega_c t$$

is applied to the reference winding. No emf is induced in the output winding when the rotor is stationary. However, as the rotor turns, a voltage of frequency ω_c is induced in the output winding. The amplitude of this voltage is proportional to the speed of rotation. In case the direction of rotation changes, the output voltage suffers a 180° phase shift.

To filter out the carrier frequency, a phase-sensitive demodulator can be used in the output which has the form of

$$e_o(t) = e(t) \sin \omega_c t$$

The forms of the modulated signal and the final filtered output are shown in Figure 5.43.

The transfer function of the ac tachometer, therefore, has the same form as that of a dc tachometer.

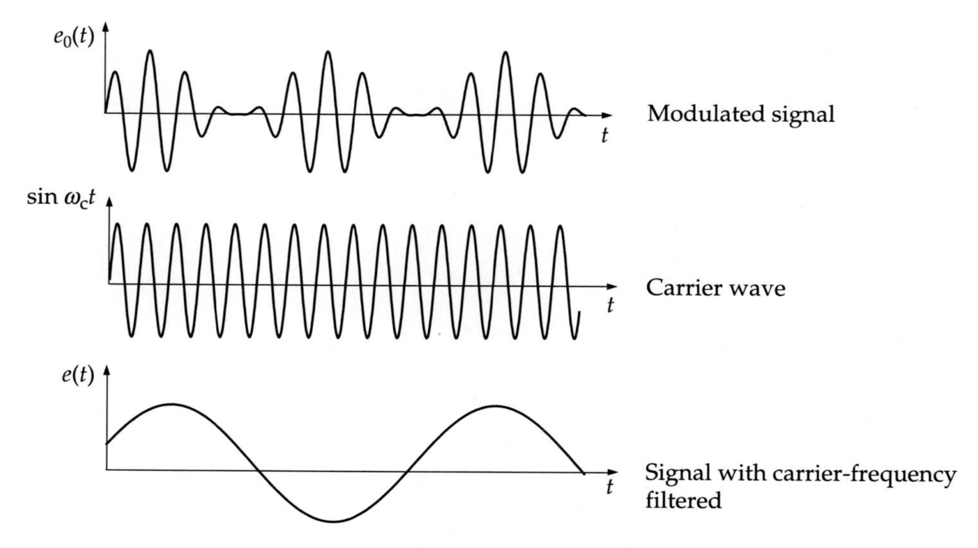

Figure 5.43 $e_o(t)$ and $e(t)$ of ac tachometer.

EXAMPLE 5.11. A tachometer has a gain of 0.05 V/(rad/s). Determine the following:

(a) The output when the shaft-speed is 20 rad/s.

(b) The shaft-speed in rad/s when the output is 1.8 V.

Solution

(a) $e(t) = K\omega(t) = 0.05 \times 20 \text{ V} = 1 \text{ V}$

(b) $\omega(t) = \dfrac{e(t)}{K} = \dfrac{1.8}{0.05} = 36 \text{ rad/s}$

EXAMPLE 5.12. Figure 5.44 represents a closed-loop control system for regulating the speed of a field-controlled dc motor. Determine the reference voltage if the speed is to be maintained at 150 rpm.

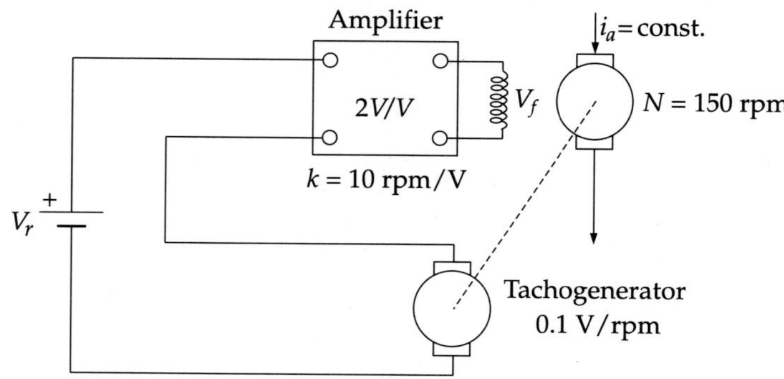

Figure 5.44 Closed-loop control of dc motor speed.

Solution The block diagram of the system looks like that given in Figure 5.45.

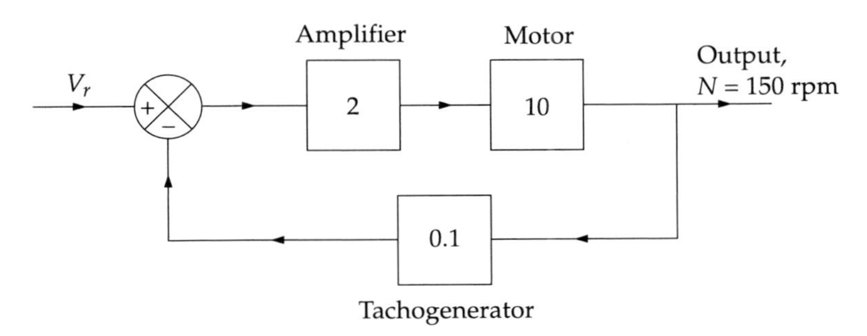

Figure 5.45 Block diagram of the system (Example 5.12).

From the block diagram we get

$$\frac{N}{V_r} = \frac{2 \times 10}{1 + 2 \times 10 \times 0.1} = \frac{20}{3}$$

Therefore,

$$V_r = \frac{3N}{20} = \frac{3 \times 160}{20} = 22.5 \text{ V}$$

Error Signal Generator

The error signal generator can be of many kinds. We consider here only two simple such devices, namely, potentiometer and synchro.

Potentiometer. The potentiometer is a simple device that converts liner or angular displacement into voltage. A fixed voltage is applied between two terminals of a fixed resistor in the form of a wound wire. A third terminal, called *wiper*, moves over this resistance and generates a variable voltage (see Figure 5.46).

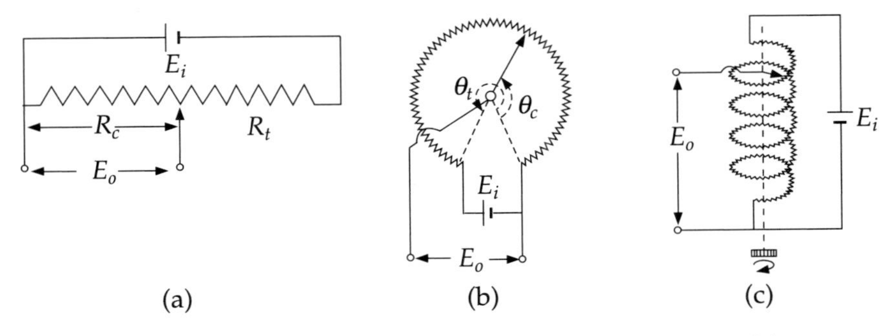

Figure 5.46 Potentiometers: (a) linear; (b) angular-displacement; (c) helical.

It is easy to see from the figure that the following relations hold good for linear and angular potentiometers:

$$\text{Linear:} \qquad \frac{R_c}{R_t} = \frac{E_o}{E_i}$$

$$\text{Angular:} \qquad \frac{\theta_c}{\theta_t} = \frac{E_o}{E_i}$$

where the subscripts c, t, o and i indicate control, total, output and input, respectively. For an N-turn angular potentiometer, $\theta_t = 2\pi N$.

EXAMPLE 5.13. A 3-turn, 100 kΩ potentiometer with 1% linearity uses a 30 V supply. Find the following:

(a) The potentiometer constant in V/turn
(b) The range of voltages at mid-point setting
(c) The voltage at the mid-point when the potentiometer is loaded with 500 kΩ (assume the potentiometer is perfectly linear)

Solution

(a) The potentiometer constant K_P is given by

$$K_P = \frac{30}{3} = 10 \text{ V/turn}$$

(b) For the mid-point setting, the resistance is 50 kΩ. Therefore,

$$\text{Voltage at the mid-point} = 10 \times 1.5 \text{ (turn)} = 15 \text{ V}$$

Given linearity = 1%. Then,

$$\text{Voltage error} = \pm 30 \times 0.01 \text{ V} = \pm 0.3 \text{ V}$$

Therefore,

$$\text{Voltage range} = 15 \pm 0.3 \text{ V} = 14.7 \text{ to } 15.3 \text{ V}$$

(c) From Figure 5.47,

$$R = 50 \text{ k}\Omega \parallel 500 \text{ k}\Omega = \frac{50 \times 500}{50 + 500} = 45.45 \text{ k}\Omega$$

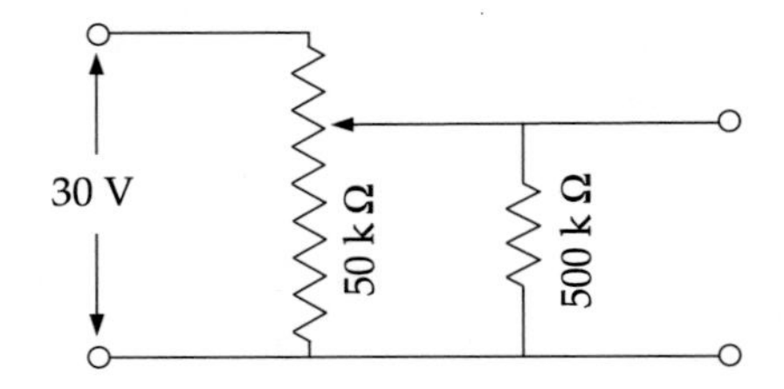

Figure 5.47 Potentiometer (Example 5.13).

By voltage division,

$$\text{Voltage} = \frac{30}{50 + 45.45} \times 45.45 \text{ V} = 14.3 \text{ V}$$

EXAMPLE 5.14. (a) Calculate the output voltage E_o if the excitation voltage is 2 V and the setting ratio is 0.4 (see Figure 5.48).

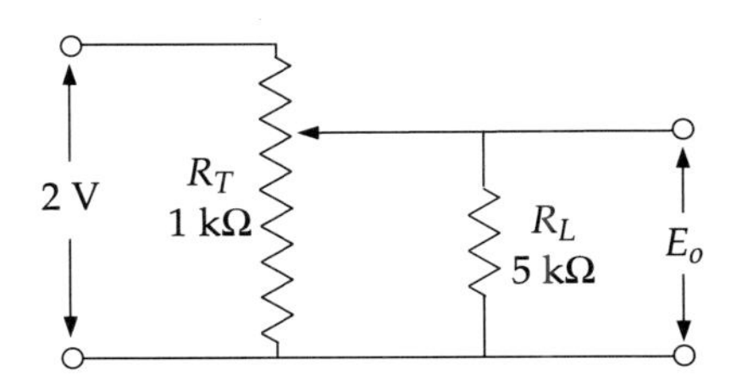

Figure 5.48 Potentiometer (Example 5.14).

(b) Also calculate the loading error.

(c) If the potentiometer is a multi-turn unit having a shaft rotation of 1080°, find the positional error associated with loading.

Solution (a) Let α = setting ratio. The resistance due to the load is

$$R = (1 - \alpha)R_T + (\alpha R_T \parallel R_L)$$

$$= 0.6 + \frac{(0.4)(1)(5)}{(0.4)(1) + 5} = 0.97 \text{ k}\Omega$$

Therefore,

$$E_o = \frac{(\alpha R_T \parallel R_L)}{R} E_i$$

$$= \frac{0.37}{0.97} \times 2 = 0.763 \text{ V}$$

(b) The error due to loading is given by

$$\text{Error} = \alpha E_i - E_o$$

$$= (0.4)(2) - 0.763 = 0.037 \text{ V}$$

(c) $\text{Positional error} = \dfrac{\text{maximum angle of rotation}}{\text{applied voltage}} \times \text{error} = \dfrac{1080°}{2} \times 0.037 = 20°.$

EXAMPLE 5.15. A helical 5-turn potentiometer has a resistance of 50 kΩ and 10,000 winding turns. Find the linearity when the measured resistance at its mid-point is 25.5 kΩ. Find the resolution of the potentiometer in V/turn and potentiometer constant in V/rad if 100 V is applied to it.

Solution The mid-point should have a resistance of 25 kΩ. But the measured value is 25.5 kΩ. So, the deviation is 0.5 kΩ. Therefore,

$$\text{Linearity} = \frac{\text{Deviation}}{\text{Desired value}} \times 100 = \frac{0.5}{25} \times 100 = 2\%$$

$$\text{Resolution} = \frac{100}{10000} = 10 \text{ mV/turn}$$

$$\text{Potentiometer constant} = \frac{E_i}{2\pi N} = \frac{100}{2\pi \times 5} = 3.183 \text{ V/rad}$$

Synchro. Basically electromechanical devices based on variation of reluctance, synchros measure angles or perform functions related to angle measurement such as remote control of angle or computation of rectangular component of vectors. They are widely used as detectors owing to their ruggedness in construction and reliability of performance. Though they find various applications, their uses as transmitter and as control transformer are of interest to us here.

Synchro transmitter. A synchro transmitter consists of a wound rotor on a dumb-bell-shaped magnetic substance and a wound stator which are arranged concentrically so that the motion of the rotor produces a variable mutual inductance between the two windings [Figure 5.49(a)]. The rotor is laminated, has a single-phase winding, and is connected to supply lines through precision slip-rings, while the stator winding is three-phase, the phases being displaced by 120° in space.

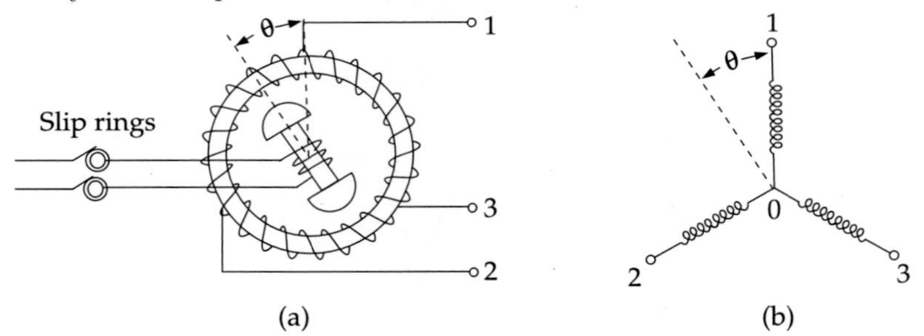

Figure 5.49 Schematic diagram of: (a) synchro transmitter; (b) its electric circuit.

Let the rotor of the synchro transmitter be excited by an ac voltage e_r given by

$$e_r = V_r \sin \omega t$$

At the *electric zero*, i.e. when the angle θ between the rotor and stator winding 1 equals 0°, the voltage induced between terminal 1 and the neutral 0 [see Figure 5.49(b)] is given by

$$e_{10}(t) = AV_r \sin \omega t$$

where A is a constant. Then the voltages induced between terminals (2 and 0) and (3 and 0) are

$$e_{20}(t) = AV_r \sin \omega t \cos 120° = -\frac{1}{2} AV_r \sin \omega t$$

$$e_{30}(t) = AV_r \sin\omega t \cos 240° = -\frac{1}{2}AV_r \sin\omega t$$

Consequently, the voltages developed between the terminals will be

$$e_{12}(t) = e_{10}(t) - e_{20}(t) = \frac{3}{2}AV_r \sin\omega t \tag{5.36}$$

$$e_{23}(t) = e_{20}(t) - e_{30}(t) = 0 \tag{5.37}$$

$$e_{31}(t) = e_{30}(t) - e_{10}(t) = -\frac{3}{2}AV_r \sin\omega t \tag{5.38}$$

Voltages shown by Eqs. (5.36)–(5.38) demonstrate that, despite having its three-phase construction, only single-phase voltages develop in the terminals.

Now, consider the case when the rotor subtends an angle θ with respect to the electric zero configuration. Then voltages developed in each stator coil with respect to the neutral 0 are

$$e_{10}(t) = AV_r \sin\omega t \cos\theta$$

$$e_{20}(t) = AV_r \sin\omega t \cos\,(120° - \theta)$$

$$e_{30}(t) = AV_r \sin\omega t \cos\,(240° - \theta)$$

Therefore, the voltages developed between terminals are

$$\begin{aligned}
e_{12}(t) &= e_{10}(t) - e_{20}(t) = AV_r \sin\omega t[\cos\theta - \cos(120° - \theta)] \\
&= AV_r \sin\omega t \left[\cos\theta + \left(\frac{1}{2}\cos\theta - \frac{\sqrt{3}}{2}\sin\theta\right)\right] \\
&= AV_r \sin\omega t \left(\frac{3}{2}\cos\theta - \frac{\sqrt{3}}{2}\sin\theta\right) \\
&= \sqrt{3}AV_r \sin\omega t \sin(\theta + 120°)
\end{aligned}$$

$$\begin{aligned}
e_{23}(t) &= e_{20}(t) - e_{30}(t) = AV_r \sin\omega t[\cos(120° - \theta) - \cos(240° - \theta)] \\
&= AV_r \sin\omega t \left(-\frac{1}{2}\cos\theta + \frac{\sqrt{3}}{2}\sin\theta + \frac{1}{2}\cos\theta + \frac{\sqrt{3}}{2}\sin\theta\right) \\
&= \sqrt{3}AV_r \sin\omega t \sin\theta
\end{aligned}$$

$$\begin{aligned}
e_{31}(t) &= e_{30}(t) - e_{10}(t) = AV_r \sin\omega t[\cos(240° - \theta) - \cos\theta] \\
&= AV_r \sin\omega t \left[\left(-\frac{1}{2}\cos\theta - \frac{\sqrt{3}}{2}\sin\theta\right) - \cos\theta\right] \\
&= AV_r \sin\omega t \left(-\frac{3}{2}\cos\theta - \frac{\sqrt{3}}{2}\sin\theta\right) \\
&= \sqrt{3}AV_r \sin\omega t \sin(\theta + 240°)
\end{aligned}$$

Figure 5.50 is a plot of these voltages vs. the rotor position. We observe that for each angular position of the rotor the set of voltages generated by the three terminals is unique. So, a measurement of this set of voltages determines the angular position of the rotor.

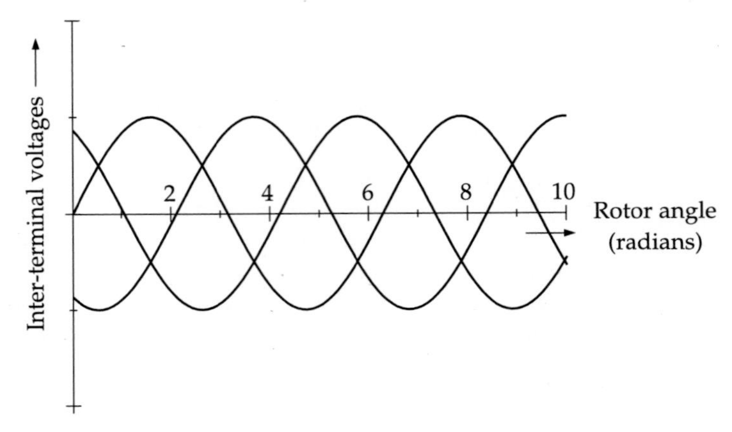

Figure 5.50 Inter-terminal voltages vs. rotor angular position plot for the synchro transmitter.

Synchro control transformer. A synchro control transformer is used to detect an error between two shaft positions and convert it to an electric signal. Since a comparison between two is involved, synchro control transformer is used in conjunction with a synchro transmitter. Their constructions are mostly similar except that the control transformer rotor is of cylindrical shape [see Figure 5.51(a)]. This is necessary to maintain a uniform flux in the air-gap when the rotor rotates. The uniform air-gap flux, in turn, provides a uniform impedance for the rotor output terminals which are often connected to a control device through an amplifier. The idea is that the amplifier should see a constant input impedance.

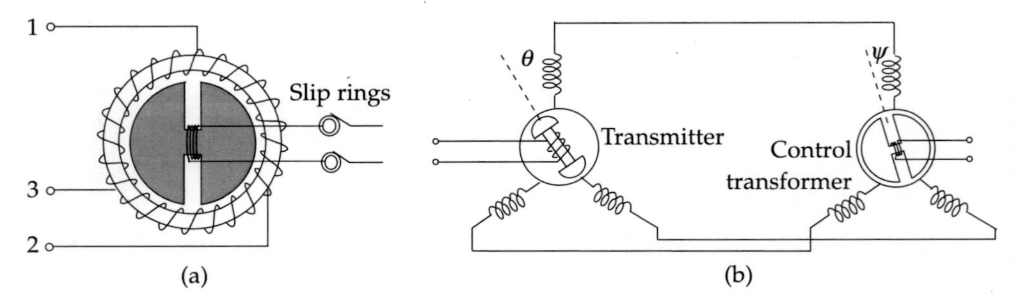

Figure 5.51 Synchro control transformer: (a) construction; (b) error detection operation.

In the error detection operation, the output of stator windings of the sychro transmitter is connected to that of the synchro control transformer, thus ensuring the same current flow through the circuits [see Figure 5.51(b)]. The voltage induced between rotor terminals of the control transformer is

$$e_0(t) = K_c V_r \sin \omega t \cos \phi \qquad (5.39)$$

where

$$e_0 = \text{error voltage}$$
$$K_c = \text{sensitivity of the control transformer}$$
$$\phi = \text{angle between the two rotors}$$

When $\phi = 90°$, the error voltage is zero. This position is known as the *electrical zero* for the control transformer. Consider the case when θ is the transmitter rotor angle and ψ is the control transformer rotor angle. Then,

$$\phi = 90° - (\theta - \psi)$$

Substituting this value in Eq. (5.39), the error voltage is

$$e_0(t) = K_c V_r \sin \omega t \sin (\theta - \psi) \tag{5.40}$$

Equation (5.40) shows that the error voltage generated by the rotor of the control transformer is not linear with respect to the relative angular positions of the two shafts because it is a sine function of the same. However, since a sine function remains almost linear within 0° to 15° angle variation, for small relative angular displacements between the two shafts

$$e_0(t) = K_c V_r (\theta - \psi) \sin \omega t$$

This situation suits well as far as error detection is concerned.

5.3 SYSTEMS

Now that we have discussed mathematical modelling of physical systems and control system components, we may discuss a few elementary control systems, namely

1. Servomechanism
2. Temperature control
3. Liquid level control

Servomechanism

By servomechanism or simply *servo*, we mean control of position—be it linear or angular—and its derivative, speed. The former is known as *positional servomechanism*, and the latter, *rate servomechanism.*

Positional servomechanism. A simple position control system is shown in Figure 5.52. Here, the input indicates the desired position of the load. The input and feedback constitute a bridge in which if the input is changed, an error voltage e_E appears at the amplifier input. The amplifier output drives the motor to make the feedback pointer assume an identical position on the potentiometer in order that $e_E = 0$.

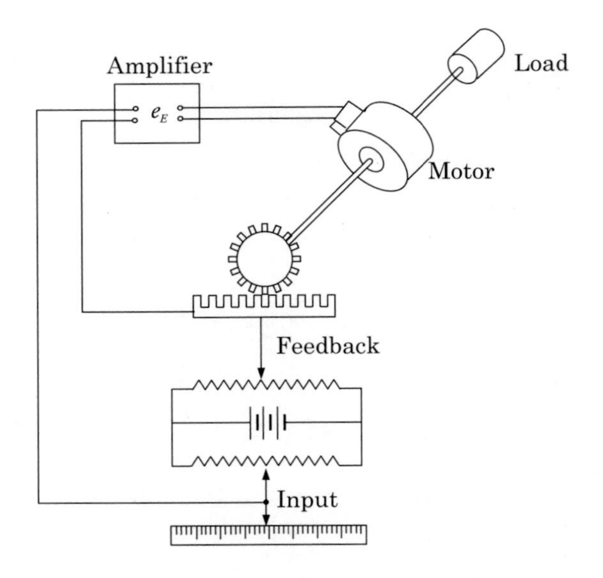

Figure 5.52 A position control system.

Figure 5.53 is a block diagrammatic representation of a position control system.

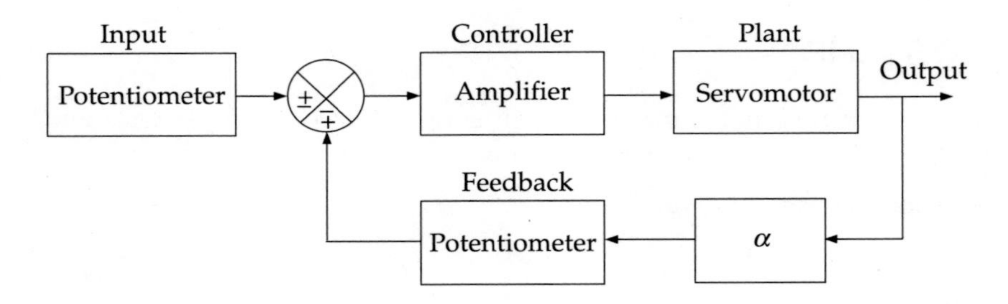

Figure 5.53 Block diagram of position control system.

Let K_p = potentiometer constant
K_a = amplifier gain
K_m = motor gain constant
J_m = moment of inertia of the motor shaft and the load
D_m = viscous friction of the motor shaft and the load
n_m = number of teeth of the motor gear
J_f = moment of inertia of the feedback gear train
D_f = viscous friction of the feedback gear train
n_p = number of teeth of the potentiometer

$\theta_r(t)$ = reference angular position

$\theta_m(t)$ = angular position of the motor shaft

$e_r(t)$ = input reference voltage corresponding to the marked position of the load $= K_p\theta_r(t)$

$e_f(t)$ = feedback voltage corresponding to feedback gear position $= K_p\theta_m(t)$

Then, the error signal

$$e(t) = e_r(t) - e_f(t) = K_p\theta_r(t) - K_p\theta_m(t) \tag{5.41}$$

The Laplace transform of Eq. (5.41) is

$$E(s) = K_p[\Theta_r(s) - \Theta_m(s)]$$

The voltage input $e_c(t)$ to the motor is given by

$$e_c(t) = K_a e(t) = K_a K_p[\theta_r(t) - \theta_m(t)] \tag{5.42}$$

The Laplace transform of Eq. (5.42) is

$$E_c(s) = K_a K_p[\Theta_r(s) - \Theta_m(s)] \tag{5.43}$$

The gear train ratio α to the feedback is given by

$$\alpha = \frac{n_m}{n_p}$$

We saw in Eq. (5.10) that the reflected inertia and viscous friction of the motor due to feedback gear train can be written as

$$J = J_m + \alpha^2 J_f$$
$$D = D_m + \alpha^2 D_f$$

Also, we know from Eq. (5.33) that

$$\frac{\Theta_m(s)}{E_c(s)} = \frac{K_m}{s(sJ + 2D)} \tag{5.44}$$

Combining Eqs. (5.43) and (5.44), we get

$$\Theta_m(s) = \frac{K_m}{s(sJ+2D)} E_c(s) = \frac{K_m K_a K_p}{s(sJ+2D)}[\Theta_r(s) - \Theta_m(s)]$$
$$= \frac{K\Theta_r(s)}{s(sJ+2D)} - \frac{K\Theta_m(s)}{s(sJ+2D)} \tag{5.45}$$

where $K = K_m K_a K_p$. On rearranging the terms in Eq. (5.45), we obtain

$$\Theta_m(s)\left[1 + \frac{K}{s(sJ+2D)}\right] = \frac{K\Theta_r(s)}{s(sJ+2D)}$$

Thus, the transfer function of the system is given by

$$G(s) = \frac{\Theta_m(s)}{\Theta_r(s)} = \frac{K}{s(sJ+2D)+K} = \frac{K}{s^2J + 2sD + K}$$

Rate servomechanism. Here, the derivative of position, i.e. the speed, is to be held constant. So, the controlled variable is the speed of rotation of a motor and hence the feedback element may be a tachometer. The corresponding block diagram is given in Figure 5.54.

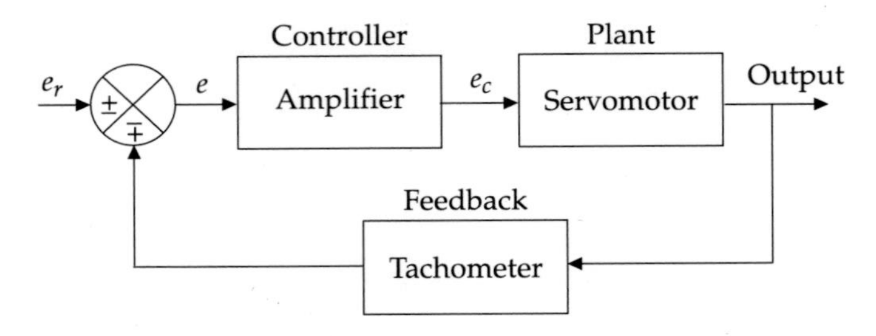

Figure 5.54 Block diagram of rate servomechanism.

For the dc tachometer, we know [see Eq. (5.35)]

$$e_f(t) = K_t \omega_t(t)$$

where K_t is the tachometer constant and ω_t, its angular velocity. So, the error signal is

$$e(t) = e_r(t) - e_f(t) = e_r(t) - K_t \omega_t(t)$$

The voltage input to the motor is

$$e_c(t) = K_a e(t) = K_a[e_r(t) - K_t \omega_t(t)] = K_a \left[e_r(t) - K_t \frac{d\theta}{dt}\right] \tag{5.46}$$

If this voltage is fed to an ac servomotor at rest, the generated stall torque τ_0 is given by [see Eq. (5.27)]

$$\tau_0 = K e_c$$

where K is a constant. Also, from mechanical considerations,

$$J \frac{d^2\theta}{dt^2} + D \frac{d\theta}{dt} = \tau_0 = K e_c$$

or

$$\left(\frac{1}{K}\right) \frac{d^2\theta}{dt^2} + \left(\frac{D}{K}\right) \frac{d\theta}{dt} = e_c \tag{5.47}$$

Substituting the value of e_c from Eq. (5.46) in Eq. (5.47), we get

$$\left(\frac{1}{K}\right) \frac{d^2\theta}{dt^2} + \left(\frac{D}{K}\right) \frac{d\theta}{dt} = K_a \left[e_r(t) - K_t \frac{d\theta}{dt}\right]$$

or

$$e_r = \left(\frac{1}{KK_a}\right) \frac{d^2\theta}{dt^2} + \left(\frac{D}{KK_a} + K_t\right) \frac{d\theta}{dt}$$

At constant speed, $\dfrac{d^2\theta}{dt^2} = 0$. Therefore,

$$\frac{d\theta}{dt} = \frac{e_r}{\dfrac{D}{KK_a} + K_t} = \frac{e_r K K_a}{D + K K_a K_t} = \frac{e_r \left(\dfrac{K}{D}\right) K_a}{1 + \left(\dfrac{K}{D}\right) K_a K_t}$$

$$= \frac{e_r K_m K_t}{1 + K_m K_a K_t}$$

where $K/D = K_m$ is called the *velocity gain constant* of the motor.

Thermal System: Temperature Control

Let us consider a temperature control system in which there is an equal inflow and outflow of liquid, and a steady temperature of the liquid in the bath is maintained by the arrangement shown in Figure 5.55.

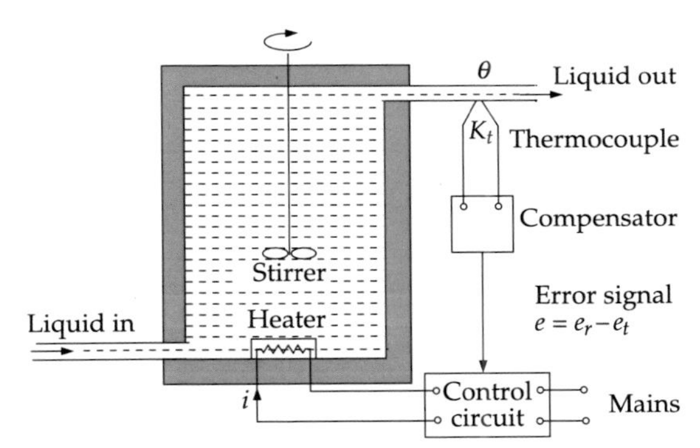

Figure 5.55 Temperature controller.

Before building a mathematical model of the system, we need to make the following simplifying assumptions:

1. The bath is perfectly insulated, i.e. there is no heat loss either by conduction or by radiation from the bath.
2. The stirrer thoroughly mixes the liquid so that the temperature within the bath is uniform everywhere.
3. The control circuit is a zero-order instrument, i.e.

$$i = K_c \tag{5.48}$$

where K_c is a constant.

Now, if

m = mass of the liquid in the tank

θ_i = temperature of the liquid flowing in

θ = temperature of the liquid in the tank

c = specific heat of the liquid in the tank

ρ = density of the liquid in the tank

R = resistance of the heating element

V = volume of liquid flowing per second

Then, from the condition of heat balance,

$$i^2 R = mc\frac{d\theta}{dt} + V\rho c(\theta - \theta_i) \tag{5.49}$$

Suppose the present values of the variables are connected to their zero-subscripted steady-state values as $e = e_0 + \Delta e$, $\theta = \theta_0 + \Delta\theta$, $\theta_l = \theta_{i0} + \Delta\theta_i$.

Then, substituting Eq. (5.48) in Eq. (5.49) and utilizing the above conditions, we get

$$K_c^2(e_0 + \Delta e)R = mc\frac{d\Delta\theta}{dt} + V\rho c(\theta_0 + \Delta\theta - \theta_{i0} - \Delta\theta_i) \tag{5.50}$$

Equating all incremental values to zero, the steady-state equation is obtained from Eq. (5.50), as

$$K_c^2 e_0^2 R = V\rho c(\theta_0 - \theta_{i0}) \tag{5.51}$$

Subtracting Eq. (5.51) from Eq. (5.50) and neglecting the term containing $(\Delta e)^2$, we have

$$2K_c^2 e_0\, \Delta e\, R = mc\frac{d\Delta\theta}{dt} + \frac{1}{R_t}(\Delta\theta - \Delta\theta_i), \qquad R_t = \frac{1}{V\rho c}$$

or

$$K\Delta e = \tau\frac{d\Delta\theta}{dt} + (\Delta\theta - \Delta\theta_i) \tag{5.52}$$

where $K = 2K_c^2 e_0\, \Delta e\, RR_t$ and $\tau = R_t mc$. R_t can be called the *thermal resistance of the heat transfer process* and τ, the time-constant.

We note that the thermocouple converts temperature to an emf e_t, which is fed to the comparator. The comparator subtracts this emf from a reference emf e_r and generates an error emf. Therefore, we can write the following equations for the incremental changes:

$$\Delta e_t = K_t \Delta\theta \tag{5.53}$$

$$\Delta e = \Delta e_r - \Delta e_t \tag{5.54}$$

The Laplace transforms of Eqs. (5.52) to (5.54) yield

$$K\Delta E(s) = \tau s\, \Delta\Theta(s) + \Delta\Theta(s) - \Delta\Theta_i(s)$$

$$\Rightarrow \qquad \Delta\Theta(s) = \frac{K\Delta E(s)}{\tau s + 1} + \frac{\Delta\Theta_i(s)}{\tau s + 1} \tag{5.55}$$

$$\Delta E_t(s) = K_t \Delta\Theta(s) \tag{5.56}$$

$$\Delta E(s) = \Delta E_r(s) - \Delta E_t(s) \tag{5.57}$$

Equations (5.55) to (5.57) help us draw the block diagram of Figure 5.56 for the system. Here, the forward-path transfer function and the disturbance transfer function are given by

$$G(s) = \frac{K}{\tau s + 1}$$

$$D(s) = \frac{\Delta\Theta_i(s)}{\tau s + 1}$$

The disturbance is caused by the change in temperature of the inflowing liquid.

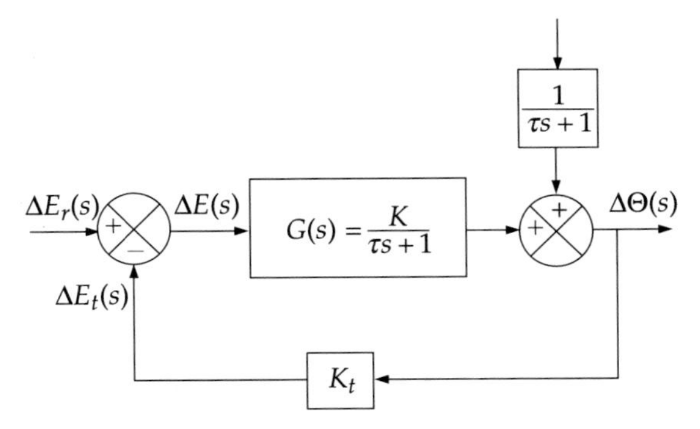

Figure 5.56 Block diagram for temperature control system.

If there is no change in temperature of the inflowing liquid, the disturbance signal is zero. Then, from Figure 5.56, we get

$$\begin{aligned}\Delta\Theta(s) &= \Delta E(s) G(s)\\ &= [\Delta E_r(s) - \Delta E_t(s)]G(s)\\ &= \frac{K}{\tau s + 1} \cdot \Delta E_r(s) - \frac{KK_t}{\tau s + 1} \cdot \Delta\Theta(s)\end{aligned}$$

or

$$\Delta\Theta(s)\left(1 + \frac{KK_t}{\tau s + 1}\right) = \frac{K}{\tau s + 1} \cdot \Delta E_r(s)$$

or

$$\frac{\Delta\Theta(s)}{\Delta E_r(s)} = \frac{K}{\tau s + 1 + KK_t}$$

If the reference signal is changed by a step, the corresponding change in the temperature of the tank liquid can be obtained from

$$\Delta\Theta(s) = \frac{1}{s} \cdot \frac{K}{\tau s + 1 + KK_t} \tag{5.58}$$

The steady state value of $\theta(t)$ can be obtained by taking the inverse Laplace transform of Eq. (5.58) and then putting $t \to \infty$. Alternatively, using the final value theorem[5] it can be derived straightaway from Eq. (5.58) as

$$\begin{aligned}\Delta\theta(t)_{\text{ss}} &= \lim_{s\to 0} s\Delta\Theta(s) \\ &= \lim_{s\to 0}\left(s \cdot \frac{1}{s} \cdot \frac{K}{\tau s + 1 + KK_t}\right) \\ &= \frac{K}{1 + KK_t}\end{aligned}$$

We note that the feedback loop reduces the temperature change by a factor of $1/(1 + KK_t)$ if there is an unwanted change in the reference voltage.

It can be found by the same method that any fluctuation in the temperature of the inflowing liquid cannot disturb in a big way the temperature of the liquid in the tank because of the reduction by the same factor. So, by controlling the loop-gain, the influence of unwanted changes can be kept under control.

Liquid System: Level Control

Control of liquid levels is an important feature in many industrial processes. We consider a simple system, shown in Figure 5.57, where the level is being maintained by a controller which actuates a pump to supply liquid at a rate of Q_i m^3/s. Let Q_o be the liquid outflow rate and h, the liquid head in the tank. Of course, a steady liquid-level can be maintained by making

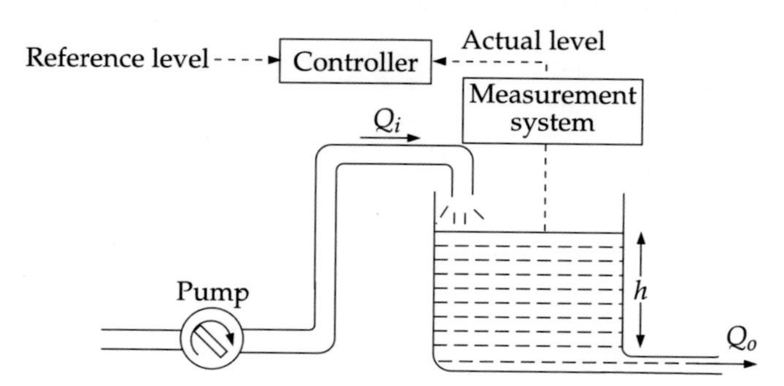

Figure 5.57 A liquid-level control system.

[5] See Section 1.2 at page 6.

$Q_i = Q_o$. However, there may be fluctuations in the supply rate due to many factors and, therefore, a level control measure is necessary.

The well-known Poiseuille's equation[6] states that for a laminar flow of liquid through a horizontal pipe,

$$\Delta p = \frac{8\mu l}{\pi a^4} Q_o = R' Q_o$$

where

μ = viscosity of the fluid flowing through the pipe
l = length of the pipe
a = inner radius of the pipe
Q_o = rate of fluid flowing out of the pipe
Δp = pressure head that causes the flow

$R' = 8\mu l/(\pi a^4)$ is called the *fluid resistance*. Since $\Delta p = h\rho g$, where g is the acceleration due to gravity and h is the height of the liquid in the tank, we can write

$$h = RQ_o$$

where $R = R'/(\rho g)$.

The rate of storage of liquid in the tank is the difference between the inflow rate and outflow rate of the liquid. Hence, if C is the cross-sectional area of the tank, then

$$C\frac{dh}{dt} = Q_i - Q_o = Q_i - \frac{h}{R}$$

or

$$RC\frac{dh}{dt} + h = RQ_i \tag{5.59}$$

Also, if

r_l = reference level of liquid in the tank
h = actual level of the liquid in the tank
e = error in liquid level = $r_l - h$
K = gain of the controller

then we have

$$RQ_i = Ke \tag{5.60}$$

The Laplace transforms of Eqs. (5.59) and (5.60) give

$$(\tau s + 1)H(s) = KE(s), \qquad \tau = RC$$

The system may be represented by the block diagram of Figure 5.58.

[6]See, for example, *The General Properties of Matter* by CJ Smith, Edward Arnold, London.

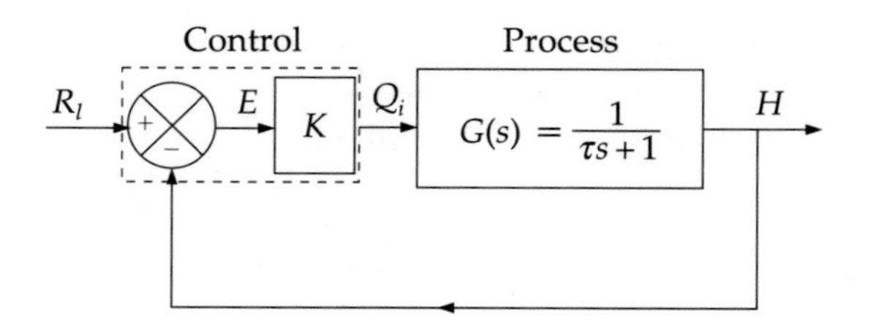

Figure 5.58 Block diagram of liquid level control system.

The block diagram gives the following relations between the error and the reference signals:

$$\frac{H(s)}{R_l(s)} = \frac{KG(s)}{1+KG(s)} = \frac{K}{\tau s+K+1}$$

$$E(s) = R_l(s) - H(s) = R_l(s)\left(1 - \frac{K}{\tau s+K+1}\right)$$

$$= R_l(s) \cdot \frac{1+\tau s}{\tau s+K+1}$$

or

$$\frac{E(s)}{R_l(s)} = \frac{1+\tau s}{\tau s+K+1}$$

For a step change in the reference signal,

$$\Delta E(s) = \frac{1}{s} \cdot \frac{\tau s+1}{\tau s+K+1}$$

the steady-state value of which is given by

$$\Delta e(t)_{\text{ss}} = \lim_{s\to 0} s\,\Delta E(s) = \lim_{s\to 0} \frac{\tau s+1}{\tau s+1+K} = \frac{1}{1+K} \tag{5.61}$$

Equation (5.61) suggests that to decrease the error in the level of the liquid, one has to increase K. This, in effect, means to use a high power pump. However, one has to consider the cost of the control at the same time and so the question of optimization arises.

5.4 TRANSPORTATION LAG IN SYSTEMS

We know that transfer functions represent the ratio of Laplace transforms of output and input in linear time-invariant systems. Generally, they are ratios of two polynomials in the s-space. When they are derived, apart from assuming that all initial conditions are set to zero, there is another tacit assumption that there is no time delay for the error signal generation. But, in many cases, this assumption is not valid. Even computer controlled systems have time delays, or transportation lags, because the computer takes some time to perform numerical operations before it produces an error signal. For mechanical, pneumatic or chemical systems, the time delay is no doubt greater.

Suppose a signal generated at a distance d travels with a velocity v to reach the controller. If T be the corresponding time lag, it is given by

$$T = \frac{d}{v}$$

Because of this time delay, the output lags behind the controller by T seconds. Thus, the monitored feedback signal $b(t)$ and output $c(t)$ are related as

$$b(t) = c(t - T) \tag{5.62}$$

The Laplace transform of Eq. (5.62) yields

$$B(s) = e^{-sT}C(s) \tag{5.63}$$

Equation (5.63) gives the transfer function between the feedback signal and the output as

$$\frac{B(s)}{C(s)} = e^{-sT} \tag{5.64}$$

The function e^{-sT}, called the *transportation lag* or *input delay* or *dead time*, is very useful in handling control problems. We may recall that we encountered this function in Example 3.12 at page 99. We will consider others in Examples 10.15 at page 337 and 11.27 at page 394 .

REVIEW QUESTIONS

5.1 Choose the correct answers:

(i) In position control system, the device used for rate feedback voltage is called
   (a) potentiometer (b) synchrotransmitter
   (c) tachogenerator (d) servomotor

(ii) In the force-voltage analogy, mass is analogous to
   (a) charge (b) current (c) inductance (d) resistance

(iii) A rotational system is described by the differential equation

$$2\frac{d\omega}{dt} + \omega = 10$$

where the RHS represents the constant torque acting on the system and ω represents the angular velocity. The solution for ω is
   (a) $10(1 - e^{-t/2})$ (b) $10e^{-t/2}$ (c) $10(1 + e^{-t/2})$ (d) $5(1 - e^{-t/2})$

(iv) An ac servomotor is basically a
   (a) universal motor (b) single-phase induction motor
   (c) two-phase induction motor (d) three-phase induction motor

(v) In force-voltage analoous system, displacement is equivalent to
   (a) current (b) flux (c) charge (d) inductance

5.2 Derive the transfer function of an armature-controlled dc motor.

5.3 Draw the typical torque-speed characteristics of a two-phase induction motor. How is this achieved and why? Show that the transfer function of a two-phase induction motor can be written in the form

$$T(s) = \frac{\Theta_m(s)}{V_z(s)} = \frac{\text{Motor shaft displacement}}{\text{Control phase voltage}} = \frac{K_m}{s(1 + s\tau_m)}$$

What are the expressions for K_m and τ_m and what are they called?

5.4 (a) Determine the transfer function of a dc servomotor in the field controlled mode.

(b) Derive the transfer function of a two-phase, 50 Hz servomotor with the following specifications:

Starting torque	: 0.16 N-m
Inertia of motor	: 1×10^{-5} kg-m^2
Supply voltage	: 115 V
No-load angular velocity	: 304 rad/s

Assume a straight line characteristic and zero viscous friction.

5.5 Consider the mechanical system shown in the following figure. $F(t)$ is the external force applied to the system and $y(t)$ the displacement of the mass element M from the equilibrium position. Draw the free-body diagram of the mass element and write down the ordinary differential equation that describes the system dynamics.

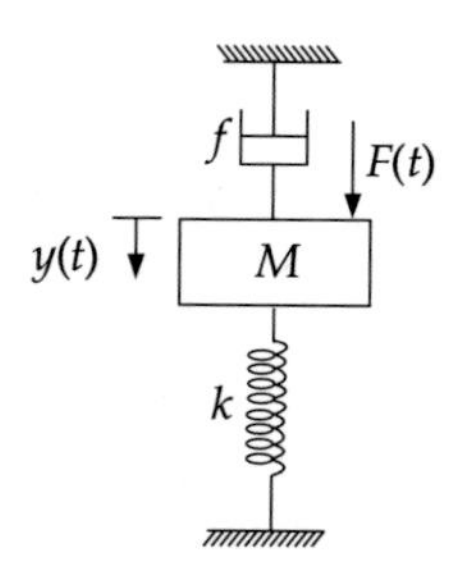

5.6 For the mechanical system shown in the following figure, write the differential equations of performance with equivalent mechanical network.

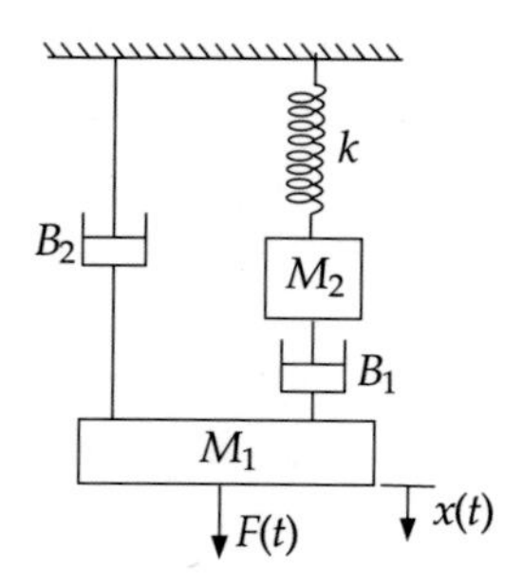

5.7 Derive the system equations and find the value of $X_2(s)/F(s)$ for the system shown in the following figure.

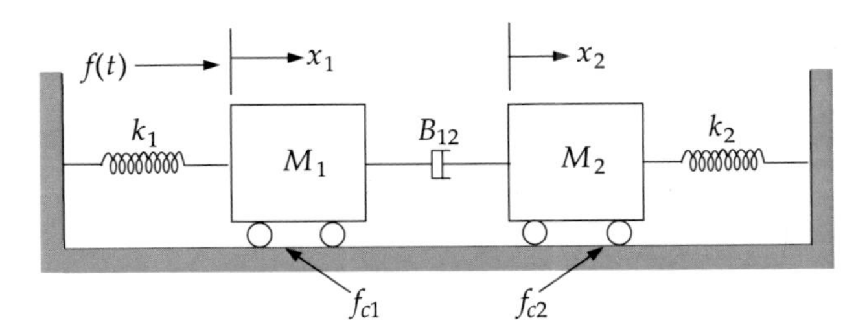

5.8 For the mechanical system shown in the following figure, (a) draw the equivalent mechanical network, (b) write performance equations, and (c) draw the analogous electrical network.

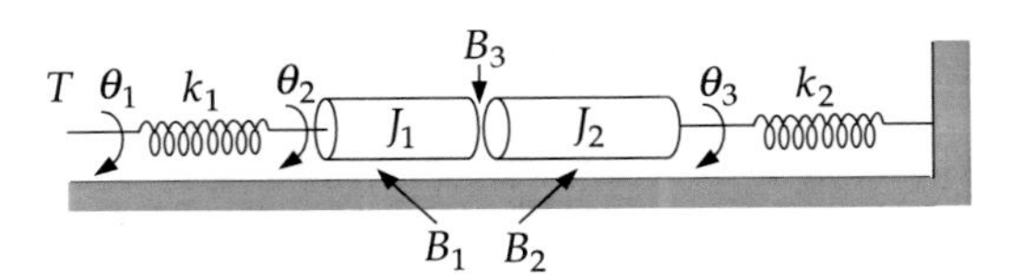

5.9 (a) Is there any special constructional features for the motors used in servo system?
(b) Find the transfer function of the motor given in the following figure:

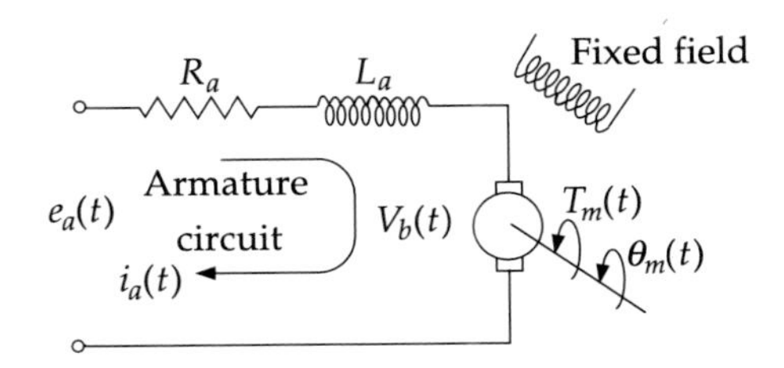

(c) What is the order of the transfer function?
(d) Sketch the torque-speed curve of the motor.

5.10 Write short notes on the following:

(a) The ac tachogenerator
(b) Tachometer
(c) Synchro transmitter control transformer
(d) Synchro pair
(e) Position and speed control of an automobile
(f) Liquid level controller in the two-tank process
(g) A temperature control system.

Answers to Selected Questions

5.1. (i) (c); (ii) (c); (iii) (a); (iv) (c); (v) (c)

5.4. (b) $\dfrac{\Theta(s)}{E_c(s)} = \dfrac{139.1}{s(s+105.3)}$

5.6. $M_1\dfrac{d^2x}{dt^2} + B_2\dfrac{dx}{dt} + B_1\dfrac{d}{dt}(x - x_1) = F(t), \quad M_2\dfrac{d^2x_1}{dt^2} + kx_1 = B_1\dfrac{d}{dt}(x - x_1)$

5.7. $f(t) = M_1\dfrac{d^2x_1}{dt^2} + f_{c1}\dfrac{dx_1}{dt} + k_1x_1 + B_{12}\dfrac{d}{dt}(x_1 - x_2)$

$$B_{12}\frac{d}{dt}(x_1 - x_2) = M_2\frac{d^2x_2}{dt^2} + f_{c2}\frac{dx_2}{dt} + k_2x_2$$

$$\frac{X_2(s)}{F(s)} = \frac{sB_{12}}{\left[s^2M_1 + s\left(f_{c1} + B_{12}\right) + k_1\right]\left[s^2M_2 + s\left(f_{c2} + B_{12}\right) + k_2\right] + s^2B_{12}^2}$$

5.8. (b) $J_1\dfrac{d^2\theta_1}{dt^2} + B_1\dfrac{d\theta_1}{dt} + B_3\dfrac{d}{dt}(\theta_1 - \theta_2) + k_1\theta_1 = T$

$$J_2\frac{d^2\theta_1}{dt^2} + B_2\frac{d\theta_1}{dt} + k_2\theta_2 = B_3\frac{d}{dt}(\theta_1 - \theta_2)$$

(c) The analogous (f–i analogy) electrical equations are

$$L_1\frac{di_1}{dt} + R_1i_1 + R_3(i_1 - i_2) + \frac{\int i_1dt}{C_1} = V$$

$$L_2\frac{di_2}{dt} + R_2i_2 + \frac{\int i_2dt}{C_2} = R_3(i_1 - i_2)$$

The circuit may be drawn from these equations.

CHAPTER 6

Feedback Control Characteristics

6.1 INTRODUCTION

At the outset we would like to mention that the feedback can be of two types:

- Degenerative or negative feedback
- Regenerative or positive feedback

Unless otherwise specified, normally the word 'feedback' in control systems means the degenerative feedback.

In Chapter 4, we have seen how to obtain mathematical models of components of control systems. Also, in Chapter 2 we discussed feedback control systems. In this chapter, we will study the following:

1. Transfer functions of open-loop and closed-loop systems
2. The effect of parameter variation in open-loop and closed-loop systems
3. The effects of feedback on
   (a) Sensitivity
   (b) Time-constant
   (c) Overall gain
   (d) Stability
   (e) Disturbance
4. Steady-state error of a control system

6.2 TRANSFER FUNCTION

Open-loop System

The block diagram and signal flow graph of an open-loop control system are shown in Figure 6.1. The output of the open-loop system is

$$C(s) = G(s)R(s)$$

Therefore, though $G(s)$ should be called the open-loop transfer function, the convention is to define the open-loop transfer function (OLTF) as the signal from the feedback path. Thus,

$$\text{OLTF} = G(s)H(s)$$

Unless otherwise specified, the value of $H(s)$ is usually assumed to be 1.

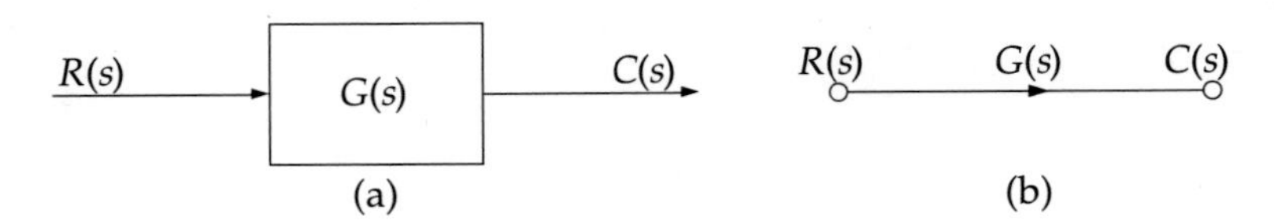

Figure 6.1 Open-loop system: (a) block diagram; (b) signal flow graph.

Closed-loop System

The block diagram and signal flow graph of a closed-loop system are shown in Figure 6.2.

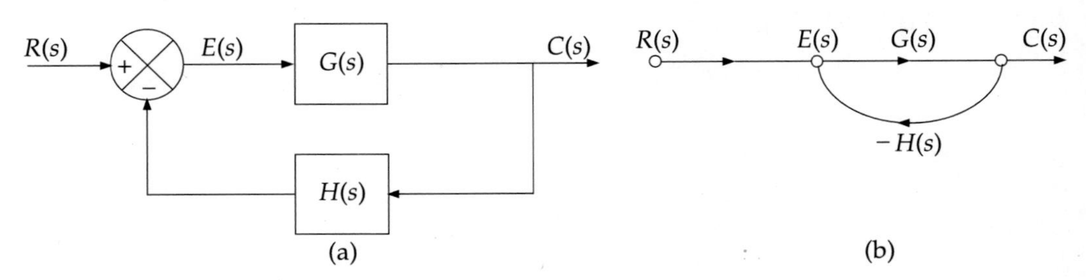

Figure 6.2 Closed-loop system: (a) block diagram; (b) signal flow graph.

The basic difference between open-loop and closed-loop systems is the generation of *error signal*, $E(s)$, in the latter. The output of the closed-loop system is, therefore,

$$C(s) = E(s)G(s) = [R(s) - H(s)]G(s)$$

The closed-loop transfer function (CLTF) is, thus,

$$\text{CLTF} = \frac{C(s)}{R(s)} = \frac{G(s)}{1 + G(s)H(s)} \tag{6.1}$$

6.3 EFFECT OF PARAMETER VARIATION

Open-loop system. Let, $\Delta G(s)$ = change in $G(s)$ due to parameter variation, and $\Delta C(s)$ = corresponding change in the output. Then,

$$C(s) + \Delta C(s) = [G(s) + \Delta G(s)]R(s) = G(s)R(s) + \Delta G(s)R(s)$$
$$= C(s) + \Delta G(s)R(s)$$

$$\therefore \qquad \Delta C(s) = \Delta G(s)R(s) \tag{6.2}$$

Closed-loop system. From Eq. (6.1),

$$C(s) = \left[\frac{G(s)}{1 + G(s)H(s)}\right] R(s)$$

Therefore,

$$C(s) + \Delta C(s) = \left[\frac{G(s) + \Delta G(s)}{1 + [G(s) + \Delta G(s)]H(s)}\right] R(s)$$

$$= \left[\frac{G(s) + \Delta G(s)}{1 + G(s)H(s)}\right] R(s) \qquad [\text{neglecting } \Delta G(s)H(s)]$$

$$= \frac{G(s)}{1 + G(s)H(s)} R(s) + \frac{\Delta G(s)}{1 + G(s)H(s)} R(s)$$

$$= C(s) + \frac{\Delta G(s)}{1 + G(s)H(s)} R(s)$$

Thus,

$$\Delta C(s) = \frac{\Delta G(s)}{1 + G(s)H(s)} R(s) \tag{6.3}$$

Comparing Eqs. (6.2) and (6.3), we observe that the closed-loop variation in output is $1/[1 + G(s)H(s)]$ times that of the open-loop system. Usually, $G(s)H(s)$ is high, which means that the feedback lowers the variation in output with respect to parameter variation.

6.4 EFFECTS OF FEEDBACK

We know that feedback exists only for closed-loop systems. For open-loop systems, there is no feedback. However, for the sake of comparison about the effect of feedback, we have to consider the corresponding situations in open-loop systems as well.

On Sensitivity

If K is a parameter of the system and T is the transfer function of the system, then the sensitivity of T with respect to K is defined as

$$S_K^T = \frac{\partial T/T}{\partial K/K}$$

Open-loop system. Here, $T = G$. Hence,

$$S_G^T = \frac{\partial G/G}{\partial G/G} = 1$$

Therefore, the sensitivity of an open-loop system to variation of G is 100%.

Closed-loop system. Here,

$$T = \frac{G}{1 + GH}$$

$$\frac{\partial T}{\partial G} = \frac{1}{1 + GH} - \frac{GH}{(1 + GH)^2} \tag{6.4a}$$

$$\frac{\partial T}{\partial H} = -\frac{G^2}{(1 + GH)^2} \tag{6.4b}$$

Therefore,

$$\begin{aligned} S_G^T &= \frac{\partial T}{\partial G} \cdot \frac{G}{T} \\ &= \left[\frac{1}{1+GH} - \frac{GH}{(1+GH)^2}\right](1+GH) \qquad \text{[using Eq. (6.4a)]} \\ &= \frac{1}{1+GH} \\ &< 1 \qquad \text{(since } GH \text{ is high)} \end{aligned}$$

Thus, the sensitivity for a closed-loop system is reduced by a factor. This essentially means that the closed-loop system is much less sensitive to G variation.

Now,

$$\begin{aligned} S_H^T &= \frac{\partial T}{\partial H} \cdot \frac{H}{T} \\ &= -\frac{G^2}{(1+GH)^2} \cdot \frac{H(1+GH)}{G} \qquad \text{[using Eq. (6.4b)]} \\ &= -\frac{GH}{1+GH} \\ &= 1 \qquad \text{(since } GH \text{ is high)} \end{aligned}$$

So, a feedback loop is highly sensitive to H variation.

Note: (a) Since a closed-loop system is highly sensitive to H variation, it is necessary to use robust feedback elements that are *not* affected by environmental factors.

(b) G is normally made of power elements while H comprises measuring/monitoring elements that operate at low power levels. So, it is easier and less costly to maintain the constancy of H.

On Time-constant

Let us consider an elementary system where the process is of first order and the negative feedback is of zero order[1] .

Open-loop system. Here,

$$G(s) = \frac{C(s)}{R(s)} = \frac{K}{1+sT} \tag{6.5}$$

For a unit impulse input, $R(s) = 1$. Therefore,

$$C(s) = \frac{K}{T} \cdot \frac{1}{s+(1/T)}$$

Thus,

$$c(t) = \frac{K}{T} \exp\left[-(t/T)\right] \tag{6.6}$$

[1]Means order of the transfer function of the feedback element.

Closed-loop system. Here,

$$T(s) = \frac{G(s)}{1 + G(s)H(s)}$$

If $H(s) = K'$, then

$$T(s) = \frac{\dfrac{K}{1+sT}}{1 + \dfrac{KK'}{1+sT}} = \frac{K/T}{s + \dfrac{1+KK'}{T}} \tag{6.7}$$

Therefore,

$$c(t) = \frac{K}{T} \exp\left[-(1/T')\right] \tag{6.8}$$

where

$$T' = \frac{T}{1+KK'}$$

Now, $T' < T$ because $KK' \gg 1$. Thus, a comparison of Eq. (6.8) with Eq. (6.6) reveals that a closed-loop system has a lower time-constant, or faster response, than an open-loop system.

On Overall Gain

In the case of an open-loop system,

$$\text{Overall gain } = G(s)$$

while, for a closed-loop system,

$$\text{Overall gain } = \frac{G(s)}{1 + G(s)H(s)}$$

So, the overall gain of a closed-loop system is less than that of an open-loop system possessing the same process gain.

On Stability

Consider a first order system with zero order feedback.

Open-loop system. From Eq. (6.5),

$$G(s) = \frac{K/T}{s + (1/T)}$$

Therefore, the pole is at $-\dfrac{1}{T}$.

Closed-loop system. From Eq. (6.7),

$$T(s) = \frac{K/T}{s + [(1+KK')/T]}$$

Therefore, the pole is at $-\dfrac{1+KK'}{T}$.

The poles are plotted in Figure 6.3. It may be seen from the figure that for increasing values of K', the location of the pole shifts more towards the left in the s-plane. This means that, with proper feedback, the stability of a system may be increased. However, we will see later that, in general, open-loop systems, which are simpler in construction, are more stable than closed-loop systems which are rather complex in design.

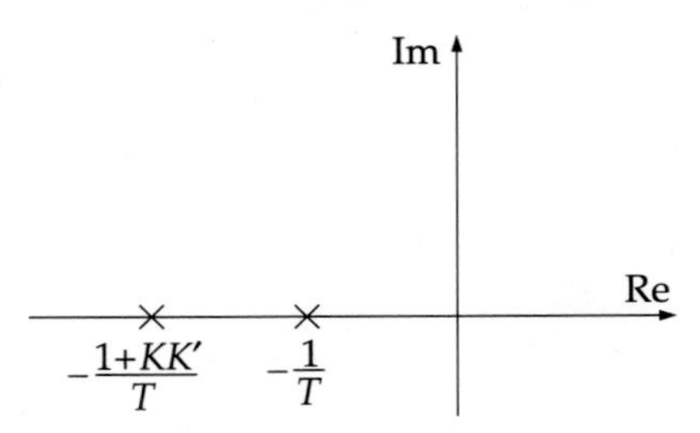

Figure 6.3 Location of poles for open-loop and closed-loop systems.

On Disturbance

Open-loop system. Here, if the disturbance is on the input side, it gets multiplied by G while, if it is on the output side, it remains as such. So, in an open-loop system, there is no question of controlling the disturbance.

Closed-loop system. Consider Figure 6.4, where D_1 and D_2 indicate disturbances.

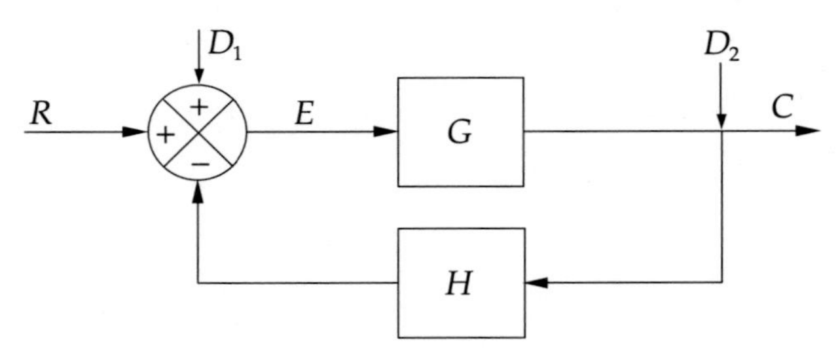

Figure 6.4 Closed-loop system with disturbances.

Here,

$$C \pm D_2 = (E \pm D_1)G = (R - CH \pm D_1)Gr$$

$$C\left(\frac{1}{G} + H\right) = R \pm D_1 \mp \frac{D_2}{G}r$$

$$C = \frac{G}{1+GH}R \pm \frac{G}{1+GH}D_1 \mp \frac{1}{1+GH}D_2$$

$$= \frac{R}{H} \pm \frac{D_1}{H} \mp \frac{D_2}{GH} \qquad \text{for } GH \gg 1$$

This result shows that the value of D_2 gets considerably reduced, whereas the value of D_1 is not much reduced. Thus, a closed-loop system considerably reduces noise (or disturbance) at process output, but not so effectively at the process input.

EXAMPLE 6.1. A feedback system with high gain K is shown in the following figure:

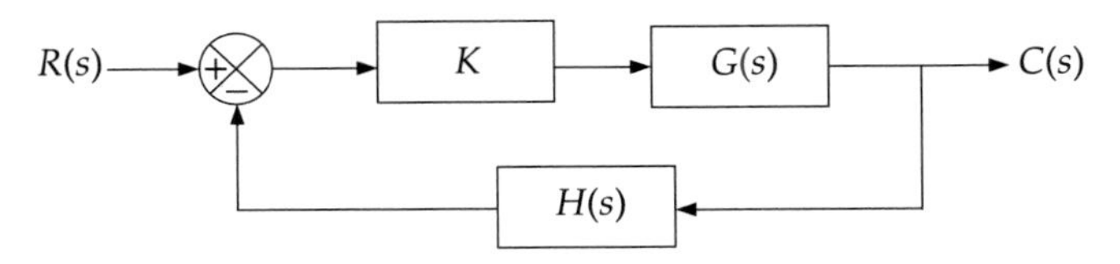

The the closed-loop transfer function function is

(a) sensitive to perturbations in $G(s)$ and $H(s)$

(b) sensitive to perturbations in $G(s)$ but not to perturbations in $H(s)$

(c) sensitive to perturbations in $H(s)$ but not to perturbations in $G(s)$

(d) insensitive to perturbations in $G(s)$ and $H(s)$

Solution The closed-loop transfer function is given by

$$T = \frac{KG}{1 + KGH}$$

Determination of the sensitivity of T to variations in G:

$$S_G^T = \frac{\partial T}{\partial G} \cdot \frac{G}{T} \tag{i}$$

Now,

$$\frac{\partial T}{\partial G} = \frac{K(1 + KGH) - K^2GH}{(1 + KGH)^2} = \frac{K}{(1 + KGH)^2} \tag{ii}$$

$$\frac{G}{T} = \frac{1 + KGH}{K} \quad \text{[from Eq. (i)]} \tag{iii}$$

Therefore,

$$S_G^T = \frac{K}{(1 + KGH)^2} \cdot \frac{1 + KGH}{K} \quad \text{[from Eq. (ii) and Eq. (iii)]}$$

$$= \frac{1}{1 + KGH} \approx \frac{1}{KGH}$$

$$\ll 1 \tag{iv}$$

Determination of the sensitivity of T to variations in H:

$$S_H^T = \frac{\partial T}{\partial H} \cdot \frac{H}{T} \tag{v}$$

Now,

$$\frac{\partial T}{\partial H} = -\frac{KG \cdot KG}{(1 + KGH)^2} \tag{vi}$$

$$\frac{H}{T} = H \cdot \frac{1+KGH}{KG} \qquad \text{[from Eq. (v)]} \tag{vii}$$

Therefore,
$$S_H^T = -\frac{K^2G^2}{(1+KGH)^2} \cdot \frac{H(1+KGH)}{KG} \qquad \text{[from Eqs. (vi) and (vii)]}$$

$$= -\frac{KGH}{1+KGH} \approx -\frac{KGH}{KGH}$$

$$= -1 \tag{viii}$$

From Eqs. (iv) and (viii), we observe that T is sensitive to perturbations in H, but not to perturbations in G.
Ans: (c).

6.5 STEADY-STATE ERROR

The error $e(t)$ is defined as

$$e(t) = r(t) - c(t)$$

Hence,
$$E(s) = R(s) - C(s)$$

The steady-state error e_{ss} is defined as

$$e_{ss} = e(t)|_{t\to\infty}$$

which, on applying the final value theorem[2], turns out to be

$$e_{ss} = \lim_{s\to 0} sE(s) \tag{6.9}$$

Now, let us examine how the steady-state errors for a step input to an open-loop and a closed-loop systems compare.

Open-loop system. Here,

$$E(s) = R(s) - R(s)G(s) = [1 - G(s)]R(s) \tag{6.10}$$

$R(s)$ being equal to $\frac{1}{s}$, from Eqs. (6.9) and (6.10), we have

$$e_{ss} = \lim_{s\to 0} [1 - G(s)] = 1 - G(0)$$

Closed-loop system. Here, for a unity feedback system,

$$E(s) = R(s) - R(s)\frac{G(s)}{1+G(s)} = \frac{1}{1+G(s)}R(s) \tag{6.11}$$

[2]See Section 1.2 at page 6.

Therefore, for a step input, from Eqs. (6.9) and (6.11),

$$e_{ss} = \lim_{s \to 0} \frac{1}{1+G(s)} = \frac{1}{1+G(0)}$$

Since $G(0)$ is high, the value of steady-state error is much less in a closed-loop system. In other words, a closed-loop system gives much better accuracy than an open-loop system.

Table 6.1 summarizes the differences between an open-loop and a closed-loop system.

Table 6.1 Distinction between an Open-loop and a Closed-loop System

Property	*Open-loop*	*Closed-loop*	*Comments*
Transfer function and gain	$G(s)H(s)$	$\dfrac{G(s)}{1+G(s)H(s)}$	Indicates that the open-loop gain is higher because $G(s)$ is normally high.
Parameter variation	$\Delta C(s) = \Delta G(s)H(s)$	$\Delta C(s) = \dfrac{\Delta G(s)}{1+G(s)H(s)}$	Variation in output with parameter variation is lower in a closed-loop.
Sensitivity	$S_G^T = 1$	$S_G^T = \dfrac{1}{1+GH}$	Closed-loop is much less sensitive to G variation, while open-loop sensitivity is 100%.
Time-constant	For a first order system with impulse input $c(t) = \dfrac{K}{T} \exp\,[-(t/T)]$	For the same system with the same input $c(t) = \dfrac{K}{T} \exp[-(t/T')]$ where $T' < T$	Since a closed-loop system has a lower time constant, its response is faster.
Stability	For a first order system, the pole is at $s = -\dfrac{1}{T}$	For the same system, the pole is at $s = -\dfrac{1+KK'}{T}$	For the closed-loop system, the location of the pole may be shifted to a more stable position by adjusting K'.
Disturbance	The question of control of disturbance does not arise.	Noise or disturbance is considerably reduced at the output.	Closed-loop systems are able to control disturbances.
Steady-state error	For a unit step input $e_{ss} = 1 - G(0)$	For the same input $e_{ss} = \dfrac{1}{1-G(0)}$	Error is much less in closed-loop systems.

EXAMPLE 6.2. Calculate the sensitivity of the system given in the following figure with respect to (a) the forward path transfer gain and (b) feedback path transfer function at $\omega = 1.3$ rad/s.

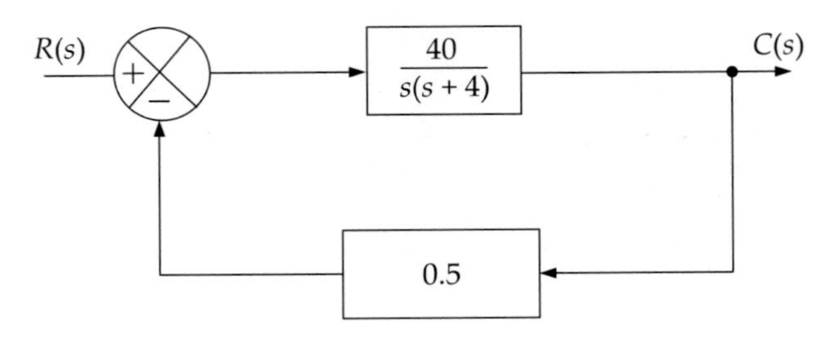

Solution Given: $G(s) = \dfrac{40}{s(s+4)}$, $H(s) = 0.5$

(a) $S_G^T = \dfrac{1}{G(s)H(s)} = \dfrac{s^2+4s}{s^2+4s+20}$

$$S_G^T\Big|_{j\omega} = \frac{-\omega^2 + 4j\omega}{V+20} = \frac{-1.69 + j5.2}{-1.69 + j5.2 + 20} = \frac{-1.69 + j5.2}{18.31 + j5.2}$$

Therefore,
$$|S_G^T| = \sqrt{\frac{(-1.69)^2 + (5.2)^2}{(18.31)^2 + (5.2)^2}} = 0.287$$

(b) $S_H^T = \dfrac{-G(s)H(s)}{1+G(s)H(s)} = -\dfrac{20}{s^2+4s+20}$

$$S_H^T\Big|_{j\omega} = -\frac{20}{18.31 + j5.2}$$

$$|S_H^T| = \frac{20}{\sqrt{(18.31)^2 + (5.2)^2}} = 1.051$$

Therefore,
$$S_H^T = -1.051$$

EXAMPLE 6.3.

$$G(s) = \frac{\omega_n^2}{s^2 + 2\zeta\omega_n s + \omega_n^2}$$

where $\omega_n = 8\pi/T$, $T = 6.28$ s and $\zeta = 0.3$. Calculate the open-loop and closed-loop sensitivities for changes in A and H, where A is the gain and H is the feedback path transfer function.

Solution

$$\omega_n = \frac{8\pi}{6.28} = 4 \text{ rad/s}$$

$$G(s) = \frac{4^2}{s^2 + 2(0.3)(4) + 4^2} = \frac{16}{s^2 + 2.4s + 16}$$

For open-loop sensitivity,

$$T(s) = AG(s)$$

Hence,

$$\frac{\partial T}{\partial A} = G(s), \qquad \frac{A}{T} = \frac{1}{G(s)}$$

Therefore,

$$S_A^T = \frac{\partial T}{\partial A} \cdot \frac{A}{T} = 1$$

There is no question of having S_H^T here.

For closed-loop sensitivity,

$$T(s) = \frac{AG(s)}{1 + AG(s)H} = \frac{16A}{s^2 + 2.4s + 16(AH + 1)}$$

$$\frac{\partial T}{\partial A} = \frac{16A}{s^2 + 2.4s + 16(AH + 1)}\left[1 - \frac{16AH}{s^2 + 2.4s + 16(AH + 1)}\right]$$

$$S_A^T = 1 - \frac{16AH}{s^2 + 2.4s + 16(AH + 1)} = \frac{s^2 + 2.4s + 16}{s^2 + 2.4s + 16(AH + 1)}$$

$$\frac{\partial T}{\partial H} = -\frac{(16A)^2}{[s^2 + 2.4s + 16(AH + 1)]^2}$$

$$S_H^T = \frac{\partial T}{\partial H} \cdot \frac{H}{T} = -\frac{(16A)^2}{[s^2 + 2.4s + 16(AH + 1)]^2} \cdot \frac{H}{T}$$

$$= -\frac{16AH}{s^2 + 2.4s + 16(AH + 1)}$$

EXAMPLE 6.4. Refer to the control system given in Figure 6.5. $G = 210$ and $H = 0.12$.

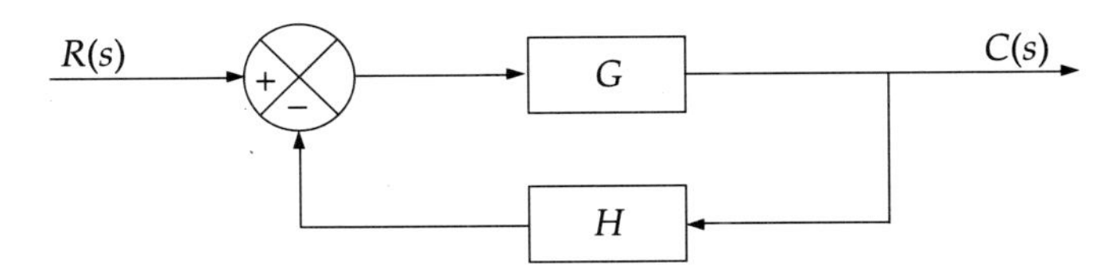

Figure 6.5 Control system (Example 6.4).

Find (a) reference voltage E_r when $E_o = 240$ V, and (b) percentage change in the generator terminal when the forward path gain has been reduced by 12%.

Solution

$$T(s) = \frac{E_o(s)}{E_r(s)} = \frac{G(s)}{1 + G(s)H(s)} = \frac{210}{1 + (210)(0.12)} = 8.015$$

(a) $E_r = \dfrac{E_o}{T} = \dfrac{240}{8.015} = 29.24$ V

(b) $S_G^T = \dfrac{\partial T}{\partial G} \cdot \dfrac{G}{T} = \left[\dfrac{1}{1+GH} - \dfrac{GH}{(1+GH)^2}\right] \cdot (1+GH) = \dfrac{1}{1+GH} = \dfrac{1}{26.2}$

Now, $$S_G^T = \frac{\partial T/T}{\partial G/G} = \frac{\%\text{ change in } T}{\%\text{ change in } G} = \frac{1}{26.2}$$

Therefore, $$\%\text{ change in } T = \frac{12}{26.2} = 0.458$$

6.6 REGENERATIVE FEEDBACK

As already mentioned, in a regenerative feedback system, the fed back signal is added to the reference signal rather than the conventional practice of subtracting it. This is mostly not desired in a control system.

For example, if one desires to control the temperature of a room at 25°C, the control system continuously measures the room temperature, subtracts the feedback signal from the reference, i.e. 25°C, and if the result is negative, which means the room temperature is higher, the cooling mechanism is switched on. So, in a negative feedback system, a negative error signal means cooling is required, and a positive error signal means heating is required. Now, if the feedback is positive, which means the signal is added to the reference, the result is always positive, and hence, the cooling mechanism is never switched on.

For similar reasons, the degenerative or negative feedback is mostly resorted to. From the mathematical viewpoint, a positive feedback has the transfer function

$$\frac{C(s)}{R(s)} = \frac{G(s)}{1 - G(s)H(s)}$$

The negative sign in the denominator indicates that there is a possibility that $G(s)H(s) = 1$ when the system gives an infinite output for a finite input.

However, in some special cases the regenerative or positive feedback is needed to increase the loop gain of the feedback system. Consider the feedback system as depicted in the block diagram of Figure 6.6.

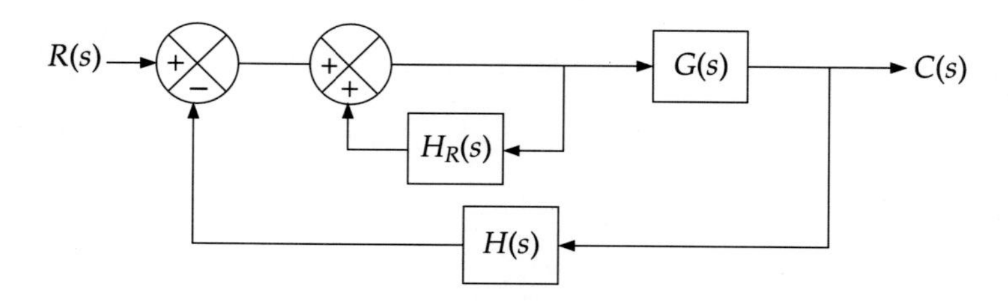

Figure 6.6 Use of regenerative feedback in a control system.

Here, the inner loop reduces to $\dfrac{1}{1 - H_R(s)}$, making the total transfer function

$$\frac{C(s)}{R(s)} = \frac{\dfrac{G(s)}{1 - H_R(s)}}{1 + \dfrac{G(s)H(s)}{1 - H_R(s)}} = \frac{G(s)}{1 - H_R(s) + G(s)H(s)} \tag{6.12}$$

In this system, if $H_R(s) \approx 1$, Eq. (6.12) reduces to

$$\frac{C(s)}{R(s)} \approx \frac{1}{H(s)} \tag{6.13}$$

Equation (6.13) shows that by a proper choice of the regenerative feedback $H_R(s)$, the closed-loop transfer function can be made insensitive to $G(s)$.

REVIEW QUESTIONS

6.1 Choose the correct answer:

(i) The output of a feedback control system should be a function of
   (a) input (b) feedback signal
   (c) reference and output (d) all of above

(ii) In a closed-loop control system the error signal is
   (a) the difference between the measured value and the reference value
   (b) the sum of the measured value and the reference value
   (c) ratio of the measured value and the reference value
   (d) none of the above.

(iii) Consider the following characteristics:
   1. sensitivity is 100 2. disturbance can be controlled
   3. steady-state error is much less 4. gain is higher.
   Which of these characteristics are true for a closed-loop system?
   (a) 1 and 2 (b) 2 and 3 (c) 3 and 4 (d) 1 and 4

(iv) Which of the characteristics stated in (iii) above are true for an open-loop system?
   (a) 1 and 2 (b) 2 and 3 (c) 3 and 4 (d) 1 and 4

(v) The effect of negative feedback is to
   (a) reduce the overall gain
   (b) increase the sensitivity of parameter variation in forward path
   (c) slow the dynamic response
   (d) decrease the bandwidth.

6.2 (a) Draw the block diagram of a closed-loop control system and indicate the following on it:

(i) plant
(ii) command input
(iii) controlled output
(iv) actuating signal
(v) feedback element and control element

(b) Mention the important features of closed-loop control systems.

6.3 (a) Define sensitivity of a system variable T with respect to another parameter X.

(b) Find the sensitivity of the overall transfer function of the closed-loop system shown in Figure 6.7 with respect to (i) forward path transfer function, and (ii) feedback path transfer function at the frequency 1.4 rad/s.

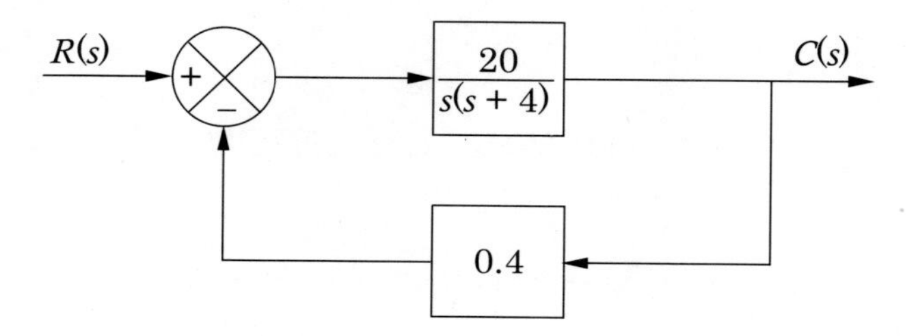

Figure 6.7 System [Question 6.3(b)].

6.4 (a) What is regenerative feedback? What is its application?

(b) Show that the use of negative feedback in a control system reduces the sensitivity of the system to parameter variation and disturbance signal.

Answers to Selected Questions

6.1 (i) (a); (ii) (a); (iii) (b); (iv) (d); (v) (a)

6.3 (b) (i) $S_G^T = 0.72$; (ii) $S_H^T = -0.97$

CHAPTER 7

Time-domain Performance of Control Systems

7.1 INPUT SIGNALS

Control systems are basically dynamic systems. So, studying their behaviour with time is important. Indeed, the study of variation of output of a control system is important from many perspectives including its stability. However, the actual input to the system is usually unknown and, therefore, some standard test input signals are used in the study. We discussed such signals in Section 3.3 and their Laplace transforms. Here, we summarize them in Table 7.1.

Table 7.1 Standard Test Signals

Signal	$r(t)$	$R(s)$
Unit impulse	$\delta(t) = \begin{cases} \lim\limits_{\varepsilon \to 0} \dfrac{1}{\varepsilon} & \text{for } 0 \le t \le \varepsilon \\ 0 & \text{for } t > \varepsilon \end{cases}$	$\mathscr{L}\{\delta(t)\} = 1$
Unit step	$u(t) = \begin{cases} 0 & \text{for } t \le 0 \\ 1 & \text{for } t > 0 \end{cases}$	$\mathscr{L}\{u(t)\} = \dfrac{1}{s}$
Unit ramp	$r(t) = 1$	$\mathscr{L}\{r(t)\} = \dfrac{1}{s^2}$
Unit parabolic	$p(t) = \dfrac{t^2}{2}$	$\mathscr{L}\{p(t)\} = \dfrac{1}{s^3}$

It may be noted from the Laplace transforms that each successive signal is the integral of the preceding one.

7.2 RESPONSES

Once an input is given to a system, it will produce an output or response. The response can be

(i) Transient

(ii) Steady-state

The transient response is that part of the response which approaches zero as time tends to infinity. The part of the response that remains after the transients die out, is called the *steady-state response.* The parameter, which is important in steady state, is the steady-state error, e_{ss}. It is defined as

$$e_{ss} = \lim_{t \to \infty} e(t) = \lim_{t \to \infty} [r(t) - c(t)] \tag{7.1}$$

where $e(t)$, being the difference between reference input $r(t)$ and output $c(t)$, is the error.

7.3 TIME RESPONSE ANALYSIS

First Order System

If the dynamic relation between the reference input $r(t)$ and output $c(t)$ of a system is of the form

$$a_1 \frac{dc(t)}{dt} + a_0 c(t) = b_0 r(t)$$

then it can be written as

$$\tau \frac{dc(t)}{dt} + c(t) = Kr(t) \tag{7.2}$$

where $\tau = a_1/a_0$ is the *time constant* and $K = b_0/a_0$ is the *static sensitivity* of the system. Taking Laplace transform of Eq. (7.2) and assuming all initial conditions = 0, we have

$$(s\tau + 1)C(s) = KR(s) \tag{7.3}$$

Equation (7.3) yields the transfer function

$$G(s) \equiv \frac{C(s)}{R(s)} = \frac{K}{s\tau + 1}$$

The order of the transfer function is 1. So, such systems are called *first order systems.* The block diagram and signal flow graph of a first order system are given in Figure 7.1.

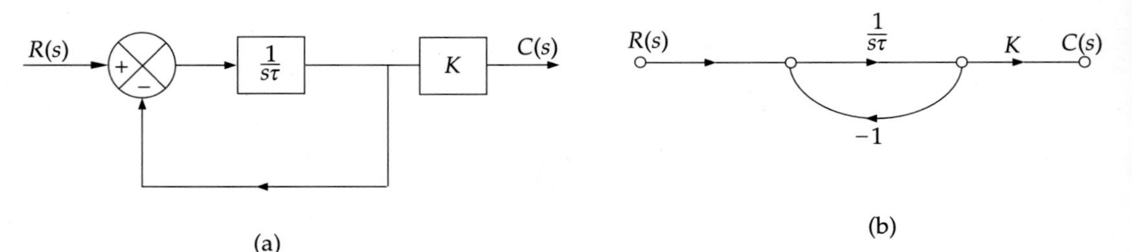

Figure 7.1 First order system: (a) block diagram; (b) signal flow graph.

Now, we check the response of the system for different standard inputs. For convenience, let us assume that $K = 1$ so that the transfer function is

$$G(s) = \frac{1}{s\tau + 1}$$

Unit impulse input. Since, for this input $R(s) = 1$, we have

$$C(s) = \frac{1}{s\tau + 1} = \frac{1}{\tau} \cdot \frac{1}{s + \dfrac{1}{\tau}}$$

On taking inverse Laplace transform, we get

$$c(t) = \frac{1}{\tau} e^{-t/\tau}$$

So, as $t \to 0$ $\quad c(t) \to \dfrac{1}{\tau}$

as $t \to \infty$ $\quad c(t) \to 0$

The time-response curve is given in Figure 7.2.

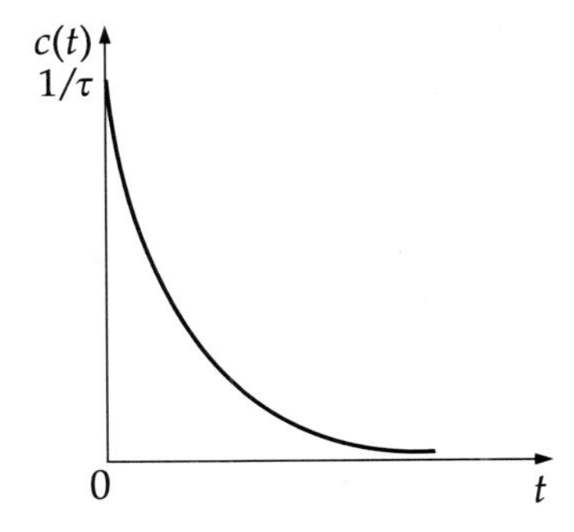

Figure 7.2 Time-response curve for unit impulse input to a first order system.

Unit step input. In the case of unit step input, $R(s) = 1/s$. Therefore,

$$C(s) = \frac{1}{s} \cdot \frac{1}{s\tau + 1} = \frac{1}{s} - \frac{1}{s + \dfrac{1}{\tau}} \tag{7.4}$$

which, on inverse Laplace transform, yields

$$c(t) = 1 - e^{-t/\tau} \tag{7.5}$$

It may be observed from Eq. (7.5) that

$$\text{Initial slope of the curve} = \left.\frac{dc(t)}{dt}\right|_{t=0} = \left.\frac{1}{\tau} e^{-t/\tau}\right|_{t=0} = \frac{1}{\tau} \tag{7.6}$$

We see from Eq. (7.6) that if the value of τ is high, the initial slope is low and vice versa. Thus, the lower the τ, the faster the response. Also the lower the τ, the higher the stability. This is clear from a plot of its poles [see Figure 7.3(a), where a plot is shown for two values of τ with $\tau_1 > \tau_2$]. Here, the transfer function may be obtained from Eq. (7.4) as

$$G(s) = \frac{C(s)}{R(s)} = \frac{1}{s\tau + 1}$$

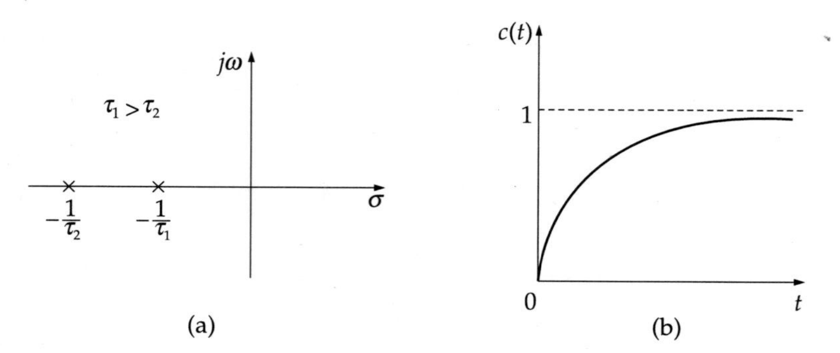

Figure 7.3 First order system: (a) plot of poles of closed-loop system; (b) time-response curve for unit step input.

Poles are obtained by solving the characteristic equation $s\tau + 1 = 0$, yielding $s = -1/\tau$. The steady-state error is given as

$$e_{\text{ss}} = \lim_{t \to \infty} [r(t) - c(t)] = 1 - 1 = 0$$

This result is obtained from the time response of the system. Without finding out the time response, i.e., without doing inverse Laplace transform, we can find out the steady-state error by applying the final value theorem. We will discuss the method later (see Section 7.4 at page 259).

An interesting result may be obtained from Table 7.2, where values of the output are given against t/τ. It may be seen that the step response of the first order system attains 99.3% of the final value for $t = 5\tau$. The time-response curve is shown in Figure 7.3(b).

Table 7.2 Response with Time for Unit Step Input

t/τ	1	2	3	4	5
$c(t)$ %	63.2	86.0	95.0	98.2	99.3

EXAMPLE 7.1. A first order process has a static gain of 1, a time constant of 1 s, and a transport delay of 0.1 s. Its transfer function is

(a) $\dfrac{e^{-s}}{1+0.1s}$ (b) $\dfrac{e^{-10s}}{1+s}$ (c) $\dfrac{e^{-0.1s}}{1+s}$ (d) $\dfrac{e^{-0.1s}}{1+0.1s}$

Solution Since this is a first order system with a time delay, let the transfer function be

$$G(s) = \frac{Ke^{-s\tau_d}}{1+\tau s}$$

where the terms have their usual significance. Given $K = 1$, $\tau = 1$ s, $\tau_d = 0.1$ s. Then the transfer function is

$$G(s) = \frac{e^{-0.1s}}{1+s}$$

Ans: (c).

EXAMPLE 7.2. A unit step is applied at $t = 0$ to a first order system without time delay. The response has a value of 1.264 units at $t = 10$ min and 2 units at steady state. What is the transfer function of the system?

Solution Since it is a first order system, let the transfer function be given by

$$G(s) = \frac{K}{1 + \tau s} \tag{i}$$

We know from Eq. (7.13) that for a unit step input, the output is given by

$$c(t) = 1 - e^{-t/\tau}$$

However, since the steady state output is 2 units, $K = 2$, and the output is given by

$$c(t) = 2\left(1 - e^{-t/\tau}\right) \tag{ii}$$

Now, from the given condition at $t = 10$ min $= 600$ s, we get from Eq. (ii),

$$1.264 = 2\left(1 - e^{-600/\tau}\right)$$

or

$$1 - e^{-600/\tau} = \frac{1.264}{2}$$

or

$$e^{-600/\tau} = 1 - 0.632 = 0.368$$

or

$$\frac{-600}{\tau} = \ln 0.368 = -1$$

Therefore,

$$\tau = 600 \text{ s}$$

Thus, we get the required transfer function from Eq. (i) as

$$G(s) = \frac{2}{1 + 600s}$$

EXAMPLE 7.3. A system with transfer function of $1/(\tau s + 1)$, subjected to a step input takes 10 seconds to reach 50% of the step height. The value of τ is

(a) 6.9 s (b) 10 s (c) 14.4 s (d) 20 s

Solution For an input of step A, the response of a first order system is given by

$$c(t) = A\left(1 - e^{-t/\tau}\right)$$

Given, when $t = 10$ s, $c(t) = A/2$. So, we have

$$\frac{A}{2} = A\left(1 - e^{-10/\tau}\right)$$

$\Rightarrow$

$$e^{-10/\tau} = 1 - \frac{1}{2} = \frac{1}{2}$$

$\Rightarrow$

$$-\frac{10}{\tau} = \ln 0.5 = -0.6931$$

$\Rightarrow$

$$\tau = \frac{10}{0.6931} = 14.4 \text{ s}$$

Ans: (c).

EXAMPLE 7.4. In the feedback scheme in the following figure, the time-constant of the closed-loop system is

(a) $A\beta\tau$ (b) $1 + A\beta\tau$ (c) τ (d) $\dfrac{\tau}{1 + A\beta}$

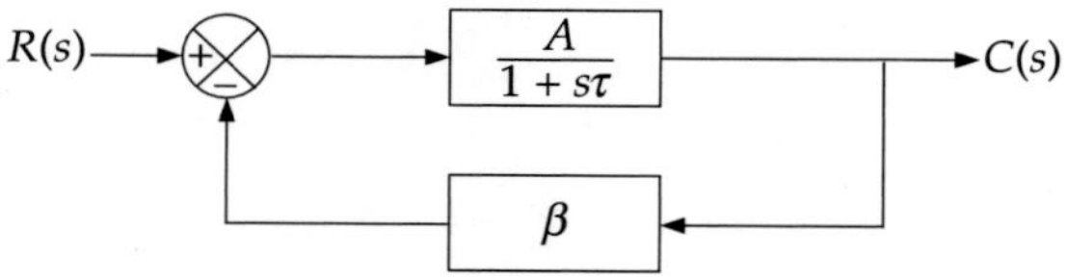

Solution The closed-loop transfer function is

$$\frac{C(s)}{R(s)} = \frac{\dfrac{A}{1+s\tau}}{1 + \dfrac{A\beta}{1+s\tau}} = \frac{A}{s\tau + 1 + A\beta}$$

$$= \frac{A/(1+A\beta)}{s \cdot \dfrac{\tau}{1+A\beta} + 1} \equiv \frac{K}{s\tau' + 1}$$

So, the time constant $= \tau' = \dfrac{\tau}{1 + A\beta}$.

Ans: (d)

Unit ramp input. For the unit ramp input, $R(s) = \dfrac{1}{s^2}$. Therefore,

$$C(s) = \frac{1}{s^2} \cdot \frac{1}{s\tau + 1} = \frac{A}{s^2} + \frac{B}{s} + \frac{C}{s\tau + 1} \text{ (say)}$$

Then[1],

$$A = \left[s^2 C(s)\right]_{s=0} = \left(\frac{1}{s\tau + 1}\right)_{s=0} = 1$$

$$B = \left[\frac{d}{ds}\left\{s^2 C(s)\right\}\right]_{s=0} = \left[\frac{d}{ds}\left(\frac{1}{s\tau + 1}\right)\right]_{s=0}$$

$$= -\left.\frac{\tau}{(s\tau + 1)^2}\right|_{s=0} = -\tau$$

$$C = \left[(s\tau + 1)C(s)\right]_{s=-(1/\tau)} = \left.\frac{1}{s^2}\right|_{s=-(1/\tau)} = \tau^2$$

[1]See Section 1.2 at page 9 for the method.

Thus,

$$C(s) = \frac{1}{s^2} - \frac{\tau}{s} + \frac{\tau^2}{s\tau + 1} = \frac{1}{s^2} - \frac{\tau}{s} + \frac{\tau}{s + (1/\tau)} \tag{7.7}$$

which, on inverse Laplace transform, yields

$$c(t) = t - \tau + \tau e^{-t/\tau} = t - \tau\left(1 - e^{-t/\tau}\right) \tag{7.8}$$

The time-response curve is given in Figure 7.4. The steady-state error is

$$\begin{aligned} e_{ss} &= \lim_{t \to \infty} [r(t) - c(t)] \\ &= t - (t - \tau) \qquad \text{[from Eq. (7.8)]} \\ &= \tau \end{aligned}$$

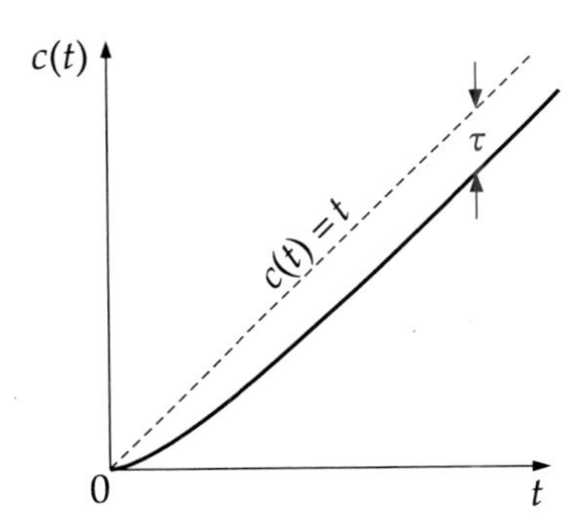

Figure 7.4 Time-response curve for the unit ramp input. - - -, input; — output.

We find from Table 7.3 that the interrelation of different inputs also holds good for the corresponding responses.

Table 7.3 Mutual Relationship between Input and Response

	Input, r(t)		*Response, c(t)*	
Function	*Form*	*Relationship*	*Form*	*Relationship*
Ramp	$r(t) = t$		$t - \tau + \tau e^{-t/\tau}$	
Step	$u(t) = 1$	$\frac{dr(t)}{dt}$	$1 - e^{-t/\tau}$	$\left.\frac{d}{dt}c(t)\right\vert_{\text{ramp}}$
Impulse	$\delta(t)$	$\left.\frac{d}{dt}u(t)\right\vert_{t=0}$	$\frac{1}{\tau}e^{-t/\tau}$	$\left.\frac{d}{dt}c(t)\right\vert_{\text{step}}$

Second Order Systems

If the dynamic input-output relationship of a system is represented by a differential equation of the form

$$a_2 \frac{d^2c(t)}{dt^2} + a_1 \frac{dc(t)}{dt} + a_0 c(t) = b_0 r(t) \tag{7.9}$$

it is called a *second order system* because we will presently see that the transfer function of the system has an order of 2.

Equation (7.9) can be written as

$$\frac{1}{\omega_n^2} \frac{d^2c(t)}{dt^2} + \frac{2\zeta}{\omega_n} \frac{dc(t)}{dt} + c(t) = Kr(t) \tag{7.10}$$

where

$$\frac{1}{\omega_n^2} = \frac{a_2}{a_0}, \qquad \frac{2\zeta}{\omega_n} = \frac{a_1}{a_0}, \qquad K = \frac{b_0}{a_0}$$

ω_n, ζ and K are called *natural frequency of oscillation*, *damping ratio* and *static sensitivity* (or *gain*), respectively. Taking Laplace transform of Eq. (7.10) and assuming it to be a relaxed system (i.e., all initial conditions = 0), we get

$$\left(\frac{s^2}{\omega_n^2} + \frac{2\zeta s}{\omega_n} + 1\right) C(s) = KR(s) \tag{7.11}$$

On rearranging Eq. (7.11), we get the transfer function for the second order system as

$$G(s) \equiv \frac{C(s)}{R(s)} = \frac{K}{\dfrac{s^2}{\omega_n^2} + \dfrac{2\zeta s}{\omega_n} + 1} = \frac{K\omega_n^2}{s^2 + 2\zeta\omega_n s + \omega_n^2} \tag{7.12}$$

This can also be written in the form

$$\frac{C(s)}{R(s)} = \frac{K}{s(s\tau + 1) + K} = \frac{K/\tau}{s^2 + (s/\tau) + (K/\tau)} \tag{7.13}$$

where

$$\begin{aligned} \frac{K}{\tau} &= \omega_n^2 \quad \Rightarrow \quad \omega_n = \sqrt{\frac{K}{\tau}} \\ 2\zeta\omega_n &= \frac{1}{\tau} \quad \Rightarrow \quad \zeta = \frac{1}{2\omega_n \tau} \end{aligned} \tag{7.14}$$

The results given in Eq. (7.14) are important because they indicate how the undamped natural frequency and damping ratio behave with the gain. With the increase in gain, the natural frequency increases which, in turn, reduces the damping ratio.

The block diagram and signal flow graph representations of a second order system are given in Figure 7.5. From Eq. (7.12), we see that the characteristic equation for the system is

$$s^2 + 2\zeta\omega_n s + \omega_n^2 = 0$$

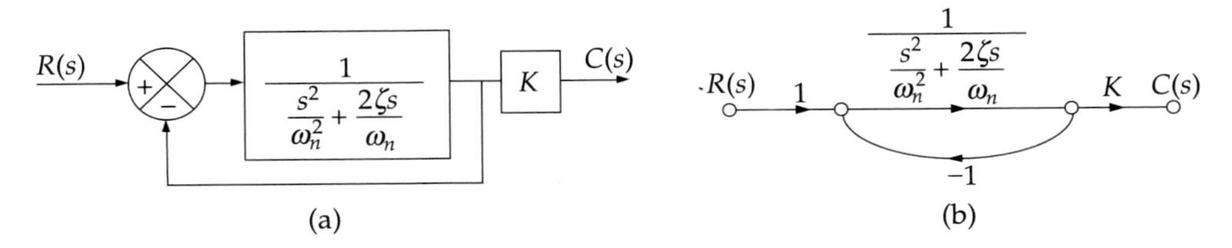

Figure 7.5 Second order system: (a) block diagram; (b) signal flow graph.

Therefore, its roots are given by

$$s_1,\ s_2 = -\zeta\omega_n \pm \omega_n\sqrt{\zeta^2 - 1} \tag{7.15}$$

Depending on the value of ζ, three distinct cases present themselves from Eq. (7.15) as shown in Table 7.4.

Table 7.4 Three Distinct Cases of Second Order Systems

Case	*Damping ratio value*	*Roots of characteristic equation*	*System specification*
1	$\zeta > 1$	$s_1 \neq s_2$, and both are real	Overdamped
2	$\zeta = 1$	$s_1 = s_2$, and both are real	Critically damped
3	$\zeta < 1$	s_1 and s_2 are complex, one being the conjugate of the other	Underdamped

We will consider only the unit step input for the second order system.

Underdamped ($0 < \zeta < 1$) system. For the second order underdamped system, we have

$$C(s) = \frac{\omega_n^2}{s^2 + 2\zeta\omega_n s + \omega_n^2} \cdot R(s) = \frac{\omega_n^2}{s^2 + 2\zeta\omega_n s + \omega_n^2} \cdot \frac{1}{s} \tag{7.16}$$

$$= \frac{\omega_n^2}{s(s + \zeta\omega_n - \jmath\omega_n\sqrt{1-\zeta^2})(s + \zeta\omega_n + \jmath\omega_n\sqrt{1-\zeta^2})}$$

$$\equiv \frac{A}{s} + \frac{B}{(s + \zeta\omega_n - \jmath\omega_n\sqrt{1-\zeta^2})} + \frac{C}{(s + \zeta\omega_n + \jmath\omega_n\sqrt{1-\zeta^2})} \tag{7.17}$$

Then,
$$A = sC(s)|_{s=0} = \left.\frac{\omega_n^2}{s(s + \zeta\omega_n - \jmath\omega_n\sqrt{1-\zeta^2})(s + \zeta\omega_n + \jmath\omega_n\sqrt{1-\zeta^2})}\right|_{s=0}$$

$$= \frac{\omega_n^2}{\zeta^2\omega_n^2 + \omega_n^2(1-\zeta^2)} = 1 \tag{7.18}$$

$$B = (s + \zeta\omega_n - \jmath\omega_n\sqrt{1-\zeta^2})C(s)\Big|_{s=-\zeta\omega_n+\jmath\omega_n\sqrt{1-\zeta^2}}$$

$$= \frac{\omega_n^2}{s(s + \zeta\omega_n + \jmath\omega_n\sqrt{1-\zeta^2})}\Bigg|_{s=-\zeta\omega_n+\jmath\omega_n\sqrt{1-\zeta^2}}$$

$$= \frac{1}{(-\zeta + \jmath\sqrt{1-\zeta^2})(2\jmath\sqrt{1-\zeta^2})}$$

$$= \frac{-\zeta - \jmath\sqrt{1-\zeta^2}}{(-\zeta - \jmath\sqrt{1-\zeta^2})(-\zeta + \jmath\sqrt{1-\zeta^2})(2\jmath\sqrt{1-\zeta^2})} = -\frac{\zeta}{2\jmath\sqrt{1-\zeta^2}} - \frac{1}{2} \tag{7.19}$$

$$C = (s + \zeta\omega_n + \jmath\omega_n\sqrt{1-\zeta^2})C(s)\Big|_{s=-\zeta\omega_n-\jmath\omega_n\sqrt{1-\zeta^2}}$$

$$= \frac{\omega_n^2}{s(s + \zeta\omega_n - \jmath\omega_n\sqrt{1-\zeta^2})}\Bigg|_{s=-\zeta\omega_n-\jmath\omega_n\sqrt{1-\zeta^2}}$$

$$= \frac{1}{(-\zeta - \jmath\sqrt{1-\zeta^2})(2\jmath\sqrt{1-\zeta^2})} = \frac{\zeta}{2\jmath\sqrt{1-\zeta^2}} - \frac{1}{2} \tag{7.20}$$

Substituting the values of A, B and C in Eq. (7.17), we get

$$C(s) = \frac{1}{s} - \left(\frac{\zeta}{2\jmath\sqrt{1-\zeta^2}} + \frac{1}{2}\right)\frac{1}{s + \zeta\omega_n - \jmath\omega_n\sqrt{1-\zeta^2}} + \left(\frac{\zeta}{2\jmath\sqrt{1-\zeta^2}} - \frac{1}{2}\right)\frac{1}{s + \zeta\omega_n + \jmath\omega_n\sqrt{1-\zeta^2}} \tag{7.21}$$

The inverse Laplace transform of Eq. (7.21) yields

$$c(t) = 1 - \left(\frac{\zeta}{2\jmath\sqrt{1-\zeta^2}} + \frac{1}{2}\right)e^{-\left(\zeta-\jmath\sqrt{1-\zeta^2}\right)\omega_n t} + \left(\frac{\zeta}{2\jmath\sqrt{1-\zeta^2}} - \frac{1}{2}\right)e^{-\left(\zeta+\jmath\sqrt{1-\zeta^2}\right)\omega_n t} \tag{7.22}$$

Equation (7.22) yields, on simplification, the relation

$$c(t) = 1 - e^{-\zeta\omega_n t}\left[\frac{\zeta}{\sqrt{1-\zeta^2}}\sin\omega_d t + \cos\omega_d t\right] \tag{7.23}$$

where $\omega_d = \omega_n\sqrt{1-\zeta^2}$. Thus,

$$c(t) = 1 - \frac{e^{-\zeta\omega_n t}}{\sqrt{1-\zeta^2}}\sin(\omega_d t + \phi) \tag{7.24}$$

where $\tan\phi = \dfrac{\sqrt{1-\zeta^2}}{\zeta}$.

The plot of Eq. (7.24) shows a sine wave decaying in an exponential envelope (see Figure 7.6).

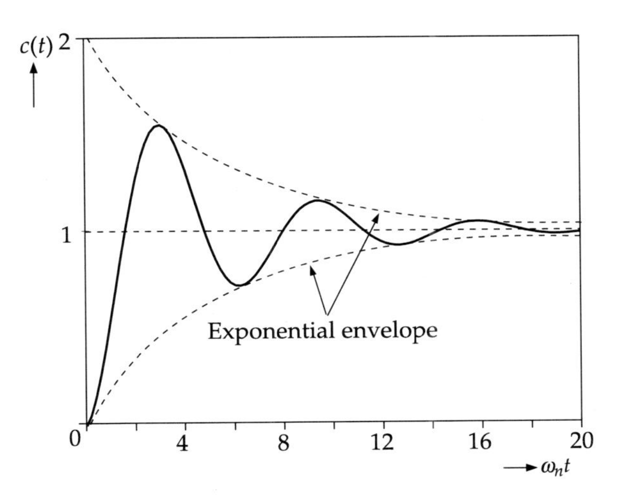

Figure 7.6 Plot of Eq. (7.24).

We observe that Eq. (7.16) consists of a real pole at the origin and two complex poles. The exponentially decaying amplitude of the response is caused by the real pole whereas the sinusoidal oscillation is caused by the complex poles. The time constant of the exponential decay is actually reciprocal of the real part of the system pole.

Damping ratio. In this context, let us see what the physical definition of the damping ratio ζ is. Often it is defined as the ratio of the exponential decay frequency to the natural frequency of the system. That is,

$$\zeta = \frac{\zeta\omega_n}{\omega_n} = \frac{\text{Exponential decay frequency}}{\text{Natural frequency (rad/s)}} \tag{7.25}$$

Also,

$$\text{Exponential decay frequency} = \frac{1}{\text{Exponential time constant}}$$

$$\text{Natural frequency (rad/s)} = \frac{2\pi}{\text{Natural period (s)}}$$

Therefore, from Eq. (7.25), we have

$$\zeta = \frac{1}{2\pi} \cdot \frac{\text{Natural period (s)}}{\text{Exponential time constant}}$$

The exponential decay frequency σ_D is thus defined as

$$\sigma_D = \zeta\omega_n$$

Critically damped ($\zeta = 1$) system. For the critically damped second order system,

$$C(s) = \frac{\omega_n^2}{s(s+\omega_n)^2} = \frac{1}{s} - \frac{\omega_n}{(s+\omega_n)^2} - \frac{1}{s+\omega_n} \tag{7.26}$$

Therefore,

$$c(t) = 1 - \omega_n t\, e^{-\omega_n t} - e^{-\omega_n t} = 1 - (1 + \omega_n t)\, e^{-\omega_n t} \tag{7.27}$$

Overdamped ($\zeta > 1$) system. In the case of overdamped second order system, the response works out to be

$$c(t) = 1 - \frac{\zeta + \sqrt{\zeta^2 - 1}}{2\sqrt{\zeta^2 - 1}} e^{\left(-\zeta + \sqrt{\zeta^2 - 1}\right)\omega_n t} + \frac{\zeta - \sqrt{\zeta^2 - 1}}{2\sqrt{\zeta^2 - 1}} e^{\left(-\zeta - \sqrt{\zeta^2 - 1}\right)\omega_n t} \tag{7.28}$$

The response curves for critically damped and overdamped systems are given in Figure 7.7.

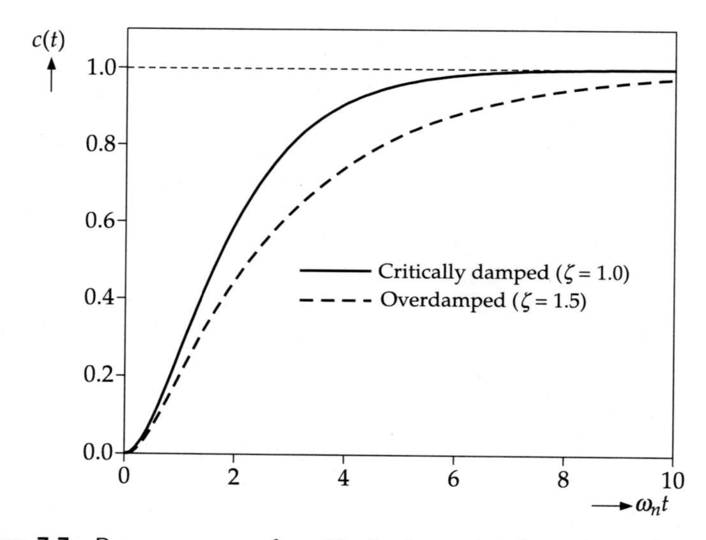

Figure 7.7 Response curves for critically damped and overdamped systems.

It is apparent from the curves that the higher the value of ζ , the more the rise time and, therefore, the settling time. Also, it can be seen from the curves that a critically damped second order system attains the step input value at the time constant value of nearly 8 while this value was nearly 5 for the first order system.

Performance Parameters for Transient Response

The performance parameters for transient response are defined in Table 7.5. These parameters are indicated in Figure 7.8 which shows the step response of a second order system. Now, let us discuss these parameters one by one.

Table 7.5 Performance Parameters for Transient Response

Parameter	*Definition*
Delay time, t_d	The time required for the step response of the system to reach 50% of the final value.
Rise time, t_r	The time required for the step response to reach the final output for the first time in underdamped systems or grow from 10% to 90% of the value in overdamped systems.
Peak time, t_p	The time required by the step response of the system to reach the first peak which is the maximum value of the output.
Maximum overshoot, M_p	The maximum amount by which the output deviates from the input. It is generally expressed as per cent of the input.
Settling time, t_s	The time the system takes to settle within a given per cent of the final value.

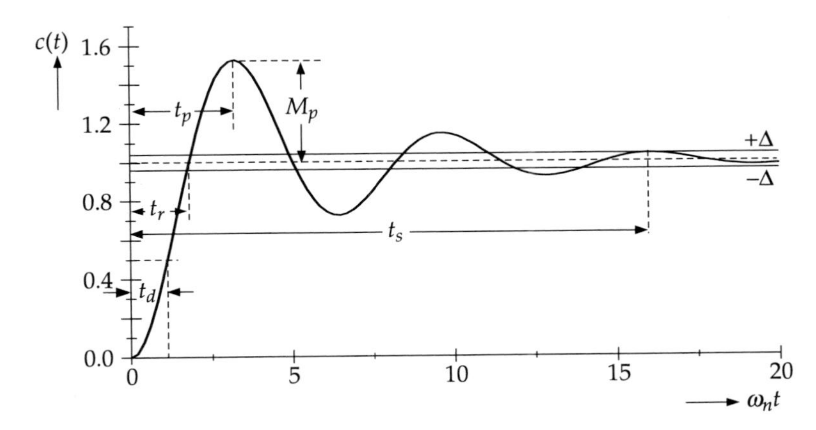

Figure 7.8 Graphical presentation of performance parameters of an underdamped second order system. - - - - input; — output.

Delay time. According to the definition, $c(t)|_{t=t_d} = \frac{1}{2}$. So, from Eqs. (7.24), (7.27) and (7.28), we get

$$\frac{e^{-\zeta x}}{\sqrt{1-\zeta^2}}\sin(\sqrt{1-\zeta^2}x+\phi) = \frac{1}{2} \qquad \text{for } 0<\zeta<1$$

$$(1+x)e^{-x} = \frac{1}{2} \qquad \text{for } \zeta = 1 \tag{7.29}$$

$$\frac{\zeta+\sqrt{\zeta^2-1}}{2\sqrt{\zeta^2-1}}e^{(-\zeta+x\sqrt{\zeta^2-1})x} + \frac{\zeta-\sqrt{\zeta^2-1}}{2\sqrt{\zeta^2-1}}e^{(-\zeta-x\sqrt{\zeta^2-1})x} = \frac{1}{2} \qquad \text{for } \zeta>1$$

where $x = \omega_n t_d$. Solutions to Eq. (7.29) can be obtained for different values of ζ and plotted to draw a graph of ζ vs. x. A linear approximation to this graph is given by the relation

$$x = 1 + 0.7\zeta \tag{7.30}$$

Equation (7.30) yields the approximate expression for t_d as

$$t_d = \frac{1 + 0.7\zeta}{\omega_n} \tag{7.31}$$

Rise time. We consider the underdamped system for simplicity. If t_r = time required to attain the steady-state value, we note that the sine factor of Eq. (7.24) should have a zero value to make $c(t) = 1$. Thus, from Eq. (7.24),

$$\sin\,(\omega_d t_r + \phi) = \sin N\pi \tag{7.32}$$

For the first time, i.e., for $N = 1$, we have, from Eq. (7.32),

$$\omega_n t_r \sqrt{1-\zeta^2} + \phi = \pi$$

or

$$t_r = \frac{\pi - \phi}{\omega_n \sqrt{1-\zeta^2}} \tag{7.33}$$

where

$$\phi = \tan^{-1} \frac{\sqrt{1-\zeta^2}}{\zeta}$$

Peak time. If t_p is the time required to reach the first peak, we can find an expression for it from Eq. (7.24) using the condition that $c(t)$ will be maximum here. Thus,

$$\left.\frac{dc(t)}{dt}\right|_{t=t_p} = 0$$

$$\Rightarrow \quad -\frac{e^{\zeta\omega_n t_p}}{\sqrt{1-\zeta^2}}\omega_d \cos\,(\omega_d t_p + \phi) + \frac{\zeta\omega_n e^{\zeta\omega_n t_p}}{\sqrt{1-\zeta^2}}\omega_d \sin\,(\omega_d t_p + \phi) = 0$$

$$\Rightarrow \quad \omega_d \cos\,(\omega_d t_p + \phi) - \zeta\omega_n \sin\,(\omega_d t_p + \phi) = 0$$

$$\Rightarrow \quad \sqrt{1-\zeta^2} \cos\,[(\omega_n \sqrt{1-\zeta^2})t_p + \phi] - \zeta \sin\,[(\omega_n \sqrt{1-\zeta^2})t_p + \phi] = 0$$

$$\Rightarrow \quad \sin\phi \cos\,[(\omega_n \sqrt{1-\zeta^2})t_p + \phi] - \cos\phi \sin\,[(\omega_n \sqrt{1-\zeta^2})t_p + \phi] = 0$$

$$\Rightarrow \quad \sin\,[(\omega_n \sqrt{1-\zeta^2})t_p + \phi - \phi] = 0 = \sin N\pi \tag{7.34}$$

For the first peak, $N = 1$. Therefore, from Eq. (7.34),

$$t_p = \frac{\pi}{\omega_n \sqrt{1-\zeta^2}} \tag{7.35}$$

Maximum overshoot. Since it is the overshoot value, at the peak time,

$$M_p = c(t)\big|_{t=t_p} - 1 = -\frac{e^{\zeta\omega_n t_p}}{\sqrt{1-\zeta^2}} \sin\left[(\omega_n\sqrt{1-\zeta^2})t_p + \phi\right] \tag{7.36}$$

Substituting t_p value from Eq. (7.35) in Eq. (7.36), after a little algebraic manipulation, we get

$$M_p = e^{-\frac{\pi\zeta}{\sqrt{1-\zeta^2}}} \tag{7.37}$$

The maximum per cent overshoot can be obtained by multiplying the RHS of Eq. (7.37) by 100. It may be observed that M_p is a function of ζ only. Its behaviour with ζ is shown in Figure 7.9.

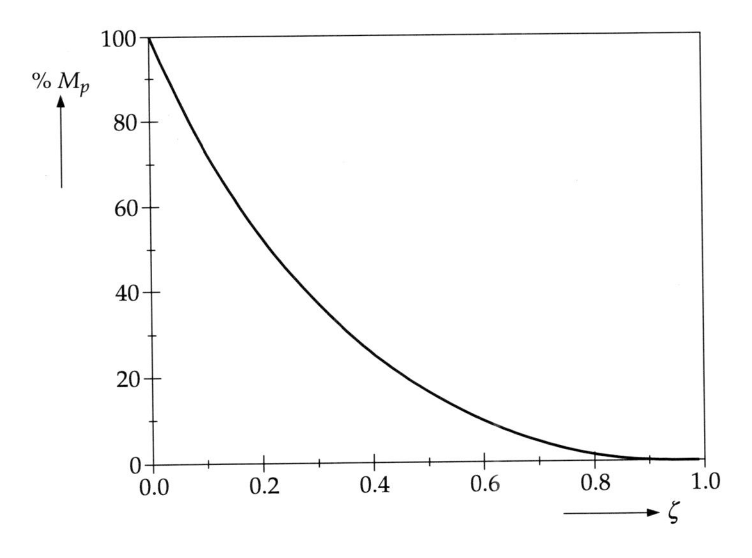

Figure 7.9 ζ vs. $\%M_p$ plot.

Settling time. If t_s is the time for settling within 2% of the final value, then

$$t_s = \frac{4}{\zeta\omega_n} \tag{7.38}$$

This approximate relation can be arrived at from the following considerations. Suppose we want that the output settles within a tolerance band of $\pm\Delta$. That means the amplitude of the oscillating sine factor should equal Δ. Therefore,

$$\frac{e^{\zeta\omega_n t_p}}{\sqrt{1-\zeta^2}} = \Delta \tag{7.39}$$

Taking the logarithm of both sides of Eq. (7.39) and rearranging, we get

$$-\zeta\omega_n t_s = \ln(\Delta\sqrt{1-\zeta^2}) = \ln\Delta - \frac{\zeta^2}{2}$$

or

$$t_s = \frac{\zeta}{2\omega_n} - \frac{\ln\Delta}{\zeta\omega_n} \tag{7.40}$$

Since $\zeta < 1$, the second term on the RHS of Eq. (7.40) dominates and so we may write it as

$$t_s \approx -\frac{\ln\Delta}{\zeta\omega_n} \tag{7.41}$$

Substituting $\Delta = 0.02$ in Eq. (7.41), we get the approximate relation of Eq. (7.38).

Steady-state error. From the definition of error,

$$e(t) = r(t) - c(t) = \frac{e^{\zeta\omega_n t_p}}{\sqrt{1-\zeta^2}} \sin(\omega_d t + \phi) \tag{7.42}$$

Therefore,

$$e_{\text{ss}} = e(t)\big|_{t\to\infty} = 0 \tag{7.43}$$

With this background on the response of systems, we will work out a few problems and then discuss how steady-state error can be calculated without having recourse to calculation of actual response as a function of time.

EXAMPLE 7.5. A measurement system with input $x(t)$ and output $y(t)$ is described by the differential equation

$$3\frac{dy}{dx} + 5y = 8x$$

The static sensitivity of the system is

(a) 0.60 (b) 1.60 (c) 1.67 (d) 2.67

Solution The given equation can be written as

$$\frac{3}{5}\frac{dy}{dx} + y = \frac{8}{5}x$$

Comparing this equation with Eq. (7.2) at page 226, the static sensitivity is found to be $\frac{8}{5} = 1.6$.

Ans: (b).

EXAMPLE 7.6. A unity feedback system has an open-loop transfer function $G(s) = K/[s(s+3)]$. The value of K that yields a damping ratio of 0.5 for the closed-loop system is

(a) 1 (b) 3 (c) 5 (d) 9

Solution The characteristic equation is given by

$$s(s+3)+K=0$$

So, by comparison with a standard second order system, we have

$$\omega_n^2 = K$$

$$\Rightarrow \qquad \omega_n = \sqrt{K}$$

$$2\zeta\omega_n = 3$$

$$\Rightarrow \qquad \zeta = \frac{3}{2\omega_n} = \frac{3}{2\sqrt{K}} = 0.5 \quad \text{(given)}$$

or $$\sqrt{K} = 3$$

$$\Rightarrow \qquad K = 9$$

Ans: (d).

EXAMPLE 7.7. If a first order system and its time response to a unit step input are as shown in the following figure, then the gain K is

(a) 0.25 (b) 0.8 (c) 1 (d) 4

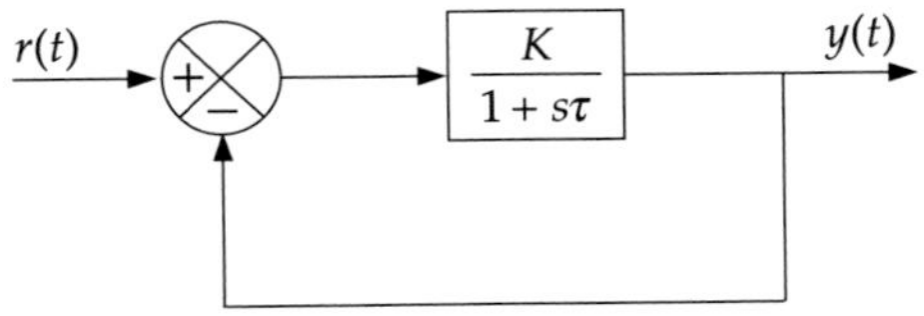

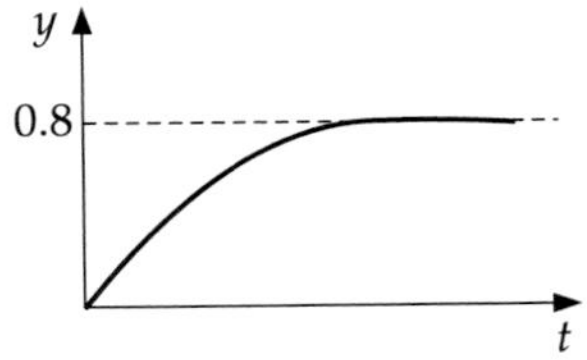

Solution Let us calculate the response as follows:

$$\frac{Y(s)}{R(s)} = \frac{K}{s\tau + K + 1}$$

$$\Rightarrow \qquad Y(s) = \frac{1}{s} \cdot \frac{K}{s\tau + K + 1} \qquad \left[\because R(s) = \frac{1}{s}\right]$$

Therefore, $$y_{ss} = \lim_{s\to 0} sY(s) = \lim_{s \to 0} s \cdot \frac{1}{s} \cdot \frac{K}{s\tau + K + 1} = \frac{K}{K+1} \tag{i}$$

From Eq. (i) and the response curve, we get

$$\frac{K}{K+1} = 0.8$$

or $$1 + \frac{1}{K} = \frac{1}{0.8}$$

or $$K = \frac{0.8}{0.2} = 4$$

Ans: (d).

EXAMPLE 7.8. A unity feedback system has open-loop transfer function $G(s) = 100/[s(s+p)]$. The time at which the response to a unit step input reaches its peak is $\pi/8$ seconds.

(i) The damping coefficient for the closed-loop system is

(a) 0.4 (b) 0.6 (c) 0.8 (d) 1

(ii) The value of p is

(a) 6 (b) 12 (c) 14 (d) 16

Solution Since $H(s) = 1$, the closed-loop transfer function is given by

$$\frac{C(s)}{R(s)} = \frac{100}{s^2 + ps + 100}$$

Therefore, we have

$$\omega_n^2 = 100 \quad \Rightarrow \quad \omega_n = 10$$

$$2\zeta\omega_n = p \quad \Rightarrow \quad \zeta = \frac{p}{20}$$

(i) Now,

$$t_p = \frac{\pi}{\omega_n\sqrt{1-\zeta^2}} = \frac{\pi}{8} \quad \text{(given)}$$

$$\Rightarrow \quad \omega_n\sqrt{1-\zeta^2} = 8$$

$$\Rightarrow \quad \sqrt{1-\zeta^2} = \frac{8}{10} = 0.8$$

$$\Rightarrow \quad \zeta^2 = 1 - 0.64 = 0.36$$

Therefore, $\zeta = 0.6$

Ans: (b).

(ii) $p = 2\zeta\omega_n = 2 \times 0.6 \times 10 = 12$

Ans: (b).

EXAMPLE 7.9. The unit step response of a unity feedback system with open-loop transfer function $G(s) = K/[(s+1)(s+2)]$ is shown in the figure.

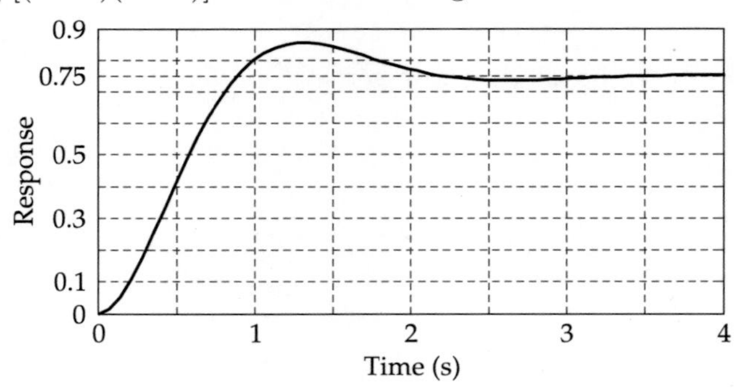

The value of K is

(a) 0.5 (b) 2 (c) 4 (d) 6

Solution From the given forward transfer function and unity feedback, the closed-loop transfer function is obtained as

$$\frac{C(s)}{R(s)} = \frac{K}{(s+1)(s+2)+K} = \frac{K}{s^2+3s+2+K}$$

or $$C(s) = \frac{K}{s^2+3s+2+K} \cdot \frac{1}{s} \qquad \left[\because \ R(s) = \frac{1}{s}\right]$$

We get from the given diagram $c_{ss} = 0.75$. So,

$$\lim_{s\to 0} s \cdot C(s) = \lim_{s\to 0} \frac{K}{s^2+3s+2+K} = 0.75$$

$$\Rightarrow \qquad \frac{K}{2+K} = \frac{3}{4}$$

$$\Rightarrow \qquad K = 6$$

Ans: (d).

EXAMPLE 7.10. For a unity feedback system with open-loop gain of $G(s) = \dfrac{9}{s(s+3)}$, the closed-loop system will have a damping ratio of

(a) 0.33 (b) 0.5 (c) 1.0 (d) 3

Solution The closed-loop transfer function is given by

$$\frac{C(s)}{R(s)} = \frac{\dfrac{9}{s(s+3)}}{1+\dfrac{9}{s(s+3)}} = \frac{9}{s^2+3s+9}$$

Comparing the closed-loop transfer function to that of a standard second order system, we have

$$\omega_n^2 = 9 \qquad \Rightarrow \qquad \omega_n = 3$$

$$2\zeta\omega_n = 3 \qquad \Rightarrow \qquad \zeta = \frac{3}{2\omega_n} = \frac{3}{2 \times 3} = 0.5$$

Ans: (b).

EXAMPLE 7.11. The transfer function of a control system is given by

$$\frac{C(s)}{R(s)} = \frac{25}{s^2+6s+25}.$$

The first maximum value of the response occurs at a time $t_{\max}$ given by

(a) $\dfrac{\pi}{8}$ (b) $\dfrac{\pi}{4}$ (c) $\dfrac{\pi}{2}$ (d) π

Solution Comparing the closed-loop transfer function to that of a standard second order system, we have

$$\omega_n^2 = 25 \quad \Rightarrow \quad \omega_n = 5$$

$$2\zeta\omega_n = 6 \quad \Rightarrow \quad \zeta = \frac{6}{2\omega_n} = \frac{6}{2 \times 5} = 0.6$$

Therefore,

$$t_{\max} = \frac{\pi}{5\sqrt{1-0.6^2}} = \frac{\pi}{4}$$

Ans: (b).

EXAMPLE 7.12. The impulse response of a system S_1 is given by $x_1(t) = 4e^{-2t}$. The step response of a system S_2 is given by $x_2(t) = 2\big(1 - e^{-3t}\big)$.

(a) Find the transfer functions of the systems S_1 and S_2.

(b) The two systems are cascaded together without any interaction. Find the unit step response of the cascaded system.

(c) Sketch the unit ramp response of the cascaded system.

Solution (a) Taking the Laplace transform of the response of S_1, we get

$$X_1(s) = \frac{4}{s+2}$$

Therefore,

$$G_1(s) = \frac{4}{s+2} \qquad [\because\ Y_1(s) = 1]$$

Taking the Laplace transform of the response of S_2, we get

$$X_2(s) = 2\left(\frac{1}{s} - \frac{1}{s+3}\right) = \frac{6}{s(s+3)}$$

$$Y_2(s) = \frac{1}{s}$$

Thus,

$$G_2(s) = \frac{6}{s(s+3)} \cdot s = \frac{6}{s+3}$$

(b) The transfer function of the cascaded system is

$$G(s) = G_1(s)G_2(s) = \frac{24}{(s+1)(s+3)}$$

The Laplace transform of unit ramp is $R(s) = \dfrac{1}{s^2}$. Therefore,

$$C(s) = \frac{24}{(s+1)(s+3)} \cdot \frac{1}{s^2}$$

$$\equiv \frac{A}{s^2} + \frac{B}{s} + \frac{C}{s+2} + \frac{D}{s+3}$$

$$A = \left.\frac{24}{(s+1)(s+3)}\right|_{s=0} = 4$$

$$B = \frac{d}{ds}\left[s^2C(s)\right]_{s=0}$$

$$= \frac{d}{ds}\left[\frac{24}{(s+1)(s+3)}\right] = -\left.\frac{24(2s+5)}{(s+2)^2(s+3)^2}\right|_{s=0}$$

$$= -\frac{10}{3}$$

$$C = \left.\frac{24}{s^2(s+3)}\right|_{s=-2} = 6$$

$$D = \left.\frac{24}{s^2(s+2)}\right|_{s=-3} = -\frac{8}{3}$$

$$C(s) = \frac{4}{s^2} - \frac{10}{3}s + \frac{6}{s+2} - \frac{8}{3}e^{-3t}$$

Therefore, $$c(t) = 4t - \frac{10}{3}u(t) + 6e^{-2t} - \frac{8}{3}e^{-3t}$$

(c) The plot of the response is shown in the following figure.

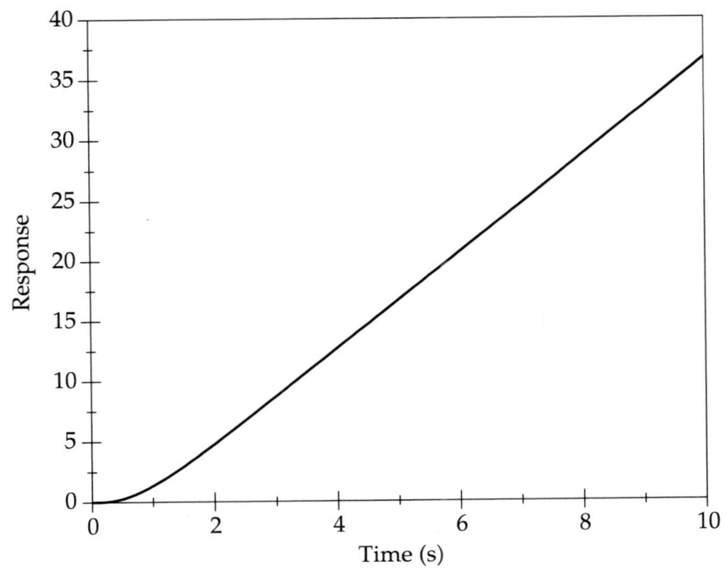

EXAMPLE 7.13. For a unity feedback system having

$$G(s) = \frac{25}{s(s+10)}$$

find (a) ω_n; (b) ζ ; (c) ω_d; (d) t_p; and (e) M_p.

Solution Since it is a unity feedback system, its overall transfer function is given by

$$T(s) = \frac{G(s)}{1+G(s)} = \frac{25}{s(s+10)+25}$$

$$= \frac{25}{s^2+10s+25} \equiv \frac{\omega_n^2}{s^2+2\zeta\omega_n s+\omega_n^2}$$

Comparing the denominator of the given transfer function with that of the standard second order system, we get

(a) $\omega_n^2 = 25$. Therefore, $\omega_n = 5$ rad/s

(b) $2\zeta\omega_n = 10$. Therefore, $\zeta = \dfrac{5}{\omega_n} = 1$ (critically damped)

(c) $\omega_d = \omega_n\sqrt{1-\zeta^2} = 0$

(d) $t_p = \dfrac{\pi}{\omega_n\sqrt{1-\zeta^2}} \to \infty$ (no peak)

(e) $M_p = 0$ since it has no peak

EXAMPLE 7.14. The open-loop transfer function of a unity feedback system is given by

$$G(s) = \frac{K}{s(1+sT)}$$

where T and K are positive constants. By what factor should the gain K be reduced such that

(a) the peak overshoot of the system to unit step input is reduced from 75% to 25%, and

(b) the damping ratio increases from 0.1 to 0.6 ?

Solution The overall transfer function of the system is given by

$$T(s) = \frac{G(s)}{1+G(s)} = \frac{K}{s(1+sT)+K}$$

$$= \frac{K/T}{s^2+\dfrac{s}{T}+\dfrac{K}{T}} \equiv \frac{\omega_n^2}{s^2+2\zeta\omega_n s+\omega_n^2}$$

By comparison, we get $\omega_n = \sqrt{K/T}$, $\quad \zeta = \dfrac{1}{2\sqrt{KT}}$.

(a) From Eq. (7.37),

$$(M_p)_1 = 100e^{-\frac{\pi\zeta_1}{\sqrt{1-\zeta_1^2}}} = 75 \qquad \text{(i)}$$

$$(M_p)_2 = 100e^{-\frac{\pi\zeta_2}{\sqrt{1-\zeta_2^2}}} = 25 \qquad \text{(ii)}$$

On taking logarithms of both the sides, we get

$$-\frac{\pi\zeta_1}{\sqrt{1-\zeta_1^2}} = \ln\frac{75}{100} = \ln 3 - \ln 4 \qquad \text{[from Eq. (i)]}$$

or
$$\frac{\pi\zeta_1}{\sqrt{1-\zeta_1^2}} = \frac{\ln 4 - \ln 3}{\pi} = 0.0916 \qquad \text{(iii)}$$

$$\frac{\pi\zeta_2}{\sqrt{1-\zeta_2^2}} = \frac{\ln 4 - \ln 1}{\pi} = 0.4413 \qquad \text{[from Eq. (ii)]} \qquad \text{(iv)}$$

Substituting $\zeta_1 = \dfrac{1}{2\sqrt{K_1T}}$ and $\zeta_2 = \dfrac{1}{2\sqrt{K_2T}}$, from Eqs. (iii) and (iv), we obtain

$$\frac{1}{\sqrt{4K_1T-1}} = 0.0916$$

or
$$K_1T = 30.045 \qquad \text{(v)}$$

$$\frac{1}{\sqrt{4K_2T-1}} = 0.4413$$

or
$$K_2T = 1.534 \qquad \text{(vi)}$$

Thus, from Eqs. (v) and (vi), we have

$$\frac{K_1}{K_2} = \frac{30.045}{1.534} \qquad \text{(vii)}$$

Equation (vii) suggests that K should be reduced nearly 20 times to lower the peak overshoot from 75% to 25%.

(b) Here,

$$\frac{\zeta_1}{\zeta_2} = \sqrt{\frac{K_2T}{K_1T}}$$

or
$$\frac{K_1}{K_2} = \frac{\zeta_2^2}{\zeta_1^2} = \frac{(0.6)^2}{(0.1)^2} = 36 \qquad \text{(viii)}$$

The import of Eq. (viii) is that K has to decrease by 1/36 of its present value to increase the damping ratio from 0.1 to 0.6.

EXAMPLE 7.15. Find the time response of the following system for input $r(t) = \sin t$:

$$T(s) = \frac{1}{(s+1)(s^2+1)}$$

Solution

$$r(t) = \sin t$$

$$\Rightarrow \qquad R(s) = \frac{1}{s^2+1}$$

Therefore,
$$C(s) = \frac{1}{(s+1)(s^2+1)^2} \equiv \frac{A}{s+1} + \frac{Bs+C}{(s^2+1)^2} + \frac{Ds+E}{s^2+1}$$

Equating the numerators of both the sides, we get from comparing the coefficients of s^6, s^5, s^4, s^3, s^2, s and the constants

$$A = \frac{1}{4}, \quad B = -\frac{1}{2}, \quad C = \frac{1}{2}, \quad D = -\frac{1}{4} \text{ and } E = \frac{1}{4}$$

Thus,

$$\begin{aligned} C(s) &= \frac{1}{4}\cdot\frac{1}{s+1} - \frac{1}{2}\cdot\frac{s-1}{(s^2+1)^2} - \frac{1}{4}\cdot\frac{s-1}{s^2+1} \\ &= \frac{1}{4}\cdot\frac{1}{s+1} - \frac{1}{2}\cdot\frac{s-1}{(s^2+1)^2} + \frac{1}{2}\cdot\frac{1}{(s^2+1)^2} - \frac{1}{4}\cdot\frac{s}{s^2+1} + \frac{1}{4}\cdot\frac{1}{s^2+1} \end{aligned}$$

On performing inverse Laplace transform, we get

$$\begin{aligned} c(t) &= \frac{1}{4}e^{-t} - \frac{1}{2}\frac{d}{dt}(t\sin t) + \frac{1}{2}t\sin t - \frac{1}{4}\cos t + \frac{1}{4}\sin t \\ &= \frac{1}{4}e^{-t} - \frac{1}{2}\sin t - \frac{1}{2}t\cos t + \frac{1}{2}t\sin t - \frac{1}{4}\cos t + \frac{1}{4}\sin t \\ &= \frac{1}{4}e^{-t} - \frac{1}{2\sqrt{2}}\sin\left(t + \frac{\pi}{4}\right) + \frac{1}{\sqrt{2}}t\sin\left(t - \frac{\pi}{4}\right) \end{aligned}$$

EXAMPLE 7.16. A step input is applied to a unity feedback system having the following open-loop transfer function:

$$G(s) = \frac{15}{s(s+3)}$$

Find (a) the closed-loop transfer function; (b) ω_n; (c) ζ; and (d) ω_d.

Solution (a) The closed-loop transfer function is

$$\frac{G(s)}{1+G(s)} = \frac{15}{s(s+3)+15} = \frac{15}{s^2+3s+15}$$

(b) Comparing the denominator of closed-loop transfer function with that of a standard second order transfer function, we get

$$\omega_n^2 = 15$$

or

$$\omega_n = \sqrt{15} = 3.873 \text{ rad/s}$$

(c) From the same comparison, we also have

$$2\zeta\omega_n = 3$$

or

$$\zeta = \frac{3}{2\omega_n} = \frac{3}{2\sqrt{15}} = 0.387$$

(d) $\omega_d = \omega_n\sqrt{1-\zeta^2} = 3.571$ rad/s

EXAMPLE 7.17. A second order control system, having $\zeta = 0.4$ and $\omega_n = 5$ rad/s, is subjected to a step input. Determine (a) closed-loop transfer function; (b) t_r; (c) t_p; (d) t_s; and (e) M_p.

Solution (a) The standard closed-loop transfer function for a second order control system is given by

$$\frac{C(s)}{R(s)} = \frac{\omega_n^2}{s^2 + 2\zeta\omega_n s + \omega_n^2}$$

Substituting the values of ζ and ω_n, we get

$$\frac{C(s)}{R(s)} = \frac{25}{s^2 + 4s + 25}$$

(b) $t_r = \dfrac{\pi - \cos^{-1}\zeta}{\omega_n\sqrt{1-\zeta^2}} = \dfrac{\pi - \cos^{-1}0.4}{5\sqrt{1-0.16}} = 0.433$ s

(c) $t_p = \dfrac{\pi}{\omega_n\sqrt{1-\zeta^2}} = \dfrac{\pi}{5\sqrt{1-0.16}} = 0.686$ s

(d) $t_s = \dfrac{4}{\zeta\omega_n} = \dfrac{4}{(0.4)(0.5)} = 2$ s

(e) $M_p = 100\exp\left(-\dfrac{\pi\zeta}{\sqrt{1-\zeta^2}}\right) = 100\exp\left(-\dfrac{0.4\pi}{\sqrt{1-0.16}}\right) = 25.4\%$

EXAMPLE 7.18. A unity feedback control system has the following open-loop transfer function:

$$G(s) = \frac{4s+1}{4s^2}$$

Find expressions for its time response when it is subjected to (a) unit impulse input and (b) unit step input.

Solution The closed-loop transfer function of the system is given by

$$T(s) = \frac{C(s)}{R(s)} = \frac{4s+1}{4s^2+4s+1} = \frac{4s+1}{(2s+1)^2}$$

(a) For a unit impulse input, $R(s) = 1$.

Therefore,
$$C(s) = \frac{4s+1}{(2s+1)^2} = \frac{s+\frac{1}{4}}{\left[s+\frac{1}{2}\right]^2} \equiv \frac{A}{\left[s+\frac{1}{2}\right]^2} + \frac{B}{s+\frac{1}{2}}$$

Then,
$$A = \left(s+\frac{1}{2}\right)^2 C(s)\bigg|_{s=-1/2} = \left(s+\frac{1}{2}\right)_{s=-1/2} = -\frac{1}{4}$$

$$B = \frac{d}{ds}\left\{\left(s+\frac{1}{2}\right)^2 C(s)\right\}_{s=-1/2} = 1$$

Thus,
$$C(s) = -\frac{1}{4\left[s+\frac{1}{2}\right]^2} + \frac{1}{s+\frac{1}{2}}$$

The inverse Laplace transform of the above equation yields

$$c(t) = e^{-t/2} - \frac{1}{4}te^{-t/2} = \left[1 - \frac{t}{4}\right]e^{-t/2}$$

(b) For a unit step input, $R(s) = \dfrac{1}{s}$. Therefore,

$$C(s) = \frac{s+\frac{1}{4}}{s\left[s+\frac{1}{2}\right]^2} \equiv \frac{A}{s} + \frac{B}{\left[s+\frac{1}{2}\right]^2} + \frac{C}{s+\frac{1}{2}}$$

$$A = sC(s)|_{s=0} = \frac{s+\frac{1}{4}}{\left[s+\frac{1}{2}\right]^2}\Bigg|_{s=0} = 1$$

$$B = \left(s+\frac{1}{2}\right)^2 C(s)\bigg|_{s=-1/2} = \frac{s+\frac{1}{4}}{s}\bigg|_{s=-1/2} = \frac{1}{2}$$

$$C = \frac{d}{ds}\left\{\left(s+\frac{1}{2}\right)^2 C(s)\right\}\bigg|_{s=-1/2} = \frac{d}{ds}\left\{\frac{s+\frac{1}{4}}{s}\right\}\bigg|_{s=-1/2}$$

$$= \left[\frac{1}{s} - \frac{s+\frac{1}{4}}{s^2}\right]_{s=-1/2} = -1$$

Thus,
$$C(s) = \frac{1}{s} + \frac{1}{2\left[s+\frac{1}{2}\right]^2} - \frac{1}{s+\frac{1}{2}}$$

Its inverse Laplace transform yields

$$c(t) = 1 + \frac{1}{2}te^{-t/2} - e^{-t/2} = 1 - \left(1 - \frac{t}{2}\right)e^{-t/2}$$

EXAMPLE 7.19. Determine the transfer function of the control system given in Figure 7.10 and derive an expression for its output if the input is a step having a magnitude of 2 units.

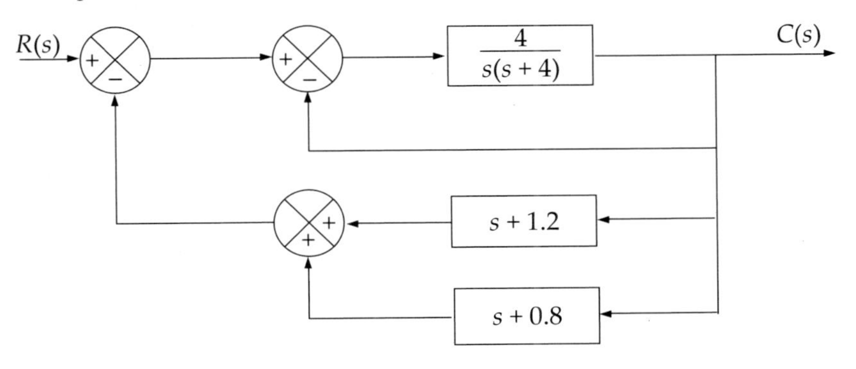

Figure 7.10 Block diagram of the control system (Example 7.19).

Solution The block diagram can be reduced as shown in Figure 7.10(a).

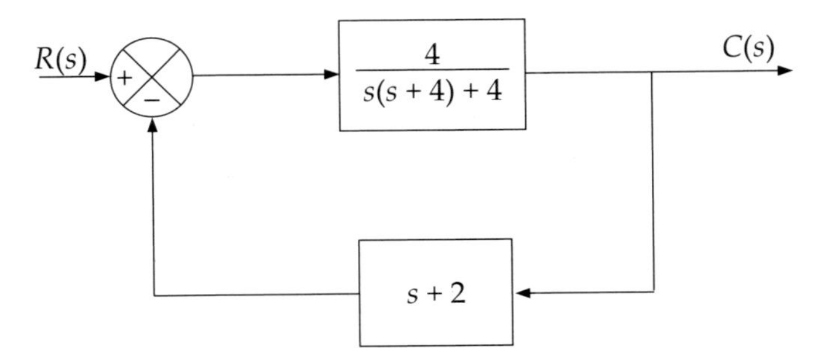

Figure 7.10(a) Reduced block diagram.

The reduced block diagram yields the closed-loop transfer function of the system as

$$T(s) = \frac{C(s)}{R(s)} = \frac{4}{s(s+4)+4+4(s+2)}$$

$$= \frac{4}{s^2+8s+12} = \frac{4}{(s+2)(s+6)}$$

Given $R(s) = \dfrac{2}{s}$. Therefore,

$$C(s) = \frac{8}{s(s+2)(s+6)} \equiv \frac{A}{s} + \frac{B}{s+2} + \frac{C}{s+6}$$

Then,

$$A = sC(s)|_{s=0} = \left.\frac{8}{(s+2)(s+6)}\right|_{s=0} = \frac{2}{3}$$

$$B = (s+2)C(s)|_{s=-2} = \left.\frac{8}{s(s+6)}\right|_{s=-2} = -1$$

$$C = (s+6)C(s)|_{s=-6} = \left.\frac{8}{s(s+2)}\right|_{s=-6} = \frac{1}{3}$$

Thus,

$$C(s) = \frac{2}{3}\cdot\frac{1}{s} - \frac{1}{s+2} + \frac{1}{3}\cdot\frac{1}{s+6}$$

Its inverse Laplace transform yields

$$c(t) = \frac{2}{3} - e^{-2t} + \frac{1}{3}e^{-6t}$$

EXAMPLE 7.20. The measurement conducted on a servomechanism shows the error response to be

$$e(t) = 1.66\ e^{-8t}\ \sin\ (6t + 37°)$$

where the input is a sudden unit displacement. Determine the natural frequency, damping ratio, and damped angular frequency of the system.

Solution Given $r(t) = 1$. Then,

$$e(t) = r(t) - c(t) = 1 - c(t)$$

Since the error is sinusoidal, it is a second order underdamped system where, for a step input,

$$c(t) = 1 - \frac{e^{-\zeta\omega_n t}}{\sqrt{1-\zeta^2}}\ \sin\left(\omega_d t + \tan^{-1}\frac{\sqrt{1-\zeta^2}}{\zeta}\right)$$

Thus,

$$e(t) = \frac{e^{-\zeta\omega_n t}}{\sqrt{1-\zeta^2}}\ \sin\left(\omega_d t + \tan^{-1}\frac{\sqrt{1-\zeta^2}}{\zeta}\right) = 1.66\ e^{-8t}\ \sin\ (6t + 37°) \quad \text{(i)}$$

Comparing the two sides of Eq. (i), we get

$$\tan^{-1}\frac{\sqrt{1-\zeta^2}}{\zeta} = 37°$$

$$\Rightarrow \qquad \frac{\sqrt{1-\zeta^2}}{\zeta} = \tan 37° = 0.7536$$

$$\Rightarrow \qquad \zeta = 0.7986$$

$$\omega_d = 6 \text{ rad/s}$$

$$\Rightarrow \qquad \omega_n = \frac{6}{\sqrt{1-\zeta^2}} = 9.969 \text{ rad/s}$$

EXAMPLE 7.21. In the block diagram shown in Figure 7.11,

$$G(s) = A/s^2, \qquad H(s) = ms + n$$

For $A = 10$, determine the values of m and n for a step input with a time constant 0.1 s which gives a peak over-shoot of 30%.

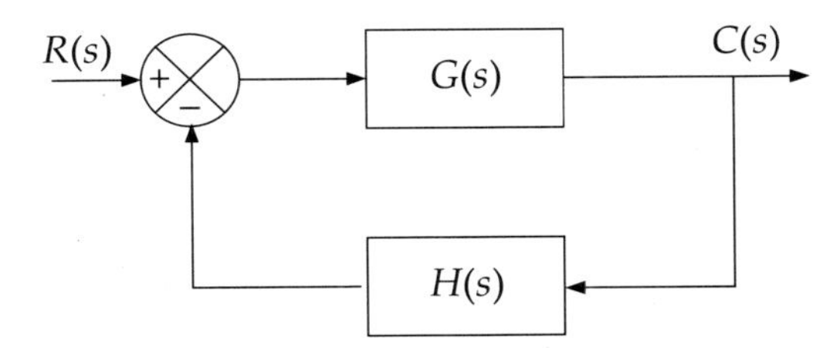

Figure 7.11 Block diagram (Example 7.21).

Solution The overall transfer function of the system is given by

$$T(s) = \frac{C(s)}{R(s)} = \frac{G(s)}{1+G(s)} = \frac{A}{s^2 + Ams + An} \equiv \frac{\omega_n^2}{s^2 + 2\zeta\omega_n s + \omega_n^2}$$

Comparing the denominators of the transfer function of the system and the denominator of the standard second order transfer function, we get

$$2\zeta\omega_n = Am = 10m \tag{i}$$

$$\omega_n^2 = An = 10n$$

From Eq. (i),

$$\zeta = \frac{10m}{2\omega_n} = \frac{10m}{2\sqrt{10n}} \tag{ii}$$

Now, $\qquad$ Time constant $\dfrac{1}{\zeta\omega_n} = \dfrac{2}{10m} = \dfrac{1}{5m} = 0.1$ s (given)

Therefore, $\qquad m = 2$

Substituting this value in Eq. (ii), we get

$$\zeta = \frac{10 \times 2}{2\sqrt{10m}} = \frac{10}{\sqrt{10n}} \tag{iii}$$

Again,

$$\text{Peak overshoot, } \%M_p = 100 \exp\left(-\frac{\pi\zeta}{\sqrt{1-\zeta^2}}\right) = 30 \text{ (given)}$$

$$\Rightarrow \qquad -\frac{\pi\zeta}{\sqrt{1-\zeta^2}} = \ln\,(0.3) = -1.204$$

$$\Rightarrow \qquad \pi^2\zeta^2 = (1.204)^2(1-\zeta^2)$$

$$\Rightarrow \qquad \zeta = \sqrt{\frac{(1.204)^2}{\pi^2 + (1.204)^2}} = 0.3579$$

Substituting the value of ζ in Eq. (iii), we get

$$\sqrt{\frac{10}{n}} = 0.3579$$

Thus,

$$n = \frac{10}{(0.3579)^2} = 78.07$$

EXAMPLE 7.22. The system and its response curve are given in Figures 7.12(a) and 7.12(b). Determine the values of M, B and K.

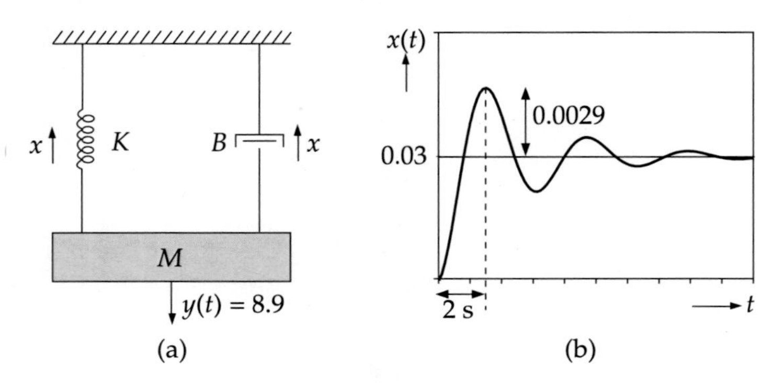

Figure 7.12 Example 7.22: (a) the system; (b) its response curve.

Solution The differential equation for the system can be written as

$$M\frac{d^2x(t)}{dt^2} + B\frac{dx(t)}{dt} + Kx(t) = y(t)$$

Its Laplace transform, for all initial conditions = 0, is

$$(s^2M + sB + K)X(s) = Y(s)$$

Therefore, the transfer function of the system is

$$T(s) = \frac{X(s)}{Y(s)} = \frac{1}{s^2M + sB + K} = \frac{1/M}{s^2 + s\dfrac{B}{M} + \dfrac{K}{M}} \qquad \text{(i)}$$

This yields the characteristic equation

$$s^2 + s\frac{B}{M} + \frac{K}{M} = 0 \equiv s^2 + 2\zeta\omega_n s + \omega_n^2$$

By comparison, we get

$$\omega_n = \sqrt{K/M}$$

$$2\zeta\omega_n = \frac{B}{M}$$

Therefore,

$$\zeta = \frac{B}{2\omega_n M} = \frac{B}{2\sqrt{KM}} \qquad \text{(ii)}$$

Now, considering the peak overshoot,

$$M_p = \exp\left(-\frac{\pi\zeta}{\sqrt{1-\zeta^2}}\right) = \frac{0.0029}{0.03} \quad \text{[from the response curve]}$$

$$\Rightarrow \qquad \frac{\pi\zeta}{\sqrt{1-\zeta^2}} = -\ln 0.0967 = 2.3365$$

$$\Rightarrow \qquad \zeta = \frac{2.3365}{\sqrt{\pi^2 + 2.3365^2}} = 0.597 \qquad \text{(iii)}$$

From Figure 7.12 (a) and Eq. (i),

$$X(s) = \frac{8.9}{s(Ms^2 + Bs + K)}$$

So,

$$x(\infty) = 0.03 = \lim_{s\to 0} sX(s) = \lim_{s\to 0} \frac{8.9}{Ms^2 + Bs + K} = \frac{8.9}{K}$$

Therefore,

$$K = \frac{8.9}{0.03} = 296.67 \text{ N/m} \qquad \text{(iv)}$$

Again,

$$t_p = \frac{\pi}{\omega_n\sqrt{1-\zeta^2}} = 2 \text{ (from response)}$$

$$\Rightarrow \qquad \omega_n = \frac{\pi}{2\sqrt{1-\zeta^2}} = \frac{\pi}{2\sqrt{1-0.597^2}} = 1.958 \text{ rad/s} \qquad \text{(v)}$$

Combining the results of Eqs. (iv) and (v), we get

$$\omega_n = \sqrt{\frac{K}{M}} = 1.958$$

$$\Rightarrow \qquad M = \frac{K}{\omega_n^2}$$

$$= \frac{296.67}{(1.958)^2} = 77.383 \text{ kg} \tag{vi}$$

And Eq. (ii), together with Eq. (vi), yields

$$\zeta = \frac{B}{2\sqrt{KM}}$$

or

$$B = 2\zeta\sqrt{KM} = 2(0.597)\sqrt{(296.67)(77.383)} = 180.91 \text{ N-s/m}$$

EXAMPLE 7.23. In the closed-loop system shown in Figure 7.13, determine the value of K and a such that the per cent peak overshoot due to the unit step input does not exceed 10. The settling time should be equal to 1 s.

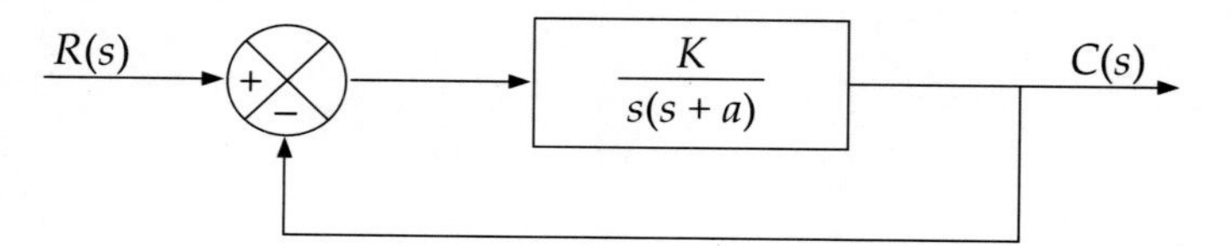

Figure 7.13 Closed-loop system (Example 7.23).

Solution The overall transfer function of the system is

$$T(s) = \frac{C(s)}{R(s)} = \frac{K}{s(s+a)+K} = \frac{K}{s^2+as+K}$$

Comparing the characteristic equation $s^2 + as + K = 0$ with that of a standard second order system, we get

$$\omega_n = \sqrt{K}, \qquad \zeta = \frac{a}{2\omega_n} = \frac{a}{2\sqrt{K}} \tag{i}$$

Now,

$$t_s = \frac{4}{\zeta\omega_n} = 1 \text{ s (given)}$$

Therefore,

$$\zeta\omega_n = 4 = \frac{a}{2}$$

or

$$a = 8 \tag{ii}$$

The peak overshoot condition gives

$$\exp\left(-\frac{\pi\zeta}{\sqrt{1-\zeta^2}}\right) = 0.1$$

$$\Rightarrow \qquad \frac{\pi\zeta}{\sqrt{1-\zeta^2}} = 2.303 \qquad \text{(iii)}$$

Substituting the values of ζ from Eq. (i) in Eq. (iii), we obtain

$$\frac{\frac{\pi a}{2\sqrt{K}}}{\sqrt{1-\frac{a^2}{4K}}} = 2.303$$

$$\Rightarrow \qquad \frac{\pi a}{\sqrt{4K-a^2}} = 2.303$$

Putting the value of a from Eq. (ii) in this equation, we have

$$\frac{\pi(8)}{\sqrt{4K-64}} = 2.303$$

or

$$\frac{64\pi^2}{2.303^2} = 4K - 64$$

or

$$K = \frac{1}{4}\left[\frac{64\pi^2}{2.303^2} + 64\right] = 45.774$$

EXAMPLE 7.24. Determine the value of controller gains K_p and K_1 for the system shown in Figure 7.14 so that its damping ratio and natural frequency of oscillation become 1 and 4 Hz, respectively.

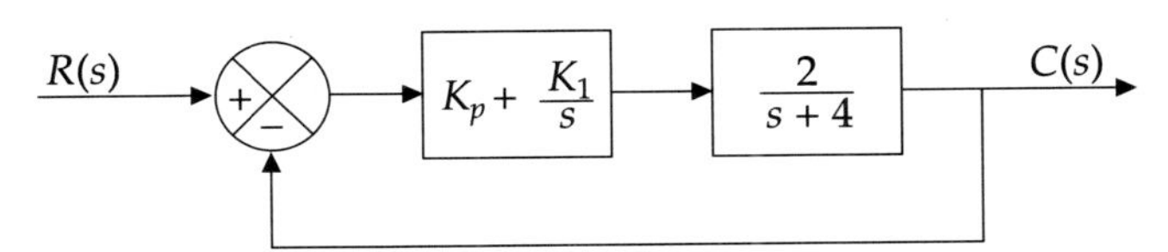

Figure 7.14 Control system (Example 7.24).

Solution Here,

$$G(s) = \frac{2(K_1 + K_p s)}{s(s+4)}$$

$$T(s) = \frac{G(s)}{1+G(s)} = \frac{2(K_1 + K_p s)}{s(s+4) + 2(K_1 + K_p s)} = \frac{2(K_1 + K_p s)}{s^2(2K_p + 4s + 2K_1)}$$

Comparing the characteristic equation $s^2(2K_p + 4s + 2K_1) = 0$ with that of a standard second order system, we get

$$\omega_n^2 = 2K_1$$

Given $\nu_n = 4$ Hz. Then,

$$\omega_n = 2\pi\nu_n = 8\pi$$

$$\Rightarrow \qquad K_1 = \frac{\omega_n^2}{2} = \frac{64\pi^2}{2} = 315.83$$

and

$$2\zeta\omega_n = 2K_p + 4$$

or

$$K_p = \frac{2\zeta\omega_n - 4}{2} = \zeta\omega_n - 2$$

$$= 8\pi - 2 \qquad [\because \zeta = 1 \text{ (given)}]$$

$$= 23.13$$

Performance parameters from location of poles. Suppose a pair of dominant poles of a system are given by $p_1 = -a + \jmath b$ and $p_2 = -a - \jmath b$ as shown in Figure 7.15.

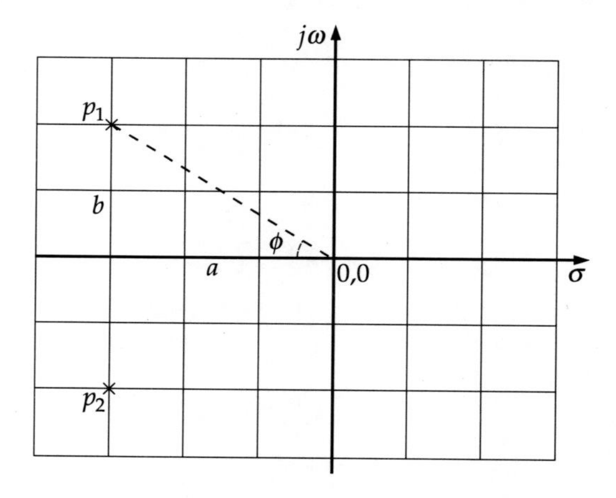

Figure 7.15 Dominant pole locations.

Then, the damping ratio is given by

$$\zeta = \cos\phi = \frac{a}{\sqrt{a^2 + b^2}} \tag{7.44}$$

The natural frequency of oscillation is obtained from the radial distance of the pole from the origin as

$$\omega_n = \sqrt{a^2 + b^2} \tag{7.45}$$

The damped oscillation frequency ω_d and the exponential decay frequency σ_D can be found from Eq. (7.44) as

$$\begin{aligned} \omega_d &\equiv \omega_n\sqrt{1 - \zeta^2} = b \\ \sigma_D &\equiv \zeta\omega_n = a \end{aligned} \tag{7.46}$$

EXAMPLE 7.25. The dominant poles of a servosystem are located at $s = (-2 \pm \jmath 2)$. The damping ratio of the system is

(a) 1 (b) 0.8 (c) 0.707 (d) 0.6

Solution From Eq. (7.44),

$$\zeta = \cos\phi = \frac{2}{\sqrt{2^2 + 2^2}} = \frac{1}{\sqrt{2}} = 0.707$$

Ans: (c).

Alternatively, Let the transfer function be

$$G(s) = \frac{K}{(s + 2 - \jmath 2)(s + 2 + \jmath 2)} = \frac{K}{s^2 + 4s + 8}$$

Then,

$$\omega_n^2 = 8 \quad \Rightarrow \quad \omega_n = \sqrt{8}$$

$$2\zeta\omega_n = 4 \quad \Rightarrow \quad \zeta = \frac{4}{2\sqrt{8}} = 0.707$$

7.4 ANALYSIS OF STEADY-STATE ERROR

We have already defined steady-state error in Eq. (7.1) at page 226 and calculated the same for first and second order systems in time-domain by a method that involves the inverse Laplace transform of responses. But this error can be calculated directly from the Laplace transformed values of the response. We will discuss the procedure in this section.

From Eq. (7.1), we know that

$$e_{ss} = \lim_{t\to\infty} e(t) \tag{7.47}$$

Applying the final value theorem[2] to Eq. (7.47), we get

$$e_{ss}(t) = \lim_{s\to 0} sE(s) \tag{7.48}$$

Consider a canonical feedback system given in Figure 7.16.

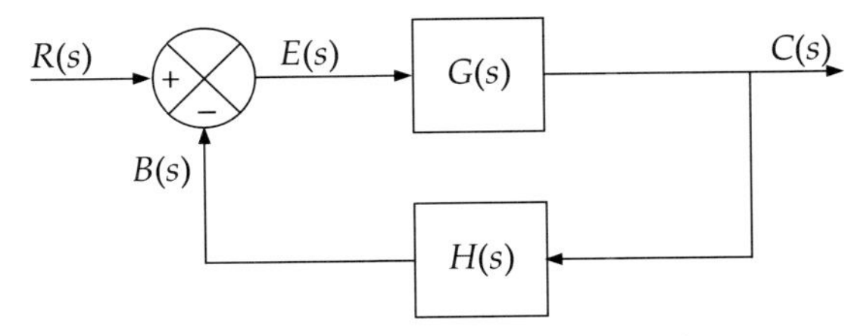

Figure 7.16 Canonical feedback control system.

[2]See Section 1.2 at page 6.

The overall transfer function for this system is given by

$$T(s) = \frac{C(s)}{R(s)} = \frac{G(s)}{1 + G(s)H(s)} \tag{7.49}$$

From Figure 7.16, we get

$$E(s) = R(s) - B(s) = R(s) - C(s)H(s) = R(s) - E(s)G(s)H(s) \tag{7.50}$$

This yields

$$E(s) = \frac{R(s)}{1 + G(s)H(s)} \tag{7.51}$$

Therefore,

$$e_{\text{ss}} = \lim_{s \to 0} sE(s) = \lim_{s \to 0} \frac{sR(s)}{1 + G(s)H(s)} \tag{7.52}$$

From Eq. (7.52), we find that, while the numerator is related to the input, the denominator, which is the characteristic polynomial, relates to the type of the system. Thus, the steady-state error of a system depends on the following:

- Input to the system
- Type of the system

Before we discuss the effects of input and type of system on the steady-state error, let us define what are called *static error coefficients* or *static error constants.*

Static Error Constants

Often called *figures of merit* of a control system, there are three static error constants:

1. Static position error constant, K_p
2. Static velocity error constant, K_v
3. Static acceleration error constant, K_a

These constants are defined in such a way that the higher the value of the constant, the less the value of e_{ss}. Note that each error constant variable is the derivative of its preceding one, i.e., velocity is the derivative of position and acceleration is the derivative of velocity. These originate from their definitions.

Input Types and Error Constants

Step input. Here,

$$R(s) = \frac{1}{s}$$

Therefore,

$$\begin{aligned} e_{\text{ss}} &= \lim_{s \to 0} \left(\frac{s}{1 + G(s)H(s)} \cdot \frac{1}{s} \right) = \lim_{s \to 0} \frac{1}{1 + G(s)H(s)} \\ &= \frac{1}{1 + \lim\limits_{s \to 0} G(s)H(s)} \equiv \frac{1}{1 + K_p} \end{aligned} \tag{7.53}$$

where
$$K_p = \lim_{s \to 0} G(s)H(s) \tag{7.54}$$

K_p is called the *static position error constant.*

Ramp input. Here,

$$R(s) = \frac{1}{s^2}$$

Therefore,

$$\begin{aligned} e_{ss} &= \lim_{s \to 0} \left[\frac{s}{1 + G(s)H(s)} \cdot \frac{1}{s^2} \right] \\ &= \lim_{s \to 0} \frac{1}{s[1 + G(s)H(s)]} \\ &= \frac{1}{\lim\limits_{s \to 0} sG(s)H(s)} \equiv \frac{1}{K_v} \end{aligned} \tag{7.55}$$

where
$$K_v = \lim_{s \to 0} sG(s)H(s) \tag{7.56}$$

K_v is called the *static velocity error constant.*

Note: There is an s in its definition. This indicates a differentiation for once in consonance with the fact that velocity is the first derivative of the position variable.

Parabolic input. Here,

$$R(s) = \frac{1}{s^3}$$

Therefore,

$$\begin{aligned} e_{ss} &= \lim_{s \to 0} \left(\frac{s}{1 + G(s)H(s)} \cdot \frac{1}{s^3} \right) \\ &= \lim_{s \to 0} \frac{1}{s^2[1 + G(s)H(s)]} \\ &= \frac{1}{\lim\limits_{s \to 0} s^2G(s)H(s)} \equiv \frac{1}{K_a} \end{aligned} \tag{7.57}$$

where
$$K_a = \lim_{s \to 0} s^2G(s)H(s) \tag{7.58}$$

K_a is called the *static acceleration constant.* Predictably, its definition contains s^2 which indicates a second derivative of the position variable.

System Types and Steady-State Errors

We defined the type of a system earlier (Section 3.2 at page 105). It is determined by the number of poles at the origin (i.e. at $s = 0$) existing in the open-loop transfer function, $G(s)H(s)$, of a system. The general form of $G(s)H(s)$ is

$$G(s)H(s) = \frac{K(1+sT_1)(1+sT_2)\cdots(1+sT_m)}{s^k(1+sT_a)(1+sT_b)\cdots(1+sT_n)} \tag{7.59}$$

In Eq. (7.59), k indicates the number of poles at the origin. Therefore, it is a type k system.

EXAMPLE 7.26. Identify the type of the system for which

$$G(s) = \frac{K(1+3s)}{s^2} \quad \text{and} \quad H(s) = \frac{1+5s}{s^2+6s+7}$$

Solution Here,

$$G(s)H(s) = \frac{K(1+3s)(1+5s)}{s^2(s^2+6s+7)}$$

Since the number of poles at the origin is 2 (indicated by s^2 in the denominator), type = 2.

EXAMPLE 7.27. Identify the type of the system, for which

$$G(s) = \frac{4}{s^2+6s+7} \quad \text{and} \quad H(s) = s+3$$

Solution Here,

$$G(s)H(s) = \frac{4(s+3)}{s^2+6s+7}$$

It has no poles at the origin. Therefore, type = 0.

With this background, we see how the static error constants, and for that matter, steady-state errors, are related to types of systems.

Type 0. Here,

$$G(s)H(s) = \frac{K(1+sT_1)(1+sT_2)\cdots(1+sT_m)}{(1+sT_a)(1+sT_b)\cdots(1+sT_n)}$$

Step input

$$K_p = \lim_{s\to 0} G(s)H(s) = \frac{K\cdot 1\cdot 1\cdots 1}{1\cdot 1\cdots 1} = K$$

Therefore,

$$e_{ss} = \frac{1}{1+K_p} = \frac{1}{1+K}$$

Ramp input

$$K_v = \lim_{s\to 0} sG(s)H(s) = \lim_{s\to 0} \frac{sK\cdot 1\cdot 1\cdots 1}{1\cdot 1\cdots 1} = 0$$

Therefore,

$$e_{ss} = \frac{1}{K_v} = \infty$$

Parabolic input

$$K_a = \lim_{s^2\to 0} s^2G(s)H(s) = \lim_{s\to 0} \frac{s^2K\cdot 1\cdot 1\cdots 1}{1\cdot 1\cdots 1} = 0$$

Therefore,

$$e_{ss} = \frac{1}{K_a} = \infty$$

Type 1. Here,

$$G(s)H(s) = \frac{K(1 + sT_1)(1 + sT_2) \cdots (1 + sT_m)}{s(1 + sT_a)(1 + sT_b) \cdots (1 + sT_n)}$$

Step input
$$K_p = \lim_{s \to 0} G(s)H(s) = \lim_{s \to 0} \frac{K \cdot 1 \cdot 1 \cdots 1}{s \cdot 1 \cdot 1 \cdots 1} = \infty$$

Therefore,
$$e_{ss} = \frac{1}{1 + K_p} = \frac{1}{1 + \infty} = 0$$

Ramp input
$$K_v = \lim_{s \to 0} sG(s)H(s) = \lim_{s \to 0} \frac{Ks \cdot 1 \cdot 1 \cdots 1}{s \cdot 1 \cdot 1 \cdots 1} = K$$

Therefore,
$$e_{ss} = \frac{1}{K_v} = \frac{1}{K}$$

Parabolic input
$$K_a = \lim_{s^2 \to 0} s^2 G(s)H(s) = \lim_{s \to 0} \frac{Ks^2 \cdot 1 \cdot 1 \cdots 1}{s \cdot 1 \cdot 1 \cdots 1} = 0$$

Therefore,
$$e_{ss} = \frac{1}{K_a} = \frac{1}{0} = \infty$$

Type 2. Here,

$$G(s)H(s) = \frac{K(1 + sT_1)(1 + sT_2) \cdots (1 + sT_m)}{s^2(1 + sT_a)(1 + sT_b) \cdots (1 + sT_n)} \tag{7.60}$$

Step input
$$K_p = \lim_{s \to 0} G(s)H(s) = \infty$$

Therefore,
$$e_{ss} = \frac{1}{1 + K_p} = 0$$

Ramp input
$$K_v = \lim_{s \to 0} sG(s)H(s) = \infty$$

Therefore,
$$e_{ss} = \frac{1}{K_v} = 0$$

Parabolic input
$$K_a = \lim_{s^2 \to 0} s^2 G(s)H(s) = K$$

Therefore,
$$e_{ss} = \frac{1}{K_a} = \frac{1}{K}$$

The results so far obtained are summarized in Table 7.6. The following points emerge from the table:

1. Ramp and parabolic inputs should not be applied to type 0 system.
2. Parabolic input should not be applied to type 1 system.
3. For types ≥ 3, $e_{ss} = 0$ for all inputs.

Table 7.6 Steady-state error for Different Types and Different Inputs

Type	*Steady-state error for*		
	Step input	*Ramp input*	*Parabolic input*
	$e_{ss} = \dfrac{1}{1+K_p}$	$e_{ss} = \dfrac{1}{K_v}$	$e_{ss} = \dfrac{1}{K_a}$
0	$\dfrac{1}{1+K}$	∞	∞
1	0	$\dfrac{1}{K}$	∞
2	0	0	$\dfrac{1}{K}$
3	0	0	0

Two more points which are not apparent from Table 7.6 are as follows:

- e_{ss} for a linear combination of inputs can be superposed.
- The method fails for sinusoidal inputs because the final value theorem cannot be applied in that case

EXAMPLE 7.28. The system $G(s) = \dfrac{0.8}{s^2+s-2}$ is subjected to a step input. The system output $y(t)$ as $t \to \infty$ is

(a) 0.8 (b) 0.4 (c) −0.4 (d) unbounded

Solution Applying the final value theorem, we get

$$c(t)\Big|_{t\to\infty} = \lim_{s\to 0} s \cdot \frac{1}{s} \cdot \frac{0.8}{s^2+s-2} = -0.4$$

Ans: (c).

EXAMPLE 7.29. A unity feedback closed-loop second order system has a transfer function of

$$\frac{81}{s^2+0.6s+9}$$

and is excited by a step input of 10 units. The steady-state error of the output is

(a) 10 (b) 0.0 (c) 1.0 (d) 0.1

Solution Given:

$$G(s)H(s) = \frac{81}{s^2+0.6s+9}, \qquad R(s) = \frac{10}{s}$$

Therefore,

$$e_{ss} = \lim_{s\to 0} s \cdot \frac{10}{s} \cdot \frac{1}{1+\dfrac{81}{s^2+0.6s+9}} = \frac{10}{1+9} = 1$$

Ans: (c)

EXAMPLE 7.30. For a unit step input, a system with a closed-loop transfer function of

$$G(s) = \frac{20}{s^2 + 2s + 5}$$

has a steady-state output of

(a) 10 (b) 5 (c) 2 (d) 4

Solution Given:

$$R(s) = \frac{1}{s}$$

Applying the final value theorem, the stead-state output is given by

$$c(t)\big|_{ss} = \lim_{s \to 0} s \cdot C(s) = \lim_{s \to 0} s \cdot G(s)R(s) = \lim_{s \to 0} s \cdot \frac{20}{s^2 + 2s + 5} \cdot \frac{1}{s} = \frac{20}{5} = 4$$

Ans: (d).

EXAMPLE 7.31. The loop transfer function of a system is given by

$$G(s)H(s) = \frac{10e^{-0.1s}}{s(s+5)(s^2 + 4s + 16)}$$

When the system is subjected to a parabolic input of $t^2/2$, the steady-state error is

(a) zero (b) 0.1 (c) infinity (d) $e^{-0.1}/0.1$

Solution The Laplace transform of the input is $R(s) = 1/s^3$. Therefore,

$$\begin{aligned} e_{ss} &= \lim_{s \to 0} \frac{sR(s)}{1 + G(s)H(s)} \\ &= \lim_{s \to 0} \frac{s}{s^3} \cdot \frac{1}{1 + \dfrac{10e^{-0.1s}}{s(s+5)(s^2 + 4s + 16)}} \\ &= \lim_{s \to 0} \frac{1}{s} \cdot \frac{(s+5)(s^2 + 4s + 16)}{s(s+5)(s^2 + 4s + 16) + 10e^{-0.1s}} \\ &= \infty \end{aligned}$$

Ans: (c).

EXAMPLE 7.32. The steady-state error due to a unit step disturbance input $D(s)$ in Figure 7.17(a) is

(a) 0 (b) 0.012 (c) 0.021 (d) 0.025

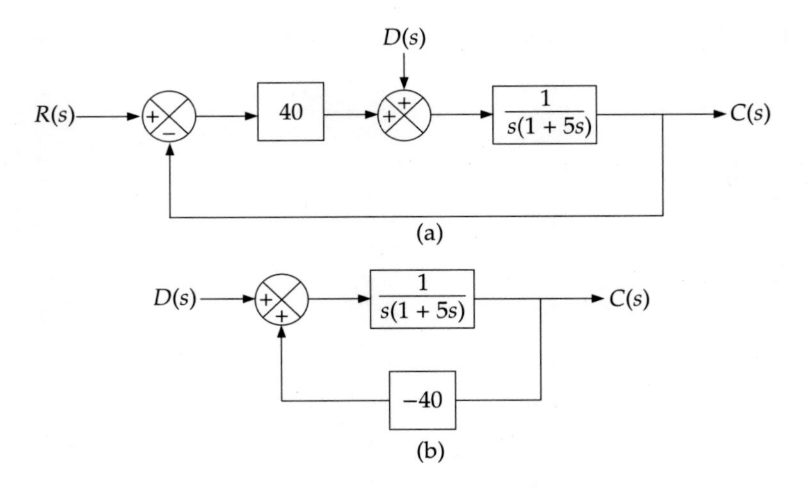

Figure 7.17 System (Example 7.32).

Solution To find the influence of $D(s) = (1/s)$, we put $R(s) = 0$ and redraw the block diagram as shown in Figure 7.17(b). Then

$$G(s)H(s) = \frac{-40}{s(1+5s)}$$

$$1 + G(s)H(s) = \frac{s(1+5s) - 40}{s(1+5s)}$$

$$e_{ss} = \lim_{s \to 0} \frac{sD(s)}{1 + G(s)H(s)}$$

$$= \lim_{s \to 0} s \cdot \frac{1}{s} \cdot \frac{s(1+5s)}{s(1+5s) - 40} = 0$$

Ans: (a).

EXAMPLE 7.33. The dynamic response of a measurement system is given by the differential equation

$$4\frac{d^2y}{dt^2} + 640\frac{dy}{dt} + 40000y = 80000x$$

where y is the output and x is the input.

(a) Determine the
  (i) natural frequency
  (ii) damping ratio, and
  (iii) static sensitivity of the system.

(b) If it is used to measure a point of frequency of 10 Hz, what would be the ratio of the indicated amplitude to the input amplitude?

Solution We write the given differential equation as

$$\frac{4}{40000}\frac{d^2y}{dt^2} + \frac{640}{40000}\frac{dy}{dt} + y = \frac{80000}{40000}$$

Comparing it with Eq. (7.10) at page 232, we get

(a) (i)
$$\frac{1}{\omega_n^2} = \frac{4}{40000}$$

$$\Rightarrow \quad \omega_n^2 = 10000$$

$$\Rightarrow \quad \omega_n = 100 \text{ rad/s}$$

(ii)
$$\frac{2\zeta}{\omega_n} = \frac{640}{40000}$$

$$\Rightarrow \quad \zeta = \frac{640}{40000} \times \frac{100}{2} = 0.8$$

(iii)
$$K = \frac{80000}{40000} = 2$$

(b) The Laplace transform of the given equation yields

$$(4s^2 + 640s + 40000)Y(s) = 80000X(s)$$

Therefore,

$$\frac{Y(s)}{X(s)} = \frac{20000}{s^2 + 160s + 10000}$$

or
$$\frac{y(\jmath\omega)}{x(\jmath\omega)} = \frac{20000}{-\omega^2 + \jmath\,160\omega + 10000}$$

or
$$\left|\frac{y(\jmath\omega)}{x(\jmath\omega)}\right|_{\omega=2\pi\times 10} = \left.\frac{20000}{\sqrt{\left(10000 - \omega^2\right)^2 + (160\omega)^2}}\right|_{\omega=20\pi}$$

$$= \frac{20000}{\sqrt{[10000 - (20\pi)^2]^2 + (160 \times 20\pi)^2}} = 1.7044$$

EXAMPLE 7.34. The open-loop transfer function of a unity feedback system is given by

$$G(s) = \frac{100}{s^2(s+4)(s^2+5s+25)}$$

Find static error coefficients and the steady-state error of the system when it is subjected to an input of $r(t) = 2 + 4t + 2t^2$.

Solution The static error coefficients are

$$K_p = \lim_{s\to 0} G(s)H(s) = \lim_{s\to 0} \frac{100}{s^2(s+4)(s^2+5s+25)} = \infty$$

$$K_v = \lim_{s\to 0} sG(s)H(s) = \lim_{s\to 0} \frac{100}{s(s+4)(s^2+5s+25)} = \infty$$

$$K_a = \lim_{s\to 0} s^2G(s)H(s) = \lim_{s\to 0} \frac{100}{(s+4)(s^2+5s+25)} = \frac{100}{(4)(25)} = 1$$

Taking the Laplace transform of the input, we get

$$R(s) = \frac{2}{s} + \frac{4}{s^2} + \frac{4}{s^3}$$

We observe that this is a combination of step, ramp and parabolic inputs. Hence the steady-state error is given by

$$e_{ss} = \frac{2}{1+K_p} + \frac{4}{K_v} + \frac{4}{K_a}$$

Substituting the values obtained for the error coefficients, we get

$$e_{ss} = \frac{2}{1+\infty} + \frac{4}{\infty} + \frac{4}{1} = 4$$

EXAMPLE 7.35. A servomechanism, the block diagram of which is given in Figure 7.18, is designed to keep a radar antenna pointed at a flying aeroplane. If the aeroplane is flying with a velocity of 600 km/h, at a range of 2 km, and the maximum tracking error is to be within 0.1°, determine the required velocity error coefficient.

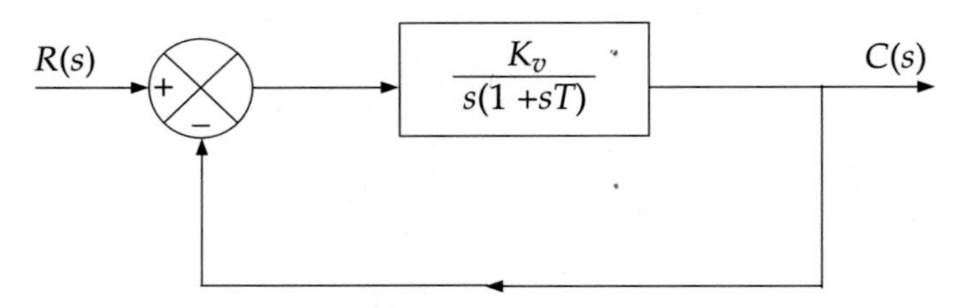

Figure 7.18 Block diagram of servomechanism (Example 7.35).

Solution Given velocity = 600 km/h = 166.67 m/s, radial distance = 2 km = 2000 m. Therefore, the angular velocity is

$$\omega = \frac{v}{r} = \frac{166.67}{2000} = 0.083 \text{ rad/s} = 4.756 \text{ deg/s}$$

The input to the tracking system is the angle swept in time t s. So,

$$r(t) = \omega t = 4.756t$$

$$\Rightarrow \qquad R(s) = \frac{4.756}{s^2}$$

The open-loop transfer function of the system is

$$G(s)H(s) = \frac{K_v}{s(1 + sT)}$$

Therefore,

$$1 + G(s)H(s) = \frac{s(1 + sT) + K_v}{s(1 + sT)}$$

Thus, the steady-state error is given by

$$\begin{aligned} e_{ss} &= \lim_{s \to 0} \frac{R(s)}{1 + G(s)H(s)} \\ &= \lim_{s \to 0} \frac{4.756}{s2} \cdot \frac{s(1 + sT)}{s(1 + sT) + K_v} \\ &= \frac{4.756}{K_v} \end{aligned}$$

To maintain steady-state tracking error at $0.1°$,

$$\frac{4.756}{K_v} = 0.1$$

$\Rightarrow$

$$K_v = 47.56 \text{ /s}$$

EXAMPLE 7.36. Find type, error coefficients and e_{ss} for the system, shown in Figure 7.19, for the inputs (a) $r(t) = 12$ (b) $r(t) = 6t$ and (c) $r(t) = 12 + 6t + \frac{1}{2}t^2$.

$$G_1(s) = \frac{25}{(s + 5)(s + 10)}$$

$$H_1(s) = 5s$$

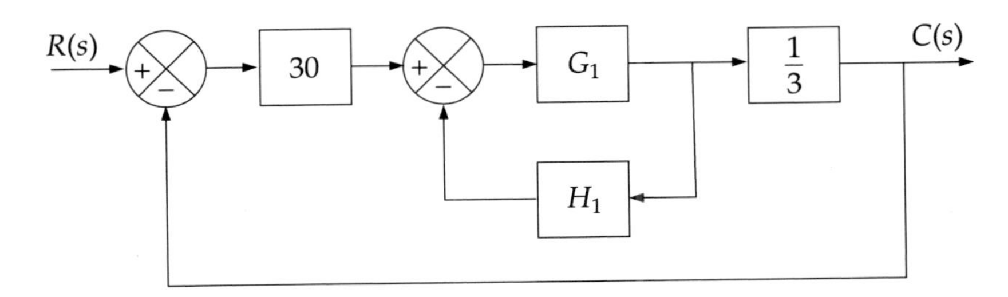

Figure 7.19 Block diagram for the control system (Example 7.36).

Solution The block diagram can be reduced to give Figure 7.20.

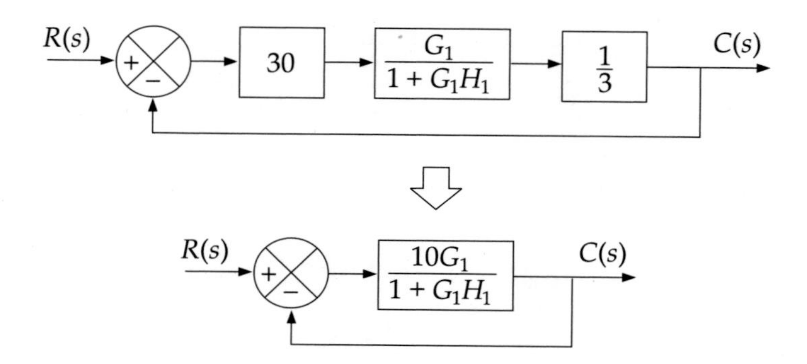

Figure 7.20 Reduced block diagram (Example 7.36).

Therefore, the open-loop transfer function and closed-loop transfer function of the system are

$$G(s)H(s) = \frac{10G_1}{1+G_1H_1} = \frac{250}{(s+5)(s+10)+125s} = \frac{250}{s^2+140s+50}$$

$$\frac{C(s)}{R(s)} = \frac{250}{s^2+140s+300}$$

The closed-loop transfer function has no poles at the origin. Therefore, it is a type 0 system. For a type 0 system, only the static position error coefficient, K_p, exists, its value being

$$K_p = \lim_{s\to 0} G(s)H(s) = \lim_{s\to 0} \frac{250}{s^2+140s+50} = 5$$

(a) Here, $R(s) = \frac{12}{s}$. Therefore,

$$e_{ss} = \lim_{s\to 0} \frac{sR(s)}{1+G(s)H(s)} = \frac{12}{1+K_p} = 2$$

(b) Here, $R(s) = \frac{6}{s^2}$. Therefore,

$$e_{ss} = \lim_{s\to 0} \frac{6}{s[1+G(s)H(s)]} = \infty$$

(c) Here, $$R(s) = \frac{12}{s} + \frac{6}{s^2} + \frac{1}{2s^2}$$

Since errors for individual inputs can be superposed and we have already seen that e_{ss} for the second part is ∞, without calculating it for the third part we can say that for this input also, $e_{ss} = \infty$.

EXAMPLE 7.37. Determine the range of values of K of the system shown in Figure 7.21 so that $e_{ss} < 0.004$ when $r(t) = 0.2t$.

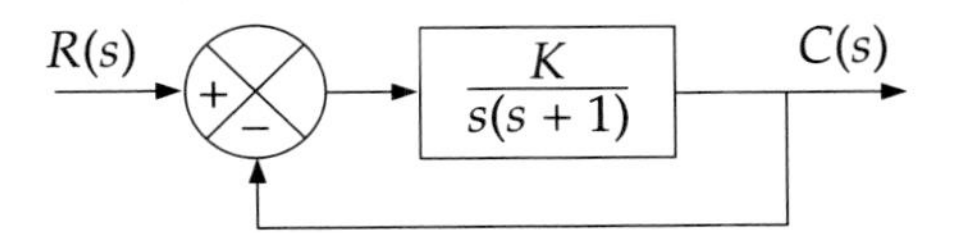

Figure 7.21 Control system (Example 7.37).

Solution The open-loop transfer function for the system is

$$G(s)H(s) = \frac{K}{s(s+1)}$$

Hence,
$$1 + G(s)H(s) = \frac{s(s+1)+K}{s(s+1)}$$

Given,
$$r(t) = 0.2t$$

⇒
$$R(s) = \frac{0.2}{s^2}$$

Thus,

$$e_{ss} = \lim_{s\to 0} \frac{sR(s)}{1+G(s)H(s)}$$
$$= \lim_{s\to 0} \left[\frac{0.2}{s^2} \cdot \frac{s^2(s+1)}{s^2+s+K}\right]$$
$$= \lim_{s\to 0} \frac{0.2(s+1)}{s^2+s+K}$$
$$= \frac{0.2}{K}$$

From the given condition
$$\frac{0.2}{K} < 0.004$$

⇒
$$K > \frac{0.2}{0.004}$$

⇒
$$K > 50$$

EXAMPLE 7.38. The transfer function of a system is $1/(1+sT)$. The input to the system is $tu(t)$. The output would track the system. Find the value of the steady-state error.

Solution Given:
$$G(s) = \frac{1}{1+sT}$$

and since the output would track the system, $H(s) = 1$.

Also,

$$r(t) = tu(t) = t \quad \text{for} \quad t > 0$$

$$\Rightarrow \qquad R(s) = \frac{1}{s^2}$$

Therefore,

$$e_{ss} = \lim_{s \to 0} \frac{sR(s)}{1 + G(s)H(s)} = \lim_{s \to 0} \frac{s(1 + sT) \cdot \frac{1}{s^2}}{2 + sT} = \infty$$

Alternative method. We can arrive at the same result from the following study in the t-space. For this case, the closed-loop transfer function is given by

$$\frac{C(s)}{R(s)} = \frac{G(s)}{1 + G(s)H(s)} = \frac{1}{2 + sT}$$

Therefore,

$$C(s) = \frac{1}{2 + sT} \cdot \frac{1}{s^2} = \frac{1/T}{s^2\left(s + \frac{2}{T}\right)} \equiv \frac{A}{s^2} + \frac{B}{s} + \frac{C}{s + \frac{2}{T}}$$

where A, B and C are constants. The constants are evaluated as follows:

$$A = s^2 C(s)\Big|_{s=0} = \left.\frac{1/T}{s + \frac{2}{T}}\right|_{s=0} = \frac{1}{2}$$

$$B = \frac{d}{ds}\left[s^2 C(s)\right]_{s=0} = \frac{1}{T} \cdot \frac{d}{ds} \left.\frac{1}{s + \frac{2}{T}}\right|_{s=0} = -\frac{1}{T} \cdot \left.\frac{1}{\left(s + \frac{2}{T}\right)^2}\right|_{s=0} = -\frac{T}{4}$$

$$C = \left(s + \frac{2}{T}\right) C(s)\Bigg|_{s=-2/T} = \left.\frac{1}{s^2 T}\right|_{s=-2/T} = \frac{T}{4}$$

Thus,

$$C(s) = \frac{1}{2s^2} - \frac{T}{4s} + \frac{T}{4\left(s + \frac{2}{T}\right)}$$

which, on inverse Laplace transform, yields

$$c(t) = \frac{t}{2} - \frac{T}{4} + \frac{T}{4} \cdot e^{-2t/T}$$

Therefore, the steady-state error is given by

$$\begin{aligned} e_{ss} &= \lim_{t \to \infty}\left[r(t) - c(t)\right] = \lim_{t \to \infty}\left[t - \left(\frac{t}{2} - \frac{T}{4} + \frac{T}{4} \cdot e^{-2t/T}\right)\right] \\ &= \lim_{t \to \infty}\left[\frac{t}{2} + \frac{T}{4}\left(1 - e^{-2t/T}\right)\right] \\ &= \infty \end{aligned}$$

Note: This is done in order to demonstrate the power as well as the convenience offered by the final value theorem.

EXAMPLE 7.39. A unit ramp input is applied to the system shown in the following figure.

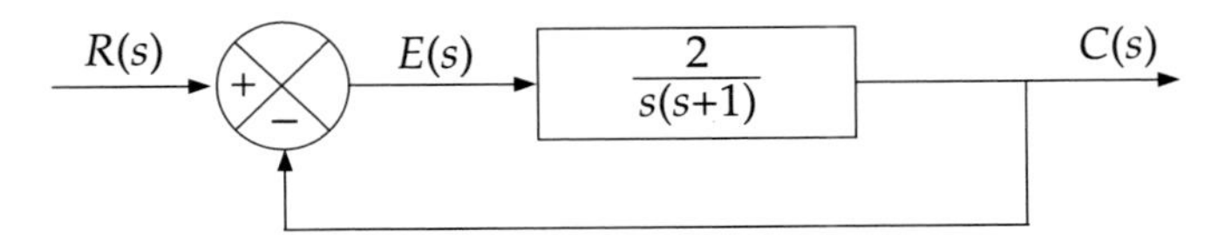

The steady-state error in its output is

(a) 0 (b) 0.5 (c) 1 (d) 2

Solution Given:

$$R(s) = \frac{1}{s^2} \qquad \text{and} \qquad G(s)H(s) = \frac{2}{s(s+1)}$$

Therefore,

$$e_{ss} = \lim_{s\to 0} sE(s) = \lim_{s\to 0} \frac{sR(s)}{1+G(s)H(s)}$$

$$= \lim_{s\to 0} s \cdot \frac{1}{s^2} \cdot \frac{s(s+1)}{s(s+1)+2} = \frac{1}{2} = 0.5$$

Ans: (b).

Dynamic Error Coefficients

In the previous section we discussed how to evaluate the steady-state error utilizing the final value theorem. The method suffers from the limitation that it allows determination of error coefficients for only three kinds of input. Also, for those permitted kinds of input, it does not show how the error varies with time. These disadvantages can be overcome by following the method of finding generalized error coefficients that are often termed *dynamic error coefficients*.

The error function, as we found in Eq. (7.51), is given by

$$\frac{E(s)}{R(s)} = \frac{1}{1+G(s)H(s)}$$

The denominator on the RHS of the equation is a function of s and it can be written as power series in s as follows:

$$\frac{E(s)}{R(s)} = \frac{1}{K_1} + \frac{1}{K_2}s + \frac{1}{K_3}s^2 + \cdots$$

$$\Rightarrow \qquad E(s) = \frac{1}{K_1}R(s) + \frac{1}{K_2}sR(s) + \frac{1}{K_3}s^2R(s) + \cdots \tag{7.61}$$

The inverse Laplace transform of Eq. (7.61) yields

$$e(t) = \frac{1}{K_1}r(t) + \frac{1}{K_2}\frac{d}{dt}r(t) + \frac{1}{K_3}\frac{d^2}{dt^2}r(t) + \cdots \tag{7.62}$$

Therefore, if the input function is known, dynamic error coefficients can be figured out through a comparison between Eq. (7.62) and the polynomial obtained from the inverse Laplace transform[3]. The following examples will make the procedure clear.

EXAMPLE 7.40. For a system having

$$G(s)H(s) = \frac{K(s+4)}{s(s^3+5s^2+6s)}$$

find

(a) Type of the system

(b) Static error constants

(c) Error due to input $(At^2/2)$

Solution

(a) $G(s)H(s) = \dfrac{K(s+4)}{s(s^3+5s^2+6s)} = \dfrac{K(s+4)}{s^2(s^2+5s+6)}$. Therefore, type = 2.

(b) Since it is a type 2 system, $K_p = K_v = \infty$. And

$$K_a = \lim_{s\to 0} s^2 G(s)H(s) = \lim_{s\to 0} \frac{K(s+4)}{s^2+5s+6} = \frac{2K}{3}$$

(c) The input is parabolic. Its Laplace transform is $R(s) = \dfrac{A}{s^3}$. Therefore,

$$e_{ss} = \frac{A}{K_a} = \frac{3A}{2K}$$

EXAMPLE 7.41. The open-loop transfer function of a servo system with unity feedback is given by

$$G(s) = \frac{500}{s(1+0.1s)}$$

(a) Evaluate the error series of the system.

(b) Determine the steady-state error of the system for the input $r(t) = 1 + 2t + t^2$.

Solution Since it is a unity feedback system, we have

$$\frac{E(s)}{R(s)} = \frac{1}{1+G(s)} = \frac{s+0.1s^2}{500+s+0.1s^2}$$

The long division of the RHS of the equation yields

$$\begin{array}{r|llll}
 & +0.002s & +0.000196s^2 & -0.000000792s^3 & \\
\hline
500+s+0.1s^2 & +s & +0.1s^2 & & \\
 & +s & +0.002s^2 & +0.0002s^3 & \\
\hline
 & & +0.098s^2 & -0.0002s^3 & \\
 & & +0.098s^2 & +0.000196s^3 & +0.0000196s^4 \\
\hline
 & & & -0.000396s^3 & -0.0000196s^4 \\
 & & & -0.000396s^3 & -0.000000792s^4 \\
\hline
\end{array}$$

[3]Some authors define error coefficients as $C_i = 1/K_i$.

Thus,

$$\frac{E(s)}{R(s)} = 0.002s + 0.000196s^2 - 0.000000792s^3 + \cdots$$

$$\Rightarrow \quad E(s) = 0.002sR(s) + 0.000196s^2R(s) - 0.000000792s^3R(s) + \cdots \qquad \text{(i)}$$

$$\equiv \frac{1}{K_1}R(s) + \frac{1}{K_2}sR(s) + \frac{1}{K_3}s^2R(s) + \frac{1}{K_4}s^3R(s) + \cdots \qquad \text{(ii)}$$

The inverse Laplace transform of Eq. (i) yields

$$e(t) = 0.002\frac{dr(t)}{dt} + 0.000196\frac{d^2r(t)}{dt^2} - 0.000000792\frac{d^3r(t)}{dt^3} + \cdots \qquad \text{(iii)}$$

Given $r(t) = 1 + 2t + t^2$. Therefore,

$$\frac{dr(t)}{dt} = 2 + 2t$$

$$\frac{d^2r(t)}{dt^2} = 2 \qquad \text{(iv)}$$

$$\frac{d^3r(t)}{dt^3} = 0$$

Substituting the results of Eqs. (iv) in Eq. (iii), we get the required error series as

$$e(t) = 0.002(2 + 2t) + 0.0000196(2) - 0.000000792(0) + \cdots$$

$$= 0.004392 + 0.004t$$

(a) The dynamic error coefficients can be calculated by comparing Eq. (i) and Eq. (ii) as follows:

$$K_1 = \frac{1}{0} = \infty$$

$$K_2 = \frac{1}{0.002} = 500$$

$$K_3 = \frac{1}{0.000196} = 5102.04$$

$$K_4 = -\frac{1}{0.000000792} = -1262626.26$$

(b) The steady-state error in time-domain is

$$\lim_{t\to\infty} e(t) = \lim_{t\to\infty} (0.004392 + 0.004t) = \infty$$

EXAMPLE 7.42. The open-loop transfer function of a unity feedback system is given by

$$G(s) = \frac{50}{s(s+10)}$$

Find out dynamic error coefficients and the error series when the input is $r(t) = 1 + 2t + t^2$.

Solution It being a unity feedback system, we have

$$\frac{E(s)}{R(s)} = \frac{10s + s^2}{50 + 10s + s^2}$$

Note: Since we want to find out a polynomial in ascending powers of s, we have to arrange the numerator and the denominator in that order, before performing the long division.

By long division, we get

$$E(s) = (0)R(s) + 0.2sR(s) - 0.02s^2R(s) + (0)s^3R(s) + 0.004s^4R(s) + \cdots \quad \text{(i)}$$

$$\equiv \frac{1}{K_1}R(s) + \frac{1}{K_2}sR(s) + \frac{1}{K_3}s^2R(s) + \frac{1}{K_4}s^3R(s) + \frac{1}{K_5}s^4R(s) + \cdots \quad \text{(ii)}$$

Comparing the coefficients of Eq. (i) and Eq. (ii), we obtain the dynamic error coefficients as

$$K_1 = \frac{1}{0} = \infty$$

$$K_2 = \frac{1}{0.2} = 5$$

$$K_3 = -\frac{1}{0.02} = -50$$

$$K_4 = \frac{1}{0} = \infty$$

$$K_5 = \frac{1}{0.004} = 250$$

Now, the inverse Laplace transform of Eq. (i) yields

$$e(t) = (0)r(t) + 0.2\frac{dr(t)}{dt} - 0.02\frac{d^2r(t)}{dt^2} + (0)\frac{d^3r(t)}{dt^3} + 0.004\frac{d^4r(t)}{dt^4} + \cdots$$

The input is $r(t) = 1 + 2t + t^2$. Hence,

$$\frac{dr(t)}{dt} = 2 + 2t$$

$$\frac{d^2r(t)}{dt^2} = 2$$

Therefore, the error series is given by

$$e(t) = 0.2(2 + 2t) - 0.02(2) = 0.36 + 0.4t$$

7.5 OTHER ERROR PARAMETERS

We now define two other important error parameters, namely,

- Integral square error (ISE)
- Integral absolute error (IAE)

Integral Square Error (ISE)

Unlike the steady-state error, the integral square error is a measure of the system error through the entire time period, i.e. from $t = 0$ to $t = \infty$. It is defined as

$$\text{ISE} = \int_0^\infty e^2(t)\, dt \tag{7.63}$$

where $e(t)$ is the error signal. For example, what is the ISE for the unit step input to an underdamped second order system? For the unit step input to an underdamped second order system, the error signal is given by[4]

$$e(t) = 1 - c(t) = \frac{e^{-\zeta\omega_n t}}{\sqrt{1-\zeta^2}} \sin\left[\left(\omega_n\sqrt{1-\zeta^2}\right)t + \phi\right] \qquad \phi = \tan^{-1}\frac{\sqrt{1-\zeta^2}}{\zeta}$$

Therefore,

$$\begin{aligned}\text{ISE} &= \int_0^\infty \frac{e^{-2\zeta\omega_n t}}{1-\zeta^2} \sin^2\left[\left(\omega_n\sqrt{1-\zeta^2}\right)t + \phi\right] dt \\ &= \frac{1}{2\omega_n}\left(\frac{1}{2\zeta} + 2\zeta\right)\end{aligned} \tag{7.64}$$

Equation (7.64) indicates that the ISE for a second order underdamped system depends on two factors, namely,

- Damping ratio
- Natural frequency

Equation (7.63) gives the magnitude of the ISE from the t-space representation $e(t)$ of the error function. It would be convenient if we could find it out from the s-space representation $E(s)$ of the error function.

A real integration theorem states that

$$\int_0^\infty f(t)dt = \lim_{t\to\infty}\int_0^t f(t)dt$$

Applying the final value theorem, we get from this equation

$$\begin{aligned}\lim_{t\to\infty}\int_0^t f(t)dt &= \lim_{s\to 0} s\mathscr{L}\left\{\int_0^t f(t)dt\right\} \\ &= \lim_{s\to 0} s\frac{F(s)}{s}\end{aligned}$$

or

$$\int_0^\infty f(t)dt = \lim_{s\to 0} F(s) \tag{7.65}$$

[4]See Eq. (7.24) at page 234.

Combining Eqs (7.63) and (7.65), we get

$$\text{ISE} = \lim_{s \to 0} E^2(s) \tag{7.66}$$

EXAMPLE 7.43. A unity feedback position control system has a forward path transfer function $G(s) = K/s$. For unit step input, compute the value of K that minimizes ISE.

Solution Given:

$$G(s) = K/s, \quad H(s) = 1$$

Therefore,

$$T(s) = \frac{C(s)}{R(s)} = \frac{G(s)}{1 + G(s)} = \frac{K}{s + K}$$

Thus,

$$C(s) = \frac{K}{s(s + K)} \qquad \left[\because R(s) = \frac{1}{s}\right]$$

$$= \frac{1}{s} - \frac{1}{s + K} \tag{i}$$

The inverse Laplace transform of Eq. (i) yields

$$c(t) = 1 - e^{-Kt}$$

$$\Rightarrow \qquad e(t) = e^{-Kt}$$

Therefore,

$$\text{ISE} = \int_0^{\infty} e^{-2Kt}\, dt = -\left.\frac{e^{-2Kt}}{2K}\right|_0^{\infty} = \frac{1}{2K}$$

This result indicates that ISE $\to 0$ as $K \to \infty$.

EXAMPLE 7.44. A unity feedback system has a forward path transfer function

$$G(s) = \frac{10}{s(s + 5)}$$

It is subjected to a unit ramp input. The integral of the squared error (ISE) is

(a) zero (b) infinity (c) always negative (d) 0.5

Solution We know that

$$E(s) = \frac{R(s)}{1 + G(s)H(s)} = \frac{1}{s^2} \cdot \frac{s^2 + 5s}{s^2 + 5s + 10} \qquad \left[\because R(s) = \frac{1}{s^2}\right]$$

$$= \frac{s + 5}{s(s^2 + 5s + 10)}$$

So, from Eq. (7.66),

$$\text{ISE} = \lim_{s \to 0} \left(\frac{s + 5}{s(s^2 + 5s + 10)}\right)^2 = \infty$$

Ans: (b).

EXAMPLE 7.45. The forward path transfer function of a unity feedback system is given by

$$G(s) = \frac{(1+5s)(1+10s)(1+2s)}{(1+s)(1+8s)(1+20s)}$$

If $e(t)$ is the error to a unit impulse input, the value of the performance index

$$J = \int_0^\infty e(t)dt$$

is equal to

(a) zero (b) infinity (c) -12 (d) 0.5

Solution

$$E(s) = \frac{1}{1+G(s)} \qquad [\because \; R(s) = 1, \; H(s) = 1]$$

$$= \frac{(1+s)(1+8s)(1+20s)}{(1+s)(1+8s)(1+20s) + (1+5s)(1+10s)(1+2s)}$$

From Eq. (7.65),

$$J = \int_0^\infty e(t)dt = \lim_{s\to 0} E(s) = \lim_{s\to 0} \frac{(1+s)(1+8s)(1+20s)}{(1+s)(1+8s)(1+20s) + (1+5s)(1+10s)(1+2s)}$$

$$= \frac{1\times 1\times 1}{1\times 1\times 1 + 1\times 1\times 1} = 0.5$$

Ans: (d).

Integral Absolute Error (IAE)

The integral absolute error is a measure of average error over a certain time period, T, and is defined as

$$\text{IAE} = \int_0^T |e(t)| \, dt \tag{7.67}$$

7.6 EFFECTS OF ADDITION OF POLES AND ZEROS ON RESPONSE OF SYSTEMS

It is interesting to study the effects of addition of poles and zeros on the response of systems. While doing that, we will restrict ourselves to second order underdamped systems and step response only. Such systems can be represented by their open-loop or closed-loop transfer functions. We will consider both separately. The other assumption is that poles and zeros will be located on the left half plane of the s-space.

Addition of Poles

As already discussed, this study can be made on open-loop and closed-loop transfer functions. To make the study meaningful, we will consider unity feedback systems for open-loop cases.

Closed-loop transfer function. Let the closed-loop transfer function be

$$T(s) = \frac{1}{s^2 + 0.6s + 1}$$

Obviously, it is a second order system with $\omega_n = 1$ and $\zeta = 0.3$. We add a pole to make it

$$T(s) = \frac{1}{s^2 + 0.6s + 1} \cdot \frac{1}{\tau_p s + 1} \tag{7.68}$$

Figure 7.22 shows step responses for the system represented by Eq. (7.68) for $\tau_p = 0$, 1, 2 and 4. Here we observe that as τ_p increases, rise time increases, but peak overshoot decreases. We add that, though observed in a particular case, the conclusion is general.

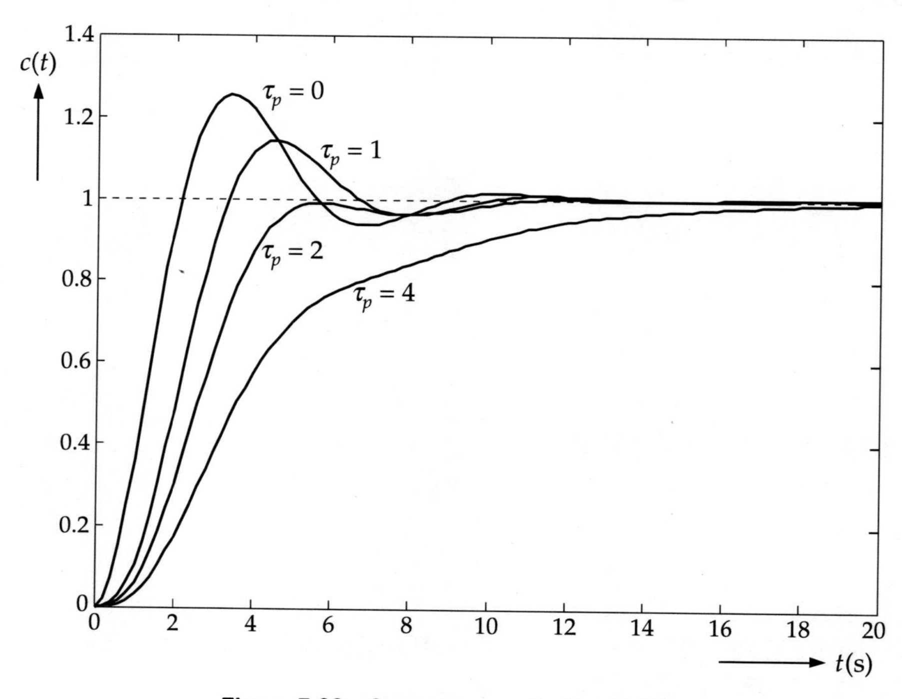

Figure 7.22 Step responses for Eq. (7.68).

Open-loop transfer function. Let the open-loop transfer function be

$$G(s) = \frac{1}{s(s+2)}$$

We add a pole at $s = -1/\tau_p$ to make it

$$G(s) = \frac{1}{s(s+2)(1+\tau_p s)}$$

The corresponding unity feedback closed-loop transfer function is

$$T(s) = \frac{G(s)}{1 + G(s)} = \frac{1}{s(s+2)(1+\tau_p s) + 1} \tag{7.69}$$

The step responses of Eq. (7.69) corresponding to $\tau_p = 0$, 1, 2 and 4 are plotted in Figure 7.23.

It may be observed from this plot that as τ_p increases, both peak overshoot and rise time increase. An increase of τ_p means that the pole comes closer to the origin in the s-plane. The conclusion, though drawn for a particular case, is in fact quite general.

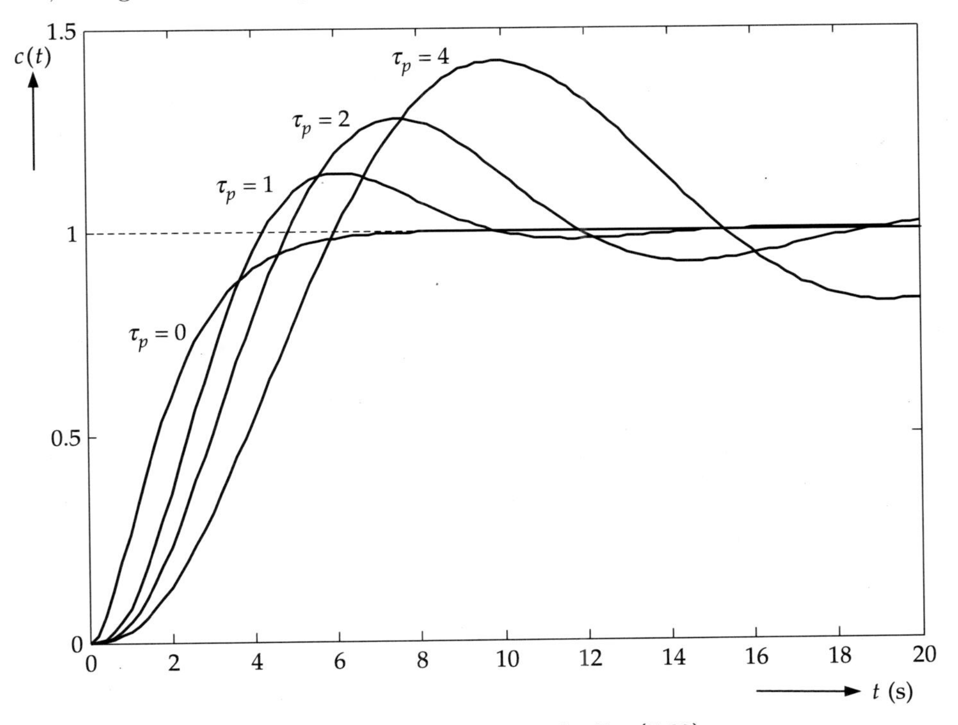

Figure 7.23 Step responses for Eq. (7.69).

Addition of Zeros

As in the case of poles, here also, we will consider open-loop and closed-loop cases separately and a unity feedback will be considered for the open-loop.

Closed-loop transfer function. Let the closed-loop transfer function be

$$T(s) = \frac{1}{s^2 + s + 1} \tag{7.70}$$

We add a zero to it at $s = -1/\tau_z$ so that

$$T(s) = \frac{1 + \tau_z s}{s^2 + s + 1} \tag{7.71}$$

Equation (7.71) can be rewritten as

$$T(s) = \frac{1}{s^2 + s + 1} + \tau_z s \cdot \frac{1}{s^2 + s + 1} \tag{7.72}$$

So, if $c(t)$ be the step response of Eq. (7.70), the step response of Eq. (7.72) is

$$c_0(t) = c(t) + \tau_z \frac{d}{dt} c(t)$$

Figure 7.24 shows the plot of $c(t)$, $\tau_z(dc(t)/dt)$ and $c_0(t)$. It is clear from the figure that the zero term decreases the rise time but increases the peak overshoot.

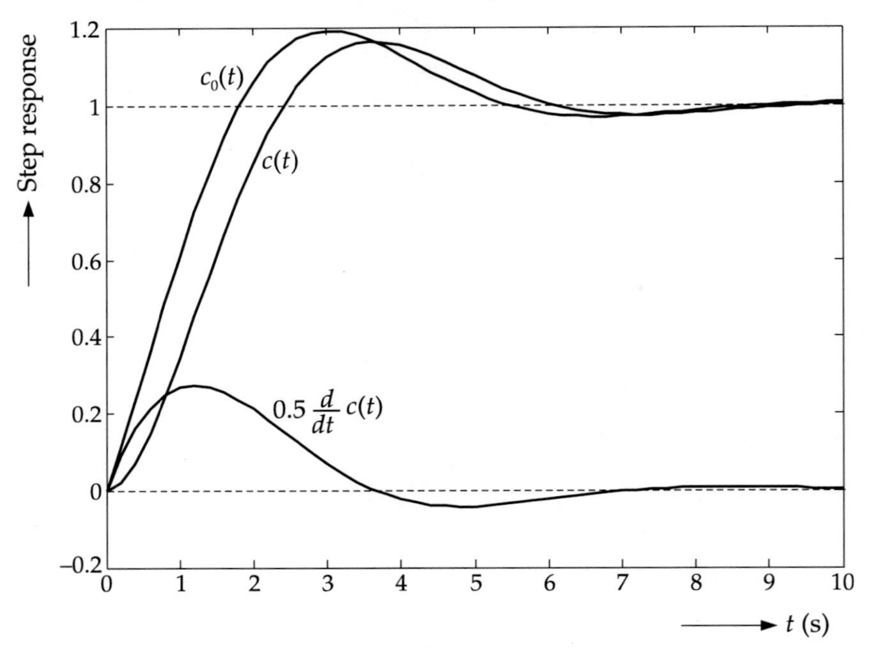

Figure 7.24 Effect of addition of zero to closed-loop transfer function.

Open-loop transfer function. Let the open-loop transfer function be

$$G(s) = \frac{1}{s(s+1)}$$

and the zero at $s = -1/\tau_z$. Then the unity feedback closed-loop transfer function becomes

$$T(s) = \frac{1 + \tau_z s}{(1 + \tau_z s)[s(s+1)]}$$

We observe here that while the term $(1 + \tau_z s)$ in the numerator increases the maximum overshoot, $\tau_z s$ term in the denominator adds to the damping term which, in turn, reduces the maximum overshoot. The combination of two counteracting effects produces the following complicated result:

1. For lower values of τ_z, addition of zero reduces maximum overshoot and improves the rise time.
2. But, after an optimum value of τ_z, the situation is reversed.

Let us consider the open-loop transfer function with a zero added to it. Thus,

$$G(s) = \frac{5(1+\tau_z)}{s(s+1)(s+2)}$$

Its unity feedback closed-loop form is given by

$$T(s) = \frac{5(1+\tau_z)}{s(s+1)(s+2)+5(1+\tau_z)} = \frac{5(1+\tau_z)}{s^3+3s^2+(2+5\tau_z s)s+5} \tag{7.73}$$

The step response of Eq. (7.73) is shown in Figure 7.25 for $\tau_z = 0$, 0.2, 0.4, 2, 4 and 8. It may be observed from the figure that the peak overshoot as well as rise time goes down as τ_z increases from 0 to 2. After that, though the rise time continues to decrease, the peak overshoot goes on increasing.

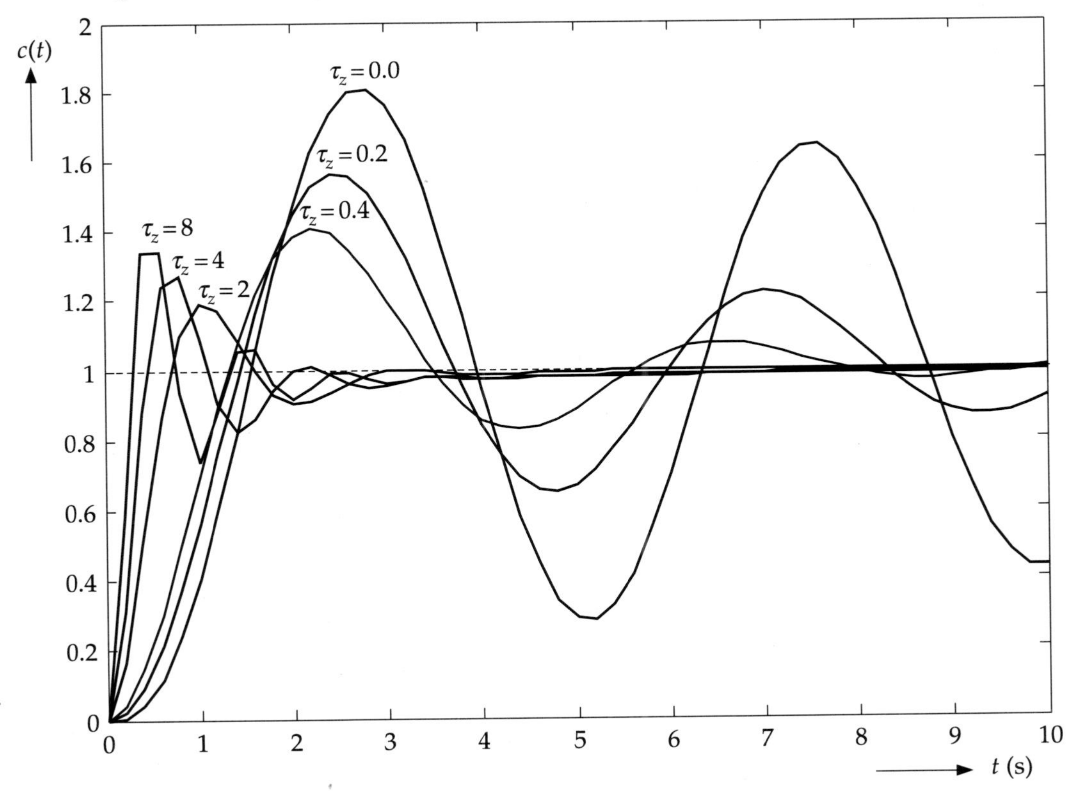

Figure 7.25 Effect of addition of zero to open-loop transfer function.

The results of addition of poles and zeros are summarized in Table 7.7.

Table 7.7 Summary of Effects of Addition of Poles and Zeros on the Step Response of Systems

Type of transfer function	*As the pole nears the origin*		*As the zero nears the origin*	
	Rise time t_r	*Peak overshoot* M_p	*Rise time* t_r	*Peak overshoot* M_p
Open-loop	Increases	Increases	Decreases	Decreases up to a certain value of τ, but increases thereafter
Closed-loop	Increases	Decreases	Decreases	Increases

7.7 MINIMUM AND NON-MINIMUM PHASE SYSTEMS

In Section 3.2 at page 107, we talked about minimum and non-minimum phase systems. Here, we want to see the difference of response that these systems offer.

We recall that minimum phase systems have only positive terms in their transfer functions. If the system is of non-minimum phase, the sweep in phase area is larger. Let us compare the unit step responses of minimum and non-minimum phase systems, the transfer functions of which are given by

$$\frac{C(s)}{R(s)} = G_{\min}(s) = \frac{s+1}{s^2+s+1}$$

$$\frac{C(s)}{R(s)} = G_{\text{nonmin}}(s) = \frac{-s+1}{s^2+s+1}$$

We present the unit step response of the systems in Figures 7.26 and 7.27, respectively.

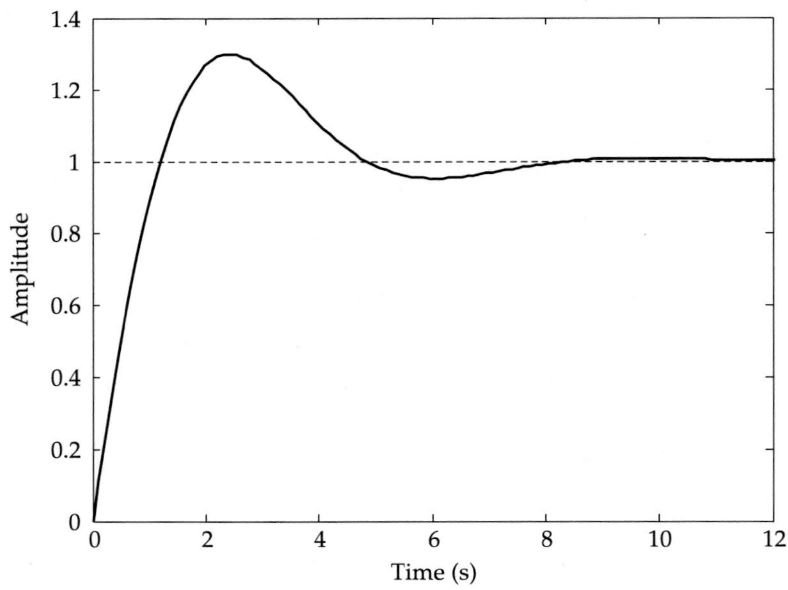

Figure 7.26 Unit step response of a minimum phase system.

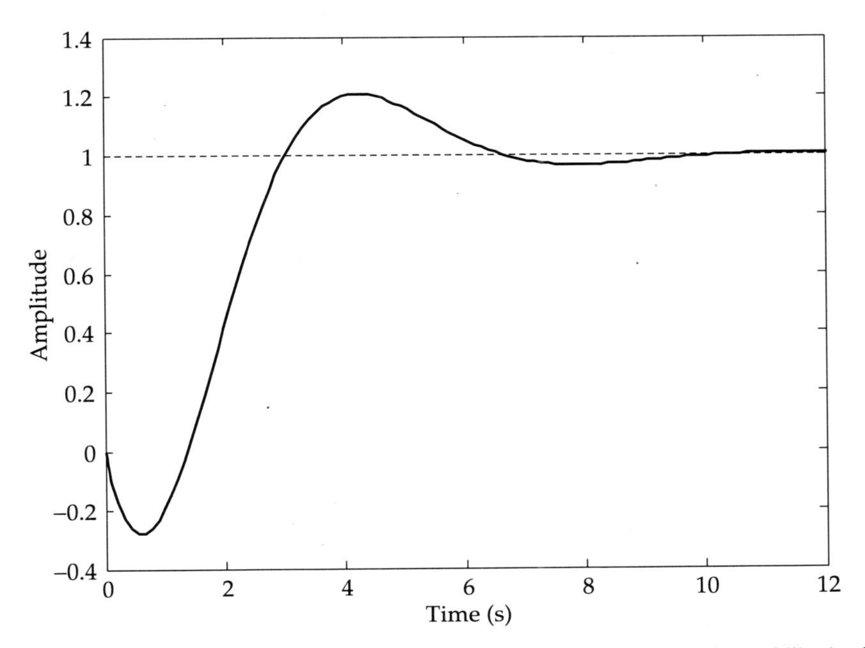

Figure 7.27 Unit step response of a non-minimum phase system. Note the initial instability in the response which is typical to a non-minimum phase system.

The difference in behaviour is apparent from Figure 7.27. The non-minimum phase system produces an initial response that is going in the opposite direction to that of the input. This is typical for such systems.

REVIEW QUESTIONS

7.1 Choose the correct answer

(i) The characteristic equation of a system is $s^2 + 2s + 2 = 0$. The system is

(a) critically damped (b) underdamped
(c) overdamped (d) none of these

(ii) For a control system to be operated in overdamped condition, the value of the damping ratio should be

(a) 0 (b) > 1 (c) 1 (d) < 1

(iii) The steady-state error for a type 3 system for a unit step input is

(a) 0 (b) ∞ (infinity) (c) 1 (d) 3

(iv) The transfer function of a system is $\dfrac{10}{1+s}$. The steady-state error due to step input is

(a) 10 (b) 0 (c) $\dfrac{1}{11}$ (d) ∞

(v) The second order system defined by the transfer function

$$\frac{C(s)}{R(s)} = \frac{25}{s^2 + 5s + 25}$$

is given a unit step input. The time taken for the output to settle within ±2% of the final value is

(a) 1.2 s (b) 1.6 s (c) 2 s (d) 0.4 s

(vi) Addition of a pole to the closed-loop transfer function

(a) increases rise time (b) decreases rise time
(c) increases overshoot (d) has no effect

(vii) The measure of average error over a specified time period is called

(a) integral absolute error (b) integral square error
(c) steady-state error (d) none of these

(viii) A second order system has a damping ratio of 0.9. The system is

(a) underdamped (b) overdamped
(c) critically damped (d) insufficient information

(ix) A system has a single pole at the origin. Its impulse response will be

(a) constant (b) ramp
(c) decaying exponential (d) oscillatory

(x) For a type 1 system, the steady-state error due to step input is

(a) $\dfrac{1}{1+K_p}$ (b) $\dfrac{1}{K_p}$ (c) 0 (d) ∞

(xi) The settling time of a second order underdamped system for settling within ±2% of the final value is given by

(a) $\dfrac{4}{\zeta\omega_n}$ (b) $\dfrac{3}{\zeta\omega_n}$ (c) $\dfrac{4\zeta}{\omega_n}$ (d) $\dfrac{2}{\zeta\omega_n}$

(xii) For a unit step input, a system with a closed-loop transfer function of $20/(s^2 + 2s + 5)$ has a steady-state output of

(a) 10 (b) 5 (c) 2 (d) 4

(xiii) For a second order system given by

$$2\frac{d^2y}{dt^2} + 4\frac{dy}{dt} + 8y = 8x$$

the damping ratio is

(a) 0.1 (b) 0.25 (c) 0.333 (d) 0.5

(xiv) The unit step response of a particular control system is $c(t) = 1 - 10e^{-t}$. The transfer function is

(a) $\dfrac{10}{s+1}$ (b) $\dfrac{s-9}{s+1}$ (c) $\dfrac{1-9s}{s+1}$ (d) $\dfrac{1-9s}{s(s+1)}$

(xv) The steady-state error for a type 2 system subjected to a unit ramp input is
(a) 2 (b) 1 (c) 0 (d) ∞

(xvi) Addition of a zero to the closed-loop transfer function
(a) increases rise time (b) decreases rise time
(c) increases overshoot (d) has no effect

7.2 Consider the following block diagram:

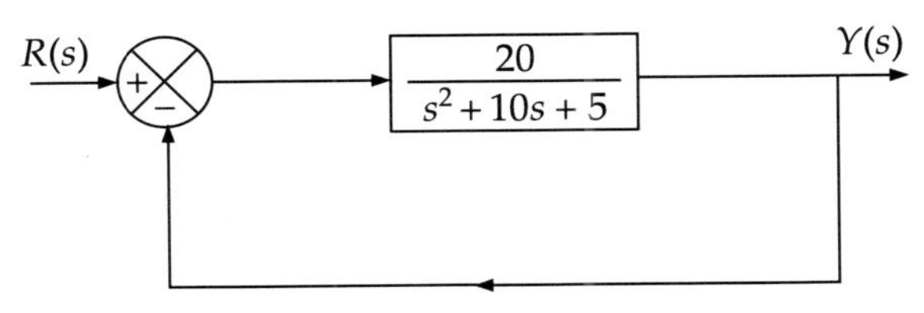

(a) Find the closed-loop transfer function $T(s) = Y(s)/R(s)$.
(b) Find the CL unit step response.
(c) Find the final value of $y(t)$.

7.3 (a) Starting from cold condition, a first order system when subjected to a step input, has a temperature rise of 25°C after an hour, and 37.5°C after 2 hours. Calculate the final steady temperature rise and the thermal time constant.

(b) Sketch the time-domain response of a typical underdamped second order system to a step input. On this sketch, indicate the following:
(i) Maximum peak overshoot, M_p
(ii) Rise time, t_r
(iii) Delay time, t_d
(iv) Settling time, t_s
(v) Steady-state error, e_{ss}

7.4 (a) A second order differential equation is given by

$$\frac{d^2x}{dt^2} + 5\frac{dx}{dt} + 7x = 7y$$

Find the following:
(i) Undamped natural frequency
(ii) Damping ratio
(iii) Damped natural frequency
(iv) Damping coefficient
(v) Time constant

(b) Explain the unit impulse response of a first order system.

7.5 Find the expression for per cent peak overshoot of a second order underdamped system when subjected to a unit step input.

7.6 Consider the unity feedback system with $G(s) = \dfrac{K}{s(s+4)}$. If $K_v = 12$, find the per cent overshoot and peak time.

7.7 (a) Find the error coefficient for the system shown in the following figure.

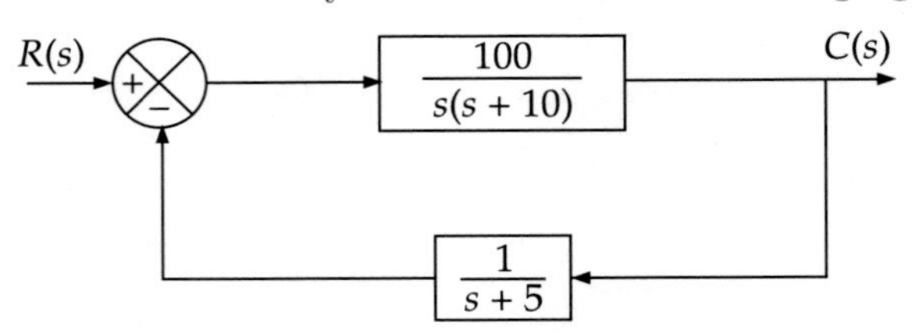

(b) What do you mean by transient response and steady-state response?

7.8 What is steady-state error? Suppose unit ramp input each is given to type 0, type 1 and type 2 systems. Compare the steady-state error for these three cases.

7.9 (a) Find out the time response analysis of a unity feedback underdamped control system having open loop gain of $\omega_n^2/(s^2 + 2\zeta\omega_n s)$ for unit step input.

(b) Define and calculate the value of peak time and settling time for the above time response equation.

7.10 Derive an expression for step response of an underdamped second order system. Sketch the response indicating all time-domain specifications.

7.11 Find the steady-state error of the unity feedback system having $G(s) = \dfrac{10}{0.1s+1}$ in following the input $r(t) = 1 + tu(t)$.

7.12 (a) Find out the response of a typical underdamped second order closed-loop system with unity feedback having unit step input. Draw the response curve.

(b) A second order servo system has poles at $-1 \pm \jmath 2$ and zero at $-1 \pm \jmath 0$. Its steady-state output for a unit step input is $\dfrac{1}{2}$. Determine the transfer function.

7.13 (a) Sketch the unit step response of a second order system indicating peak overshoot, rise time and settling time.

(b) A temperature sensing device can be modelled as a first order system with a time constant of 6 s. It is suddenly subjected to a step input of 30°C to 150°C. What temperature will be indicated in 10 s after the process has started?

7.14 (a) Define the following:

(i) Rise time (ii) Delay time

(iii) Settling time (iv) Overshoot

(b) The open-loop transfer function of a unity feedback control system is given by

$$G(s) = \frac{A}{s(1+sT)}$$

(i) By what factor should the amplifier gain A be multiplied so that the damping ratio is increased from a value of 0.2 to 0.6?

(ii) By what factor should the amplifier gain be multiplied so that the overshoot of the unit step response is reduced from 80% to 20%?

7.15 (a) Find out the response of a typical underdamped second order closed-loop system with unity feedback having step input.

(b) For a unity feedback closed loop control system, the open-loop transfer function is given by

$$G(s) = \frac{25}{s(s+5)}$$

Calculate the value of rise time (t_r) and maximum overshoot (M_p) when the system is subjected to a unit step input.

7.16 Evaluate the static error constants for a unity feedback system having a forward path transfer function $G(s) = 50/s(s+10)$. Estimate steady-state errors of the systems for the input $r(t)$ given by $r(t) = 1 + 2t + t^2$.

Answers to Selected Questions

7.1 (i) (b); (ii) (b); (iii) (a); (iv) (c); (v) (b); (vi) (a); (vii) (a); (viii) (a); (ix) (a); (x) (c); (xi) (a); (xii) (d); (xiii) (d); (xiv) (b); (xv) (c) ; (xvi) (b).

7.2 (a) $20/(s^2 + 10s + 25)$; (b) $y(t) = 0.8 - 4(t + 0.2)e^{-5t}$; (c) $y_{ss} = 0.8$.

7.3 (a) 3.79 h.

7.4 (a) (i) 2.65 rad/s; (ii) 0.95; (iii) 0.87 rad/s; (iv) 0.71; (v) 0.4 s.

7.6 38.6%, 0.47 s.

7.7 (a) $K_1 = \infty$, $K_2 = 2$, $K_3 = -10$, $K_4 = -66.67$.

7.11 ∞.

7.12 (b) $\dfrac{2(s+1)}{s^2 + 2s + 5}$.

7.13 (b) 127.3°C.

7.14 (b) (i) 0.11; (ii) 0.024.

7.15 (b) 0.484 s, 16.3%.

7.16 $K_p = \infty$, $K_v = 50$, $K_a = 0$. e_{ss} for the given input is ∞.

CHAPTER 8

Basics of Controllers

8.1 INTRODUCTION

We know that an automatic control system can be of two kinds—open-loop and closed-loop—depending on whether or not the control element receives a feedback from the output. Figure 8.1 gives a visual presentation of their difference.

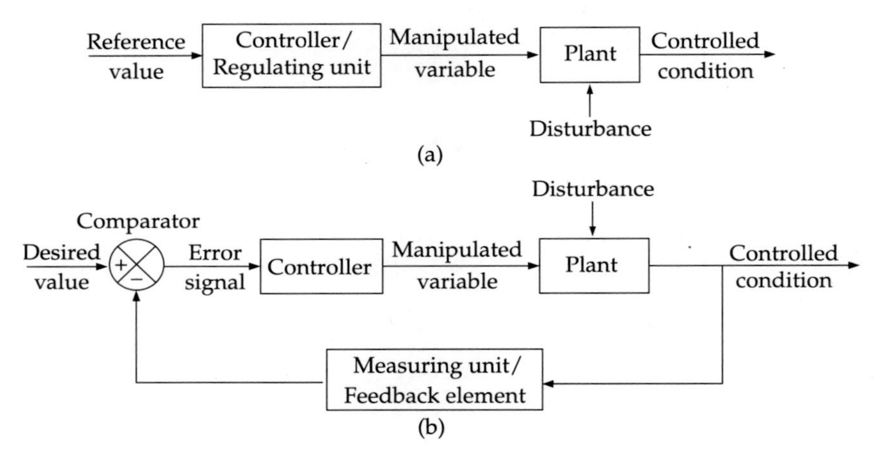

Figure 8.1 Automatic control systems: (a) open-loop; (b) closed-loop.

From the figure we see that controllers constitute an integral element in either control system. In an automatic control system, the controlled value of the variable is continuously compared with its desired value and the deviation is automatically rectified. This procedure can be accomplished through four basic modes:

1. Two-position or ON-OFF control
2. Proportional control
3. Integral control
4. Derivative control

Except the first one, various combinations of other three basic modes may be employed to suit the plant characteristics. So, controllers can be divided into following six categories depending on their mode of control:

1. Proportional (P) controller
2. Integral (I) controller
3. Derivative (D) controller
4. Proportional plus integral (PI) controller
5. Proportional plus derivative (PD) controller
6. Proportional plus integral plus derivative (PID) controller

In the following sections we will consider these controllers and evaluate their performance.

8.2 TWO-POSITION (ON-OFF) CONTROLLER

Two-position controllers are generally electric devices such as relays, solenoid operated valves, etc. Schematically, their action may be represented as in Figure 8.2.

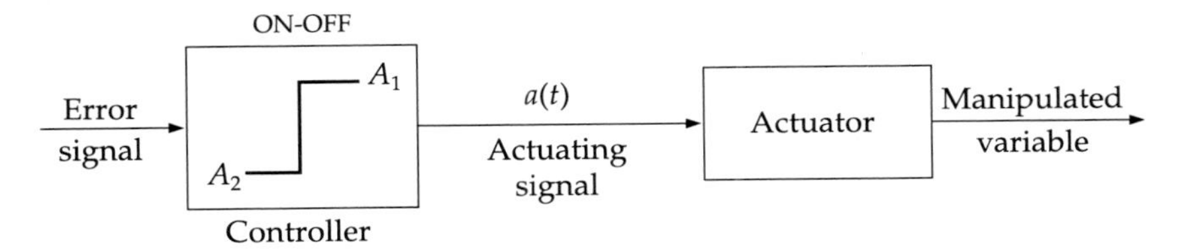

Figure 8.2 Schematic representation of a two-position controller.

If $a(t)$ is the actuating signal, then mathematically the action can be described as

$$a(t) = \begin{cases} A_1 \text{ (max or ON position)} & = a_1 e(t) > 0 \\ A_2 \text{ (min or OFF position)} & = a_2 e(t) < 0 \end{cases}$$

The range through which the error signal must pass before switching takes place is called the *differential gap* (Figure 8.3).

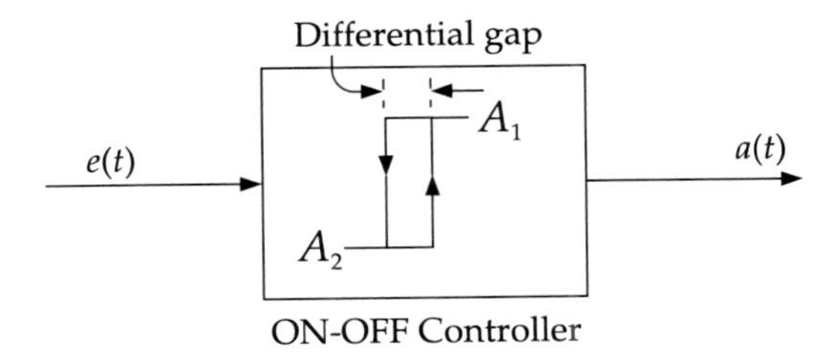

Figure 8.3 Differential gap.

The advantages and disadvantages of this type of control are listed in Table 8.1.

Table 8.1 Advantages and Disadvantages of ON-OFF Controller

Advantages	*Disadvantages*
1. Simple, economical.	1. Transient conditions and steady-state accuracy are not much improved.
2. Used in domestic appliances as well as in industry where rough control is necessary.	2. Controlled output oscillates around the set point.

8.3 PROPORTIONAL (P) CONTROLLER

In a proportional controller, the actuating signal is proportional to the error signal. Mathematically, if $a(t)$ is the actuating signal, then

$$a(t) \propto e(t) = K_p e(t) \tag{8.1}$$

where K_p is called the *controller gain.* Though generally its value is greater than 1, it can be less than 1 in some applications. Since in most of the cases it amplifies the error signal, it facilitates detection of small deviations and their eventual remedy by the actuator.

The Laplace transform of Eq. (8.1) is

$$A(s) = K_p E(s)$$

Accuracy

Consider, for example, the second order system given in Figure 8.4.

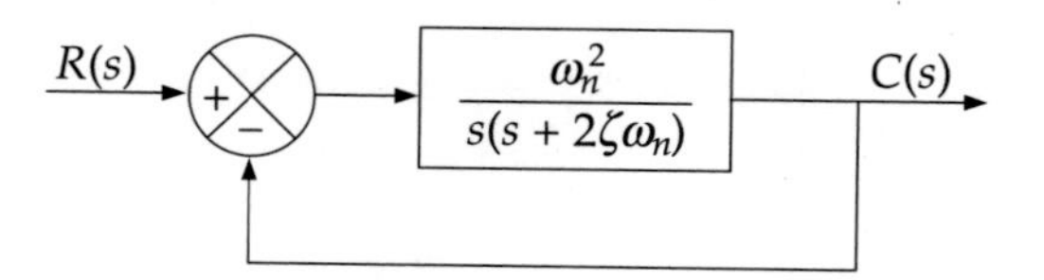

Figure 8.4 Second order system.

Here, with the addition of a proportional controller, the open-loop transfer function becomes

$$G(s)H(s) = \frac{K_p \omega_n^2}{s(s + 2\zeta\omega_n)}$$

Therefore,

$$e_{ss} = \lim_{s \to 0} \frac{sR(s)}{1 + G(s)H(s)} = \lim_{s \to 0} \frac{sR(s)}{1 + \dfrac{K_p \omega_n^2}{s(s + 2\zeta\omega_n)}}$$

$$= \lim_{s \to 0} \frac{s^2(s + 2\zeta\omega_n)}{s(s + 2\zeta\omega_n) + K_p \omega_n^2} \cdot R(s) \tag{8.2}$$

Step input. For a step input, $R(s) = 1/s$. Therefore, from Eq. (8.2),

$$e_{ss} = \lim_{s \to 0} \frac{s^2(s + 2\zeta\omega_n)}{s(s + 2\zeta\omega_n) + K_p\omega_n^2} \cdot \frac{1}{s}$$

$$= \lim_{s \to 0} \frac{s(s + 2\zeta\omega_n)}{s(s + 2\zeta\omega_n) + K_p\omega_n^2} = 0$$

Ramp input. In this case, $R(s) = 1/s^2$. So, from Eq. (8.2),

$$e_{ss} = \lim_{s \to 0} \frac{s^2(s + 2\zeta\omega_n)}{s(s + 2\zeta\omega_n) + K_p\omega_n^2} \cdot \frac{1}{s^2}$$

$$= \frac{2\zeta}{K_p\omega_n} \tag{8.3}$$

Parabolic input. In the case of parabolic input, $R(s) = 1/s^3$. Therefore, from Eq. (8.2),

$$e_{ss} = \lim_{s \to 0} \frac{s^2(s + 2\zeta\omega_n)}{s(s + 2\zeta\omega_n) + K_p\omega_n^2} \cdot \frac{1}{s^3} = \infty$$

The advantages and disadvantages of a proportional controller are listed in Table 8.2.

Table 8.2 Advantages and Disadvantages of a Proportional Controller

Advantages	*Disadvantages*
1. This type of controller increases the forward path gain. We have seen in Eq. (7.14) that an increase in the forward path gain increases the undamped natural frequency ω_n which, in turn, reduces the damping ratio ζ. Consequently, the response becomes faster for overdamped systems. For underdamped systems, however, it increases peak overshoot though reducing the rise time.	1. A proportional controller causes offset, i.e. it may cause a sustained deviation from the desired value.
2. e_{ss} is reduced. Since for this type of controller $\zeta < 1$ and $\omega_n > 1$, it can be seen from Eq. (8.3) that $e_{ss} \ll 1$. Also, as K_p increases, e_{ss} decreases. But as discussed in (1), an increase of K_p implies lowering of ζ which may make the system unstable if K_p is too high.	2. It increases the maximum overshoot of the system because it lowers ζ.
3. It is a stable control as long as K_p is not high.	

Realization

A simple op-amp realization of the proportional controller is indicated in Figure 8.5. The input-output relationship for such a circuit is given by

$$\left|\frac{e_o}{e_i}\right| = \frac{R_2}{R_1}$$

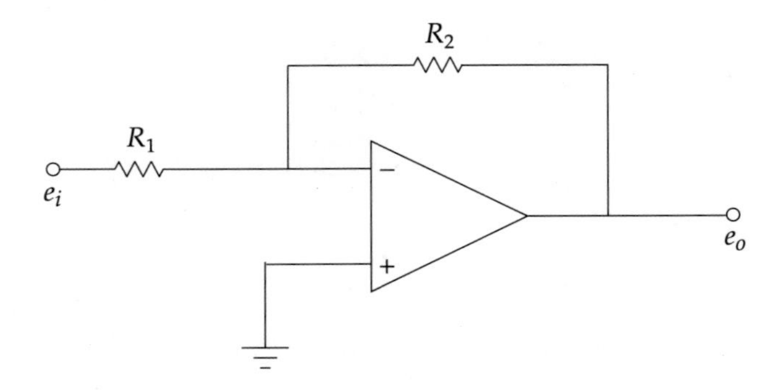

Figure 8.5 Proportional controller circuit.

8.4 INTEGRAL (I) CONTROLLER

In an integral controller, the actuating action can be mathematically put as

$$a(t) \propto \int e(t)\ dt = K_i \int e(t)\ dt \tag{8.4}$$

where K_i, a constant, is the *controller gain.*

The Laplace transform of Eq. (8.4) is

$$A(s) = \frac{K_i}{s} E(s) \tag{8.5}$$

Realization

A simple op-amp realization of the integral controller is shown in Figure 8.6. This is nothing but an integrator circuit for which the output-input relation is given by

$$e_o = \frac{1}{RC} \int |e_i|\ dt \tag{8.6}$$

The Laplace transform of Eq. (8.6) yields the transfer function of the controller as

$$G(s) = \frac{E_o(s)}{E_i(s)} = \frac{1}{s} \cdot \frac{1}{RC}$$

which has the form of Eq. (8.5).

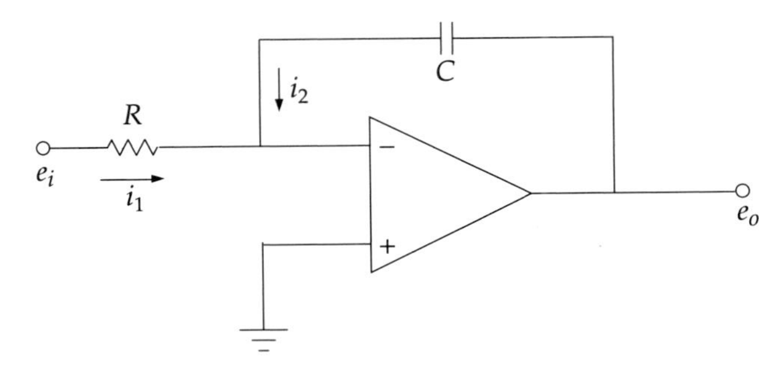

Figure 8.6 Op-amp realization of an integral controller.

8.5 DERIVATIVE (D) CONTROLLER

The derivative controller produces an actuating signal from the error signal as

$$a(t) \propto \frac{de(t)}{dt} = K_d \frac{de(t)}{dt} \tag{8.7}$$

where K_d is a constant. The Laplace transform of Eq. (8.7) is

$$A(s) = K_d s E(s)$$

Realization

The simple op-amp realization (Figure 8.7) of the derivative controller is nothing else than a differentiator for which the output is related to the input as

$$e_o = RC \frac{de_i}{dt} \tag{8.8}$$

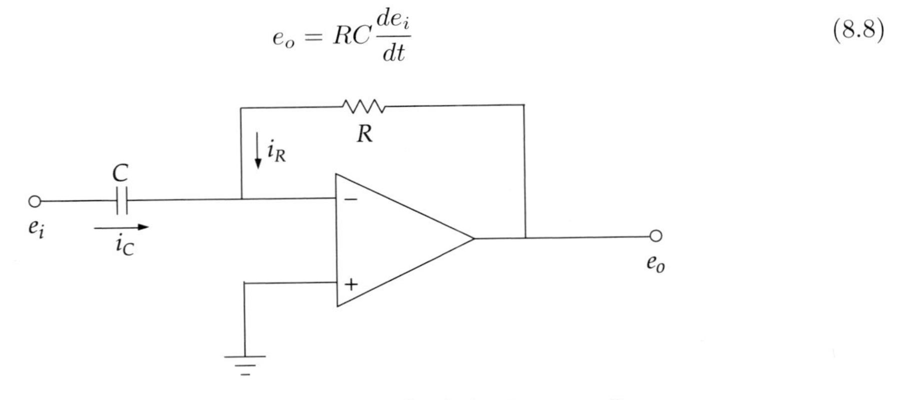

Figure 8.7 Op-amp realization of a derivative controller.

From the Laplace transform of Eq. (8.8), we get the transfer function as

$$G(s) = \frac{E_o(s)}{E_i(s)} = sRC$$

Advantages and disadvantages of derivative controllers are given in Table 8.3.

Table 8.3 Advantages and Disadvantages of a Derivative Controller

Advantages	*Disadvantages*
1. Also called the *rate controller*, a derivative controller improves the transient response of the system.	1. It amplifies the noise signal.
	2. It may cause saturation effect in the actuator.
	3. It does not improve e_{ss} because of generation of a zero in the transfer function.

The disadvantages of a derivative controller far outweigh the only one advantage it offers. This is why *it is never used alone.*

8.6 PROPORTIONAL PLUS INTEGRAL (PI) CONTROLLER

In the case of a PI controller,

$$a(t) = K_p e(t) + K_i \int e(t)\, dt \tag{8.9}$$

$$\Rightarrow \qquad A(s) = \left(1 + \frac{K_i}{s}\right) E(s)$$

if $K_p = 1$.

Control Action

The actuating action of a PI controller for error input of a unit step can be worked out from Eq. (8.9) as

$$a(t) = K_p u(t) + K_i \int u(t)\, dt = K_p + K_i t$$

Thus, a step error signal will produce a ramp actuating signal as shown in Figure 8.8.

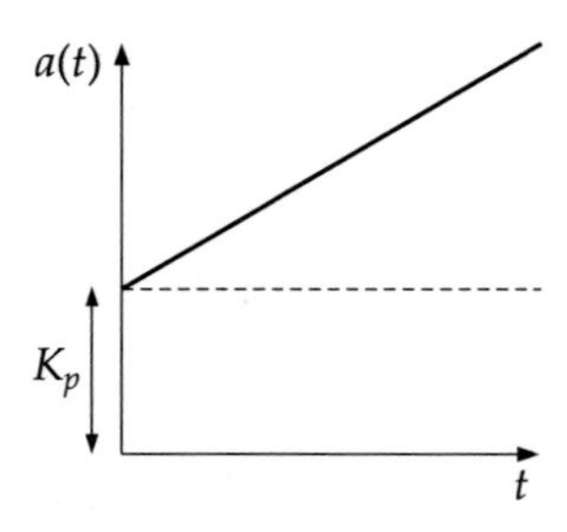

Figure 8.8 Actuating signal of a PI controller for the step error input.

Stability

To check the stability of a PI controller, let us consider a second order closed-loop system to which it is attached (see Figure 8.9).

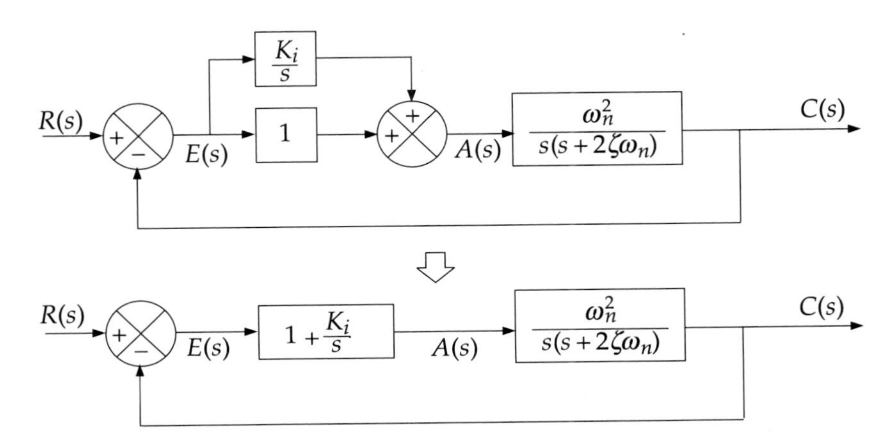

Figure 8.9 PI controller attached to second order plant.

The overall transfer function of the system is given by

$$T(s) = \frac{C(s)}{R(s)} = \frac{\left(1+\frac{K_i}{s}\right)\left[\frac{\omega_n^2}{s(s+2\zeta\omega_n)}\right]}{1+\left(1+\frac{K_i}{s}\right)\left[\frac{\omega_n^2}{s(s+2\zeta\omega_n)}\right]}$$

$$= \frac{(s+K_i)\omega_n^2}{s^3+2\zeta\omega_n s^2+(s+K_i)\omega_n^2}$$

The characteristic equation is

$$s^3+2\zeta\omega_n s^2+(s+K_i)\omega_n^2 = 0 \tag{8.10}$$

It possesses three roots. If $K_i > 2\zeta\omega_n$, then two roots have positive real parts that will make the system unstable. However, if $K_i < 2\zeta\omega_n$, then all the three roots have negative real parts indicating a stable system[1].

Now we discuss the steady-state error or, in other words, the accuracy that a PI controller offers.

[1]For a discussion on the location of poles vis-à-vis the stability of a system, see Section 9.3 at page 314. The condition can be worked out using Routh's rule.

Accuracy

The steady-state error for the system with PI controller is

$$e_{ss} = \lim_{s\to 0} sE(s) = \lim_{s\to 0} \frac{sR(s)}{1+G(s)H(s)}$$

$$= \lim_{s\to 0} \frac{s^3(s+2\zeta\omega_n)R(s)}{s^3+2\zeta\omega_n s^2+\omega_n^2 s++K_i\omega_n^2}$$

Step input. In the case of step input,

$$e_{ss} = \lim_{s\to 0} \frac{s^3(s+2\zeta\omega_n)}{s^3+2\zeta\omega_n s^2+\omega_n^2 s++K_i\omega_n^2} \cdot \frac{1}{s} = 0$$

Ramp input. In the case of ramp input,

$$e_{ss} = \lim_{s\to 0} \frac{s^3(s+2\zeta\omega_n)}{s^3+2\zeta\omega_n s^2+\omega_n^2 s++K_i\omega_n^2} \cdot \frac{1}{s^2} = 0$$

Parabolic input. Here,

$$e_{ss} = \lim_{s\to 0} \frac{s^3(s+2\zeta\omega_n)}{s^3+2\zeta\omega_n s^2+\omega_n^2 s++K_i\omega_n^2} \cdot \frac{1}{s^3} = \frac{2\zeta}{K_i\omega_n}$$

Table 8.4 reveals that the accuracy of a PI controller is superior to that of a P controller.

Table 8.4 Comparison between P and PI Controllers

Input	*Steady-state error*	
	P ($K_p = 1$)	*PI*
Step	0	0
Ramp	$\frac{2\zeta}{\omega_n}$	0
Parabolic	∞	$\frac{2\zeta}{K_i\omega_n}$

Advantages

The advantages of a PI controller are as follows:

1. e_{ss} is reduced, which means the accuracy improves.
2. The order and type of the system increase by 1. That improves damping and reduces overshoot.
3. The noise is filtered.
4. The offset is eliminated.

Realization

The simple op-amp realization of a PI controller is given in Figure 8.10.

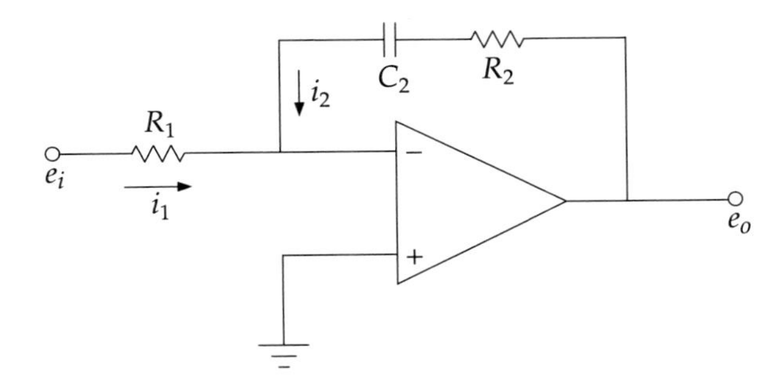

Figure 8.10 Op-amp realization of PI controller.

For the circuit given in the Figure,

$$i_1 = \frac{e_i}{R_1} = |i_2|$$

Therefore,

$$e_o = \frac{1}{C_2}\int |i_2|\, dt + i_2 R_2 = \frac{1}{R_1 C_2}\int |e_i|\, dt + \frac{R_2}{R_1} e_i \tag{8.11}$$

The Laplace transform of Eq. (8.11) yields

$$\frac{E_o(s)}{E_i(s)} = \frac{R_2}{R_1} + \frac{1}{R_1 C_2 s} \equiv K_p + \frac{K_i}{s}$$

8.7 PROPORTIONAL PLUS DERIVATIVE (PD) CONTROLLER

In a PD controller,

$$a(t) = K_p e(t) + K_d \frac{de(t)}{dt}$$

Its Laplace transform produces

$$\frac{A(s)}{E(s)} = 1 + K_d s$$

Control Action

If the error signal is a unit step, then

$$a(t) = K_p u(t) + K_d \frac{du(t)}{dt} = K_p + K_d \delta(t)$$

which indicates that the actuating signal is an impulse. However, if the error signal is a ramp, it is easy to see that the actuating signal is also a ramp (see Figure 8.11).

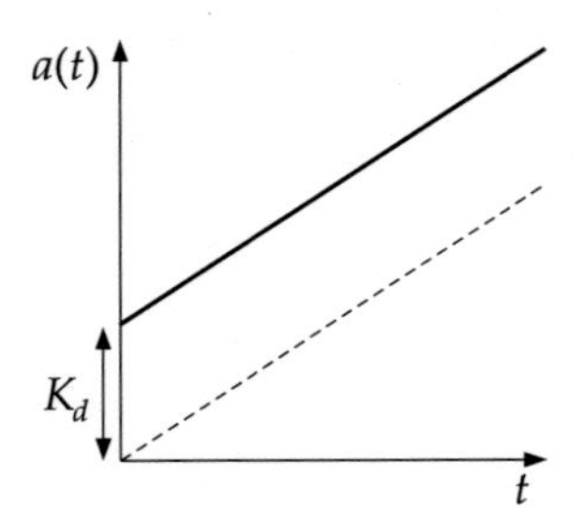

Figure 8.11 Actuating signal of a PD controller for ramp error signal.

Transient Response

Figure 8.12 gives the block diagram of the PD-controlled second order process.

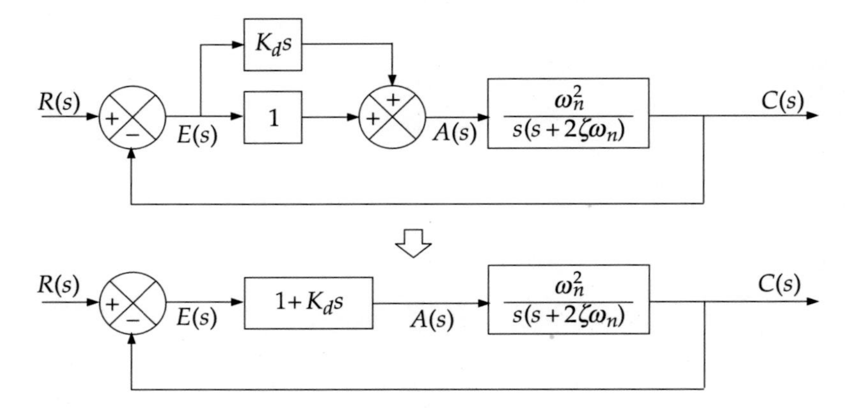

Figure 8.12 PD-controlled second order process.

Its transfer function is

$$T(s) = \frac{C(s)}{R(s)} = \frac{(1+K_d s)\omega_n^2}{s^2 + (2\zeta\omega_n + K_d\omega_n^2)s + \omega_n^2}$$

The characteristic equation

$$s^2 + (2\zeta\omega_n + K_d\omega_n^2)s + \omega_n^2 = 0$$

can be rewritten as

$$s^2 + 2\left(\zeta + \frac{K_d\omega_n}{2}\right)\omega_n s + \omega_n^2 = 0$$

$\Rightarrow$
$$s^2 + 2\zeta'\omega_n s + \omega_n^2 = 0$$

where

$$\zeta' = \zeta + \frac{K_d\omega_n}{2} \tag{8.12}$$

Equation (8.12) shows that the effective damping of the system increases by inducting a PD controller. As a result, the maximum overshoot is reduced (see Figure 7.9 for ζ vs. M_p graph).

Accuracy

Here,

$$e_{ss} = \lim_{s\to 0} \frac{sR(s)}{1+G(s)H(s)} = \lim_{s\to 0} \frac{s^2(s+2\zeta\omega_n)R(s)}{s^2+(2\zeta\omega_n+K_d\omega_n^2)s+\omega_n^2} \tag{8.13}$$

Step input. In the case of step input,

$$e_{ss} = \lim_{s\to 0} \frac{s^2(s+2\zeta\omega_n)}{s^2+(2\zeta\omega_n+K_d\omega_n^2)s+\omega_n^2} \cdot \frac{1}{s} = 0$$

Ramp input. In the case of ramp input,

$$e_{ss} = \lim_{s\to 0} \frac{s^2(s+2\zeta\omega_n)}{s^2+(2\zeta\omega_n+K_d\omega_n^2)s+\omega_n^2} \cdot \frac{1}{s^2} = \frac{2\zeta}{\omega_n} \tag{8.14}$$

Parabolic input. In the case of parabolic input,

$$e_{ss} = \lim_{s\to 0} \frac{s^2(s+2\zeta\omega_n)}{s^2+(2\zeta\omega_n+K_d\omega_n^2)s+\omega_n^2} \cdot \frac{1}{s^3} = \infty$$

Comparing Eq. (8.13) with Eq. (8.2), we find that the steady-state error is nearly the same for both P and PD controllers. But, the PD controller scores higher over the transient response because it increases the damping ratio and thus reduces the peak overshoot.

Realization

An op-amp realization of the PD controller circuit is shown in Figure 8.13.

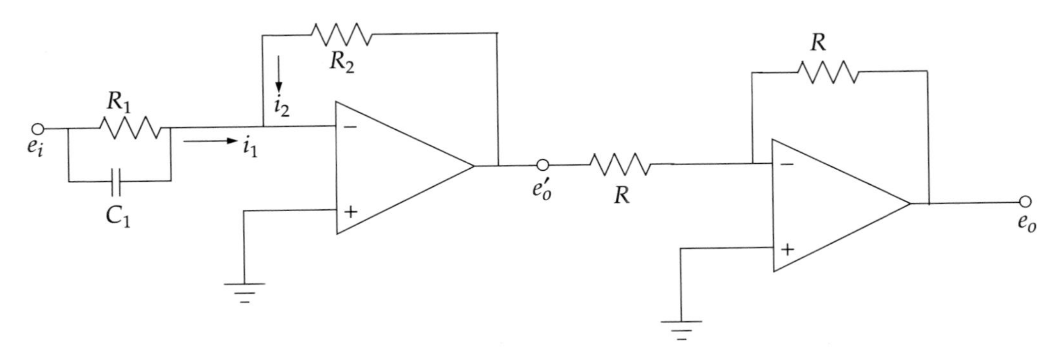

Figure 8.13 PD controller circuit.

From the circuit we get

$$i_1 = -i_2, \qquad e_o = -e_o'$$

Applying KVL to the Laplace-transformed circuit, we obtain

$$E_i(s) = \frac{R_1 \cdot \frac{1}{sC_1}}{R_1 + \frac{1}{sC_1}} \cdot I_1(s)$$

$$\Rightarrow \qquad I_1(s) = \frac{sR_1C_1 + 1}{R_1} \cdot E_i(s) \tag{8.15}$$

$$E_o' = E_o(s) = R_2 I_2(s) = R_2 |I_1(s)|$$

$$= R_2 \left| \frac{sR_1C_1 + 1}{R_1} \right| \cdot E_i(s) \quad \text{[from Eq. (8.15)]}$$

$$= \left| R_2C_1 s + \frac{R_2}{R_1} \right| \cdot E_i(s)$$

Thus,

$$\frac{E_o(s)}{E_i(s)} = \frac{R_2}{R_1} + R_2C_1 s \equiv K_p + K_d s$$

8.8 PROPORTIONAL PLUS INTEGRAL PLUS DERIVATIVE (PID) CONTROLLER

As the name suggests, in a PID controller the actuating signal has the following form:

$$a(t) = K_p e(t) + K_i \int e(t)\ dt + K_d \frac{de(t)}{dt}$$

The block diagram of the PID controller action is shown in Figure 8.14.

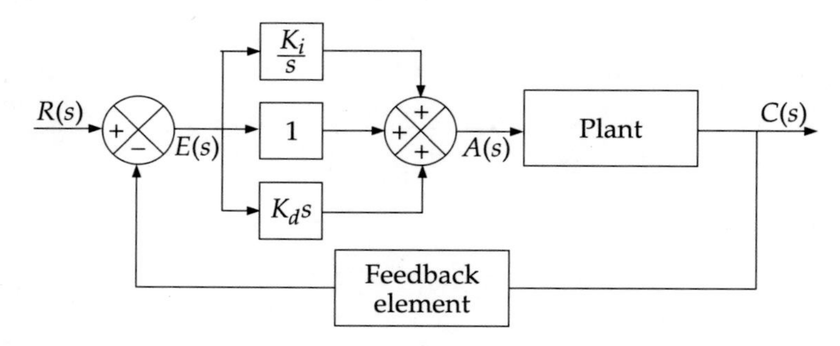

Figure 8.14 Illustration of PID control action.

Without making a detailed analysis, we may observe that the PD control improves the transient part and the PI control improves the steady-state part. Therefore, from the principle of superposition of errors, we may conclude that a combination of PI and PD control improves the overall control.

The op-amp realization of a PID controller is given in Figure 8.15.

Table 8.5 presents a summary of the salient features of three basic modes of control to help select a controller for a particular process.

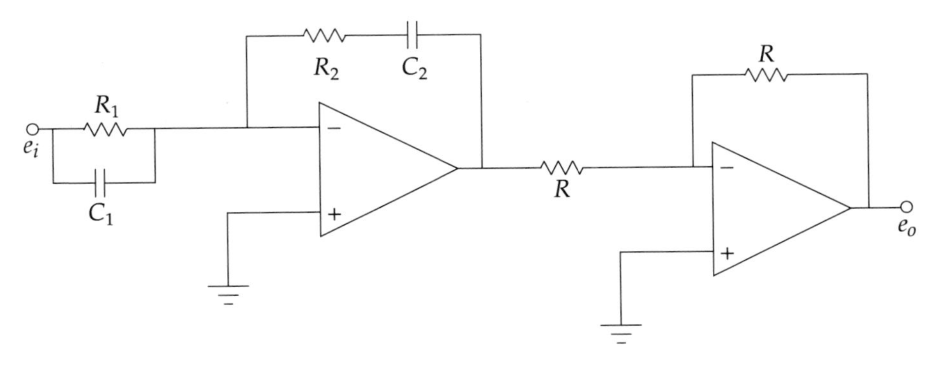

Figure 8.15 PID controller circuit.

Table 8.5 Selection of Control Action

Proportional action is chosen if:	*Integral action is chosen if:*	*Derivative action is chosen if:*
1. Load changes are small.	1. Offset is not permitted.	1. Transient response parameters such as rise time, maximum overshoot etc. are to be improved upon.
2. Offset can be tolerated.	2. High degree of accuracy is required.	2. Plant load changes have to be tracked by the controller.
3. Error signal, which is small, needs amplification.		

8.9 RATE FEEDBACK CONTROLLER

Also called *derivative feedback controller* and *output derivative controller*, this controller incorporates derivative action in the feedback path as shown in Figure 8.16.

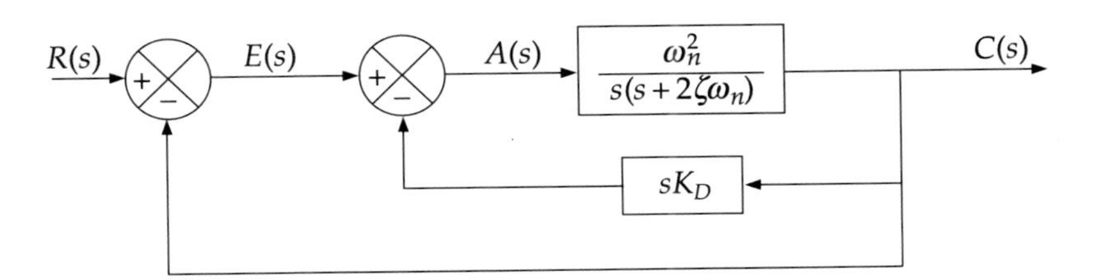

Figure 8.16 Rate feedback control action.

Here,

$$a(t) = e(t) - K_D \frac{dc(t)}{dt}$$

Thus, the forward path gain is

$$G(s) = \frac{C(s)}{R(s)} = \frac{\dfrac{\omega_n^2}{s(s+2\zeta\omega_n)}}{1+\dfrac{sK_D\omega_n^2}{s(s+2\zeta\omega_n)}} = \frac{\omega_n^2}{s^2+(2\zeta\omega_n+K_D\omega_n^2)s}$$

The overall transfer function is

$$\frac{C(s)}{R(s)} = \frac{G(s)}{1+G(s)} = \frac{\omega_n^2}{s^2+(2\zeta\omega_n+K_D\omega_n^2)s+\omega_n^2}$$

The characteristic equation is

$$s^2+(2\zeta\omega_n+K_D\omega_n^2)s+\omega_n^2 = 0 \tag{8.16}$$

Equation (8.16) can be written in the standard form of the characteristic equation of a second order system as

$$s^2+2\zeta'\omega_n s+\omega_n^2 = 0$$

where

$$\zeta' = \frac{2\zeta\omega_n+K_D\omega_n^2}{2\omega_n} = \zeta+\frac{K_D\omega_n}{2} \tag{8.17}$$

It is clear from Eq. (8.17), that the damping ratio is increased. Consequently,

(a) the maximum overshoot is decreased[2], and

(b) the rise-time is increased[3].

Now, let us check the accuracy of the rate feedback controller.

Accuracy

The steady-state error for the rate feedback controller is given by

$$e_{ss} = \lim_{s\to 0}\frac{sR(s)}{1+G(s)H(s)} = \lim_{s\to 0}\frac{s[s^2+(2\zeta\omega_n+K_D\omega_n^2)s]R(s)}{s^2+(2\zeta\omega_n+K_D\omega_n^2)s+\omega_n^2}$$

$$= \lim_{s\to 0}\frac{s^2(s+2\zeta\omega_n+K_D\omega_n^2)R(s)}{s^2+(2\zeta\omega_n+K_D\omega_n^2)s+\omega_n^2}$$

Step input. In the case of step input,

$$e_{ss} = \lim_{s\to 0}\frac{s^2(s+2\zeta\omega_n+K_D\omega_n^2)}{s^2+(2\zeta\omega_n+K_D\omega_n^2)s+\omega_n^2}\cdot\frac{1}{s} = 0$$

[2]See Figure 7.9.

[3]See Eq. (8.14).

Ramp input. Here,

$$e_{ss} = \lim_{s \to 0} \frac{s^2(s + 2\zeta\omega_n + K_D\omega_n^2)}{s^2 + (2\zeta\omega_n + K_D\omega_n^2)s + \omega_n^2} \cdot \frac{1}{s^2}$$

$$= \frac{2\zeta\omega_n + K_D\omega_n^2}{\omega_n^2}$$

$$= \frac{2\zeta}{\omega_n} + K_D \tag{8.18}$$

Parabolic input. In the case of parabolic input,

$$e_{ss} = \lim_{s \to 0} \frac{s^2(s + 2\zeta\omega_n + K_D\omega_n^2)}{s^2 + (2\zeta\omega_n + K_D\omega_n^2)s + \omega_n^2} \cdot \frac{1}{s^3} = \infty$$

A comparison between the PD controller and the rate feedback controller, both of which employ the derivative action, will be of interest here.

Comparison between PD and Rate Feedback Controllers

1. Damping ratio increases in both the cases. As a consequence, peak overshoot and settling time decrease while rise time increases in both the cases.
2. The PD controller introduces an additional zero at $s = -1/K_d$ in the closed loop transfer function while the rate feedback controller does not. We have seen in Section 7.6 that the addition of a zero reduces rise time but increases the peak overshoot. So, even if $K_d = K_D$, the responses will not be the same because of this zero factor.
3. The type of the system does not change in both the cases, which implies that the pattern of the steady-state error will remain the same for both the cases. However, by comparing Eq. (8.14) with Eq. (8.18), we find that the ramp input error is higher for the rate feedback controller.

EXAMPLE 8.1. In the system shown in Figure 8.17, the damping ratio is to be made 0.8 using derivative control. Determine the value of K_d.

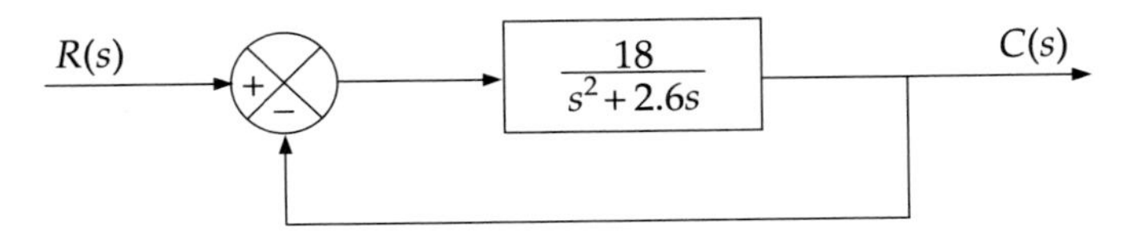

Figure 8.17 System (Example 8.1).

Solution Here,

$$T(s) = \frac{C(s)}{R(s)} = \frac{\dfrac{18}{s^2 + 2.6s}}{1 + \dfrac{18}{s^2 + 2.6s}} = \frac{18}{s^2 + 2.6s + 18}$$

Therefore, the characteristic equation is

$$s^2 + 2.6s + 18 = 0 \equiv s^2 + 2\zeta\omega_n s + \omega_n^2$$

By comparison,

$$\omega_n^2 = 18 \quad \Rightarrow \quad \omega_n = 4.243 \text{ rad/s}$$

$$2\zeta\omega_n = 2.6 \quad \Rightarrow \quad \zeta = \frac{2.6}{(2)(4.243)} = 0.3$$

Now, with the derivative control, ζ' has to be 0.8. So, from Eq. (8.17),

$$K_d = (\zeta' - \zeta)\frac{2}{\omega_n} = (0.8 - 0.3)\frac{2}{4.243} = 0.24$$

EXAMPLE 8.2. In the system shown in Figure 8.18, determine the derivative feedback constant K_D which will increase the damping factor of the system to 0.6. What is the steady-state error to unit ramp input with this setting of the derivative feedback constant?

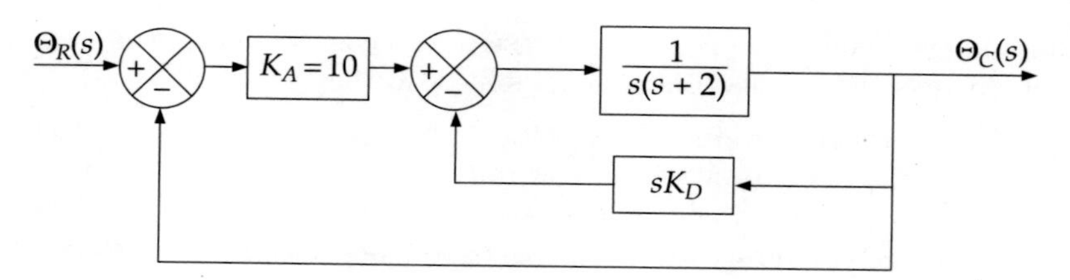

Figure 8.18 System (Example 8.2).

Solution Given

$$G_1(s) = \frac{1}{s(s+2)}, \qquad H_1(s) = sK_0$$

The transfer function of the inner feedback loop is

$$G(s) = \frac{G_1(s)}{1 + G_1(s)H_1(s)} = \frac{1}{s(s+2) + sK_D}$$

And the overall transfer function is

$$T(s) = \frac{\Theta_C(s)}{\Theta_R(s)} = \frac{K_A G(s)}{1 + K_A G(s)}$$

$$= \frac{K_A}{s(s+2) + sK_D + K_A}$$

$$= \frac{10}{s^2 + (2 + K_D)s + 10}$$

The characteristic equation is

$$s^2 + (2 + K_D)s + 10 = 0 \equiv s^2 + 2\zeta\omega_n s + \omega_n^2$$

By comparison,

$$\omega_n^2 = 10 \quad \Rightarrow \quad \omega_n = 3.162 \text{ rad/s}$$

$$2\zeta\omega_n = 2 + K_D \quad \Rightarrow \quad K_D = 2\zeta\omega_n - 2 = 2(0.6)(3.162) - 2 = 1.8$$

Therefore,

$$e_{ss} = \lim_{s\to 0} \frac{s\Theta_R(s)}{1 + K_A G(s)} = \lim_{s\to 0} \frac{s \cdot (1/s^2)}{1 + \dfrac{10}{s^2 + 2s + 1.8s}}$$

$$= \lim_{s\to 0} \frac{s + 3.8}{s^2 + 3.8s + 10}$$

$$= 0.38$$

EXAMPLE 8.3. Figure 8.19 shows the block diagram of a position control system.

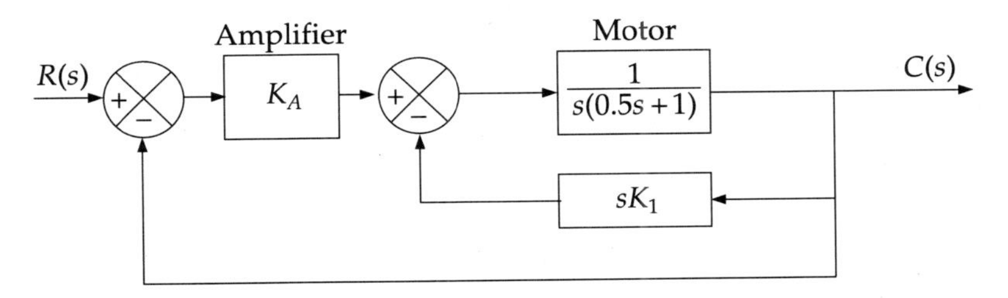

Figure 8.19 Position control system (Example 8.3).

(a) In the absence of the derivative feedback, i.e. $K_1 = 0$, determine the damping ratio of the system for amplifier gain $K_A = 5$.
(b) Find suitable values of parameters K_A and K_1 so that the damping ratio of the system is increased to 0.7 without affecting the steady state error as obtained in part (a).

Solution

(a) Here,

$$T(s) = \frac{C(s)}{R(s)} = \frac{\dfrac{s}{s(0.5s+1)}}{1 + \dfrac{s}{s(0.5s+1)}} = \frac{10}{s^2 + 2s + 10}$$

Therefore, the characteristic equation is

$$s^2 + 2s + 10 = 0 \equiv s^2 + 2\zeta\omega_n s + \omega_n^2$$

By comparison,

$$\omega_n^2 = 10 \quad \Rightarrow \quad \omega_n = 3.162 \text{ rad/s}$$

$$2\zeta\omega_n = 2 \quad \Rightarrow \quad \zeta = \frac{1}{3.162} = 0.316$$

(b) The steady-state error (for a ramp input) for part (a) is given by

$$e_{ss} = \lim_{s \to 0} \frac{sR(s)}{1 + \dfrac{5}{s(0.5s+1)}} = \lim_{s \to 0} \frac{0.5s+1}{s(0.5s+1)+5} = 0.2$$

Now, the transfer function for the inner loop is

$$G_1(s) = \frac{\dfrac{1}{s(0.5s+1)}}{1 + \dfrac{sK_1}{s(0.5s+1)}} = \frac{1}{s(0.5s+1) + sK_1}$$

So, the overall transfer function is

$$T(s) = \frac{C(s)}{R(s)} = \frac{K_A G_1(s)}{1 + K_A G_1(s)} = \frac{\dfrac{K_A}{s(0.5s+1)+sK_1}}{1 + \dfrac{K_A}{s(0.5s+1)+sK_1}}$$

$$= \frac{K_A}{0.5s^2 + (K_1+1)s + K_A}$$

$$= \frac{K_A/0.5}{s^2 + \dfrac{K_1+1}{0.5} + \dfrac{K_A}{0.5}}$$

The characteristic equation is

$$s^2 + \frac{K_1+1}{0.5} + \frac{K_A}{0.5} = 0 \equiv s^2 + 2\zeta\omega_n s + \omega_n^2$$

By comparison,

$$\omega_{n'}^2 = \frac{K_A}{0.5} \quad \Rightarrow \quad \omega_{n'} = \sqrt{\frac{K_A}{0.5}}$$

$$2\zeta\omega_{n'} = \frac{K_1+1}{0.5} \quad \Rightarrow \quad \zeta' = \frac{1}{2\omega_n'} = \frac{(K_1+1)}{0.5 \times 2}\sqrt{\frac{0.5}{K_A}} = 0.7 \text{ (given)}$$

Therefore,

$$\frac{K_1+1}{\sqrt{K_A}} = \frac{0.7}{\sqrt{0.5}} \tag{i}$$

Now,

$$e_{ss} = \lim_{s\to 0} \frac{s \times \dfrac{1}{s^2}}{1 + \dfrac{K_A}{s(0.5s+1) + sK_1}} = \lim_{s\to 0} \frac{0.5s + 1 + K_1}{s(0.5s+1+K_1) + K_A}$$

$$= \frac{1 + K_1}{K_A}$$

Thus,

$$\frac{K_1 + 1}{K_A} = 0.2 \quad \text{[determined for part (a)]} \qquad \text{(ii)}$$

Dividing Eq. (i) by Eq. (ii), we get

$$\sqrt{K_A} = \frac{0.7}{(\sqrt{0.5})(0.2)}$$

$$\Rightarrow \qquad K_A = 24.5 \qquad \text{(iii)}$$

From Eqs. (ii) and (iii),

$$K_1 = 0.2K_A - 1 = 3.9$$

REVIEW QUESTIONS

8.1 Choose the correct answer

(i) Excellent transient and steady state response are provided by
   (a) proportional action system
   (b) proportional + differential action system
   (c) proportional + integral + differential action system
   (d) none of the above

(ii) Derivative controllers are never used alone because of
   (a) their large input impedance
   (b) they amplify noise signal and cause saturation of the actuator
   (c) they worsen the transient response of the system
   (d) none of the above

(iii) Which of the following control actions is to be employed for eliminating the offset?
   (a) Proportional control
   (b) Proportional plus derivative control
   (c) Proportional plus integral control
   (d) None of these

(iv) By the use of PD control to a second order system, the rise time

(a) decreases (b) increases

(c) remains the same (d) has no effect

(v) Derivative feedback control

(a) increases rise time

(b) increases overshoot

(c) decreases the steady-state error

(d) does not affect the steady-state error

8.2 (a) Why is derivative controller never used alone?

(b) What is the effect of derivative controller on peak overshoot of a second order underdamped system?

8.3 (a) Derive the expression for PI, PD and PID controller.

(b) Write down the advantages and disadvantages of PI, PD and PID control action.

8.4 Discuss, with the help of block diagrams, PI, PD and PID controllers on transient response of a second order system.

8.5 (a) Explain the function of a PID controller enumerating the benefits of using a PID controller.

(b) Draw the circuit diagram of a PID controller using op-amp.

8.6 Describe the procedure adopted in the design of a PID controller consisting of a PI portion connected in cascade with a PD portion.

8.7 Consider the system with

$$G(s) = \frac{10^{-4}}{s^2 - 1.21}, \quad H(s) = 1$$

Design a proportional plus derivative controller $G_C(s) = K_p(1 + K_d s)$ such that the damping ratio $\zeta = 0.7$ and undamped natural frequency $\omega_n = 0.5$ rad/s for the closed loop system.

8.8 Figure 8.20 shows a feedback control system employing a proportional plus error derivative controller.

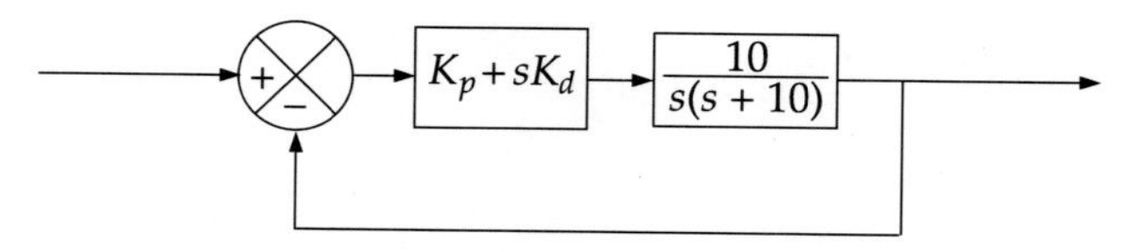

Figure 8.20 Feedback control system (Question 8.8).

(a) State the type of the plant. Determine the closed-loop transfer function and state the order of the system.

(b) Find out the values of K_p and K_d so that the following time-domain specifications are met: e_{ss} to a unit ramp ≤ 0.02 and % peak overshoot ≤ 15.

(c) How does the zero in the closed-loop transfer function (introduced by the error derivative action) improve the speed of response of the system?

Answers to Selected Questions

8.1 (i) (c); (ii) (b); (iii) (c); (iv) (b); (v) (a).

8.7 $K_p = 1.46 \times 10^4$, $K_d = 0.48$.

8.8 (a) Type 1, CLTF $= \dfrac{10(K_p + sK_d)}{s^2 + 10s(K_d + 1) + 10K_p}$, Order 2; (b) $K_p = 50$, $K_d = 1.73$.

CHAPTER 9

Stability Concepts

When a system receives an input, its output passes through an initial transient stage. After the transients die down, it should reach a steady state. The question that a designer is concerned with is whether the steady state is stable or not. In fact, the designer should be interested in testing the stability of the system, not only before activating it, but even before it is constructed.

The question is: What is the precise definition of stability?

9.1 DIFFERENT KINDS OF STABILITY

In connection with the stability of a linear time-invariant system, a few terms are generally used. These are defined in Table 9.1.

Table 9.1 Definition of Different Kinds of Stability

Kind of stability	*Definition*
Bounded input bounded output (BIBO) stability	Output of the system should be bounded, i.e. it should remain confined within a limited range when a bounded input is applied to it.
Asymptotic (or zero input) stability	In the absence of any input, there should be no output from the system.
Absolute stability	If a bounded input is applied, the system remains stable for all values of system parameters.
Relative stability	It indicates how close the system is to instability. The values of *gain margin* and *phase margin* are measures of relative stability.
Conditional stability	Stability of a system for a particular range of any/some parameter(s).

9.2 STABILITY CONDITION FOR A BIBO SYSTEM

Let us consider a transfer function defined as

$$G(s) = \frac{C(s)}{R(s)} = \frac{b_0 s^m + b_1 s^{m-1} + b_2 s^{m-2} + \cdots + b_{m-1} s^1 + b_m}{a_0 s^n + a_1 s^{n-1} + a_2 s^{n-2} + \cdots + a_{n-1} s + a_n} \tag{9.1}$$

We know, for the impulse input $R(s) = 1$, and, therefore, $C(s) = G(s)$. Taking inverse Laplace transform, $c(t) = g(t)$. This is true only for the impulse input. But it can be shown that for any input $r(t)$, the output $c(t)$ can be written in the form of convolution[1]

$$c(t) = \int_0^{\infty} g(\tau) r(t-\tau)\, d\tau \tag{9.2}$$

where $g(\tau)$ is the impulse response of the system. If $r(t)$ remains bounded (i.e., finite) and $c(t)$ has to remain bounded for all t, we find from Eq. (9.2) that it is necessary that

$$\int_0^{\infty} g(\tau)\, d\tau = \text{finite} \tag{9.3}$$

We wrote Eq. (9.3) intuitively. Mathematically, it can be derived as follows. We take absolute value on both sides of Eq. (9.2) to get

$$|c(t)| = \left| \int_0^{\infty} g(\tau) r(t-\tau)\, d\tau \right| \tag{9.4}$$

A theorem states that the absolute value of an integral is no greater than the integral of the absolute value of the integrand. Therefore, it follows from Eq. (9.4) that

$$\begin{aligned} |c(t)| &\le \int_0^{\infty} |g(\tau) r(t-\tau)|\, d\tau \\ &\le \int_0^{\infty} |g(\tau)||r(t-\tau)|\, d\tau \end{aligned} \tag{9.5}$$

If $r(t)$ is bounded, it means

$$|r(t)| \le M < \infty \quad \text{for } t \ge t_0 \tag{9.6}$$

Substitution of Eq. (9.6) in Eq. (9.5) implies

$$\begin{aligned} |c(t)| &\le \int_0^{\infty} M|g(\tau)\, d\tau \\ &\le M \int_0^{\infty} |g(\tau)\, d\tau \end{aligned} \tag{9.7}$$

Now, if $c(t)$ has to remain bounded, it means

$$|c(t)| \le N < \infty \quad \text{for } t \ge t_0 \tag{9.8}$$

[1] See Appendix C at page 665.

Combining Eqs. (9.7) and (9.8), we get

$$\int_0^{\infty} |g(\tau)\, d\tau \leq P < \infty \tag{9.9}$$

Equation (9.9) is the same as that we arrived at Eq. (9.3) from intuitive reasoning.

9.3 EFFECT OF LOCATION OF POLES ON STABILITY

We know that the overall transfer function of a system is defined as

$$T(s) = \frac{C(s)}{R(s)} \tag{9.10}$$

where the terms have their usual significance. Poles are the values of s for which $T(s) \to \infty$. That happens when $R(s) = 0$. So, by solving the characteristic equation $R(s) = 0$, we can get the poles of the transfer function of the system. We will consider the stability of systems by placing poles at different locations in the s-plane.

Single Pole at the Origin

Here, the transfer function is given by

$$\frac{C(s)}{R(s)} = \frac{1}{s}$$

Its impulse response is

$$c(t) = 1$$

The corresponding plot in the s-plane and the response are shown in Figure 9.1. Obviously, the output is bounded at all times, which means it is a stable system.

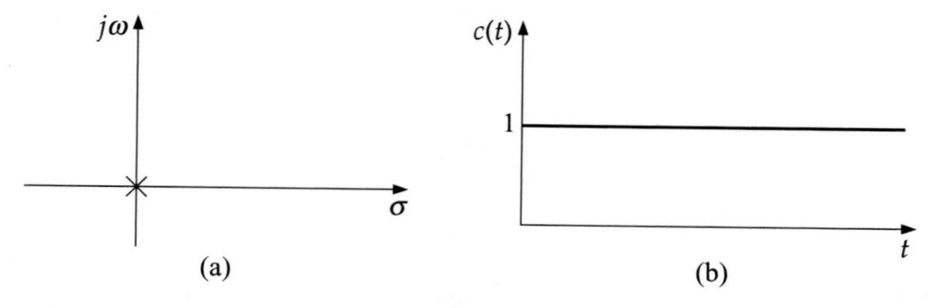

Figure 9.1 (a) Pole location; (b) impulse response for having a single pole at the origin.

A Pair of Poles at the Origin

The presence of a pair of poles at the origin implies a transfer function of the form

$$\frac{C(s)}{R(s)} = \frac{1}{s^2}$$

The corresponding impulse response is the ramp function, i.e.,

$$c(t) = t$$

The pole plot and response are shown in Figure 9.2. The response shows the output is not bounded as $t \to \infty$. So, the system is unstable.

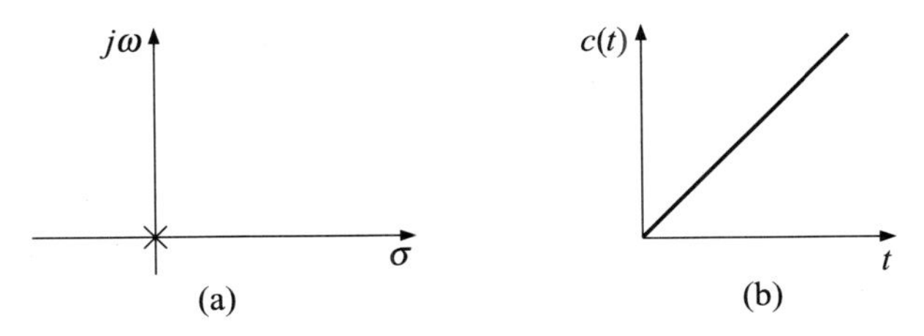

Figure 9.2 (a) Pole plot; (b) impulse response for having a pair of poles at the origin.

Single Real Pole at the Right Hand Plane (RHP)

Let us consider the simple transfer function

$$\frac{C(s)}{R(s)} = \frac{1}{s-1}$$

The impulse response, here, is

$$c(t) = e^t$$

The s-plane plot of the pole and the response are shown in Figure 9.3. The response clearly indicates that the output is not bounded as $t \to \infty$. So, the system is unstable.

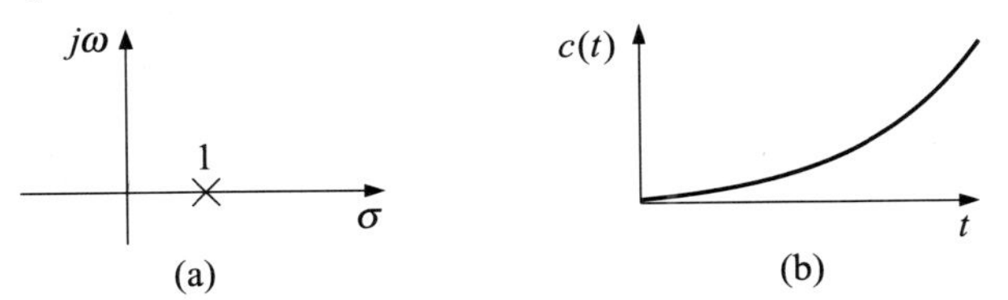

Figure 9.3 (a) Pole plot; (b) impulse response for having a single pole at the RHP.

Single Real Pole at the Left Hand Plane (LHP)

Let our transfer function be

$$\frac{C(s)}{R(s)} = \frac{1}{s+1} \tag{9.11}$$

The corresponding impulse response is

$$c(t) = e^{-t} \tag{9.12}$$

The pole plot of Eq. (9.11) and response plot of Eq. (9.12) are shown in Figure 9.4. The latter shows that the output becomes zero as $t \to \infty$. Consequently, the system is stable.

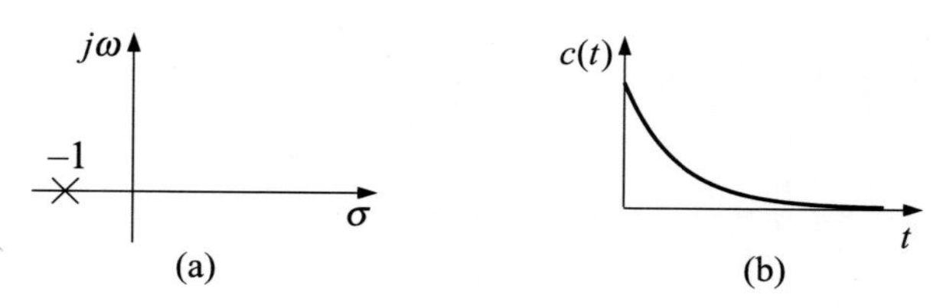

Figure 9.4 (a) Pole plot; (b) impulse response for having a single pole at the LHP.

A Pair of Poles on the Imaginary Axis

A transfer function having the attribute of a pair of poles on the imaginary axis is

$$\frac{C(s)}{R(s)} = \frac{1}{s^2+1} \tag{9.13}$$

Its impulse response is

$$c(t) = \sin t \tag{9.14}$$

The pole plot of Eq. (9.13) and the response plot of Eq. (9.14) are presented in Figure 9.5. The response, oscillating between +1 and −1, indicates that the system is marginally (or in a limited way) stable.

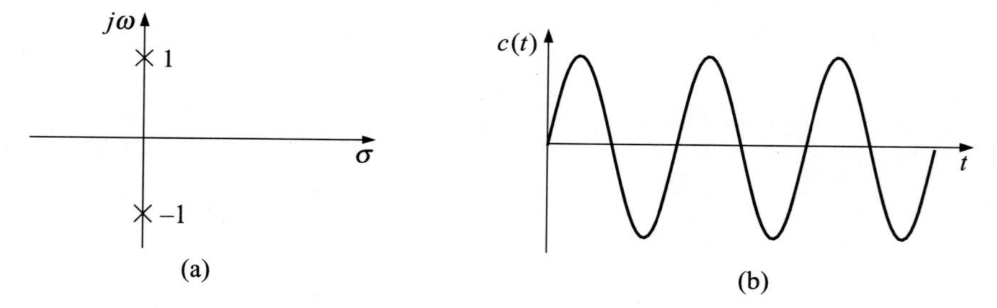

Figure 9.5 (a) Pole plot; (b) impulse response for having a pair of poles on the imaginary axis.

Two Pairs of Repeated Poles on the Imaginary Axis

Let the transfer function be

$$\frac{C(s)}{R(s)} = \frac{1}{(s^2+1)^2} \tag{9.15}$$

The impulse response is

$$c(t) = t\,\sin t \tag{9.16}$$

The plots of poles for Eq. (9.15) and response given by Eq. (9.16) are shown in Figure 9.6. It is abundantly clear from the response graph that the system is unstable if it possesses two pairs of repeated poles on the imaginary axis.

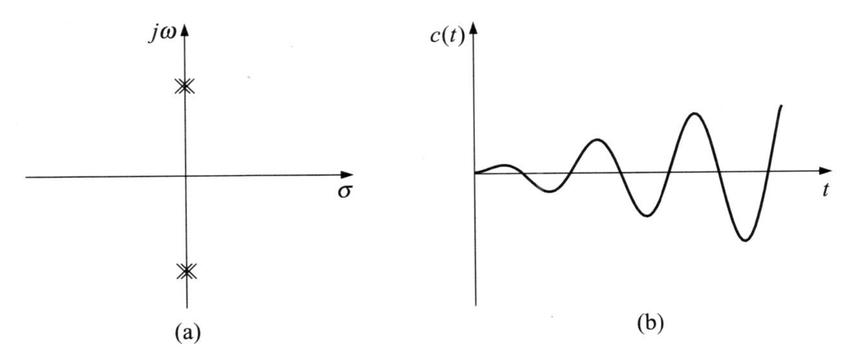

Figure 9.6 (a) Pole plot; (b) impulse response when there are two pairs of repeated poles on the imaginary axis.

A Pair of Complex Poles in the RHP

Let the corresponding transfer function be

$$\frac{C(s)}{R(s)} = \frac{1}{(s-1)^2+1} \tag{9.17}$$

Its impulse response is

$$c(t) = e^{t} \sin t \tag{9.18}$$

The poles corresponding to Eq. (9.17) and the output represented by Eq. (9.18) are shown in Figure 9.7. The output clearly shows that the system having this kind of transfer function is unstable.

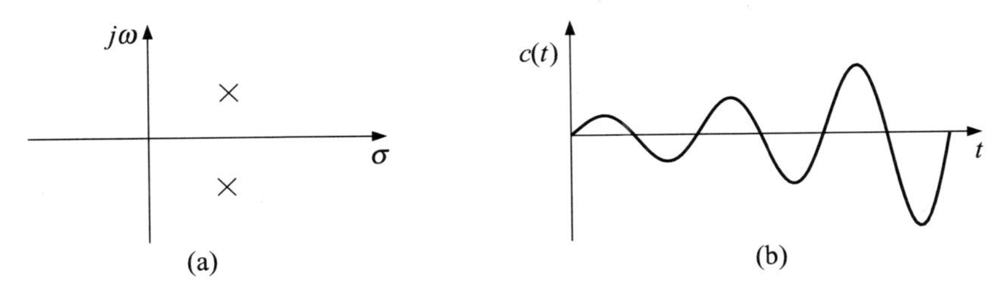

Figure 9.7 (a) Pole plot for Eq. (9.17); (b) response plot of Eq. (9.18).

A Pair of Complex Poles in the LHP

Here, we assume that the transfer function is

$$\frac{C(s)}{R(s)} = \frac{1}{(s+1)^2+1} \tag{9.19}$$

The corresponding impulse response is, therefore,

$$c(t) = e^{-t} \sin t \tag{9.20}$$

We plot the poles of Eq. (9.19) and the response given by Eq. (9.20) in Figure 9.8. From the appearance of the response there is no doubt that the system is stable.

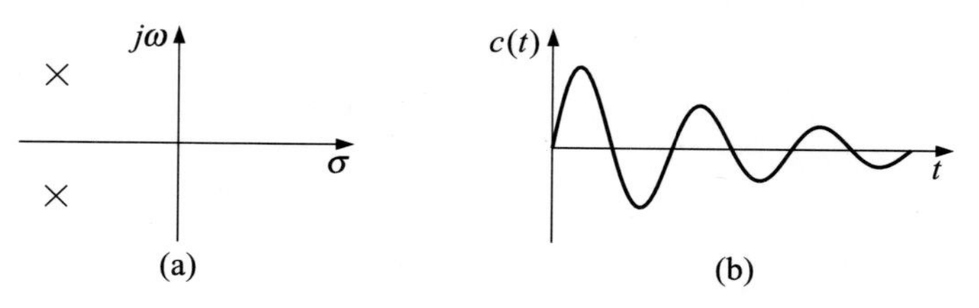

Figure 9.8 (a) Pole plot for Eq. (9.19); (b) response plot of Eq. (9.20).

Table 9.2 summarizes what we observed from the impulse response plots of transfer functions having poles at different locations of the s-plane vis-a-vis the stability of the systems.

Table 9.2 Summary of Observations

Location of poles	*Stability*
Negative (i.e. left) half of the s-plane, i.e. LHP	Stable
Positive (i.e. right) half of the s-plane, i.e. RHP	Unstable
Conjugate pair of poles on the imaginary axis	Continuous oscillation. System marginally (or limitedly) stable
Repeated poles on the imaginary axis	Unstable
Repeated poles at the origin	Unstable

Now let us examine a few stability problems in the light of these observations. And then we shall see how the stability of systems can be studied by using different methods.

EXAMPLE 9.1. Comment on the stability of the systems, the open-loop transfer functions of which are given by

(a) $G(s) = \dfrac{10(s+2)}{(s+1)(s+4)(s+5)}$

(b) $G(s) = \dfrac{5(s-3)}{(s+1)(s^2+4)}$

(c) $G(s) = \dfrac{Ks}{(s^2+1)^2(s+5)}$

(d) $G(s) = \dfrac{20}{(s-2)(s+3)}$

(e) $G(s) = \dfrac{30}{s^4+30s^3+s^2+10s}$

Solution

(a) Poles are at $s = -1,\ -4,\ -5$ and zero at $s = -2$. Stable.

(b) Poles are at $s = -1,\ \pm\jmath 2$ and zero at $s = 3$. Marginally stable.

(c) Poles are at $s = -5, \pm\jmath$ (double). Unstable.

(d) Poles are at $s = 2, -3$. Unstable.

(e) One pole is at $s = 0$. May be stable.

EXAMPLE 9.2. Match the roots in the s-plane with the impulse responses shown in the adjacent figure.

(a) Two imaginary roots

(b) Two complex roots in the LHP

(c) A single root on the negative real axis

(d) A single root at the origin

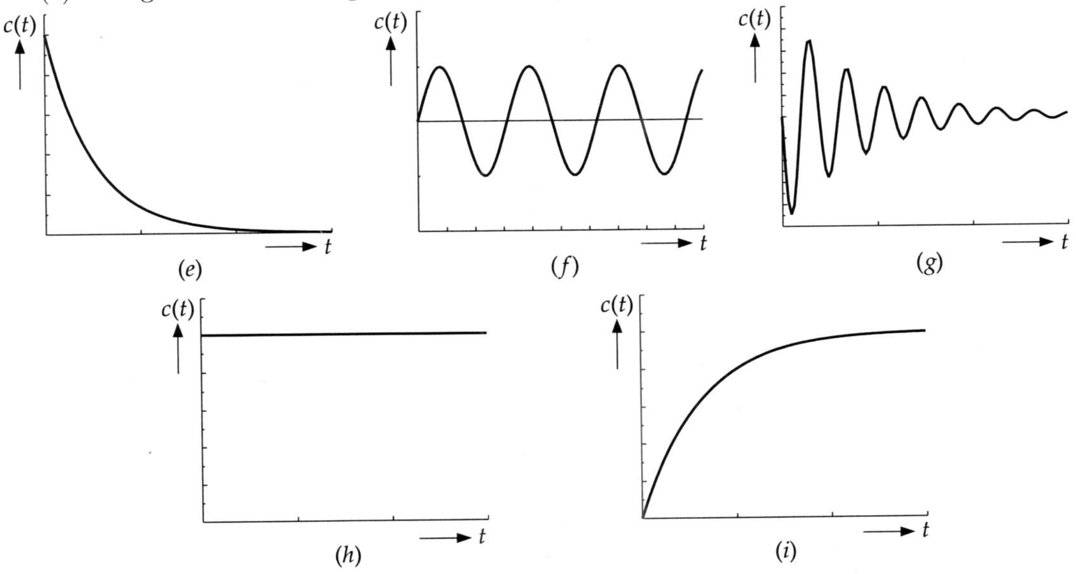

Solution $(a) \to (f)$, $(b) \to (g)$, $(c) \to (e)$, $(d) \to (h)$.

9.4 METHODS OF STUDY OF STABILITY

In this book we will consider the following five methods that are generally used to study the stability of a system:

1. Hurwitz-Routh criteria
2. Root-locus technique
3. Bode plot
4. Polar plot
5. Nyquist plot

While the first two methods deal with the study of the problem in the s-plane, the remaining methods are concerned with the study of problems in the frequency-domain (i.e. $\jmath\omega$-plane).

REVIEW QUESTIONS

9.1 Choose the correct answer:

(i) A system which has some roots with real parts equal to zero, but none with positive real parts is
   (a) relatively stable (b) marginally stable
   (c) absolutely stable (d) absolutely unstable

(ii) The system, represented by its transfer function, has some poles lying on the imaginary axis of s-plane. The system is
   (a) unconditionally stable (b) conditionally stable
   (c) stable (d) marginally stable

(iii) A system is stable if
   (a) bounded inputs produce bounded outputs
   (b) unbounded inputs produce bounded outputs
   (c) bounded inputs produce unbounded outputs
   (d) all bounded inputs produce bounded outputs

(iv) If the poles corresponding to the transfer function of a system lie on the left half of the s-plane, the system
   (a) is unstable (b) is marginally stable
   (c) achieves stability with time (d) behaviour is unpredictable

(v) If a conjugate pair of poles corresponding to the transfer function of a system lie on the imaginary axis of the s-plane, the system
   (a) is unstable (b) is marginally stable
   (c) achieves stability with time (d) behaviour is unpredictable

(vi) If repeated poles corresponding to the transfer function of a system lie on the imaginary axis of the s-plane, the system
   (a) is unstable (b) is marginally stable
   (c) achieves stability with time (d) behaviour is unpredictable

(vii) A system is always stable if in the s-plane
   (a) all poles are in the left half and zeros are in the right half
   (b) all poles and zeros are in the left half
   (c) all poles and zeros are in the right half
   (d) all poles in the right half and zeros are in the left half

(viii) The system whose characteristic equation has the following roots is marginally stable
   (a) $-j,\ j,\ -1$ (b) $-3, -2, 0$
   (c) $-2 + j3, -2 - j3, -2$ (d) $-3, -2, -1$

(ix) An example of bounded input is
   (a) e^{-4t} (b) e^{2t}
   (c) t (d) $e^{t} \sin t$

9.2 Define the following:
   (a) Absolutely stable system
   (b) Conditionally stable system
   (c) Asymptotically stable system

9.3 (a) What is the necessary condition of stability?
   (b) Define stable, unstable and limitedly stable systems.
   (c) What do you mean by bounded-input bounded-output (BIBO) stability?

9.4 Show that for a bounded input bounded output stable system $\int_0^\infty |g(\tau)|d\tau$ is finite, where $g(\tau)$ is the impulse response of the system.

Answers to Selected Questions

9.1 (i) (b); (ii) (b); (iii) (a); (iv) (c); (v) (b); (vi) (a); (vii) (b); (viii) (a); (ix) (a).

CHAPTER 10

Hurwitz-Routh Stability Criteria

We know that the closed-loop transfer function of a linear time-invariant system can be written as

$$\frac{C(s)}{R(s)} = \frac{b_0 s^m + b_1 s^{m-1} + \cdots + b_{m-1}s + b_m}{a_0 s^n + a_1 s^{m-1} + \cdots + a_{n-1}s + a_n}$$

which generates the characteristic equation

$$a_0 s^n + a_1 s^{m-1} + \cdots + a_{n-1}s + a_n = 0 \tag{10.1}$$

The roots of this equation represent closed-loop poles, on the location of which the stability of a system depends. Conditions, though not sufficient, for having no such poles in the RHP are as follows:

1. All coefficients, $a_0,\ a_1, \ldots a_n$ must have the same sign.
2. All powers of s must be present in the descending order.

Since these conditions are not sufficient, one has to solve the characteristic equation. Hurwitz and Routh developed criteria which allow one to determine the stability without solving the characteristic equation.

10.1 HURWITZ'S CRITERION

Hurwitz[1] defined certain determinants out of coefficients of the characteristic equation. The sufficient condition is that these determinants must be positive for having all roots of the characteristic equation in the LHP.

The Hurwitz determinants are formed as follows. The main determinant, D_n, is given by

$$D_n = \begin{vmatrix} a_1 & a_3 & a_5 & \cdots & a_{2n-1} \\ a_0 & a_2 & a_4 & \cdots & a_{2n-2} \\ 0 & a_1 & a_3 & \cdots & a_{2n-3} \\ 0 & a_0 & a_2 & \cdots & a_{2n-4} \\ \vdots & \vdots & \vdots & \vdots & \vdots \\ 0 & 0 & 0 & \cdots & a_n \end{vmatrix}$$

[1] Adolf Hurwitz (1859–1919): German mathematician.

Thus,

$$D_1 = |a_1|$$

$$D_2 = \begin{vmatrix} a_1 & a_3 \\ a_0 & a_2 \end{vmatrix}$$

$$D_3 = \begin{vmatrix} a_1 & a_3 & a_5 \\ a_0 & a_2 & a_4 \\ 0 & a_1 & a_3 \end{vmatrix}$$

$$D_4 = \begin{vmatrix} a_1 & a_3 & a_5 & a_7 \\ a_0 & a_2 & a_4 & a_6 \\ 0 & a_1 & a_3 & a_5 \\ 0 & a_0 & a_2 & a_4 \end{vmatrix}$$

The condition for stability is

$$D_1 > 0,\ D_2 > 0,\ D_3 > 0\ \ldots D_k > 0$$

Hurwitz's criterion is not so useful in comparison to other methods as it has the following shortcomings:

1. It is time-consuming and cumbersome to solve large determinants for higher order systems.
2. It does not yield exact number of poles located in the RHP.
3. Marginal stability prediction, by this method, becomes tough.

Nevertheless, we consider a few examples to illustrate the method.

EXAMPLE 10.1. Using Hurwitz's criterion, study the stability of unity feedback (a) first and (b) second order systems.

Solution (a) For the first order system,

$$G(s) = \frac{K}{\tau s + 1}$$

Therefore,

$$T(s) = \frac{G(s)}{1 + G(s)} = \frac{K}{\tau s + K + 1}$$

The characteristic equation is

$$\tau s + K + 1 = 0$$

Therefore,

$$a_0 = \tau$$

$$a_1 = K + 1$$

$$D_1 = |K + 1| = K + 1$$

$D_1 > 0$ for $K \geq 0$. Thus, the system is stable.

(b) For the second order system,

$$G(s) = \frac{K\omega_n^2}{s^2 + 2\zeta\omega_n s + \omega_n^2}$$

Therefore,

$$T(s) = \frac{G(s)}{1 + G(s)} = \frac{K\omega_n^2}{s^2 + 2\zeta\omega_n s + (K+1)\omega_n^2}$$

The characteristic equation is

$$s^2 + 2\zeta\omega_n s + (K+1)\omega_n^2 = 0$$

Therefore,

$$a_0 = 1$$

$$a_1 = 2\zeta\omega_n$$

$$a_2 = (K+1)\omega_n^2$$

The Hurwitz determinants are

$$D_1 = |a_1| = 2\zeta\omega_n$$

$$D_2 = \begin{vmatrix} a_1 & a_3 \\ a_0 & a_2 \end{vmatrix} = \begin{vmatrix} 2\zeta\omega_n & 0 \\ 1 & (K+1)\omega_n^2 \end{vmatrix} = 2\zeta\omega_n^2(K+1)$$

Now,

$$D_1 > 0 \quad \text{for} \quad \zeta > 0$$

$$D_2 > 0 \quad \text{for} \quad \zeta > 0;\ K \geq 0$$

The conditions are normally true. Thus, the system is stable.

EXAMPLE 10.2. Find, using Hurwitz's criterion, whether a system having the characteristic equation

$$2s^4 + s^3 + 3s^2 + 4s + 10 = 0$$

is stable or not.

Solution Here,

$$a_0 = 2, \quad a_1 = 1, \quad a_2 = 3, \quad a_3 = 4, \quad a_4 = 10$$

Therefore,

$$D_1 = |a_1| = 1$$

$$D_2 = \begin{vmatrix} a_1 & a_3 \\ a_0 & a_2 \end{vmatrix} = \begin{vmatrix} 1 & 4 \\ 2 & 3 \end{vmatrix} = -5$$

We need not work out D_3 and D_4 because already $D_2 < 0$. The system is unstable.

10.2 ROUTH'S RULE

Routh[2] gave an alternative but much simpler sufficient condition for finding the stability of a system. In this method, the coefficients of the characteristic equation [Eq. (10.1)] are arranged in an array, called the *Routh array*, as given in the following representation.

$$
\begin{array}{c|ccccc}
s^n & a_0 & a_2 & a_4 & a_6 & \cdots \\
s^{n-1} & a_1 & a_3 & a_5 & a_7 & \cdots \\
s^{n-2} & b_1 & b_2 & b_3 & b_4 & \cdots \\
s^{n-3} & c_1 & c_2 & c_3 & c_4 & \cdots \\
\vdots & \vdots & & & & \\
s^0 & a_n & & & &
\end{array}
$$

where

$$b_1 = \frac{a_1a_2 - a_0a_3}{a_1}, \qquad b_2 = \frac{a_1a_4 - a_0a_5}{a_1}$$

$$b_3 = \frac{a_1a_6 - a_0a_7}{a_1}, \qquad b_4 = \frac{a_1a_8 - a_0a_9}{a_1}$$

$$c_1 = \frac{b_1a_3 - a_1b_2}{b_1}, \qquad c_2 = \frac{b_1a_5 - a_1b_3}{b_1}$$

The elements of the first two rows of the Routh array are obtained from the characteristic equation. The elements of the subsequent rows are formed by using the above formulae. If the connectivities shown in the figure are followed, the pattern of formation of the formulae will be clear.

The stability of a system can be predicted from the array according to the following rules:

1. For the system to be stable, all the elements in the first column of the Routh array must have the same sign.
2. In case there is a change of sign, the system is unstable.
3. The number of changes of sign indicates the number of roots lying in the RHP of the CLTF.

In fact, Routh's derivation of these criteria follows from Hurwitz's formulation for the problem. This is why, these criteria are often referred to as *Hurwitz-Routh criteria.*

Let us work out a few examples here to show how Hurwitz-Routh criteria are utilized to study stability problems.

EXAMPLE 10.3. Find, using Routh's rule, whether a system having the characteristic equation

$$2s^4 + s^3 + 3s^2 + 4s + 10 = 0$$

is stable or not.

[2]Edward John Routh (1831–1907): English mathematician.

Solution The Routh array is

$$
\begin{array}{c|ccc}
s^4 & 2 & 3 & 10 \\
s^3 & 1 & 4 & 0 \\
s^2 & -5(b_1) & 10(b_2) & 0 \\
s^1 & 6(c_1) & 0(c_2) & \\
s^0 & 10(d_1) & &
\end{array}
$$

The calculation of elements from the third row onwards is as follows:

$$b_1 = \frac{(1)(3) - (2)(4)}{1} = -5, \qquad b_2 = \frac{(1)(10) - (2)(0)}{1} = 10$$

$$c_1 = \frac{(-5)(4) - (1)(10)}{-5} = 6, \qquad c_2 = \frac{(-5)(0) - (1)(0)}{-5} = 0$$

$$d_1 = \frac{(6)(10) - (-5)(0)}{6} = 10$$

There is change of sign in the first column of the Routh array. So, the system is unstable. In fact, there are two changes of sign, from + to −, and then from − to +. Thus, two roots lie in the RHP.

EXAMPLE 10.4. The characteristic equation of a closed-loop system is $s^2 + Ks + 4K - 1 = 0$. For K greater than what value is the system stable?

Solution The Routh array is given by

$$
\begin{array}{c|cc}
s^2 & 1 & 4K \\
s^1 & K & 0 \\
s^0 & 4K - 1 &
\end{array}
$$

For stability of the system,

$$4K - 1 > 0$$

or

$$K > \frac{1}{4}$$

EXAMPLE 10.5. The condition for stability of a closed-loop system with a characteristic equation $s^3 + bs^2 + cs + 1 = 0$, with positive coefficients is

(a) $b + c > 1$ (b) $bc > 1$

(c) $b = c$ (d) $b > c$

Solution The Routh array for the characteristic equation is

$$
\begin{array}{c|cc}
s^3 & 1 & c \\
s^2 & b & 1 \\
s^1 & c - \dfrac{1}{b} & 0 \\
s^0 & c &
\end{array}
$$

For stability of the system,

$$c > \frac{1}{b}$$

or

$$bc > 1$$

Ans: (b).

EXAMPLE 10.6. Consider the unity negative feedback system shown in the following figure:

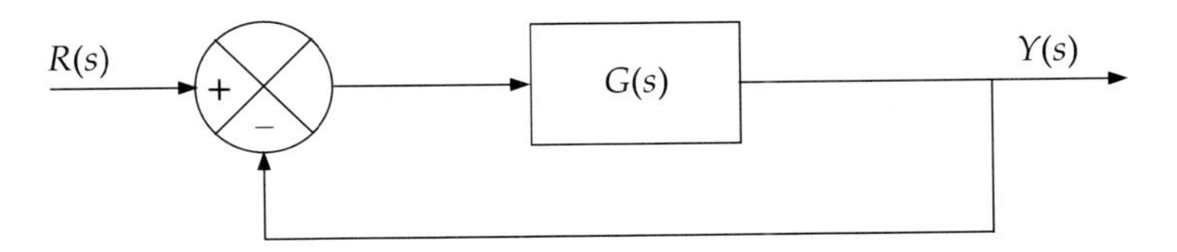

where

$$G(s) = \frac{K}{(1 + 0.5s)(1 + 0.5s + 0.25s^2)}$$

Using Routh's criterion, find the value of K for which the system is stable.

Solution Here,

$$T(s) = \frac{Y(s)}{R(s)} = \frac{G(s)}{1 + G(s)} = \frac{K}{(1 + 0.5s)(1 + 0.5s + 0.25s^2) + K}$$

The characteristic equation is

$$(1 + 0.5s)(1 + 0.5s + 0.25s^2) + K = 0$$

$$0.125s^3 + 0.5s^2 + s + (K + 1) = 0$$

The Routh array is

$$\begin{array}{c|ccc} s^3 & 0.125 & 1.0 & 0 \\ s^2 & 0.5 & K+1 & 0 \\ s^1 & 1.0 - 0.25(K+1) & 0 & \\ s^0 & K+1 & & \end{array}$$

The calculation of elements from the third row onwards is as follows:

$$b_1 = \frac{(0.5)(1.0) - (0.125)(K + 1)}{0.5} = 1.0 - 0.25(K + 1)$$

$$b_2 = \frac{(0.5)(0) - (0.125)(0)}{0.5} = 0$$

$$c_1 = \frac{[1.0 - 0.25(K + 1)](K + 1) - (0.5)(0)}{1.0 - 0.25(K + 1)} = K + 1$$

For the system to be stable, from Routh's criterion it is necessary that the third row first column factor should be positive. Thus,

$$1.0 - 0.25(K+1) > 0$$
$$0.25(K+1) < 1.0$$
$$K+1 < 4$$
$$K < 3$$

and the fourth row first column factor should be positive. This means

$$K+1 > 0 \quad \Rightarrow \quad K > -1$$

Therefore, for the stability of the system

$$3 > K > -1$$

EXAMPLE 10.7. For the feedback control system shown in the following figure, using Routh's tabulation, determine the range of scalar variable K for which the system is stable.

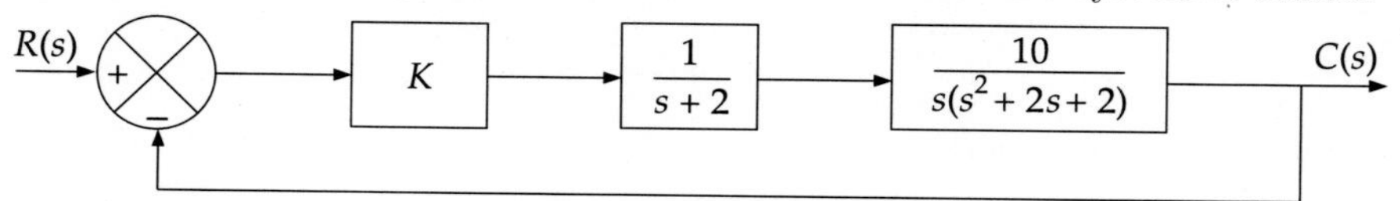

Solution The closed-loop transfer function is given by

$$\frac{10K}{(s+3)[s(s^2+2s+2)]+10K}$$

The characteristic equation is

$$s^4 + 5s^3 + 8s^2 + 6s + 10K = 0$$

Therefore, the Routh table is as follows:

s^4	1	8	$10K$
s^3	5	6	
s^2	6.8	$10K$	
s^1	$6 - 7.35K$	0	
s^0	$10K$		

For stability of the system, we have

$$6 - 7.35K > 0$$

or

$$K < \frac{6}{7.35}$$
$$< 0.816$$

and

$$10K > 0$$

or

$$K > 0$$

Thus, the required range is $0 < K < 0.816$.

EXAMPLE 10.8. The closed-loop control system (shown in the following figure) has $\tau_1 > 0$. The system will remain stable for all τ_1 in the range of

(a) $\tau_1 < 0.5$ (b) $0.5 < \tau_1 < 1.0$ (c) $\tau_1 < 1.0$ (d) $\tau_1 > 1.0$

Solution The closed-loop transfer function is given by

$$\frac{C(s)}{R(s)} = \frac{10\dfrac{[1+(1/\tau_1 s)]}{s(s+1)}}{1+10\dfrac{[1+(1/\tau_1 s)]}{s(s+1)}} = \frac{10\left(1+\dfrac{1}{\tau_1 s}\right)}{s(s+1)+10\left(1+\dfrac{1}{\tau_1 s}\right)}$$

The characteristic equation is

$$s(s+1)+10\left(1+\frac{1}{\tau_1 s}\right) = 0$$

or

$$\tau_1 s^3 + \tau_1 s^2 + 10\tau_1 s + 10 = 0$$

The Routh array is

$$\begin{array}{c|cc} s^3 & \tau_1 & 10\tau_1 \\ s^2 & \tau_1 & 10 \\ s^1 & 10\tau_1 - 10 & 0 \\ s^0 & 10 & \end{array}$$

The stability condition is

$$10\tau_1 - 10 > 0$$

or

$$\tau_1 > 1$$

Ans: (d).

10.3 SPECIAL CASES

Row of Zeros

When all the elements in any row of the Routh array are zero, it indicates that there are symmetrically located roots in the s-plane. That means there exists

- either a pair of real roots with opposite signs;
- and/or a pair of conjugate roots on the imaginary axis;
- and/or complex conjugate roots forming quadrates in the s-plane.

These are indeed indications of a marginally stable system which crosses the imaginary axis temporarily to attain stability but reverts to unstable state again.

In such cases, an auxiliary equation is to be formed by the elements of the row just above the row of zeros of the Routh array. The first derivative of the auxiliary polynomial gives the elements of the row containing zeros. We consider a few examples to show how to handle this kind of problem.

EXAMPLE 10.9. The characteristic equation of a system is given by

$$s^6 + 2s^5 + 8s^4 + 12s^3 + 18s^2 + 12s + 12 = 0$$

Find the stability of the system.

Solution The Routh array is

$$\begin{array}{c|cccc} s^6 & 1 & 8 & 18 & 12 \\ s^5 & 2 & 12 & 12 & \\ s^4 & 2 & 12 & 12 & \\ s^3 & 0 & 0 & 0 & \end{array}$$

We find that all the elements in the s^3-row are zeros. Therefore, we need to form the auxiliary polynomial from the s^4-row with the elements as coefficients. The auxiliary polynomial thus formed is

$$\begin{aligned} A(s) &= 2s^4 + 12s^2 + 12 \\ &= s^4 + 6s^2 + 6 \end{aligned}$$

Mark that we have divided all the coefficients of the polynomial by 2. This is allowed since dividing or multiplying coefficients by the same number does not affect the roots of the corresponding characteristic equation.

Now,

$$\frac{dA(s)}{ds} = 4s^3 + 12s$$

So, 4 and 12 are the elements for the s^3-row. The Routh array now becomes

$$\begin{array}{c|cccc} s^6 & 1 & 8 & 18 & 12 \\ s^5 & 1 & 6 & 6 & \\ s^4 & 1 & 6 & 6 & \\ s^3 & 4 & 12 & 0 & \\ s^2 & 3 & 6 & & \\ s^1 & 4 & & & \\ s^0 & 6 & & & \end{array}$$

The auxiliary equation is

$$s^4 + 6s^2 + 6 = 0$$

This yields

$$s^2 = \frac{-6 \pm \sqrt{(6)^2 - 4(1)(6)}}{2} = -3 \pm \sqrt{3}$$

$$= -4.732 \quad \text{and} \quad -1.268$$

Therefore,

$$s = \pm \jmath\, 2.175, \quad s = \pm \jmath\, 1.126$$

These are also among the roots of the characteristic equation. And there is no sign change in the elements of the first column of the Routh array formed with the help of the auxiliary polynomial. Because there are multiple, non-repeated roots on the imaginary-axis, we conclude that the system is marginally stable.

EXAMPLE 10.10. The partial Routh array of the characteristic equation of a system is given by

$$\begin{array}{c|ccc} s^4 & 1 & a & 8 \\ s^3 & 3 & 12 & \end{array}$$

The system oscillates with a frequency of 2 rad/s. The value of the parameter a of the system is

(a) 6 (b) 2 (c) 8 (d) 12

Solution The next row of the Routh array is

$$\begin{array}{c|cc} s^2 & a-4 & 8 \end{array}$$

Therefore, the auxiliary equation is

$$(a-4)s^2 + 8 = 0$$

or

$$a - 4 = \frac{8}{\omega^2} = \frac{8}{4} = 2 \qquad [\because \quad s^2 = -\omega^2 \text{ and } \omega = 2 \text{ (given)}]$$

or

$$a = 6$$

Ans: (a).

EXAMPLE 10.11. The Routh-Hurwitz array for a third order characteristic equation is given as

$$\begin{array}{c|cccc} s^3 & 1 & b & 0 & 0 \\ s^2 & a & c & 0 & 0 \\ s^1 & \dfrac{6-k}{3} & 0 & & \\ s^0 & k & & & \end{array}$$

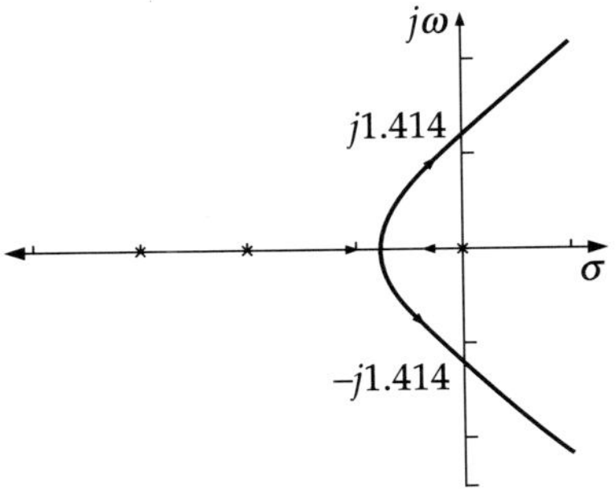

The coefficients a, b, c and k are such that $a = 3$, $b > 0$, $c > 0$, $k > 0$ and c is a function of k. The root locus for the corresponding characteristic equation is as shown in the given figure. The values of k and c for critical stability are

(a) 6 and 6 (b) 6 and 2 (c) 3 and 2 (d) 3 and 3

Solution From the Routh-Hurwitz array, we have the auxiliary equation

$$as^2 + c = 0 \tag{i}$$

The root locus intersects the $j\omega$-axis at $s = \pm j1.414 = \pm j\sqrt{2}$ from which we get

$$s = \pm j\sqrt{2}$$

or

$$s^2 + 2 = 0 \tag{ii}$$

Comparing Eqs. (i) and (ii), we obtain

$$\frac{c}{a} = 2$$

or

$$c = 6 \qquad [\because \quad a = 3 \text{ (given)}]$$

For the critical stability of the system, from the Routh-Hurwitz array, we get

$$\frac{6-k}{3} = 0$$

or

$$k = 6$$

Ans: (a).

EXAMPLE 10.12. The first two rows of Routh's tabulation of a third order equation are as follows:

$$\begin{array}{c|cc} s^3 & 2 & 2 \\ s^2 & 4 & 4 \end{array}$$

This means that there are:

(a) Two roots at $s = \pm j$ and one root in right half s-plane
(b) Two roots at $s = \pm j2$ and one root in left half s-plane
(c) Two roots at $s = \pm j2$ and one root in right half s-plane
(d) Two roots at $s = \pm j$ and one root in left half s-plane

Solution From Routh's tabulation we construct the auxiliary equation and find the roots on the $j\omega$-axis as follows:

$$4s^2 + 4 = 0$$

or

$$s = \pm j \tag{i}$$

Also, from the two rows of Routh's tabulation, we construct the characteristic equation and extract other roots therefrom as follows:

$$2s^3 + 4s^2 + 2s + 4 = 0$$

$$\Rightarrow \quad s^3 + 2s^2 + s + 2 = 0$$

$$\Rightarrow \quad s(s^2 + 1) + 2(s^2 + 1) = 0$$

$$\Rightarrow \quad (s^2 + 1)(s + 2) = 0 \qquad \text{(ii)}$$

From Eqs. (i) and (ii), we note that the choice given in (d) is true.
Ans: (d).

EXAMPLE 10.13. The open-loop transfer function of a unity feedback system is given by

$$G(s) = \frac{K}{(s+2)(s+4)(s^2+6s+25)}$$

By applying the Hurwitz-Routh criterion, discuss the stability of the closed-loop system as a function of K. Determine the values of K which will cause sustained oscillation in the closed-loop system. What are the corresponding frequencies?

Solution Here,

$$T(s) = \frac{C(s)}{R(s)} = \frac{G(s)}{1+G(s)} = \frac{K}{(s+2)(s+4)(s^2+6s+25)+K}$$

The characteristic equation is

$$(s+2)(s+4)(s^2+6s+25) + K = 0$$

or

$$s^4 + 12s^3 + 69s^2 + 198s + (200+K) = 0$$

The Routh array is

s^4	1	69	$200+K$
s^3	12	198	0
s^2	52.5	$200+K$	0
s^1	$152.3 - 0.23K$	0	
s^0	$200+K$		

$$b_1 = \frac{(12)(69) - (1)(198)}{12} = 52.5$$

$$b_2 = \frac{(12)(200+K) - (1)(0)}{12} = 200 + K$$

$$c_1 = \frac{(52.5)(198) - (12)(200+K)}{52.5} = 152.3 - 0.23K$$

$$d_1 = \frac{(152.3 - 0.23K)(200+K) - (52.5)(0)}{152.3 - 0.23K} = 200 + K$$

For stability:

$$200 + K > 0 \quad \Rightarrow \quad K > -200$$

and

$$152.3 - 0.23K > 0 \quad \Rightarrow \quad K < 662.17$$

Therefore,

$$-200 < K < 662.17$$

For marginal stability, $K = 662.17$ so that the last two rows of the Routh array become zeros. Then,

$$\begin{array}{c|ccc} s^4 & 1 & 69 & 200 + K \\ s^3 & 12 & 198 & 0 \\ s^2 & 52.5 & 200 + K & 0 \\ s^1 & 0 & 0 & 0 \\ s^0 & 0 & & \end{array}$$

The auxiliary equation is formed as

$$A(s) = 52.5s^2 + (200 + K) = 0$$

or

$$52.5s^2 = -862.17 \qquad [\because \ K = 662.17]$$

Therefore,

$$s = \pm \jmath\, 4.05 \equiv \jmath\omega$$

Hence, the frequency of oscillation ω is 4.05 rad/s or 25.45 Hz.

Zero in First Term of a Row

When the first term in any row of the Routh array is zero while the rest of the row has at least one non-zero term, the problem can be solved by one of three methods. Let us work out one example to make ourselves familiar with different methods.

EXAMPLE 10.14. Examine the stability of the system represented by the characteristic equation

$$s^5 + s^4 + 2s^3 + 2s^2 + 3s + 5 = 0$$

Solution The Routh array is

$$\begin{array}{c|ccc} s^5 & 1 & 2 & 3 \\ s^4 & 1 & 2 & 5 \\ s^3 & 0 & -2 & 0 \end{array}$$

We cannot proceed further because there is a '0' in the first column. This problem can be solved by three methods.

Method 1. We replace '0' with a small positive number ε. Then we examine the sign change by taking $\varepsilon \to 0$. The ensuing Routh array is

$$\begin{array}{c|ccc} s^5 & 1 & 2 & 3 \\ s^4 & 1 & 2 & 5 \\ s^3 & 0 = \varepsilon & -2 & 0 \\ s^2 & \displaystyle\lim_{\varepsilon \to 0} \frac{2\varepsilon + 2}{\varepsilon} = \infty & 5 & \\ s^1 & \displaystyle\lim_{\varepsilon \to 0} \frac{-4\varepsilon - 4 - 5\varepsilon^2}{2\varepsilon + 2} = -2 & 0 & \\ s^0 & 5 & & \end{array}$$

There is a change of sign in the first column. The system is, therefore, unstable.

Method 2. We substitute $s = 1/z$. Next we transform the characteristic equation, construct the Routh array, and conclude therefrom. With transformation, the characteristic equation becomes

$$\frac{1}{z^5} + \frac{1}{z^4} + \frac{2}{z^3} + \frac{2}{z^2} + \frac{3}{z} + 5 = 0$$

$$\Rightarrow \qquad 5z^5 + 3z^4 + 2z^3 + 2z^2 + z + 1 = 0$$

The corresponding Routh array is

$$\begin{array}{c|ccc} z^5 & 5 & 2 & 1 \\ z^4 & 3 & 2 & 1 \\ z^3 & -4/3 & -2/3 & \\ z^2 & 1/2 & 1 & \\ z^1 & 2 & 0 & \\ z^0 & 1 & & \end{array}$$

There is a change of sign in the first column. Thus, the system is unstable.

Method 3. We multiply the characteristic equation by a factor $(s + \alpha)$, where $\alpha > 0$. Normally, the factor $(s + 1)$ is used. Then we check the stability from the Routh array of the resulting equation. In the present example, the equation, on multiplication by $(s + 1)$, becomes

$$s^6 + 2s^5 + 3s^4 + 4s^3 + 5s^2 + 8s + 5 = 0$$

Its Routh array is

$$\begin{array}{c|cccc} s^6 & 1 & 3 & 5 & 5 \\ s^5 & 2 & 4 & 8 & 0 \\ s^4 & 1 & 1 & 5 & \\ s^3 & 2 & -2 & & \\ s^2 & 2 & 5 & & \\ s^1 & -7 & & & \\ s^0 & 5 & & & \end{array}$$

There is a change of sign in the first column. The system is, therefore, unstable.

Systems with Transportation Lag

We derived the transfer function for transportation lag in Section 5.4 at page 206. Here, we will consider how to study stability of systems with transportation lag through the Routh-Hurwitz criteria. To do that, we will have to have resort to the following approximations:

1. For small values of T, we can write from the expansion of the exponential function as

$$e^{-sT} \approx 1 - sT \tag{10.2}$$

2. Alternatively, we can use the following relation:

$$\begin{aligned} e^{-sT} &= e^{-sT/2} \cdot e^{-sT/2} \\ &= \frac{e^{-sT/2}}{e^{sT/2}} \\ &= \frac{1 - \dfrac{sT}{2} + \dfrac{1}{2} \cdot \left(\dfrac{sT}{2}\right)^2 - \ldots}{1 + \dfrac{sT}{2} + \dfrac{1}{2} \cdot \left(\dfrac{sT}{2}\right)^2 - \ldots} \\ &\approx \frac{1 - \dfrac{sT}{2}}{1 + \dfrac{sT}{2}} \qquad \text{[for small values of } T\text{]} \qquad (10.3) \\ &= \frac{2 - sT}{2 + sT} \end{aligned}$$

In fact, Eq. (10.3) is nothing but the first order Padé[3] approximant.

Equation (10.2) adds a zero in the RHP of the system transfer function while Eq. (10.3) adds a zero in the RHP and pole in the LHP. Though Eq. (10.2) is recommended by some authors, it is indeed a poor approximation because it gives a completely wrong step response of the system. Other books on classical control system theory usually provide a relation with equal degree for the numerator and the denominator as given in Eq. (10.3). This approximation is better though it generates a puzzling jump at $t = 0$. The results will be evident from Figure 10.1 where we have plotted the step response of the following open-loop transfer function for both the approximations:

$$G(s) = \frac{10e^{-s}}{s + 1} \tag{10.4}$$

We work out one example using both the approximations.

[3]Henri Eugéne Padé (1863–1953) was a French mathematician who developed approximation techniques for functions using rational functions.

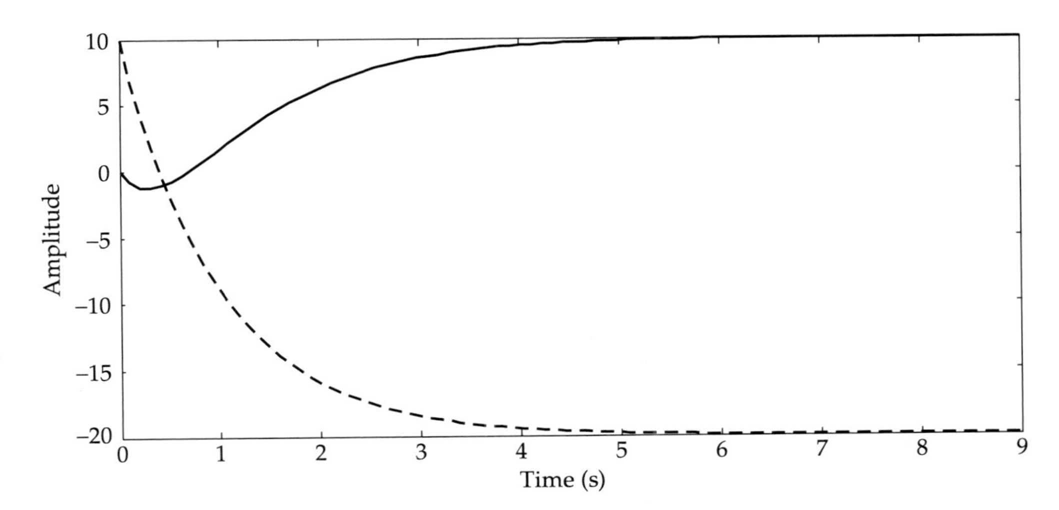

Figure 10.1 Step response of the system given by Eq. (10.4) - - -, using Eq. (10.2) for the transportation lag; —, using Eq. (10.3).

EXAMPLE 10.15. A unity feedback system has an open-loop transfer function

$$G(s)H(s) = \frac{Ke^{-s}}{s(s^2 + 5s + 9)}$$

Determine, by using the Hurwitz-Routh stability criteria, the range of K for which the closed-loop system will be stable.

Solution Here,

$$T(s) = \frac{C(s)}{R(s)} = \frac{G(s)}{1 + G(s)} = \frac{Ke^{-s}}{s(s^2 + 5s + 9) + Ke^{-s}}$$

The characteristic equation is

$$s(s^2 + 5s + 9) + Ke^{-s} = 0 \tag{i}$$

Using Eq. (10.2)*:* Equation (i) can be rewritten as

$$s(s^2 + 5s + 9) + K(1 - s) = 0$$

or

$$s^3 + 5s^2 + (9 - K)s + K = 0$$

The Routh array is

$$\begin{array}{c|ccc}
s^3 & 1 & 9 - K & 0 \\
s^2 & 5 & K & 0 \\
s^1 & \dfrac{45 - 6K}{5} & 0 & \\
s^0 & K & &
\end{array}$$

$$b_1 = \frac{5(9-K)-K}{5} = \frac{45-6K}{5}$$

The stability criteria are

$$K > 0$$

and
$$\frac{45-6K}{5} > 0$$

$\Rightarrow$
$$45 - 6K > 0$$

$\Rightarrow$
$$K < 7.5$$

Therefore,

$$0 < K < 7.5$$

Using Eq. (10.3)*:* Again, Eq. (i) can be rewritten as

$$s(s^2 + 5s + 9) + \frac{K(2-s)}{2+s} = 0$$

or
$$s^4 + 7s^3 + 19s^2 + (18-K)s + 2K = 0$$

The Routh array is

$$\begin{array}{c|ccc} s^4 & 1 & 19 & 2K \\ s^3 & 7 & 18-K & 0 \\ s^2 & \dfrac{115+K}{7} & 2K & \\ s^1 & \dfrac{2070-11K-K^2}{7} & & \\ s^0 & 2K & & \end{array}$$

The stability criteria are

$$K > 0$$

and
$$2070 - 11K - K^2 > 0$$

$\Rightarrow$
$$-(K - 40.3)(K + 51.3) > 0$$

$\Rightarrow$
$$40.3 > K > -51.3$$

Therefore,

$$40.3 > K > 0$$

Let us now consider a few additional assorted examples and then see how relative stability can be studied by Hurwitz-Routh criteria.

EXAMPLE 10.16. The system shown in Figure 10.2 has oscillation of 2.5 rad/s. Determine the values of K and p. There are no poles in the RHP.

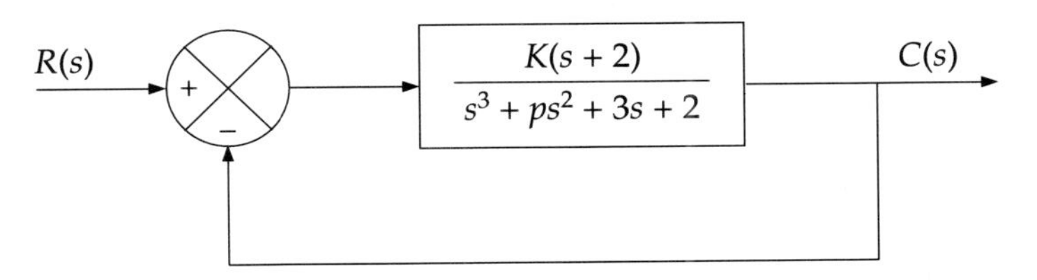

Figure 10.2 System (Example 10.16).

Solution The characteristic equation, $1 + G(s) = 0$, is

$$s^3 + ps^2 + 3s + 2 + K(s + 2) = 0$$

$$s^3 + ps^2 + (3 + K)s + (2K + 2) = 0$$

The Routh array is

$$\begin{array}{c|cc} s^3 & 1 & 3+K \\ s^2 & p & 2K+2 \\ s^1 & (3+K) - \dfrac{2K+2}{p} & 0 \\ s^0 & 2K+2 & \end{array}$$

Since there is no pole in the RHP and the system oscillates (i.e. it is marginally stable),

$$(3 + K) - \frac{2K + 2}{p} = 0$$

or
$$3 + K = \frac{2K + 2}{p} \tag{i}$$

The auxiliary equation is

$$ps^2 + 2K + 2 = 0$$

or
$$s^2 = -\frac{2K + 2}{p}$$

Therefore,
$$s = \pm j\sqrt{\frac{2K + 2}{p}} \equiv \pm j\omega \tag{ii}$$

From Eq. (ii) we get the frequency of oscillation as

$$\omega = \sqrt{\frac{2K + 2}{p}} = 2.5 \quad \text{(given)}$$

$$\frac{2K + 2}{p} = 6.25 \tag{iii}$$

Solving Eqs. (i) and (iii), we get $K = 3.25$ and $p = 1.36$.

EXAMPLE 10.17. For a system having the characteristic equation $2s^4 + 4s^2 + 1 = 0$, find the following:

(a) The number of roots in the LHP
(b) The number of roots in the RHP
(c) The number of roots on the imaginary axis

Use the Routh-Hurwitz criterion.

Solution The characteristic equation is

$$2s^4 + 0s^3 + 4s^2 + 0s + 1 = 0$$

Therefore, the Routh array is

$$\begin{array}{c|ccc} s^4 & 2 & 4 & 1 \\ s^3 & (0)\ 1 & (0)\ 1 & 0 \\ s^2 & 2 & 1 & \\ s^1 & 1/2 & 0 & \\ s^0 & 1 & & \end{array}$$

Here, the elements of the s^3-row are zero (indicated in parentheses). So, we check the auxiliary polynomial

$$2s^4 + 4s^2 + 1$$

and get its derivative

$$8s^3 + 8s \equiv s^3 + s$$

The coefficients of the derivative are the revised elements of the row, which are 1, 1.

Now, the roots of the auxiliary equation

$$2s^4 + 4s^2 + 1 = 0$$

can be found as[4]

$$\pm\jmath\,1.3065 \quad \text{and} \quad \pm\,\jmath\,0.541$$

We observe that:

(a) No root lies in the LHP.
(b) No root lies in the RHP.
(c) All four roots lie on the imaginary axis, but they are not repetitive.

Since the Routh array shows no change of sign in the first column, we conclude that the system is marginally stable.

Note: In this case, the auxiliary equation is the same as the characteristic equation. Therefore, the results could be obtained without having recourse to the Routh-Hurwitz criterion.

[4]The equation can be solved by putting $s^2 = x$.

EXAMPLE 10.18. Find the stability of a system having the characteristic equation

$$s^6 + 2s^5 + 7s^4 + 10s^3 + 14s^2 + 8s + 8 = 0$$

Solution The Routh array is

$$\begin{array}{c|cccc} s^6 & 1 & 7 & 14 & 8 \\ s^5 & 2 & 10 & 8 & 0 \\ s^4 & 2 & 10 & 8 & \\ s^3 & (0)\ 2 & (0)\ 5 & 0 & \\ s^2 & 5 & 8 & & \\ s^1 & 1.8 & 0 & & \\ s^0 & 8 & & & \end{array}$$

Since there is no change of sign in the first column of the Routh array, the system is not unstable. Here, the row of s^3 contains zero elements. So, we check the auxiliary polynomial

$$2s^4 + 10s^2 + 8$$

and its derivative

$$8s^3 + 20s \equiv 2s^3 + 5s$$

The revised elements of the s^3-row are, therefore, 2 and 5. The roots of the auxiliary equation

$$2s^4 + 10s^2 + 8 = 0$$

are

$$\pm \jmath 1 \quad \text{and} \quad \pm \jmath 2$$

However, the roots on the imaginary axis are not repetitive, which indicates that the system is marginally stable.

10.4 RELATIVE STABILITY BY HURWITZ-ROUTH CRITERIA

So far we have studied absolute stability of systems using the Hurwitz-Routh criteria. A qualitative analysis of stability that determines the location of roots of the characteristic equation (i.e. poles of the closed-loop transfer function in the s-plane) is termed *absolute stability.*

Relative stability, on the other hand, means a comparative analysis of a system stability, which determines the parameters such as maximum overshoot to a step input, damping ratio, gain margin, and phase margin of a system.

Hurwitz-Routh criteria can be utilized to study the relative stability of systems to a limited extent. This is done by shifting the origin of the s-plane to another location s' by substituting

$$s = z + s'$$

and checking by the application of Routh's rule how many poles lie on what side of the shifted origin. Figure 10.3 shows what the substitution means for $s' = -\alpha$.

The following examples will make the procedure clear.

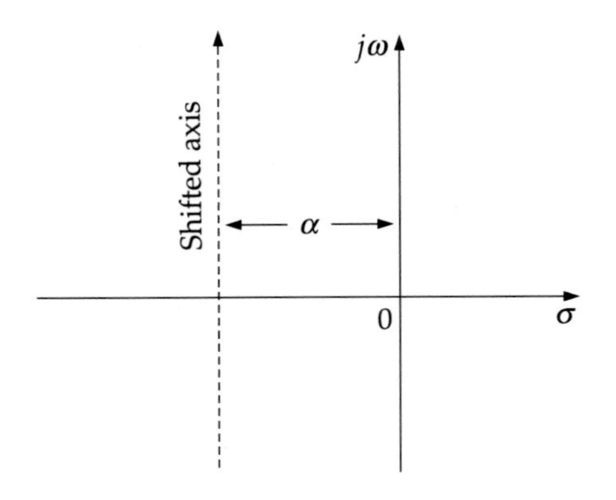

Figure 10.3 Shifting of origin of the s-plane.

EXAMPLE 10.19. Determine how many roots of the characteristic equation

$$s^3 + 10s^2 + 37s + 52 = 0$$

are more negative than $s = -1$.

Solution Substituting $s = z - 1$ in the equation, we get

$$(z-1)^3 + 10(z-1)^2 + 37(z-1) + 52 = 0$$

or

$$z^3 + 7z^2 + 20z + 24 = 0$$

Its Routh array is

z^3	1	20
z^2	7	24
z^1	16.57	
z^0	24	

There is no sign change in the first column of the Routh array. It indicates that no root lies to the right of the shifted axis of $s = -1$. Nor do they lie on the line of $s = -1$ because, in that case, there would have been a marginally stable condition. Thus, all the roots are more negative than $s = -1$.

EXAMPLE 10.20. Determine the range of values of K ($K > 0$) such that the characteristic equation

$$s^3 + 3(K+1)s^2 + (7K+5)s + (4K+7) = 0$$

has roots more negative than $s = -1$. Apply the Routh-Hurwitz criterion.

Solution Substituting $s = z - 1$ in the equation, we get

$$(z-1)^3 + 3(K+1)(z-1)^2 + (7K+5)(z-1) + (4K+7) = 0$$

or

$$z^3 + 3Kz^2 + (K+2)z + 4 = 0$$

Its Routh array is given by

$$\begin{array}{c|cc} z^3 & 1 & K+2 \\ z^2 & 3K & 4 \\ z^1 & \dfrac{3K(K+2)-4}{3K} & \\ z^0 & 4 & \end{array}$$

For stability of the system,

$$3K > 0$$
$$3K(K+2) - 4 = 0$$

The second condition is a quadratic equation in K which yields

$$K = -2.63 \quad \text{and} \quad 0.53$$

Applying the first condition to these solutions, we get the acceptable value for $K = 0.53$. Therefore, the required range is $K > 0.53$.

EXAMPLE 10.21. The open-loop transfer function of a unity feedback system is given by

$$G(s) = \frac{K}{(s+1)(s+3)(s+5)}$$

Find the limiting value of K such that all the closed-loop poles are on the left of the straight line $s = -1$ in the s-plane.

Solution The characteristic equation of the system is

$$(s+1)(s+3)(s+5) + K = 0 \tag{i}$$

Substituting $s = z - 1$ in Eq. (i), we obtain

$$z(z+2)(z+4) + K = 0$$

or

$$z^3 + 6z^2 + 8z + K = 0 \tag{ii}$$

The Routh array constructed from Eq. (ii) is

$$\begin{array}{c|cc} z^3 & 1 & 8 \\ z^2 & 6 & K \\ z^1 & 8 - \dfrac{K}{6} & 0 \\ z^0 & K & \end{array}$$

The stability conditions yield

$$K > 0,$$

$$8 - \frac{K}{6} > 0$$

or

$$8 > \frac{K}{6}$$

or

$$48 > K$$

Therefore, the required limiting values are $0 < K < 48$.

EXAMPLE 10.22. A unity feedback control system is characterized by the open-loop transfer function

$$G(s) = \frac{K(s+13)}{s(s+3)(s+7)}$$

(a) Calculate the range of values of K for the stability of the system. Use the Hurwitz-Routh criteria.

(b) For $K = 1$, check if all the roots of the characteristic equation have a damping ratio greater than 0.5.

Solution Since this is a unity feedback system, the characteristic equation is given by

$$1 + G(s) = 0$$

$$\Rightarrow \quad 1 + \frac{K(s+13)}{s(s+3)(s+7)} = 0$$

$$\Rightarrow \quad s(s+3)(s+7) + K(s+13) = 0$$

or

$$s^3 + 10s^2 + (21+K)s + 13K = 0$$

The Routh array is, therefore,

$$\begin{array}{c|cc} s^3 & 1 & 21+K \\ s^2 & 10 & 13K \\ s^1 & \dfrac{210-3K}{10} & \\ s^0 & 13K & \end{array}$$

(a) For stability of the system,

$$13K > 0 \quad \Rightarrow \quad K > 0$$

and

$$\frac{210-3K}{10} > 0 \quad \Rightarrow \quad K < 70$$

Thus

$$0 < K < 70$$

(b) At $K = 1$, the characteristic equation turns out to be

$$s^3 + 10s^2 + 22s + 13 = 0 \quad \text{(i)}$$

We need to find out the roots of Eq. (i) to check the damping ratio. It being a cubic equation, we examine a trial solution $s = -1$. We find that the left-hand polynomial becomes zero for this s value. So, this is one root. To find out the other two roots, we divide the left-hand polynomial by $s + 1$ to get the quadratic equation

$$s^2 + 9s + 13 = 0$$

whose solutions are

$$\frac{-9 \pm \sqrt{9^2 - 4(13)}}{2} = -7.1926, \; -1.8074$$

We observe that all the three roots are lying on the negative side of the real axis. Therefore, all of them subtend an angle of $\theta = 180°$. Now, the damping ratio is

$$\left| \frac{\sqrt{1-\zeta^2}}{\zeta} \right| = \tan\theta = \tan 180° = 0$$

$\Rightarrow$

$$\zeta = 1$$

which is > 0.5 for all the roots.

REVIEW QUESTIONS

10.1 Choose the correct answer:

(i) If the Routh array for a system is given by

s^4	1	2	5
s^3	1	3	0
s^2	−1	5	
s^1	−8		
s^0	5		

the system is

(a) stable (b) marginally stable
(c) unstable (d) conditionally stable

(ii) The Routh array for a system is given by

s^4	1	3	3
s^3	1	2	0
s^2	1	3	
s^1	−1		
s^0	3		

The number of poles lying in the RHP is

(a) 0 (b) 1 (c) 2 (d) 3

(iii) The number of sign changes in the first column of the Routh array indicates
 (a) the number of open-loop zeros in the RHP
 (b) the number of closed-loop zeros in the RHP
 (c) the number of open-loop poles in the RHP
 (d) the number of closed-loop poles in the RHP

(iv) The Routh array for a system is

$$\begin{array}{c|cc} s^3 & 0.25 & 1 \\ s^2 & 1 & 4 \\ s^1 & 0 & 0 \\ s^0 & 0 & \end{array}$$

The auxiliary equation for this array is
 (a) $4s^2 + 1 = 0$ (b) $0.25s^3 + s = 0$
 (c) $s^2 + 4 = 0$ (d) $s^3 + 0.25s = 0$

(v) If the changes of signs of the elements of the first column of Routh array are two for the fifth order characteristic equation, the location of roots in the s-plane are as
 (a) 2 in the left half and 3 in the right half
 (b) 2 in the right half and 3 in the left half
 (c) all 5 in the left half
 (d) all 5 in the right half

(vi) Which of the systems represented by their characteristic equations is stable? A, B, C, D are positive constants.
 (a) $As^2 + Bs + C = 0$ (b) $As^4 + Bs^2 + Cs + D = 0$
 (c) $-As^2 + Bs - C = 0$ (d) $As^2 - Bs - C = 0$

(vii) The Routh-Hurwitz criterion gives
 (a) relative stability (b) absolute stability
 (c) gain margin (d) phase margin

10.2 Find the possible range of K given in the following figure for the closed-loop system to be stable.

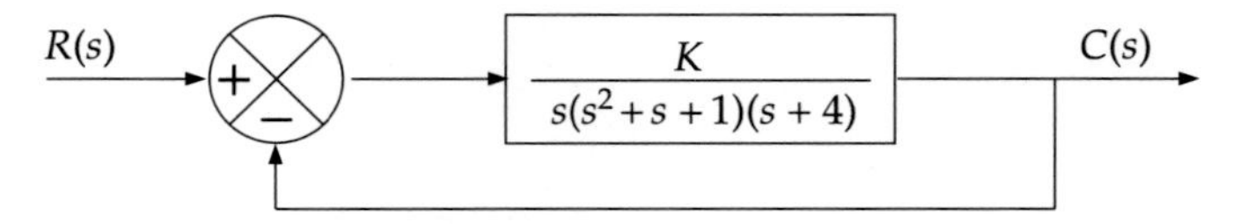

10.3 (a) With the help of Routh-Hurwitz stability criterion, determine the range of values of K $(K > 0)$ such that the characteristic equation $s^3 + 3(K+1)s^2 + (7K+5)s + (4K+7) = 0$ has roots more negative than $s = -1$.

(b) Using Routh-Hurwitz criteria, check whether the system represented by the following characteristic equation is stable or not.

$$s^4 + 2s^3 + 6s^2 + 8s + 8 = 0$$

Comment on the location of roots.

10.4 Check the stability of the system given in the following figure using Routh-Hurwitz stability criterion.

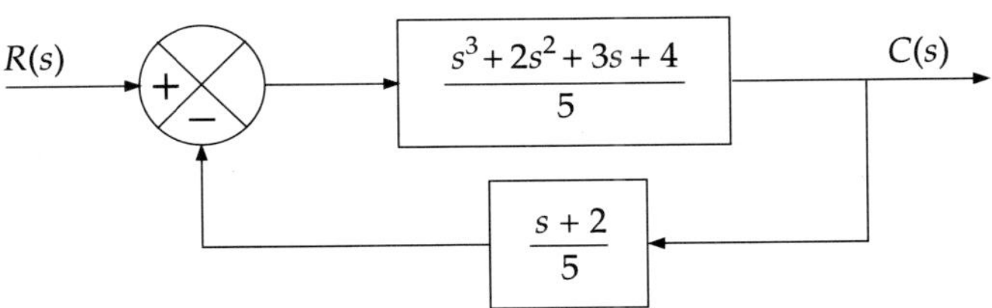

10.5 Determine the range of K such that the feedback system having characteristic equation

$$s(s^2+s+1)(s+4)+K=0$$

will be stable.

10.6 (a) State and explain Routh's stability criterion.

(b) Consider the characteristic equation

$$s^4+Ks^3+s^2+s+1=0$$

Using Routh's stability criterion, determine the range of K for stability.

10.7 A unity feedback system is characterized by the open-loop transfer function

$$G(s)=\frac{K(s+13)}{s(s+3)(s+7)}$$

Check whether for $K=1$, all roots of the characteristic equation are more negative than $s=-2$.

10.8 The open-loop transfer function of a control system is given by

$$G(s)H(s)=\frac{Ke^{-s}}{s(s^2+2s+1)}$$

Determine the maximum value of K for the closed-loop stability.

Answers to Selected Questions

10.1 (i) (c); (ii) (c); (iii) (d); (iv) (c); (v) (b); (vi) (a); (vii) (a)

10.2 $0 < K < 3.36$

10.3 (a) $K > 0.53$; (b) Marginally stable. Roots are at $s = 0 \pm \jmath 2,\ -1 \pm \jmath 1$

10.4 Stable

10.5 $0 < K < 0.96$

10.6 (b) $K > 3$

10.7 All roots

10.8 $K = \frac{2}{3}$, assuming $e^{-s} \approx (1-s)$

CHAPTER 11

Root Locus Technique

11.1 INTRODUCTION

We have seen that the location of poles in the s-plane determines the stability as well as transient response of a closed-loop system. In the root locus technique, introduced by WR Evans[1] in 1948, the loci of the roots of the characteristic equation, i.e. poles, are plotted by varying a system parameter, generally the system gain-factor K, from 0 to ∞. The resulting graph, called the *root locus*, helps us study the stability of the system. Root locus is indeed a powerful graphical method for analysis and design of control systems. Given a feedback control system, the root loci illustrate how the poles of the closed-loop system vary with the closed-loop gain. This chapter describes the root locus, its properties, and a step-by-step procedure for constructing a root locus. Examples are included to illustrate how to use the method to study the stability of a control system.

11.2 DEPENDENCE OF ROOT LOCUS ON *K* VARIATION

Consider the Single Input Single Output (SISO) system represented by Figure 11.1.

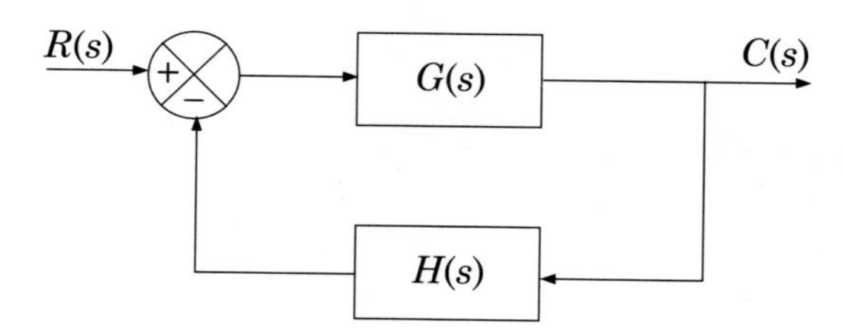

Figure 11.1 Block diagram of a SISO.

The overall transfer function of the system is

$$T(s) = \frac{C(s)}{R(s)} = \frac{G(s)}{1 + G(s)H(s)}$$

[1]Walter R Evans (1920–1999) was an American engineer.

So, the characteristic equation is

$$1 + G(s)H(s) = 0 \tag{11.1}$$

or

$$G(s)H(s) = -1 \tag{11.2}$$

Since s is generally a complex quantity, we can write Eq. (11.2) as

$$|G(s)H(s)|\angle G(s)H(s) = |1|\angle \pm (2q+1)180^\circ \qquad q = 0,\ 1,\ 2\ldots$$

Thus,

$$|G(s)H(s)| = |1| \tag{11.3}$$

$$\angle G(s)H(s) = \pm(2q+1)180^\circ \qquad q = 0,\ 1,\ 2\ldots \tag{11.4}$$

Equations (11.3) and (11.4) are called the *magnitude and angle condition* of Evans. We know that the transfer function can be expressed as

$$G(s)H(s) = \frac{K(s-b_1)(s-b_2)\cdots(s-b_m)}{(s-a_1)(s-a_2)\cdots(s-a_n)} \tag{11.5}$$

where K is the gain-factor, m is the number of zeros, and n is the number of poles. Now, from Eqs. (11.1) and (11.5),

$$1 + \frac{K(s-b_1)(s-b_2)\cdots(s-b_m)}{(s-a_1)(s-a_2)\cdots(s-a_n)} = 0 \tag{11.6}$$

or

$$(s-a_1)(s-a_2)\cdots(s-a_n) + K(s-b_1)(s-b_2)\cdots(s-b_m) = 0 \tag{11.7}$$

Normally, $n > m$. Then the order of the characteristic equation is n and it will possess n roots. It is also apparent from Eq. (11.7) that the value of roots will depend on the value of K. As the value of K changes, roots alter positions in the s-plane. So, the variation of K from 0 to ∞ will generate loci of roots in the s-plane.

We note from Eq. (11.7) that

- When $K = 0$: The second term of Eq. (11.7) does not exist and hence the roots of the resulting characteristic equation are simply the poles of the open-loop transfer function $G(s)H(s)$.

- When $K \to \infty$: The first term of Eq. (11.7) can be neglected and the roots of the resulting characteristic equation are simply the zeros of the open-loop transfer function $G(s)H(s)$.

Therefore, we observe that the loci of the roots of the characteristic equation *begin at the poles* and *end at the zeros* of open-loop transfer function $G(s)H(s)$ as K goes from 0 to ∞.

Root locus construction. As regards the root locus construction procedure, we may note that the exact loci can be calculated by varying K from 0 to ∞ and finding the roots of the characteristic equation. These positions of roots, when plotted on the s-plane, give the exact root locus plot of the system. That is what is done by MatLab. However, this procedure is cumbersome for manual plotting of the root locus.

For manual drawing of the loci, a few rules are followed to find information such as the following:

- *Part of the Re-axis* belonging to the root locus
- *Angles of asymptotes* that the loci will follow when K values are high
- *Centroid* from where the asymptotes will diverge
- *Breakaway point* for loci originating from poles facing each other
- *Angle of departure* for loci originating from complex poles
- *Point of intersection* on the Im-axis for complex loci.

Once these pieces of information are collected, approximate root loci are drawn manually. One such drawing is shown in Figure 11.2 to show the cardinal points as mentioned above.

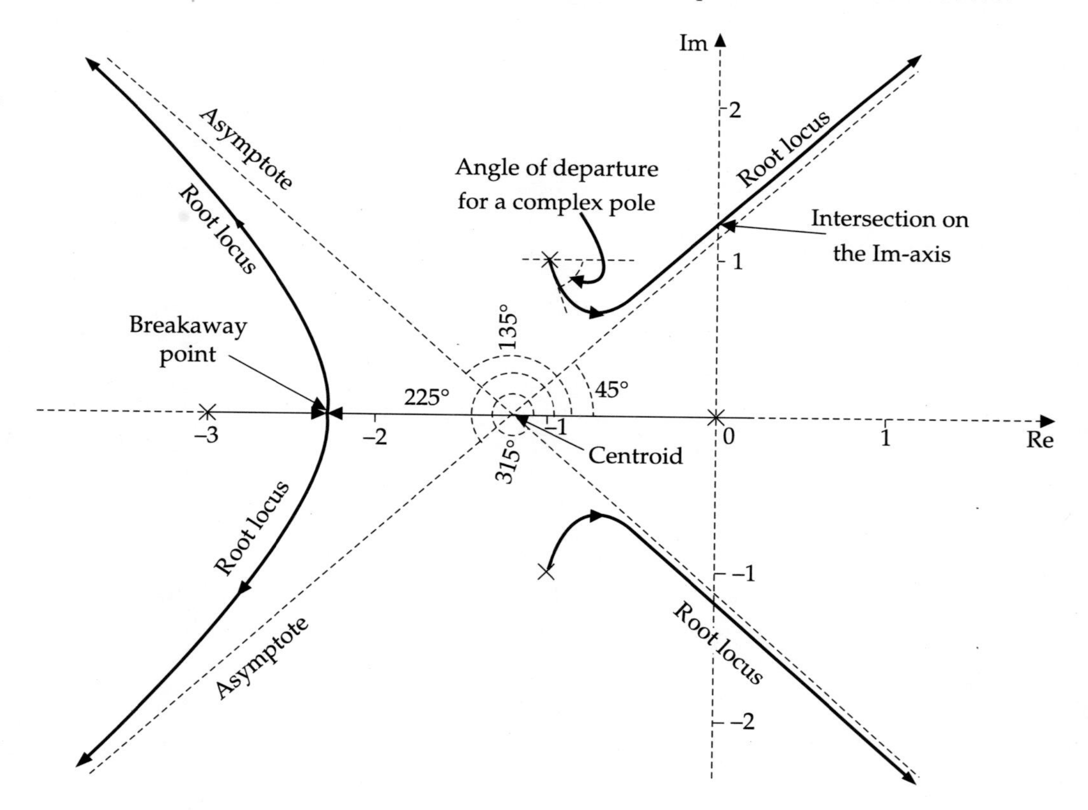

Figure 11.2 A root locus diagram showing cardinal points and asymptotes which help us to draw it.

11.3 ROOT LOCUS CONSTRUCTION PROCEDURE

To plot a root locus, we must have the open-loop transfer function $G(s)H(s)$ with K as the variation parameter. Then we follow the steps as laid down here to construct the root locus.

Step 1. *Find all poles and zeros of the system and plot them on the s-plane. Then find which part(s) of the Re-axis will belong to the root locus.* The following rules are to be followed while figuring out the part(s) of the Re-axis constituting part(s) of the root locus:

Rule 1: The root locus is always symmetrical about the real axis. That is obvious because complex roots occur in conjugate pairs.

Rule 2: All root loci start from poles because there, $K = 0$, and terminate at zeros, since there, $K \to \infty$. If there are t more poles than zeros, then t loci will terminate at ∞ as $K \to \infty$.

Rule 3: Generally, the transfer function is proper and, therefore, $n > m$. Then, the number of root loci is n

Rule 4: A point on the real axis will lie on the root locus if and only if the sum of the number of poles and zeros *to its right* is *odd* when $K > 0$ and *even* when $K < 0$.

Note: This rule owes its origin to Eq. (11.4) which indicates that the sum of angles contributed by poles and zeros should be odd multiples of 180°. The sum of angles contributed by poles and zeros to the left of the point is zero because phasors originating from the poles and zeros on the left of the point concerned are directed towards the Re-axis (therefore, angle = 0°) while those originating from poles and zeros on the right are directed towards the opposite of the Re-axis (therefore, angle = 180°). We observe that in Figure 11.3, the phasors from poles p_1, p_2 and zero z_1 directed towards the pole p_3 subtend a total angle of $3 \times 180°$. Similarly, the angle subtended by the phasor from p_1 to p_2 is 180°. Therefore, portions of the Re-axis between (p_2 and p_1) and (p_3 and z_1) constitute parts of the root locus.

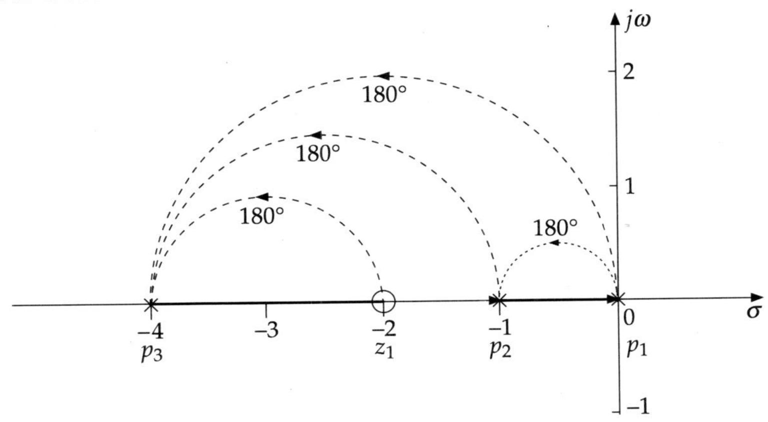

Figure 11.3 Angle contribution from poles and zeros at p_3 and p_1.

Thus, the angle contributed by poles and zeros to the right of the point has to be odd multiple of 180° to satisfy Eq. (11.4). That is possible only when their number is odd when $K > 0$ and even when $K < 0$.

In the plot of Figure 11.4, segments I and III are not part of the root locus because $(m+n)$ for them are 4 and 2, respectively, while segments II and IV, for which $(m+n)$ are 3 and 1, respectively, will be parts of the root locus.

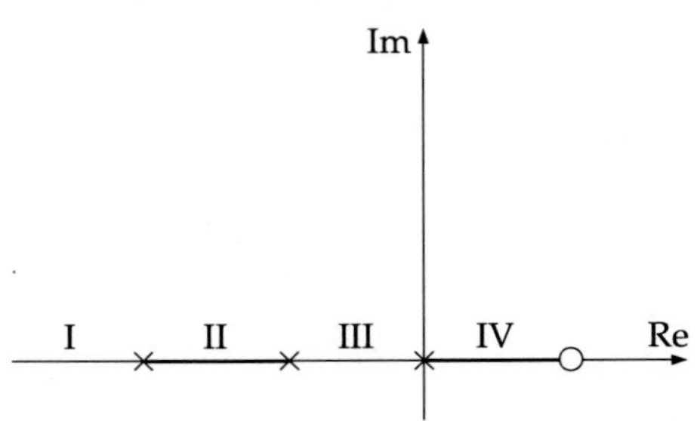

Figure 11.4 Plot showing allowable part of the root locus on the real axis.

Step 2. Find the number of asymptotes[2]*, their angles, and centroids (or centres of gravity*[3]*). Draw results.* The centroid always lies on the real axis.

Rule 5: The number of asymptotes is $n - m$.

Rule 6: Angles of asymptotes are given by

$$\theta = \begin{cases} \dfrac{2q+1}{n-m} \times 180^\circ & \text{for } K > 0 \\ \dfrac{2q}{n-m} \times 180^\circ & \text{for } K < 0 \end{cases}$$

where $q = 0,\ 1,\ 2,\ \ldots,\ (n - m)$.

Rule 7: The centroid (or centre of gravity) of asymptotes will lie on the Re-axis, and its coordinate χ_c is calculated from the relation

$$\chi_c = \frac{\Sigma \text{ Re (poles)} - \Sigma \text{ Re (zeros)}}{n - m}$$

where Σ Re (poles) means the sum of the real parts of coordinates (i.e., s-values) of poles.

EXAMPLE 11.1. Calculate the angles of asymptotes and their centroid for the system having

$$G(s)H(s) = \frac{K(s+2)}{s(s+1)(s+4)} \qquad K > 0$$

[2] *Asymptote* is a line which touches the root locus at ∞.

[3] *Centroid* (or *centre of gravity*) is the point on the σ-axis (i.e., real axis) from where all asymptotes start (i.e., where they are centred).

Solution Here, the poles are at $s = 0,\ -1$ and -4, $n = 3$. Zero is at $s = -2$, $m = 1$. The number of asymptotes $= n - m = 3 - 1 = 2$. Their angles are

$$\theta = \frac{2(0)+1}{2} \times 180^\circ = 90^\circ \qquad \text{for} \quad q = 0$$

$$= \frac{2(1)+1}{2} \times 180^\circ = 270^\circ \qquad \text{for} \quad q = 1$$

Centroid:

$$\chi_c = \frac{(-1-4)-(-2)}{2} = -1.5$$

The asymptotes, etc. are shown in Figure 11.5.

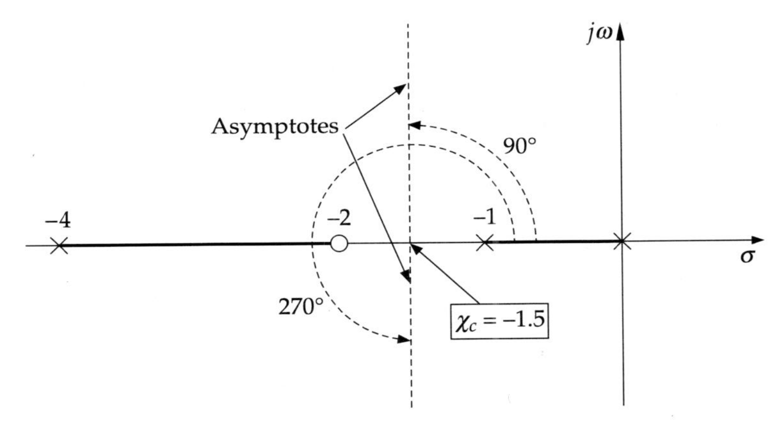

Figure 11.5 Plot (Example 11.1).

EXAMPLE 11.2. Draw the asymptotes for the system whose open-loop transfer function is given as

$$G(s)H(s) = \frac{s+1}{s(s+4)(s^2+6s+10)}$$

Solution The transfer function can be written as

$$G(s)H(s) = \frac{s+1}{s(s+4)(s+3-j1)(s+3+j1)}$$

Therefore, the poles are at $s = 0,\ -4,\ -3+j1$ and $-3-j1$. $n = 4$. Zero is at $s = -1$. $m = 1$. The number of asymptotes $= 4 - 1 = 3$. Their angles are

$$\theta = \frac{2q+1}{n-m} \times 180^\circ \qquad \text{for} \quad q = 0,\ 1,\ 2 \quad (\because\ K = 1)$$

$$= 60^\circ,\ 180^\circ,\ 300^\circ$$

Centroid

$$\chi_c = \frac{(-4-3-3)-(-1)}{3} = -3$$

The relevant plot is given in Figure 11.6.

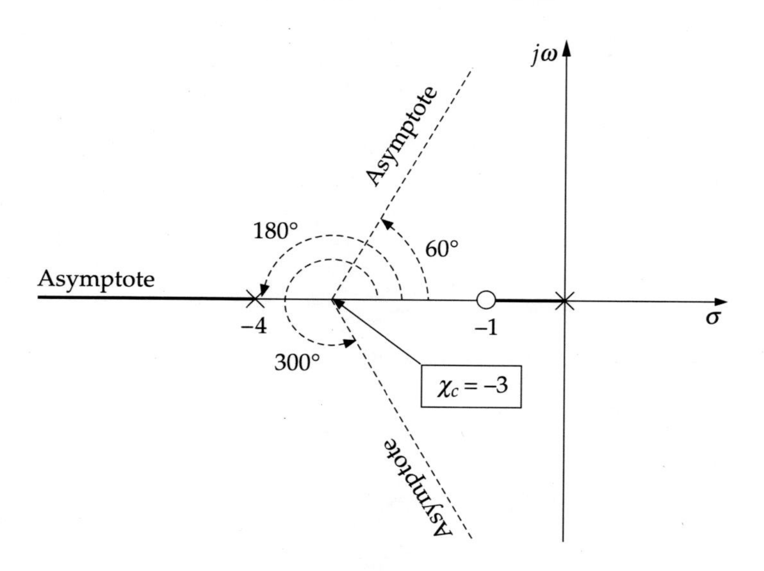

Figure 11.6 Plot (Example 11.2).

Step 3. Determine Breakaway/Break-in point, if present, and mark the point on the graph.

Breakaway point: It is the point at which the root locus breaks away from (i.e., comes out of) the real axis. The root locus leaves the real axis where there is a multiplicity of poles, typically two poles. It can be shown that K is an extremum at the breakaway point. So, the roots of the equation

$$\frac{dK}{ds} = 0$$

are the breakaway points.

Proof. Let us write Eq. (11.6) as

$$1 + K\frac{Z(s)}{P(s)} = 0 \tag{11.8}$$

Then Eq. (11.5) can be rewritten as

$$G(s)H(s) = \frac{KZ(s)}{P(s)} \tag{11.9}$$

At the breakaway point, it is necessary that

$$\frac{d}{ds}[G(s)H(s)] = 0 \tag{11.10}$$

Differentiating Eq. (11.9), we get

$$\frac{d}{ds}[G(s)H(s)] = K\left[\frac{P(s)Z'(s) - P'(s)Z(s)}{[P(s)]^2}\right] = 0$$

or

$$P(s)Z'(s) - P'(s)Z(s) = 0 \tag{11.11}$$

Again, from Eq. (11.8),

$$K\frac{Z(s)}{P(s)} = -1$$

or

$$K = -\frac{P(s)}{Z(s)}$$

Therefore,

$$\frac{dK}{ds} = \frac{P(s)Z'(s) - P'(s)Z(s)}{[Z(s)]^2} \tag{11.12}$$

Substituting the result of Eq. (11.11) in Eq. (11.12), we get

$$\frac{dK}{ds} = 0$$

It is necessary that K is real at the breakaway point. If a is the number of branches that approach the breakaway point, the locus leaves the point at an angle of $\pm 180°/a$.

Break-in point: It is the point at which the root locus enters (breaks in) the real axis. Table 11.1 indicates where to expect breakaway or break-in points.

Table 11.1 Diagnostics of Breakaway/Break-in Points

Indication	*Figure*	*Inference*
Two adjacent poles on the real axis and the segment of the real axis is a part of the root locus.	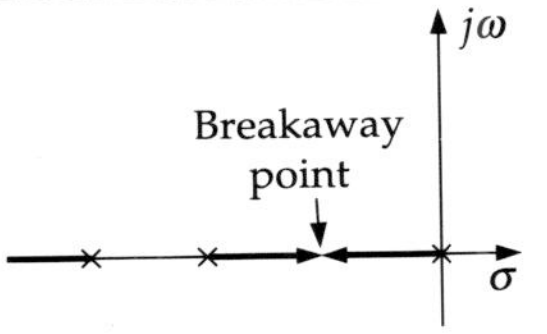 $j\omega$; Breakaway point; σ	At least one *breakaway point* exists between the poles.
Two adjacent zeros on the real axis and the segment of the real axis is a part of the root locus.	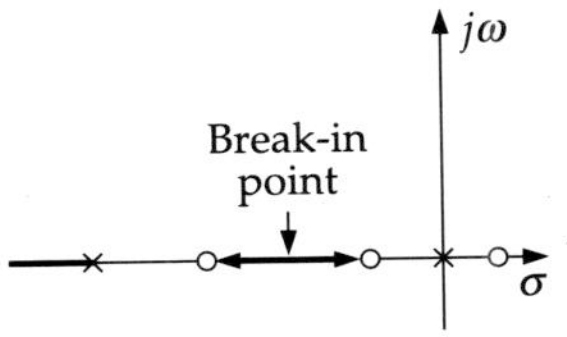 $j\omega$; Break-in point; σ	At least one *break-in point* exists between the zeros.
No poles and zeros to the left of a zero on the real axis and the segment of the real axis is a part of the root locus.	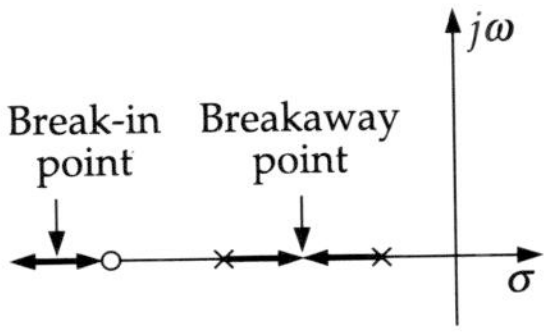 $j\omega$; Break-in point; Breakaway point; σ	At least one *break-in point* exists there.

EXAMPLE 11.3. For

$$G(s)H(s) = \frac{K}{s(s+2)(s+3)}$$

determine the coordinates of valid breakaway/break-in points.

Solution The characteristic equation is

$$1 + G(s)H(s) = 0$$

$\Rightarrow$

$$s(s+2)(s+3) + K = 0$$

or

$$K = -s^3 - 5s^2 - 6s \tag{i}$$

Therefore,

$$\frac{dK}{ds} = -3s^2 - 10s - 6 = 0$$

The roots are

$$s = \frac{-10 \pm \sqrt{100 - 72}}{6} = -0.7847, \ -2.5485$$

The corresponding values of K, namely 2.113 and -0.631 [from Eq. (i)], are real. The poles are located at $s = 0, -2$ and -3. There is no zero (see Figure 11.7). Segment I, where one breakaway point ($s = -2.55$) could be located, is not a part of the root locus, because an even number of poles exists on its right. Hence, the breakaway point is at $s = -0.785$. The number of branches approaching the breakaway point is 2. Therefore, the loci leave the real axis at angles $\pm 180°/2 = \pm 90°$.

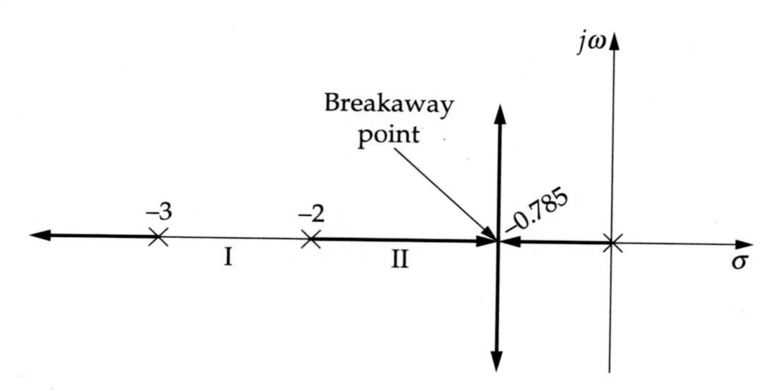

Figure 11.7 Breakaway point and the root locus.

Note: In Example 11.3, we needed to solve a quadratic equation to locate the breakaway point. But there may be cases when the relevant equation will be of higher order, making it difficult to solve. In such cases, a graphical method of locating the breakaway point may be employed. Example 11.4 demonstrates this.

EXAMPLE 11.4. Find the breakaway point for a system represented by

$$G(s)H(s) = \frac{K(s+1)}{s(s+2)(s+3)}$$

Solution The poles of the system are at $s = 0, -2$ and -3. There is a zero at $s = -1$. These are plotted in Figure 11.8(a). The plot shows that the segment between $s = -2$ and $s = -3$ is a part of the root locus. So, a breakaway point exists in between.

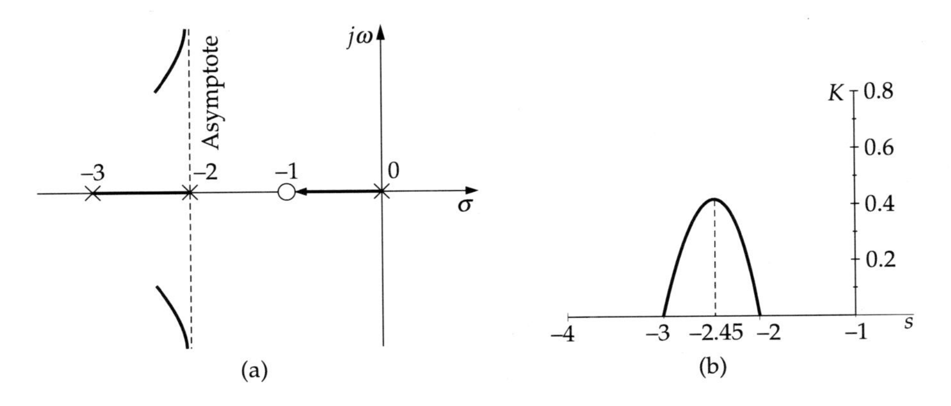

Figure 11.8 (a) Root locus in the real axis; (b) breakaway point evaluation.

The characteristic equation is given by

$$s(s+2)(s+3) + K(s+1) = 0$$

or
$$K = \frac{-s(s+2)(s+3)}{(s+1)} \tag{i}$$

$$\frac{dK}{ds} = \frac{d}{ds}\left[\frac{-s(s+2)(s+3)}{(s+1)}\right] = \frac{(s^3+5s^2+6s)-(s+1)(3s^2+10s+6)}{(s+1)^2} = 0$$

$\Rightarrow$
$$2s^3 + 8s^2 + 10s + 6 = 0 \tag{ii}$$

Equation (ii) is rather difficult to solve. But since we know that the breakaway point lies between $s = -2$ and $s = -3$, we can evaluate Eq. (i) for different values of s to find values of K in this interval and graphically obtain the maximum value at $s = -2.45$ [see below the value of K for different values of s and Figure 11.8(b)].

s	-2.00	-2.40	-2.45	-2.50	-2.60	-3.0
K	0	0.412	0.420	0.417	0.380	0

Step 4. Find the angle of departure (θ_d) from a pole and angle of arrival (θ_a) at a zero for complex poles and zeros.

This can be done either *analytically* or *graphically*. We now describe both the methods with a typical example.

EXAMPLE 11.5. Find the angle of departure for

$$G(s)H(s) = \frac{K(s+3)}{(s+2+j1)(s+2-j1)}$$

where $K > 0$.

Solution *Analytical method:* The relevant equations are:

$$\theta_d = \begin{cases} 180^\circ + \arg|G(s)H(s)| & \text{for poles} \\ 180^\circ - \arg|G(s)H(s)| & \text{for zeros} \end{cases}$$

where $\arg|G(s)H(s)|$ is the angle of $G(s)H(s)$ excluding the *pole* or *zero*, where the angle is to be calculated.

For the pole at $s = -2 + \jmath 1$,

$$\arg\left|G(s)H(s)\right|_{s=-2+\jmath 1} = \arg\left|\frac{K(s+3)}{(s+2+\jmath 1)}\right|_{s=-2+\jmath 1} = \arg\left|\frac{K(1+\jmath 1)}{2\jmath 1}\right| = \frac{\angle 45^\circ}{\angle 90^\circ} = \angle - 45^\circ$$

Therefore,
$$\theta_d = 180^\circ - 45^\circ = 135^\circ$$

For the other pole, it will obviously be -135°.

Graphical method: The relevant formula is

$$\theta_d = 180^\circ - \{\Sigma \text{ angles to a complex pole from other poles} \\ - \Sigma \text{ angles to a complex pole from other zeros}\} \quad \text{(i)}$$

Here, poles are at $s = -2 + \jmath 1$ and $s = -2 - \jmath 1$. Zero is at $s = -3$. To calculate θ_d at the pole at $s = -2 + \jmath 1$, angles subtended by the pole at $s = -2 - \jmath 1$ and the zero at $s = -3$ are plotted in Figure 11.9. Therefore, according to Eq. (i),

$$\theta_d = 180^\circ - (90^\circ - 45^\circ) = 135^\circ$$

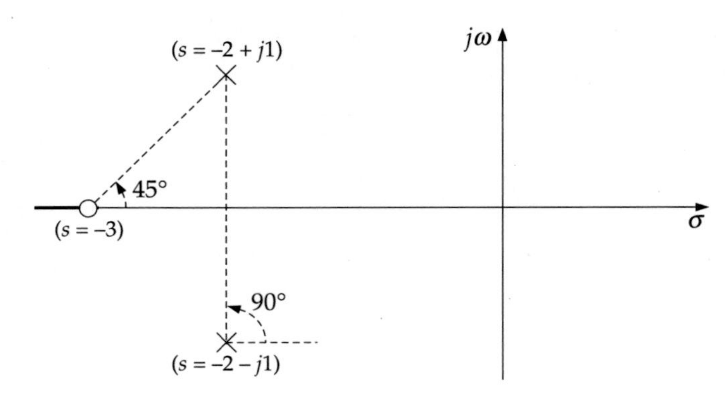

Figure 11.9 Pole, zero and their angle measurements (Example 11.5).

Step 5. Find the intersection of the root loci with the imaginary axis.

Since the roots lying on the imaginary axis will make the system marginally stable, these roots can be found with the help of the Routh array as already discussed. To recapitulate, the steps are as follows:

1. Form the Routh array.
2. Determine $K = K_{\text{marginal}}$ which creates a row of zeros in the array.
3. Solve the auxiliary equation to get the required s-values.

EXAMPLE 11.6. Find the points of intersection of the root locus with the imaginary axis for the system

$$G(s)H(s) = \frac{K}{s(s+4)(s^2+4s+13)}$$

Solution The characteristic equation is

$$1 + G(s)H(s) = 0$$

$$\Rightarrow \quad s(s+4)(s^2+4s+13) + K = 0$$

$$\Rightarrow \quad s^4 + 8s^3 + 29s^2 + 52s + +K = 0$$

The Routh array is

s^4	1	29	K
s^3	8	52	
s^2	22.5	K	
s^1	$52 - 0.35556K$	0	
s^0	K		

To determine K_{marginal}, we put

$$52 - 0.35556\, K\big|_{K=K_{\text{marginal}}} = 0$$

Therefore,

$$K_{\text{marginal}} = \frac{52}{0.35556} = 146.25$$

The auxiliary equation is

$$22.5s^2 + 146.25 = 0$$

$$\Rightarrow \quad s = \pm\, \jmath\, 2.55$$

Therefore, the intersection points on the imaginary axis are $\pm\, \jmath\, 2.55$.

These are all the steps to follow for finding out cardinal points related to drawing root loci manually. Obviously, these steps *will not give all the points to draw graphs of loci.* However, these steps will give enough *hints for making educated guesses* about the forms of the loci. With this background knowledge, let us solve a few full-fledged problems to learn the root locus technique.

EXAMPLE 11.7. The root locus plot of a unity feedback system is shown in the following figure.

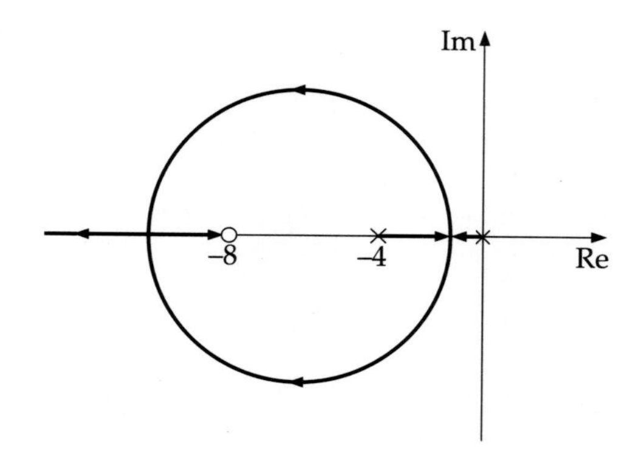

The system transfer function has the form

(a) $\dfrac{K}{s(s+4)(s+8)}$ (b) $\dfrac{K(s+8)}{s(s+4)}$ (c) $\dfrac{K(s+4)}{s(s+8)}$ (d) $\dfrac{Ks}{(s+4)(s+8)}$

Solution From the pole-zero map, we observe that the transfer function has poles at $s = 0$ and -4, and a zero at $s = -8$. Of the given choices, (b) conforms to this observation. Ans: (b).

EXAMPLE 11.8. The breakaway point of the root locus from the real axis for a closed-loop system with a loop gain

$$G(s)H(s) = \frac{K(s+10)}{(s+2)(s+5)}$$

lies

(a) between -2 and origin
(b) between -2 and -5
(c) between -10 and $-\infty$
(d) at $-\infty$

Solution The pole-zero map, shown in the following figure, indicates that the part of the real axis between $s = -2$ and $s = -5$ belongs to the root locus.

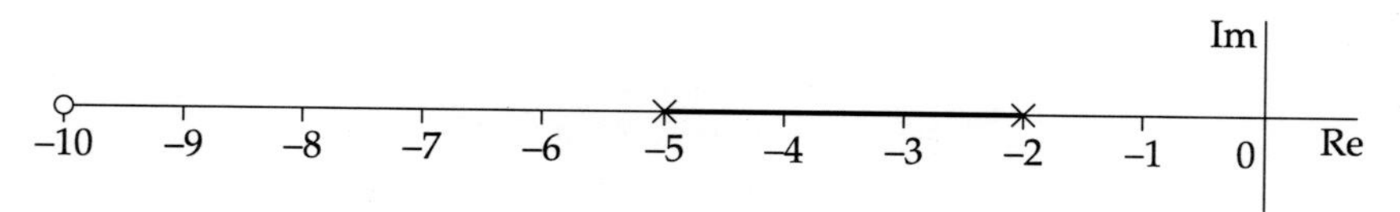

Obviously, the breakaway point must exist between $s = -2$ and $s = -5$.
Ans: (b).

EXAMPLE 11.9. The characteristic equation of a closed-loop system is $s(s+1)(s+3)+k(s+2) = 0,\ k > 0$. Which of the following statements is true?

(a) Its roots are always real.
(b) It cannot have a breakaway point in the range $-1 < \text{Re}[s] < 0$.
(c) Two of its roots tend to infinity along the asymptotes $\text{Re}[s] = -1$.
(d) It may have complex roots in the right half plane.

Solution The open-loop transfer function corresponding to the given characteristic equation is

$$G(s)H(s) = \frac{Ks(s+2)}{(s+1)(s+3)}$$

It consists of three poles at $s = 0, -1, -3$, and a zero at $s = -2$. Therefore, there are two asymptotes at the angles

$$\theta_0 = \frac{2 \times 0 + 1}{2} \cdot 180° = 90°$$

$$\theta_1 = \frac{2 \times 1 + 1}{2} \cdot 180° = 270°$$

The centroid is at $s = \dfrac{-1-3+2}{2} = -1$. The pole-zero plot along with asymptotes is shown in the following figure.

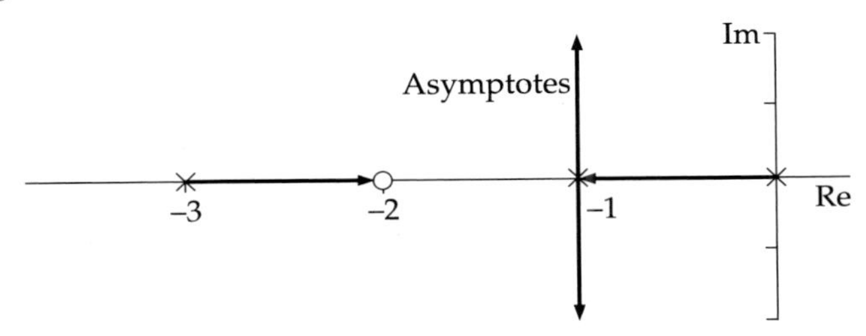

It is obvious from the plot that the two adjacent poles at $s = 0$ and -1 break away and tend to ∞ along asymptotes at $s = -1$. So, the statement (c) is true.
Ans: (c).

EXAMPLE 11.10. A unity feedback system has the transfer function

$$\frac{K(s+b)}{s^2(s+20)}$$

The value of b for which the loci of all the three roots of the closed-loop characteristic equation meet at a single point is

(a) $\dfrac{10}{9}$ (b) $\dfrac{20}{9}$ (c) $\dfrac{30}{9}$ (d) $\dfrac{40}{9}$

Solution Since $H(s) = 1$, the characteristic equation is given by

$$1 + \frac{K(s+b)}{s^2(s+20)} = 0$$

or

$$s^2(s+20) + K(s+b) = 0$$

Therefore,

$$K = -\frac{s^2(s+20)}{s+b}$$

If all loci meet at a point, then it must be a breakaway point. So, we have

$$\frac{dK}{ds} = \frac{s^2(s+20) - (3s^2+40s)(s+b)}{(s+b)^2} = 0$$

or

$$s^2(s+20) - s(3s+40)(s+b) = 0$$

or

$$2s^2 + (3b+20)s + 40b = 0 \tag{i}$$

Equation (i) yields

$$s = \frac{-(3b+20) \pm \sqrt{(3b+20)^2 - 320b}}{4} \tag{ii}$$

The condition for s to be a single point is

$$(3b+20)^2 = 320b$$

or

$$9b^2 - 200b + 400 = 0 \tag{iii}$$

Equation (iii) yields two values for b, namely,

$$b = 20 \text{ and } \frac{20}{9}$$

For $b = 20$, the breakaway point turns out to be at $s = -20$ from Eq. (ii). This value is discarded because, from the pole-zero position of the open-loop transfer function, we expect a breakaway point between 0 and -20. So, the second value, i.e. $b = 20/9$, which yields a breakaway point at $s = -6.67$ from Eq. (ii), is accepted.

Ans: (b).

EXAMPLE 11.11. Sketch the root locus of

$$G(s)H(s) = \frac{K}{s(s+1+\jmath 1)(s+1-\jmath 1)}, \qquad K > 0$$

Solution Here, poles are at $s = 0, \ -1-\jmath 1, \ -1+\jmath 1$. Therefore, $n = 3$. No zero. Therefore, $m = 0$.

Step 1. The real axis locus is present for $-\infty < s < 0$.

Step 2. *Asymptotes:* $n - m = 3$. Therefore, the number of asymptotes = 3 and $q = 0,\ 1,\ 2$. The angles of asymptotes are

$$\theta_0 = \frac{1}{3} \times 180^\circ = 60^\circ, \qquad \theta_1 = \frac{3}{3} \times 180^\circ = 180^\circ, \qquad \theta_2 = \frac{5}{3} \times 180^\circ = 300^\circ$$

The centroid coordinate is

$$\chi_c = \frac{-1-1}{3} = -\frac{2}{3} = -0.67$$

Step 3. The *breakaway point* calculation is not necessary since the loci do not break away from the real axis.

Step 4. *Angle of departure at* $s = -1 + \jmath 1$:

Analytical calculation:

$$\arg\left|G(s)H(s)\right|_{s=-1+\jmath 1} = \arg\left|\frac{K}{s(s+1+\jmath 1)}\right|_{s=-1+\jmath 1} = \arg\left|\frac{K}{(-1+\jmath 1)(2\jmath 1)}\right|$$

$$= \frac{1}{\angle(\pi - 45^\circ) + (\angle 90^\circ)} = \angle - 225^\circ$$

Therefore, $$\theta_d = 180^\circ - 225^\circ = -45^\circ$$

Graphical calculation: As shown in Figure 11.10, the calculation can be done as follows:

$$\begin{aligned}\theta_d =& 180^\circ - [90^\circ(\text{angle subtended by the other complex pole})\\ &- (\pi - 45^\circ)(\text{angle subtended by the pole at the origin})]\\ =& -45^\circ\end{aligned}$$

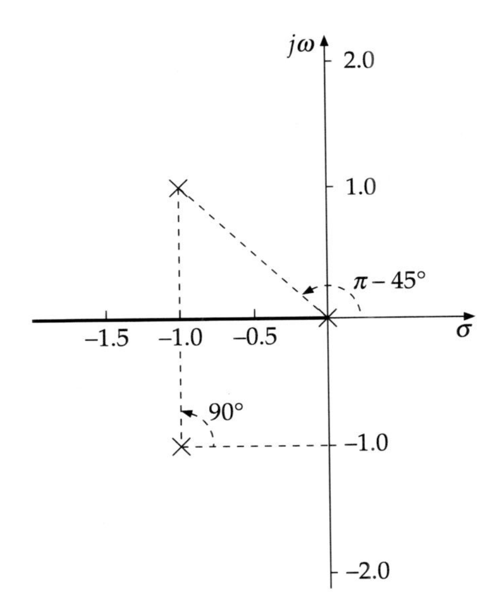

Figure 11.10 Graphical calculation of the angle of departure (Example 11.11).

Step 5. Intersection at the $\jmath\omega$-axis

$$1 + G(s)H(s) = 0$$

$$\Rightarrow \quad s(s+1+\jmath 1)(s+1-\jmath 1) + K = 0$$

$$\Rightarrow \quad s^3 + 2s^2 + 2s + K = 0$$

The Routh array is

$$\begin{array}{c|cc} s^3 & 1 & 2 \\ s^2 & 2 & K \\ s^1 & 2-\dfrac{K}{2} & 0 \\ s^0 & K & \end{array}$$

For marginal stability,

$$2-\frac{K_{\text{mar}}}{2}=0$$

$\Rightarrow$ $$K_{\text{mar}}=4$$

The auxiliary equation

$$2s^2+4=0$$

$\Rightarrow$ $$s=\pm\, j\,1.414$$

With the data that we obtained in steps 1–5, the root locus is drawn and is as shown in Figure 11.11.

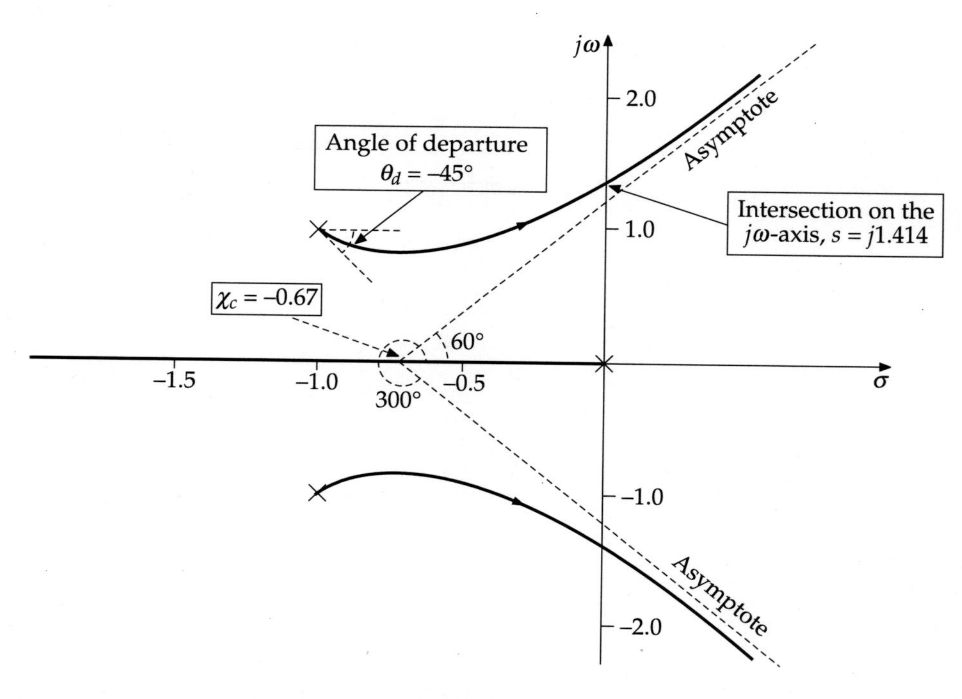

Figure 11.11 Root locus (Example 11.11).

EXAMPLE 11.12. Draw the root locus of the system having the open-loop transfer function as

$$G(s)H(s) = \frac{K}{s(s+3)(s^2+2s+2)}$$

Solution The open-loop transfer function can be written as

$$G(s)H(s) = \frac{K}{s(s+3)(s+1-\jmath 1)(s+1+\jmath 1)}$$

The poles are at $s = 0,\ -3,\ -1+\jmath 1,\ -1-\jmath 1$. Therefore, $n = 4$. No zero. Thus, $m = 0$. The real axis locus is present for $0 < s < -3$.

Asymptotes: The number of asymptotes $= (4-0) = 4$. Therefore, $q = 0,\ 1,\ 2,\ 3$.
Angles of asymptotes are:

$$\theta_0 = \frac{1}{4} \times 180° = 45° \qquad \theta_1 = \frac{3}{4} \times 180° = 135°$$

$$\theta_2 = \frac{5}{4} \times 180° = 225° \qquad \theta_3 = \frac{7}{4} \times 180° = 315°$$

Centroid:

$$\chi_c = \frac{-3-1-1}{4} = -1.25$$

Breakaway point: Since there are two adjacent poles at $s = 0$ and -3, and this segment of the real axis is a part of the root locus, a breakaway point is expected in this segment. To calculate the point, we proceed from the characteristic equation as follows:

$$s(s+3)(s^2+2s+2) + K = 0$$

$$K = -s^4 - 5s^3 - 8s^2 - 6s$$

Therefore,

$$\frac{dK}{ds} = 4s^3 + 15s^2 + 16s + 6 = 0 \tag{i}$$

Equation (i) is a cubic equation which has a root between -2 and -3 (the values of the polynomial are 2 and -15 at these s-values). So, we plot K vs. s between $s = -2$ and -3, and observe that K possesses a maximum at around $s = -2.29$ as will be evident from the following data points:

s	-2	-2.2	-2.25	-2.27	-2.28	-2.29	-2.3	-2.4	-2.5
K	4	4.2944	4.3242	4.3298	4.3312	4.3316	4.3309	4.2624	4.0625

It can be further refined by trial and error to $s = -2.289$ (approx). The other two roots of Eq. (i) being complex[4], we do not expect breakaway points in the complex plane; in fact, the breakaway point is at $s = -2.289$. Since two poles are involved, the angle at which their loci breakaway from the Re-axis is $\pm 180° \div 2 = \pm 90°$.

[4] MatLab gives the three roots as -2.2886, $-0.7307 + \jmath 0.3486$, $-0.7307 - \jmath 0.3486$.

Angle of departure for the complex pole at $s = -1 + \jmath 1$.

Analytical calculation:

$$\arg |G(s)H(s)|_{s=-1+\jmath 1} = \arg \left| \frac{K}{s(s+3)(s+1+\jmath 1)} \right|_{s=-1+\jmath 1}$$

$$= \arg \left| \frac{K}{(-1+\jmath 1)(-1+\jmath 1+3)(2\jmath 1)} \right|$$

$$= \frac{1}{\tan^{-1}\frac{1}{-1} + \tan^{-1}\frac{2}{1} + \tan^{-1}\frac{1}{0}}$$

$$= \frac{1}{\angle(\pi - 45^\circ) + \angle 26.56^\circ + \angle 90^\circ} = \angle -251.56^\circ$$

Therefore,
$$\theta_d = 180^\circ - 251.56^\circ = -71.56^\circ$$

Graphical calculation:

$$\begin{aligned}\theta_d =& 180^\circ - [90^\circ (\text{angle subtended by the other complex pole})\\ &+ (\pi - 45^\circ)(\text{angle subtended by the pole at the origin})\\ &+ 26.56^\circ (\text{angle subtended by the pole at } s = -3)]\\ =& -71.56^\circ\end{aligned}$$

Intersection on the $\jmath\omega$-axis: The characteristic equation is

$$s^4 + 5s^3 + 8s^2 + 6s + K = 0$$

Therefore, the Routh array is

$$\begin{array}{c|ccc} s^4 & 1 & 8 & K \\ s^3 & 5 & 6 & \\ s^2 & 6.8 & K & \\ s^1 & 6 - 0.735K & 0 & \\ s^0 & K & & \end{array}$$

Thus,
$$K_{\text{mar}} = \frac{6}{0.735} = 8.16$$

From the auxiliary equation

$$6.8s^2 + 8.16 = 0$$

$\Rightarrow$
$$s = \pm \jmath 1.1$$

Having obtained these data, we plot the root locus as shown in Figure 11.12.

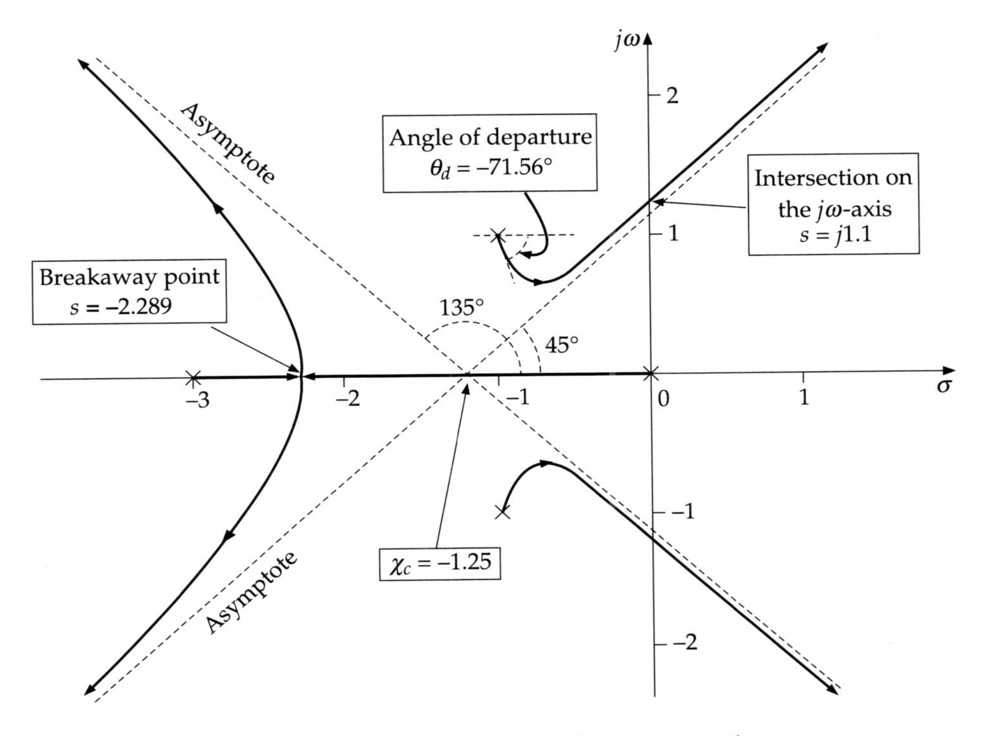

Figure 11.12 Root locus plot (Example 11.12).

EXAMPLE 11.13. The system is given by the following figure.

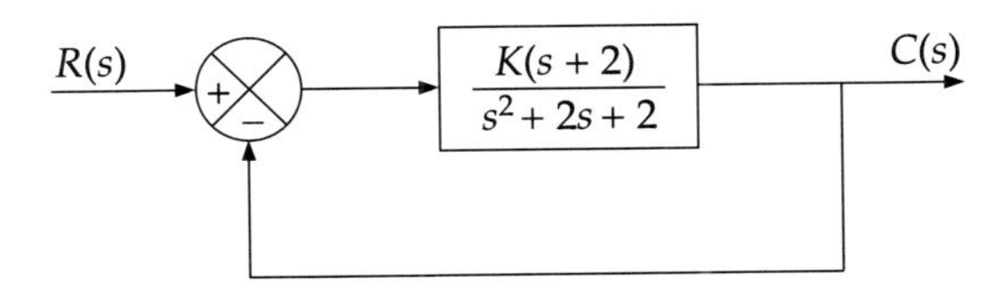

Draw the root locus and find the gain factor K at the breakaway point.

Solution The open-loop transfer function can be written as

$$G(s)H(s) = \frac{K(s+2)}{(s+1+\jmath 1)(s+1-\jmath 1)}$$

The poles are at $s = -1+\jmath 1,\ -1-\jmath 1$. Therefore, $n = 2$. Zero is at $s = -2$. Thus, $m = 1$.

Two root loci will start from two poles. One will terminate at zero (i.e., $s = -2$) and the other at $s \to -\infty$. So, the real axis locus is present for $-\infty > s > -2$.

Asymptotes: The number of asymptotes is $n - m = 1$. Hence, its angle is $180°$ $(q = 0)$.

Breakaway point: The characteristic equation is

$$s^2 + 2s + 2 + K(s+2) = 0$$

$$K = -\frac{s^2 + 2s + 2}{s+2} \tag{i}$$

$$\frac{dK}{ds} = \frac{(s^2 + 2s + 2) - (2s+2)(s+2)}{(s+2)^2} = 0$$

The roots of the equation are obtained as

$$(s^2 + 2s + 2) - (2s+2)(s+2) = 0$$

$$\Rightarrow \qquad s^2 + 4s + 2 = 0$$

Therefore, $\qquad s = -0.586 \text{ and } -3.414$

The former point does not fall on the root locus on the real axis. So, the latter is the break-in point.

Note: This is a *break-in* rather than *breakaway* point.

Angle of departure at $s = -1 + \jmath 1$*:*

$$\arg\left|\frac{K(s+2)}{s+1+\jmath 1}\right|_{s=-1+\jmath 1} = \arg\left|\frac{K(1+\jmath 1)}{2\jmath 1}\right| = \frac{\angle 45^\circ}{\angle 90^\circ} = \angle -45^\circ$$

Therefore, $\qquad \theta_d = 180^\circ - 45^\circ = 135^\circ$

To figure out the locus in the complex plane, let us choose a complex point $s = \sigma + \jmath\omega$ on the root locus and invoke the angle condition as follows:

$$\angle\{G(\sigma+\jmath\omega)H(\sigma+\jmath\omega)\} = \arg\left|\frac{\sigma+\jmath\omega+2}{(\sigma+\jmath\omega+1+\jmath 1)(\sigma+\jmath\omega+1-\jmath 1)}\right| = \pm\pi$$

$$\Rightarrow \qquad \frac{\tan^{-1}\dfrac{\omega}{\sigma+2}}{\tan^{-1}\dfrac{\omega+1}{\sigma+1} + \tan^{-1}\dfrac{\omega-1}{\sigma+1}} = -\pi$$

$$\Rightarrow \qquad \tan^{-1}\frac{\omega}{\sigma+2} - \tan^{-1}\frac{\omega+1}{\sigma+1} - \tan^{-1}\frac{\omega-1}{\sigma+1} = -\tan^{-1}(0)$$

$$\Rightarrow \qquad \tan^{-1}(0) + \tan^{-1}\frac{\omega}{\sigma+2} = \tan^{-1}\frac{\omega+1}{\sigma+1} + \tan^{-1}\frac{\omega-1}{\sigma+1}$$

$$\Rightarrow \qquad \tan^{-1}\frac{0 + \dfrac{\omega}{\sigma+2}}{1-(0)\dfrac{\omega}{\sigma+2}} = \tan^{-1}\frac{\dfrac{2\omega}{\sigma+1}}{1-\dfrac{\omega^2-1}{(\sigma+1)^2}}$$

Eliminating $\tan^{-1}$ from both sides, we get, by diagonal multiplication, the relation

$$\frac{2\omega}{\sigma+1} = \frac{\omega}{\sigma+2}\left[1 - \frac{\omega^2-1}{(\sigma+1)^2}\right]$$

$$\Rightarrow \qquad 2(\sigma+1)(\sigma+2) = (\sigma+1)^2 - \omega^2 + 1$$

$$\Rightarrow \qquad \sigma^2 + 4\sigma + 2 + \omega^2 = 0$$

$$\Rightarrow \qquad (\sigma+2)^2 + (\omega+0)^2 = 2 \tag{ii}$$

Equation (ii) represents a circle of radius $\sqrt{2}$, having its centre at $(-2, j0)$. Having obtained these data, we draw the root locus as given in Figure 11.13.

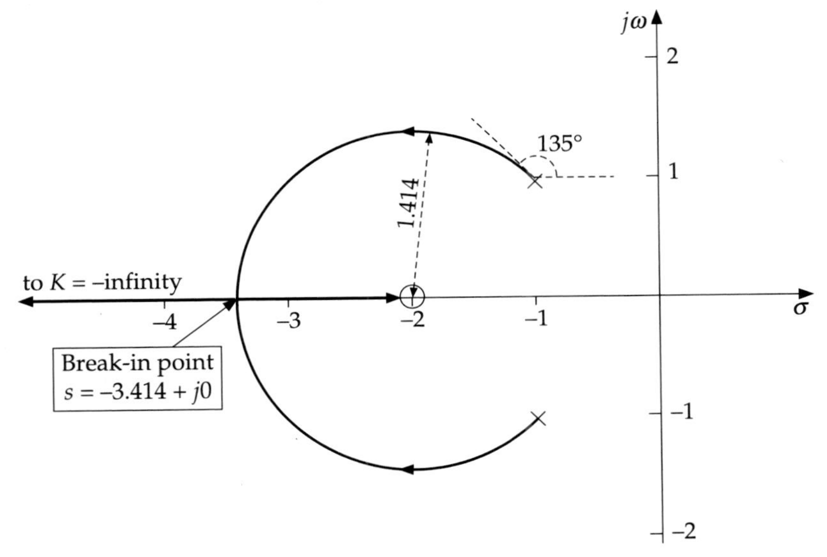

Figure 11.13 Root locus (Example 11.13).

Value of K at the break-in point:

Analytical calculation: From Eq. (i)

$$K = -\frac{(-3.414)^2 + 2(-3.414) + 2}{-3.414 + 2} = 4.828$$

Graphical calculation:

$$K = \frac{\prod \text{vector lengths from poles to break-in point}}{\prod \text{vector lengths from zeros to break-in point}}$$

$$= \frac{\left(\sqrt{2.414^2 + 1^2}\right)\left(\sqrt{2.414^2 + 1^2}\right)}{1.414}$$

$$= 4.828$$

EXAMPLE 11.14. Plot the root locus for

$$G(s)H(s) = \frac{K}{s(s^2 + 6s + 8)}$$

Then find the value of K:

(a) For marginal stability

(b) At the breakaway point

(c) For damping ratio $\zeta = 0.5$

Solution The open-loop transfer function can be written as

$$G(s)H(s) = \frac{K}{s(s+2)(s+4)}$$

Therefore, the poles are at $s = 0,\ -2,\ -4$, and $n = 3$. No zero. Therefore, $m = 0$.

The root locus exists on the $j\omega$-axis between $s = 0$ and -2, and $s = -4$ and $-\infty$. The root loci start from poles and end at ∞ since there is no zero.

Asymptotes: Number $= 3 - 0 = 3$. Therefore, $q = 0,\ 1,\ 2$. The angles of asymptotes are

$$\theta_0 = \frac{1}{3} \times 180° = 60° \qquad \theta_1 = \frac{3}{3} \times 180° = 180° \qquad \theta_3 = \frac{5}{3} \times 180° = 300°$$

and the centroid co-ordinate is

$$\chi_c = \frac{-2-4}{3} = -2$$

Breakaway point: It should exist between $s = 0$ and -2. The characteristic equation is

$$s^3 + 6s^2 + 8s + K = 0$$

$$\Rightarrow \qquad K = -s^3 - 6s^2 - 8s \tag{i}$$

$$\frac{dK}{ds} = 3s^2 + 12s + 8 = 0$$

$$\Rightarrow \qquad s = -0.845, \ -3.154$$

The point-3.154 is ruled out because it does not lie on the root locus. Therefore, the breakaway point is at $s = -0.845$.

Intersection at the $j\omega$-axis: The Routh array is

$$\begin{array}{c|cc} s^3 & 1 & 8 \\ s^2 & 6 & K \\ s^1 & 8 - \dfrac{K}{6} & \\ s^0 & K & \end{array}$$

For marginal stability,

$$8 - \frac{K_{\text{mar}}}{6} = 0$$

$$\Rightarrow \qquad K_{\text{mar}} = 48$$

The auxiliary equation is

$$6s^2 + 48 = 0$$

$$\Rightarrow \qquad s = \pm\, j2.83$$

With the help of these available data, the root locus, as shown in Figure 11.14, is drawn.

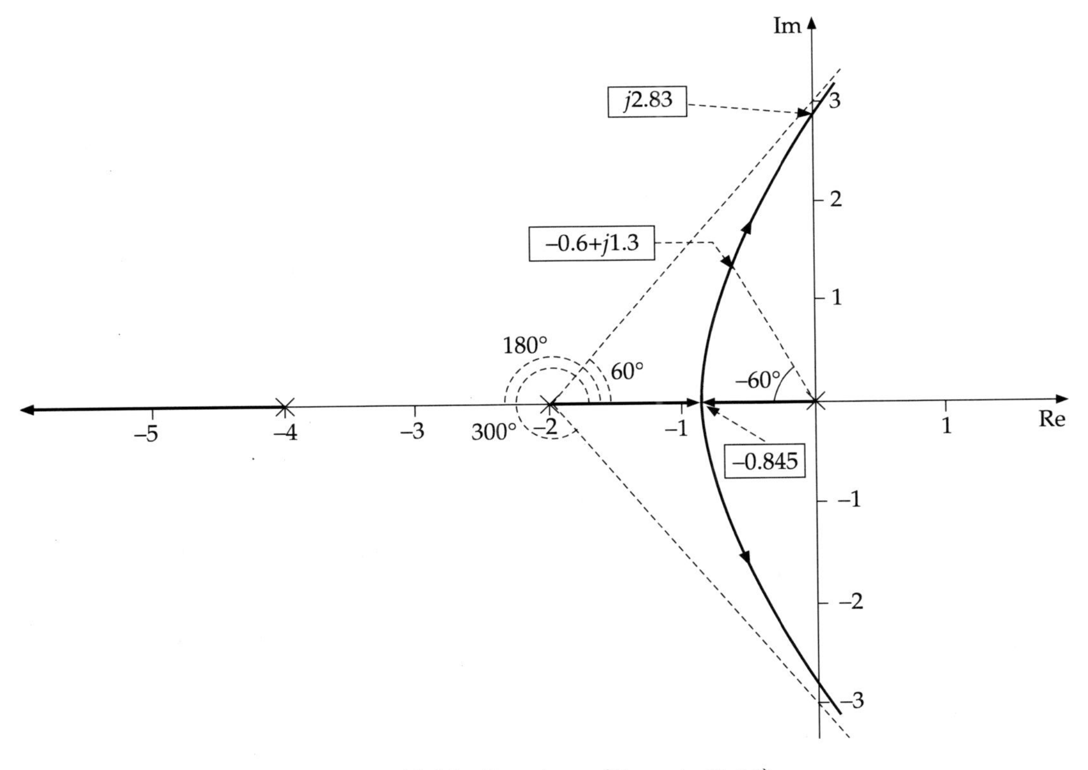

Figure 11.14 Root locus (Example 11.14).

Thus,

(a) The value of K for marginal stability is 48.

(b) The value of K at the breakaway point can be calculated as follows:
Analytically: From Eq. (i),

$$K|_{s=-0.845} = -(-0.845)^3 - 6(-0.845)^2 - 8(-0.845) = 3.079$$

Graphically:

$$K = \frac{\prod \text{vector lengths from poles to breakaway point}}{\prod \text{vector lengths from zeros to breakaway point}}$$

$$= (0.845)(2 - 0.845)(4 - 0.845) = 3.079$$

(c) For damping ratio $\zeta = 0.5$, the phase angle

$$\phi = \tan^{-1} \frac{\sqrt{1-(1/2)^2}}{1/2} = \tan^{-1} \sqrt{3} = 60^\circ$$

We draw a line in the second quadrant at an angle of 60° with the negative part of the real axis from the origin. It intercepts the locus at $-0.6 + \jmath 1.3$. The K value at this point is

$$K = \frac{\prod \text{vector lengths from poles to } (-0.6 + \jmath 1.3)}{\prod \text{vector lengths from zeros to } (-0.6 + \jmath 1.3)}$$

$$= \left[\sqrt{(0+0.6)^2 + 1.3^2}\right] \left[\sqrt{(-2+0.6)^2 + 1.3^2}\right] \left[\sqrt{(-4+0.6)^2 + 1.3^2}\right] = 9.957$$

EXAMPLE 11.15. The root locus plot for a system with transfer function

$$\frac{2}{s(s+1)(s+2)}$$

is shown in Figure 11.15. A unity feedback proportional control system is built using this system.

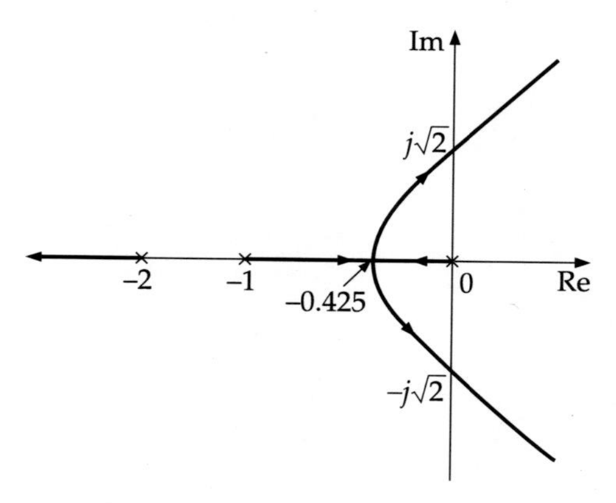

Figure 11.15 Root locus plot (Example 11.15).

(a) The maximum possible controller gain, for which the unity feedback system is stable, is approximately,

(i) 6.0 (ii) 3.0 (iii) 0.4 (iv) 0.2

(b) The maximum possible controller gain, for which the unity feedback system exhibits a non-oscillatory response to a unit step, is
(i) 6.0 (ii) 3.0 (iii) 0.4 (iv) 0.2

Solution (a) The closed-loop transfer function is given by

$$\frac{C(s)}{R(s)} = \frac{K}{1 + \dfrac{2K}{s(s+1)(s+2)}} = \frac{K}{s(s+1)(s+2) + 2K}$$

From the characteristic equation

$$s(s+1)(s+2) + 2K = 0$$

we get
$$|K| = \frac{s(s+1)(s+2)}{2}$$

Therefore,
$$|K|_{s=\jmath\sqrt{2}} = \frac{\sqrt{2}\sqrt{1+2}\sqrt{2^2+2}}{2} = 3$$

Ans: (ii).

(b) The root locus breaks away from the Re-axis at the point $s = -0.425$ and enters the complex space when its response becomes oscillatory. So the maximum gain when the system exhibits a non-oscillatory response occurs at $s = -0.425$. Therefore,

$$|K|_{s=0.425} = \left|\frac{-0.425(-0.425+1)(-0.425+2)}{2}\right| \approx 0.2$$

Ans: (iv).

EXAMPLE 11.16. The root locus of a plant is given in the following figure. The root locus crosses imaginary axis at $\omega = 4\sqrt{2}$ rad/s with gain $K = 384$. It is observed that the point $s = -1.5 + \jmath 1.5$ lies on the root locus. The gain K at $s = -1.5 + \jmath 1.5$ is computed as

(a) 11.3 (b) 21.2 (c) 41.3 (d) 61.2

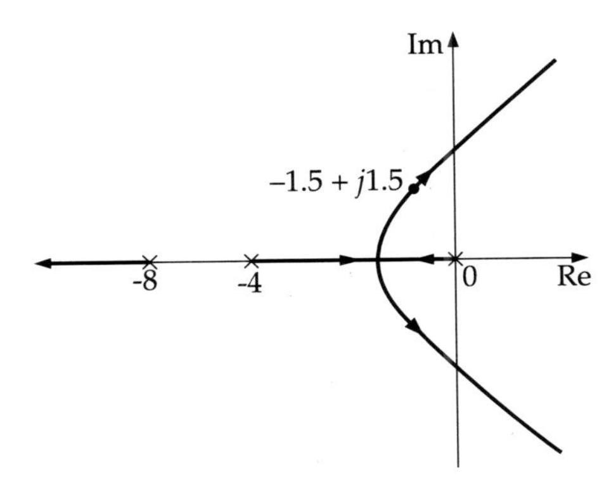

Solution From the pole-zero map, we can write the open-loop transfer function as

$$G(s)H(s) = \frac{K'}{s(s+4)(s+8)}$$

The gain at $s = \jmath\omega = \jmath 4\sqrt{2}$ is given to be 384. We know this gain is calculated as

$$\begin{aligned}
K = 384 = K' &\frac{\prod \text{distance of all poles from } s = 0 + \jmath 4\sqrt{2}}{\prod \text{distance of all zeros from } s = 0 + \jmath 4\sqrt{2}} \\
&= K'(4\sqrt{2})\left(\sqrt{4^2 + (4\sqrt{2})^2}\right)\left(\sqrt{8^2 + (4\sqrt{2})^2}\right) \\
&= K'(384)
\end{aligned}$$

Therefore, $\quad K' = 1$

Thus, the gain at $s = -1.5 + \jmath 1.5$ is

$$\begin{aligned}
K\big|_{s=-1.5+\jmath 1.5} &= \left(\sqrt{1.5^2 + 1.5^2}\right)\left(\sqrt{(4-1.5)^2 + 1.5^2}\right)\left(\sqrt{(8-1.5)^2 + 1.5^2}\right) \\
&= 41.2568
\end{aligned}$$

Ans: (c).

EXAMPLE 11.17. Sketch the root locus for the system given by

$$G(s)H(s) = \frac{K}{s(s+2)(s^2+2s+5)}$$

Solution The open-loop transfer function can be written as

$$G(s)H(s) = \frac{K}{s(s+2)(s+1-\jmath 2)(s+1+\jmath 2)}$$

The poles are at $s = 0,\ -2,\ -1+\jmath 2$ and $-1-\jmath 2$. Therefore, $n = 4$. No zero. Therefore, $m = 0$. Part of σ-axis in the root locus is between $s = 0$ and -2.

Asymptotes: Number $= 4 - 0 = 4$. Therefore, $q = 0,\ 1,\ 2,\ 3$.
Angles of asymptotes:

$$\theta_0 = \frac{1}{4} \times 180^\circ = 45^\circ \qquad \theta_1 = \frac{3}{4} \times 180^\circ = 135^\circ$$

$$\theta_2 = \frac{5}{4} \times 180^\circ = 225^\circ \qquad \theta_3 = \frac{7}{4} \times 180^\circ = 315^\circ$$

Centroid:

$$\chi_c = \frac{-2-1-1}{4} = -1$$

Breakaway points: The characteristic equation

$$s(s+2)(s^2+2s+5)+K=0$$

$$\Rightarrow \quad s^4+4s^3+9s^2+10s+K=0 \quad \text{(i)}$$

Therefore,

$$K=-s^4-4s^3-9s^2-10s$$

$$\frac{dK}{ds}=-4s^3-12s^2-18s-10=0$$

$$\Rightarrow \quad 2s^3+6s^2+9s+5=0 \quad \text{(ii)}$$

Since $s=-1$ satisfies Eq. (ii), one of its roots is $s=-1$. Dividing Eq. (ii) by $s+1$, we get

$$2s^2+4s+5=0$$

Therefore,

$$s=-1\pm \jmath 1.225$$

Thus, there are three breakaway points at $s=-1$ and $s=-1\pm \jmath 1.225$.

Angle of departure for pole at $s=-1+\jmath 2$:

Analytical calculation:

$$\arg|G(s)H(s)|_{s=-1+\jmath 2}=\arg\left|\frac{K}{s(s+2)(s+1+\jmath 2)}\right|_{s=-1+\jmath 2}=\arg\left|\frac{K}{(-1+\jmath 2)(1+\jmath 2)(\jmath 4)}\right|$$

$$=\frac{1}{(\pi-\tan^{-1}2)+\tan^{-1}2+90°]}=\angle -270°$$

Therefore,

$$\theta_d=180°-270°=-90°$$

Graphical calculation:

$$\theta_d=180°-[(\pi-\tan^{-1}2)+\tan^{-1}2+90°]=-90°$$

Note: The loci from the complex poles are heading towards the Re-axis because their angles of departure are $\mp 90°$, whereas two loci, breaking away from the Re-axis are coming at the same angles along the same line $s=-1$. So, these loci have to break away again at complex points $s=-1\pm \jmath 1.225$.

Intersection on the $\jmath\omega$-axis: From the characteristic equation [Eq. (i)], the Routh array can be constructed as

s^4	1	9	K
s^3	4	10	
s^2	6.5	K	
s^1	$10-0.615K$		
s^0	K		

For marginal stability,

$$10-0.615K_{\text{mar}}=0$$

$$\Rightarrow \quad K_{\text{mar}}=16.26$$

The auxiliary equation is

$$6.5s^2 + 16.26 = 0$$

Therefore, $$s = \pm j1.58$$

With the help of these data, we construct the root locus of Figure 11.16.

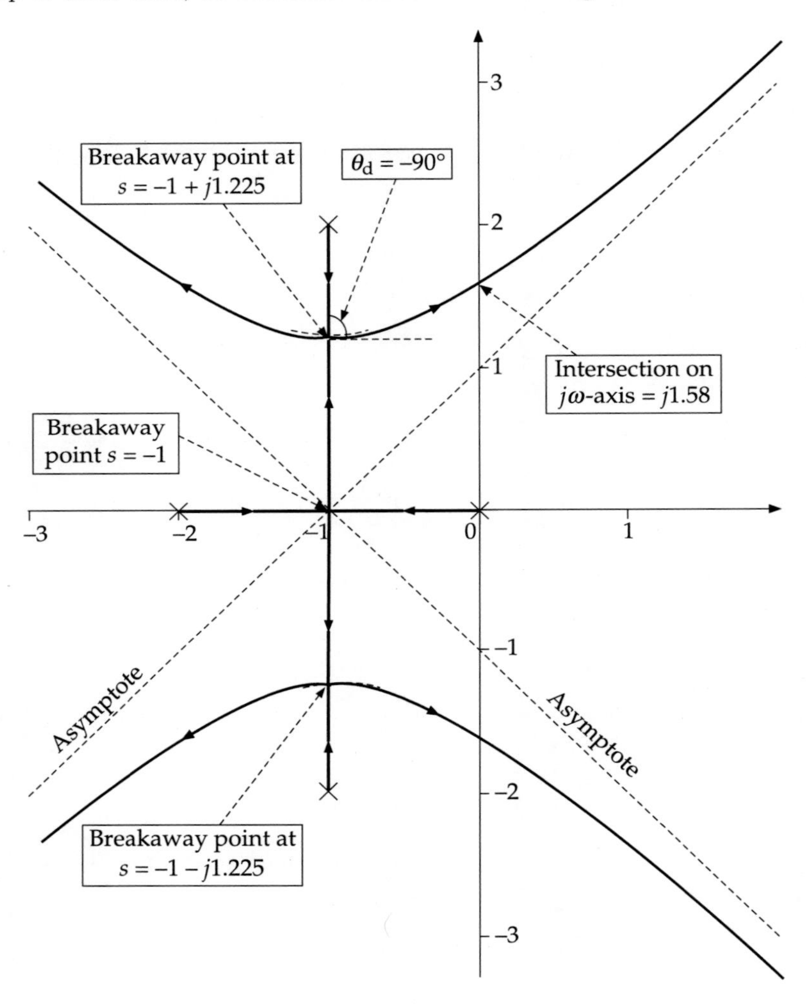

Figure 11.16 Root locus plot (Example 11.17).

EXAMPLE 11.18. Sketch the root locus of a unity feedback system with an open-loop transfer function of

$$G(s) = \frac{K}{s(s^2 + 4s + 13)}$$

Solution Since this is a unity feedback system, $H(s) = 1$. Therefore, we have

$$G(s)H(s) = \frac{K}{s(s^2 + 4s + 13)} = \frac{K}{s(s + 2 - \jmath 3)(s + 2 + \jmath 3)}$$

The poles are at $s = 0, -2 + \jmath 3$ and $-2 - \jmath 3$. Therefore, $n = 3$. No zero. Thus, $m = 0$. Part of the real axis on the root locus is from $s = 0$ to $-\infty$.

Asymptotes: Number = 3. At angles 60°, 180° and 300°.
Centroid:

$$\chi_c = \frac{-2 - 2}{3} = -1.33$$

Angle of departure for the pole at $s = -2 + \jmath 3$:

Analytical calculation:

$$\arg |G(s)H(s)|_{s=-2+\jmath 3} = \arg \left| \frac{1}{s(s + 2 + \jmath 3)} \right|_{s=-2+\jmath 3} = \arg \left| \frac{1}{(-2 + \jmath 3)(\jmath 6)} \right|$$

$$= \frac{1}{\angle\{\pi - \tan^{-1}(3/2)\}^\circ + (\angle 90^\circ)}$$

$$= \angle - 213.7^\circ$$

Therefore,

$$\theta_d = 180^\circ - 213.7^\circ = -33.7^\circ$$

Graphical calculation:

$$\theta_d = 180^\circ - \left(90^\circ + \pi - \tan^{-1}\frac{3}{2}\right) = -33.7^\circ$$

Intersection on the $\jmath\omega$-axis: The characteristic equation is

$$s^3 + 4s^2 + 13s + K = 0$$

The Routh array is, therefore,

$$\begin{array}{c|cc} s^3 & 1 & 13 \\ s^2 & 4 & K \\ s^1 & 13 - 0.25K & 0 \\ s^0 & K & \end{array}$$

For marginal stability,

$$13 - 0.25K_{\text{mar}} = 0$$

or

$$K_{\text{mar}} = 52$$

From the auxiliary equation,

$$4s^2 + 52 = 0$$

or

$$s = \pm\, \jmath 3.61$$

Having obtained these data, we sketch the root locus as in Figure 11.17.

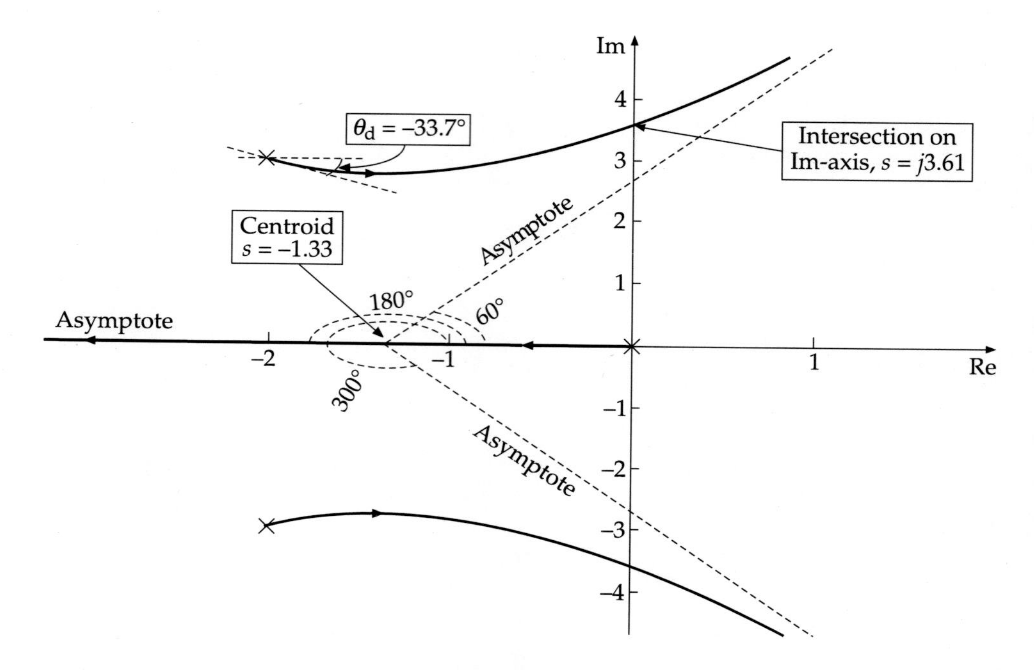

Figure 11.17 Root locus (Example 11.18).

EXAMPLE 11.19. Sketch the root locus of the closed-loop system having open-loop transfer function

$$G(s)H(s) = \frac{K}{s(s+4)(s+6)}$$

Solution Poles are at $s = 0, \; -4$ and -6. Hence, $n = 3$. No zero. Therefore, $m = 0$. Parts of the real axis that will be on the root locus are $-4 < s < 0$ and $-\infty < s < -6$.

Asymptotes: Number $= n - m = 3$. Therefore, $q = 0, \; 1, \; 2$. The angles are $60°, \; 180°, \; 300°$.

Centroid:

$$\chi_c = \frac{-4-6}{3} = -3.33$$

Breakaway point: It should exist in the interval $-4 < s < 0$. The characteristic equation is

$$s(s+4)(s+6) + K = 0$$

$$s^3 + 10s^2 + 24s + K = 0 \qquad \text{(i)}$$

$$\frac{dK}{ds} = -(3s^2 + 20s + 24) = 0$$

Therefore, $$s = -1.57, \; -5.097$$

The former point is acceptable, but the latter is not, because it does not lie on the root locus.

Intersection on the $j\omega$-axis: The Routh array is constructed from Eq. (i) as

$$\begin{array}{c|cc} s^3 & 1 & 24 \\ s^2 & 10 & K \\ s^1 & 24-0.1K & 0 \\ s^0 & K & \end{array}$$

For marginal stability,

$$24 - 0.1K_{\text{mar}} = 0$$

or

$$K_{\text{mar}} = 240$$

Therefore, from the auxiliary equation, we get

$$10s^2 + 240 = 0$$

Therefore,

$$s = \pm j4.899$$

Having obtained these data, we plot the root locus of Figure 11.18.

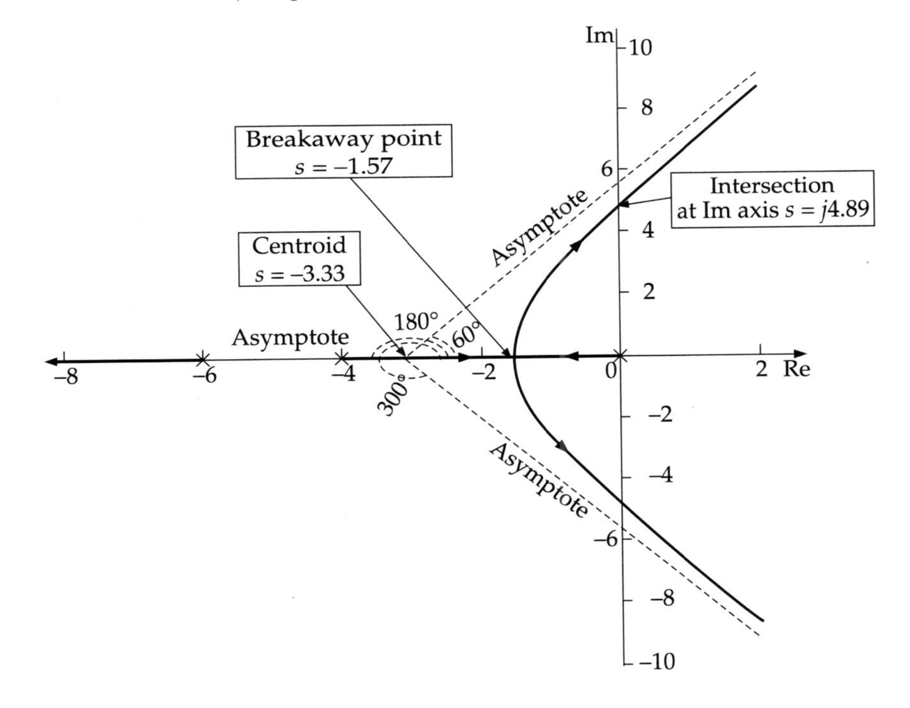

Figure 11.18 Root locus plot (Example 11.19).

EXAMPLE 11.20. Sketch the root locus for a system having

$$G(s)H(s) = \frac{K(s+3)}{(s+2+j\sqrt{5})(s+2-j\sqrt{5})}$$

Solution The poles are at $s = -2 + \jmath\sqrt{5}$ and $-2 - \jmath\sqrt{5}$. Therefore, $n = 2$. Zero at $s = -3$. Hence, $m = 1$. Part of the real axis on the root locus: $-\infty < s < -3$.

Asymptotes: Number $= 2 - 1 = 1$, angle $= 180°$.

Break-in point: It is expected at the interval $-\infty < s < -3$.
The characteristic equation is

$$(s + 2 + \jmath\sqrt{5})(s + 2 - \jmath\sqrt{5}) + K(s + 3) = 0$$

$$s^2 + 4s + 9 + K(s + 3) = 0$$

$$\frac{dK}{ds} = \frac{d}{ds}\left[-\frac{s^2 + 4s + 9}{s + 3}\right] = -\frac{(2s + 4)}{(s + 3)} + \frac{s^2 + 4s + 9}{(s + 3)^2} = -\frac{s^2 + 6s + 3}{(s + 3)^2} = 0$$

Therefore,
$$s = -0.55, \ -5.45$$

The first point is rejected because it does not lie on the root locus.

Angle of departure for $s = -2 + \jmath\sqrt{5}$:

Graphical calculation:

$$\theta_d = 180° - \left(90° - \tan^{-1}\frac{\sqrt{5}}{1}\right) = 155.91°$$

Analytical calculation:

$$\arg|G(s)H(s)|_{s=-2+\jmath\sqrt{5}} = \arg\left|\frac{K(s + 3)}{s + 2 + \jmath\sqrt{5}}\right|_{s=-2+\jmath\sqrt{5}} = \arg\left|\frac{K(1 + \jmath\sqrt{6})}{2\jmath\sqrt{5}}\right| = \frac{\angle 65.91°}{\angle 90°}$$

$$= \angle - 24.09°$$

Therefore,
$$\theta_d = 180° - 24.09° = 155.91°$$

To figure out the locus in the complex plane, let us choose a complex point $s = \sigma + \jmath\omega$ on the root locus and invoke the angle condition as follows:

$$\angle\{G(\sigma + \jmath\omega)H(\sigma + \jmath\omega)\} = \arg\left|\frac{\sigma + \jmath\omega + 3}{(\sigma + \jmath\omega + 2 + \jmath\sqrt{5})(\sigma + \jmath\omega + 2 - \jmath\sqrt{5})}\right| = \pm\pi$$

$$\Rightarrow \quad \frac{\tan^{-1}\dfrac{\omega}{\sigma + 3}}{\tan^{-1}\dfrac{\omega + \sqrt{5}}{\sigma + 2} + \tan^{-1}\dfrac{\omega - \sqrt{5}}{\sigma + 2}} = -\pi$$

$$\Rightarrow \quad \tan^{-1}\frac{\omega}{\sigma + 3} - \tan^{-1}\frac{\omega + \sqrt{5}}{\sigma + 2} - \tan^{-1}\frac{\omega - \sqrt{5}}{\sigma + 2} = -\tan^{-1}(0)$$

$$\Rightarrow \qquad \tan^{-1}(0) + \tan^{-1}\frac{\omega}{\sigma+3} = \tan^{-1}\frac{\omega+\sqrt{5}}{\sigma+2} + \tan^{-1}\frac{\omega-\sqrt{5}}{\sigma+2}$$

$$\Rightarrow \qquad \tan^{-1}\frac{0+\dfrac{\omega}{\sigma+3}}{1-(0)\dfrac{\omega}{\sigma+3}} = \tan^{-1}\frac{\dfrac{2\omega}{\sigma+2}}{1-\dfrac{\omega^2-5}{(\sigma+2)^2}}$$

Eliminating $\tan^{-1}$ from both sides, we get, by diagonal multiplication, the relation

$$\frac{2\omega}{\sigma+2} = \frac{\omega}{\sigma+3}\left[1-\frac{\omega^2-5}{(\sigma+2)^2}\right]$$

$$\Rightarrow \qquad 2(\sigma+2)(\sigma+3) = (\sigma+2)^2 - \omega^2 + 5$$

$$\Rightarrow \qquad \sigma^2 + 6\sigma + 3 + \omega^2 = 0$$

$$\Rightarrow \qquad (\sigma+3)^2 + (\omega+0)^2 = 6 = 2.45^2 \qquad \text{(i)}$$

Equation (i) represents a circle of radius 2.45, having its centre at $(-3, \jmath 0)$.

Intersection of root locus at the $\jmath\omega$-axis: It does not arise since the root loci from complex poles depart to the other direction to break in at $s = -5.45$.

With these available data, the complete root locus is drawn as in Figure 11.19.

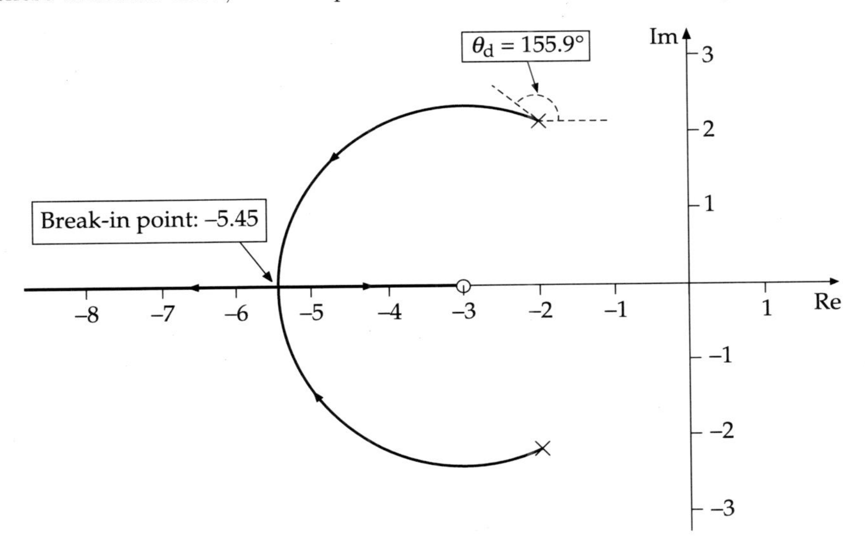

Figure 11.19 Complete root locus plot (Example 11.20).

EXAMPLE 11.21. A unity feedback system has an open loop transfer function of

$$G(s) = \frac{K}{s(s+4)(s^2+4s+20)}$$

Sketch the root locus of the system by determining the following:

(a) Number and angles of asymptotes, centroid
(b) Angle of departure of root locus from poles
(c) Breakaway point
(d) The value of K and the frequency at which the root locus crosses the $j\omega$-axis
(e) Hence find the value of K so that the system has a damping factor of 0.707

Solution The transfer function can be written as

$$G(s)H(s) = \frac{K}{s(s+4)(s+2+j4)(s+2-j4)}$$

So, the number of poles, $n = 4$. No zero, $m = 0$.

(a) *Number of asymptotes:* $n - m = 4$. *Angles of asymptotes:*

$$\theta_0 = \frac{1}{4} \times 180^\circ = 45^\circ \qquad \theta_1 = \frac{3}{4} \times 180^\circ = 135^\circ$$

$$\theta_2 = \frac{5}{4} \times 180^\circ = 225^\circ \qquad \theta_3 = \frac{7}{4} \times 180^\circ = 315^\circ$$

Centroid:

$$\chi_c = \frac{-4-2-2}{4} = -2$$

(b) *Angle of departure:*

(i) *Analytical calculation:*

$$\arg\,[G(s)H(s)]_{s=-2+j4} = \arg\left|\frac{K}{s(s+4)(s+2+j4)}\right|_{s=-2+j4}$$

$$= \arg\left|\frac{K}{(-2+j4)(2+j4)(2j4)}\right|$$

$$= \frac{1}{\tan^{-1}\frac{4}{-2} + \tan^{-1}\frac{4}{2} + \tan^{-1}\frac{4}{0}}$$

$$= \frac{1}{\pi - \tan^{-1}2 + \tan^{-1}2 + 90^\circ}$$

$$= \angle -270^\circ$$

Therefore, $$\theta_d = 180^\circ - 270^\circ = -90^\circ$$

(ii) *Graphical calculation:*

$$\theta_d = 180^\circ - [(\pi - \tan^{-1}2) + 90^\circ + \tan^{-1}2] = -90^\circ$$

(c) *Breakaway points:* The characteristic equation is

$$s(s+4)(s^2+4s+20)+K=0$$

$$\Rightarrow \quad s^4+8s^3+36s^2+80s+K=0 \quad \text{(i)}$$

Therefore,

$$K=-s^4-8s^3-36s^2-80s$$

$$\frac{dK}{ds}=-4s^3-24s^2-72s-80=0$$

$$\Rightarrow \quad s^3+6s^2+18s+20=0$$

$$\Rightarrow \quad (s+2)(s^2+4s+10)=0 \quad \text{(ii)}$$

From Eq, (ii), we get the following coordinates of the breakaway points:

$$s=-2,\ -2\pm \jmath\, 2.45$$

(d) *Intersection on the $\jmath\omega$-axis:* The Routh array is constructed from Eq. (i) as follows:

s^4	1	36	K
s^3	8	80	0
s^2	26	K	
s^1	$80-\frac{8}{26}K$		
s^0	K		

For marginal stability,

$$80-\frac{8}{26}=0$$

$$\Rightarrow \quad K=80\times\frac{26}{8}=260$$

Auxiliary equation

$$26s^2+260=0$$

$$\Rightarrow \quad s=\pm\jmath\sqrt{10}=\pm\jmath\, 3.16$$

With these data, we construct the root locus diagram as shown in Figure 11.20.

Note: There are breakaway points with complex values of s, where the two loci, breaking away from the Re-axis, again break away from two loci coming from $s=-2\pm\jmath 4$. Their angles of departure are $-90°$ and $+90°$, respectively.

To find the value of K at $\zeta=0.707=1/\sqrt{2}$, we note that the corresponding angle ϕ on the negative side of the Re-axis is

$$\phi=\tan^{-1}\frac{\sqrt{1-\zeta^2}}{\zeta}=\tan^{-1}\frac{1/\sqrt{2}}{1/\sqrt{2}}=45°$$

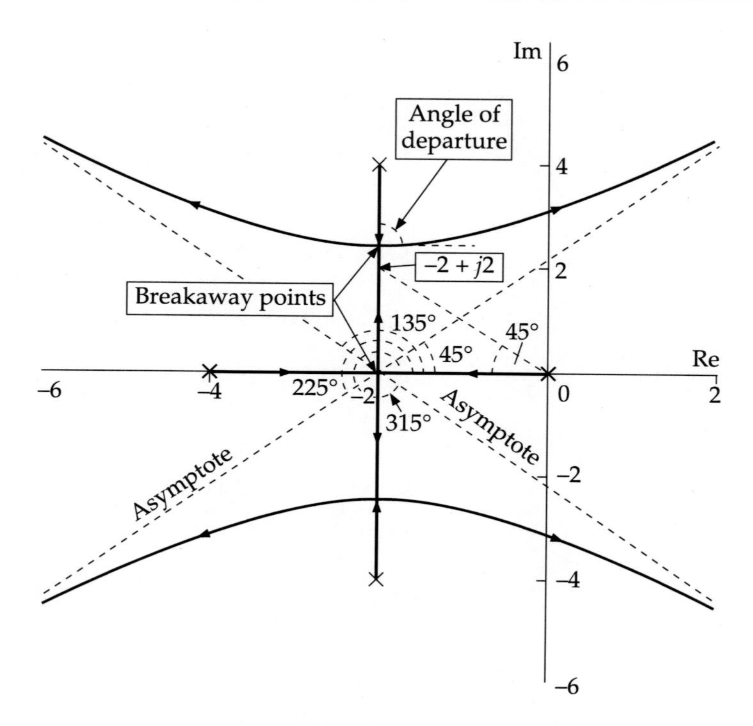

Figure 11.20 Root locus diagram (Example 11.21).

This line intersects the root locus at $s = -2 + \jmath 2$. The K value there is calculated from Eq. (i) as follows:

$$\begin{aligned}K &= -(-2+\jmath 2)(-2+\jmath 2+4)[(-2+\jmath 2)^2+4(-2+\jmath 2)+20] \\ &= -(-4-4)(4-4-4\jmath 2-8+4\jmath 2+20) = 96\end{aligned}$$

Graphically:

$$K = (4+2)(4-2)\left(\sqrt{2^2+2^2}\right)\left(\sqrt{2^2+2^2}\right) = 96$$

EXAMPLE 11.22. Sketch the root locus of the unity feedback control system whose open loop transfer function is given by

$$G(s)H(s) = \frac{K}{s(s+2)\left(s^2+4s+13\right)}$$

Find

(a) the number, angle and centroid of asymptotes
(b) the angle of departure
(c) the breakaway point
(d) the condition for marginal stability
(e) the value of K so that the system has a damping factor of 0.5

Solution The open-loop transfer function can be written as

$$G(s)H(s) = \frac{K}{s(s+2)(s+2+\jmath 3)(s+2-\jmath 3)}$$

The pole-zero map shows two poles on the Re-axis at $s = 0$ and $s = -2$. So part of the Re-axis between them is part of the root locus.

(a) *Asymptotes:* The number of poles $n = 4$. No zero, $m = 0$. Therefore, number of asymptotes; $n - m = 4$

$$\theta_0 = \frac{1}{4} \times 180^\circ = 45^\circ \qquad \theta_1 = \frac{3}{4} \times 180^\circ = 135^\circ$$

$$\theta_2 = \frac{5}{4} \times 180^\circ = 225^\circ \qquad \theta_3 = \frac{7}{4} \times 180^\circ = 315^\circ$$

Centroid:

$$\chi_c = \frac{-2-2-2}{4} = -1.5$$

(b) *Angle of departure:*

(i) *Analytical calculation:*

$$\arg\,[G(s)H(s)]_{s=-2+\jmath 3} = \arg\left|\frac{K}{(-2+\jmath 3)(\jmath 3)(2\jmath 3)}\right|$$

$$= \frac{1}{(\pi - \tan^{-1}\frac{3}{2}) + 90^\circ + 90^\circ}$$

$$= \angle -303.7^\circ$$

Therefore,

$$\theta_d = 180^\circ - 303.7^\circ = -123.7^\circ$$

(ii) *Graphical calculation:*

$$\theta_d = 180^\circ - \left[90^\circ + 90^\circ + \left(\pi - \tan^{-1}\frac{3}{2}\right)\right] = -123.7^\circ$$

(c) *Breakaway point:* Starting from the characteristic equation

$$s(s+2)(s^2+4s+13) + K = 0 \qquad \text{(i)}$$

$$\Rightarrow \qquad K = -s^4 - 6s^3 - 21s^2 - 26s$$

$$\frac{dK}{ds} = 4s^3 + 18s^2 + 42s + 26 = 0$$

$$\Rightarrow \qquad s = -0.89, \quad -1.8 \pm \jmath 2$$

(d) *Intersection on $\jmath\omega$-axis:* From Eq. (i), we get

$$s^4 + 6s^3 + 21s^2 + 26s + K = 0$$

So, the Routh array is constructed as follows:

$$\begin{array}{c|ccc} s^4 & 1 & 21 & K \\ s^3 & 6 & 26 & \\ s^2 & 16.67 & K & \\ s^1 & 26-(6/16.67)K & & \\ s^0 & K & & \end{array}$$

For marginal stability,

$$26 - \frac{6}{16.67}K = 0$$

Therefore,

$$K = 26 \times \frac{16.67}{6} = 72.24$$

From the auxiliary equation

$$16.67s^2 + 26 \times \frac{16.67}{6} = 0$$

$\Rightarrow$

$$s = \sqrt{-\frac{26}{6}} = \pm \jmath 2.08$$

With these available data, the root locus, as shown in Figure 11.21, is drawn.

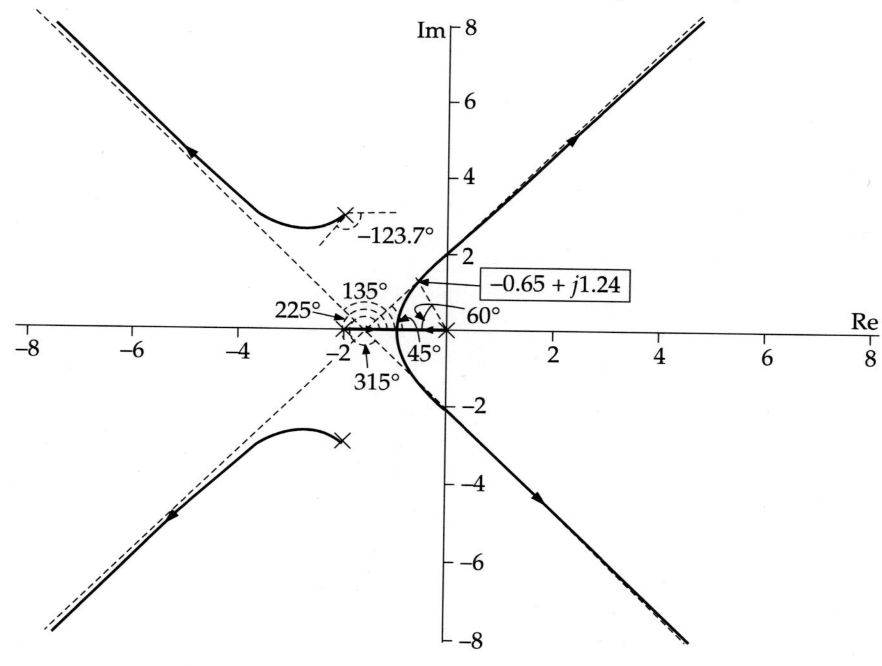

Figure 11.21 Root locus (Example 11.22).

To determine the value of K where $\zeta = 0.5$, we draw a line at an angle $\tan^{-1}\frac{\sqrt{1-0.5^2}}{0.5} = \tan^{-1}\sqrt{3} = 60°$ on the negative Re-axis. The line cuts the root locus at $s = -0.65 + \jmath 1.24$. The value of K at this point is calculated from Eq. (i) as follows:

$$\begin{aligned}
K &= |-0.65 + \jmath 1.24||-0.65 + \jmath 1.24 + 2||-0.65 + \jmath 1.24 + 2 + \jmath 3||-0.65 + \jmath 1.24 + 2 - \jmath 3| \\
&= \sqrt{0.65^2 + 1.24^2} \times \sqrt{1.35^2 + 1.24^2} \times \sqrt{1.35^2 + 4.24^2} \times \sqrt{1.35^2 + 1.76^2} \\
&= (1.4)(1.833)(4.4497)(2.2181) = 25.3
\end{aligned}$$

EXAMPLE 11.23. Draw the root locus for the unity feedback system whose open-loop transfer function is

$$G(s) = \frac{K(s+1)}{(s-1)(s+2)(s+4)}$$

Find the range of K for which the system is stable. Show all relevant steps of calculation.

Solution Parts of the Re-axis which belong to the root locus are between $s = -4$ and -2, and $s = -1$ and 1. The number of poles: $n = 3$; number of zeros: $m = 1$. Therefore, number of asymptotes: $n - m = 2$.

Step 1. Asymptotes:

$$\theta_0 = \frac{180°}{3-1} = 90° \qquad \theta_1 = \frac{180° \times 3}{3-1} = 270°$$

Centroid:

$$\chi_c = \frac{1-2-4-1}{3-1} = -2$$

Step 2. Breakaway point: We start from the characteristic equation

$$(s-1)(s+2)(s+4) + K(s+1) = 0 \tag{i}$$

$$K = -\frac{(s-1)(s+2)(s+4)}{s+1} = -\frac{s^3 + 5s^2 + 2s - 8}{s+1}$$

$$\frac{dK}{ds} = -\left[\frac{(s+1)(3s^2 + 10s + 2) - (s^3 + 5s^2 + 2s - 8)}{(s+1)^2}\right] = 0$$

$$\Rightarrow \qquad 2s^3 + 8s^2 + 10s + 10 = 0$$

$$\Rightarrow \qquad s^3 + 4s^2 + 5s + 5 = 0$$

$$\Rightarrow \qquad s = -2.864, \quad -0.568 \pm \jmath 1.193$$

Of these, only $s = -2.864$ is acceptable because from the pole-zero plot, we do not expect any complex breakaway point.

Step 3. Intersection on the $j\omega$-axis: From the characteristic equation given by Eq. (i), we get

$$s^3 + 5s^2 + (2 + K)s + (-8 + K) = 0 \tag{ii}$$

We construct the Routh array from Eq. (ii) as follows:

$$\begin{array}{c|cc} s^3 & 1 & 2+k \\ s^2 & 5 & -8+K \\ s^1 & \dfrac{2+6K}{5} & \\ s^0 & -8+K & \end{array}$$

For marginal stability

$$-8 + K = 0$$

$$\Rightarrow \qquad K = 8$$

So, the stability condition is $K > 8$. The auxiliary equation is

$$5s^2 - 8 + 8 = 0$$

$$\Rightarrow \qquad s = 0$$

This means that the root loci do not intersect the $j\omega$-axis.

With these available data, the root locus plot, as shown in Figure 11.22, is drawn.

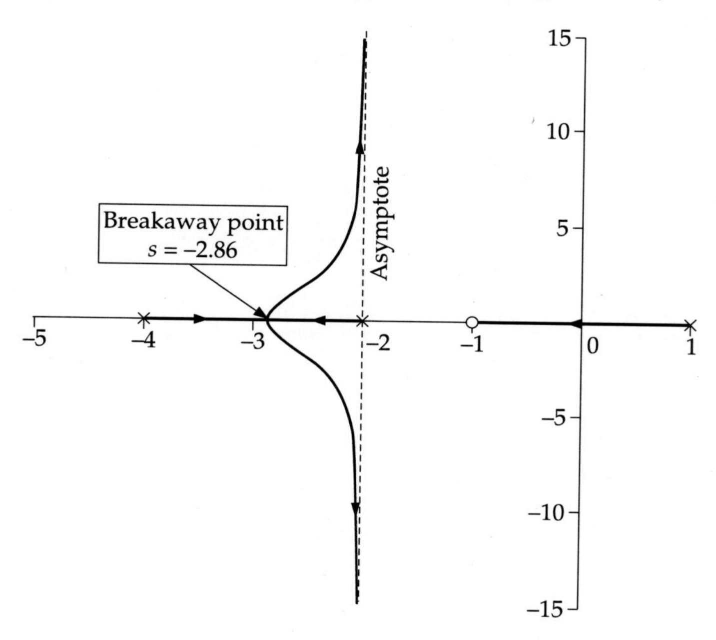

Figure 11.22 Root locus plot (Example 11.23).

EXAMPLE 11.24. The open-loop transfer function of a feedback control system is given by

$$G(s)H(s) = \frac{K(s+6)}{s(s+4)}$$

(a) Sketch the root locus with K as a variable parameter and show that the loci of complex roots are part of a circle.
(b) Determine the breakaway/break-in points, if any.
(c) Determine the range of K for which the system is underdamped.
(d) Determine the value of K for critical damping.
(e) Determine the minimum value of the damping ratio.

Solution

(a) The number of poles $n = 2$, and the number of zero $m = 1$. The pole-zero map (see Figure 11.23) shows that the part of the Re-axis between $s = 0$ and $s = -4$ belongs to the root locus. Since $n - m = 1$, there exists only one asymptote at 180°.

To show that the loci of complex roots are part of a circle, let $s = \sigma + \jmath\omega$ be a complex point on the locus. The angle criterion of Evans must be satisfied by this point since it is on the root locus. That means,

$$\angle G(\sigma + \jmath\omega) = \arg\left|\frac{K(\sigma + \jmath\omega + 6)}{(\sigma + \jmath\omega)(\sigma + \jmath\omega + 4)}\right| = \pm\pi$$

$$\tan^{-1}\frac{\omega}{\omega+6} - \tan^{-1}\frac{\omega}{\sigma} - \tan^{-1}\frac{\omega}{\sigma+4} = -\pi$$

or $$\pi + \tan^{-1}\frac{\omega}{\omega+6} = \tan^{-1}\frac{\omega}{\sigma} + \tan^{-1}\frac{\omega}{\sigma+4}$$

or $$\tan^{-1}(0) + \tan^{-1}\frac{\omega}{\omega+6} = \tan^{-1}\frac{\omega}{\sigma} + \tan^{-1}\frac{\omega}{\sigma+4} \qquad (\because \pi = \tan^{-1} 0)$$

or $$\tan^{-1}\frac{0 + \dfrac{\omega}{\sigma+6}}{1 - (0)\dfrac{\omega}{\sigma+6}} = \tan^{-1}\frac{\dfrac{\omega}{\sigma} + \dfrac{\omega}{\sigma+4}}{1 - \dfrac{\omega^2}{\sigma(\sigma+4)}} \tag{i}$$

Eliminating $\tan^{-1}$ from both sides and multiplying the sides diagonally, we get from Eq. (i), the relation

$$\frac{\omega}{\sigma} + \frac{\omega}{\sigma+4} = \frac{\omega}{\sigma+6}\left[1 - \frac{\omega^2}{\sigma(\sigma+4)}\right]$$

$$\Rightarrow \qquad \frac{2\sigma+4}{\sigma(\sigma+4)} = \frac{\sigma^2 + 4\sigma - \omega^2}{\sigma(\sigma+4)(\sigma+6)}$$

$$\Rightarrow \qquad (2\sigma+4)(\sigma+6) = \sigma^2 + 4\sigma - \omega^2 \tag{ii}$$

On rearrangement, Eq. (ii) yields

$$\sigma^2 + 12\sigma + 24 + \omega^2 = 0 \tag{iii}$$

Converting Eq. (iii) to the form

$$(\sigma + 6)^2 + (\omega + 0)^2 = 36 - 24 = 12$$

we observe that this is an equation of a circle with a radius of $\sqrt{12}$ having its centre at $(-6, \jmath 0)$.

(b) *Breakaway/break-in points:* Starting from the characteristic equation, we get

$$1 + \frac{K(s+6)}{s(s+4)} = 0$$

or

$$K = -\frac{s(s+4)}{s+6} \tag{iv}$$

$$\frac{dK}{ds} = \frac{(s^2 + 4s) - (s+6)(2s+4)}{(s+6)^2} = 0$$

or

$$s^2 + 12s + 24 = 0$$

Therefore,

$$s = \frac{-12 \pm \sqrt{48}}{2} = -2.54, \ -9.46$$

From the pole-zero plot, it is clear that the breakaway point is $s = -2.54$, and the break in point, $s = -9.46$. With these data, we draw the root locus as presented in Figure 11.23.

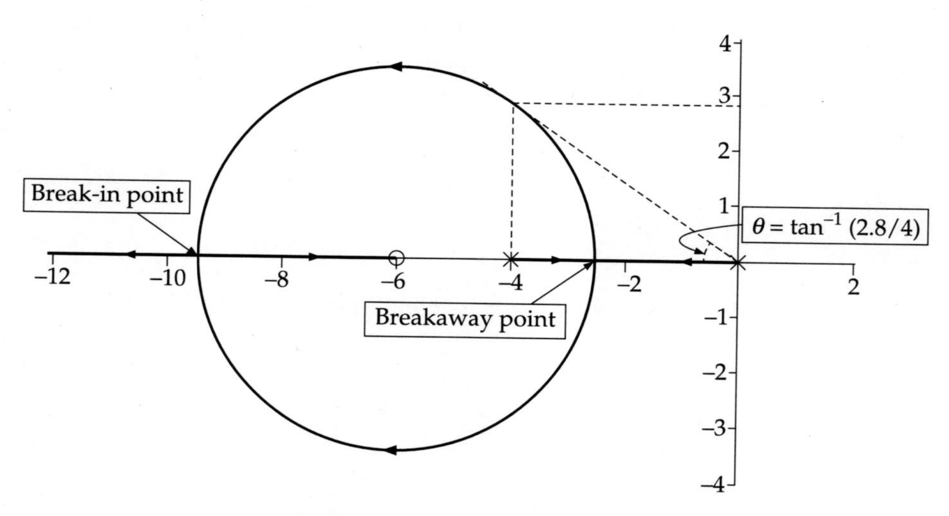

Figure 11.23 Root locus (Example 11.24).

(c) *Value of K for which the system is underdamped:* We know from Eq. (7.15) that if the system is underdapmed, i.e. $\zeta < 1$, s assumes complex values. We observe from the root locus that s assumes complex values between the breakaway and break-in points. The K values at these points can be calculated as follows.

Analytical calculation: From Eq. (iv), we get

$$K_{\text{breakaway}} = -\left.\frac{s(s+4)}{s+6}\right|_{s=-2.54} = -\frac{-2.54(-2.54+4)}{-2.54+6} = 1.072$$

$$K_{\text{break-in}} = -\left.\frac{s(s+4)}{s+6}\right|_{s=-9.46} = -\frac{-9.46(-9.46+4)}{-9.46+6} = 14.928$$

Graphical calculation:

$$K_{\text{breakaway}} = \frac{\prod \text{distance of poles from the breakaway point}}{\prod \text{distance of zeros from the breakaway point}} = \frac{(2.54-0)(4-2.54)}{6-2.54} = 1.072$$

$$K_{\text{break-in}} = \frac{\prod \text{distance of poles from the break-in point}}{\prod \text{distance of zeros from the break-in point}} = \frac{(9.46-0)(9.46-4)}{9.46-6} = 14.928$$

Therefore, the system remains underdamped in the interval when $14.928 > K > 1.072$.

(d) *Value of K for the critical damping:* For critical damping, $\theta = \tan^{-1}\frac{\sqrt{1-1^2}}{1} = 0°$. The root locus except its detour from the Re-axis between the breakaway and break-in points satisfies this criterion. So, for critical damping, $K < 1.072$ and $K > 14.928$.

(e) *Minimum value of the damping ratio:* Since $\frac{\sqrt{1-\zeta^2}}{\zeta} = \tan\theta$, the minimum value of the damping ratio occurs when θ is the largest. From Figure 11.23, we observe that

$$\theta_{\text{max}} = \tan^{-1}\frac{2.8}{4} = 34.992°$$

Hence,

$$\frac{\sqrt{1-\zeta_{\text{min}}^2}}{\zeta_{\text{min}}} = \tan 34.992° = 0.7$$

$$\Rightarrow \quad 1.49\,\zeta_{\text{min}}^2 = 1$$

$$\Rightarrow \quad \zeta_{\text{min}} = 0.82$$

EXAMPLE 11.25. The open-loop transfer function of a unity feedback system is given by

$$G(s) = \frac{K(s+0.5)}{s^2(s+4.5)}$$

Sketch the root locus.

Solution Poles are at $s = 0$ (repeated) and $s = -4.5$. Therefore, $n = 3$. Zero at $s = -0.5$. Therefore, $m = 1$. The pole-zero map (see Figure 11.24) shows that the part of the Re-axis between $s = -4.5$ and $s = -0.5$ belongs to the root locus.

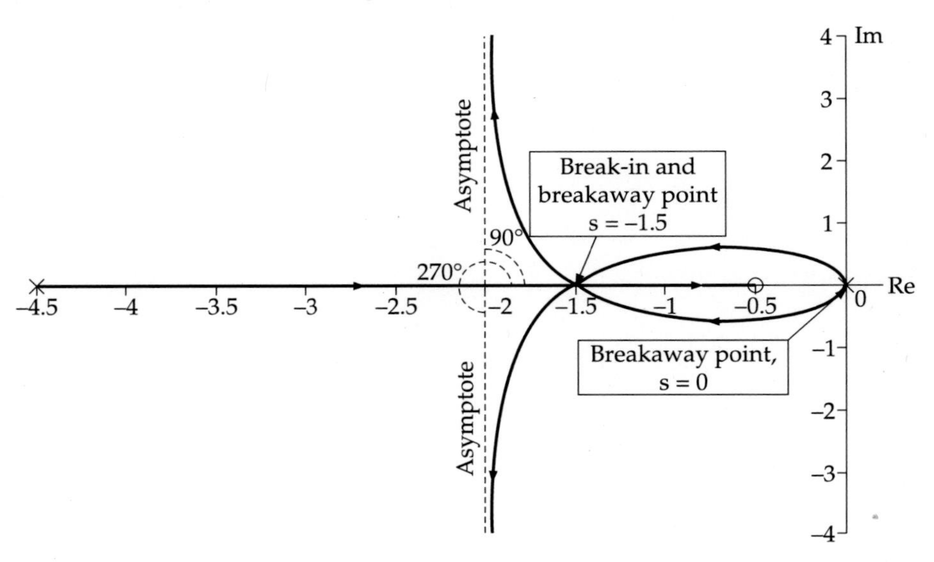

Figure 11.24 Root locus (Example 11.25).

Asymptotes: Number $= n - m = 2$.

$$\theta_0 = \frac{180^\circ}{3-1} = 90^\circ \qquad \theta_1 = \frac{180^\circ \times 3}{3-1} = 270^\circ$$

Centroid:

$$\chi_c = \frac{0.5 - 4.5}{2} = -2$$

Breakaway/Break-in points: The characteristic equation is

$$1 + \frac{K(s+0.5)}{s^2(s+4.5)} = 0 \tag{i}$$

$$s^3 + 45s^2 + Ks + 0.5K = 0 \tag{ii}$$

From Eq. (i),

$$K = -\frac{s^2(s+4.5)}{s+0.5}$$

$$\frac{dK}{ds} = -\left[\frac{(s+0.5)(3s^2+9s) - (s^3+4.5s^2)}{(s+0.5)^2}\right] = -\frac{2s\left(s^2+3s+\dfrac{9}{4}\right)}{(s+0.5)^2} = -\frac{2s\left(s+\dfrac{3}{2}\right)^2}{(s+0.5)^2} = 0$$

$$\therefore \qquad s = 0,\ -\frac{3}{2},\ -\frac{3}{2}$$

Here, $s = 0$, where two poles exist, is a breakaway point and $s = -\frac{3}{2}$ has to be a break-in as well as breakaway point in order that the loci follow the asymptotes. The two poles at the origin break away at an angle of $\frac{180°}{2} = 90°$. The plotted root locus is shown in Figure 11.24.

The root locus does not cross the Im-axis though it touches it at $s = 0$, where, we find from Eq. (ii), $K = 0$. So, the system is marginally stable at $K = 0$, and stable for all other values of K.

EXAMPLE 11.26. Consider the following root locus diagram of a system and the following statements:

1. The open-loop system is a second order system
2. The system is overdamped for $k > 1$
3. The system is absolutely stable for all values of k

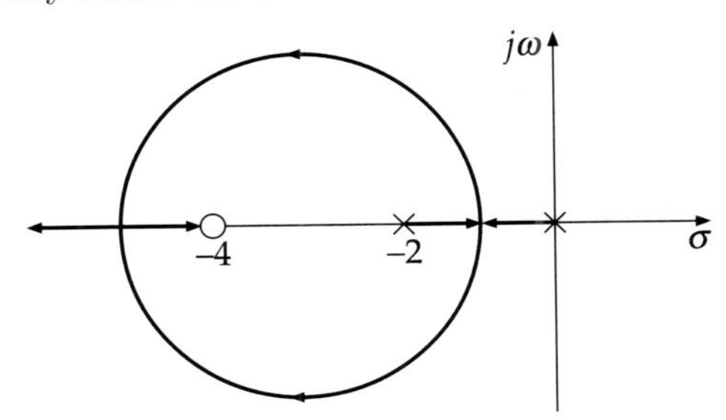

Of the above statements:

(a) 1, 2 and 3 are correct (b) 1 and 3 are correct

(c) 1 and 2 are correct (d) 2 and 3 are correct

Solution From the pole-zero plot, the open-loop transfer function is written as

$$G(s)H(s) = \frac{k(s+4)}{s(s+2)}$$

It is obvious that the open-loop system is of second order. Secondly, since the root locus lies on the LHP of the Im-axis, the system is absolutely stable. Now, the question is whether the system is overdamped for $k > 1$. To determine that, we observe that the characteristic equation is given by

$$1 + G(s)H(s) = 0$$

or

$$s(s+2) + k(s+4) = 0$$

or

$$s^2 + (2+k)s + 4k = 0 \tag{i}$$

From Eq. (i),

$$\omega_n^2 = 4k$$

$\Rightarrow$

$$\omega_n = 2\sqrt{k}$$

$$2\zeta\omega_n = 2 + k$$

$$\Rightarrow \qquad \zeta = \frac{2+k}{2\omega_n} = \frac{2+k}{4\sqrt{k}} \qquad \text{(ii)}$$

The k vs. ζ plot, as obtained from Eq. (ii), is shown in the following figure:

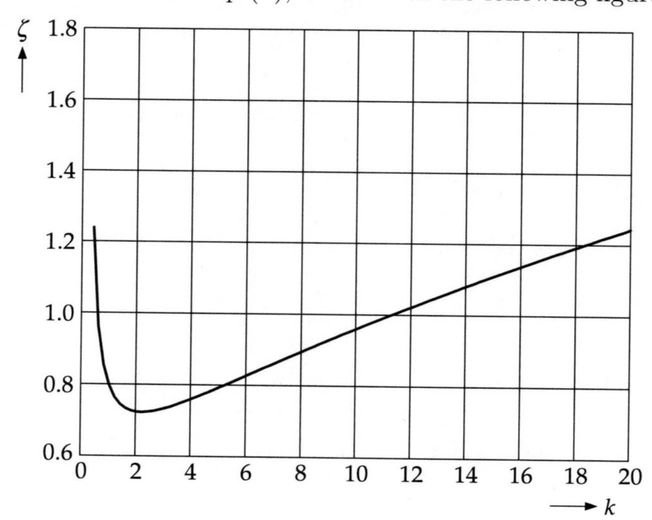

It is clear from the plot that the system is critically damped for $k \approx 1$ and 12. In fact, these are k-values at the breakaway ($s = -1.172$) and break-in ($s = -6.828$) points. For $k < 1$ and $k > 12$, the system is overdamped. So, assuming $k > 1$ but $k < 12$, the system is not overdamped. This means that the 2nd statement is wrong. Thus, the answer is (b).

Systems with Transportation Lag

We considered the stability of systems with transportation lag through Routh-Hurwitz formalism in Section 10.3 at page 336. In the following example, we study the same through the root locus technique.

EXAMPLE 11.27. The open-loop transfer functionof a unity feedback system is given by

$$G(s) = \frac{Ke^{-1.5s}}{6s+1}$$

Sketch its root locus.

Solution Using Eq. (10.3), the given open-loop transfer function can be written as

$$G(s)H(s) = \frac{Ke^{-1.5s}}{6s+1} = \frac{K(2-1.5s)}{(2+1.5s)(6s+1)} = \frac{-K(1.5s-2)}{(1.5s+2)(6s+1)} \qquad \text{(i)}$$

The poles are at $s = -1.33$ and $s = -0.17$. Hence, $n = 2$. Zero is at $s = 1.33$. Therefore, $m = 1$.

We note that the gain of the open-loop transfer function is negative, i.e. $K < 0$. Therefore, part of the Re-axis lying between $s = -1.33$ and $s = -0.17$ belongs to the root locus because the number of poles and zeros to the right of $s = -1.33$ is *even*.

Breakaway point: The characteristic equation is given by

$$1 + G(s)H(s) = 1 - \frac{K(1.5s - 2)}{(1.5s + 2)(6s + 1)} = 0 \tag{ii}$$

$$\Rightarrow \qquad 9s^2 + (13.5 - 1.5K)s + (2 + 2K) = 0 \tag{iii}$$

Therefore,

$$K = \frac{(1.5s + 2)(6s + 1)}{1.5s - 2}$$

$$= \frac{9s^2 + 13.5s + 2}{1.5s - 2} \qquad \text{[from Eq. (ii)]}$$

$$\frac{dK}{ds} = \frac{18s + 13.5}{1.5s - 2} - \frac{(9s^2 + 13.5s + 2)1.5}{(1.5s - 2)^2} = 0$$

$$\Rightarrow \qquad \frac{27s^2 - 15.8s - 27 - 13.5s^2 - 20.25s - 3}{(1.5s - 2)^2} = 0$$

$$13.5s^2 - 36s - 30 = 0$$

or

$$s = 3.33, \ -0.67$$

Of the two points, $s = -0.67$ is the breakaway point and $s = 3.33$ is both the break-in and breakaway point.

Intersection on the $j\omega$-axis: From Eq. (iii), we construct the Routh array as follows:

s^2	9	$2 + 2K$
s^1	$13.5 - 1.5K$	0
s^0	$2 + 2K$	

For marginal stability,

$$2 + 2K = 0 \quad \Rightarrow \quad K = -1$$

$$13.5 - 1.5K = 0 \quad \Rightarrow \quad K = 9$$

When $K = 9$, we get, from the auxiliary equation, the relation

$$9s^2 + 20 = 0$$

or

$$s^2 = -\frac{20}{9}$$

Therefore,

$$s = \pm j1.49$$

Shape of the loci in the complex plane: For this purpose we choose a point $s = \sigma + j\omega$ and apply the angle condition for $K < 0$ to the open-loop transfer function as follows:

$$\angle G(s)H(s) = 2q(180^\circ) = \arg\left|\frac{-K[1.5(\sigma + \jmath\omega) - 2]}{[1.5(\sigma + \jmath\omega) + 2][6(\sigma + \jmath\omega) + 1]}\right| \qquad \text{[from Eq. (i)]}$$

$$= \arg\left|\frac{-K[(1.5\sigma - 2) + \jmath 1.5\omega]}{[1.5\sigma + 2) + \jmath 1.5\omega][(6\sigma + 1) + \jmath\, 6\omega]}\right|$$

$$= \tan^{-1}\frac{1.5\omega}{1.5\sigma - 2} - \left[\tan^{-1}\frac{1.5\omega}{1.5\sigma + 2} + \tan^{-1}\frac{6\omega}{6\sigma + 1}\right]$$

$$0^\circ{}_{\text{when } q=0} = \tan^{-1}\frac{1.5\omega}{1.5\sigma - 2} - \left\{\tan^{-1}\frac{[1.5\omega/(1.5\sigma + 2)] + [6\omega/(6\sigma + 1)]}{1 - [1.5\omega/(1.5\sigma + 2)]\cdot[6\omega/(6\sigma + 1)]}\right\}$$

or $$\tan^{-1}\frac{1.5\omega}{1.5\sigma - 2} = \tan^{-1}\frac{(18\sigma + 13.5)\omega}{9\sigma^2 + 13.5\sigma + 2 - 9\omega^2}$$

Eliminating $\tan^{-1}$ and multiplying the quantities diagonally, we get

$$27\sigma^2 + 20.25\sigma - 36\sigma - 27 = 13.5\sigma^2 + 20.25\sigma + 3 - 13.5\omega^2$$

$$\Rightarrow \quad 13.5\sigma^2 - 36\sigma + 13.5\omega^2 - 30 = 0$$

$$\Rightarrow \quad \sigma^2 - 2.67\sigma + (1.33)^2 - (1.33)^2 + \omega^2 - 2.22 = 0$$

$$\Rightarrow \quad (\sigma - 1.33)^2 + (\omega - 0)^2 = 2^2 \qquad \text{(iv)}$$

Equation (iv) represents a circle of radius 2, having its centre at $(1.33, \jmath 0)$.

With these data, the root locus is drawn and presented in Figure 11.25.

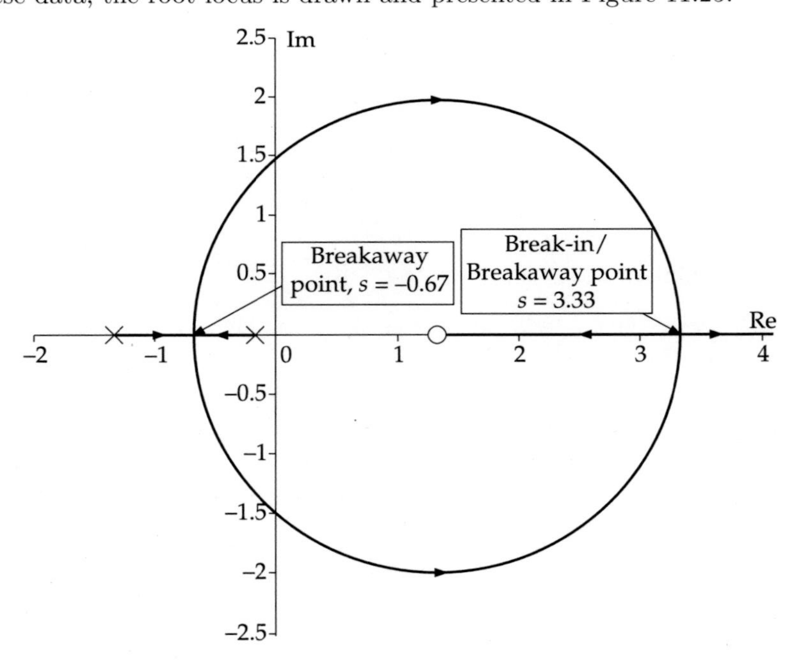

Figure 11.25 Root locus for system with transportation lag (Example 11.27).

11.4 POLE ZERO CANCELLATION RULE

The pole zero cancellation rule is:

> Poles of $G(s)$ cannot be cancelled by zeros of $H(s)$ though poles of $H(s)$ can be cancelled by zeros of $G(s)$.

We explain the reason behind the rule through examples. Suppose we have

$$G(s) = \frac{1}{(s+a)(s+b)}$$

$$H(s) = \frac{s(s+a)}{(s+c)}$$

If we allow pole zero cancellation, the characteristic equation becomes

$$1 + G(s)H(s) = 0$$

$$\Rightarrow \qquad 1 + \frac{s\cancel{(s+a)}}{\cancel{(s+a)}(s+b)(s+c)} = (s+b)(s+c) + s = 0 \tag{11.13}$$

whereas the closed-loop transfer function is

$$\frac{C(s)}{R(s)} = \frac{G(s)}{1+G(s)H(s)}$$

$$= \frac{\dfrac{1}{(s+a)(s+b)}}{1 + \dfrac{s(s+a)}{(s+b)(s+c)(s+a)}}$$

$$= \frac{(s+c)}{(s+a)[s+(s+b)(s+c)]}$$

And the characteristic equation from the closed-loop transfer function is given by

$$(s+a)[s+(s+b)(s+c)] = 0 \tag{11.14}$$

Comparing Eq. (11.14) with Eq. (11.13), we find that the closed-loop pole at $s = -a$ is missing in the latter because of pole zero cancellation. Therefore, this cancellation is not permitted.

Next, let us consider

$$G(s) = \frac{(s+c)}{(s+a)(s+b)}$$

$$H(s) = \frac{s}{(s+c)}$$

Allowing zero pole cancellation, the characteristic equation becomes

$$1 + G(s)H(s) = 0$$

$$\Rightarrow \qquad 1 + \frac{s\cancel{(s+c)}}{(s+a)(s+b)\cancel{(s+c)}} = (s+a)(s+b) + s = 0 \tag{11.15}$$

Here, the closed-loop transfer function is

$$\frac{C(s)}{R(s)} = \frac{G(s)}{1+G(s)H(s)} = \frac{\dfrac{(s+c)}{(s+a)(s+b)}}{1+\dfrac{s(s+c)}{(s+a)(s+b)(s+c)}}$$

$$= \frac{(s+c)}{(s+c)[s+(s+a)(s+b)]}$$

$$= \frac{1}{s+(s+a)(s+b)}$$

Consequently, the characteristic equation is

$$s + (s+a)(s+b) = 0 \tag{11.16}$$

We find Eq. (11.16) perfectly tallies with Eq. (11.15) which was derived allowing zero pole cancellation.

EXAMPLE 11.28. Sketch the root locus for a system having

$$G(s) = \frac{K}{s+1}$$

$$H(s) = \frac{s+1}{s^2+4s+5}$$

Solution We note that if we simply multiply $G(s)$ by $H(s)$, one pole of $G(s)$, namely that at $s = -1$, gets cancelled by a zero of $H(s)$. This is not permitted. So, to keep the characteristic equation intact, we restore the pole and write the open-loop transfer function as

$$G(s)H(s) = \frac{K}{(s+1)(s^2+4s+5)}$$

$$= \frac{K}{(s+1)(s+2-j1)(s+2+j1)}$$

Poles are at $s = -1,\ -2+j1$ and $-2-j1$. Therefore, $n = 3$. No zero. Thus, $m = 0$. Part of the real axis on the root locus: $-\infty < s < -1$.

Asymptotes: Number = 3, angles = $60°,\ 180°,\ 300°$.

Centroid:

$$\chi_c = \frac{-1-2-2}{3} = -1.67$$

Angle of departure for pole at $s = -2 + \jmath 1$:

Analytical calculation:

$$\arg |G(s)H(s)|_{s=-2+\jmath 1} = \arg \left| \frac{K}{(s+1)(s+2+\jmath 1)} \right|_{s=-2+\jmath 1}$$

$$= \arg \left| \frac{K}{(\jmath 1 - 1)(2\jmath 1)} \right|$$

$$= \frac{1}{[\angle(\pi - 45^\circ)] + (\angle 90^\circ)}$$

$$= \angle - 225^\circ$$

Therefore,

$$\theta_d = 180^\circ - 225^\circ = -45^\circ$$

Graphical calculation:

$$\theta_d = 180^\circ - [90^\circ + (\pi - 45^\circ)] = -45^\circ$$

Intersection at the $\jmath\omega$*-axis*: The characteristic equation is

$$s^3 + 5s^2 + 9s + 5 + K = 0$$

The Routh array is

s^3	1	9
s^2	5	$5 + K$
s	$8 - 0.2K$	0
s^0	K	

For marginal stability,

$$8 - 0.2K_{\text{mar}} = 0$$

or

$$K_{\text{mar}} = 40$$

From the auxiliary equation

$$5s^2 + 45 = 0$$

we get

$$s = \pm \jmath 3$$

With the help of these data, we plot the root locus of Figure 11.26.

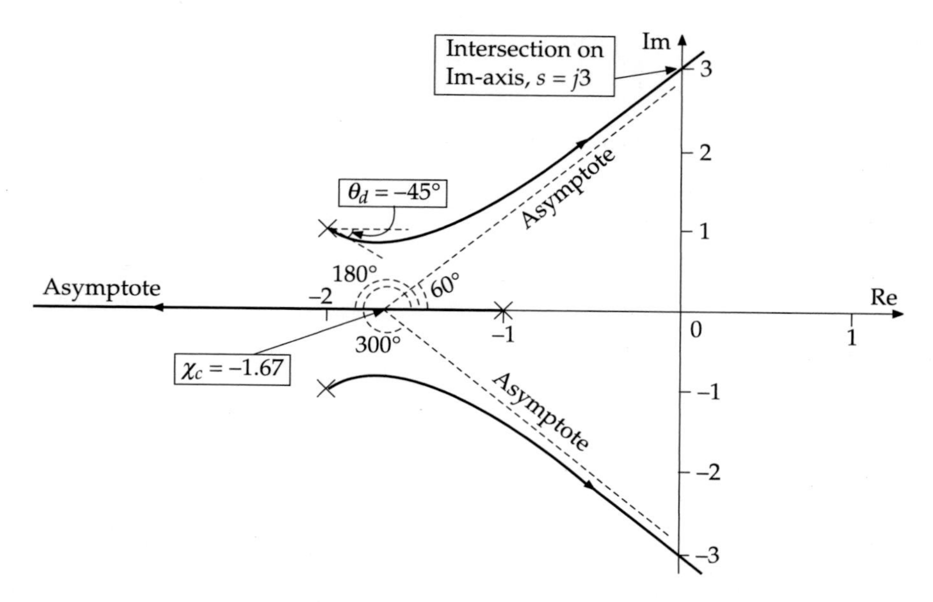

Figure 11.26 Root locus plot (Example 11.28).

11.5 STABILITY MARGINS THROUGH ROOT LOCUS

The relative stability of a control system is very often determined from a study of two parameters, called *gain margin* and *phase margin.*

Gain Margin

The maximum allowable loop gain in decibels before the system becomes unstable is called the *gain margin.*

We know that a CLTF pole placed on the RHP of the $j\omega$-axis makes the system unstable. In a root locus study, the locus of a root is plotted by increasing the value of the gain factor K. While doing so, once the locus touches the $j\omega$-axis for a certain value of K, the system becomes marginally stable. Beyond that value of K, the system crosses the Im-axis and becomes unstable.

So, the gain margin can be calculated from the root locus by finding out the ratio of K value for marginal stability (i.e. at the crossover of the imaginary axis) to that required by the design. Thus,

$$\text{Gain margin (Gm)} = 20 \log \frac{K_{\text{mar}}}{K_{\text{required}}} \text{ dB} \tag{11.17}$$

In case the root locus does not cross the $j\omega$-axis, an infinite value is assigned to K.

Phase Margin

The maximum allowable increment of phase angle of $G(s)H(s)$ in degrees up to which the system remains stable is called the *phase margin.*

A $-180°$ phase angle makes a system unstable. So, to determine the phase margin, we have to calculate by what angle the phase of a given system falls short of $-180°$. Thus, the phase margin of system can be calculated from the K value by following the procedure:

- For the given K value, we calculate ω_g for which $|G(j\omega_g)H(j\omega_g)| = 1$.
- Then we calculate the corresponding $\arg |G(j\omega_g)H(j\omega_g)|$.

The phase margin is then calculated from the relation

$$\text{Phase margin (Pm)} = 180° + \arg |G(j\omega_g)H(j\omega_g)|$$

However, it may not be easy to follow this procedure for all systems. It is rather easy to calculate phase margin from Bode plots which we will discuss later.

EXAMPLE 11.29. Sketch the root locus plot for the unity feedback system having open-loop transfer function

$$G(s) = \frac{K}{s(s+1)(s+3)}$$

and find

(a) K for critical damping
(b) K for zero damping
(c) Gain margin for gain = 5

Solution The poles are at $s = 0, -1, -3$. Therefore, $n = 3$. No zero. Thus, $m = 0$. Part of σ-axis (i.e., Re-axis) in the root locus: between $s = 0$ and -1, and $s = -3$ and $-\infty$.

Asymptotes: Number = 3, therefore, $q = 0, 1, 2$. Their angles and the centroid are as follows:

$$\theta_0 = \frac{1}{3} \times 180° = 60° \qquad \theta_1 = \frac{3}{3} \times 180° = 180°$$

$$\theta_2 = \frac{5}{3} \times 180° = 300° \qquad \chi_c = \frac{-1-3}{3} = -1.33$$

Breakaway point: It is expected between $s = 0$ and -1 and calculated from the characteristic equation as follows:

$$s(s+1)(s+3) + K = 0$$

$$s^3 + 4s^2 + 3s + K = 0 \tag{i}$$

$$\Rightarrow \qquad K = -s^3 - 4s^2 - 3s \tag{ii}$$

$$\frac{dK}{ds} = 3s^2 + 8s + 3 = 0$$

Therefore,

$$s = -0.45, \ -2.22$$

Its location at $s = -2.22$ is ruled out because the point does not lie on the root locus. Therefore, the breakaway point is at $s = -0.45$.

Intersection at the $j\omega$-axis: We construct the Routh array from the characteristic equation [Eq. (i)] as follows:

$$\begin{array}{c|cc} s^3 & 1 & 3 \\ s^2 & 4 & K \\ s & 3-0.25K & 0 \\ s^0 & K & \end{array}$$

For marginal stability,

$$3 - 0.25K_{\text{mar}} = 0$$

or

$$K_{\text{mar}} = 12$$

The auxiliary equation is

$$4s^2 + 12 = 0$$

Therefore,

$$s = \pm j1.73$$

With the help these data, we plot the root locus as shown in Figure 11.27.

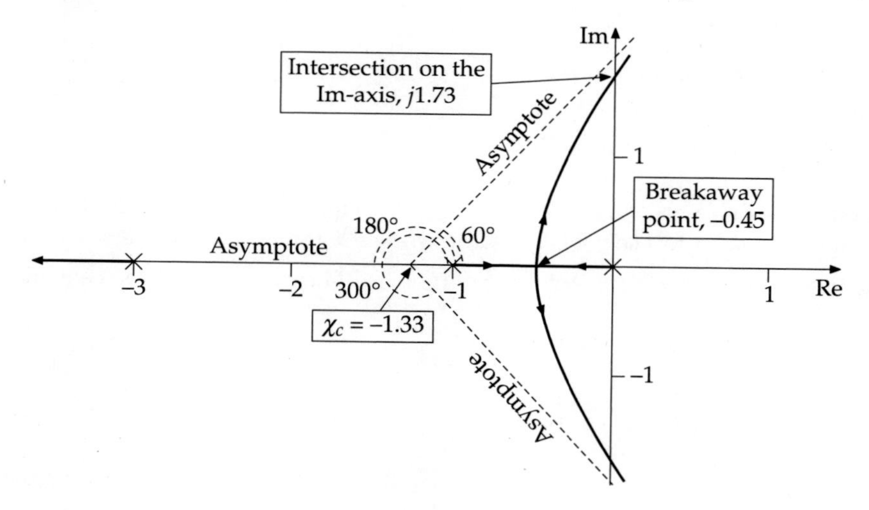

Figure 11.27 Root locus plot (Example 11.29).

(a) For critical damping, $\zeta = 1$. Therefore, $\phi = \tan^{-1}\dfrac{\sqrt{1-\zeta^2}}{\zeta} = 0°$. The root locus intersection at 0° from the origin is at $s = -0.45$. From Eq. (ii),

$$K\big|_{s=-0.45} = -(-0.45)^3 - 4(-0.45)^2 - 3(-0.45) = 0.63$$

Graphically also,

$$K\big|_{s=-0.45} = (0.45)(1-0.45)(3-0.45) = 0.63$$

(b) For $\zeta = 0$, $K = K_{\text{mar}} = 12$.

(c) From Eq. (11.17),

$$\text{Gm} = 20 \log \frac{K_{\text{mar}}}{K_{\text{required}}}$$

We have already found that $K_{\text{mar}} = 12$. Graphically also,

$$\begin{aligned} K &= \prod (\text{vector lengths from poles to } \jmath\sqrt{3}) \\ &= (\sqrt{3})\left(\sqrt{1^2 + (\sqrt{3})^2}\right)\left(\sqrt{3^2 + (\sqrt{3})^2}\right) = \sqrt{3} \times 2 \times 2\sqrt{3} \\ &= 12 \end{aligned}$$

Therefore, the required gain margin is

$$20 \log \left(\frac{12}{5}\right) = 7.6 \text{ dB}$$

EXAMPLE 11.30. The open-loop transfer function for a system is given by

$$G(s)H(s) = \frac{K}{s(s+4)^2}$$

Calculate the phase margin for the system for $K_{\text{design}} = 25$.

Solution Substituting $\jmath\omega$ for s, we get

$$|G(\jmath\omega)H(\jmath\omega)| = \left| \frac{25}{\jmath\omega(\jmath\omega + 4)^2} \right| = 1$$

or

$$\left|\jmath\omega(\jmath\omega + 4)^2\right| = 25$$

$$\Rightarrow \qquad \omega(\omega^2 + 16) = 25$$

$$\Rightarrow \qquad \omega^3 + 16\omega - 25 = 0$$

The real root of this cubic equation[5] is given by

$$\omega = 1.3934 \equiv \omega_g$$

The corresponding argument of $G(\jmath\omega)H(\jmath\omega)$ is

$$\begin{aligned} \arg \left|G(\jmath\omega_g)H(\jmath\omega_g)\right| &= \frac{\angle 0^\circ}{\angle 90^\circ + 2\left(\tan^{-1} \dfrac{1.3934}{4}\right)} \\ &= -[90^\circ + 2(19.21^\circ)] \\ &= -128.41^\circ \end{aligned}$$

Therefore, the phase margin is

$$180^\circ - 128.41^\circ = 51.59^\circ$$

[5] Nowadays, scientific pocket calculators provide this facility.

EXAMPLE 11.31. Draw the root locus for the open-loop transfer function

$$G(s)H(s) = \frac{K}{s(s+2)(s+4)}$$

and determine the gain margin and phase margin for $K = 12$.

Solution The root locus has been worked out in Example 11.14. From there we get $K_{\text{mar}} = 48$. Therefore,

$$\text{Gain margin} = 20\log\frac{K_{\text{mar}}}{K_{\text{required}}} = 20\log\frac{48}{12} = 20\log 4 = 12.04 \text{ dB}$$

For phase margin calculation, we put

$$|G(\jmath\omega)H(\jmath\omega)| = \left|\frac{12}{\jmath\omega(\jmath\omega+2)(\jmath\omega+4)}\right| = 1$$

$$\Rightarrow \qquad \frac{12}{\omega\sqrt{\omega^2+4}\sqrt{\omega^2+16}} = 1$$

$$\Rightarrow \qquad \omega\sqrt{\omega^4+20\omega^3+64} = 12$$

$$\Rightarrow \qquad \omega^6 + 20\omega^4 + 64\omega^2 - 144 = 0 \qquad \text{(i)}$$

Substituting $x = \omega^2$ in Eq. (i), we get the cubic equation

$$x^3 + 20x^2 + 64x - 144 = 0$$

Its three roots are

$$x_1 = 1.49713$$
$$x_2 = -15.14716$$
$$x_3 = -6.34997$$

Evidently, the first root yields real values for ω. Since frequency cannot be negative, $\omega = 1.2236$ rad/s is the gain crossover frequency. The phase margin is, therefore,

$$\text{Pm} = 180° + \frac{\angle 0°}{\angle 90° + \tan^{-1}(1.2236/2) + \tan^{-1}(1.2236/4)}$$

$$= 180° - (90° + 31.46° + 17°)$$

$$= 41.54°$$

11.6 EFFECTS OF ADDITION OF POLES AND ZEROS ON ROOT LOCUS

We will study the effects by considering one example. Suppose our open-loop transfer function is given by

$$G(s)H(s) = \frac{K}{s(s+3)} \qquad (11.18)$$

The root locus plot of Eq. (11.18) is given in Figure 11.28.

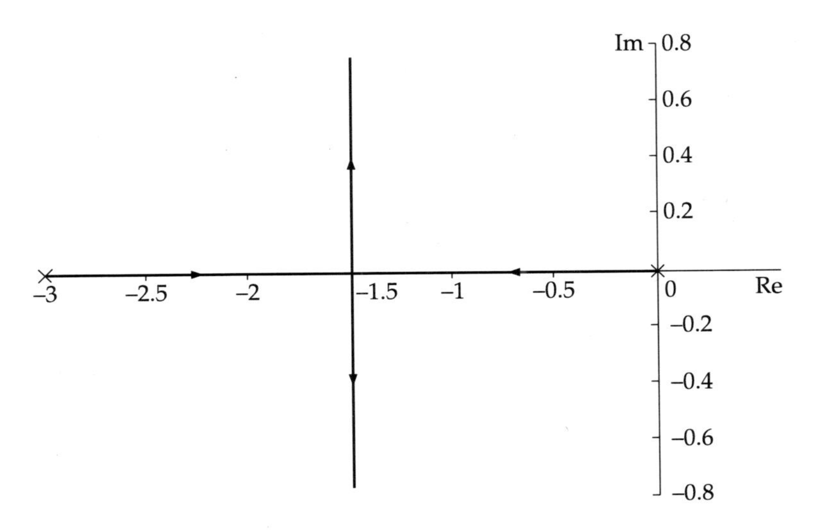

Figure 11.28 Root locus plot of Eq. (11.18).

Addition of Poles

Now we add one pole at $s = -5$ so that

$$G(s)H(s) = \frac{K}{s(s+3)(s+5)} \tag{11.19}$$

The root locus plot of Eq. (11.19) is given in Figure 11.29.

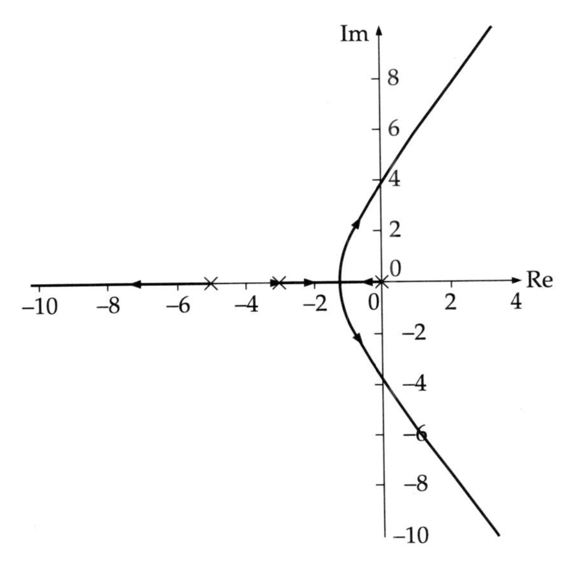

Figure 11.29 Root locus plot of Eq. (11.19).

It is clear from Figures 11.28 and 11.29 that before the addition of pole, the system was perfectly stable. After the addition of the pole:

1. The complex part of the root loci bends toward the RHP.
2. The angles of asymptotes change from $\pm 90°$ to $\pm 60°$.
3. The breakaway point moves to the right.
4. The system becomes unstable for values of K when the loci cross the Im-axis.

Dominant poles and zeros. In this context, we would like to mention that those transfer function poles having maximum effect on the transient response of the system are called *dominant poles*. The same definition applies for dominant zeros as well.

The location of poles in the s-plane has a lot of significance as far as the transient response of the system is concerned. LHP poles located near the Im-axis are found to affect the transient response more than those that are located far away from it. Indeed, these poles give rise to transient responses that decay much more slowly than the far-away ones. Owing to this influence, poles in the proximity of the Im-axis are called dominant poles. Of course, closeness is a relative term and much depends on the structure of the transfer function.

Addition of Zeros

Now let us consider the case when we add a zero in the LHP of Eq. (11.18) so that

$$G(s)H(s) = \frac{K(s+5)}{s(s+3)} \tag{11.20}$$

The root locus plot of Eq. (11.20) is given in Figure 11.30.

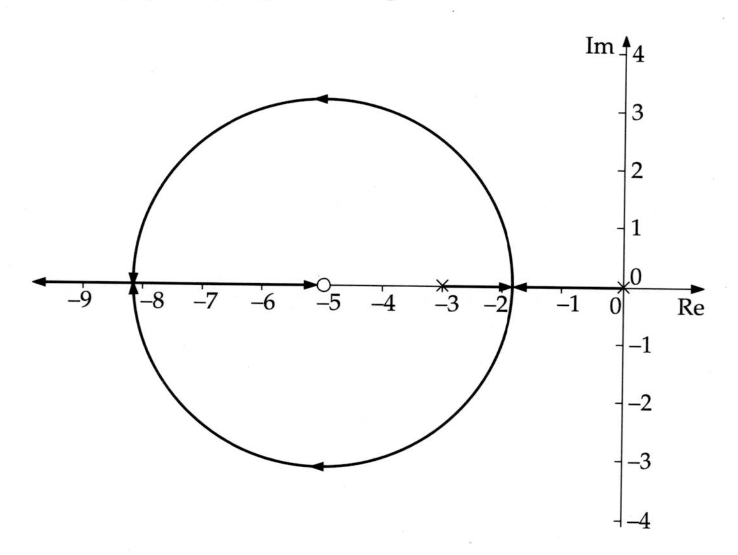

Figure 11.30 Root locus plot of Eq. (11.20).

Comparing Figure 11.30 with Figure 11.28, we find that the addition of a zero bends the root loci towards the left to form a circle. As a result, the relative stability of the system improves. So, we may infer that the addition of zeros moves root loci more towards the left half of the s-plane.

Although inferred from a particular case, the results are quite general.

REVIEW QUESTIONS

11.1 Choose the correct answer:

(i) The open-loop transfer function of a unity feedback system is given by

$$G(s) = \frac{Ks}{(s+1)(s+2)(s+3)}$$

The number of asymptotes of the root-locus plot that tend to infinity is

(a) 1 (b) 2 (c) 0 (d) 3

(ii) Root loci start from

(a) open-loop poles (b) open-loop zeros
(c) closed-loop poles (d) none of the above

(iii) Root loci terminate at

(a) open-loop poles (b) open-loop zeros
(c) closed-loop poles (d) none of the above

(iv) The open-loop transfer function of a unity feedback system is given by

$$G(s) = \frac{K(s+4)}{(s+1)(s+2)(s+3)}$$

The centroid of asymptotes is located at

(a) $s = -1$ (b) $s = -2$ (c) $s = -3$ (d) $s = -4$

(v) The open-loop transfer function of a unity feedback system is given by

$$G(s) = \frac{Ks}{(s+1)(s+2)(s+3)} \quad (K > 0)$$

The angle(s) of asymptote(s) is/are

(a) 60°, 180°, 270° (b) 90°, 180° (c) 180° (d) 0°

(vi) If there are two adjacent poles on the real axis and the segment of the real axis is a part of the root locus, then at least

(a) one break-in point exists between the poles
(b) one breakaway point exists between the poles
(c) two break-in points exist between the poles
(d) two breakaway points exist between the poles

(vii) At the breakaway point, the gain factor K should be

(a) real (b) imaginary (c) complex (d) none of these

(viii) The root locus diagram is

(a) always symmetric about the real axis
(b) always symmetric about the imaginary axis
(c) never symmetric about the real axis
(d) always symmetric about both the real and imaginary axes

(ix) The root locus diagram can be used to determine

(a) absolute stability (b) relative stability
(c) conditional stability (d) none of the above

(x) The root loci of a system have three asymptotes. The system can have

(a) five poles and two zeros (b) three poles and one zero
(c) four poles and two zeros (d) six poles and four zeros

(xi) The open-loop transfer function for a unity feedback system is

$$\frac{K}{(s+1)(s+2)(s+3)}$$

When K varies from 0 to ∞, the breakaway point in the root locus plot will be between

(a) 0 and -1 (b) -2 and -3
(c) -1 and-2 (d) beyond -3

(xii) A system having three zeros and four poles has root loci equal to

(a) 3 (b) 4 (c) 1 (d) 7

(xiii) Root locus is a plot of

(a) closed-loop poles (b) closed-loop zeros
(c) open-loop zeros (d) open-loop poles

11.2 A point on the real axis lies on the root locus if the number of open-loop poles plus zeros on the real axis to the right of the point is odd. Justify the statement.

11.3 "The breakaway points of the root locus are the solutions of $(dK/ds) = 0$". Justify.

11.4 Consider the open-loop transfer function

$$G(s)H(s) = \frac{K}{s(0.2s+1)(0.05s+1)}$$

Draw the root locus and determine for what value of gain K the phase margin will be 40°.

11.5 The open-loop transfer function of a system is given by

$$G(s)H(s) = \frac{5(s+\alpha)}{s(s+2)}$$

Draw the root locus and find the value of α, for which the damping ratio is $1/\sqrt{2}$.

11.6 Obtain a root locus of the unity feedback system whose forward transfer function is

$$\frac{K(s+2)(s+3)}{s(s+1)}$$

Determine the break-in and breakaway points and the values of K at those points. State whether the system is underdamped or overdamped for the following values of K: (i) 0.05, (ii) 5, and (iii) 15. Determine the range of K for which the system is stable.

11.7 For the system with the open-loop transfer function

$$G(s)H(s) = \frac{K}{s(s+1)(s+2)}, \qquad 0 \le K \le \infty$$

determine whether the following points are on the root locus: (i) $s = -0.5$, (ii) $s = 0.5$, (iii) $s = \jmath 1.732$, and (iv) $s = -1 + \jmath$.

11.8 Plot the root locus of a unity feedback control system whose transfer function is given by

$$G(s) = \frac{K}{s(s+2)(s+3)}$$

11.9 (a) What is a root locus?

(b) Explain why generally each branch of a root locus originates from an open-loop pole and terminates to an open-loop zero as K increases from zero to infinity.

(c) Consider a feedback system with the characteristic equation given by

$$1 + \frac{K}{s(s^2+2s+4)} = 0$$

Roughly draw the root locus of the system.

11.10 A unity feedback control system has

$$G(s) = \frac{K}{s(s+2)(s+5)}$$

Sketch the root locus and show on it a line for $\zeta = 0.5$ and the value of K for this damping.

11.11 The open-loop transfer function of a unity feedback control system is given by

$$G(s) = \frac{K(s+3)}{s(s^2+2s+2)(s+5)(s+6)}$$

(a) Draw the root locus as a function of K.

(b) Determine the value of K which makes the relative damping ratio of the closed-loop complex pole equal to 0.4.

11.12 The open-loop transfer function of a unity feedback control system is given by

$$G(s) = \frac{K}{s(s^2 + 8s + 32)}$$

Sketch the root-locus plot for the system by determining

(a) the number, angle and the centroid of the asymptotes;
(b) angle of departure;
(c) the breakaway point; and
(d) the intersection of imaginary axis by the root locus

11.13 (a) State the rules of construction of root loci.
(b) Draw the root loci for the open-loop transfer function

$$G(s)H(s) = \frac{K}{s(s + 1)(s + 2)}$$

(c) Find the critical value of K for which the system becomes unstable and the critical frequency thereof.

11.14 A unity feedback control system has an open-loop transfer function

$$G(s) = \frac{K}{s(s^2 + 4s + 13)}$$

Sketch the root locus plot of the system by determining the following:

(a) The centroid, number and angle of asymptotes
(b) The angle of departure of root loci from the poles
(c) The breakaway point, if any
(d) The value of K and frequency at which the root loci cross the $j\omega$-axis.

11.15 A control system is characterized by an open-loop transfer function of

$$G(s)H(s) = \frac{K}{s(s + 2)(s^2 + 4s + 8)}$$

(a) Draw the root locus of the characteristic equation.
(b) Find the value of K for which the closed-loop system has a pair of poles on the imaginary axis.
(c) Find out the closed-loop poles for the value of K thus obtained.

11.16 The open-loop transfer function of a feedback control system is given by

$$G(s)H(s) = \frac{K}{(s + 4)(s^2 + 4s + 8)}$$

(a) Draw the root locus plot of the system by determining the centroid, number and angle of asymptotes, angle of departure of root locus branches from the open loop poles, breakaway points, if any, and the intersection points of the root locus branches with the imaginary axis.

(b) Find out the value of K for which the dominant pole pair will have a damping ratio of 0.5.

(c) Determine the closed-loop poles for the value of K as computed in part (b).

Answers to Selected Questions

11.1 (i) (b); (ii) (a); (iii) (b); (iv) (a); (v) (b); (vi) (b); (vii) (a); (viii) (a); (ix) (b); (x) (a); (xi) (c); (xii) (b); (xiii) (d)

11.4 5.22

11.5 4.9

11.6 Break-in at $s = -2.37$, breakaway at $s = -0.63$: (i) critically damped, (ii) underdamped; and (iii) critically damped $-5 < K < 0$.

11.7 (i) Yes; (ii) No; (iii) No; (iv) No

11.10 11.4

11.11 (b) 11.8

11.12 (a) Number = 3, angles 60°, 180°, 240°, centroid at $s = -2.67$; (b) 135°; (c) No breakaway point; (d) $s = \pm j5.66$

11.13 (c) $K = 6$, $\omega = 1.41$ rad/s.

11.14 (a) Centroid at $s = -1.33$, number = 3, angles 60°, 180°, 240°; (b) 146.3°; (c) No breakaway point; (d) $K = 52$, $\omega = 3.6$ rad/s.

11.15 (b) 35.56; (c) $s_{1,2} = -3 \pm j2.08$, $s_{3,4} = 0 \pm j1.63$.

11.16 (a) Centroid at $s = -2.67$, number = 3, angles 60°, 180°, 240°, angle of departure = 65°, no breakaway point, intersection points at the Im-axis at $s = \pm j4.9$; (b) 12.9; (c) $s_1 = -4.99$, $s_{2,3} = -1.5 \pm j2.59$.

CHAPTER 12

Frequency Domain Analysis

12.1 WHY FREQUENCY DOMAIN?

While talking about the methods of analysis of stability of a control system in Section 9.4, we mentioned that three of the widely used methods are based on analysis in the frequency domain. In this chapter we will discuss what the frequency domain analysis means and what its advantages and disadvantages are.

On previous occasions, while testing the response of a system, we used four standard inputs: (i) impulse, (ii) step, (iii) ramp, and (iv) parabolic inputs. Now, suppose we give a sinusoidal input such as

$$x(t) = A \sin \omega t \tag{12.1}$$

to a control system. What happens can of course be studied by the previously laid down procedure in Chapter 7. But a general analysis shows that we need not go into such an elaborate analysis for a sinusoidal input because, for a sinusoidal input, after the transients die out:

- The output remains sinusoidal with the same frequency.
- Only the amplitude and phase of the input change after passing through any LTI system.

In fact, one of the most commonly used test functions for a circuit or system is the sine wave. This is not because sine waves are a particularly common signal. Indeed, they are quite rare—the transmission of electricity (a 50 Hz sine wave) is one example.

The importance of sine waves comes from the Fourier[1] theory. As we know, any signal going into a circuit can be represented by a sum of sine and cosine waves of varying frequency and amplitude (often an infinite sum). As a simple example, consider the graphs shown in Figure 12.1. The top graph on the left [Figure 12.1(a)] shows a ramp function. The top graph on the right [Figure 12.1(b)] shows a combination of a constant plus sine and cosine waves of varying amplitude. The bottom graph [Figure 12.1(c)] shows the sum of the waves shown in Figure 12.1(b) (dashed curve) and the original ramp function (solid curve).

Evidently, even just one pair of sine and cosine waves are sufficient in this case to closely approximate the original function. Fourier's theorem states that any function (with some very minor restrictions) can be represented in this way. This is why sine waves are important—not

[1] Fourier (pronounced *Fooryay*), Jean Baptiste Joseph, Baron (1768–1830) was born in France. He was both a mathematician and a physicist.

because they are common, but because we can represent arbitrarily complex functions using only this rather simple function.

HW Bode[2] and H Nyquist[3] developed procedures for studying stability of feedback systems in the frequency domain. In Chapters 13 and 15, we will discuss those procedures.

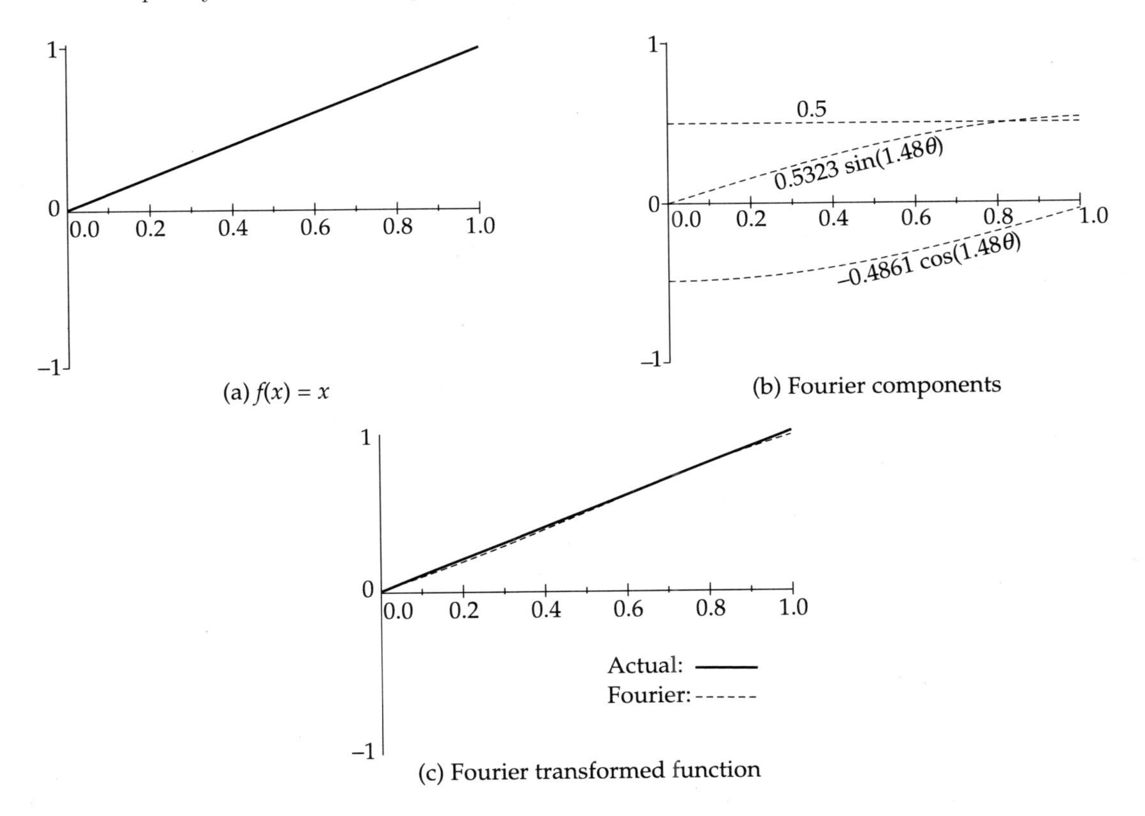

Figure 12.1 Demonstration of the Fourier transform: (a) a ramp function; (b) Fourier components — 0.5, 0.5323 sin (1.48θ) and –0.4861 cos (1.48θ) — of the ramp function; (c) comparison of the Fourier-transformed representation the original ramp function.

12.2 CONSEQUENCES OF SINUSOIDAL INPUT

The question is, what do we observe if a time-domain analysis of response to a sinusoidal input is made?

Suppose we give an input $x(t)$ to a system and the corresponding output is $y(t)$. If the Laplace transform of the input and output are $X(s)$ and $Y(s)$, respectively, and the transfer

[2]Hendrik Wade Bode (1905–1982) developed his technique in 1938. A pioneer of modern control theory and electronic instrumentation, he was an American engineer of Dutch ancestry.

[3]Harry Nyquist (1889–1976), an American (born Swedish) physicist, developed his method in 1934.

function is $G(s)$, we know that they are related as

$$G(s) = \frac{Y(s)}{X(s)}$$

In general terms, the numerator and denominator can be expressed as product of m and n factors, respectively [see, for example, Eq. (3.3)]. For the following study, we treat the numerator as one factor and write the transfer function as

$$G(s) = \frac{Y(s)}{X(s)} = \frac{N(s)}{(s+a_1)(s+a_2)\ldots(s+s_n)} \tag{12.2}$$

The Laplace transform of Eq. (12.1) gives

$$X(s) = \frac{A\omega}{s^2+\omega^2}$$

So, from Eq. (12.2),

$$Y(s) = \frac{N(s)}{(s+a_1)(s+a_2)\ldots(s+s_n)} \cdot \frac{A\omega}{s^2+\omega^2} \tag{12.3}$$

The RHS of Eq. (12.3) can be expanded in terms of partial fractions as

$$Y(s) = \left(\frac{A_1}{s+a_1} + \frac{A_2}{s+a_2} + \ldots \frac{A_n}{s+a_n}\right) + \frac{B_1}{s+\jmath\omega} + \frac{B_2}{s-\jmath\omega} \tag{12.4}$$

The inverse Laplace transform of Eq. (12.4) yields

$$y(t) = \underbrace{A_1 e^{-a_1 t} + A_2 e^{-a_2 t} + \ldots + A_n e^{-a_n t}}_{\text{transients}} + \underbrace{B_1 e^{-\jmath\omega t} + B_2 e^{\jmath\omega t}}_{\text{steady state}} \tag{12.5}$$

As $t \to \infty$, the terms marked 'transients' in Eq. (12.5) will die out, but those marked 'steady state' will endure. So, we can write

$$y_{\text{ss}}(t) = B_1 e^{-\jmath\omega t} + B_2 e^{\jmath\omega t} \tag{12.6}$$

The coefficients can be found out as

$$B_1 = (s+\jmath\omega)G(s) \cdot \left.\frac{A\omega}{s^2+\omega^2}\right|_{s=-\jmath\omega} = G(s) \cdot \left.\frac{A\omega}{s-\jmath\omega}\right|_{s=-\jmath\omega}$$

$$= -G(-\jmath\omega)\frac{A}{2\jmath} \tag{12.7}$$

$$B_2 = (s-\jmath\omega)G(s) \cdot \frac{A\omega}{s^2+\omega^2}\Bigg| s = \jmath\omega = G(s) \cdot \left.\frac{A\omega}{s+\jmath\omega}\right|_{s=\jmath\omega}$$

$$= G(\jmath\omega)\frac{A}{2\jmath} \tag{12.8}$$

Since the complex quantity $G(j\omega)$ can be written in the form of $G(j\omega) = |G(j\omega)|e^{j\phi}$, using Eqs. (12.7) and (12.8), we have from Eq. (12.6)

$$y_{ss}(t) = A\,|G(j\omega)| \left[\frac{e^{j(\omega t+\phi)} - e^{-j(\omega t+\phi)}}{2j} \right]$$

$$= A\,|G(j\omega)| \sin(\omega t + \phi) \tag{12.9}$$

Comparing the input [Eq. (12.1)] and the steady-state output [Eq. (12.9)], we find that

1. The frequency of the output remains the same as the input.
2. The amplitude and phase of the output differ from those of the input.

The results are important and they make the frequency-domain study much simpler.

12.3 SYSTEM OUTPUT WHEN INPUT AND TRANSFER FUNCTION ARE GIVEN

Given that sine waves are important, how can we analyze the response of a circuit or system to sinusoidal inputs?

The answer is that there are many ways of doing this, depending on the mathematical sophistication. Let us demonstrate a fairly basic technique that uses phasors[4].

Using complex impedances, it is possible to find the transfer function of a circuit. For example, we consider the circuit of Figure 12.2(a).

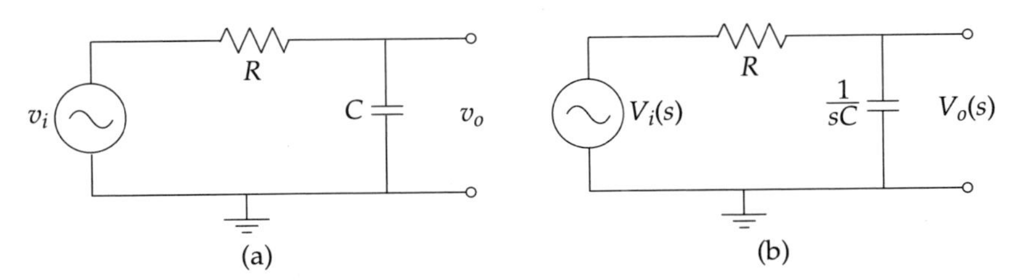

Figure 12.2 (a) The circuit; (b) its Laplace transformed components.

Its transfer function can be obtained from the Laplace transformed components [see Figure 12.2(b)] as

$$V_o(s) = \frac{V_i(s) \times (1/sC)}{R + (1/sC)} = \frac{V_i(s)}{1 + sRC}$$

Therefore,

$$G(s) = \frac{V_o(s)}{V_i(s)} = \frac{1}{1 + sRC} \tag{12.10}$$

In the frequency domain, $s \equiv j\omega$.

Consider the case when $R = 1$ and $C = 0.1$. Then,

$$G(s) = \frac{1}{1 + 0.1s} \equiv \frac{1}{1 + j0.1\omega}$$

[4]Description of phasors can be found in almost any circuit theory textbook.

Generally, we know the input V_i and want to find the output V_o. We can do this by a simple multiplication:

$$V_o = V_i \cdot G(s) = V_i \cdot \frac{1}{1 + \jmath 0.1\omega}$$

If we have a phasor representation for the input and the transfer function, the multiplication is simple (multiply magnitudes and add phases). Finding the output becomes easy. We now give a few examples.

EXAMPLE 12.1. Find the output of the system given by Eq. (12.10) when the input is

(a) $3\sin(30t)$

(b) $3\sin(30t + 40°)$

(c) $3\sin(10t + 40°)$

(d) $3\cos(10t)$

Solution (a) Here, the input is $V_i = 3\sin(30t)$, and the transfer function is

$$G(\jmath\omega) = \frac{1}{1 + \jmath 0.1\omega} = \frac{1}{1 + \jmath 3}$$

Therefore,

	Input	*Transfer characteristic*
Amplitude	$M_i = 3$	$M_t = \dfrac{1}{\sqrt{1^2 + 3^2}} = 0.316$
Phase	$\phi_i = 0°$	$\phi_t = \dfrac{1}{\tan^{-1}(3/1)} = -71.6°$

From the principle of phasor calculation,

$$\text{Output magnitude} = M_i \times M_t = 3 \times 0.316 = 0.95$$

$$\text{Output phase} = \phi_i - \phi_t = 0° - 71.6° = -71.6°$$

Thus $$V_o = 0.95\sin(30t - 71.6°)$$

So, while transmitting the signal, the circuit reduces the amplitude from 3 to 0.95 and changes the phase from 0° to −71.6°.

(b) Here, the input is $V_i = 3\sin(30t + 40°)$, the transfer function remaining the same. We note that only the phase has changed from the previous case, the magnitude and frequency remaining the same. Therefore,

	Input	*Transfer characteristic*
Phase	$\phi_i = 40°$	$\phi_t = \dfrac{1}{\tan^{-1}(3/1)} = -71.6°$

From the principle of phasor calculation,

$$\text{Output phase} = \phi_i - \phi_t = 40^\circ - 71.6^\circ = -31.6^\circ$$

Thus,
$$V_o = 0.95 \sin (30t - 31.6^\circ)$$

(c) Here, the input is $V_i = 3 \sin (10t + 40^\circ)$, and the transfer function is

$$G(\jmath\omega) = \frac{1}{1 + \jmath 0.1\omega} = \frac{1}{1 + \jmath 1}$$

Therefore,

	Input	*Transfer characteristic*
Amplitude	$M_i = 3$	$M_t = \dfrac{1}{\sqrt{1^2 + 1^2}} = 0.707$
Phase	$\phi_i = 40^\circ$	$\phi_t = \dfrac{1}{\tan^{-1}(1/1)} = -45^\circ$

From the principle of phasor calculation,

$$\text{Output magnitude} = M_i \times M_t = 3 \times 0.707 = 2.12$$

$$\text{Output phase} = \phi_i - \phi_t = 40^\circ - 45^\circ = -5^\circ$$

Therefore,
$$V_o = 2.12 \sin (30t - 5^\circ)$$

(d) Here, the input is $V_i = 3 \cos (10t) = 3 \sin (10t + 90^\circ)$, the transfer function remains the same as in (c). We note that the transfer characteristics have remained unaltered, though the input phase has changed. Therefore, from the principle of phasor calculation,

$$\text{Output phase} = \phi_i - \phi_t = 90^\circ - 45^\circ = 45^\circ$$

Thus,
$$V_o = 2.12 \sin(30t + 45^\circ)$$

12.4 FREQUENCY RESPONSE SPECIFICATIONS

The frequency response of a control system is generally characterized in terms of the following parameters:

1. Resonant peak, M_r
2. Resonant frequency, ω_r
3. Bandwidth, Bw
4. Gain crossover frequency, ω_g
5. Phase crossover frequency, ω_p
6. Cut-off frequency, ω_c
7. Gain margin, Gm
8. Phase margin, Pm

We consider these eight parameters in detail now.

1. Resonant peak, M_r. The closed-loop transfer function in the frequency domain is written as

$$\frac{G(\jmath\omega)}{1+G(\jmath\omega)H(\jmath\omega)} \equiv |M(\jmath\omega)|\angle\phi$$

The maximum value of $|M(\jmath\omega)|$, as ω is varied, is called the *resonant peak* and is denoted as M_r.

M_r value gives an indication about the relative stability of a system. A large M_r indicates a large maximum overshoot, M_p, in step response of the system. Normally, M_r lies between 1.1 and 1.5.

2. Resonant frequency, ω_r. The resonant frequency is the frequency at which the resonant peak occurs.

3. Bandwidth, Bw. As ω is varied, the value of $M(\jmath\omega)$ also varies. Ultimately, its level may drop down by 3 dB below[5] its value at zero frequency which is normally 0 dB level. The range of frequency between 0 dB level and −3 dB level of $M(\jmath\omega)$ is called the *bandwidth*.

While a large Bw may enhance the system response, it makes the system susceptible to noise as well. On the other hand, a low Bw indicates a sluggish system response.

4. Gain crossover frequency, ω_g. This is the frequency at which the gain of the system is unity.

5. Phase crossover frequency, ω_p. This is the frequency at which the phase angle of $M(\jmath\omega)$ is 180°.

6. Cut-off frequency, ω_c. It is the frequency at which the power level falls to half of its value, i.e. magnitude of $M(\jmath\omega)$ equals $1/\sqrt{2}$.

7. Gain margin, Gm. We have already defined[6] that the factor by which the gain can be increased before the system becomes unstable is called its *gain margin* (Gm).

The gain margin of a system can be determined from its open-loop frequency response. It is the reciprocal of the open-loop gain at the phase crossover frequency.

8. Phase margin, Pm. The gain margin alone is not sufficient to express the relative stability of a system. A system may have a very good gain margin, but it may require a small additional phase lag to render the system unstable. Hence, a further criterion, termed *phase margin* (Pm), is necessary to assess the stability of systems of this type. It is defined as the additional phase lag that will make the system marginally stable[7].

The phase margin is determined from the open-loop frequency response. It is the clockwise angle through which the unit vector must be rotated before it lies on the negative real axis[8]. Thus, phase margin is the difference of the phase shift of the system at the gain crossover frequency plus 180°.

[5] It corresponds to $1/\sqrt{2}$ (= 0.707) times the value of M at zero frequency.

[6] See Section 11.5 at page 400.

[7] See also Section 11.5 at page 400.

[8] The reason why it is so defined will be clear when we will discuss the Nyquist plot.

Utility of gain and phase margins. Control system designers frequently use gain and phase margins to be sure about the frequency response of a system. Of course, these margins relate to the open-loop systems only.

A good gain margin is desirable from the point of view of robust stability of the system. However, one should not select an arbitrarily large gain margin because the dynamic performance characteristics of the system are related to it and the response may become unacceptably overdamped and sluggish as a result.

To provide a good overall stability to a system, the rule of thumb is that the gain and phase margins should be around 12 dB and 45° to 60°, respectively. In most of the cases, a good Gm automatically ensures a good Pm and vicé versa. But there are exceptions to this rule as well.

12.5 FREQUENCY DOMAIN RELATIONS

We study these relations for a second order system where they are relevant. For a unity feedback second order system,

$$G(s) = \frac{C(s)}{R(s)} = \frac{\omega_n^2}{s^2 + 2\zeta\omega_n s + \omega_n^2} \qquad H(s) = 1$$

Therefore,

$$G(\jmath\omega) = \frac{C(\jmath\omega)}{R(\jmath\omega)} = \frac{\omega_n^2}{-\omega^2 + \jmath 2\zeta\omega_n\omega + \omega_n^2}$$

$$\equiv \frac{1}{1 + \jmath 2\zeta u - u^2} \qquad [\text{where } u = (\omega/\omega_n)]$$

If we express $G(\jmath\omega)$ as

$$G(\jmath\omega) = |G(\jmath\omega)|\angle G(\jmath\omega) \equiv M\angle\phi$$

then,

$$M = \frac{1}{\sqrt{(1-u^2)^2 + (2\zeta u)^2}} \tag{12.11}$$

$$\phi = -\tan^{-1}\frac{2\zeta u}{1-u^2} \tag{12.12}$$

We may plot M vs. u and ϕ vs. u as shown in Figure 12.3, Eqs. (12.11) and (12.12) giving us the values of Table 12.1 at once.

Table 12.1 Magnitude and Phase Values

u	M	ϕ
0	1	0
1	$1/(2\zeta)$	$-\pi/2$
∞	0	$-\pi$

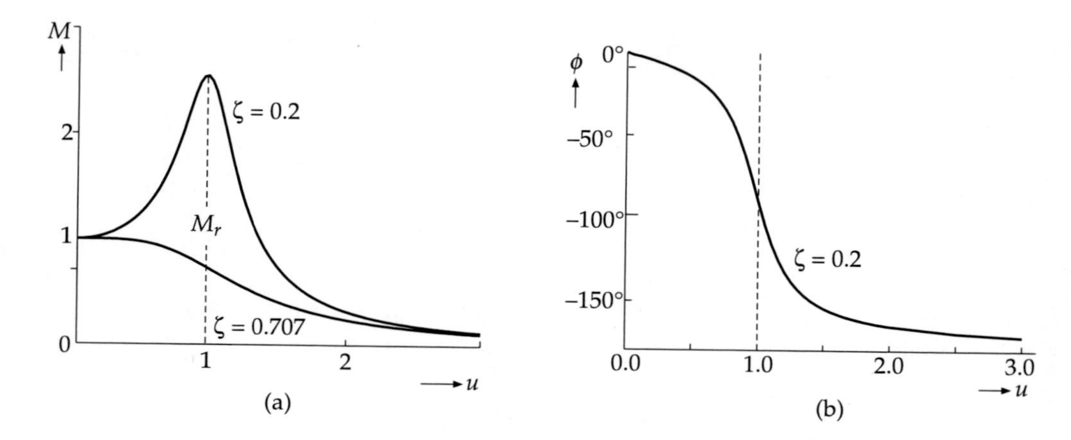

Figure 12.3 (a) Magnitude plot; (b) phase plot.

Resonant frequency, ω_r. For resonant frequency,

$$\left.\frac{dM(\omega)}{d\omega}\right|_{\omega_r} = \left.\frac{dM(u)}{du}\right|_{u_r} = 0$$

$$\Rightarrow \qquad -\frac{1}{2}\cdot\frac{2(1-u_r^2)(-2u_r)+8\zeta^2u_r}{[(1-u_r^2)^2+4\zeta^2u_r^2]^{3/2}} = 0$$

$$\Rightarrow \qquad u_r^2 = 1-2\zeta^2$$

$$\Rightarrow \qquad u_r = \sqrt{1-2\zeta^2} \tag{12.13}$$

Therefore,

$$\omega_r = \omega_n\sqrt{1-2\zeta^2} \tag{12.14}$$

Resonant peak, M_r. From Eqs. (12.13), (12.11) and (12.12),

$$M_r = \frac{1}{2\zeta\sqrt{1-\zeta^2}} \tag{12.15}$$

$$\phi_r = -\tan^{-1}\frac{\sqrt{1-2\zeta^2}}{\zeta} \tag{12.16}$$

It is interesting to note from Eqs. (12.14) and (12.15) that

1. As $\zeta \to 0$, $M_r \to \infty$ and $\omega_r \to \omega_n$. This is the resonance condition.
2. For $\zeta = 1/\sqrt{2} = 0.707$, $\omega_r = 0$ and $M_r = 1$. So, for $0 < \zeta < 0.707$, $\omega_r < \omega_n$ and $M_r > 1$.
3. No resonant peak occurs for $\zeta > 0.707$.

Bandwidth, Bw. To calculate the Bw, we find the u_c-value corresponding to $M = 1/\sqrt{2}$.

Then,

$$\frac{1}{\sqrt{(1-u_c^2)^2+4\zeta^2u_c^2}}=\frac{1}{\sqrt{2}}$$

$$\Rightarrow \qquad u_c^2 = 1-2\zeta^2+\sqrt{(2\zeta^2-1)^2+1}$$

$$\Rightarrow \qquad u_c = \sqrt{1-2\zeta^2+\sqrt{(2\zeta^2-1)^2+1}}$$

$$\Rightarrow \qquad \omega_c = \omega_n\sqrt{1-2\zeta^2+\sqrt{(2\zeta^2-1)^2+1}} \tag{12.17}$$

This cut-off frequency is the measure of the bandwidth. Figure 12.4(a) gives a graphic idea about the bandwidth. If we define normalized bandwidth as ω_c/ω_n, then the graph given in Figure 12.4(b) indicates what it looks like against ζ values. The graph as well as Eq. (12.17) shows that for $\zeta = 1/\sqrt{2}$, the normalized bandwidth equals 1.

From Eq. (12.17), we note the following:

- Bw $= \omega_c$
- Bw decreases as ζ increases [see Figure 12.4(b)].

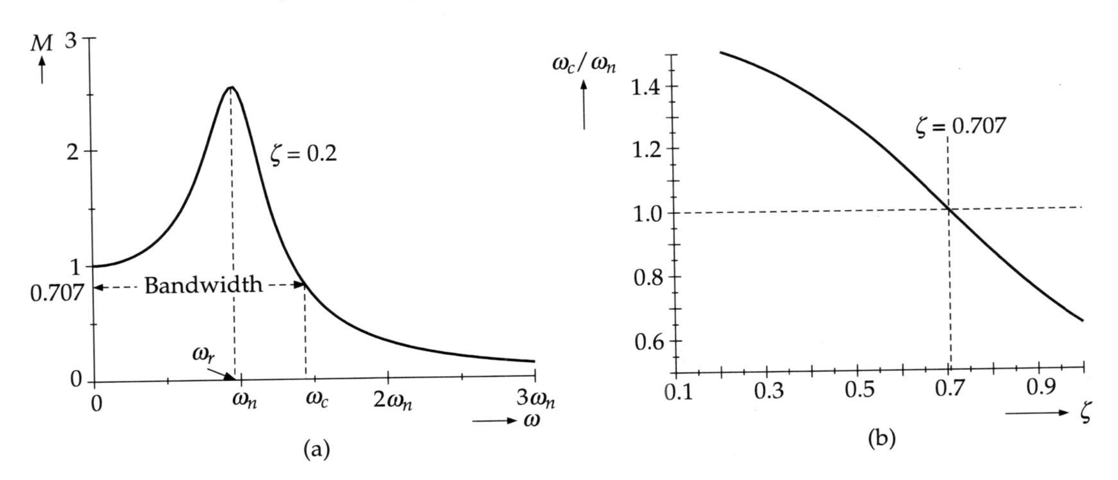

Figure 12.4 (a) Magnitude curve and bandwidth; (b) ζ vs. normalized Bw.

Comparison between Time Domain and Frequency Domain Parameters

For the sake of comparison, we better choose a second order system. In the frequency domain, a system is mainly characterized by its resonant frequency, ω_r and resonant peak, M_r, while in the time domain the damped frequency, ω_d and the peak overshoot, M_p in response to a step input are major characteristic parameters.

The resonant frequency of the frequency domain and the damped frequency of the time domain are both related to the natural frequency of oscillation, ω_n and the damping ratio, ζ as follows:

$$\omega_r = \omega_n\sqrt{1-2\zeta^2}$$

$$\omega_d = \omega_n\sqrt{1-\zeta^2}, \qquad 0 \le \zeta \le 1$$

Therefore,

$$\frac{\omega_r}{\omega_d} = \sqrt{\frac{1-2\zeta^2}{1-\zeta^2}}$$

The interrelation of ω_r and ω_d can be visualized from Figure 12.5(a).

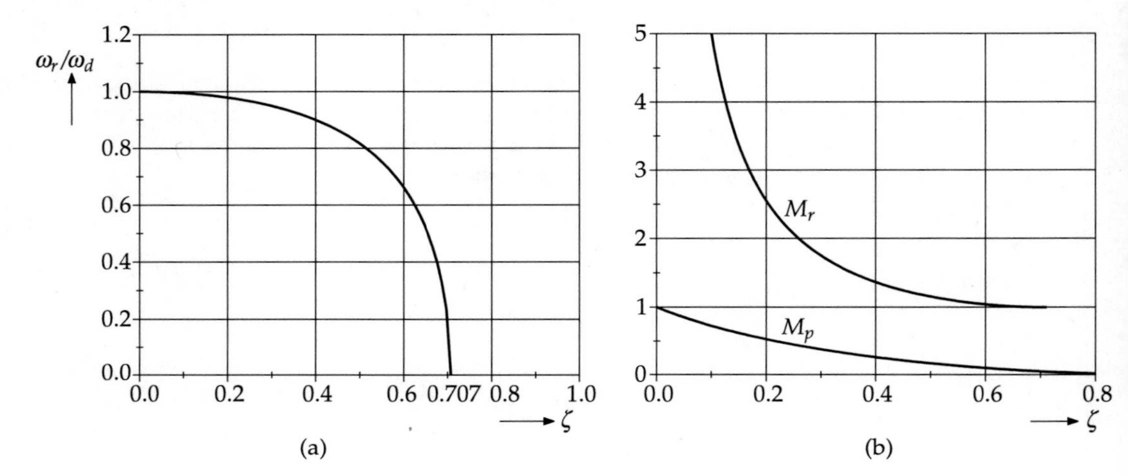

Figure 12.5 (a) Ratio of ω_r and ω_d vs. the damping ratio ζ; (b) plots of M_r and M_p vs. ζ.

Similarly, we observe that the resonant peak, M_r in the frequency domain and the peak overshoot, M_p in the time domain are also ζ-dependent as can be seen from the following formulae:

$$M_r = \frac{1}{2\zeta\sqrt{1-\zeta^2}}$$

$$M_p = \exp\left(-\frac{\pi\zeta}{\sqrt{1-\zeta^2}}\right)$$

The visual picture of their interrelation is shown in Figure 12.5(b). Of course, this relation exists up to $\zeta = 0.707$ beyond which the resonant peak does not exist.

EXAMPLE 12.2. Consider the following systems:

System 1: $G(s) = \dfrac{1}{2s+1}$

System 2: $G(s) = \dfrac{1}{5s+1}$

The true statement regarding the systems is:

(a) Bandwidth of System 1 > bandwidth of System 2
(b) Bandwidth of System 1 < bandwidth of System 2
(c) Bandwidths of both the systems are the same
(d) Bandwidths of both the systems are infinite

Solution In the frequency domain, the amplitudes of the systems are

$$M_1(j\omega) = \frac{1}{j2\omega_1 + 1} = \frac{1}{\sqrt{4\omega_1^2 + 1}}$$

$$M_2(j\omega) = \frac{1}{j5\omega_2 + 1} = \frac{1}{\sqrt{25\omega_2^2 + 1}}$$

To determine bandwidth, we put $M = \frac{1}{\sqrt{2}}$. So, the cut-off frequencies are determined as follows:

$$\frac{1}{\sqrt{4\omega_{c1}^2 + 1}} = \frac{1}{\sqrt{2}} \qquad \frac{1}{\sqrt{25\omega_{c2}^2 + 1}} = \frac{1}{\sqrt{2}}$$

or

$$\omega_{c1} = \frac{1}{2} \qquad \omega_{c2} = \frac{1}{5}$$

Obviously, the bandwidth of System 1 > bandwidth of System 2.
Ans: (a).

EXAMPLE 12.3. A linear time-invariant causal system has a frequency response given in the polar form as

$$\frac{1}{\sqrt{1+\omega^2}}\angle - \tan^{-1}\omega$$

For input $x(t) = \sin t$, the output is

(a) $\frac{1}{\sqrt{2}}\cos t$ (b) $\frac{1}{\sqrt{2}}\cos\left(t - \frac{\pi}{4}\right)$ (c) $\frac{1}{\sqrt{2}}\sin t$ (d) $\frac{1}{\sqrt{2}}\sin\left(t - \frac{\pi}{4}\right)$

Solution From the given response and the input of frequency 1, we find the following magnitude M and phase ϕ of the output:

$$M = \left.\frac{1}{\sqrt{1+\omega^2}}\right|_{\omega=1} = \frac{1}{\sqrt{2}}$$

$$\phi = -\tan^{-1} 1 = -45^\circ$$

Therefore, the output will be

$$y(t) = \frac{1}{\sqrt{2}}\sin\left(t - \frac{\pi}{4}\right)$$

Ans: (d).

EXAMPLE 12.4. The open-loop transfer function of a feedback control system is $1/(s-1)^3$. The gain margin of the system is

(a) 16 (b) 8 (c) 4 (d) 2

Solution At the outset, we note that the Gm value is not in dB. Now, the open-loop transfer function in the frequency domain yields

$$G(\jmath\omega)H(\jmath\omega) = \frac{1}{(\jmath\omega+1)^3} \equiv \frac{1}{\left(\sqrt{\omega^2+1}\right)}\angle -3\tan^{-1}\omega$$

To calculate the phase crossover frequency ω_p, the gain at ω_p and the gain margin, we have

$$-3\tan^{-1}\omega_p = -180°$$

$$\Rightarrow \quad \tan^{-1}\omega_p = 60°$$

$$\Rightarrow \quad \omega_p = \tan 60° = \sqrt{3} \text{ rad/s}$$

$$\text{Gain}\Big|_{\omega_p} = \frac{1}{\sqrt{\left(\sqrt{3}\right)^2+1}} = 0.125$$

Therefore,

$$\text{Gm} = \frac{1}{0.125} = 8$$

Ans: (b).

EXAMPLE 12.5. The loop transfer function of a system is given by

$$G(s)H(s) = \frac{10e^{-Ls}}{s}$$

The phase crossover frequency is 5 rad/s. The value of the dead time L is

(a) $\pi/20$ (b) $\pi/10$ (c) $-\pi/20$ (d) zero

Solution From the given transfer function, we get in the frequency domain

$$G(\jmath\omega)H(\jmath\omega) = \frac{10e^{-\jmath L\omega}}{\jmath\omega}$$

$$\Rightarrow \quad \equiv \frac{10}{\omega}\angle(-L\omega - 90°)$$

For phase crossover,

$$-L\omega_p - 90° = -180°$$

or

$$L\omega_p = 90° = \frac{\pi}{2}$$

or

$$L = \frac{\pi}{2\omega_p} = \frac{\pi}{2\times 5} = \frac{\pi}{10}$$

Ans: (b).

EXAMPLE 12.6. Determine the frequency domain specifications for the unity feedback system given by

$$G(s) = \frac{256}{s(s+8)}$$

Solution The closed-loop transfer function of the system is

$$T(s) = \frac{C(s)}{R(s)} = \frac{G(s)}{1+G(s)H(s)} = \frac{256}{s^2+8s+256} \quad \text{(i)}$$

The standard transfer function of a second order system is given by

$$\frac{C(s)}{R(s)} = \frac{K\omega_n^2}{s^2+2\zeta\omega_n s+\omega_n^2} \quad \text{(ii)}$$

Comparing the denominators of Eq. (i) and (ii), we get

$$\omega_n^2 = 256 \qquad \text{or} \qquad \omega_n = \sqrt{256} = 16 \text{ rad/s}$$

$$2\zeta\omega_n = 8 \qquad \text{or} \qquad \zeta = \frac{8}{2(16)} = 0.25$$

Therefore,

$$M_r = \frac{1}{2\zeta\sqrt{1-\zeta^2}} = \frac{1}{2(0.25)\sqrt{1-(0.25)^2}}$$

$$= 2.066$$

$$\omega_r = \omega_n\sqrt{1-2\zeta^2} = (16)\sqrt{1-2(0.25)^2}$$

$$= 14.97 \text{ rad/s}$$

$$\text{Bw} = \omega_c = \omega_n\sqrt{1-2\zeta^2+\sqrt{(2\zeta^2-1)^2+1}}$$

$$= (16)\sqrt{1-2(0.25)^2+\sqrt{[2(0.25)^2-1]^2+1}}$$

$$= 35.26 \text{ rad/s}$$

EXAMPLE 12.7. Determine the frequency domain specifications for the system whose closed-loop transfer function is given by

$$\frac{C(s)}{R(s)} = \frac{50}{s^2+10s+100}$$

Solution Since this is a closed-loop transfer function, the characteristic equation is given by

$$s^2+10s+100 = 0$$

Comparing this equation, with that of a standard second order system, we get

$$\omega_n^2 = 100 \qquad \text{or} \qquad \omega_n = \sqrt{100} = 10 \text{ rad/s}$$

$$2\zeta\omega_n = 10 \qquad \text{or} \qquad \zeta = \frac{10}{2(10)} = 0.5$$

Therefore,

$$M_r = \frac{1}{2\zeta\sqrt{1-\zeta^2}} = \frac{1}{2(0.5)\sqrt{1-(0.5)^2}}$$

$$= 1.154$$

$$\omega_r = \omega_n\sqrt{1-2\zeta^2} = (10)\sqrt{1-2(0.5)^2}$$

$$= 7.07 \text{ rad/s}$$

$$\text{Bw} = \omega_c = \omega_n\sqrt{1-2\zeta^2+\sqrt{(2\zeta^2-1)^2+1}}$$

$$= (10)\sqrt{1-2(0.5)^2+\sqrt{[2(0.5)^2-1]^2+1}}$$

$$= 12.72 \text{ rad/s}$$

EXAMPLE 12.8. A control system is represented by Figure 12.6. Determine K and α values so that $M_r = 2.0$ and $\omega_r = 10$. Find the corresponding bandwidth and the settling time.

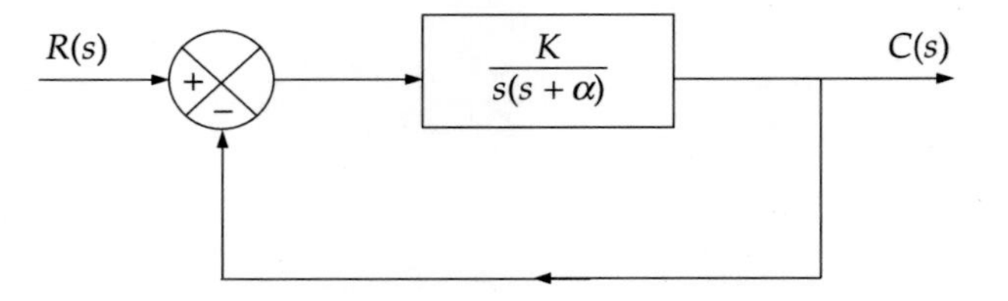

Figure 12.6 System (Example 12.8).

Solution The closed-loop transfer function of the system is

$$\frac{C(s)}{R(s)} = \frac{G(s)}{1+G(s)} = \frac{K}{s(s+\alpha)+K} = \frac{K}{s^2+\alpha s+K} \tag{i}$$

Comparing the denominator of Eq. (i) with that of a standard second order transfer function, we get

$$\omega_n^2 = K \qquad \text{or} \qquad \omega_n = \sqrt{K} \tag{ii}$$

$$2\zeta\omega_n = \alpha \qquad \text{or} \qquad \zeta = \frac{\alpha}{2\sqrt{K}} \tag{iii}$$

Now,

$$M_r = \frac{1}{2\zeta\sqrt{1-\zeta^2}} = \frac{1}{\dfrac{\alpha}{\sqrt{K}}\sqrt{1-\dfrac{\alpha^2}{4K}}} = 2 \text{ (given)}$$

This yields

$$\frac{2K}{\alpha\sqrt{4K-\alpha^2}} = K$$

or

$$\alpha\sqrt{4K-\alpha^2} = K$$

or

$$K^2 - 4\alpha^2 K + \alpha^4 = 0$$

$$K = \frac{4\alpha^2 \pm \sqrt{16\alpha^4 - 4\alpha^4}}{2}$$

$$= 3.732\alpha^2,\ 0.268\alpha^2 \qquad \text{(iv)}$$

Secondly,

$$\omega_r = \omega_n\sqrt{1-2\zeta^2} = (\sqrt{K})\sqrt{1-2\left(\frac{\alpha}{2\sqrt{K}}\right)^2} = 10 \text{ (given)}$$

This yields

$$\sqrt{\frac{2K-\alpha^2}{2}} = 10$$

$\Rightarrow$

$$2K - \alpha^2 = 200 \qquad \text{(v)}$$

Of the two values of K, as obtained in Eq. (iv), only $3.732\alpha^2$ yields a real value for α when calculated from Eq. (v) as

$$2(3.732\alpha^2) - \alpha^2 = 200$$

or

$$\alpha = \sqrt{\frac{200}{2(3.732)-1}} = 5.5624$$

A real value for α is necessary to make K positive. With this value of α, from Eq. (iv), K turns out to be

$$K = 3.732(5.5624)^2 = 115.47$$

Substituting these values of α and K in Eqs. (ii) and (iii), we get

$$\omega_n = \sqrt{115.47} = 10.75 \text{ rad/s}$$

$$\zeta = \frac{5.5624}{2\sqrt{115.47}} = 0.26$$

Therefore, the settling time is

$$t_s = \frac{4}{\zeta\omega_n} = \frac{4}{(0.26)(10.75)} = 1.43 \text{ s}$$

and the bandwidth is

$$\begin{aligned}\text{Bw} = \omega_c &= \omega_n\sqrt{1 - 2\zeta^2 + \sqrt{(2\zeta^2 - 1)^2 + 1}} \\ &= (10.75)\sqrt{1 - 2(0.26)^2 + \sqrt{[2(0.26)^2 - 1]^2 + 1}} \\ &= 15.9 \text{ rad/s}\end{aligned}$$

EXAMPLE 12.9. A unity feedback system has an open-loop transfer function of

$$G(s) = \frac{e^{-0.1s}}{1 + 0.5s}$$

(a) Show that the phase crossover frequency of the system is approximately 16.9 rad/s.

(b) What is the gain margin of the system?

Solution Since $H(s) = 1$, the open-loop transfer function in frequency domain is

$$\begin{aligned}G(j\omega)H(j\omega) &= \frac{e^{-j0.1\omega}}{1 + j0.5\omega} \\ &\equiv \frac{|1|\angle - 0.1\omega}{\sqrt{1 + 0.25\omega^2}\angle \tan^{-1} 0.5\omega} \\ &= \frac{1}{\sqrt{1 + 0.25\omega^2}}\angle(-0.1\omega - \tan^{-1} 0.5\omega)\end{aligned}$$

(a) Since at the phase crossover frequency ω_p the phase equals $-180°$, we have

$$-0.1\omega_p - \tan^{-1} 0.5\omega_p = -\pi$$

or

$$0.1\omega_p + \left(\frac{\pi}{2} - \frac{1}{0.5\omega_p}\right) = \pi \qquad \left[\because \tan^{-1} x \cong \frac{\pi}{2} - \frac{1}{x} \text{ for } x > 1\right]$$

or

$$0.1\omega_p - \frac{1}{0.5\omega_p} = \frac{\pi}{2}$$

or

$$\frac{0.05\omega_p^2 - 1}{0.5\omega_p} = \frac{\pi}{2} = 1.571$$

or

$$0.05\omega_p^2 - 0.785\omega_p - 1 = 0 \tag{i}$$

Equation (i) yields two values of ω_p, namely, 16.9 rad/s and -1.18 rad/s. The latter value is rejected because the frequency cannot be negative.

(b) The gain margin is calculated as

$$\text{Gm} = -20\log\frac{1}{\sqrt{1+0.25\omega_p^2}} = -20\log\frac{1}{\sqrt{1+0.25\times 16.9^2}} = 18.6\text{ dB}$$

EXAMPLE 12.10. A unity feedback control loop with an open-loop transfer function of the form $K/[s(s+a)]$ has a gain crossover frequency of 1 rad/s and a phase margin of 60°. If an element having a transfer function $(s-\sqrt{3})/(s+\sqrt{3})$ is inserted into the loop, what will be the phase margin?

Solution In the frequency domain, the given transfer function can be written as

$$\frac{K}{\omega\sqrt{\omega^2+a^2}}\angle\left(-90^\circ - \tan^{-1}\frac{\omega}{a}\right)$$

The given value of phase margin is 60° and the gain crossover frequency is $\omega_g = 1$ rad/s. So, we have

$$180^\circ - 90^\circ - \tan^{-1}\frac{1}{a} = 60^\circ \qquad [\because \omega_g = 1]$$

$$\Rightarrow \qquad \tan^{-1}\frac{1}{a} = 30^\circ$$

$$\Rightarrow \qquad \frac{1}{a} = \tan 30^\circ = \frac{1}{\sqrt{3}}$$

$$\Rightarrow \qquad a = \sqrt{3}$$

Now, at the gain crossover frequency, the gain is 1 so that in dB it is 0. Therefore,

$$\left.\frac{K}{\omega\sqrt{\omega^2+a^2}}\right|_{\omega=1} = 1$$

$$\Rightarrow \qquad \frac{K}{\sqrt{1+a^2}} = 1$$

$$\Rightarrow \qquad K = \sqrt{a^2+1} = \sqrt{3+1} = 2$$

So, the original open-loop transfer function is given by

$$G(s)H(s) = \frac{2}{s\left(s+\sqrt{3}\right)}$$

With the insertion of a new element of transfer function $\dfrac{s-\sqrt{3}}{s+\sqrt{3}}$, the new open-loop transfer function becomes

$$G'(s)H(s) = \frac{2}{s\left(s+\sqrt{3}\right)}\cdot\frac{s-\sqrt{3}}{s+\sqrt{3}} = \frac{2\left(s-\sqrt{3}\right)}{s\left(s+\sqrt{3}\right)^2}$$

To determine the Pm of this open-loop transfer function, we first find the gain crossover frequency as follows:

$$\frac{2\sqrt{\omega_g^2+3}}{\omega_g\left(\omega_g^2+3\right)} = 1$$

or

$$\frac{2}{\omega_g\sqrt{\omega_g^2+3}} = 1$$

or

$$\frac{4}{\omega_g^2+3} = \omega_g^2$$

or

$$\omega_g^4 + 3\omega_g^2 - 4 = 0 \tag{i}$$

Substituting $\omega_g^2 = x$, the two values of x that we get from Eq. (i) are 1 and -4. The latter is rejected because a frequency cannot be imaginary. So, $\omega_g = 1$. That ω_g remains unchanged is expected because a zero and a pole have been added to the transfer function on both sides of the Im-axis. Anyway, the phase margin is now calculated as follows:

$$\begin{aligned}\text{Pm} &= 180^\circ + \left(\pi - \tan^{-1}\frac{\omega_g}{\sqrt{3}}\right) - 90^\circ - 2\tan^{-1}\frac{\omega_g}{\sqrt{3}} \\ &= 270^\circ - 3\tan^{-1}\frac{1}{\sqrt{3}} \\ &= 270^\circ - 3\times 30^\circ = 180^\circ\end{aligned}$$

EXAMPLE 12.11. Consider a unity feedback system with open-loop transfer function

$$G(s) = \frac{1+6s}{s^2(1+s)(1+2s)}$$

(i) The phase crossover frequency of the system in radians per second is

(a) 0.125 (b) 0.25 (c) 0.5 (d) 1

(ii) The gain margin of the system is

(a) 0.125 (b) 0.25 (c) 0.5 (d) 1

Solution In the frequency domain the open-loop transfer function can be written as

$$\begin{aligned}G(j\omega)H(j\omega) &= \frac{1+j6\omega}{(j\omega)^2(1+j\omega)(1+j2\omega)} \\ &\equiv \frac{\sqrt{1+36\omega^2}}{\omega^2\sqrt{1+\omega^2}\sqrt{1+4\omega^2}}\angle\left(\tan^{-1}6\omega - 180^\circ - \tan^{-1}\omega - \tan^{-1}2\omega\right)\end{aligned}$$

(i) To find out the phase crossover frequency ω_p, it is necessary that

$$\tan^{-1} 6\omega_p - 180^\circ - \tan^{-1} \omega_p - \tan^{-1} 2\omega_p = -180^\circ$$

$$\Rightarrow \qquad \tan^{-1} 6\omega_p = \tan^{-1} \omega_p + \tan^{-1} 2\omega_p = \tan^{-1} \frac{3\omega_p}{1 - 2\omega_p^2}$$

Taking the tangents of both sides, we get

$$1 - 2\omega_p^2 = \frac{3\omega_p}{6\omega_p} = \frac{1}{2}$$

$$\Rightarrow \qquad 2\omega_p^2 = \frac{1}{2}$$

$$\Rightarrow \qquad \omega_p = \frac{1}{2} = 0.5$$

Ans: (c).

(ii) To calculate the Gm, we observe that

$$\text{Gain}\Big|_{\omega_p} = \left.\frac{\sqrt{1 + 36\omega^2}}{\omega^2\sqrt{1 + \omega^2}\sqrt{1 + 4\omega^2}}\right|_{\omega=0.5}$$

$$= \frac{\sqrt{1 + \frac{36}{4}}}{\frac{1}{4}\sqrt{\left(1 + \frac{1}{4}\right)\left(1 + \frac{4}{4}\right)}} = \frac{\sqrt{10}}{\frac{1}{4}\sqrt{\frac{10}{4}}}$$

$$= 8$$

Therefore,

$$\text{Gm} = \frac{1}{8} = 0.125$$

Ans: (a).

With this background on the frequency domain, we move to the study of stability of systems in this domain via Bode, polar and Nyquist plots.

REVIEW QUESTIONS

12.1 Choose the correct answer:

(i) The phase crossover frequency is the frequency at which the phase is

(a) 0° (b) 90° (c) -180° (d) 180°

(ii) The gain crossover frequency is at which the gain is

(a) 0 (b) 1 (c) 10 (d) 100

(iii) At the cut-off frequency the magnitude $|M(j\omega)|$ equals

(a) $\frac{1}{2}$ (b) $\frac{1}{\sqrt{2}}$ (c) 1 (d) 0

(iv) The bandwidth is equal to

(a) gain crossover frequency (b) phase crossover frequency
(c) cut-off frequency (d) resonant frequency

(v) If the damping ratio increases, the bandwidth

(a) decreases (b) increases
(c) remains the same (d) none of the above is correct

(vi) No resonant peak occurs for

(a) $\zeta < 0.707$ (b) $\zeta > 0.707$ (c) $\zeta < 0.5$ (d) $\zeta > 0.5$

(vii) At resonance

(a) $M_r \to 0$ (b) $M_r \to 1$ (c) $M_r \to \frac{1}{\sqrt{2}}$ (d) $M_r \to \infty$

(viii) A closed-loop system is stable for

(a) Gm = 15 dB and Pm = 70° (b) Gm = −15 dB and Pm = 70°
(c) Gm = 15 dB and Pm = −70° (d) Gm =−15 dB and Pm = −70°

(ix) The loop transfer function of a system is given by $G(s) = \frac{e^{-0.1s}}{s}$. The phase crossover frequency is given by

(a) $\frac{\pi}{2}$ (b) $\frac{\pi}{10}$ (c) $\frac{\pi}{0.2}$ (d) $\frac{\pi}{4}$

(x) The open-loop transfer function of a unity feedback system is given by $G(s) = \frac{e^{-0.1s}}{s}$. The gain margin of this system is

(a) 11.95 dB (b) 17.67 dB (c) 21.33 dB (d) 23.9 dB

(xi) A system has a gain margin of −4. The system is

(a) stable (b) unstable
(c) marginally stable (d) asymptotically stable

(xii) Gain margin is a measure of

(a) relative stability (b) controllability
(c) observability (d) absolute stability

12.2 Define the following:

(a) Gain margin
(b) Phase margin
(c) Phase crossover frequency
(d) Gain crossover frequency

12.3 Determine frequency domain specifications for a unity feedback control system having

$$G(s) = \frac{144}{s(s+1)}$$

12.4 A control system is represented by the following figure. Determine the values of K and α so that $M_r = 1.05$ and $\omega_r = 12$ rad/s. Find out the corresponding bandwidth.

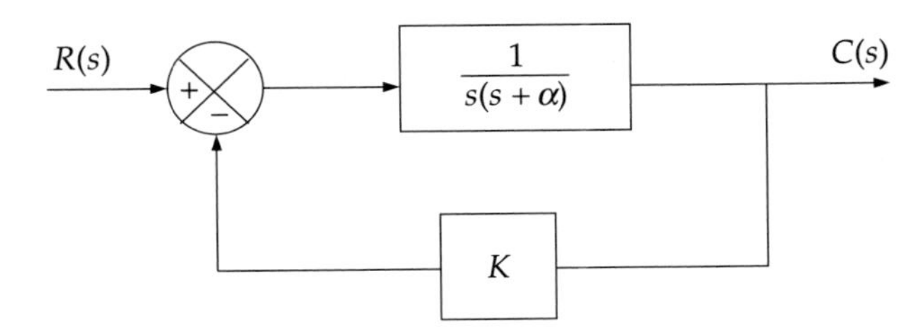

Answers to Selected Questions

12.1 (i) (c); (ii) (b); (iii) (b); (iv) (c); (v) (a); (vi) (b); (vii) (d); (viii) (a); (ix)(c); (x) (d); (xi) (b); (xii) (a).

12.3 $\omega_r = 11.98$ rad/s; $M_r = 12.51$; Bw $= 18.62$ rad/s.

12.4 $K = 472.7$; $\alpha = 25.64$; Bw $= 25.25$ rad/s.

CHAPTER 13

Bode Plot

13.1 INTRODUCTION

In Chapter 12, we showed that the study of sinusoidal inputs is important and how phasor representations of the input and the system transfer function can be used to easily determine the system output. This is referred to as the *frequency domain behaviour of a system.*

Bode[1] plots are a very useful way of representing the gain and phase of a system as a function of frequency. These plots are nothing but a way of representing the transfer function with two plots—magnitude vs. frequency and phase vs. frequency. Plots are logarithmic on the frequency axis, and the magnitude is plotted in dBs (deciBels[2]).

The question arises: Why is logarithmic scale used? Drawing the phase is fairly simple. We can draw each phase term separately, and then simply add them. The magnitude term is not so straightforward because of the fact that the magnitude terms are *multiplied.* It would be much easier if they were added. Then we could draw each term on a graph and just add them.

One way to transform multiplication into addition is by using the logarithm. Instead of using a simple logarithm, Bode used deciBel or dB. The relationship between a quantity Q and its dB representation X is given by

$$X = 20 \log_{10} Q \tag{13.1}$$

Table 13.1 gives an idea of how X values vary with Q.

Table 13.1 dB Values for Common Quantities

Quantity (Q)	1/100	1/10	1/2	$1/\sqrt{2}$	1	$\sqrt{2}$	2	3.16	5	10	100
dB value (X)	−40	−20	−6	−3	0	3	6	10	14	20	40

13.2 EQUIVALENCE OF SCALES IN LOGARITHMIC PLOT

A frequency domain transfer function is a complex quantity and hence it can be represented as a phasor, i.e. in magnitude and angle form, as follows:

$$G(j\omega) = |G(j\omega)|\angle G(j\omega) = M(\omega)\angle\phi$$

[1] Pronounced Boh-dee in English, Boh-dah in Dutch.

[2] Named after Alexander Graham Bell (1847–1922), the American who invented the telephone.

If the gain M is expressed in dB, then[3]

$$\text{Gain in dB} = 20\log M(\omega) \tag{13.2}$$

Now, suppose

$$x = \log\omega$$

Therefore,

$$x_2 - x_1 = \log\omega_2 - \log\omega_1 = \log\frac{\omega_2}{\omega_1}$$

$x_2 - x_1$ gives an interval between two frequencies in logarithmic scale. If the interval is an *octave*, i.e. if $\omega_2 = 2\omega_1$, using Eq. (13.2), we find

$$\text{Gain in dB} = 20\log 2 = 6$$

So, it is a 6 dB gain. But if the interval is a decade, i.e. if $\omega_2 = 10\omega_1$, it is a 20 dB gain. Now, since

$$\frac{6}{\log 2} = 20 = \frac{20}{\log 10}$$

we have

$$\frac{6\text{ dB}}{\text{octave}} = \frac{20\text{ dB}}{\text{decade}}$$

when the frequency is plotted in a logarithmic scale.

13.3 BODE PLOT REQUIREMENTS

The open-loop transfer function of a system of type N and gain-factor K can be written as

$$G(s)H(s) = \frac{K(1+sT_1)(1+sT_2)\cdots\omega_n^2}{s^N(1+sT_a)\cdots(s^2+2\zeta\omega_n s+\omega_n^2)}$$

Obviously, it contains first order terms in the numerator and (first and second) order terms in the denominator. The magnitude and phase of its frequency domain representation can be rewritten in terms of dB as

$$\begin{aligned}|M| = 20\log|G(\jmath\omega)H(\jmath\omega)| &= 20\log K + 20\log|1+\jmath\omega T_1| + \cdots \\ &- 20N\log|\jmath\omega| - 20\log|1+\jmath\omega T_a| - \cdots \\ &- 20\log\left|\frac{\omega_n^2-\omega^2}{\omega_n^2} + \jmath\frac{2\zeta\omega_n\omega}{\omega_n^2}\right|\end{aligned} \tag{13.3}$$

$$\begin{aligned}\phi = \angle G(\jmath\omega)H(\jmath\omega) &= \angle 0^\circ + \tan^{-1}\omega T_1 + \tan^{-1}\omega T_2 + \cdots \\ &- N\angle 90^\circ - \tan^{-1}\omega T_a - \cdots \\ &- \tan^{-1}\left[\frac{2\zeta\omega_n\omega}{\omega_n^2-\omega^2}\right]\end{aligned} \tag{13.4}$$

A scrutiny of Eqs. (13.3) and (13.4) reveals that the expressions for magnitude and phase consist of the following factors:

[3] We will write 'log' to indicate $\log_{10}$ and 'ln' to indicate $\log_e$ or natural logarithm.

1. Constant: K
2. Zeros or poles at the origin: $(j\omega)^{\pm 1}$
3. Simple zeros or poles: $(1 + j\omega T)^{\pm 1}$
4. Quadratic zeros or poles: $(1 + j2\zeta u - u^2)^{\pm 1}$, where $u = \omega/\omega_n$

Let us examine the characteristics of Bode plots for each factor individually.

Plot for Constant K

The magnitude plot is a horizontal line through $20 \log |K|$ and the phase plot is a horizontal line through 0° (see Figure 13.1).

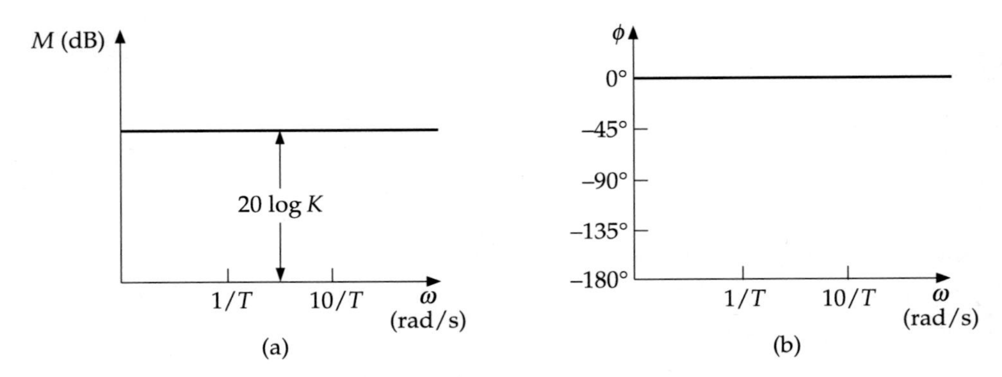

Figure 13.1 (a) Magnitude plot for constant; (b) phase plot for constant.

Plots for Zero and Pole at Origin

(a) Zero at origin: $j\omega^N$. Here,

$$|M| = 20N \log |\omega|$$

$$\phi = (+N) \times 90^\circ$$

For $\omega = 10$ and $N = 1$, $|M| = 20$ dB. Hence, for a single zero at the origin, the magnitude plot is a straight line of slope 20 dB/decade[4] passing through 0 dB at $\omega = 1$. For a single zero, the phase plot is a horizontal line through 90° (see Figure 13.2).

(b) Pole at origin: $j\omega^{-N}$. For obvious reasons, plots for the pole at the origin are mirror images of those for zero (see Figure 13.3).

[4]Or 6 dB/octave.

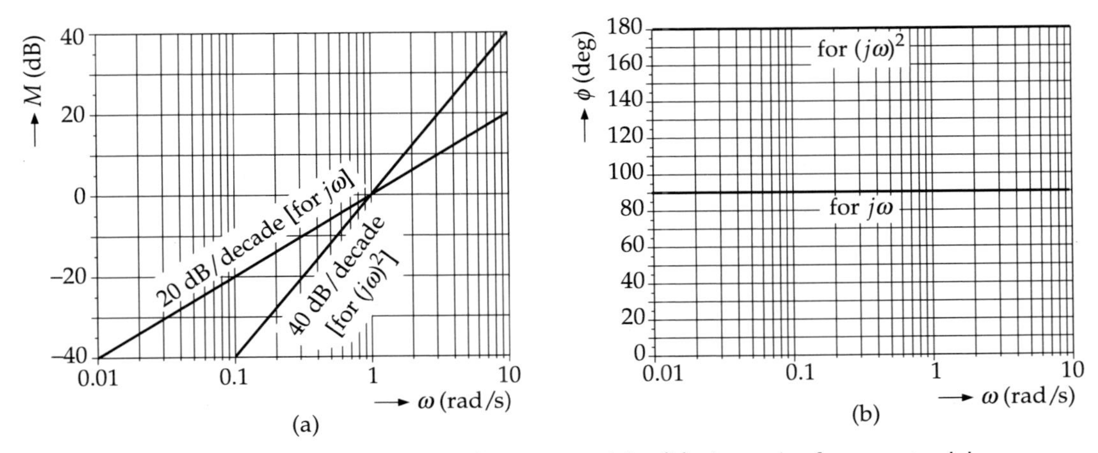

Figure 13.2 (a) Magnitude plot for zero at origin; (b) phase plot for zero at origin.

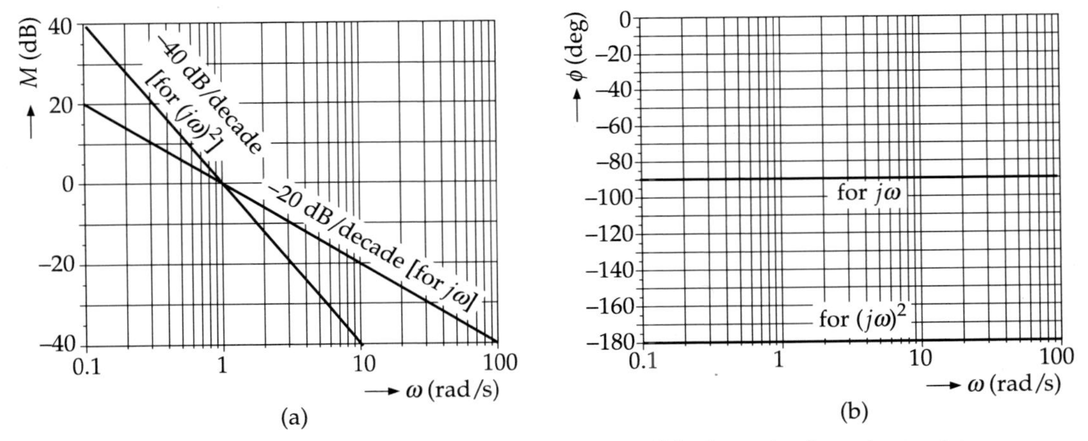

Figure 13.3 (a) Magnitude plot for pole at origin; (b) phase plot for pole at origin.

Plots for Simple Zero and Simple Pole

(a) Simple zero: $(1 + \jmath\,\omega T)$.

Magnitude. Here,

$$|M| = 20 \log \sqrt{1 + \omega^2 T^2} \tag{13.5}$$

At low frequencies,

$$\omega T \ll 1$$

Therefore,

$$|M| = 0 \text{ dB}$$

At high frequencies,

$$\omega T \gg 1$$

Therefore, $$|M| = 20 \log \omega T \text{ dB} \tag{13.6}$$

For $\omega T = 10$, $|M| = 20$ dB. Thus, a simple zero is represented by a horizontal line through 0 dB and then a straight line of slope 20 dB/decade through the *corner* (or *break*) *frequency*.

Corner (or *Break*) *frequency* is the intersection point of low and high frequency asymptotes. So, if ω_c is the corner frequency[5], we can equate the $|M|$-value at $\omega_c T$ to that at $\omega T = 1$. Thus,

$$20 \log \omega_c T = 20 \log 1$$

$$\Rightarrow \qquad \omega_c = \frac{1}{T} \tag{13.7}$$

The plot drawn for simple zero, using Eqs. (13.6) and (13.7), is called *asymptotic*. The actual curve can be obtained by using Eq. (13.5) [see Figure 13.4(a)]. The error in the asymptotic value at $\omega_c T = 1$ is

$$\text{Actual} - \text{asymptotic} = 20 \log \sqrt{1+1} - 20 \log 1 = 3.01 \text{ dB}$$

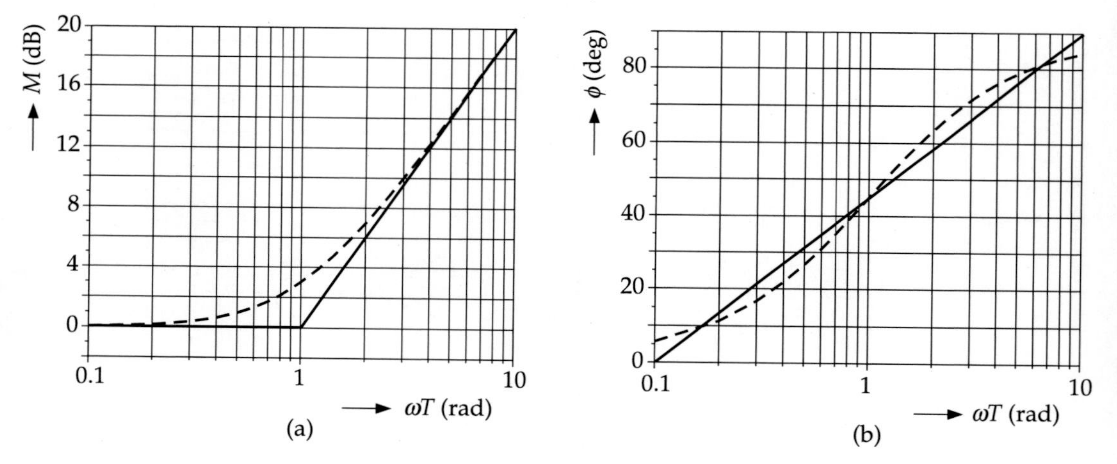

Figure 13.4 Plot for a simple zero: (a) magnitude; (b) phase. —, asymptotic; - - -, actual.

Table 13.2 lists the error for other frequencies. It can be seen that the error is maximum at the corner frequency.

Table 13.2 Error at Frequencies Near the Corner Frequency

Frequency	$0.5/T$	$1/T$	$2/T$
Approx. error	1 dB	3 dB	1 dB

[5]It should not be confused with the *cut-off frequency* which is also written as ω_c.

Phase. The exact relation is

$$\phi = \tan^{-1} \omega T$$

Approximately, it is 45°/decade because $\phi|_{\omega T=0.1} = 0°$ and $\phi|_{\omega T=1} = 45°$. Table 13.3 as well as Figure 13.4(b) shows that the maximum error in approximate plot is about 6°.

Table 13.3 Exact and Asymptotic Phase Values for Simple Zero

ω	$0.1/T$	$1/T$	$2/T$	$10/T$
$\phi_{\text{asymptotic}}$	0°	45°	60°	90°
ϕ_{exact}	5.7°	45°	64°	84.3°

(b) Simple pole: $(1 + \jmath\omega T)^{-1}$. For obvious reasons, plots for a simple pole (Figure 13.5) are mirror images of those for zero.

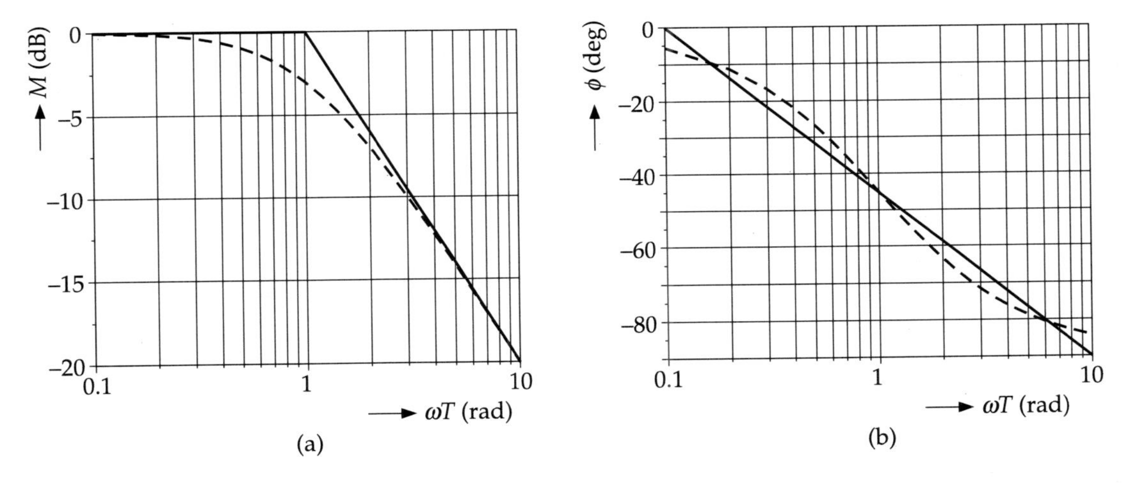

Figure 13.5 Plot for a simple pole: (a) magnitude; (b) phase. —, asymptotic; - - -, actual.

Plots for Quadratic Zero and Pole

(a) Quadratic zero

Magnitude. Here,

$$|M| = 20 \log \left| \left(1 - \frac{\omega^2}{\omega_n^2}\right) + \jmath\, 2\zeta \frac{\omega}{\omega_n} \right|$$

$$= 20 \log \sqrt{\left(1 - \frac{\omega^2}{\omega_n^2}\right)^2 + 4\zeta^2 \frac{\omega^2}{\omega_n^2}} \text{ dB} \tag{13.8}$$

For low frequencies, i.e., at $\omega \ll \omega_n$,

$$|M| = 20 \log 1 = 0 \text{ dB} \tag{13.9}$$

That means the $\log \omega$ asymptote is a horizontal line through 0 dB at low frequencies. For high frequencies, i.e., at $\omega \gg \omega_n$,

$$|M| = 20 \log \sqrt{\left(\frac{\omega^2}{\omega_n^2}\right)^2} = 40 \log \frac{\omega}{\omega_n} \tag{13.10}$$

For a decade, i.e. $\omega/\omega_n = 10$, $|M| = 40$ dB from Eq. (13.10). This means that the $\log \omega$ asymptote is a straight line of 40 dB/decade (12 dB/octave) at frequencies above the corner frequency. The corner frequency, as can be seen by equating Eqs. (13.9) and (13.10), is $\omega_c = \omega_n$. For $\omega = \omega_n$, the exact value of $|M|$, as determined from Eq. (13.8), is

$$|M|_{\text{exact}} = 20 \log \sqrt{4\zeta^2} = 20 \log 2\zeta \tag{13.11}$$

$$\text{Error} = |M|_{\text{exact}} - |M|_{\text{approx}} = 20 \log 2\zeta \tag{13.12}$$

Equation (13.12) shows that the error at the corner frequency is a function of ζ. Table 13.4 gives an idea of the error at different damping ratios.

Table 13.4 Error vs. Damping Ratio at the Corner Frequency

ζ	0.1	0.3	0.5	0.7	0.9	1.0
Error (dB)	−13.98	−4.44	0	2.92	5.10	6.02

The complete asymptotic plot, along with the exact plot, of the magnitude of a quadratic zero is given in Figure 13.6(a).

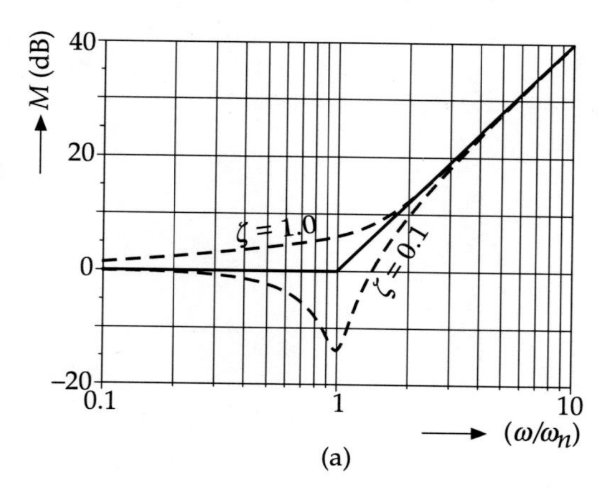

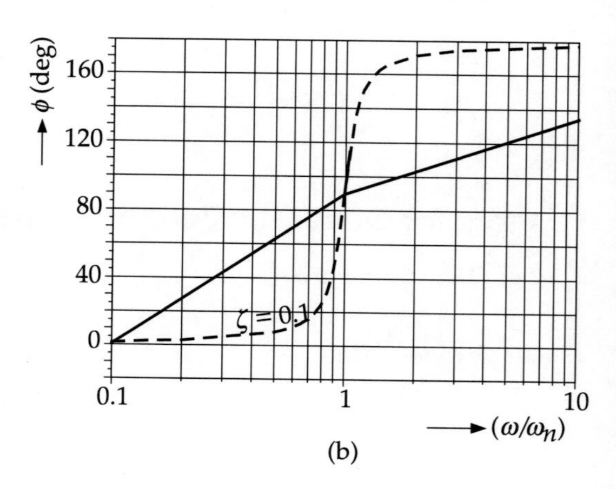

Figure 13.6 Plot for a quadratic zero: (a) magnitude; (b) phase. —, asymptotic; - - -, actual.

Phase. For a quadratic zero, the phase is given by

$$\phi = \tan^{-1}\left[\frac{2\zeta \cdot (\omega/\omega_n)}{-1 - (\omega/\omega_n)^2}\right] \tag{13.13}$$

From Eq. (13.13), we can quickly obtain the following values of frequency and phase:

ω	0	ω_n	∞
ϕ	$0°$	$90°$	$180°$

However, Eq. (13.13) indicates negative values for ϕ for $\omega > \omega_n$. To correct it, we need to add 180° to such values. The graph [Figure 13.6] shows that, although asymptotic, Bode plots for the magnitudes at higher ζ are not too bad. The asymptotic plot for the phase of quadratic zero is far from acceptable.

(b) Quadratic pole. Plots for a quadratic pole (Figure 13.7) are mirror images of those for zero.

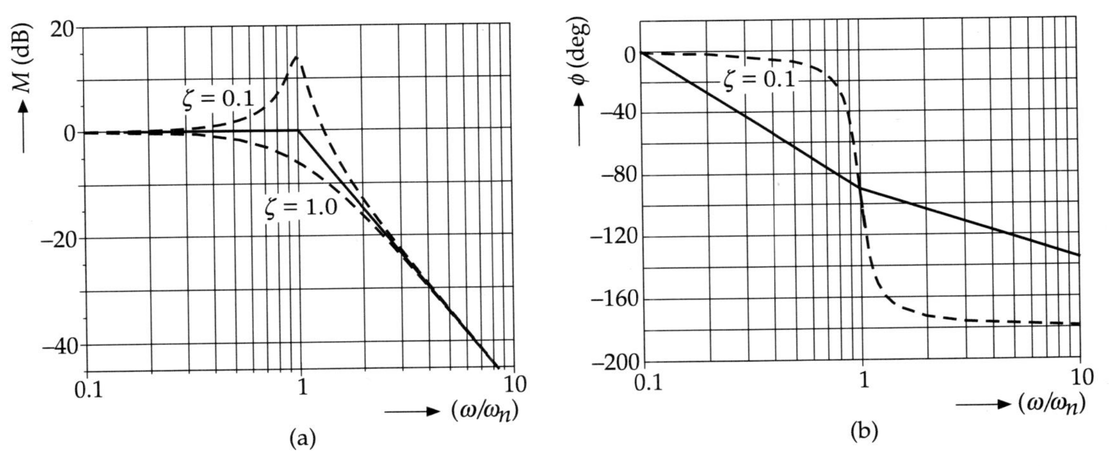

Figure 13.7 Plot for a quadratic pole: (a) magnitude; (b) phase. —, asymptotic; - - -, actual.

13.4 DRAWING BODE PLOTS

To draw a Bode diagram for an open-loop transfer function, we have to take the following steps:

Step 1. We rewrite the transfer function in a form such that first order factors look like $1 + s/a$ and the second order factors look like $1 + bs + s^2/c$, where a, b and c are constants. Then we convert the function to frequency domain form by substituting $s = \jmath\omega$. We consider an example to show how to do it.

$$G(s)H(s) = 30 \cdot \frac{s+12}{(s+2)(s+60)}$$

$$= 30 \cdot \frac{12\left(1+\dfrac{s}{12}\right)}{2\left(1+\dfrac{s}{2}\right)60\left(1+\dfrac{s}{60}\right)} = 3 \cdot \frac{1+\dfrac{s}{12}}{\left(1+\dfrac{s}{2}\right)\left(1+\dfrac{s}{60}\right)}$$

$$\equiv 3 \cdot \frac{1+\dfrac{j\omega}{12}}{\left(1+\dfrac{j\omega}{2}\right)\left(1+\dfrac{j\omega}{60}\right)}$$

Step 2. We find corner frequencies for different factors of the open-loop transfer function. In the above example, corner frequencies are 2, 12 and 60 rad/s. We have to remember that, when we plot the graph, the frequency in the semi-log graph paper[6] must start from a lower value than the lowest of corner frequencies.

Step 3. For the phase plot, we have to estimate the total variation of angle using Eq. (13.4) and accordingly choose the axis in the graph paper. In the above example, the frequency should start from 0.2 rad/s, which is 1/10th of the lowest corner frequency 2, and it should extend to 600 rad/s, which is ten times the highest corner frequency 60.

Step 4. We plot magnitude and phase for each factor using Table 13.5.

Step 5. We add results obtained from Step 4 to draw the overall Bode diagram.

Prescriptions for making Bode plots are summed up in Table 13.5. It has to be kept in mind that these prescriptions pertain to *asymptotic Bode plots* and *not exact ones.* Drawing exact Bode plots entails evaluation of magnitude and phase values for the entire open-loop transfer functionat many frequencies. MatLab[7] precisely does that job. Nevertheless, asymptotic plots are useful for approximate calculations and, therefore, find wide use in the design of control systems.

Note:

1. For quadratic poles and zeros, peaks exist for $0 < \zeta < 0.707$ $(= 1/\sqrt{2})$ and peak frequency is typically very near the corner frequency.
2. For quadratic poles and zeros, if $\zeta < 0.02$, draw phase vertically from $0°$ to $-180°$ at corner frequency.
3. For nth order pole or zero, make asymptotes, peaks and slopes n times higher (i.e., second order asymptote is -40 dB/decade and phase goes from $0°$ to $-180°$). *Do not change frequencies*, plot only values and slopes.

[6]The semi-log graph paper can be downloaded from http://www.intmath.com/downloads/Graph-paper.php.

[7]A software produced by The Mathworks Inc., Natick(Massachusetts). Appendix D at page 671 provides a MATLAB primer.

Table 13.5 Prescriptions for Making Asymptotic Bode Plots

Term	*Factor*	*Prescription*
Magnitude plot		
Constant	K	Horizontal line at $20 \log \lvert K \rvert$ dB
Zero at origin	$j\omega$	+20 dB/decade; through 0 dB at $\omega = 1$
Pole at origin	$\dfrac{1}{j\omega}$	−20 dB/decade; through 0 dB at $\omega = 1$
Simple zero	$1 + \dfrac{j\omega}{\omega_c}$	+20 dB/decade line starting at corner frequency extending to infinity
Simple pole	$\dfrac{1}{1 + \dfrac{j\omega}{\omega_c}}$	−20 dB/decade line starting at corner frequency extending to infinity
Quadratic zero	$1 + j2\zeta\left(\dfrac{\omega}{\omega_n}\right) - \left(\dfrac{\omega}{\omega_n}\right)^2$	+40 dB/decade line from $\omega = \omega_n$. If $\zeta < 0.7$, draw *dip* at frequency $\omega_r = \omega_n\sqrt{1-2\zeta^2}$ with amplitude $\lvert G(j\omega)H(j\omega)\rvert = +20 \log\left(2\zeta\sqrt{1-\zeta^2}\right)$
Quadratic pole	$\dfrac{1}{1 + j2\zeta\left(\dfrac{\omega}{\omega_n}\right) - \left(\dfrac{\omega}{\omega_n}\right)^2}$	−40 dB/decade line from $\omega = \omega_n$. If $\zeta < 0.7$, draw *peak* at frequency $\omega_r = \omega_n\sqrt{1-2\zeta^2}$ with amplitude $\lvert G(j\omega)H(j\omega)\rvert = -20 \log\left(2\zeta\sqrt{1-\zeta^2}\right)$
Phase plot		
Constant	K	For $K > 0$: Horizontal line through 0° For $K < 0$: Horizontal line through ± 180°.
Zero at origin	$j\omega$	Horizontal line through +90°
Pole at origin	$\dfrac{1}{j\omega}$	Horizontal line through −90°
Simple zero	$1 + \dfrac{j\omega}{\omega_c}$	Line starting at $(0.1\omega_c,\ 0^\circ)$ and ending at $(10\omega_c,\ +90^\circ)$
Simple pole	$\dfrac{1}{1 + \dfrac{j\omega}{\omega_c}}$	Line starting at $(0.1\omega_c,\ 0^\circ)$ and ending at $(10\omega_c,\ -90^\circ)$
Quadratic zero	$-\left(\dfrac{\omega}{\omega_n}\right)^2 + j2\zeta\left(\dfrac{\omega}{\omega_n}\right) + 1$	Line starting at $\left[\omega = \dfrac{\omega_n}{2}\log(2/\zeta),\ 0^\circ\right]$ and ending at $\left[\omega = \dfrac{2\omega_n}{\log(2/\zeta)},\ +180^\circ\right]$
Quadratic pole	$\dfrac{1}{1 + j2\zeta\left(\dfrac{\omega}{\omega_n}\right) - \left(\dfrac{\omega}{\omega_n}\right)^2}$	Line starting at $\left[\omega = \dfrac{\omega_n}{2}\log(2/\zeta),\ 0^\circ\right]$ and ending at $\left[\omega = \dfrac{2\omega_n}{\log(2/\zeta)},\ -180^\circ\right]$

EXAMPLE 13.1. Draw the asymptotic Bode diagram for the unity feedback system having

$$G(s) = \frac{10(s+20)}{s^2 + 210s + 2000}$$

Solution

Step 1. Rewrite the transfer function to a suitable form as discussed earlier. Thus,

$$G(s)H(s) = \frac{10^3(s+20)}{s^2+210s+2000} = \frac{10^3(s+20)}{(s+10)(s+200)}$$

$$= 10^3 \frac{20\left(1+\frac{s}{20}\right)}{10\left(1+\frac{s}{10}\right)200\left(1+\frac{s}{200}\right)} = \frac{10\left(1+\frac{s}{20}\right)}{\left(1+\frac{s}{10}\right)\left(1+\frac{s}{200}\right)}$$

Therefore, $$G(j\omega)H(j\omega) = \frac{10\left(1+\frac{j\omega}{20}\right)}{\left(1+\frac{j\omega}{10}\right)\left(1+\frac{j\omega}{200}\right)}$$

Step 2. The open-loop transfer function consists of the following components:

(a) A constant gain $K = 10$
(b) A simple zero at $\omega_c = 20$
(c) A pair of simple poles at $\omega_c = 10$ and $\omega_c = 200$

Step 3. Before starting the plot, we need to determine the scales, especially the frequency scale which should be common for both the magnitude and the phase plots. Here, the phase is the determining factor. We observe that the lowest and highest corner frequencies are 10 and 200, respectively. So, for the phase plot we need a span of 1 (0.1×10) to 2000 (10×200). Therefore, we choose a span of 1 to 10000 for the frequency.

For the magnitude we choose a span of $+20$ to -20 dB to cover both the constant and the poles. For phase, we need to select a span of $+90°$ to $-90°$.

Magnitude plot

(a) For $K = 10$, $|M| = 20 \log 10 = 20$ dB. So, a horizontal line is drawn at 20 dB.
(b) For the simple zero, a 20 dB/decade line is drawn at $\omega_c = 20$.
(c) For the two simple poles, two -20 dB/decade lines are drawn at $\omega_c = 10$ and $\omega_c = 200$.

Phase plot

(a) A horizontal line at 0° is drawn for the constant K.
(b) A line joining 0° at $\omega = 2$ ($0.1\times$ corner frequency of zero) and 90° at $\omega = 200$ ($10\times$ corner frequency of zero) is drawn for the zero at $\omega_c = 20$.
(c) A line joining 0° at $\omega = 1$ ($0.1\times$ corner frequency of zero) and $-90°$ at $\omega = 100$ ($10\times$ corner frequency of zero) is drawn for the pole at $\omega_c = 10$.
(d) A line joining 0° at $\omega = 20$ ($0.1\times$ corner frequency of zero) and $-90°$ at $\omega = 2000$ ($10\times$ corner frequency of zero) is drawn for the pole at $\omega_c = 200$.

Step 4. Individual contributions are added to obtain the resultant asymptotic plots for magnitude and phase [see Figures 13.8(a) and 13.8(b)]. The question is, how do we add the individual components?

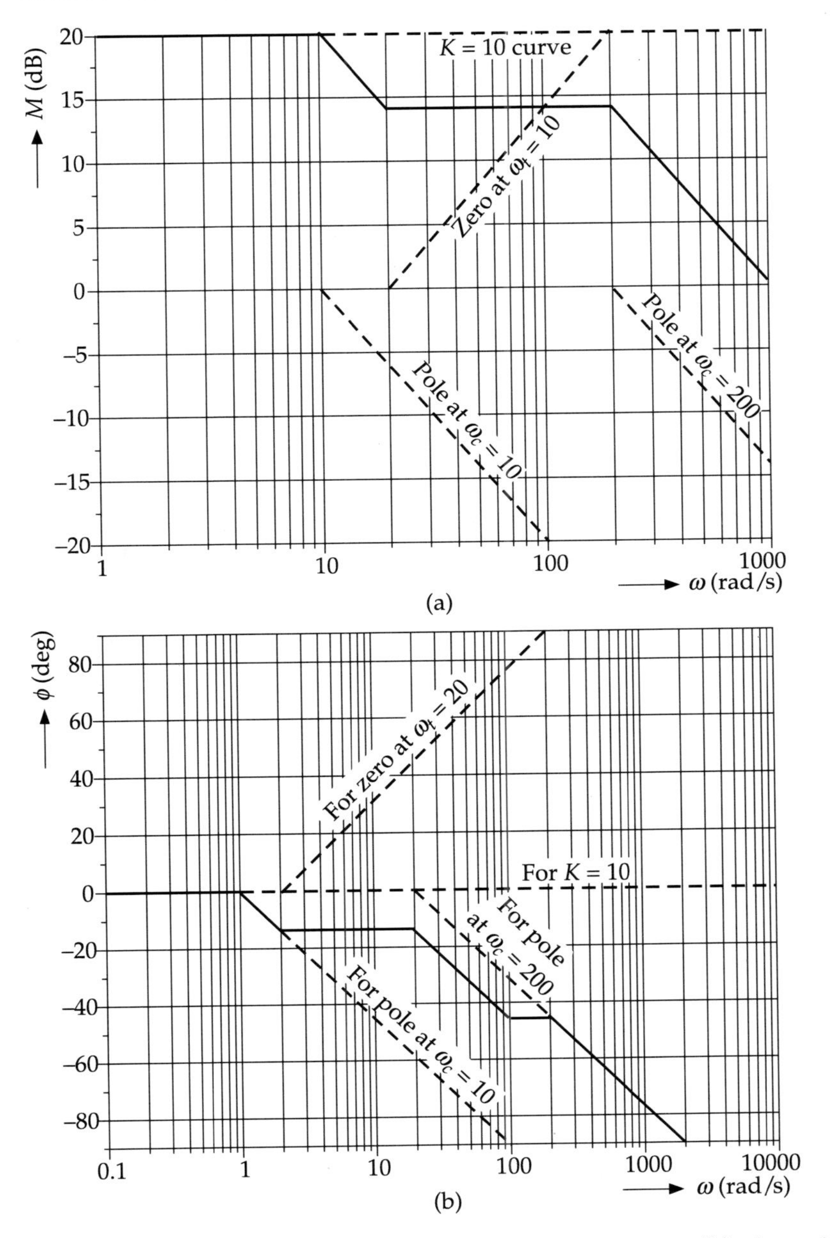

Figure 13.8 Asymptotic Bode diagram (Example 13.1): (a) magnitude plot; (b) phase plot. - - - lines indicate plots of components while solid lines indicate the resultant asymptotic plots.

For the magnitude plot, we note that the only component that exists till $\omega = 10$ is the 20 dB contribution from the constant. Then the -20 dB/decade contribution from the simple pole at $\omega_c = 10$ starts. So the the resultant curve assumes the same slope till $\omega = 20$ when it encounters a $+20$ dB/decade contribution from the simple zero. The contributions from the zero and pole cancel each other till the resultant curve reaches $\omega = 200$ when the contribution from another simple pole sets in. So, from this point onwards the resultant curve continues with a slope of -20 dB/decade.

Contributions from the phase components are added in a similar fashion.

EXAMPLE 13.2. Draw the Bode diagram for the system having the open-loop transfer function as

$$G(s)H(s) = \frac{2500(s+10)}{s(s+2)(s^2+30s+2500)}$$

Solution Comparing the second order term in the denominator to its standard form, we get

$$\omega_n^2 = 2500 \qquad \text{or} \qquad \omega_n = 50$$

$$2\zeta\omega_n = 30 \qquad \text{or} \qquad \zeta = \frac{30}{2 \times 50} = 0.3$$

The transfer function can be written in the Bode form as

$$G(s)H(s) = 2500 \cdot \frac{10\left(1+\frac{s}{10}\right)}{s(2)\left(1+\frac{s}{2}\right)(2500)\left(\frac{s^2}{50^2}+\frac{2 \times 15}{50}\zeta\frac{s}{50}+1\right)}$$

Therefore,
$$G(\jmath\omega)H(\jmath\omega) = \frac{5\left(1+\frac{\jmath\omega}{10}\right)}{\jmath\omega\left(1+\frac{\jmath\omega}{2}\right)\left[1+0.6\left(\frac{\jmath\omega}{50}\right)+\left(\frac{\jmath\omega}{50}\right)^2\right]}$$

The transfer function comprises the following factors:

1. A constant gain $K = 5$
2. A zero at $\omega_c = 10$
3. Poles at $\omega_c = 0$ and 2
4. A pair of quadratic (or complex) poles at $\omega_n = 50$

Therefore, the frequency span chosen is from 0.1 to 100.

Magnitude plot

1. The constant gain is a horizontal line at $20 \log 5 = 14$ dB.
2. The asymptote for zero at $\omega_c = 10$ is drawn with a slope of 20 dB/decade.
3. The asymptote for the pole at the origin extends from 0.1 to ∞. Passing through $\omega = 1$ at 0 dB, it has a slope of -20 dB/decade.
4. The asymptote for pole at $\omega_c = 2$ is drawn with a slope of -20 dB/decade.

5. The asymptote for the quadratic pole at $\omega_n = 50$ has a slope of -40 dB/decade. Since $\zeta = 0.3$, corrections are necessary to determine peak frequency and peak magnitude, which are given as

$$\omega_r = \omega_n\sqrt{1-\zeta^2} = 45.3 \text{ rad/s}$$

$$|G(\jmath\omega_r)H(\jmath\omega_r)| = -20\log(2\zeta\sqrt{1-\zeta^2}) = 4.85 \text{ dB}$$

6. Individual contributions are added to draw the final plot.

Phase plot

1. The phase of the constant gain is 0°.
2. Zero at $\omega_c = 10$ gives a straight line connecting 0° at $\omega = 1.0$ and 90° at $\omega = 100$.
3. Pole at the origin gives a horizontal line at 90°.
4. Pole at $\omega_c = 2$ gives a straight line connecting 0° at $\omega = 0.2$ and $-90°$ at $\omega = 20$.
5. Though the asymptotic phase characteristics for the quadratic pole can be drawn by drawing a straight line connecting 0° at $\omega = (\omega_n/2)\log(2/\zeta) = 20.6$ and $-180°$ at $\omega = 2\omega_n/\log(2/\zeta) = 121.4$, it is better to calculate the actual phase variation between $\omega = 5$ and 100 with the help of Eq. (13.13)[8].
6. Individual contributions are added to obtain the resultant plot.

The plots are given in Figure 13.9.

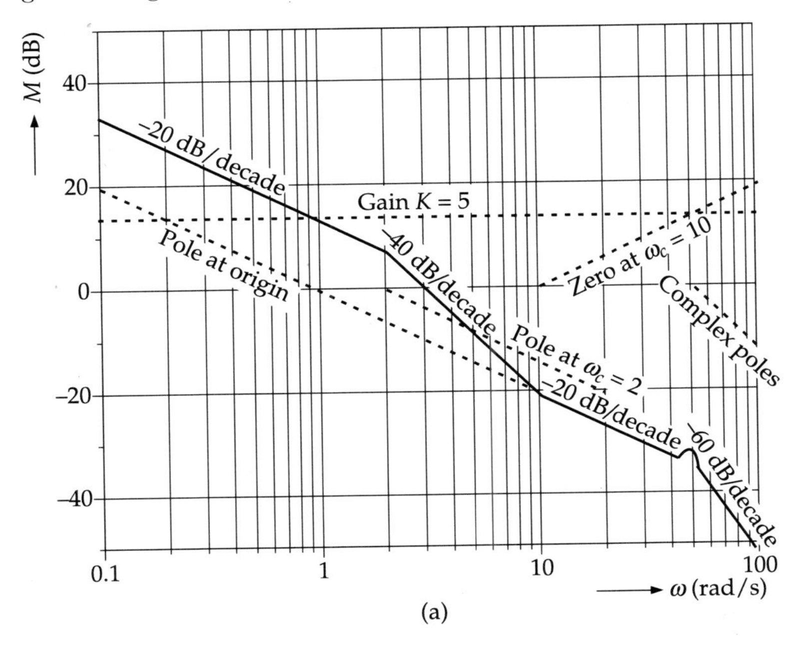

(a)

[8]With a negative sign.

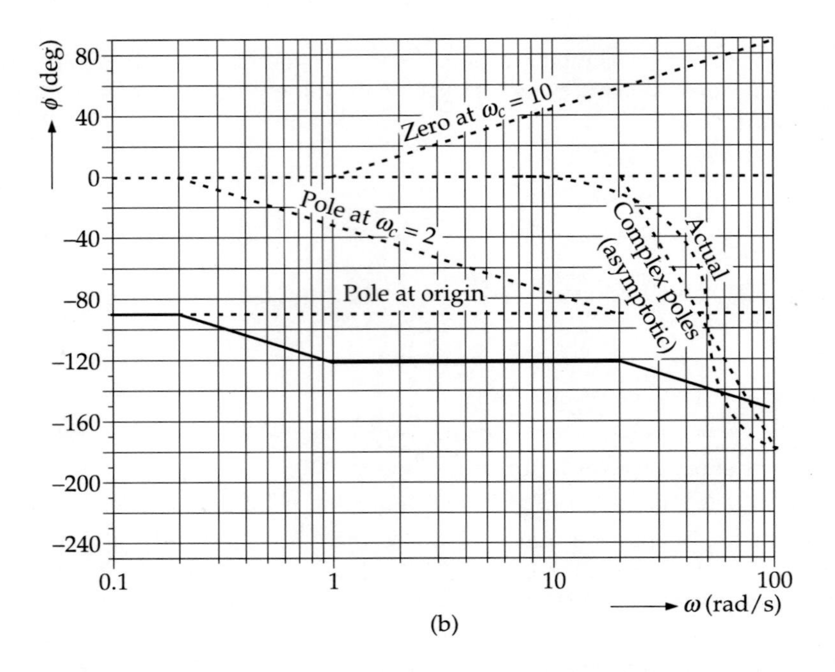

Figure 13.9 Asymptotic Bode diagram (Example 13.2): (a) magnitude plot; (b) phase plot.

13.5 STABILITY STUDY FROM BODE PLOT

The parameters which give an idea about the stability of a system are

1. Gain crossover frequency, ω_g
2. Phase crossover frequency, ω_p
3. Gain margin, Gm
4. Phase margin, Pm

We have already defined these parameters in Section 12.4 at page 417. Here, we discuss how these parameters can be found out from Bode plots. In fact, Bode plots are eminently suitable for finding out these parameters.

Gain crossover frequency, ω_g. It is the frequency at which the Bode magnitude plot crosses the 0 dB axis (see Figure 13.10).

Phase crossover frequency, ω_p. It is the frequency at which the Bode phase plot crosses the $-180°$ axis (see Figure 13.10).

Gain margin, Gm. If ω_p is the phase crossover frequency, then the reciprocal of gain at this frequency is the gain margin. Quantitatively, it is defined as

$$\text{Gm} = |M|^{-1}_{\omega=\omega_p} = -20\log\left|G(j\omega_p)H(j\omega_p)\right| \text{ dB}$$

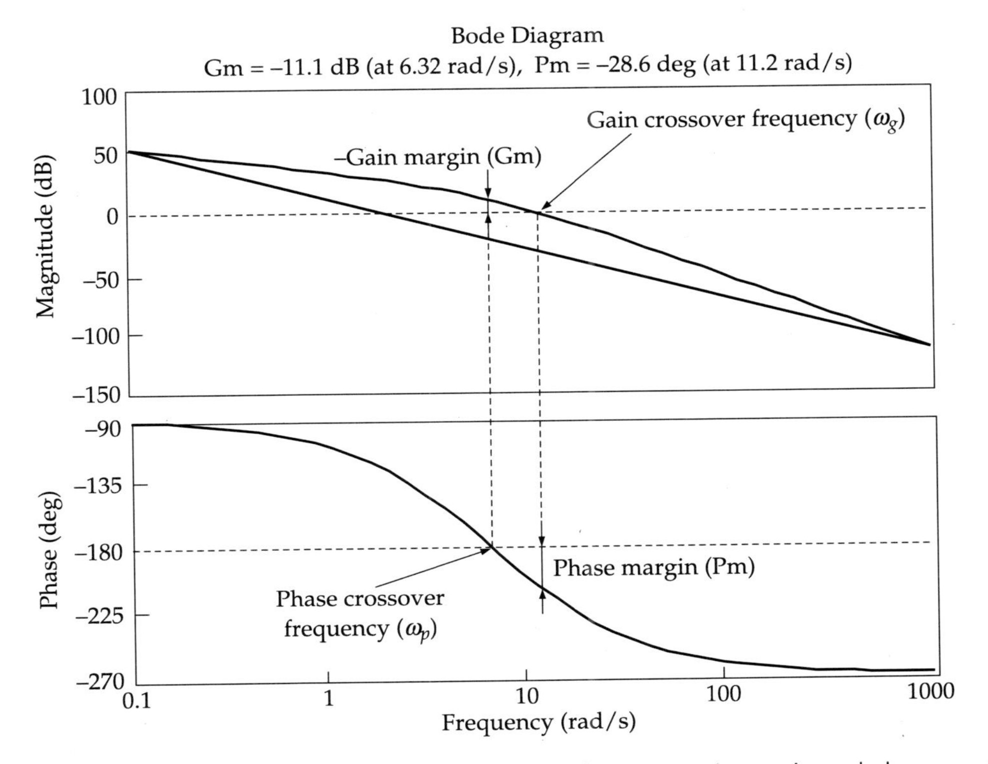

Figure 13.10 Gain crossover frequency, phase crossover frequency, gain margin, and phase margin.

From Bode plots,

1. We determine the frequency where the phase plot crosses the $-180°$ line. This frequency is the phase crossover frequency, ω_p.
2. Corresponding to ω_p, we find out the value of the gain from the magnitude plot. This value of the gain, with its sign reversed, is the gain margin. That means, if the value of the gain is found negative, the gain margin is positive and vice versa.

Phase margin, Pm. If ϕ is the phase angle at the gain crossover frequency ω_g, then the phase margin is given by $180° + \phi$. Symbolically,

$$\text{Pm} = 180° + \phi|_{\omega=\omega_g} \text{ degree}$$

To determine the phase margin from Bode plots,

1. In the magnitude plot, we find out the frequency where the curve crosses the 0 dB line. That frequency is the gain crossover frequency, ω_g.
2. Corresponding to ω_g, we find out the phase value from the phase plot. This phase value, added to $180°$, gives the phase margin.

Obviously, to determine the stability margins, both the magnitude and the phase plot have to be made in the same frequency range and one on top of the other.

The stability conditions are as follows:

Gain margin	*Phase margin*	*Frequency*	*System*
>0	>0	$\omega_p > \omega_g$	Stable
0	0	$\omega_p = \omega_g$	Marginally stable
<0	≥ 0	$\omega_p < \omega_g$	Unstable
≥ 0	<0	$\omega_p > \omega_g$	Unstable

The gist of what has been stated in the table is that

- The system is marginally stable if Gm = Pm = 0.
- The system is unstable if either Gm or Pm is negative.

EXAMPLE 13.3. A unity feedback system has the following open-loop frequency response:

ω (rad/s)	2	3	4	5	6	8	10
$\lvert G(j\omega)\rvert$	7.5	4.8	3.15	2.25	1.70	1.00	0.64
$\angle G(j\omega)$	$-118°$	$-130°$	$-140°$	$-150°$	$-157°$	$-170°$	$-180°$

The gain and phase margins of the system are:

(a) 0 dB, $-180°$ (b) 3.88 dB, $-170°$
(c) 0 dB, $10°$ (d) 3.88 dB, $10°$

Solution From the given table we find that the phase crossover frequency of the system is 10 rad/s, where the phase angle is $-180°$. At that frequency the gain is 0.64. Hence,

$$\text{Gm} = -20 \log 0.64 = 3.88 \text{ dB}$$

Also, we find from the table that the gain crossover frequency, where the gain is 1.00 (so that gain = 0 dB), is 8 rad/s. At that frequency, the phase reads $-170°$. Therefore,

$$\text{Pm} = 180° - 170° = 10°$$

Ans: (d).

EXAMPLE 13.4. Sketch the Bode plot for the system having its open-loop transfer function as

$$G(s)H(s) = \frac{20}{s(s+1)(s+4)}$$

Find the gain margin and phase margin, and comment on the stability of the system.

Solution The transfer function can be written in Bode form as

$$G(s)H(s) = \frac{5}{s\left(1+\dfrac{s}{1}\right)\left(1+\dfrac{s}{4}\right)}$$

$$G(j\omega)H(j\omega) = \frac{5}{j\omega\left(1+\dfrac{j\omega}{1}\right)\left(1+\dfrac{j\omega}{4}\right)}$$

The transfer function comprises the following components:

1. A constant $K = 5$
2. A pole at the origin
3. Two simple poles at $\omega_c = 1$ and 4

Magnitude plot

1. For the constant, a horizontal line is drawn at 20 log 5 = 14 dB.
2. For the pole at the origin, a −20 dB/decade line that passes through $\omega = 1$ at 0 dB is drawn.
3. For two simple poles, two −20 decade/dB lines are drawn at $\omega_c = 1$ and 4.
4. Individual contributions are added to get the resultant curve.

Phase plot

1. For the constant, the phase is 0°.
2. For the pole at the origin, the phase is a horizontal line at −90°.
3. For the pole at $\omega_c = 1$, a line between 0° at $\omega = 0.1$ and −90° at $\omega = 10$ is drawn. For the other pole, a parallel line is drawn between $\omega = 0.4$ and $\omega = 40$.
4. Individual contributions are added to get the final curve.

ω_p, ω_g, *Gm* and *Pm*

From the asymptotic phase plot (Figure 13.11), we find that the phase curve cuts the −180° line at $\omega = 2.0$ rad/s. This is the phase crossover frequency ω_p.

From the magnitude plot, we observe that the magnitude curve cuts the 0 dB line at $\omega = 2.2$ rad/s. This is the gain crossover frequency ω_g.

The $\omega = 2.0$ rad/s line, when extended to the magnitude plot, intersects it at 2 dB. Therefore, gain margin = −2 dB.

The $\omega = 2.2$ rad/s line is extended to the phase plot. It intersects the phase plot at −182°. Therefore, phase margin = 180° − 182° = −2°.

Since both Gm and Pm are negative, the system is unstable.

Note: The actual plots, drawn through the MatLab, show $\omega_p = \omega_g = 2.0$ rad/s and Gm = Pm = 0, which indicate a marginally stable system. This variation in conclusion occurs owing to the approximate nature of asymptotic plots.

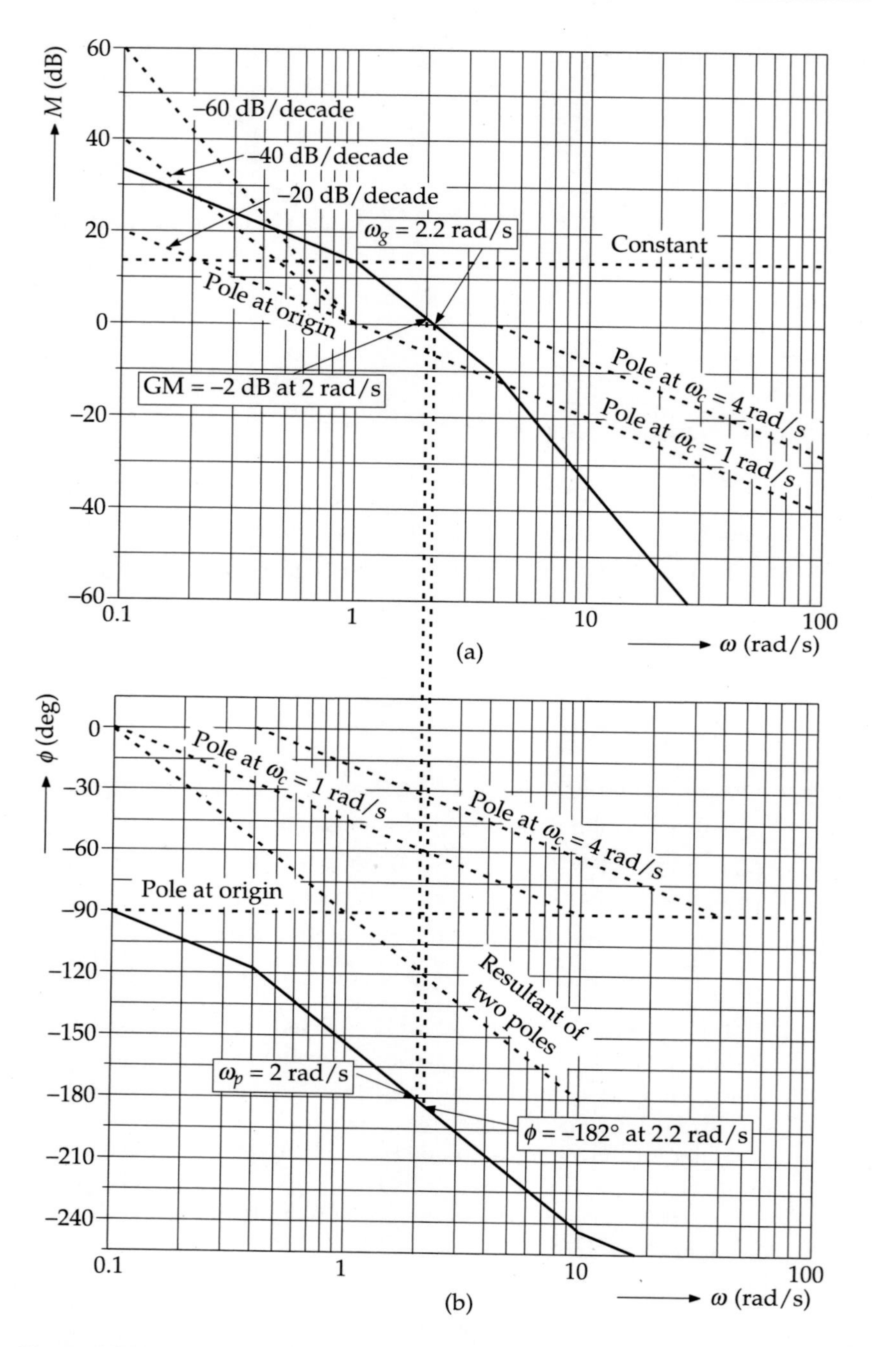

Figure 13.11 Bode plots (Example 13.4): (a) magnitude plot; (b) phase plot.

EXAMPLE 13.5. Draw the Bode plot of the system whose open-loop transfer function is given by

$$G(s) = \frac{64(s+2)}{s(s+0.5)}$$

Find the gain margin and phase margin from the plot.

Solution The transfer function can be written in the Bode form as

$$G(s) = 64 \cdot \frac{2}{0.5} \cdot \frac{1+\dfrac{s}{2}}{s\left(1+\dfrac{s}{0.5}\right)} = \frac{256\left(1+\dfrac{s}{2}\right)}{s\left(1+\dfrac{s}{0.5}\right)}$$

$$G(j\omega) = \frac{256\left(1+\dfrac{j\omega}{2}\right)}{j\omega\left(1+\dfrac{j\omega}{0.5}\right)}$$

The transfer function comprises the following components:

1. A constant $K = 256$
2. A pole at the origin
3. A simple zero at $\omega_c = 2$
4. A simple pole at $\omega_c = 0.5$

Magnitude plot

1. For the constant, a horizontal line is drawn at 20 log 256 = 48.2 dB.
2. For the pole at the origin, a −20 dB/decade line that passes through $\omega = 1$ at 0 dB is drawn.
3. For the simple zero, a 20 dB/decade line is drawn at $\omega_c = 2$.
4. For the simple pole, a −20 dB/decade line is drawn at $\omega_c = 0.5$.
5. Individual contributions are added to get the final curve.

Phase plot

1. For the constant, the phase is 0°.
2. For the pole at the origin, the phase is a horizontal line at −90°.
3. For the zero at $\omega_c = 2$, a line between 0° at $\omega = 0.2$ and 90° at $\omega = 20$ is drawn.
4. For the pole at $\omega_c = 0.5$, a line between 0° at $\omega = 0.05$ and −90° at $\omega = 5$ is drawn.
5. Individual contributions are added to get the final curve.

From the plots (Figure 13.12), we find that $\omega_p = \infty$, $\omega_g = 64$ and phase at $\omega_g = -90°$. So, gain margin = ∞, phase margin = 180° − 90° = 90°. Therefore, the system is stable.

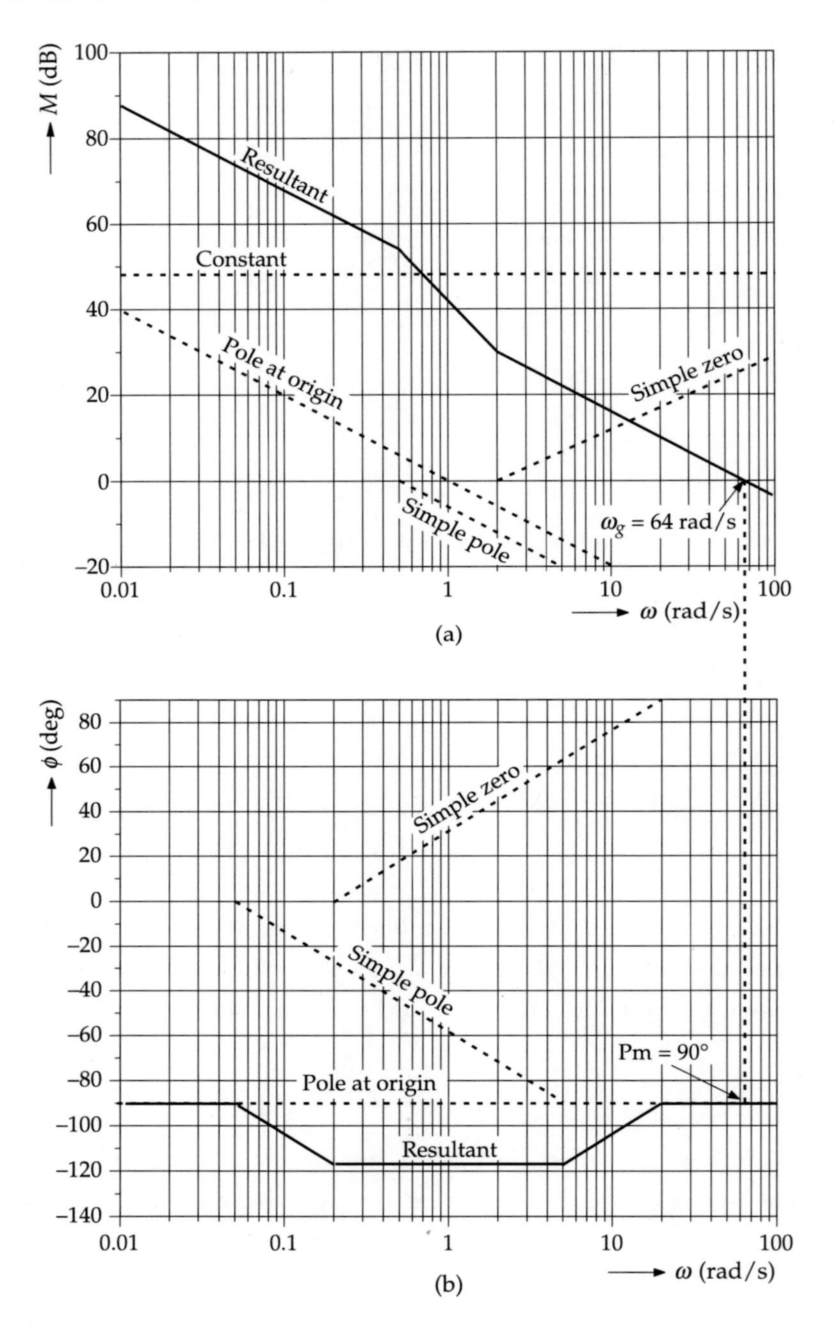

Figure 13.12 Bode plots: (a) magnitude plot; (b) phase plot (Example 13.5).

EXAMPLE 13.6. The open-loop transfer function of a control system is given by

$$G(s)H(s) = \frac{K}{s(s+4)(s+10)}$$

From its Bode plot, find the value of K when the system is marginally stable.

Solution The open-loop transfer function can be written in the Bode form as

$$G(s)H(s) = \frac{K}{4 \times 10} \cdot \frac{1}{s\left(1+\frac{s}{4}\right)\left(1+\frac{s}{10}\right)} = \frac{K'}{s\left(1+\frac{s}{4}\right)\left(1+\frac{s}{10}\right)}, \quad K' = \frac{K}{40}$$

Therefore, $$G(j\omega)H(j\omega) = \frac{K'}{j\omega\left(1+\frac{j\omega}{4}\right)\left(1+\frac{j\omega}{10}\right)}$$

The transfer function comprises the following:

1. A constant K'
2. A pole at the origin
3. Two simple poles at $\omega_c = 4$ and 10

Magnitude plot

1. The constant is an unknown quantity. So, we ignore it for the time being.
2. For the pole at the origin, a -20 dB/decade line that passes $\omega = 1$ at 0 dB is drawn.
3. For the simple poles, two -20 dB/decade lines are drawn $\omega_c = 4$ and 10.
4. Individual contributions are added to get the final curve.

Phase plot

1. For the constant, the phase is $0°$.
2. For the pole at the origin, the phase is a horizontal line at $-90°$.
3. For the pole at $\omega_c = 4$, a line between $0°$ at $\omega = 0.4$ and $-90°$ at $\omega = 40$ is drawn. For the second pole at $\omega_c = 10$, another line between $0°$ at $\omega = 1$ and $-90°$ at $\omega = 100$ is drawn.
4. Individual contributions are added to get the final curve.

Bode magnitude and phase plots are given in Figure 13.13. It may be seen from the phase plot, which is not affected by the unknown value of the constant (because the constant contributes nothing to the phase), that the phase crossover frequency is 7.5 rad/s. Therefore, to make the system marginally stable, the gain crossover frequency has to be made equal to 7.5 rad/s. The magnitude plot shows that at ω_p the gain is -20 dB. So, if we raise the magnitude curve by 20 dB, the gain becomes 0 dB (i.e. unity) to make $\omega_g = \omega_p$.

Now,

$$20 \log K' = 20 \text{ dB}$$

or $$K' = 10 = \frac{K}{40}$$

or $$K = 400$$

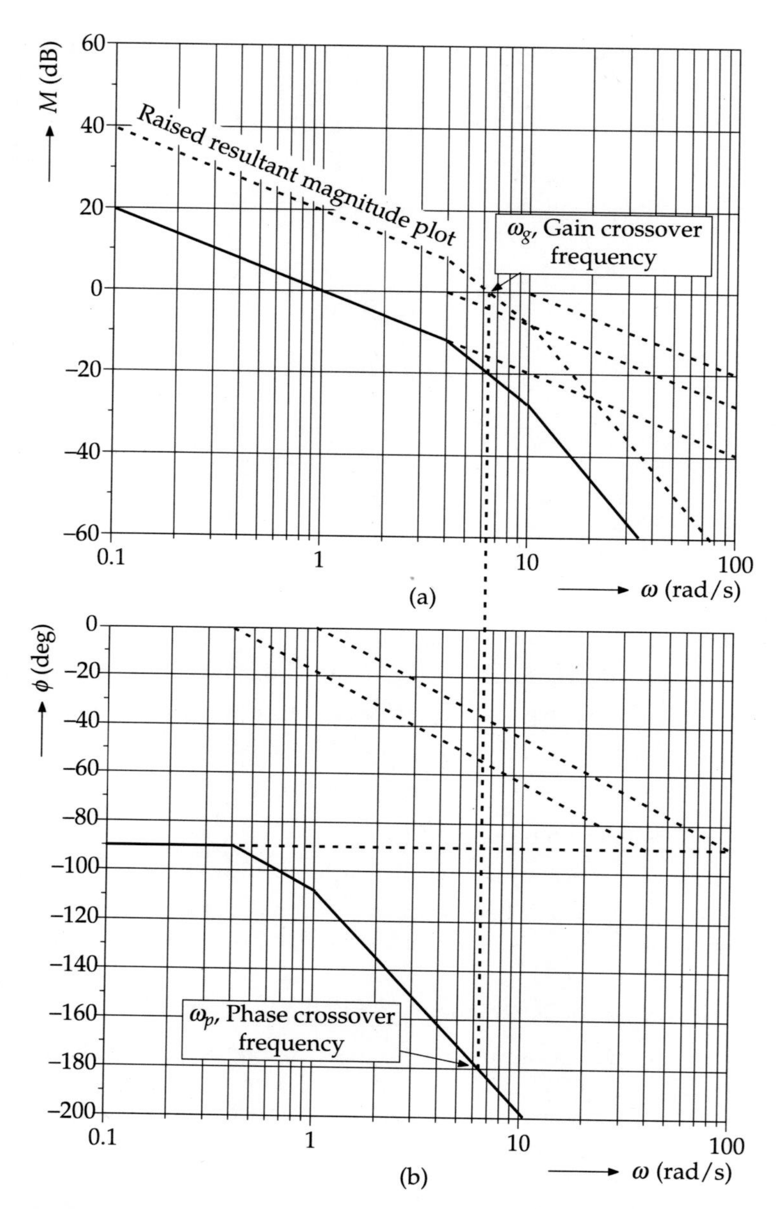

Figure 13.13 Bode plots: (a) magnitude plot; (b) phase plot (Example 13.6).

EXAMPLE 13.7. Sketch the Bode plot (both gain and phase) for the transfer function given below and determine the gain crossover frequency.

$$\frac{100(1+0.5s)}{s(s^2+16s+100)}$$

Solution Comparing the second order term in the denominator to its standard form, we get

$$\omega_n^2 = 100 \qquad \text{or} \qquad \omega_n = 10$$

$$2\zeta\omega_n = 16 \qquad \text{or} \qquad \zeta = \frac{16}{2 \times 10} = 0.8$$

The transfer function can be written in the Bode form as

$$G(s)H(s) = \frac{100\left(1 + \dfrac{s}{2}\right)}{s(100)\left(1 + \dfrac{2 \times 8}{10} \cdot \dfrac{s}{10} + \dfrac{s^2}{10^2}\right)}$$

$$G(j\omega)H(j\omega) = \frac{\left(1 + \dfrac{j\omega}{2}\right)}{j\omega\left[1 + 2 \times 0.8\left(\dfrac{j\omega}{10}\right) + \left(\dfrac{j\omega}{10}\right)^2\right]}$$

$$\equiv \frac{\left(1 + \dfrac{j\omega}{2}\right)}{j\omega\left[1 + 2j\zeta\left(\dfrac{\omega}{\omega_n}\right) - \left(\dfrac{\omega}{\omega_n}\right)^2\right]}$$

The transfer function comprises the following factors:

1. A zero at $\omega_c = 2$
2. A pole at $\omega_c = 0$
3. A quadratic pole (or, a pair of complex poles) at $\omega_n = 10$

Magnitude plot

1. The asymptote for zero at $\omega_c = 2$ is drawn with a slope of 20 dB/decade.
2. The asymptote for the pole at the origin extends from 0.1 to ∞. Passing through $\omega = 1$ at 0 dB, it has a slope of -20 dB/decade.
3. The asymptote for the quadratic pole at $\omega_n = 10$ has a slope of -40 dB/decade. Since $\zeta = 0.8$, which is > 0.707, it has no peak. Therefore, it is not necessary to determine peak frequency and peak magnitude and make corrections to the curve.
4. Individual contributions are added to draw the final plot.

Phase plot

1. The zero at $\omega_c = 2$ gives a straight line connecting 0° at $\omega = 0.2$ and 90° at $\omega = 20$.
2. The pole at the origin gives a horizontal line at 90°.
3. The asymptotic phase characteristics for the quadratic pole can be drawn by drawing a straight line connecting

$$0° \text{ at } \omega = \frac{\omega_n}{2}\log\left(\frac{2}{\zeta}\right) = \frac{10}{2}\log\left(\frac{2}{0.8}\right) = 1.99$$

and

$$-180° \text{ at } \omega = 2\omega_n \div \log\left(\frac{2}{0.8}\right) = 50.26$$

4. Individual contributions are added to obtain the resultant plot.

The plots are given in Figure 13.14. From the magnitude plot [Figure 13.14(a)], we find that the frequency corresponding to 0 dB gain is 1. Thus, the gain crossover frequency is 1 rad/s.

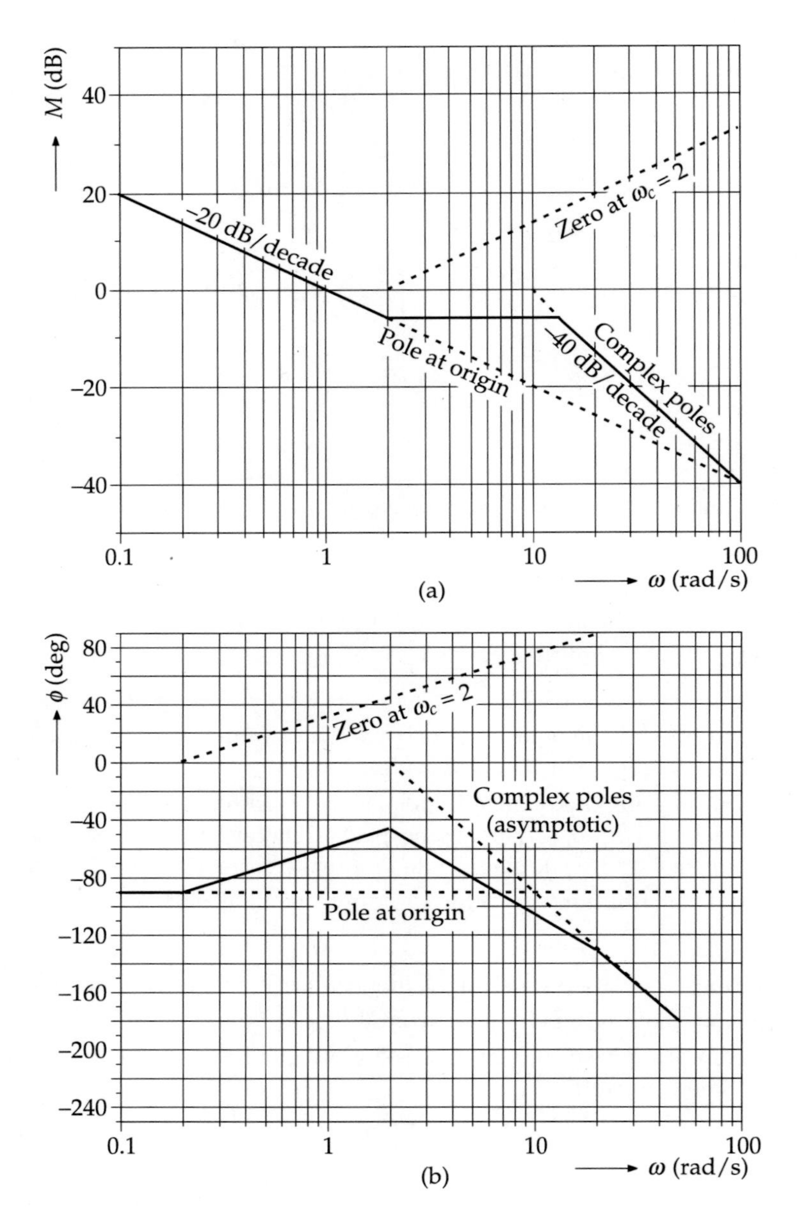

Figure 13.14 Bode plots (Example 13.7): (a) amplitude plot; (b) phase plot.

EXAMPLE 13.8. Sketch the asymptotic Bode plot for the open loop transfer function

$$G(s)H(s) = \frac{2(s+0.25)}{s^2(s+1)(s+0.5)}$$

Is the system stable?

Solution The transfer function can be written in Bode form as

$$G(s)H(s) = \frac{1+\dfrac{s}{0.25}}{s^2\left(1+\dfrac{s}{1}\right)\left(1+\dfrac{s}{0.5}\right)}$$

$$G(j\omega)H(j\omega) = \frac{1+\dfrac{j\omega}{0.25}}{(j\omega)^2\left(1+\dfrac{j\omega}{1}\right)\left(1+\dfrac{j\omega}{0.5}\right)}$$

The open-loop transfer function comprises the following components:

1. A zero at $\omega_c = 0.25$
2. Two poles at the origin
3. Two simple poles at $\omega_c = 0.5$ and 1

Magnitude plot

1. For the zero, a 20 decade/dB line is drawn at $\omega_c = 0.25$.
2. For two poles at the origin, a -40 dB/decade line that passes through $\omega = 1$ is drawn.
3. For two simple poles, two -20 decade/dB lines are drawn at $\omega_c = 0.5$ and 1.
4. Individual contributions are added to get the final curve.

Phase plot

1. For the zero at $\omega_c = 0.25$, a line between $0°$ at $\omega = 0.025$ and $90°$ at $\omega = 2.5$ is drawn.
2. For two poles at the origin, the phase is a horizontal line at $-180°$.
3. For the pole at $\omega_c = 0.5$, a line between $0°$ at $\omega = 0.05$ and $-90°$ at $\omega = 5$ is drawn. For the other pole a parallel line is drawn between $\omega = 0.1$ and $\omega = 10$.
4. Individual contributions are added to get the final curve.

From the asymptotic Bode plots (Figure 13.15), we find that $\omega_p = 0.2$ rad/s, whereas $\omega_g = 1.3$ rad/s. Here, gain margin $= -23$ dB and phase margin $= 180° - 210° = -30°$. Therefore, the system is unstable.

However, actual plots, done by the MatLab, show $\omega_g = 1.12$ rad/s, $\omega_p = 0.354$ rad/s, Gm $= -20.6$ dB and Pm $= -36.7°$. This variation in values occurs owing to the approximate nature of our asymptotic plots.

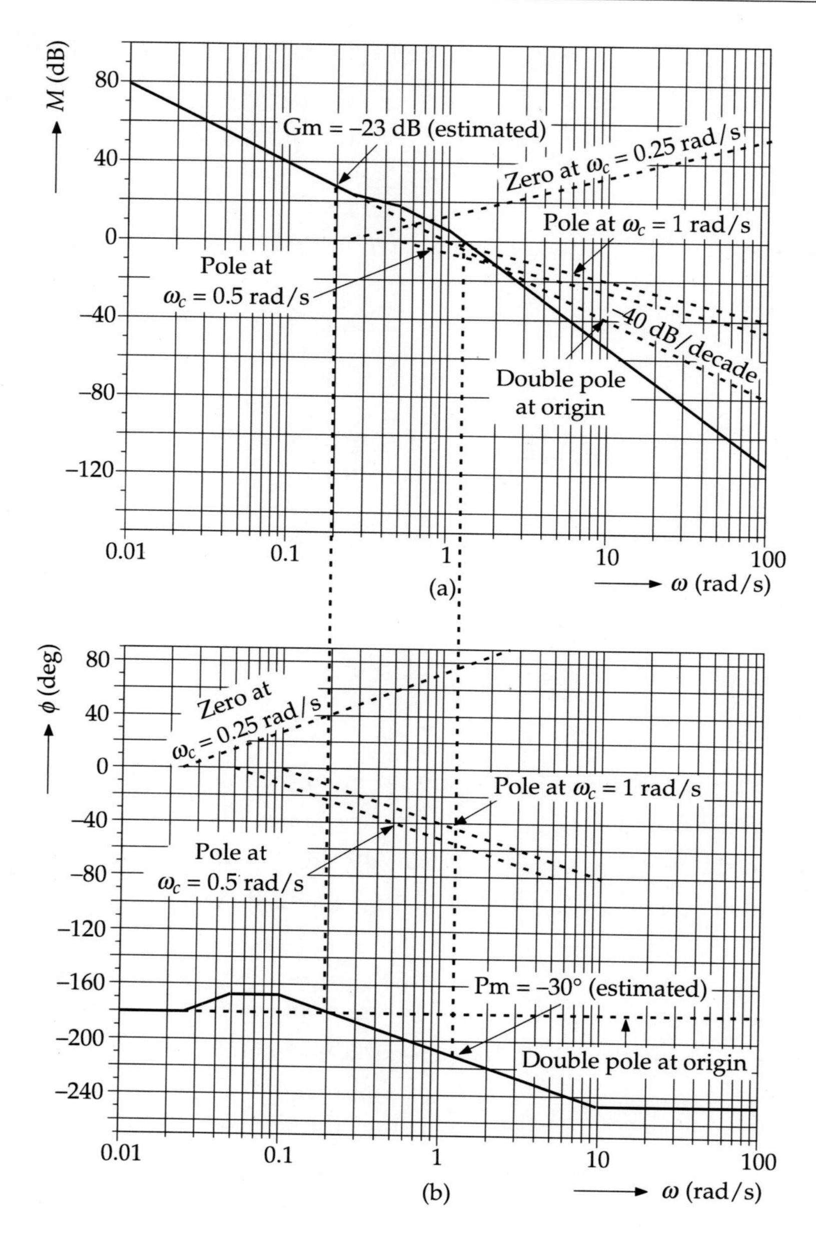

Figure 13.15 Bode plots: (a) magnitude plot; (b) phase plot (Example 13.8).

EXAMPLE 13.9. Sketch the Bode plot for a unity feedback system whose open-loop transfer function is given by

$$G(s) = \frac{500}{s(s+1)(s+50)}$$

From the Bode plot, determine the following:

(a) Gain and phase cross-over frequencies
(b) Gain and phase margin
(c) Stability of the closed loop system

Solution The transfer function can be written in the Bode form as

$$G(s)H(s) = \frac{10}{s\left(1+\frac{s}{1}\right)\left(1+\frac{s}{50}\right)}$$

$$G(\jmath\omega)H(\jmath\omega) = \frac{10}{\jmath\omega\left(1+\frac{\jmath\omega}{1}\right)\left(1+\frac{\jmath\omega}{50}\right)}$$

The transfer function comprises the following components:

1. A constant $K = 10$
2. A pole at the origin
3. Two simple poles at $\omega_c = 1$ and 50

Magnitude plot

1. For the constant, a horizontal line is drawn at 20 log 10 = 20 dB.
2. For the pole at the origin, a −20 dB/decade line that passes through $\omega = 1$ at 0 dB is drawn.
3. For two simple poles, two −20 decade/dB lines are drawn at $\omega_c = 1$ and 50.
4. Individual contributions are added to get the final curve.

Phase plot

1. For the constant, the phase is 0°.
2. For the pole at the origin, the phase is a horizontal line at −90°.
3. For the pole at $\omega_c = 1$, a line between 0° at $\omega = 0.1$ and −90° at $\omega = 10$ is drawn. For the other pole, a parallel line is drawn between $\omega = 5$ and $\omega = 500$.
4. Individual contributions are added to get the final curve.
   (a) From the asymptotic Bode plots (Fig 13.16), we find that $\omega_p = 9.0$ rad/s whereas $\omega_g = 3.1$ rad/s.
   (b) Here, gain margin = 19 dB and phase margin = 180° – 150° = 30°.
   (c) Therefore, the system is stable.

Note: Actual plots, drawn by the MatLab, show ω_g =3.08 rad/s, ω_p = 7.07 rad/s and Gm = 14.2 dB and Pm = 14.5°, which indicate a stable system. The variation in values occurs owing to the approximate nature of asymptotic plots.

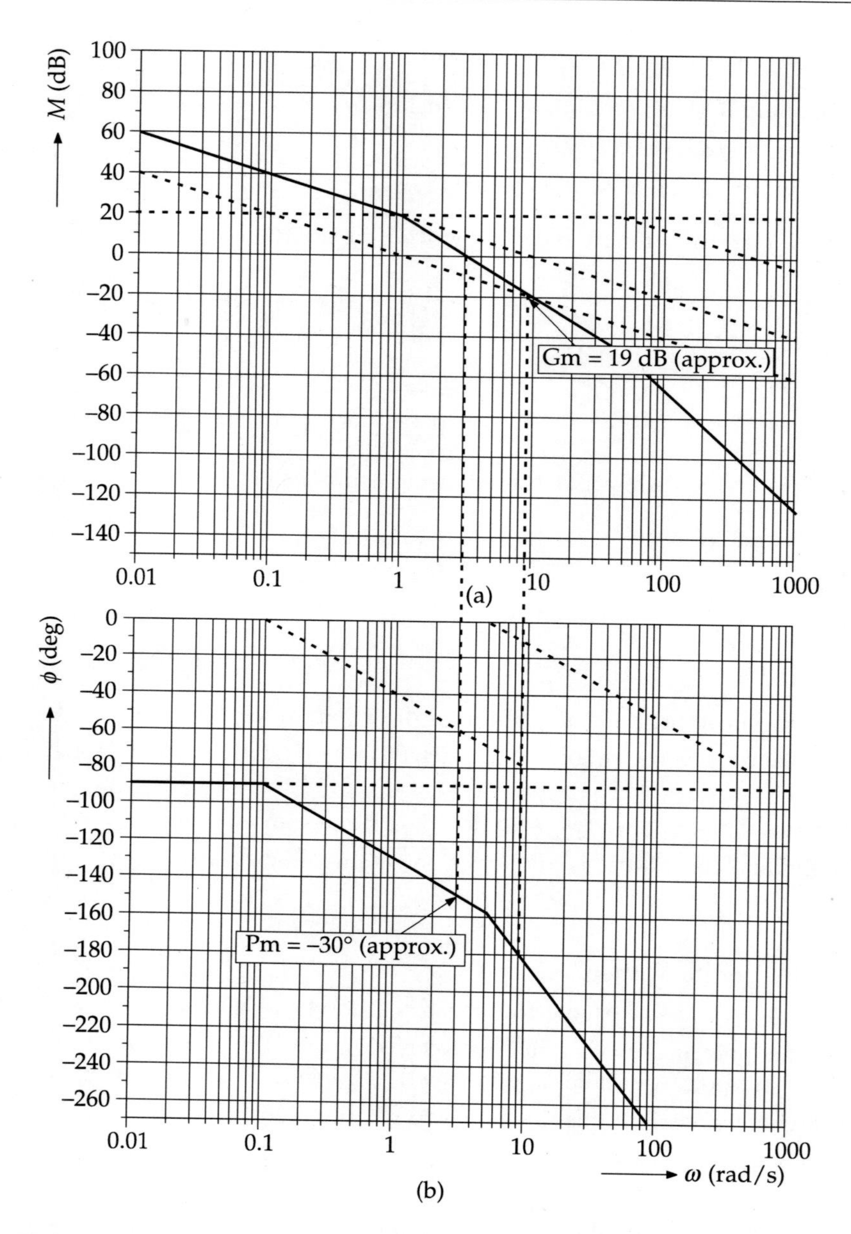

Figure 13.16 Bode plot: (a) magnitude plot; (b) phase plot (Example 13.9).

EXAMPLE 13.10. Draw the Bode plot of the open-loop transfer function

$$G(s) = \frac{50}{s(1 + 0.25s)(1 + 0.1s)}$$

From the graph determine

(a) Gain crossover frequency
(b) Phase crossover frequency
(c) Gain and phase margins
(d) Stability of the system

Solution The given open-loop transfer function can be written in the Bode form as

$$G(s) = \frac{50}{s\left(1+\frac{s}{4}\right)\left(1+\frac{s}{10}\right)}$$

$$G(j\omega) = \frac{50}{j\omega\left(1+\frac{j\omega}{4}\right)\left(1+\frac{j\omega}{10}\right)}$$

The transfer function consists of the following components:

1. A constant $K = 50$
2. A pole at the origin
3. Two simple poles at $\omega_c = 4$ and 10

Magnitude plot

1. For the constant, a horizontal line is drawn at $20\log 50 = 34$ dB.
2. For the pole at the origin, a -20 dB/decade line that passes through $\omega = 1$ at 0 dB is drawn.
3. For the two simple poles, two -20 dB/decade lines are drawn from $\omega_c = 4$ and 10.
4. Individual contributions are added to get the final magnitude plot.

Phase plot

1. For the pole at the origin, the phase is a horizontal line at $-90°$.
2. For the pole at $\omega_c = 4$ rad/s, a line is drawn connecting 0° at $\omega = 0.4$ rad/s and $-90°$ at $\omega = 40$ rad/s.
3. For the pole at $\omega_c = 10$ rad/s, a line is drawn connecting 0° at $\omega = 1$ rad/s and $-90°$ at $\omega = 100$ rad/s.
4. Individual contributions are added to get the final phase plot.

The Bode plot is presented in Figure 13.17. From the plot, we get the following data:

(a) $\omega_g = 12$ rad/s

(b) $\omega_p = 7.5$ rad/s

(c) Gm $= -10$ dB, Pm $= -20°$

(d) From the Gm and Pm values, it is obvious that the system is unstable.

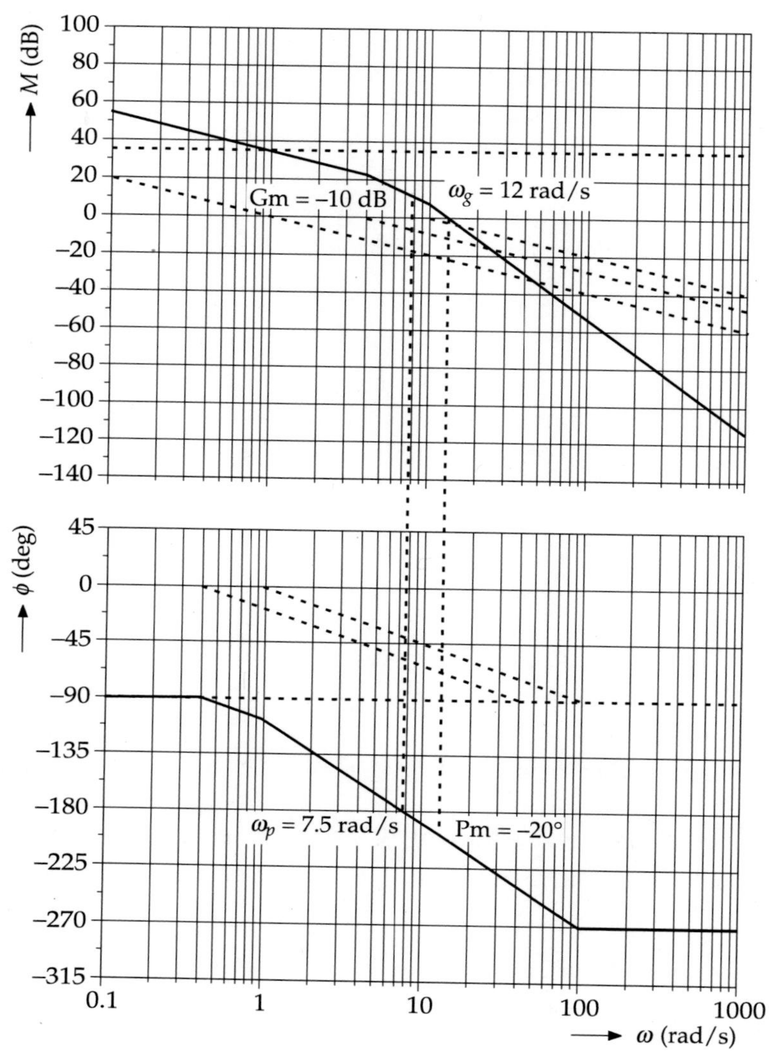

Figure 13.17 Bode plot (Example 13.10).

EXAMPLE 13.11. For the open-loop transfer function of a unity feedback system given by

$$G(s) = \frac{K(20+s)}{(s+1)(s+2)(s+10)}$$

construct the Bode plot for $K = 10$. Determine

(a) Gain crossover frequency
(b) Phase crossover frequency
(c) Gain margin
(d) Phase margin

Solution The given open-loop transfer function can be written in the Bode form as

$$G(s) = \frac{10(20)\left(1+\dfrac{s}{20}\right)}{(1+s)(2)\left(1+\dfrac{s}{2}\right)(10)\left(1+\dfrac{s}{10}\right)}$$

$$= \frac{10\left(1+\dfrac{s}{20}\right)}{(1+s)\left(1+\dfrac{s}{2}\right)\left(1+\dfrac{s}{10}\right)}$$

$$G(j\omega) = \frac{10\left(1+\dfrac{j\omega}{20}\right)}{(1+j\omega)\left(1+\dfrac{j\omega}{2}\right)\left(1+\dfrac{j\omega}{10}\right)}$$

The transfer function consists of the following components:

1. A constant $K = 10$
2. A zero at $\omega_c = 20$
3. Three simple poles at $\omega_c = 1,\ 2$ and 10.

Magnitude plot

1. For the constant, a horizontal line is drawn at $20 \log 10 = 20$ dB.
2. For the three simple poles, three -20 dB/decade lines are drawn from $\omega_c = 1, 2$ and 10.
3. Individual contributions are added to get the final magnitude plot.

Phase plot

1. For the zero at $\omega_c = 20$ rad/s, a line is drawn connecting $0°$ at $\omega = 2$ rad/s and $90°$ at $\omega = 200$ rad/s.
2. For the pole at $\omega_c = 1$ rad/s, a line is drawn connecting $0°$ at $\omega = 0.1$ rad/s and $-90°$ at $\omega = 10$ rad/s.
3. For the pole at $\omega_c = 2$ rad/s, a line is drawn connecting $0°$ at $\omega = 0.2$ rad/s and $-90°$ at $\omega = 20$ rad/s.
4. For the pole at $\omega_c = 10$ rad/s, a line is drawn connecting $0°$ at $\omega = 1$ rad/s and $-90°$ at $\omega = 100$ rad/s.
5. Individual contributions are added to get the final phase plot.

The Bode plot is presented in Figure 13.18. From the plot, we get the following data:

(a) $\omega_g = 4.5$ rad/s (b) $\omega_p = 10$ rad/s (c) Gm = 15 dB (d) Pm = $30°$

Since both Gm and Pm values are positive, the system is stable.

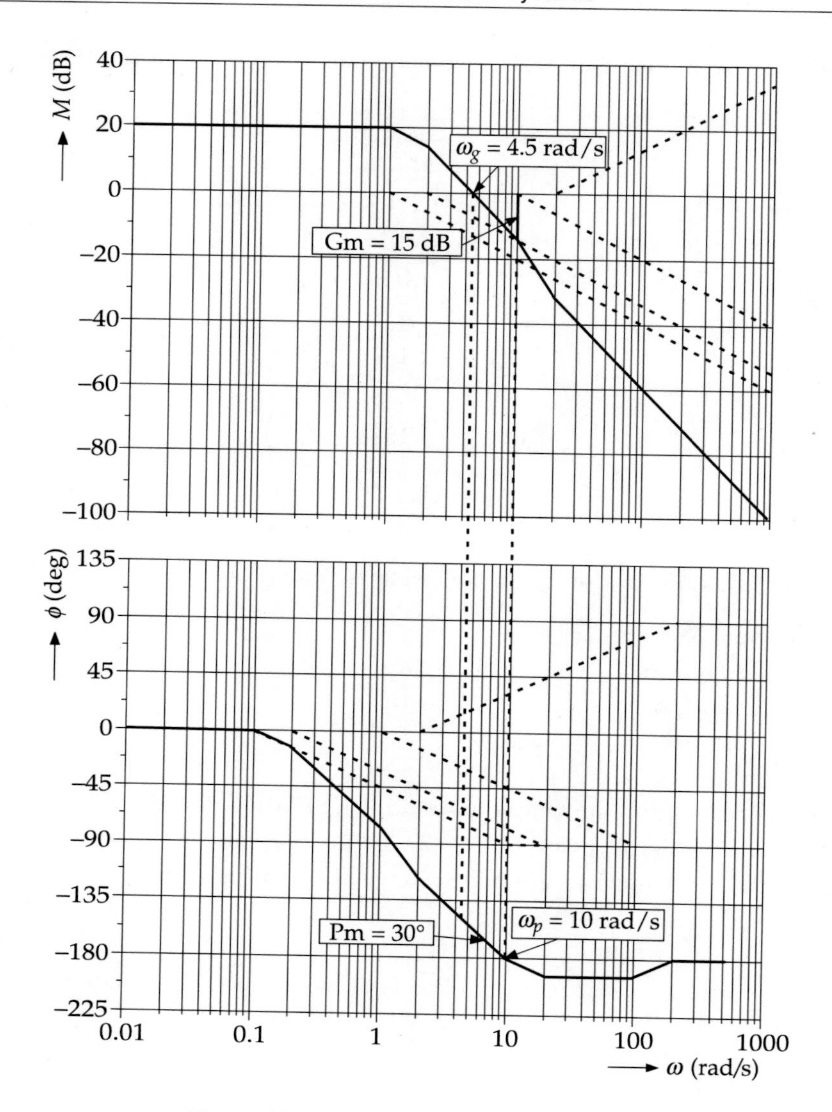

Figure 13.18 Bode plot (Example 13.11).

EXAMPLE 13.12. Construct the Bode plot for the unity feedback control system having

$$G(s) = \frac{10(s+10)}{s(s+2)(s+5)}$$

(a) From the plot, obtain gain margin, phase margin, gain crossover frequency, and phase crossover frequency

(b) Comment on the stability of the system

Solution The given open-loop transfer function can be written in the Bode form as

$$G(s) = \frac{10(10)\left(1+\dfrac{s}{10}\right)}{s(2)\left(1+\dfrac{s}{2}\right)(5)\left(1+\dfrac{s}{5}\right)} = \frac{10\left(1+\dfrac{s}{10}\right)}{s\left(1+\dfrac{s}{2}\right)\left(1+\dfrac{s}{5}\right)}$$

$$G(j\omega) = \frac{10\left(1+\dfrac{j\omega}{10}\right)}{j\omega\left(1+\dfrac{j\omega}{2}\right)\left(1+\dfrac{j\omega}{5}\right)}$$

The transfer function consists of the following components:

1. A constant $K = 10$
2. A zero at $\omega_c = 10$
3. A pole at the origin
4. Two simple poles at $\omega_c = 2$ and 5

Magnitude plot

1. For the constant, a horizontal line is drawn at $20 \log 10 = 20$ dB.
2. For the pole at the origin, a -20 dB/decade line that passes through $\omega = 1$ at 0 dB is drawn.
3. For the two simple poles, two -20 dB/decade lines are drawn from $\omega_c = 2$ and 5.
4. Individual contributions are added to get the final magnitude plot.

Phase plot

1. For the zero at $\omega_c = 10$ rad/s, a line is drawn connecting 0° at $\omega = 1$ rad/s and 90° at $\omega = 100$ rad/s.
2. For the pole at the origin, the phase is a horizontal line at $-90°$.
3. For the pole at $\omega_c = 2$ rad/s, a line is drawn connecting 0° at $\omega = 0.2$ rad/s and $-90°$ at $\omega = 20$ rad/s.
4. For the pole at $\omega_c = 5$ rad/s, a line is drawn connecting 0° at $\omega = 0.5$ rad/s and $-90°$ at $\omega = 50$ rad/s.
5. Individual contributions are added to get the final phase plot.

The Bode plot is presented in Figure 13.19. From the plot, we get the following data:

(a) Gm = 10 dB; Pm = 10°; $\omega_g = 4.5$ rad/s; $\omega_p = 7$ rad/s

(b) Since both Gm and Pm values are positive, the system is stable.

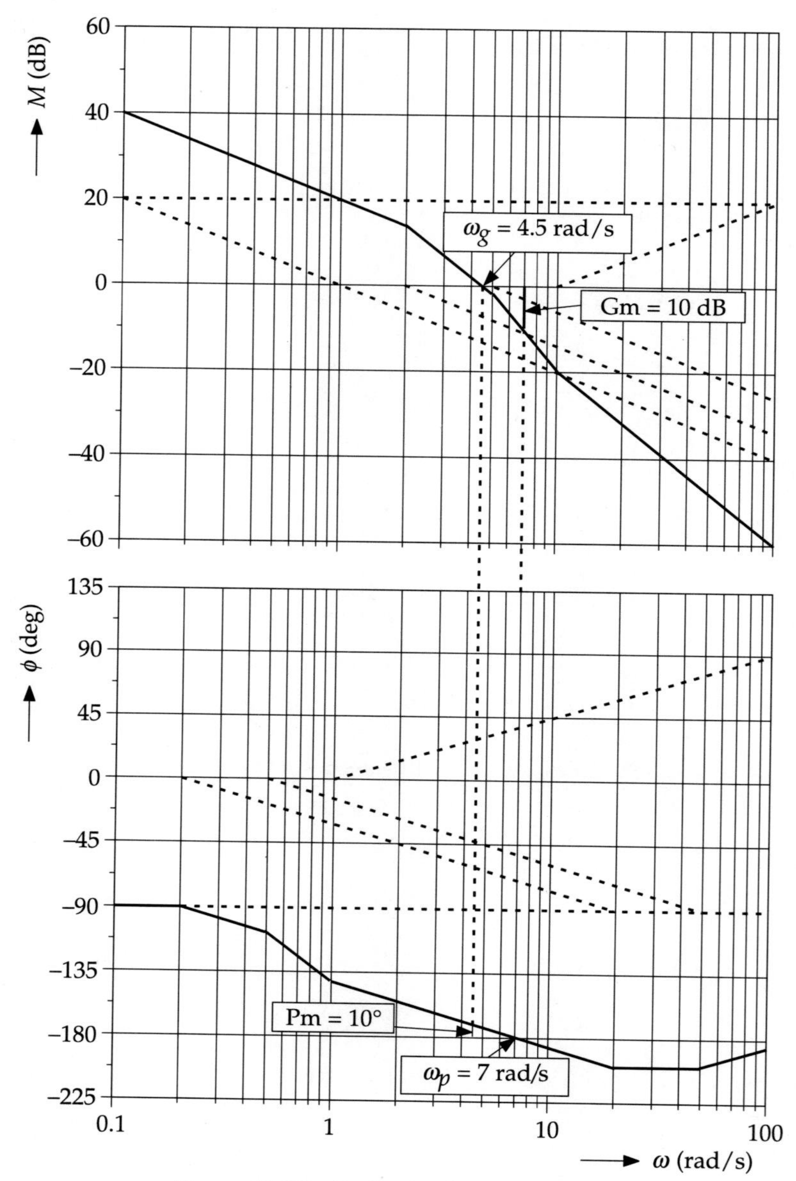

Figure 13.19 Bode plot (Example 13.12).

EXAMPLE 13.13. Sketch the plot showing the magnitude in dBs and phase angle in degrees as a function of frequency in logarithm scale for the transfer function given by

$$G(s) = \frac{10}{s(1+0.5s)(1+0.1s)}$$

Hence determine the gain margin and phase margin of the system. Comment on the stability of the system.

Solution The given open-loop transfer function can be written in the Bode form as

$$G(s) = \frac{10}{s\left(1+\frac{s}{2}\right)\left(1+\frac{s}{10}\right)}$$

$$G(\jmath\omega) = \frac{10}{\jmath\omega\left(1+\frac{\jmath\omega}{2}\right)\left(1+\frac{\jmath\omega}{10}\right)}$$

The transfer function consists of the following components:

1. A constant $K = 10$
2. A pole at the origin
3. Two simple poles at $\omega_c = 2$ and 10

Magnitude plot

1. For the constant, a horizontal line is drawn at $20\log 10 = 20$ dB.
2. For the pole at the origin, a -20 dB/decade line that passes through $\omega = 1$ at 0 dB is drawn.
3. For the two simple poles, two -20 dB/decade lines are drawn from $\omega_c = 2$ and 10.
4. Individual contributions are added to get the final magnitude plot.

Phase plot

1. For the pole at the origin, the phase is a horizontal line at $-90°$.
2. For the pole at $\omega_c = 2$ rad/s, a line is drawn connecting $0°$ at $\omega = 0.2$ rad/s and $-90°$ at $\omega = 20$ rad/s.
3. For the pole at $\omega_c = 10$ rad/s, a line is drawn connecting $0°$ at $\omega = 1$ rad/s and $-90°$ at $\omega = 100$ rad/s.
4. Individual contributions are added to get the final phase plot.

The Bode plot is presented in Figure 13.20. From the plot, we get the following data:

$$\text{Gm} = \text{Pm} \approx 0 \qquad [\because \omega_g \approx \omega_p]$$

Since both Gm and Pm values are zeros, the system is marginally stable.

Note: The MatLab shows a stable system with Gm $= 1.58$ dB at $\omega_p = 4.47$ rad/s and Pm $= 10.7°$ at $\omega_g = 4.08$ rad/s. Our crossover frequencies are pretty close, but owing to the approximate plots, we have arrived at a wrong conclusion about the stability of the system.

Normally, the asymptotic plots have large errors near corner frequencies which may invert the conclusion. To investigate that, we calculated the actual magnitude and phase for a few frequencies from the relations

$$M = 20\log\left(\frac{10}{\omega\sqrt{1+0.25\omega^2}\sqrt{1+0.01\omega^2}}\right)$$

$$\phi = -90° - \tan^{-1} 0.5\omega - \tan^{-1} 0.1\omega$$

The results obtained are as follows:

ω (rad/s)	M (dB)	ϕ (deg)	ω (rad/s)	M (dB)	ϕ (deg)
0.1	20	−93	20	−33.15	−238
0.3	30.36	−100	30	−43.1	−248
0.5	25.75	−107	40	−50.45	−253
1.0	18.98	−122	50	−55.9	−256
1.5	5.27	−135	60	−60.8	−259
2	10.8	−146	70	−64.8	−260
5	−3.6	−185	80	−68.2	−261
10	−17.1	−214	100	−74.0	−263

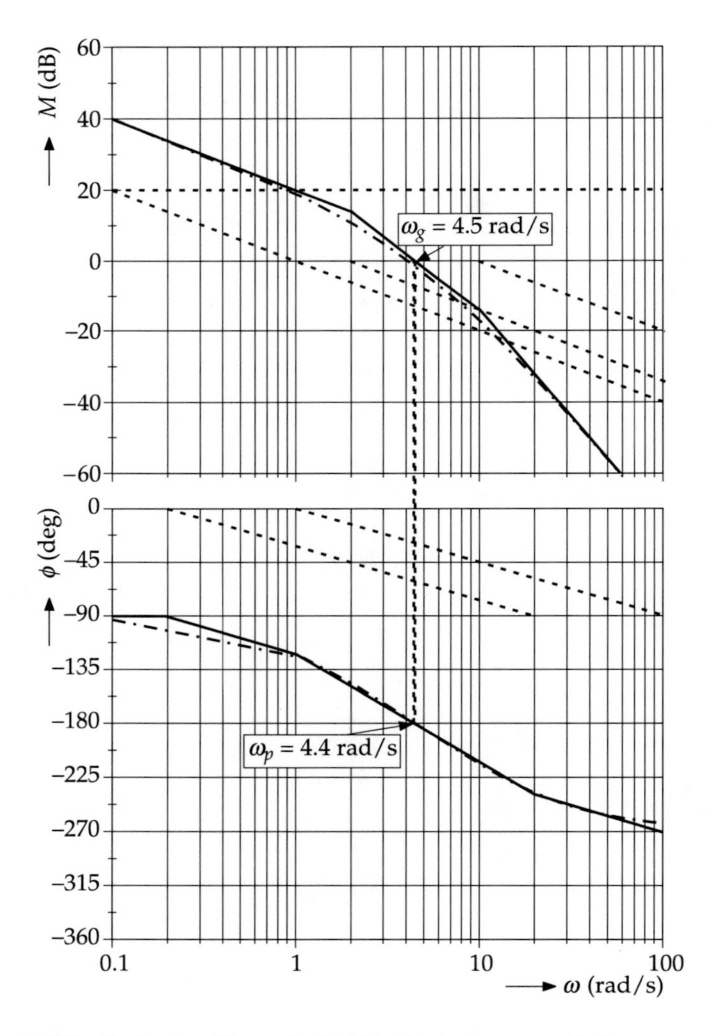

Figure 13.20 Bode plot (Example 13.13). Dash-dot curves indicate exact plots.

The exact curves obtained from these values are shown by dash-dot curves in Figure 13.20. It can be seen from the figure that a small change in the vital area of the magnitude plot has made $\omega_g > \omega_p$ in the asymptotic plots, whereas the situation is opposite in the exact plots.

EXAMPLE 13.14. The closed-loop transfer function of a position control system is given by

$$T(s) = \frac{9}{s^2 + 1.2s + 9}$$

(a) Draw the Bode plot of the system.

(b) Find out the resonant frequency ω_r and resonant peak M_r from the plot. Also find out the phase lag at ω_r.

(c) From the plot find out the 3 dB bandwidth of the system.

Solution

(a) We note that this is a closed-loop transfer function though Bode plots are made for open-loop transfer function. But here, instead of finding the stability of the system, we are asked to find some frequency domain parameters for the system through the Bode-type plots. So, we go for exact magnitude and phase plots in the Bode scheme and do not look for asymptotic plots because the denominator consists of complex poles. To do that, we have to write the transfer function in magnitude and phase form as

$$T(s) = \frac{9}{-\omega^2 + 1.2\jmath\omega + 9}$$

$$= \left|\frac{9}{\sqrt{(9-\omega^2)^2 + (1.2\omega)^2}}\right| \angle -\tan^{-1}\frac{1.2\omega}{9-\omega^2}$$

or

$$M \equiv 20\log\left|\frac{9}{\sqrt{(9-\omega^2)^2 + (1.2\omega)^2}}\right| \text{ dB}$$

and

$$\phi \equiv -\left(\frac{180}{\pi}\cdot\tan^{-1}\frac{1.2\omega}{9-\omega^2}\right) \text{ degree}$$

Now, we have to calculate magnitude in dB and phase in degrees. By comparison with the standard second order transfer function, we note that the undamped natural frequency of the system is 3 rad/s. So, the resonant frequency will be close to it and, therefore, while tabulating data, we have to find magnitude and phase at close intervals near 3 rad/s. The calculated magnitude and phase are given in Table 13.6. These values are presented in Bode plot form in Figure 13.21.

(b) From the magnitude and phase plot, we get

1. Resonant frequency = 2.9 rad/s
2. Resonant peak = 8.1 dB
3. Phase lag at $\omega_r = -80°$

(c) From the magnitude plot, we find that the -3 dB bandwidth is 4.53 rad/s.

Table 13.6 Calculated Values of Magnitude and Phase (Illustration of Example 13.14)

ω (rad/s)	M (dB)	ϕ (degree)	ω (rad/s)	M (dB)	ϕ (degree)
1.0	0.9268	−8.5308	3.0	7.9588	−90.0
2.0	4.2048	−25.641	3.1	7.5589	−99.3124
2.5	6.8937	−47.4896	3.2	7.2459	−107.896
2.6	7.3962	−54.3236	3.3	6.2393	−115.514
2.7	7.8067	−62.1759	3.5	4.5819	−127.733
2.8	8.0689	−70.7534	4.0	0.5094	−145.561
2.85	8.1274	−75.6095	5.0	−5.5687	−159.444
2.9	8.1301	−80.3776			

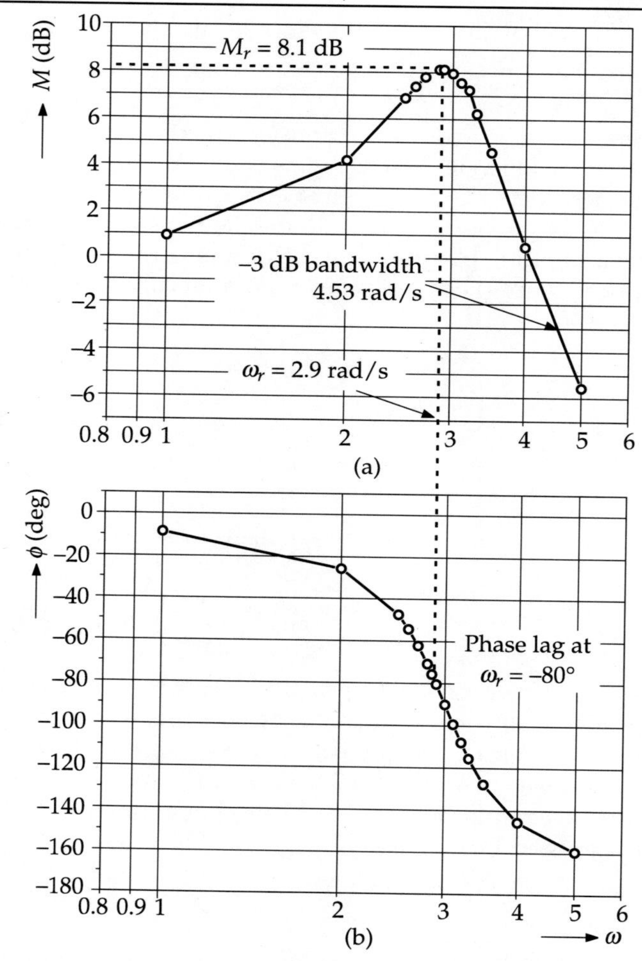

Figure 13.21 Bode plot of (a) magnitude; (b) phase (Example 13.14).

13.6 BODE PLOT FOR SYSTEMS WITH TRANSPORTATION LAG

We have seen in Eq. (5.64) that the transfer function for transportation lag can be written as

$$G(s) = e^{-sT}$$

where T is the time delay in seconds. In the frequency domain, this transfer function becomes

$$G(j\omega) = e^{-j\omega T} \tag{13.14}$$

Putting Eq. (13.14) in magnitude and phase form, we obtain

$$\begin{aligned} G(j\omega) &= \cos\omega T - j\sin\omega T \\ &= \left|\sqrt{\cos^2\omega T + \sin^2\omega T}\right| \angle -\tan^{-1}\frac{\sin\omega T}{\cos\omega T} \\ &= |1| \angle -\omega T \end{aligned}$$

In dB, the magnitude turns out to be 0 while the phase angle, which varies linearly with the frequency ω for a given T, is in degrees

$$\phi = -57.3\,\omega T \text{ degree}$$

EXAMPLE 13.15. Draw the Bode plot for

$$G(s)H(s) = \frac{e^{-0.4s}}{s(s+2)}$$

and find gain and phase margins.

Solution Written in the Bode form, the open-loop transfer function is

$$G(j\omega)H(j\omega) = \frac{0.5e^{-j0.4\omega}}{j\omega\left(1 + \dfrac{j\omega}{2}\right)}$$

We observe that the transfer function comprises the following components:

1. A constant = 0.5
2. A transportation lag of 0.4 s
3. A pole at the origin
4. A pole at $\omega_c = 2$.

Magnitude plot

1. For the constant, a horizontal line is drawn at 20 log 0.5 = −6 dB.
2. The transportation lag has no contribution to magnitude.
3. For the pole at the origin, a −20 dB/decade line that passes through $\omega = 1$ at 0 dB is drawn.
4. For the simple pole, −20 dB/decade line is drawn at $\omega_c = 2$.
5. Individual contributions are added to get the final curve.

Phase plot

Because of the transportation lag contribution, it is inconvenient to add individual contributions of different components in the asymptotic plot. Therefore, we calculate the phase angle from the equation

$$\begin{aligned}\angle e^{-j0.4\omega} &= -57.3(0.4\omega)\\ &= -22.92\ \omega\end{aligned}$$

$$\begin{aligned}\phi &= \angle G(j\omega)H(j\omega)\\ &= -90° - \tan^{-1}\frac{\omega}{2} - 22.92\ \omega\end{aligned}$$

and then draw the actual Bode phase diagram from the values given in Table 13.7.

Table 13.7 Data for Actual Bode Phase Plot (Example 13.7)

ω (rad/s)	$\angle j\omega$ (deg)	$\angle(1 + j\omega/2) = \tan^{-1}(\omega/2)$ (deg)	$\angle e^{-j0.4\omega} = -22.92\omega$ (deg)	ϕ (deg)
0.1	−90	−2.862	−2.292	−95.154
0.2	−90	−5.711	−4.584	−100.295
0.5	−90	−14.036	−11.460	−115.496
1.0	−90	−26.565	−22.920	−139.485
2.0	−90	−45.0	−45.84	−180.84
5.0	−90	−68.199	−114.6	−272.799
10.0	−90	−78.690	−229.2	−397.89

From the asymptotic plot of magnitude [Figure 13.22(a)] and actual plot of phase [Figure 13.22(b)], we find that $\omega_g = 5$ rad/s, $\omega_p = 2$ rad/s, Gm = 14 dB and Pm = (180° − 118°) = 62°. Therefore, the system is stable.

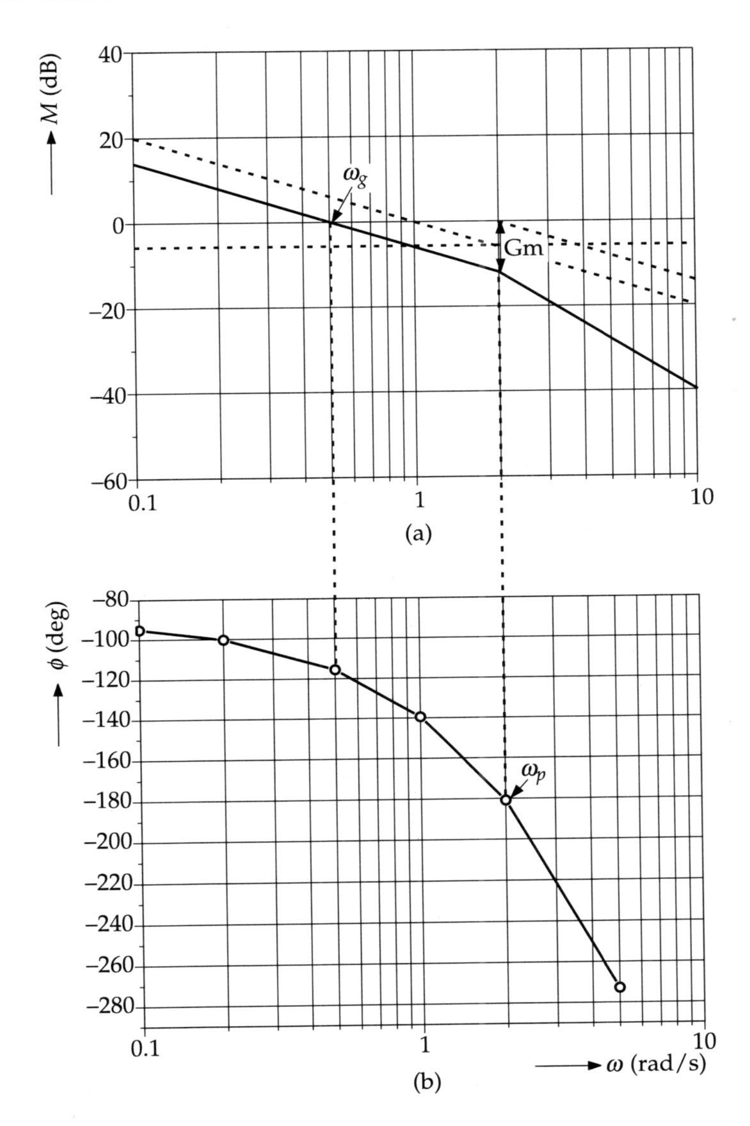

Figure 13.22 Bode plot: (a) asymptotic magnitude plot; (b) actual phase plot (Example 13.15).

13.7 TRANSFER FUNCTION FROM BODE PLOT

If the system is of minimum phase type[9], the transfer function of a system can be derived from the Bode magnitude plot by following the inverse procedure. We consider a few examples to see how it is done.

[9]See Section 3.2 at page 107.

EXAMPLE 13.16. The Bode asymptotic magnitude plot of a transfer function is shown in the following diagram.

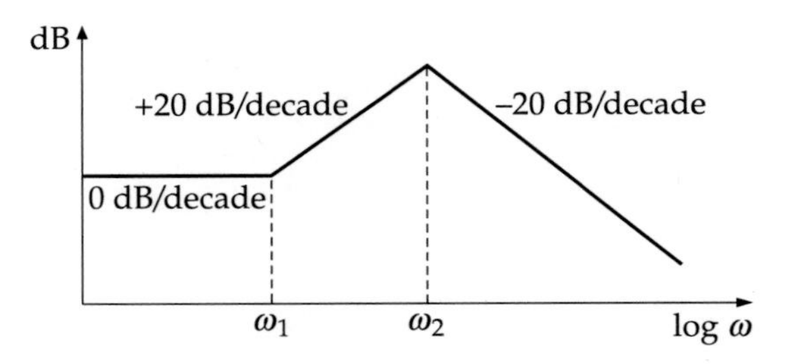

In the frequency range shown, the transfer function has

(a) 3 poles and 1 zero (b) 1 pole and 2 zeros
(c) 2 poles and 1 zero (d) 2 poles and 2 zeros

Solution Since the origin of the graph is not indicated and that the initial portion of the plot shows a horizontal straight line, it may be the resultant of a simple pole and a simple zero or caused by a constant gain. The second part is created by a simple zero. The third part, which reverses the slope, must have been caused by the existence of two poles or a quadratic pole. So, the nearest answer is (a), assuming that the initial part is caused by a constant and that there is no quadratic pole in the third part.

EXAMPLE 13.17. The asymptotic approximation of the log-magnitude vs. frequency plot of a system containing only real poles and zeros is shown in the following diagram

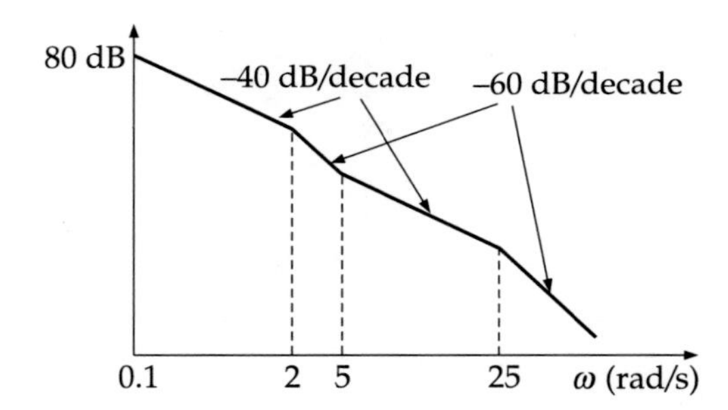

Its transfer function is

(a) $\dfrac{10(s+5)}{s(s+2)(s+25)}$ (b) $\dfrac{1000(s+5)}{s^2(s+2)(s+25)}$

(c) $\dfrac{100(s+5)}{s(s+2)(s+25)}$ (d) $\dfrac{80(s+5)}{s^2(s+2)(s+25)}$

Solution The initial slope of -40 dB/decade suggests the existence of two simple poles at the origin. The slope gets enhanced to -60 dB/decade at the corner frequency 2 rad/s, which suggests the existence of another simple pole there. The slope improves to -40 dB/decade at the corner frequency 5 rad/s owing obviously to the existence of a zero there. Finally, at the corner frequency 25 rad/s, the slope again becomes −40 dB/decade as a consequence of the existence of another simple pole there. So, let our open-loop transfer function be

$$G(s)H(s) = \frac{K(s+5)}{s^2(s+2)(s+25)}$$

or
$$G(\jmath\omega)H(\jmath\omega) = \frac{K(\jmath\omega+5)}{(\jmath\omega)^2(\jmath\omega+2)(\jmath\omega+25)}$$

$$\equiv \frac{K\sqrt{\omega^2+25}}{\omega^2\sqrt{\omega^2+4}\sqrt{\omega^2+625}}\angle\left(\tan^{-1}\frac{\omega}{5} - 180^\circ - \tan^{-1}\frac{\omega}{2} - \tan^{-1}\frac{\omega}{25}\right)$$

From the given figure, we observe that the magnitude is 80 dB at $\omega = 0.1$ rad/s. Therefore, we have

$$20\log\left.\frac{K\sqrt{\omega^2+25}}{\omega^2\sqrt{\omega^2+4}\sqrt{\omega^2+625}}\right|_{\omega=0.1} = 80$$

or
$$\log\frac{K\sqrt{0.01+25}}{0.01\sqrt{0.01+4}\sqrt{0.01+625}} = 4$$

or
$$\log\frac{5K}{0.01\times 2\times 25} = 4 \qquad [\because 4,\ 25,\ 625 \gg 0.01]$$

or
$$\frac{K}{0.1} = \text{antilog } 4 = 10000$$

or
$$K = 1000$$

Ans: (b).

EXAMPLE 13.18. The asymptotic Bode plot for the gain magnitude of a minimum phase system $G(s)$ is shown in the following figure:

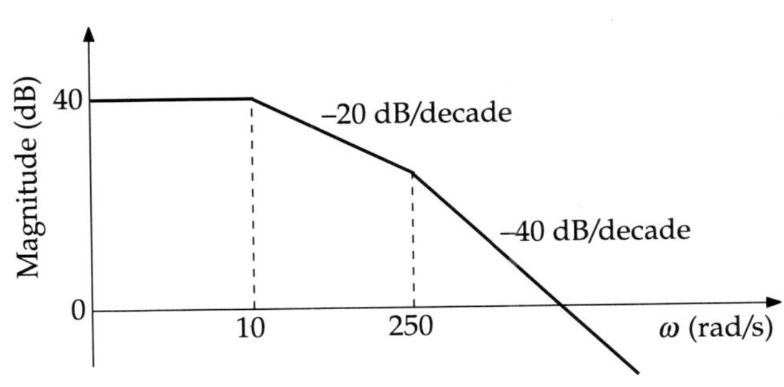

The transfer function $G(s)$ is given by

(a) $\dfrac{100}{[1+(s/10)][1+(s/250)]}$

(b) $\dfrac{40}{s(s+250)}$

(c) $\dfrac{100}{(s+10)(s+250)}$

(d) $\dfrac{100s}{(s+10)(s+250)}$

Solution From the given magnitude plot we get the following data: There are two simple poles at corner frequencies 10 and 250 rad/s, and there is a gain factor K which makes the initial part of the plot a straight line parallel to the ω-axis. To find out the value of K, we have

$$20 \log K = 40 \quad \Rightarrow \quad \log K = 2 \quad \Rightarrow \quad K = 100$$

So,

$$G(s) = \frac{100}{\left(1 + \dfrac{s}{10}\right)\left(1 + \dfrac{s}{250}\right)}$$

Ans: (a).

EXAMPLE 13.19. The asymptotic Bode magnitude plot of an open-loop system $G(S)$ with $K > 0$ and all poles and zeros on the left-hand side of the s-plane is shown in the following figure. It is completely symmetric about ω_g.

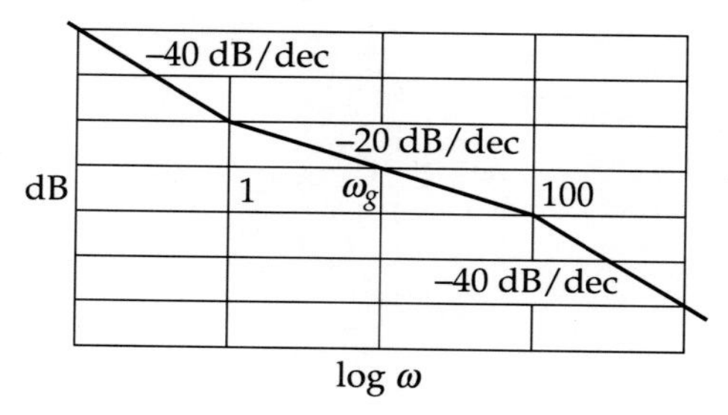

The maximum absolute phase angle contribution by $G(s)$ is given by

(a) $-78.6°$ (b) $-90°$ (c) $-101.4°$ (d) $-180°$

Solution We note that the initial slope of the magnitude plot is -40 dB/decade. This indicates the existence of a double pole at the origin. Then, at the corner frequency 1, the slope becomes -20 dB/decade indicating the presence of a zero there. Thereafter, at the corner frequency 100, the slope again becomes -40 dB/decade which suggests the presence of a pole at that frequency. Since we need to determine the phase, we need not bother about the constant factor because a constant contributes nothing to the phase. Our transfer function is, therefore,

$$G(s) = \frac{K(1+s)}{s^2\left(1 + \dfrac{s}{100}\right)}$$

In the frequency domain, its phase factor ϕ is given by

$$\phi = \tan^{-1}\omega - 180° - \tan^{-1}\frac{\omega}{100}$$

The phase of a transfer function , where a zero is symmetrically sandwiched between two poles, will obviously have a bell shape in the middle, i.e. at ω_g. The phase will have a maximum there. Since $\omega_g = 10$ (as can be seen from the plot), the maximum value of phase is

$$\phi_{\max} = \tan^{-1} 10 - 180° - \tan^{-1} 0.1 = -101.4°$$

Ans: (c).

EXAMPLE 13.20. The approximate Bode magnitude plot of a lead network with its pole and zero on the left half of the s-plane is shown in the following figure.

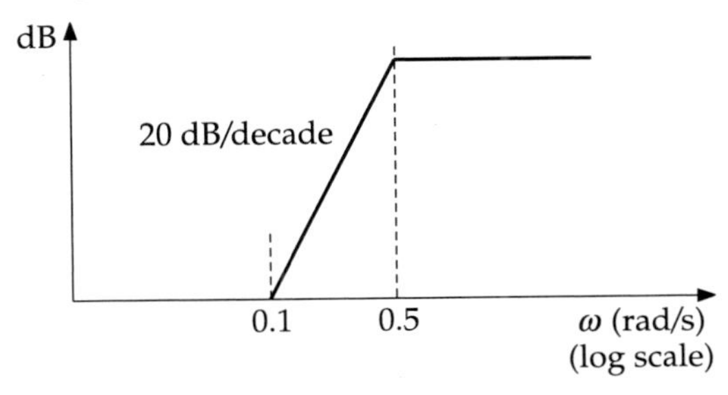

The frequency at which the phase angle of the network is maximum (in rad/s) is

(a) $\dfrac{3}{\sqrt{10}}$ (b) $\dfrac{1}{\sqrt{20}}$ (c) $\dfrac{1}{20}$ (d) $\dfrac{1}{30}$

Solution The shape of the magnitude curve suggests a zero at the corner frequency 0.1 rad/s and a pole at the corner frequency 0.5 rad/s. Hence, let the transfer function be given by

$$G(s)H(s) = \frac{K\left(1 + \dfrac{s}{0.1}\right)}{1 + \dfrac{s}{0.5}} = \frac{K(0.1 + s)}{0.5 + s}$$

The phase factor of this function in the frequency domain is given by

$$\phi = \tan^{-1}\frac{\omega}{0.1} - \tan^{-1}\frac{\omega}{0.5} = \tan^{-1} 10\omega - \tan^{-1} 2\omega$$

To maximize phase, it is necessary that

$$\frac{d\phi}{d\omega} = \frac{d}{d\omega}\left(\tan^{-1} 10\omega\right) - \frac{d}{d\omega}\left(\tan^{-1} 2\omega\right) = 0$$

or

$$\frac{10}{1 + 100\omega^2} - \frac{2}{1 + 4\omega^2} = 0$$

or

$$\frac{5}{1 + 100\omega^2} = \frac{1}{1 + 4\omega^2}$$

or

$$1 + 100\omega^2 = 5 + 20\omega^2$$

or

$$80\omega^2 = 4$$

or

$$\omega = \frac{1}{\sqrt{20}}$$

Ans: (b).

EXAMPLE 13.21. The asymptotic approximation to the log-magnitude versus frequency plot (Bode plot) of a unity feedback control system is shown in Figure 13.23. The system is a minimum phase system.

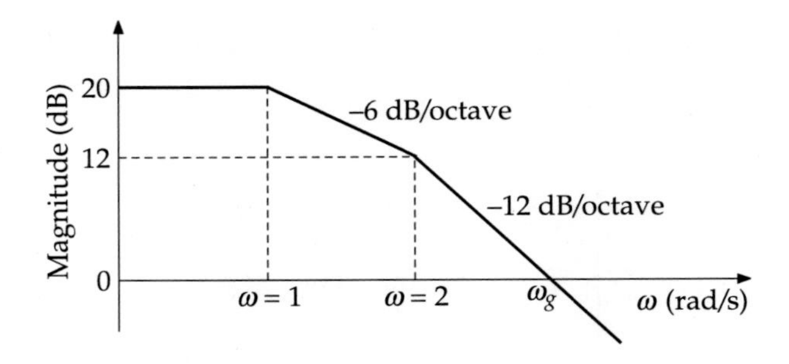

Figure 13.23 Log-magnitude versus frequency plot (Example 13.21).

Determine

(a) Gain crossover frequency, in rad/s

(b) Gain margin, in dB

(c) Phase margin, in degrees

Solution Since the plot begins with a constant magnitude of 20 dB, there exists a dc gain of anitlog (20/20) = 10 in the transfer function of the system. -6 dB/octave at $\omega = 1$ and -12 dB/octave at $\omega = 2$ indicate the existence of two simple poles at corner frequencies $\omega = 1$ and $\omega = 2$. Therefore, the open-loop transfer function is given by

$$G(s)H(s) = \frac{10}{(1+s)[1+(s/2)]}$$

or

$$G(\jmath\omega)H(\jmath\omega) = \frac{10}{(1+\jmath\omega)[1+(\jmath\omega/2)]}$$

$$\equiv \frac{10}{\sqrt{1+\omega^2}\sqrt{1+(\omega^2/4)}}\angle\left(-\tan^{-1}\omega - \tan^{-1}\frac{\omega}{2}\right)$$

(a) To find the gain crossover frequency ω_g, we equate the magnitude part to 0 dB. Thus,

$$20\log\frac{10}{\sqrt{1+\omega_g^2}\sqrt{1+(\omega_g^2/4)}} = 20\log 1$$

or

$$\sqrt{1+\omega_g^2}\sqrt{1+(\omega_g^2/4)} \doteq 10$$

or

$$(1+\omega^2)\left(1+\frac{\omega^2}{4}\right) = 100$$

or

$$4 + 5\omega_g^2 + \omega_g^4 = 400 \qquad \text{(i)}$$

Equation (i) yields two values for ω_g^2, namely, 17.56 and -22.56. The latter value is discarded because the square of a frequency cannot be negative. So, $\omega_g = \sqrt{17.56}$ $= 4.19$ rad/s.

(b) To determine the gain margin, we need to know the phase crossover frequency ω_p. We do that by equating the phase factor to $-180°$. Thus,

$$-\tan^{-1}\omega_p - \tan^{-1}\frac{\omega_p}{2} = -180°$$

or

$$\tan^{-1}\frac{\omega_p + (\omega_p/2)}{1 - \omega_p \cdot (\omega_p/2)} = 180°$$

or

$$\frac{(3/2)\omega_p}{1 - (\omega_p^2/2)} = \tan 180° = 0 \qquad \text{(ii)}$$

The RHS of Eq. (ii) becomes 0 if $\omega_p = 0$ or $\omega_p = \infty$. The former value, being impractical, is rejected. So, $\omega_p = \infty$ is accepted. This frequency is substituted in the magnitude factor to find the gain and then we need to convert its inverse to dB to find the gain margin. Now,

$$|M| = \left.\frac{10}{\sqrt{1+\omega_p^2}\sqrt{1+(\omega_p^2/4)}}\right|_{\omega_p=\infty} = \frac{1}{\infty} = 0$$

Therefore,

$$\frac{1}{|M|} = \infty$$

Thus,

$$\text{Gm} = 20\log\infty = \infty$$

(c) The phase margin is calculated as

$$\phi\Big|_{\omega_g=4.19} = -\tan^{-1}4.19 - \tan^{-1}\frac{4.19}{2} = -141.06°$$

Therefore,

$$\text{Pm} = 180° - 141.06° = 38.9°$$

EXAMPLE 13.22. Figure 13.24 shows the Bode magnitude and phase plots of a system (not to scale). Determine from the plot the transfer function of the system.

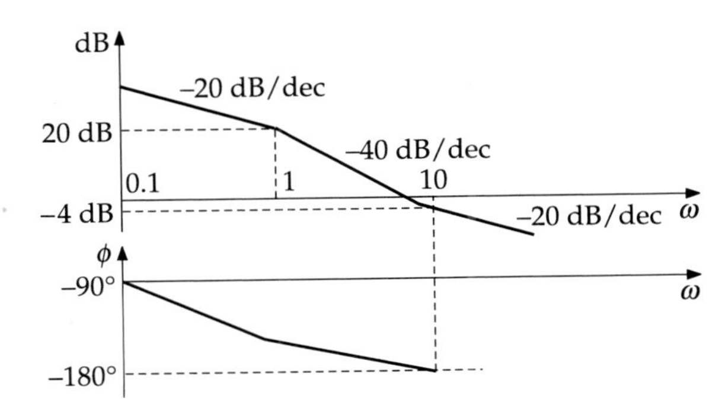

Figure 13.24 Bode magnitude and phase plots of a system (Example 13.22).

Solution The magnitude plot starts with a -20 dB/decade slope. The slope increases to -40 dB/decade at the corner frequency 1 and then again decreases to -20 dB/decade at the unknown corner frequency x. Obviously, a zero is present at the corner frequency x. So, let us write the open-loop transfer function as

$$G(s)H(s) = \frac{K\left(1 + \dfrac{s}{x}\right)}{s\left(1 + \dfrac{s}{1}\right)}$$

Since the value of the magnitude is 20 dB at $\omega = 1$ rad/s, we have

$$20 \log K = 20$$

or

$$\log K = 1$$

or

$$K = 10$$

The plot indicates that the phase crossover frequency line intersects the magnitude plot at -4 dB. Hence, Gm $= 4$ dB at $\omega_p = 10$ rad/s. Therefore,

$$\text{Gm} = 4 \text{ dB} = -20 \log |G(\jmath\omega)H(\jmath\omega)|\Big|_{\omega=10}$$

$$= -20 \log \left.\frac{10\sqrt{1 + \omega^2/x^2}}{\omega\sqrt{1+\omega^2}}\right|_{\omega=10}$$

or

$$\text{antilog}\left(-\frac{4}{20}\right) = \frac{10\sqrt{1 + 100/x^2}}{10\sqrt{1+100}}$$

or

$$0.6310 = \sqrt{\frac{1 + 100/x^2}{101}}$$

or

$$\frac{1 + 100/x^2}{101} = 0.6310^2 = 0.3982$$

or

$$1 + \frac{100}{x^2} = 101 \times 0.3982 = 40.2182$$

or

$$\frac{100}{x^2} = 39.2182$$

or

$$x = \sqrt{\frac{100}{39.2182}} = 1.5968$$

The required transfer function is, therefore, given by

$$G(s)H(s) = \frac{10[1 + (s/1.5968)]}{s(1+s)} = \frac{10(1 + 0.6262s)}{s(1+s)}$$

EXAMPLE 13.23. Obtain the transfer function of a system whose Bode magnitude plot is given in Figure 13.25.

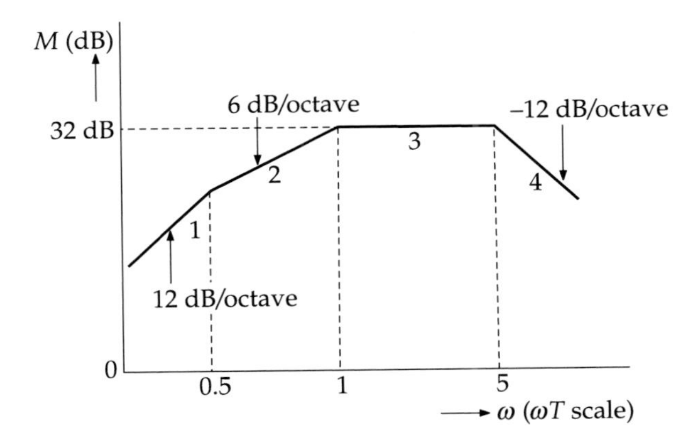

Figure 13.25 Bode magnitude plot (Example 13.23).

Solution The curve contains four segments which we have numbered as 1, 2, 3 and 4. We will designate transfer functions corresponding to these segments as G_1, G_{12}, meaning thereby the transfer function up to that segment of the curve.

At $\omega T = 0$, the slope is 12 dB/octave, i.e. 40 dB/decade. The slope is positive which indicates there are two zeros there because we know that each zero contributes a slope of 20 dB/decade. Therefore,

$$G_1 = Ks^2$$

where K is the gain.

At $\omega T = 0.5$, the slope is 6 dB/octave, i.e. 20 dB/decade. So, the change in slope is -20 dB/decade, indicating the existence of a simple pole at that corner frequency. Therefore,

$$G_{12} = \frac{Ks^2}{1 + \dfrac{s}{0.5}}$$

At $\omega T = 1$, the slope becomes zero which indicates the existence of another simple pole that nullifies the existing 6 dB/octave slope. Therefore,

$$G_{123} = \frac{Ks^2}{\left(1 + \dfrac{s}{0.5}\right)(1 + s)}$$

Now we observe that at $\omega T = 1$

1. The resultant magnitude plot is horizontal.
2. Its magnitude is 32 dB.

This indicates that the magnitude 32 dB comes from the contribution of the constant—poles and zeros having annulled their contributions here. Thus,

$$20 \log K = 32$$

or

$$K = 39.8$$

At $\omega T = 5$, the slope becomes -12 dB/octave, i.e. -40 dB/decade. This indicates the existence of a double pole there. Combining all these, we get

$$G_{1234} = \frac{39.8s^2}{\left(1+\frac{s}{0.5}\right)(1+s)\left(1+\frac{s}{5}\right)^2}$$

$$= (0.5 \times 25) \cdot \frac{39.8s^2}{(s+0.5)(s+1)(s+5)^2}$$

$$= \frac{497.6s^2}{(s+0.5)(s+1)(s+5)^2} \qquad \text{(i)}$$

Therefore, the required transfer function is given by Eq. (i).

EXAMPLE 13.24. From the asymptotic magnitude (in dB) versus frequency (log scale) plot of Figure 13.26, find the associated transfer function. Assume that no right-half plane poles or zeros are present.

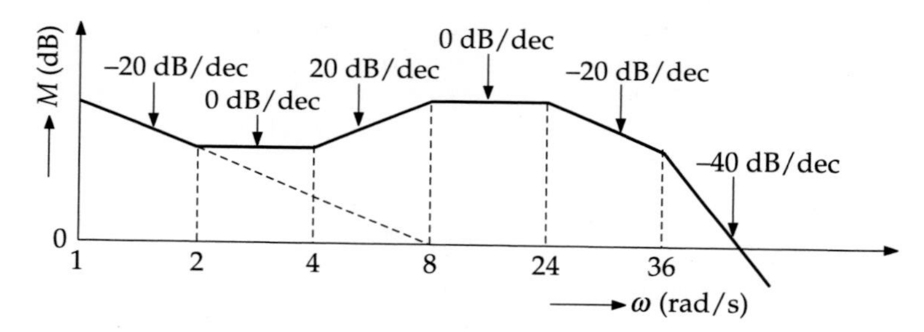

Figure 13.26 Bode magnitude plot (Example 13.24).

Solution At $\omega_c = 1$, the slope is -20 dB/decade. This indicates there is a pole at the origin. Therefore,

$$G_1 = \frac{K}{s}$$

At $\omega_c = 2$, the slope changes to 0 dB/decade, indicating the existence of a simple zero at that corner frequency. Therefore,

$$G_{12} = \frac{K\left(1+\frac{s}{2}\right)}{s}$$

At $\omega_c = 4$, the slope changes to + 20 dB/decade, which indicates the existence of another simple zero there. Therefore,

$$G_{123} = \frac{K\left(1+\frac{s}{2}\right)\left(1+\frac{s}{4}\right)}{s}$$

Now we observe that at $\omega_c = 8$ the resultant magnitude plot is horizontal. This indicates the existence of a pole at $\omega_c = 8$, which neutralizes the + 20 dB/decade slope. Therefore,

$$G_{1234} = \frac{K\left(1+\dfrac{s}{2}\right)\left(1+\dfrac{s}{4}\right)}{s\left(1+\dfrac{s}{8}\right)}$$

At $\omega_c = 24$, the slope becomes -20 dB/decade. This indicates the existence of a pole there. Therefore,

$$G_{12345} = \frac{K\left(1+\dfrac{s}{2}\right)\left(1+\dfrac{s}{4}\right)}{s\left(1+\dfrac{s}{8}\right)\left(1+\dfrac{s}{24}\right)}$$

Finally, at $\omega_c = 36$, the slope becomes -40 dB/decade, indicating the existence of another pole there. Therefore,

$$G_{123456} = \frac{K\left(1+\dfrac{s}{2}\right)\left(1+\dfrac{s}{4}\right)}{s\left(1+\dfrac{s}{8}\right)\left(1+\dfrac{s}{24}\right)\left(1+\dfrac{s}{36}\right)} \quad \text{(i)}$$

Evaluation of K: We observe that for the pole at origin, the -20 dB/decade line should have passed through 0 dB at $\omega_c = 1$. Instead, it does so at $\omega_c = 8$. That could happen because of the contribution of 20 log K dB contribution of the constant. The -20 dB/decade line has a drop of 20 log 8 dB from $\omega_c = 1$ to $\omega_c = 8$. Therefore,

$$20 \log K = 20 \log 8 \text{ dB}$$

or

$$K = 8$$

Substituting this value of K in Eq. (i) yields, on simplification, the transfer function

$$G(s) = \frac{6912(2+s)(4+s)}{s(8+s)(24+s)(36+s)}$$

EXAMPLE 13.25. From the asymptotic magnitude (in dB) versus frequency (log scale) plot of Figure 13.27, find the associated transfer function. Assume that the system is of minimum phase type.

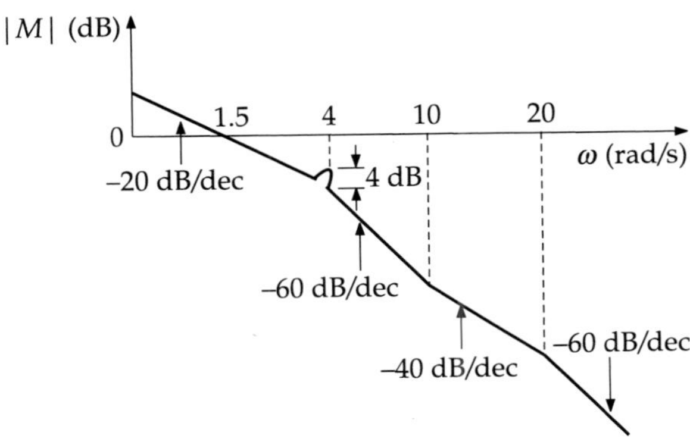

Figure 13.27 Bode magnitude plot (Example 13.25).

Solution Since the system is of minimum phase type, there is no pole or zero in the RHP.

The starting line on the left possesses a slope of −20 dB/decade, which indicates the existence of a pole at the origin. But this line, instead of passing through 0 dB at $\omega_c = 1$, passes through $\omega_c = 1.5$. This indicates a contribution from the gain K. So, our initial transfer function is

$$G_1 = \frac{K}{s}$$

At $\omega_c = 1.5$, the value of the −20 dB/decade line is 20 log 1.5. This should equal the contribution from the constant to make the resultant 0 dB. Therefore,

$$20 \log K = 20 \log 1.5$$

or

$$K = 1.5$$

Now, we observe that

1. There exists a peak at $\omega_c = 4$ with a peak height of 4 dB.
2. The slope of the magnitude plot has changed by −40 dB/decade.

These indicate the existence of a quadratic pole at that corner frequency. We know from Eq. (13.11) that for a quadratic pole[10], $\omega_c = \omega_n$ and

$$|M| = 20 \log \left(\frac{1}{2\zeta}\right) = 4 \text{ dB (given)}$$

$$\therefore \qquad \zeta = 0.315$$

So, the transfer function now becomes

$$\begin{aligned} G_{12} &= \frac{K\omega_n^2}{s(s^2 + 2\zeta\omega_n s + \omega_n^2)} \\ &= \frac{(1.5)(4)^2}{s[s^2 + 2(0.315)(4)s + 4^2]} \\ &= \frac{24}{s(s^2 + 2.52s + 16)} \end{aligned}$$

Next, we observe that the slope changes from −60 dB/decade to −40 dB/decade, indicating the existence of a zero at the corner frequency of 10 rad/s. So, the transfer function now becomes

$$G_{123} = \frac{24\left(1 + \dfrac{s}{10}\right)}{s(s^2 + 2.52s + 16)}$$

Finally, the slope again changes to −60 dB/decade at 20 rad/s, indicating the existence of another pole there. Hence the transfer function is

$$G_{1234} = \frac{24\left(1 + \dfrac{s}{10}\right)}{s\left(1 + \dfrac{s}{20}\right)(s^2 + 2.52s + 16)}$$

[10]Note that Eq. (13.11) pertains to a quadratic zero. For a quadratic pole, the inverse of the term under the radical sign in Eq. (13.8) has to be considered.

On simplification, the required transfer function turns out to be

$$G(s) = \frac{48(s+10)}{s(s+20)(s^2+2.52s+16)}$$

13.8 STATIC ERROR CONSTANTS FROM BODE PLOT

The static error constants K_p, K_v and K_a can be determined from the Bode plot. We will show how to do that with simple examples.

Let an open-loop transfer function be defined as

$$G(s)H(s) = \frac{1}{s^N}$$

In frequency domain, its magnitude in dB is written as

$$|M| = 20\log|G(\jmath\omega)H(\jmath\omega)| = 20\log\left|\frac{K}{(\jmath\omega)^N}\right|$$

$$= 20\log K - 20N\log\omega \tag{13.15}$$

Static Position Error Constant

Let us consider a type 0 system with $N = 0$. Substituting this value in Eq. (13.15), we find

$$|M| = 20\log K$$

Now we recall that the static position error constant is defined as

$$K_p = \lim_{s\to 0} G(s)H(s)$$

which in the frequency domain becomes

$$K_p = \lim_{s\to 0} G(\jmath\omega)H(\jmath\omega)$$

Thus the initial magnitude of the Bode plot, corresponding to the magnitude of the constant gain K equals the static position error constant.

Static Velocity Error Constant

Let us consider a type 1 system having $N = 1$. Putting this value in Eq. (13.15), we get, at the intersection of the Bode magnitude plot with the 0 dB line, the relation

$$0 = 20\log K - 20\log\omega$$

$$\Rightarrow \qquad K = \omega$$

The static velocity error constant is defined as

$$K_v = \lim_{s\to 0} sG(s)H(s)$$

For a type 1 system, the s-factor in the numerator and denominator cancel out yielding just the constant K at $\omega \rightarrow 0$. In other words,

$$K_v = K = \omega$$

Thus, for a type 1 system, the value of ω at the 0 dB intersection of the initial -20 dB/decade line gives the value of the static velocity error constant. A type 1 system has an initial slope of -20 dB/decade. We need to extend that line to 0 dB and find the corresponding ω which gives the value of K_v (see Figure 13.28a).

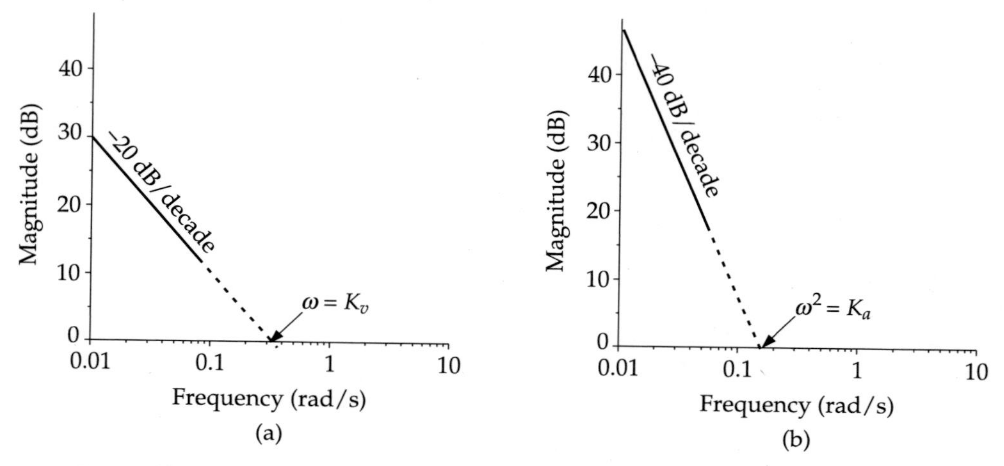

Figure 13.28 Static error constant determination: (a) velocity; (b) acceleration.

Static Acceleration Error Constant

Consider a type 2 system having $N = 2$. As before, putting this value in Eq. (13.15), we get, at the intersection of the Bode magnitude plot with the 0 dB line, the relation

$$0 = 20 \log K_a - 20 \log \omega^2$$

or

$$K_a = \omega^2$$

So, following the same argument that we put forward for the velocity error constant, we can find out the static acceleration constant by extending the initial -40 dB/decade line (a type 2 system has that slope initially) to the 0 dB line (see Figure 13.28b) and figuring out the square of the corresponding frequency value.

13.9 MINIMUM AND NON-MINIMUM PHASE SYSTEMS

We know systems having no poles or zeros in the right-half of the s-plane (RHP) of their transfer functions are called *minimum phase systems*, while those having poles and/or zeros in the RHP are called *non-minimum phase systems*. The modulus of the phase response for a non-minimum phase system is then always larger than that for a system with minimum phase behaviour though both may have the same amplitude response.

Let us consider, for example, a minimum phase system having the transfer function as

$$G(s) = \frac{1 + sT}{1 + sT_1} \tag{13.16}$$

and a non-minimum phase system having a transfer function of

$$G(s) = \frac{1 - sT}{1 + sT_1} \tag{13.17}$$

where $T_1 > T > 0$. The pole-zero plots of the systems are shown in Figure 13.29.

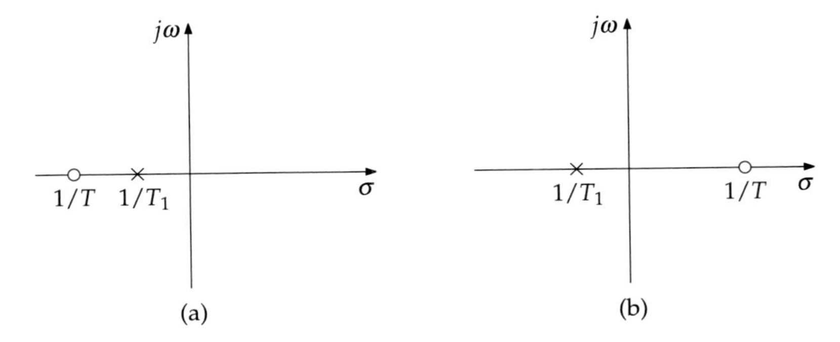

Figure 13.29 Pole-zero plot of (a) minimum; (b) non-minimum phase systems.

The frequency domain transfer functions of Eqs. (13.16) and (13.17) yield the same amplitude of

$$|M| = \frac{\sqrt{1 + \omega^2 T^2}}{\sqrt{1 + \omega^2 T_1^2}}$$

while the phases, which are different, are given by

$$\phi_{\text{min}} = \tan^{-1} \omega T - \tan^{-1} \omega T_1 = -\tan^{-1} \frac{\omega(T_1 - T)}{1 + \omega^2 T_1 T}$$

$$\phi_{\text{non-min}} = -\tan^{-1} \omega T - \tan^{-1} \omega T_1 = -\tan^{-1} \frac{\omega(T_1 + T)}{1 - \omega^2 T_1 T}$$

respectively.

Their plots, shown in Figure 13.30, depict a huge phase sweep for the non-minimum phase system. In fact, given a particular magnitude of response, the minimum phase system offers less phase shift than all other systems. A non-minimum phase system transfer function, such as that given by Eq. (13.17), can always be represented as a cascaded minimum phase and a pure phase shift element (also called *all-pass* system) as follows:

$$G(s) = \frac{1 + sT}{1 + sT_1} \cdot \frac{1 - sT}{1 + sT}$$

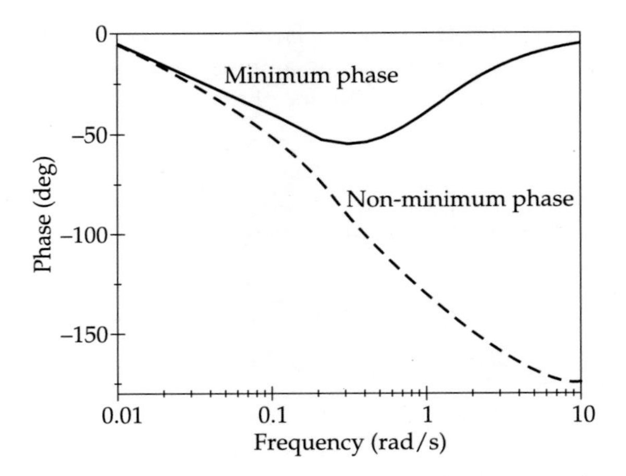

Figure 13.30 Comparison of phase shifts of minimum and non-minimum phase.

This phase-shift element, viz.,

$$\frac{1 - sT}{1 + sT}$$

which consists of reflected pole-zero in the s-plane, has a magnitude of unity, but a phase shift of

$$\phi = -2 \tan^{-1} \omega T$$

Thus, this all-pass element, which introduces a phase shift of 0° to $-180°$, increases the phase sweep of non-minimum phase systems.

Because of the high phase shift, non-minimum phase systems are slow in response and so they are not desirable in a control system.

So far, we have centred our discussion on first order systems. In general, if m and n denote degrees of polynomials of the numerator and denominator of a transfer function, for minimum phase systems,

$$\phi|_{\omega \to \infty} = -90°(n - m)$$

where ϕ is the phase angle. This criterion does not hold good for non-minimum phase systems, though for both,

$$M|_{\omega \to \infty} = -20 \text{ dB/decade}$$

where M is the magnitude. So, from the behaviour of the phase at $\omega \to \infty$, a non-minimum phase system can be recognized.

REVIEW QUESTIONS

13.1 Choose the correct answer:

(i) 20 dB/decade corresponds to

(a) 3 dB/octave
(b) 6 dB/octave
(c) 9 dB/octave
(d) none of these

(ii) A system has 4 poles and 1 zero. Its high frequency asymptote in magnitude plot has a slope of

(a) -100 dB/decade (b) 100 dB/decade
(c) -60 dB/decade (d) 60 dB/decade

(iii) Slope of asymptotes in Bode plot for a second order system is

(a) 18 dB/octave (b) -12 dB/octave
(c) 6 dB/octave (d) -3 dB/octave

(iv) The frequency 2 decade above a frequency of 2 rad/sec is

(a) 22 rad/sec (b) 102 rad/sec (c) 200 rad/sec (d) 32 rad/sec

(v) The gain of a system is 10. In terms of dB it is

(a) 0 dB (b) 1 dB (c) 20 dB (d) 100 dB

(vi) The figure shows the Bode magnitude plot of a system.

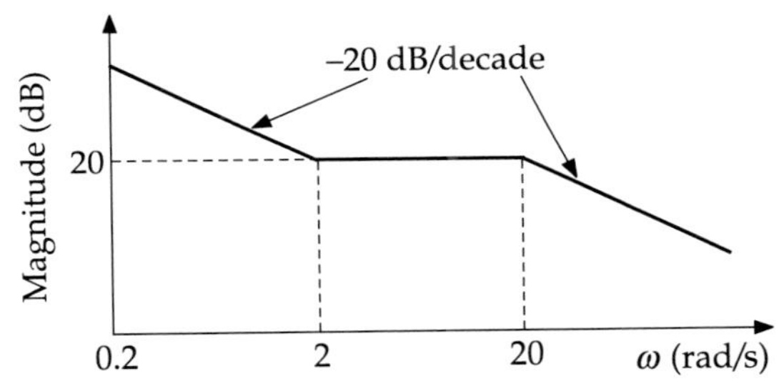

The minimum-phase transfer function of the system is given by

(a) $\dfrac{80\left(\frac{s}{2}+1\right)}{s\left(\frac{s}{20}+1\right)}$ (b) $\dfrac{20\left(\frac{s}{2}+1\right)}{s\left(\frac{s}{20}+1\right)}$

(c) $\dfrac{8(s+2)}{s(s+20)}$ (d) $\dfrac{80(s+2)}{s(s+20)}$

(vii) The Bode plot of the transfer function $G(s) = s$ is

(a) zero magnitude and zero phase shift
(b) 20 dB/decade and phase shift π
(c) 20 dB/decade and phase shift $\pi/2$
(d) -20 dB/decade and phase shift $-\pi/2$

(viii) The initial slope of Bode plot for a transfer function having simple pole at origin is

(a) 20 dB/dec (b) -40 dB/dec (c) 40 dB/dec (d) -20 dB/dec

(ix) The error at the corner frequency of an asymptotic Bode plot for the term $(1 + \jmath\omega T)^N$ is

(a) $\pm 5N$ dB (b) ± 3 dB (c) ± 6 dB (d) $\pm 3N$ dB

13.2 (a) Draw the Bode plot for a system having

$$G(s)H(s) = \frac{100}{s(s+1)(s+2)}$$

(b) For the above problem, find graphically gain margin, phase margin, gain crossover frequency, and phase crossover frequency.

(c) Comment on the stability of the system.

13.3 (a) Define gain margin and phase margin. How are they used to determine the amount of relative stability of a closed-loop system?

(b) How can you evaluate the Gm and Pm of a control system from its Bode plot?

13.4 (a) Sketch the asymptotic Bode plot for the following open-loop transfer function with unity feedback:

$$G(s) = \frac{10}{s(s+1)(s+5)}$$

(b) Calculate the gain margin and phase margin from the Bode plot. Determine the gain cross-over frequency.

13.5 Sketch the Bode plots showing the magnitude in decibels and phase angle in degrees as a function of log frequency for the transfer function

$$G(s)H(s) = \frac{10(s+1)}{s(s+2)(s+10)}$$

13.6 Draw Bode plots for a control system whose loop transfer function is

$$G(s)H(s) = \frac{80}{s(1+0.02s)(1+0.05s)}$$

Determine the gain margin and phase margin and thus check the stability.

13.7 The transfer function of a control system is

$$G(s) = \frac{1000(s+3)}{s(s+12)(s+50)}$$

Draw the Bode plot of amplitude versus log ω and phase versus log ω.

Answers to Selected Questions

13.1 (i) (b); (ii) (c); (iii) (b); (iv) (a); (v) (c); (vi) (d); (vii) (c); (viii) (d); (ix) (d)

13.2 (b) Gm $= -24$ dB, Pm $= -53°$, $\omega_g = 4.5$ rad/s, $\omega_p = 1.4$ rad/s; (c) Unstable

13.4 (b) Gm $= 9.5$ dB, Pm $= 25°$, $\omega_g = 1.2$rad/s

13.6 Gm $= -1.2$ dB, Pm $= -3.4°$; Unstable.

CHAPTER 14

Polar Plots

A frequency domain transfer function is a complex quantity. Like any other complex quantity, the complex transfer function can be expressed in magnitude and phase (angle) form, and can be plotted in polar coordinates. The polar plot is, therefore, a plot of the magnitude of $G(j\omega)$ vs. angle of $G(j\omega)$ in polar coordinates as ω is varied from 0 to ∞.

In polar plots:

1. Angles are positive if they are measured from the positive real axis in the counter-clockwise direction and vice versa.
2. The positive imaginary axis is the line at $+90°$, and the negative imaginary axis, at $-90°$.
3. The positive real axis is the line at $0°$ (or $360°$) and the negative real axis, at $+180°$ (or $-180°$).

To familiarize ourselves with this kind of plots, let us draw plots of a few standard functions of control systems.

14.1 POLAR PLOTS OF STANDARD FUNCTIONS

Integral Factor

The s-plane representation of an integral factor is given by

$$G(s) = \frac{1}{s}$$

The magnitude and angle form of this factor is given by

$$G(j\omega) = \frac{1}{j\omega} \equiv \frac{1}{\omega}\angle -\tan^{-1}\frac{1}{0} = \frac{1}{\omega}\angle -90° \tag{14.1}$$

The magnitude of Eq. (14.1) varies with ω, as shown in the following table, but the angle remains constant at $-90°$.

ω	0.01	0.1	1	∞
$1/\omega$	100	10	1	0

So, the polar plot [Figure 14.1(a)] is the negative part of the imaginary axis.

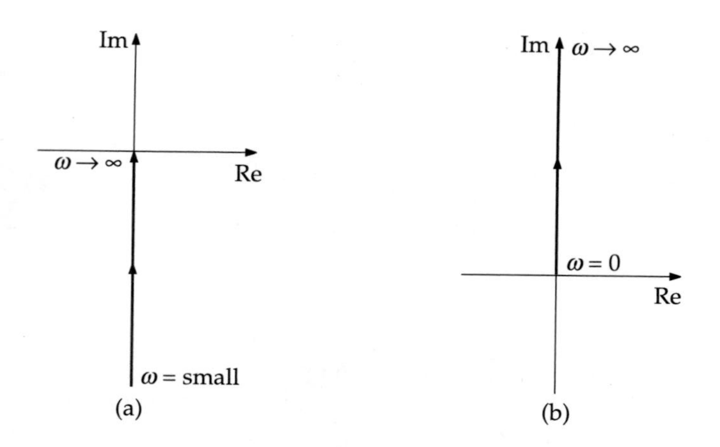

Figure 14.1 Polar plots of: (a) integral; (b) derivative factors.

Derivative Factor

The derivative factor is written in the s-plane as

$$G(s) = s$$

Its magnitude and angle form is as follows:

$$G(\jmath\omega) = \jmath\omega \equiv \omega\angle 90^\circ$$

Obviously, the positive part of the imaginary axis represents the plot [Figure 14.1(b)].

Simple Pole

In the s-plane, a simple pole is represented by the following transfer function:

$$G(s) = \frac{1}{1 + sT}$$

In the frequency domain, its magnitude and angle form is given by

$$G(\jmath\omega) = \frac{1}{1 + \jmath\omega T} \equiv \frac{1}{\sqrt{1 + \omega^2 T^2}}\angle - \tan^{-1}(\omega T)$$

In this case, both the magnitude and angle of $G(\jmath\omega)$ vary with ω. Their values at some points are as follows:

ω	$\lvert G(\jmath\omega)\rvert$	$\angle G(\jmath\omega)$
0	1	0°
$1/T$	$1/\sqrt{2}$	-45°
∞	0	-90°

The polar plot [Figure 14.2(a)], is a semi-circle of radius 0.5 having its centre at (0.5, 0).

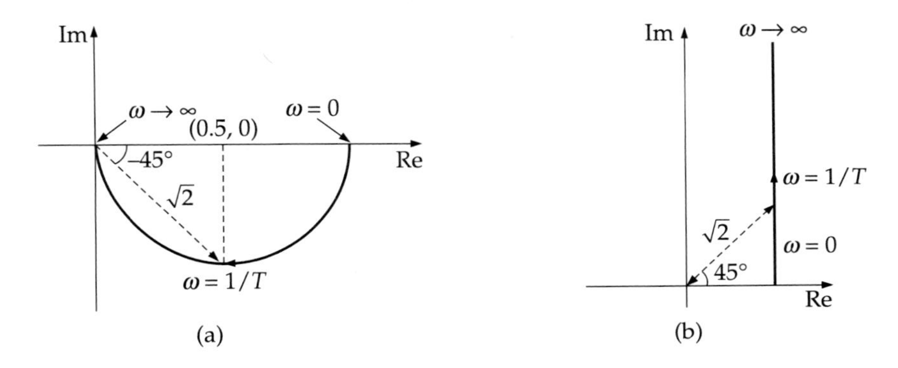

Figure 14.2 Polar plots of: (a) simple pole; (b) simple zero.

Simple Zero

The transfer function of a simple zero is given by

$$G(s) = (1 + sT)$$

In the frequency domain, its magnitude and angle form is given by

$$G(\jmath\omega) = 1 + \jmath\omega T \equiv \sqrt{1 + \omega^2 T^2} \angle \tan^{-1} (\omega T)$$

The values of magnitude and angle for a few ω values are as follows:

ω	$\lvert G(\jmath\omega)\rvert$	$\angle G(\jmath\omega)$
0	1	0°
$1/T$	$\sqrt{2}$	45°
∞	∞	–90°

The polar plot [Figure 14.2(b)] is a straight line parallel to the positive part of the imaginary axis, originating from (1, 0).

Quadratic Pole

The standard transfer function of a quadratic pole is given by

$$G(s) = \frac{\omega_n^2}{s^2 + 2\zeta\omega_n s + \omega_n^2} \qquad \zeta > 0$$

Its magnitude and angle form can be derived from its complex form as follows:

$$G(\jmath\omega) = \frac{1}{1 + \jmath 2\zeta(\omega/\omega_n) - (\omega/\omega_n)^2}$$

$$\equiv \frac{1}{\sqrt{[1 - (\omega/\omega_n)]^2 + 4\zeta^2(\omega/\omega_n)^2}} \angle - \tan^{-1} \frac{2\zeta(\omega/\omega_n)}{1 - (\omega/\omega_n)^2}$$

The values of magnitude and angle for three cardinal ω values are as follows:

ω	$\lvert G(j\omega)\rvert$	$\angle G(j\omega)$
0	1	$0°$
ω_n	$1/2\zeta$	$-90°$
∞	0	$-180°$

The polar plot is shown in Figure 14.3(a).

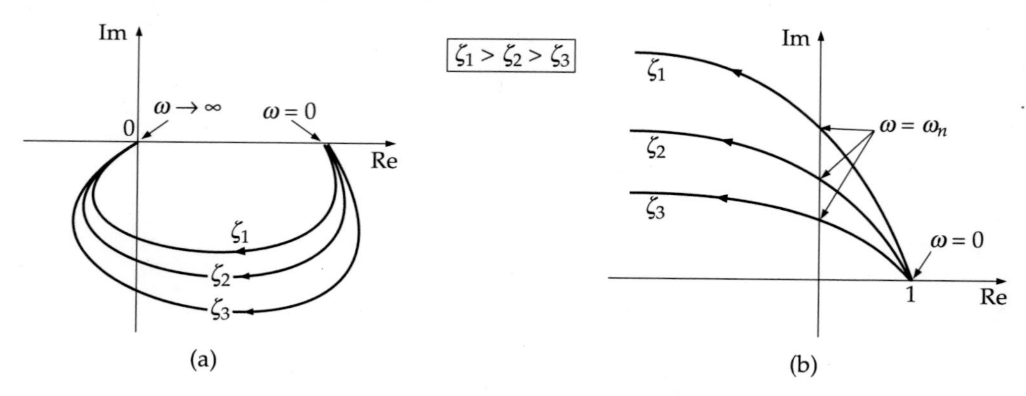

Figure 14.3 Polar plots of: (a) quadratic poles; (b) quadratic zeros for different ζ.

Quadratic Zero

In the s-plane, a quadratic zero is given by

$$G(s) = \frac{s^2 + 2\zeta\omega_n s + \omega_n^2}{\omega_n^2}, \qquad \zeta > 0$$

Its polar form is obtained as follows:

$$G(j\omega) = 1 + j2\zeta\left(\frac{\omega}{\omega_n}\right) - \left(\frac{\omega}{\omega_n}\right)^2$$

$$\equiv \sqrt{\left(1 - \frac{\omega}{\omega_n}\right)^2 + 4\zeta^2\left(\frac{\omega}{\omega_n}\right)^2} \angle \tan^{-1}\frac{2\zeta(\omega/\omega_n)}{1-(\omega/\omega_n)^2}$$

The values of magnitude and angle for three cardinal ω values are as follows:

ω	$\lvert G(j\omega)\rvert$	$\angle G(j\omega)$
0	1	$0°$
ω_n	2ζ	$90°$
∞	∞	$180°$

The polar plot is shown in Figure 14.3(b).

Transportation Lag

We have seen in Section 5.4 at page 206 that the transfer function for the transportation lag is given by

$$G(s) = e^{-sT}$$

Its polar representation is obtained as follows:

$$G(\jmath\omega) = e^{-\jmath\omega T} = \cos\omega T - \jmath\sin\omega T \equiv 1\angle - \omega T$$

Obviously, its polar plot (Figure 14.4) is a circle because its magnitude is constant for all angles.

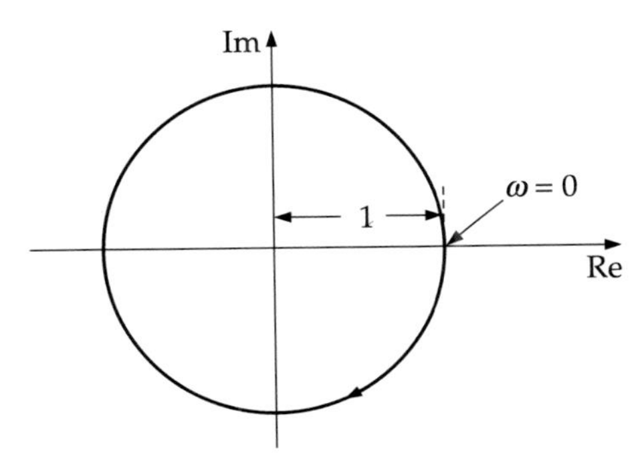

Figure 14.4 Polar plot of transportation lag.

14.2 DRAWING POLAR PLOTS

Based on what we have done so far to draw polar plots of a few standard simple transfer functions, we now delineate the steps for drawing them in general.

Step 1. Convert the transfer function to frequency domain form by replacing s with $\jmath\omega$.

Step 2. Convert the frequency domain transfer function to magnitude and angle form.

Step 3. Magnitudes $|G(\jmath\omega)|$ are found at $\omega = 0$ and $\omega = \infty$ by calculating $\lim_{\omega\to 0} |G(\jmath\omega)|$ and $\lim_{\omega\to\infty} |G(\jmath\omega)|$. In the same way, phase angles are calculated at $\omega = 0$ and $\omega = \infty$.

Step 4. Break the complex transfer function $G(\jmath\omega)$ into real and imaginary parts through rationalization.

Step 5. Equate the real part to zero and find out the corresponding frequency. Evaluate the imaginary part of $|G(\jmath\omega)|$ at this value of the frequency. This gives the intercept of the plot on the imaginary axis.

Step 6. Equate the imaginary part, obtained in *Step 4*, to zero and find out the corresponding frequency. Evaluate the real part of $|G(\jmath\omega)|$ at this value of the frequency. This gives the intercept of the plot on the real axis.

Step 7. From the data obtained so far, draw the polar plot.

It is true that we have to have an educated guess to draw the entire plot because not many points will be obtained in the foregoing steps. But with a little bit of practice, it will not be a difficult task to do.

We will draw a few polar plots before we discuss how we can utilize them in our stability studies.

Type 0 System

Let the transfer function for a type 0 system be

$$G(s) = \frac{1}{(1+sT_1)(1+sT_2)}$$

Step 1. In the frequency domain the transfer function is as follows:

$$G(\jmath\omega) = \frac{1}{(1+\jmath\omega T_1)(1+\jmath\omega T_2)}$$

Step 2. Its magnitude and angle form is given by

$$G(\jmath\omega) = \frac{1}{\sqrt{1+(\omega T_1)^2}\sqrt{1+(\omega T_2)^2}} \angle(-\tan^{-1}\omega T_1 - \tan^{-1}\omega T_2)$$

Step 3. The limiting values of the magnitude and phase are

$$\lim_{\omega\to 0}|G(\jmath\omega)| = \lim_{\omega\to 0}\frac{1}{\sqrt{1+(\omega T_1)^2}\sqrt{1+(\omega T_2)^2}} = 1$$

$$\lim_{\omega\to \infty}|G(\jmath\omega)| = \lim_{\omega\to \infty}\frac{1}{\sqrt{1+(\omega T_1)^2}\sqrt{1+(\omega T_2)^2}} = 0 \tag{14.2}$$

$$\lim_{\omega\to 0}\angle G(\jmath\omega) = \lim_{\omega\to 0}(-\tan^{-1}\omega T_1 - \tan^{-1}\omega T_2) = 0°$$

$$\lim_{\omega\to \infty}\angle G(\jmath\omega) = \lim_{\omega\to \infty}(-\tan^{-1}\omega T_1 - \tan^{-1}\omega T_2) = -180° \tag{14.3}$$

Step 4. Now we split the complex transfer function into real and imaginary parts as follows:

$$\begin{aligned} G(\jmath\omega) &= \frac{1}{(1+\jmath\omega T_1)(1+\jmath\omega T_2)} \cdot \frac{(1-\jmath\omega T_1)(1-\jmath\omega T_2)}{(1-\jmath\omega T_1)(1-\jmath\omega T_2)} \\ &= \frac{1-\omega^2 T_1 T_2 - \jmath\omega(T_1+T_2)}{(1+\omega^2 T_1^2)(1+\omega^2 T_2^2)} \\ &= \frac{1-\omega^2 T_1 T_2}{(1+\omega^2 T_1^2)(1+\omega^2 T_2^2)} - \jmath\frac{\omega(T_1+T_2)}{(1+\omega^2 T_1^2)(1+\omega^2 T_2^2)} \end{aligned} \tag{14.4}$$

Step 5. Equating the real part to zero, we obtain the corresponding frequency as follows:

$$\text{Re}\,\{G(\jmath\omega)\} = \frac{1-\omega^2 T_1 T_2}{(1+\omega^2 T_1^2)(1+\omega^2 T_2^2)} = 0$$

$$\Rightarrow \qquad 1 - \omega^2 T_1 T_2 = 0$$

$$\Rightarrow \qquad \omega = \frac{1}{\sqrt{T_1 T_2}} \tag{14.5}$$

Substituting ω from Eq. (14.5) in the imaginary part of Eq. (14.4), we get

$$\text{Im}\,\{G(j\omega)\} = -j\,\frac{\dfrac{1}{\sqrt{T_1T_2}}(T_1+T_2)}{\left(1+\dfrac{T_1^2}{T_1T_2}\right)\left(1+\dfrac{T_2^2}{T_1T_2}\right)} = -j\,\frac{\dfrac{1}{\sqrt{T_1T_2}}(T_1+T_2)}{\left(\dfrac{T_1+T_2}{T_2}\right)\left(\dfrac{T_1+T_2}{T_1}\right)}$$

$$= -j\,\frac{\sqrt{T_1T_2}}{T_1+T_2} \equiv \frac{\sqrt{T_1T_2}}{T_1+T_2}\angle -90^\circ$$

Step 6. We now equate the imaginary part obtained in *Step 4* and find out the corresponding frequency as follows:

$$\text{Im}\,\{G(j\omega)\} = \frac{\omega(T_1+T_2)}{(1+\omega^2T_1^2)(1+\omega^2T_2^2)} = 0$$

$$\Rightarrow \qquad \omega = 0,\ \infty$$

We already know the magnitude and angle values at these frequencies from Eqs. (14.2) and (14.3). So far we have got the following data for the polar plot:

ω	$\lvert G(j\omega)\rvert$	$\angle G(j\omega)$
0	1	0°
∞	0	−180°
$\dfrac{1}{\sqrt{T_1T_2}}$	$\left\lvert\dfrac{\sqrt{T_1T_2}}{T_1+T_2}\right\rvert$	−90°

Step 7. The polar plot is presented in Figure 14.5.

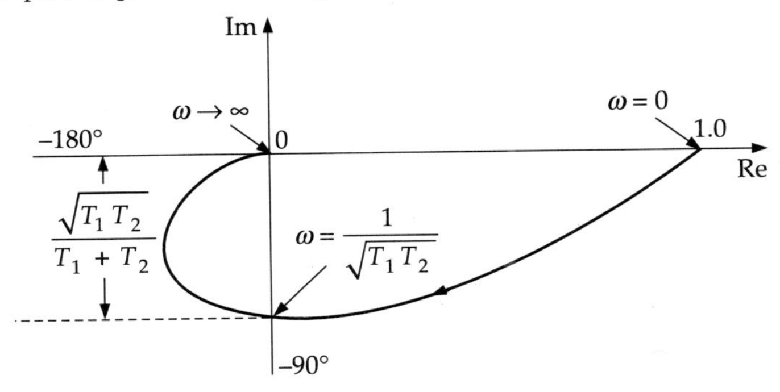

Figure 14.5 Polar plot for the type 0 system.

Type 1 System

Let the transfer function for the type 1 system be

$$G(s) = \frac{1}{s(1+sT_1)(1+sT_2)}$$

We now follow the steps, as shown before, to make its polar plot.
Step 1.

$$G(j\omega) = \frac{1}{j\omega(1+j\omega T_1)(1+j\omega T_2)}$$

Step 2.

$$G(j\omega) = \frac{1}{\omega\sqrt{1+(\omega T_1)^2}\sqrt{1+(\omega T_2)^2}}\angle(-90^\circ - \tan^{-1}\omega T_1 - \tan^{-1}\omega T_2)$$

Step 3.

$$\lim_{\omega\to 0}|G(j\omega)| = \lim_{\omega\to 0}\frac{1}{\omega\sqrt{1+(\omega T_1)^2}\sqrt{1+(\omega T_2)^2}} = \infty$$

$$\lim_{\omega\to \infty}|G(j\omega)| = \lim_{\omega\to \infty}\frac{1}{\omega\sqrt{1+(\omega T_1)^2}\sqrt{1+(\omega T_2)^2}} = 0$$

$$\lim_{\omega\to 0}\angle G(j\omega) = \lim_{\omega\to 0}\angle(-90^\circ - \tan^{-1}\omega T_1 - \tan^{-1}\omega T_2) = -90^\circ$$

$$\lim_{\omega\to \infty}\angle G(j\omega) = \lim_{\omega\to \infty}\angle(-90^\circ - \tan^{-1}\omega T_1 - \tan^{-1}\omega T_2) = -270^\circ \tag{14.6}$$

Step 4.

$$\begin{aligned} G(j\omega) &= \frac{1}{j\omega(1+j\omega T_1)(1+j\omega T_2)}\cdot\frac{-j(1-j\omega T_1)(1-j\omega T_2)}{-j(1-j\omega T_1)(1-j\omega T_2)} \\ &= \frac{-j[1-\omega^2T_1T_2 - j\omega(T_1+T_2)]}{\omega(1-\omega^2T_1^2)(1-\omega^2T_2^2)} \\ &= \frac{-\omega(T_1+T_2)}{\omega(1-\omega^2T_1^2)(1-\omega^2T_2^2)} - j\frac{1-\omega^2T_1T_2}{\omega(1-\omega^2T_1^2)(1-\omega^2T_2^2)} \end{aligned} \tag{14.7}$$

Step 5.

$$\text{Re}\,\{G(j\omega)\} = \frac{-\omega(T_1+T_2)}{\omega(1-\omega^2T_1^2)(1-\omega^2T_2^2)} = 0$$

$$\Rightarrow \qquad \omega = \infty$$

This means that the polar plot does not intercept the imaginary axis. The angle at $\omega = \infty$ is known to us from Eq. (14.6).

Step 6.

$$\text{Im}\,\{G(j\omega)\} = \frac{1-\omega^2 T_1 T_2}{\omega(1-\omega^2 T_1^2)(1-\omega^2 T_2^2)} = 0$$

$$\Rightarrow \quad 1-\omega^2 T_1^2 = 0$$

$$\Rightarrow \quad \omega = \frac{1}{\sqrt{T_1 T_2}} \tag{14.8}$$

Substituting ω from Eq. (14.8) in the real part of Eq. (14.7), we get

$$\text{Re}\,\{G(j\omega)\} = -\frac{T_1+T_2}{\left(1+\dfrac{T_1^2}{T_1T_2}\right)\left(1+\dfrac{T_2^2}{T_1T_2}\right)} = -\frac{T_1+T_2}{\left(\dfrac{T_1+T_2}{T_2}\right)\left(\dfrac{T_1+T_2}{T_1}\right)}$$

$$= -\frac{T_1T_2}{T_1+T_2} \equiv \frac{T_1T_2}{T_1+T_2}\angle 0^\circ$$

So far we have obtained the following data for the polar plot:

ω	$\|G(j\omega)\|$	$\angle G(j\omega)$
0	∞	90°
∞	0	-270°
$\dfrac{1}{\sqrt{T_1T_2}}$	$\dfrac{\sqrt{T_1T_2}}{T_1+T_2}$	-0°

Based on these data, the polar plot is drawn as in Figure 14.6.

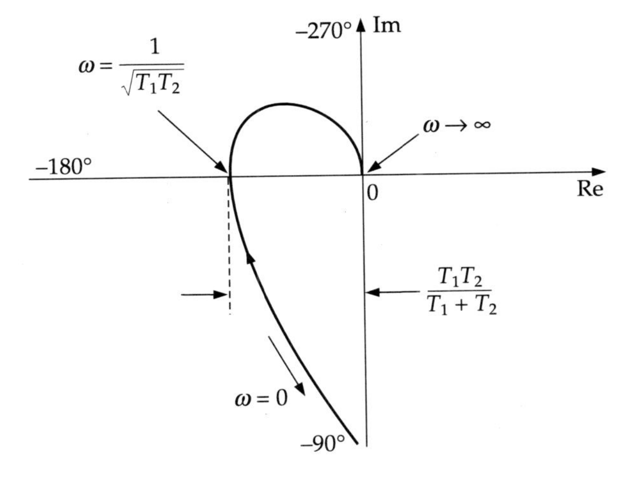

Figure 14.6 Polar plot of type 1 system.

Type 2 System

Let the transfer function of a type 2 system be given by

$$G(s) = \frac{1}{s^2(1+sT)}$$

We follow the steps to obtain the indicative data points for drawing the polar plot.

Step 1.

$$G(\jmath\omega) = \frac{1}{(\jmath\omega)^2(1+\jmath\omega T)}$$

Step 2.

$$G(\jmath\omega) = \frac{1}{\omega^2\sqrt{1+(\omega T)^2}}\angle(-180^\circ - \tan^{-1}\omega T)$$

Step 3.

$$\lim_{\omega\to 0} |G(\jmath\omega)| = \lim_{\omega\to 0} \frac{1}{\omega^2\sqrt{1+(\omega T)^2}} = \infty$$

$$\lim_{\omega\to \infty} |G(\jmath\omega)| = \lim_{\omega\to \infty} \frac{1}{\omega^2\sqrt{1+(\omega T)^2}} = 0$$

$$\lim_{\omega\to 0} \angle G(\jmath\omega) = \lim_{\omega\to 0} \angle(-180^\circ - \tan^{-1}\omega T) = -180^\circ$$

$$\lim_{\omega\to \infty} \angle G(\jmath\omega) = \lim_{\omega\to \infty} \angle(-180^\circ - \tan^{-1}\omega T) = -270^\circ \quad (14.9)$$

Step 4.

$$\begin{aligned} G(\jmath\omega) &= \frac{1}{-\omega^2(1+\jmath\omega T)}\cdot\frac{(1-\jmath\omega T)}{(1-\jmath\omega T)} \\ &= \frac{(1-\jmath\omega T)}{-\omega^2(1+\jmath\omega T)} \\ &= -\frac{1}{\omega^2(1+\omega^2T^2)} - \jmath\frac{\omega T}{\omega^2(1+\omega^2T^2)} \end{aligned}$$

Step 5.

$$\text{Re}\,\{G(\jmath\omega)\} = -\frac{1}{\omega^2(1+\omega^2T^2)} = 0$$

$\Rightarrow$

$$\omega = \infty$$

This means that the polar plot does not intercept the imaginary axis. The angle at $\omega = \infty$ is known to us from Eq. (14.9).

Step 6.

$$\text{Im}\,\{G(j\omega)\} = \frac{\omega T}{\omega^2(1+\omega^2 T^2)} = 0$$

$$\Rightarrow \qquad \omega = \infty$$

This again means that the polar plot intercepts the real axis only at the origin. The angle at $\omega = \infty$ is known to us from Eq. (14.9). So far we have obtained the following data for the polar plot:

ω	$\lvert G(j\omega)\rvert$	$\angle G(j\omega)$
0	∞	$-180°$
∞	0	$-270°$

Step 7. Based on these data, the polar plot, as shown in Figure 14.7, is drawn.

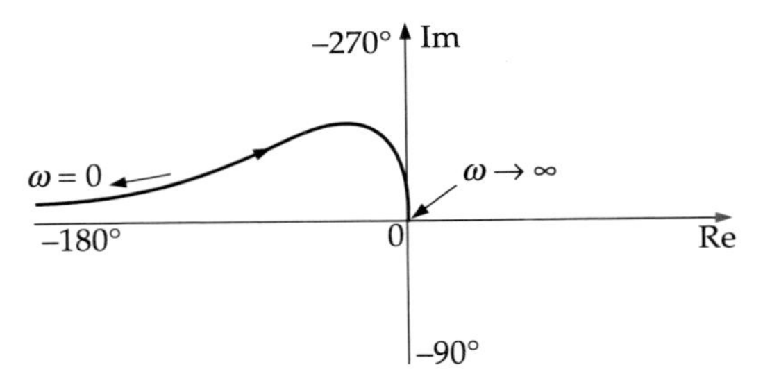

Figure 14.7 Polar plot of type 2 system.

We now work out a couple of examples to familiarize ourselves with polar plots.

EXAMPLE 14.1. The figure below shows the polar plot of a system.

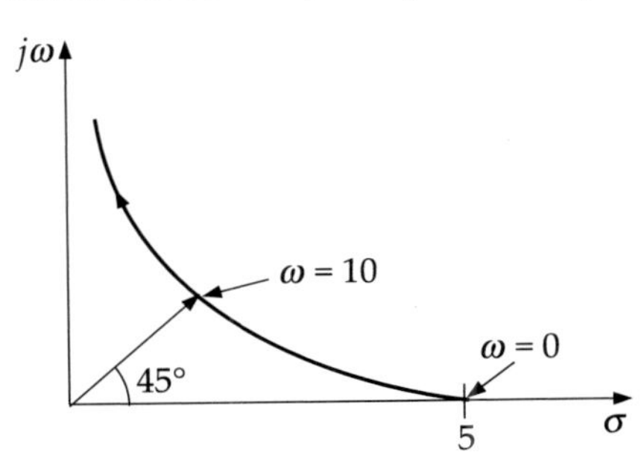

The transfer function of the system is

(a) $5(1 + 0.1s)$ (b) $1 + 0.5s$ (c) $5(1 + 10s)$ (d) $5(1 + s)$

Solution Let the transfer function be given by

$$G(s) = K(1 + sT)$$

Then, in the frequency domain,

$$G(\jmath\omega) = K(1 + \jmath\omega T) \equiv K\sqrt{1+\omega^2T^2}\angle \tan^{-1}\omega T$$

$$\phi = \tan^{-1} 10T = 45^\circ \qquad \text{[from the given diagram]}$$

$\Rightarrow$ $$10T = \tan 45^\circ = 1$$

$\Rightarrow$ $$T = 0.1$$

Now, $$K\sqrt{1+0} = 5 \qquad \text{[from the given diagram]}$$

$\Rightarrow$ $$K = 5$$

Therefore, $$G(s) = 5(1 + 0.1s)$$

Ans: (a).

EXAMPLE 14.2. Sketch the polar plot for the following transfer function:

$$G(s) = \frac{20}{s(s+1)}$$

Solution We substitute $s \equiv \jmath\omega$ and convert the transfer function to magnitude and angle form as follows:

$$G(\jmath\omega) = \frac{20}{\jmath\omega(\jmath\omega+1)} \equiv \frac{20}{\omega\sqrt{\omega^2+1}}\angle - (90^\circ + \tan^{-1}\omega)$$

Evaluating the magnitude and angle at $\omega = 0$ and ∞, we get

$$\lim_{\omega\to 0} |G(\jmath\omega)| = \lim_{\omega\to 0} \frac{20}{\omega\sqrt{\omega^2+1}} = \infty$$

$$\lim_{\omega\to \infty} |G(\jmath\omega)| = \lim_{\omega\to \infty} \frac{20}{\omega\sqrt{\omega^2+1}} = 0$$

$$\lim_{\omega\to 0} \angle G(\jmath\omega) = \lim_{\omega\to 0} \angle - (90^\circ + \tan^{-1}\omega) = -90^\circ$$

$$\lim_{\omega\to \infty} \angle G(\jmath\omega) = \lim_{\omega\to \infty} \angle - (90^\circ + \tan^{-1}\omega) = -180^\circ$$

Now, we check if there are intersections on the real and/or imaginary axes. To do that, we have to break the transfer function into real and imaginary parts as follows:

$$G(\jmath\omega) = \frac{20}{\jmath\omega(\jmath\omega+1)} \cdot \frac{-\jmath\omega(\jmath\omega-1)}{-\jmath\omega(\jmath\omega-1)}$$

$$= \frac{20\omega^2}{\omega^2(\omega^2+1)} + \jmath\frac{20\omega}{\omega^2(\omega^2+1)}$$

Equating the real part to zero, we obtain

$$\text{Re}\,\{G(\jmath\omega)\} = \frac{20\omega^2}{\omega^2(\omega^2+1)} = 0$$

$\Rightarrow$ $$\omega = \infty$$

This indicates that the polar plot does not intersect the imaginary axis. Next we equate the imaginary part to zero to find out the intersection at the real axis, i.e.

$$\text{Im}\,\{G(\jmath\omega)\} = \frac{20\omega}{\omega^2(\omega^2+1)} = 0$$

$$\Rightarrow \qquad \omega = \infty$$

The result indicates that the plot intersects the real axis only at the origin. From these data we construct the polar plot as given in Figure 14.8.

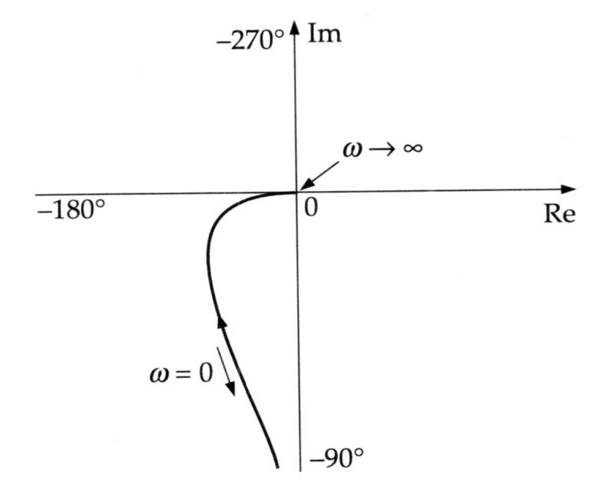

Figure 14.8 Polar plot (Example 14.2).

EXAMPLE 14.3. Sketch the polar plot for the following transfer function:

$$G(s) = \frac{20}{s(s+1)(s+2)}$$

Solution We substitute $s \equiv \jmath\omega$ and convert the transfer function to the magnitude and angle form as follows:

$$G(\jmath\omega) = \frac{20}{\jmath\omega(\jmath\omega+1)(\jmath\omega+2)} = \frac{20}{\omega\sqrt{\omega^2+1}\sqrt{\omega^4+4}}\angle\left(-90^\circ - \tan^{-1}\omega - \tan^{-1}\frac{\omega}{2}\right)$$

Evaluating the magnitude and angle at $\omega = 0$ and ∞, we get

$$\lim_{\omega\to 0}|G(\jmath\omega)| = \lim_{\omega\to 0}\frac{20}{\omega\sqrt{\omega^2+1}\sqrt{\omega^4+4}} = \infty$$

$$\lim_{\omega\to \infty}|G(\jmath\omega)| = \lim_{\omega\to \infty}\frac{20}{\omega\sqrt{\omega^2+1}\sqrt{\omega^4+4}} = 0$$

$$\lim_{\omega\to 0}\angle G(\jmath\omega) = \lim_{\omega\to 0}\angle\left(-90^\circ - \tan^{-1}\omega - \tan^{-1}\frac{\omega}{2}\right) = -90^\circ$$

$$\lim_{\omega\to \infty}\angle G(\jmath\omega) = \lim_{\omega\to \infty}\angle\left(-90^\circ - \tan^{-1}\omega - \tan^{-1}\frac{\omega}{2}\right) = -270^\circ$$

Now, we check if there are intersections on the real and/or imaginary axes. To do that, we have to break the transfer function into real and imaginary parts as follows:

$$G(\jmath\omega) = \frac{20}{\jmath\omega(\jmath\omega+1)(\jmath\omega+2)} \cdot \frac{-\jmath\omega(\jmath\omega-1)(\jmath\omega-2)}{\jmath\omega(\jmath\omega-1)(\jmath\omega-2)}$$

$$= \frac{-60\omega^2}{\omega^2(\omega^2+1)(\omega^2+4)} + \jmath\frac{20\omega(\omega^2-2)}{\omega^2(\omega^2+1)(\omega^2+4)} \tag{i}$$

Equating the real part to zero, we get

$$\text{Re}\{G(\jmath\omega)\} = \frac{-60\omega^2}{\omega^2(\omega^2+1)(\omega^2+4)} = 0$$

$\Rightarrow$

$$\omega = \infty$$

This indicates that the polar plot does not intersect the imaginary axis. Next we equate the imaginary part to zero to find out the intersection at the real axis:

$$\text{Im}\{G(\jmath\omega)\} = \frac{20\omega(\omega^2-2)}{\omega^2(\omega^2+1)(\omega^2+4)} = 0$$

$\Rightarrow$

$$\omega = \pm\sqrt{2}$$

The result indicates that the plot intersects the real axis at $\omega = \sqrt{2}$ rad/s, the negative frequency value being meaningless. Substituting this value in Eq. (i), we obtain

$$G(\jmath\sqrt{2}) = \frac{-60}{(2+1)(2+4)} = -3.33$$

From these data we construct the polar plot as given in Figure 14.9.

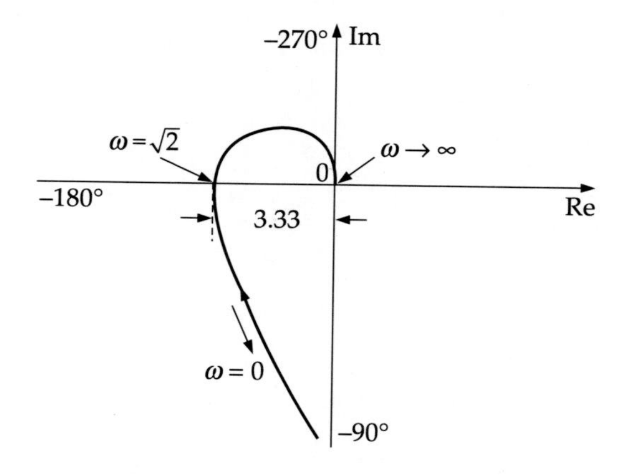

Figure 14.9 Polar plot (Example 14.3).

14.3 ALTERNATIVE METHOD FOR POLAR PLOT

This method is based on plotting the tip of the vector $M(\omega)e^{j\phi(\omega)}$ in the G-plane by varying ω from 0 to ∞. The following example will make the procedure clear.

EXAMPLE 14.4. Sketch the polar plot for

$$G(s) = \frac{10(s+1)}{s+10}$$

Solution We substitute $s \equiv j\omega$ and convert the transfer function to the magnitude and angle form as follows:

$$\begin{aligned} G(j\omega) &= \frac{10(j\omega+1)}{j\omega+10} \\ &\equiv \frac{10\sqrt{\omega^2+1}}{\sqrt{\omega^2+100}} \angle \left(\tan^{-1}\omega - \tan^{-1}\frac{\omega}{10} \right) \end{aligned} \tag{i}$$

We generate the data of Table 14.1 from Eq. (i).

Table 14.1 Data for Polar Plot of Example 14.4

ω (rad/s)	M	ϕ (deg)	$M\cos\phi + jM\sin\phi$
0	1.0	0.0	$1 + j0$
1	1.407	39.289	$1.089 + j0.891$
2	2.193	52.125	$1.346 + j1.731$
3	3.029	54.866	$1.743 + j2.477$
4	3.828	54.162	$2.241 + j3.103$
5	4.561	52.125	$2.800 + j3.600$
7	5.793	46.878	$3.960 + j4.210$
10	7.106	39.289	$5.500 + j4.500$
15	8.339	29.876	$7.231 + j4.154$
20	8.955	23.703	$8.200 + j3.600$
35	9.619	14.309	$9.321 + j2.377$
50	9.808	10.164	$9.654 + j1.729$
100	9.951	5.138	$9.911 + j0.891$
200	9.988	2.576	$9.978 + j0.449$

Then we plot $M\cos\phi$ and $M\sin\phi$ along the real and imaginary axes of Figure 14.10.

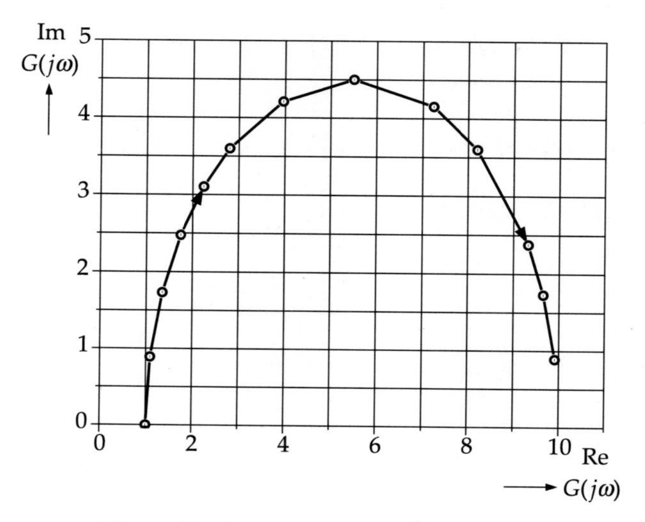

Figure 14.10 Polar plot (Example 14.4).

14.4 STABILITY STUDY FROM POLAR PLOT

A polar plot of the open-loop transfer function can be utilized to assess the stability of a system by studying the gain and phase margins.

Gain and Phase Crossover Frequencies

In the polar plot shown in Figure 14.11, the plot crosses the negative Re-axis at A and the unit circle at B. A and B denote phase crossover and gain crossover points respectively, because at A the plot crosses $-180°$ phase and at B, the unit gain.

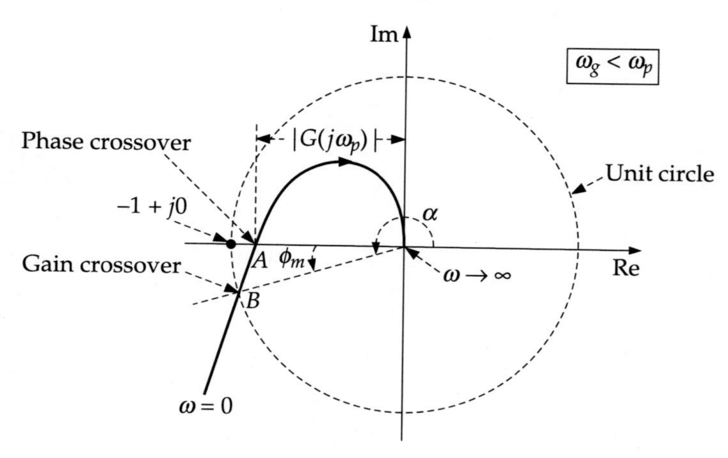

Figure 14.11 Gain and phase crossover frequencies, and gain and phase margins in a polar plot.

Gain Margin

If $|G(\jmath\omega)|$ denotes the gain at the phase crossover frequency ω_p, then according to the definition

$$\text{Gm} = 20\log\frac{1}{|G(\jmath\omega)|} = -20\log|G(\jmath\omega)| \text{ dB} \tag{14.10}$$

Four cases may emerge from polar plots of systems. The plot characteristic, gain margin, and the stability of a system corresponding to the cases are given in Table 14.2.

Table 14.2 Polar Plots vis-a-vis Gain Margins and Stabilities

Case	*Plot characteristic*	*Gain margin*	*Stability of the system*
1	The plot may not cross the negative Re-axis at all except at $\omega = \infty$, where $\lvert G(\jmath\omega)\rvert = 0$.	$\text{Gm} = \infty$	Stable
2	The plot may cross the negative Re-axis between the origin and $-1 + \jmath 0$. Figure 14.11 is such a case. Here, since $\lvert G(\jmath\omega)\rvert$ is a fraction, its logarithm will be negative.	$\text{Gm} \geq 0$	Stable
3	The plot may cross the negative Re-axis beyond $-1 + \jmath 0$ [see Figure 14.12(a)]. Here the point $-1 + \jmath 0$ lies to the right of the curve as we move with ω from 0 to ∞ in a clockwise direction. It is said the plot *encloses* $-1 + \jmath 0$ (see Figure 14.12(b) to know the region of enclosure). Obviously, here, $\lvert G(\jmath\omega)\rvert > 1$.	$\text{Gm} < 0$	Unstable
4	The plot may pass through $-1 + \jmath 0$. So, a slight increase in $\lvert G(\jmath\omega)\rvert$ will make it positive.	$\text{Gm} = 0$	Marginally stable

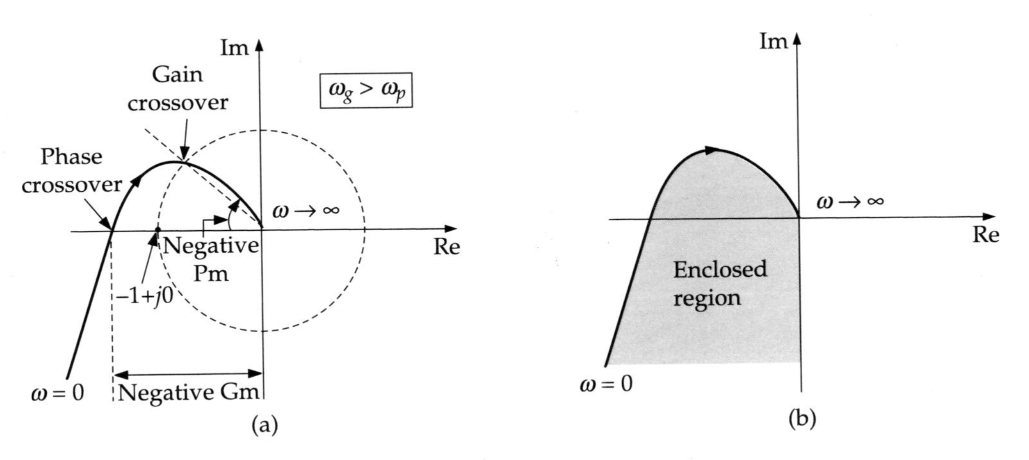

Figure 14.12 Polar plots showing: (a) unstable condition; (b) meaning of enclosed region.

Phase Margin

By definition, the phase margin is

$$\text{Pm} = 180° + \alpha \tag{14.11}$$

However (see Figure 14.11),

$$\alpha = 180° + \phi_m \tag{14.12}$$

Combining Eqs. (14.11) and (14.12), we get

$$\text{Pm} = 180° + 180° + \phi_m = \phi_m \tag{14.13}$$

So, the angle measured in the counter-clockwise direction from the negative Re-axis (i.e. 180°) to the gain crossover point is the measure of the phase margin.

We consider a couple of examples to see how stability conditions can be worked out from polar plots.

EXAMPLE 14.5. Sketch the polar plot for the unity feedback system given by

$$G(s) = \frac{K}{s(s+2)(s+5)}$$

and determine the range of K for the stability of the system.

Solution Since it is a unity feedback system, the open-loop transfer function in the frequency domain is

$$G(\jmath\omega)H(\jmath\omega) = \frac{K}{\jmath\omega(\jmath\omega+2)(\jmath\omega+5)}$$

$$\equiv \frac{K}{\omega\sqrt{\omega^2+4}\sqrt{\omega^2+25}} \angle -\left(90° + \tan^{-1}\frac{\omega}{2} + \tan^{-1}\frac{\omega}{5}\right) \tag{i}$$

From Eq. (i), we get the following data:

ω	$\lvert G(\jmath\omega)H(\jmath\omega)\rvert$	$\angle G(\jmath\omega)H(\jmath\omega)$
0	∞	$-90°$
∞	0	$-270°$

The sketch of the polar plot is given in Figure 14.13. To find the gain margin, we have to calculate the phase crossover frequency ω_p at A.

At point A, where $\omega = \omega_p$, we have

$$\phi = -180° = -90° - \tan^{-1}\frac{\omega_p}{2} - \tan^{-1}\frac{\omega_p}{5}$$

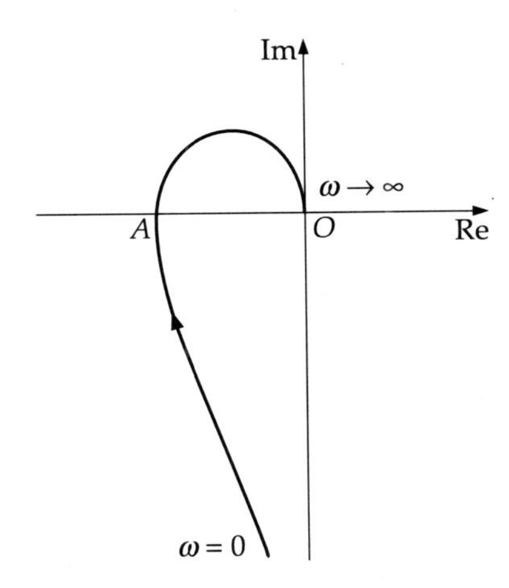

Figure 14.13 Polar plot (Example 14.5).

which gives

$$90^\circ = \tan^{-1}\frac{\omega_p}{2} + \tan^{-1}\frac{\omega_p}{5} = \tan^{-1}\frac{\dfrac{\omega_p}{2} + \dfrac{\omega_p}{5}}{1 - \dfrac{\omega_p}{2}\cdot\dfrac{\omega_p}{5}}$$

$$\Rightarrow \qquad \tan 90^\circ = \tan\left(\tan^{-1}\frac{7\omega_p}{10 - \omega_p^2}\right)$$

$$\Rightarrow \qquad \infty = \frac{7\omega_p}{10 - \omega_p^2} \tag{ii}$$

Equation (ii) indicates

$$10 - \omega_p^2 = 0$$

or

$$\omega_p = 3.162 \text{ rad/s}$$

Therefore,

$$|G(j\omega)H(j\omega)|_A = |G(j\omega)H(j\omega)|_{\omega=\omega_p}$$

$$= \frac{K}{(3.162)\sqrt{(3.162)^2 + 4\sqrt{(3.162)^2 + 25}}}$$

or

$$OA = \frac{K}{70}$$

For the system to be marginally stable, $OA = 1$. Hence,

$$1 = \frac{K}{70}$$

or

$$K = 70 \tag{iii}$$

From Eq. (iii), we get the following stability conditions:

$$\text{For } K \begin{cases} < 70 & \text{the system is stable} \\ = 70 & \text{the system is marginally stable} \\ > 70 & \text{the system is unstable} \end{cases}$$

EXAMPLE 14.6. Sketch the polar plot of the unity feedback system given by

$$G(s) = \frac{1}{(s+1)^2}$$

and find the phase margin therefrom.

Solution The open-loop transfer function in the frequency domain is

$$\begin{aligned} G(\jmath\omega)H(\jmath\omega) &= \frac{1}{(\jmath\omega+1)^2} \\ &\equiv \frac{1}{\omega^2+1}\angle -2\tan^{-1}\omega \end{aligned}$$

We get the following values to draw the polar plot:

ω	$\lvert G(\jmath\omega)H(\jmath\omega)\rvert$	$\angle G(\jmath\omega)H(\jmath\omega)$
0	1	$-0°$
∞	0	$-180°$

The polar plot is presented in Figure 14.14. To figure out the phase margin from the plot, we need to draw the unit circle and find at what angle from the 180°-line (i.e. negative real axis) the polar plot intercepts it. It is clear from the above table as well as Figure 14.14 that the unit gain occurs at $\omega = 0°$, where the phase angle is 0° and consequently the phase margin is 180°.

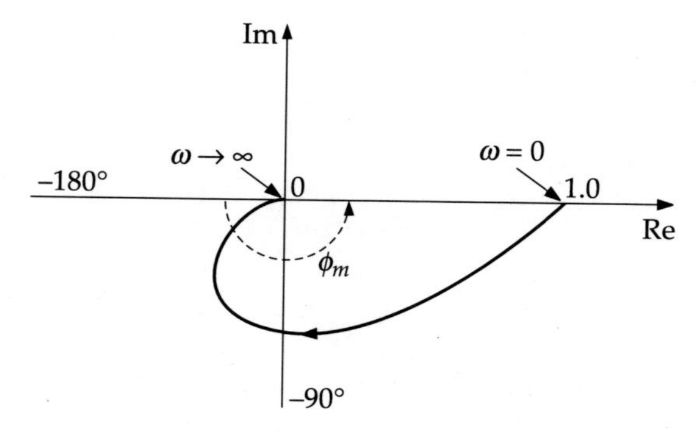

Figure 14.14 Polar plot (Example 14.6).

With this background in polar plots, we now move to Nyquist plots.

REVIEW QUESTIONS

14.1 Choose the correct answer:

A polar plot is a plot of

(a) magnitude and phase of frequency domain transfer function in the G-plane when ω varies from 0 to ∞

(b) pole and zero plot in the s-plane

(c) path traversed by poles when gain varies from 0 to ∞

(d) magnitude (in dB) and phase of frequency domain open-loop transfer function

14.2 Consider the transfer function

$$G(j\omega) = \frac{1}{j\omega(1 + j\omega T)}$$

Draw the polar plot of the function. Graphically find the gain crossover frequency and phase margin of the transfer function.

14.3 A feedback control system is characterized by the open-loop transfer function

$$G(s)H(s) = \frac{100}{s(s+4)(s+20)}$$

(a) Draw the polar plot of the open-loop system. Calculate the magnitude and phase at 10 equi-spaced frequency points (including the point where the plot intersects the imaginary axis) and use them to construct the plot.

(b) From the plot determine the gain margin, phase margin, gain crossover frequency and phase crossover frequency, and comment on the system stability.

Answers to Selected Questions

14.1. (a)

14.3. (b) Gm $= 25.7$ dB, Pm $= 69.9°$, $\omega_g = 1.2$ rad/s, $\omega_p = 8.9$ rad/s. Stable

CHAPTER 15

Nyquist Plot

15.1 INTRODUCTION

A Nyquist plot combines the two types of Bode plot—magnitude and phase—on a single graph, with frequency as a parameter along the curve. It is useful for assessing the stability of a closed-loop system.

The Nyquist plot shows the amplification/attenuation and phase-shift of the transfer function in the complex plane. The phase-shift of frequency-domain transfer function is represented by the argument and the magnitude is represented by the length of a vector from the origin in the direction described by the argument.

The method is essentially based on the following procedures and principle:

- Drawing a Nyquist path for the open-loop transfer function
- Mapping of the path to $G(s)H(s)$-plane following the polar plot scheme
- Applying the *principle of argument*

We have had some idea of polar plots from Chapter 14. Now we consider the other two aspects of Nyquist plots before going to construct the plots themselves.

15.2 MAPPING IN THE $G(S)H(S)$-PLANE

Consider an open-loop transfer function

$$G(s)H(s) = s^2 + 2$$

Suppose we consider a point $S_1 = 2 + \jmath 3$ in the s-plane [see Figure 15.1(a)]. The corresponding value of $G(s)H(s)$ is

$$G(s)H(s) = (2 + \jmath 3)^2 + 2 = -3 + \jmath 12 \tag{15.1}$$

So, if we consider a $G(s)H(s)$-plane, the point G_1 [see Figure 15.1(b)] corresponds to Eq. (15.1). It is said that the point S_1 of the s-plane maps onto the point G_1 of the $G(s)H(s)$-plane.

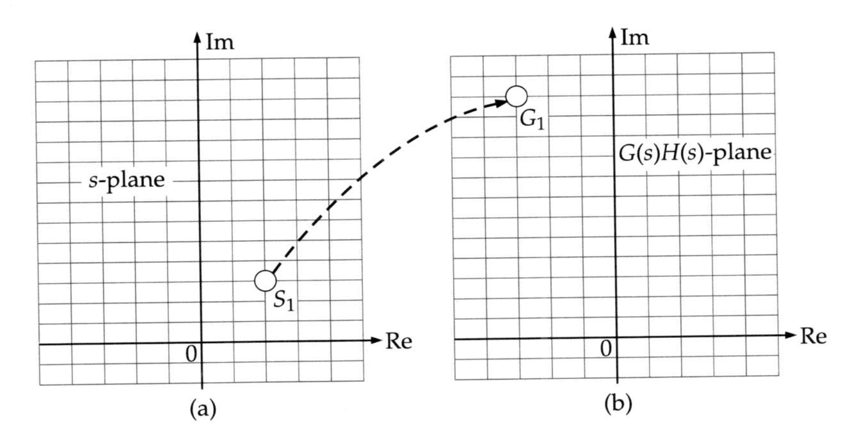

Figure 15.1 Mapping of a point in (a) s-plane to (b) $G(s)H(s)$-plane.

15.3 PRINCIPLE OF ARGUMENT

Let us consider an analytic[1] and single-valued[2] function $F(s)$ which can be written as

$$F(s) = K\frac{(s-z_1)(s-z_2)\cdots(s-z_m)}{(s-p_1)(s-p_2)\cdots(s-p_n)} \qquad m < n \tag{15.2}$$

Let us draw the plot of zeros z_1, z_2 and poles p_1, p_2 of $F(s)$ in the s-plane. Now, we draw in this plane a closed arbitrary contour which does not enclose any of the poles or zeros [Figure 15.2(a)]. As $\mathbf{s}$ moves in the clockwise (CW) direction along the contour, phasors of zeros and poles, namely, $(\mathbf{s}-\mathbf{z_1})$, $(\mathbf{s}-\mathbf{z_2})$, $(\mathbf{s}-\mathbf{p_1})$, $(\mathbf{s}-\mathbf{p_2})$, describe separate paths, phasor-heads always remaining on the contour, but none of the phasors describes a net angle. So, the total contribution of angle to $F(s)$ is zero.

Next we consider the situation when the s-plane contour encloses a zero, namely, z_1 [see Figure 15.2(b)]. Then as the contour completes its path in the CW-direction, the phasor $(\mathbf{s}-\mathbf{z_1})$ describes a net angle of -2π radians[3], while other zeros and poles of $F(s)$, as in the previous case, do not contribute any net angle to $F(s)$.

Now, these poles and zeros are related to $F(s)$ through Eq. (15.2) from which the point $F(s)$ can be written as

$$|F(s)| = K\frac{|s-z_1||s-z_2|\ldots|s-z_m|}{|s-p_1||s-p_2|\ldots|s-p_n|}$$

$$\angle F(s) = \angle(s-z_1) + \angle(s-z_2) + \cdots + \angle(s-z_m)$$
$$- \angle(s-p_1) - \angle(s-p_2) - \cdots - \angle(s-p_n)$$

---

[1] A function, whose value and the value of its derivative are defined in the prescribed domain, is called an *analytic function* within the domain.

[2] The function $F(s) = \sqrt{2}$ is not single valued while $F(s) = s$ is.

[3] The angle is negative because we are moving in the clockwise direction.

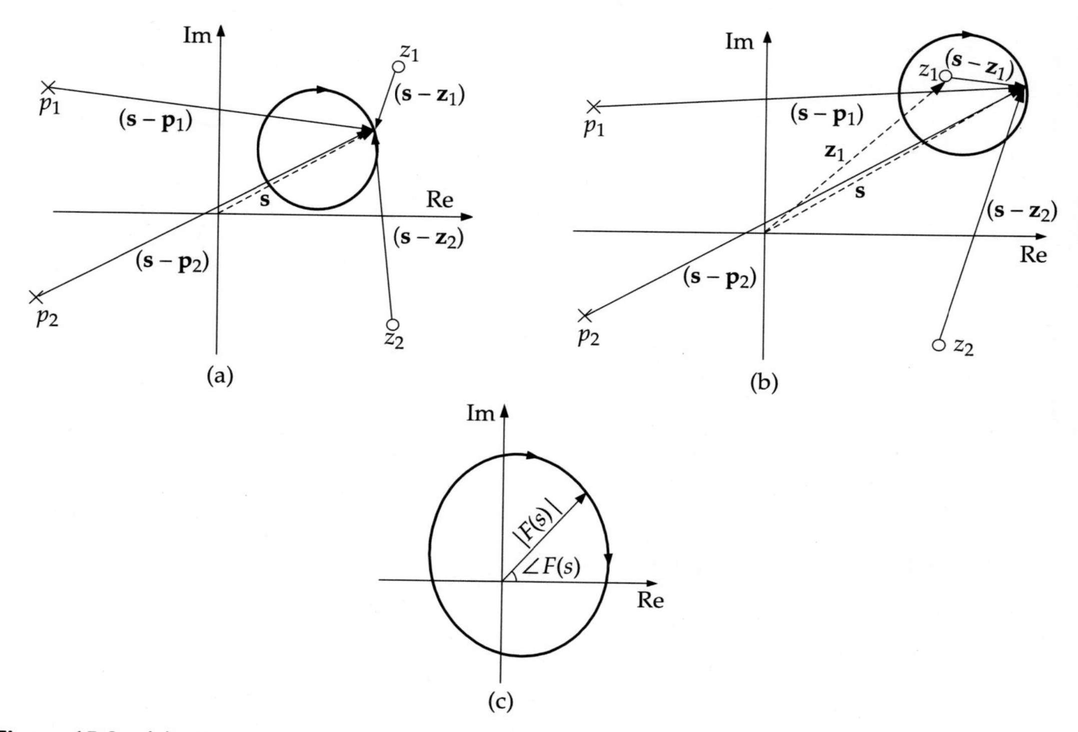

Figure 15.2 (a) When no zero or pole is enclosed in the s-plane; (b) when a zero is enclosed; (c) its mapping in the $1 + G(s)H(s)$ plane.

So, if along with the rotation of the **s** vector in the s-plane we simultaneously rotate the **s** vector in the $F(s)$-plane, an angle of -2π radians will be described *around the origin there*, **s** having started from the origin.

Thus, this enclosure of a zero by the s-plane contour will generate a CW-encirclement of the origin in the $F(s)$-plane [see Figure 15.2(c)]. Let us call the contour in the s-plane Γ_s and that in the $F(s)$-plane, $\Gamma_{F(s)}$. Then we can say that the CW Γ_s contour enclosing a zero of $F(s)$ maps to a CW encirclement of the origin by the $\Gamma_{F(s)}$ contour. If two zeros are enclosed by the Γ_s contour, there will result two encirclements of the origin by the $\Gamma_{F(s)}$ contour because, then the angles described by phasors $(\mathbf{s} - \mathbf{z_1})$ and $(\mathbf{s} - \mathbf{z_2})$ will add up to -4π radians.

What happens if, instead of a zero, a pole is enclosed by the Γ_s contour? This time, too, the $(\mathbf{s} - \mathbf{p_1})$ phasor will describe a net angle of -2π radians. But the corresponding angle contribution to $F(s)$ will be 2π radians because $(\mathbf{s} - \mathbf{p_1})$ lies in the denominator of $F(s)$.

In other words, the CW enclosure of a pole by the Γ_s contour maps to a counter-clockwise (CCW) encirclement of the origin by the $\Gamma_{F(s)}$ contour. As in the case of zeros, enclosure of two poles will generate two such encirclements, and so on.

Thus, if

N = number of clockwise encirclements of the origin of the $F(s)$-plane

Z = number of zeros of $F(s)$ enclosed by the clockwise Γ_s contour

P = number of poles enclosed by the clockwise Γ_s contour then

$$N = Z - P$$

This relation between the enclosure of zeros and poles by contours in the s-plane and encirclements of the origin in the $F(s)$-plane by the $\Gamma_{F(s)}$ contour is commonly known as the *principle of argument* derived from a theorem of Cauchy[4]. The Nyquist plot utilizes this principle to determine the stability of control systems. But before we proceed to consider such plots, let us have a look at the relation between the principle of argument and the stability of control systems. To do that, in turn, we have to know what are

1. Nyquist criterion
2. Nyquist path

15.4 NYQUIST CRITERION

We know that the open-loop transfer function of a control system can be written as

$$G(s)H(s) = K\frac{(s-z_1)(s-z_2)\cdots(s-z_m)}{(s-p_1)(s-p_2)\cdots(s-p_n)} \qquad m < n \tag{15.3}$$

The corresponding closed-loop transfer function is given by

$$\frac{C(s)}{R(s)} = \frac{G(s)}{1+G(s)H(s)} \tag{15.4}$$

and the characteristic equation is given by

$$1 + G(s)H(s) = 0 \tag{15.5}$$

We know that the roots of the characteristic equation are the poles of the closed-loop transfer function.

We have seen before that for the stability of a control system, the poles of the CLTF, i.e. the roots of the characteristic equation, must not lie in the RHP of the s-plane.

Now, combining Eqs. (15.3) and (15.5), we get

$$\begin{aligned} F(s) \equiv 1 + G(s)H(s) &= 1 + K\frac{(s-z_1)(s-z_2)\cdots(s-z_m)}{(s-p_1)(s-p_2)\cdots(s-p_n)} \qquad m < n \\ &= \frac{(s-p_1)(s-p_2)\cdots(s-p_n) + K(s-z_1)(s-z_2)\cdots(s-z_m)}{(s-p_1)(s-p_2)\cdots(s-p_n)} \\ &= \frac{(s-z_1')(s-z_2')\cdots(s-z_m')}{(s-p_1)(s-p_2)\cdots(s-p_n)} \end{aligned} \tag{15.6}$$

Comparing Eqs. (15.3) and (15.6), we observe that

- The poles of $F(s)$ are the same as poles of the open-loop transfer function
- But zeros of $F(s)$ are indeed the poles of the closed-loop transfer function if we calculate it by substituting the results of Eq. (15.6) in Eq. (15.4).

---

[4] Augustin Louis Cauchy (1789–1857) was a French mathematician.

The stability condition dictates that though a pole of the open-loop transfer function in the RHP of the s-plane may not destablize the system per se, a pole of the closed-loop transfer function, and hence, a zero of $F(s)$, must not lie in the RHP for stability. This probably indicates that an open-loop unstable system may turn out to be stable through feedback.

Therefore, to check the stability of a control system, we need to map the $F(s)$-plane by an s-plane contour which encompasses the entire RHP and check for encirclements of the origin by the $F(s)$-plane contour. While drawing the s-plane contour—we call it Γ_s contour—we need to make small detours if poles or zeros of $F(s)$ lie on the imaginary-axis. Such a contour is called the *Nyquist path.*

15.5 NYQUIST PATH

Nyquist proposed a contour in the s-plane that goes in the following manner:

- A path travelling up the $\jmath\omega$-axis, from $(0 - \jmath\infty)$ to $(0 + \jmath\infty)$.
- A semicircular arc, with radius $R \to \infty$, that starts at $(0 + \jmath\infty)$ and travels clockwise to $(0 - \jmath\infty)$.

The contour as specified above is called the *Nyquist path.* The path encloses all poles and zeros (i.e. poles of CLTF) of the $1+G(s)H(s)$ that are located in the RHP. The path should not pass through any pole [see Figure 15.3(a)], but it may contain them [as shown in Figure 15.3(b)]. Mathematical representations corresponding to the plots are given in Table 15.1.

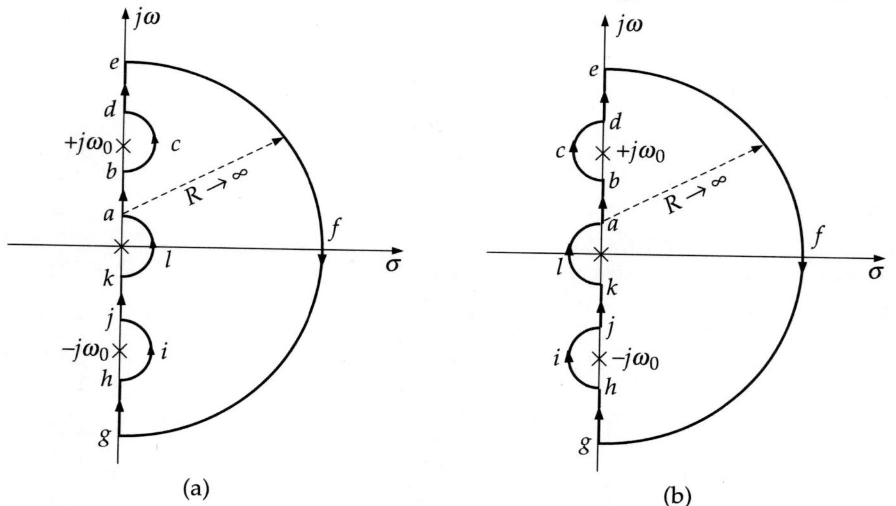

Figure 15.3 General Nyquist paths: (a) bypassing poles; (b) including poles.

Applying Cauchy's theorem to the control problem,

- If a closed path in a *specified region* is mapped to the $1 + G(s)H(s)$-plane,
- If the number of zeros (i.e. poles of CLTF) is Z and the number of poles enclosed in the RHP of the s-plane is P

then the change in the argument (i.e. phase) is given by

$$\Delta \arg = -2\pi(Z - P) \text{ rad}$$

Table 15.1 Mathematical Representation of Segments of Nyquist Paths

Segment	*s-value*	*Limits*
ab	$j\omega$	$\omega = 0$ to ω_0
bcd	$\lim_{\rho\to 0} (\rho e^{j\theta}) + j\omega_0$	$\theta = -90^\circ$ to $+90^\circ$
de	$j\omega$	$\omega = \omega_0$ to ∞
efg	$\lim_{R\to\infty} Re^{j\theta}$	$\theta = +90^\circ$ to -90°
gh	$j\omega$	$\omega = -\infty$ to $-\omega_0$
hij	$\lim_{\rho\to 0} (\rho e^{j\theta}) - j\omega_0$	$\theta = -90^\circ$ to $+90^\circ$
jk	$j\omega$	$\omega = -\omega_0$ to 0
kla	$\lim_{\rho\to 0} (\rho e^{j\theta})$	$\theta = -90^\circ$ to $+90^\circ$

Three terms, namely, specified region, enclosure, and encirclement, require clarification in this context.

Specified region. This is the right-hand side of the s-plane (i.e. RHP), where the existence of roots makes the system unstable.

Enclosure. The term *enclosed in the clockwise direction* means if a path is followed in this direction and a point is found to lie *to the right* of the path, the point is enclosed by the path. The point A in Figure 15.4(a) is enclosed by the path BO moving in the clockwise direction.

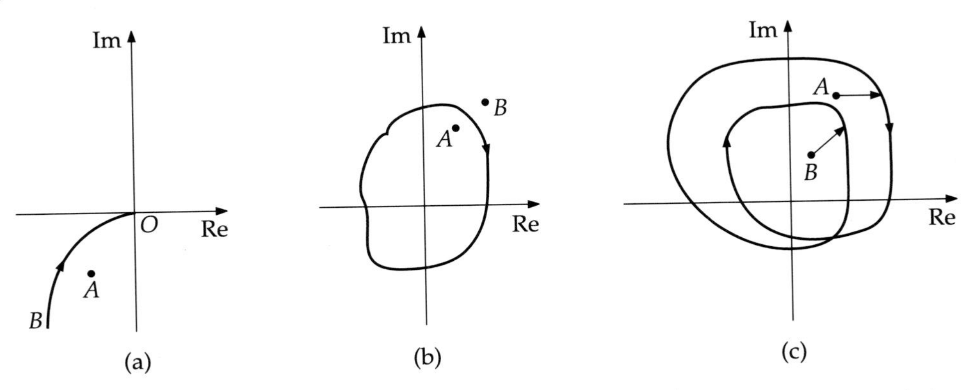

Figure 15.4 (a) Clockwise enclosure of point; (b) encirclement of point; (c) counting encirclements.

Encirclement. We have already used the term *encirclement* which needs elaboration now. The encirclement of a point by a prescribed path means that the point is found within the closed path. In Figure 15.4(b), the point A is encircled by the closed path, but the point B is not.

Positive and negative encirclements. We will consider it a positive encirclement if the path moves in a clockwise direction and negative encirclement if it is a counter-clockwise path.

Counting encirclements. To determine the number of encirclements, we can draw a vector from the point to the path and observe how many revolutions the vector suffers while moving along the path until it returns to its starting point. Following this method, we find the point A in Figure 15.4(c) is encircled once while the point B is encircled twice.

15.6 STABILITY ANALYSIS

If we map the Nyquist path to the $G(s)H(s)$-plane, it will include roots of $G(s)H(s) = 0$, and *not* of the characteristic equation $1 + G(s)H(s) = 0$. So, such a mapping can determine the stability of the open-loop transfer function and *not* the closed-loop transfer function. However, we note that

$$1 + G(s)H(s) = 0$$

or

$$G(s)H(s) = -1$$

Critical Point

The above equations suggest that an encirclement around (0, 0) in the $G(s)H(s)$-mapping for the OLTF is equivalent to an encirclement around $(-1 + \jmath 0)$ in the $G(s)H(s)$-mapping for the CLTF. The point $(-1 + \jmath 0)$ is often called the *critical point*[5].

Stability

Given a Nyquist contour Γ_s in the s-plane, the mapped contour $\Gamma_{F(s)}$ in the $F(s)$-plane shall, for a stable feedback system, encircle the point $(-1 + \jmath 0)$ in the CW-direction a number of times N such that

$$N = Z - P \tag{15.7}$$

where P is the number of poles of $F(s)$ enclosed by Γ_s, and Z is the number of zeros of $F(s)$ (i.e. the poles of CLTF) enclosed by Γ_s. The three cases that may emerge are shown in Table 15.2.

Stated in clear terms, the Nyquist stability criterion[6] states that

A closed-loop control system will be stable iff $Z = 0$, as calculated from Eq. (15.7).

---

[5]Sometimes this is written as 'Q-point'.

[6]Some texts claim that any encirclement of the point $(-1 + \jmath 0)$ causes the system to become unstable. This is not strictly accurate. Given the equation $N = Z - P$, only Z must be zero to ensure that the system is stable. An OLTF may have a pole in the RHP, without compromising stability.

Table 15.2 The Three Cases

Case	*Condition*	*Stability*
1	$N = 0$, i.e. no CW encirclement of the critical point.	This may arise owing to either of the following two possibilities: 1. $Z = P$, where $Z > 0$ and $P > 0$. Since in this case zeros (i.e. CLTF poles) exist in the RHP, this is an unstable condition for the control system. 2. $P = 0$, which automatically means $Z = 0$. Obviously, then the system is stable because there is no zero (i.e. CLTF pole) in the RHP.
2	$N > 1$, i.e. 1 or more CW encirclements of the critical point.	This may arise owing to either of the following two possibilities: 1. $Z > P,\ P > 0$. Obviously, then the system is unstable. 2. $Z > 0,\ P = 0$. Then also, the system is unstable.
3	$N < 1$, i.e. 1 or more CCW encirclements of the critical point.	This may arise owing to either of the following two possibilities: 1. $Z < P$, but $P > 1,\ Z \geq 1$. This indicates an unstable situation owing to existence of zero(s) in the RHP. 2. $Z = 0$, and hence $N = P$. This indicates a stable condition.

EXAMPLE 15.1. The loop transfer function of a system has a pole in the right half of the s-plane. The Nyquist plot makes one clockwise encirclement of the $(-1,\ 0)$ point. The closed-loop system is

(a) unstable
(b) stable
(c) marginally stable
(d) asymptotically stable

Solution Since the open-loop transfer functionhas a pole in the RHP, $P = 1$. Now the Nyquist plot makes one CW-encirclement of the critical point. Hence, $N = 1$. Therefore, from Eq. (15.7), we find

$$1 = Z - 1 \quad \text{or} \quad Z = 2$$

This means that there are two poles in the RHP of the closed-loop transfer function and, therefore, the system is unstable.
Ans: (a).

EXAMPLE 15.2. The open-loop part of a unity feedback control system $G(s)$ is unstable and has two poles on the right-hand side of the s-plane. However, the closed-loop system is stable. The number of encirclements made by the Nyquist plot of the $(-1,\ 0)$ point in the $G(s)$-plane is

(a) 2 (b) 0 (c) -3 (d) -1

Solution From the given data, $P = 2$. However, the encirclement queries do not mention if they are CW or CCW. Assuming, CCW encirclement, for the stability of the system

$$-N = Z - P = 0 - 2$$

Ans: (a).

15.7 DRAWING A NYQUIST PLOT

A step-by-step procedure for drawing Nyquist plots is as follows:

Step 1. Find from the OLTF the number of poles at the origin or on the $\jmath\omega$-axis.

Step 2. Draw the Nyquist path in the s-plane.

Step 3. Utilizing Table 15.2, map the path from a to e onto the $G(s)H(s)$-plane.

Step 4. Draw a mirror image of the map obtained in *Step 3* and connect the two.

Step 5. Count the number of CW encirclements about $(-1 + \jmath 0)$.

Step 6. Determine stability using the criterion stated in Section 15.6.

We consider two examples to familiarize ourselves with the procedure.

EXAMPLE 15.3. Draw the Nyquist plot for

$$G(s)H(s) = \frac{1}{s+2}$$

and determine the stability of the closed-loop system.

Solution

Step 1. There is only one pole at $s = -2$; no pole at the origin or on the $\jmath\omega$-axis.

Step 2. The Nyquist path is, therefore, as given in Figure 15.5(a).

Step 3. The $G(s)H(s)$-plane mapping can be done as follows:

For ab-segment: $s = \jmath\omega \quad 0 < \omega < \infty$

$$G(\jmath\omega)H(\jmath\omega) = \frac{1}{\jmath\omega + 2} \equiv \frac{1}{\sqrt{4+\omega^2}} \angle - \tan^{-1}\left(\frac{\omega}{2}\right)$$

which gives

ω	$\|G(\jmath\omega)H(\jmath\omega)\|$	$\angle G(\jmath\omega)H(\jmath\omega)$
$+0$	0.5	$0°$
$+\infty$	0	$-90°$

With the help of these data, we plot the lower segment in the Nyquist diagram [Figure 15.5(b)].

Step 4. A mirror image (upper segment) is drawn and the two segments are connected.

Step 5. Since there is no encirclement about the critical point, $N = 0$.

Step 6. There is no pole in the RHP of the open-loop transfer function. Therefore, $P = 0$. Thus, from the Nyquist criterion, $Z = 0$, and the system is stable.

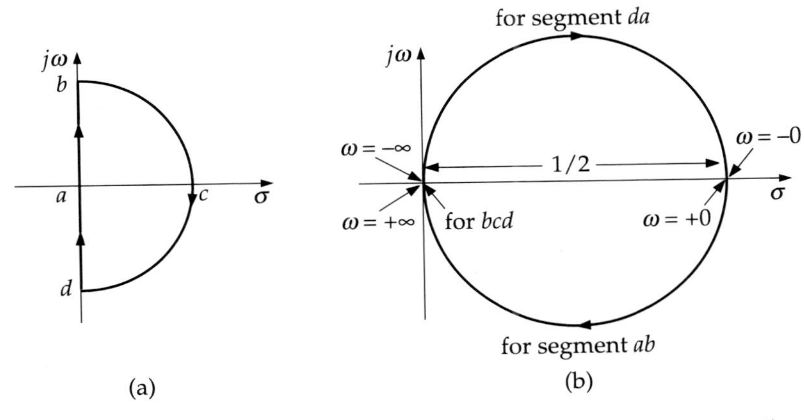

Figure 15.5 (a) Nyquist path; (b) $G(s)H(s)$-mapping (Example 15.3).

EXAMPLE 15.4. Draw Nyquist plot for

$$G(s)H(s) = \frac{1}{s(s+2)}$$

and determine stability.

Solution

Step 1. Two poles exist at $s = 0, -2$. So, there exists one pole at the origin and none on the Im-axis.

Step 2. The Nyquist path is given in Figure 15.6(a).

Step 3. The mapping is as follows:

(i) *For ab-segment:* $s \equiv \jmath\omega \quad 0 < \omega < \infty$

$$G(\jmath\omega)H(\jmath\omega) = \frac{1}{\jmath\omega(\jmath\omega+2)} \equiv \left|\frac{1}{\omega\sqrt{\omega^2+4}}\right| \angle\left(-90° - \tan^{-1}\frac{\omega}{2}\right)$$

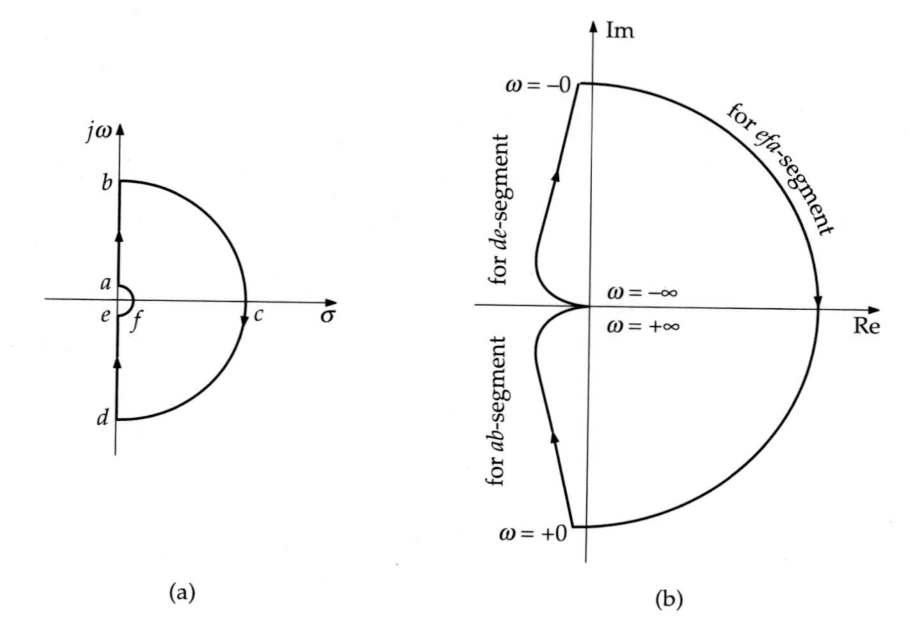

Figure 15.6 Nyquist: (a) path; (b) plot (Example 15.4).

This yields the following table:

ω	$\lvert G(j\omega)H(j\omega)\rvert$	$\angle G(j\omega)H(j\omega)$
0	∞	$-90°$
∞	0	$-180°$

(ii) *For de-segment:* A mirror image of the plot for the ab-segment.

(iii) *For bcd-segment:* $s \equiv \lim\limits_{R\to\infty} Re^{j\theta} \quad +90° < \theta < -90°$

$$|G(s)H(s)| = \lim_{R\to\infty} \frac{1}{Re^{j\theta}\left(Re^{j\theta}+2\right)} = 0$$

Therefore, it maps onto a point at the origin.

(iv) *For efa-segment:* $s \equiv \lim\limits_{\rho\to 0} \rho e^{j\theta} \quad -90° < \theta < +90°$

$$|G(s)H(s)| = \lim_{\rho\to 0}\left\{\frac{1}{\rho e^{j\theta}\left(\rho e^{j\theta}+2\right)}\right\} = \infty$$

To determine $\angle G(s)H(s)$, we observe that

$$\angle \lim_{\rho\to 0} \rho e^{j\theta} = \angle \lim_{\rho\to 0} \rho(\cos\theta + j\sin\theta) = \lim_{\rho\to 0}\left(\tan^{-1}\frac{\rho\sin\theta}{\rho\cos\theta}\right) = \tan^{-1}\tan\theta = \theta$$

$$\angle \lim_{\rho\to 0}\left(\rho e^{j\theta}+2\right) = \angle \lim_{\rho\to 0}[\rho(\cos\theta + j\sin\theta)+2] = \lim_{\rho\to 0}\frac{\rho\sin\theta}{2+\rho\cos\theta} = 0$$

Therefore, the total angle in the denominator $= \theta + 0 = \theta$. When brought to the numerator the angle is $-\theta$. Since θ varies from $-90°$ to $+90°$, the phase of the magnitude varies from $+90°$ to $-90°$. The complete Nyquist plot is given in Figure 15.6(b).

Step 4. No encirclement about $(-1 + \jmath 0)$. Therefore, $N = 0$.

Step 5. No OLTF poles in the RHP. Therefore, $P = 0$.

Step 6. Since $N = 0,\ P = 0, Z = 0$. Therefore, the system is stable.

Closing Nyquist Plots

For $s = -\jmath 0$ to $s = +\jmath 0$, ω varies from $\omega = -0$ to $\omega = +0$ since $s = \jmath\omega$ in the frequency domain. This region is marked by the indented portion [*efa* in Figure 15.6(a)] of the Nyquist path. We showed in Example 15.4 how to calculate directly the magnitude and phase of its mapping in the $G(s)H(s)$-plane so as to complete or close plots generated through mapping of *ab* and *de* segments of the Nyquist path.

However, without doing this exercise every time, a scheme may be developed as to how to close the plots. It is observed from the last two examples that the closure or completion of the plots depends on the *type of the open-loop transfer function*, i.e. the number of poles at the origin. Suppose the type of the open-loop transfer function is n and the transfer function is as follows:

$$G(s)H(s) = \frac{K(sT_a + 1)(sT_b + 1)\cdots}{s^n(sT_p + 1)(sT_q + 1)\cdots}$$

For $s = \lim\limits_{\rho\to 0} \rho e^{\jmath\theta}$, the factor $sT_a + 1$ becomes

$$\begin{aligned} sT_a + 1 &= \lim_{\rho\to 0} \rho e^{\jmath\theta} T_a + 1 \\ &= \lim_{\rho\to 0} \rho(\cos\theta T_a + \jmath \sin\theta T_a) + 1 \\ &\equiv \lim_{\rho\to 0} \left|\sqrt{(\rho\cos\theta T_a + 1)^2 + (\rho\sin\theta T_a)^2}\right| \angle \tan^{-1} \frac{\rho\sin\theta T_a}{\rho\cos\theta T_a + 1} \\ &= |1|\angle 0° \end{aligned}$$

Therefore, for $s = \lim\limits_{\rho\to 0} \rho e^{\jmath\theta}$, all the terms $(sT_a + 1),\ (sT_b + 1), \cdots (sT_p + 1),\ (sT_q + 1)\cdots$ contribute magnitudes of 1 and phases of 0°.

However, for the term s^n, we find

$$\begin{aligned} s^n &= \lim_{\rho\to 0} \rho^n e^{\jmath n\theta} \\ &= \lim_{\rho\to 0} (\rho^n \cos n\theta + \jmath\, \rho^n \sin n\theta) \\ &= \lim_{\rho\to 0} \left|\sqrt{\rho^{2n}(\cos^2 n\theta + \sin^2 n\theta)}\right| \angle \tan^{-1} \frac{\rho^n \sin n\theta}{\rho^n \cos n\theta} \\ &= |0|\angle \tan^{-1} \tan n\theta = |0|\angle n\theta \end{aligned}$$

if θ variation is positive, i.e. in the counterclockwise direction as shown in Figure 15.3(a).

However, if θ variation is negative, as shown in Figure 15.3(b), the angle will be $-n\theta$.

Now, if the variation of s from $-\jmath 0$ to $+\jmath 0$ takes place in the indented portion of the Nyquist path in the positive (CCW) direction and it is represented by the rotation of a vector $\rho \to 0$ along a semicircular arc that describes angles from $-\pi/2$ to $+\pi/2$, the corresponding phase variation in the $G(s)H(s)$-plane will be

$$n\theta = \left(-\frac{n\pi}{2}\right) - \left(+\frac{n\pi}{2}\right) = -n\pi$$

The negative sign indicates that the phase variation of $G(s)H(s)$ takes place in the *clockwise* direction.

If the Nyquist path included the pole at the origin and, therefore, it were indented to the left, s would have varied in the negative (CW) direction and would be represented by the rotation of a vector $\rho \to 0$ along a semicircular arc that describes angles from $-\pi/2$ to $+\pi/2$ in the clockwise direction, the corresponding phase variation in the $G(s)H(s)$-plane will be

$$-n\theta = -\left[\left(-\frac{n\pi}{2}\right) - \left(+\frac{n\pi}{2}\right)\right] = +n\pi$$

The positive sign indicates that the phase variation of $G(s)H(s)$ would take place in the *counterclockwise* direction.

Therefore, we conclude that

1. For a type n system, the Nyquist plots will close from $\omega = -0$ to $\omega = +0$
   (a) in the *clockwise* direction for the Nyquist path excluding poles on the path
   (b) in the *counterclockwise* direction for the Nyquist path including poles on the path

   with an angle of $n\pi$ and an arc of infinite radius.
2. For a type 0 system, the points $\omega = -0$ and $\omega = +0$ will always coincide.

Henceforward, we will follow this rule to close Nyquist plots generated by mapping of *ab* and *de* segments [see Figure 15.6(a)] of the Nyquist path.

15.8 STABILITY MARGINS

In addition to determining the absolute stability of a system, the Nyquist diagrams provide qualitative information as to the degree of its stability. The $(-1+\jmath 0)$ point plays the same role in the Nyquist diagram as the imaginary axis does in the root locus diagram. The comparison is given in Figure 15.7.

1. Consider the root locus of Figure 15.7(a). The crossing of imaginary axis tells that for $K < K_{\text{mar}}$, the system is stable. The corresponding Nyquist plot of Figure 15.7(b) does not encircle the $(1+\jmath 0)$ point.
2. When $K = K_{\text{mar}}$, the system is marginally stable. The corresponding Nyquist curve of Figure 15.7(c) passes through the $(1+\jmath 0)$ point.
3. Finally, when $K > K_{\text{mar}}$, the system is unstable, the corresponding Nyquist curve of Figure 15.7(d) encloses the $(1+\jmath 0)$ point.

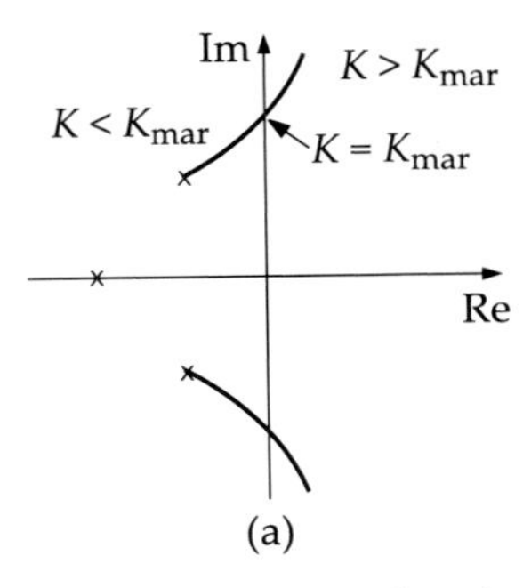

(a)

Root locus diagram showing stable ($K < K_{mar}$), marginally stable($K = K_{mar}$) and unstable ($K > K_{mar}$) conditions.

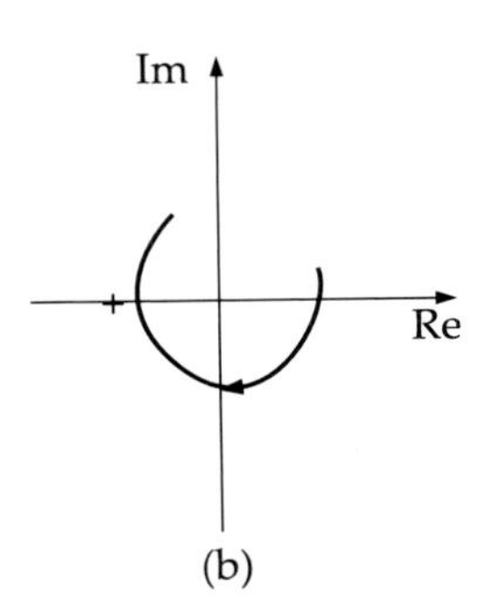

(b)

Nyquist plot (arbitrary) showing no CW encirclement of the $(-1 + j0)$ point. This corresponds to $K < K_{mar}$ condition of the root locus.

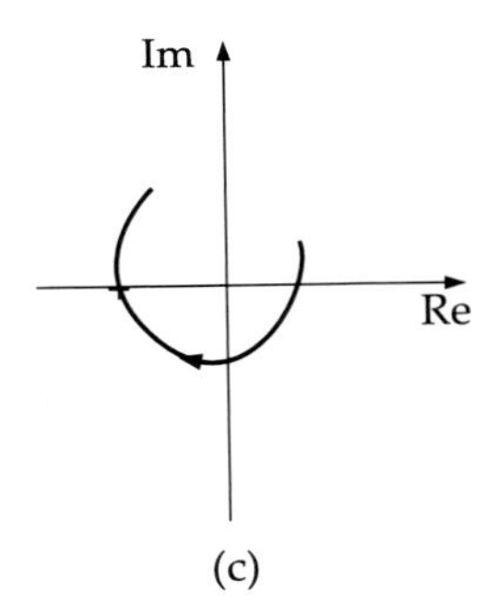

(c)

Nyquist plot (arbitrary) where the plot passes through the $(-1 + j0)$ point. This corresponds to $K = K_{mar}$ condition of the root locus.

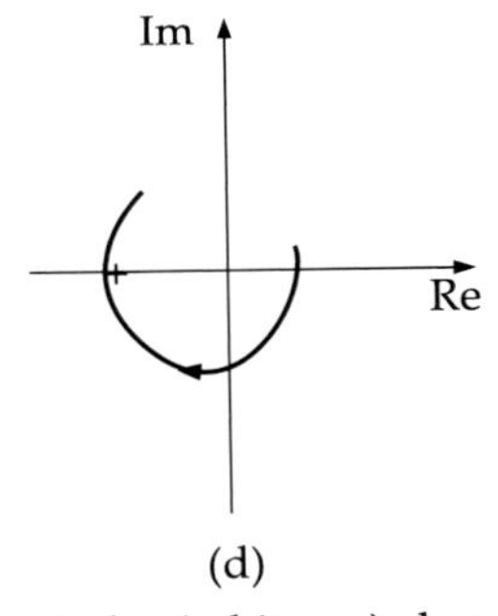

(d)

Nyquist plot (arbitrary) showing CW encirclement of the $(-1 + j0)$ point. This corresponds to $K > K_{mar}$ condition of the root locus.

Figure 15.7 Comparison of the role of Im-axis in root locus and $(-1 + j0)$ point in Nyquist plot.

We know that the quantitative measures of relative stability are provided by gain and phase margins.

Gain Margin

By definition, the gain margin is the reciprocal of the open-loop gain at the frequency where phase is $-180°$ (known as phase crossover frequency, ω_p). In a typical Nyquist diagram (Figure 15.8), A is the point on the plot where the phase is $-180°$. The corresponding open-loop gain is OA. Therefore, Gm $= 1/OA$. Expressed in dB, the gain margin is given by

$$\text{Gm} = -20 \log (OA) \text{ dB}$$

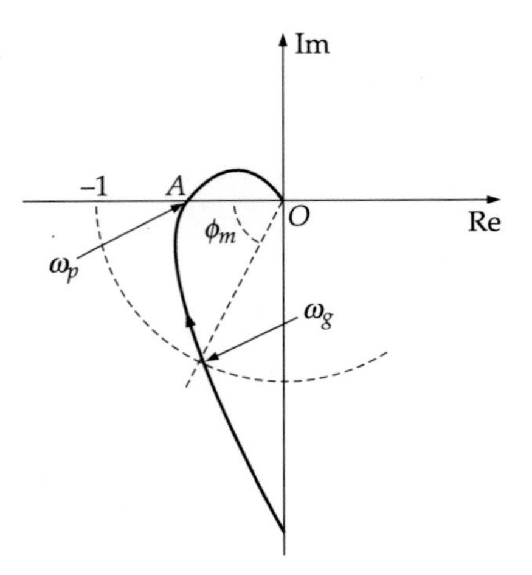

Figure 15.8 Gain and phase margins from Nyquist plot.

Phase Margin

By definition, it is the difference of the phase shift of the system at the gain crossover frequency (i.e. where the gain is unity) plus 180°. So, to determine the phase margin, a circle of radius unity is drawn. The phase margin is given by the subtended angle ϕ_m in Figure 15.8 [see Eq. (14.13)], where the unit circle cuts the Nyquist plot because there the gain is unity.

The following examples will make the procedure clear.

EXAMPLE 15.5. Draw a Nyquist plot, decide stability and calculate the gain margin for the system characterized by its open-loop transfer function

$$G(s)H(s) = \frac{12}{s(s+1)(s+2)}$$

Solution The OLTF shows one pole at the origin (Type 1 system) and none on the $j\omega$-axis. Therefore, the Nyquist path is as given in Figure 15.9(a).

Mapping: (i) *For ab-segment:* $s \equiv j\omega \quad 0 < \omega < \infty$

$$G(j\omega)H(j\omega) = \frac{12}{j\omega(j\omega+1)(j\omega+2)} \equiv \left|\frac{12}{\omega\sqrt{\omega^2+1}\sqrt{\omega^2+4}}\right| \angle\left(-90^\circ - \tan^{-1}\omega - \tan^{-1}\frac{\omega}{2}\right)$$

This equation generates values of magnitude and angle of the frequency-domain open-loop transfer function, which are listed in the following table:

ω	$\lvert G(j\omega)H(j\omega)\rvert$	$\angle G(j\omega)H(j\omega)$
$+0$	∞	-90°
$+\infty$	0	-270°

The mapping of the segment is shown in Figure 15.9(b) with an appropriate label.

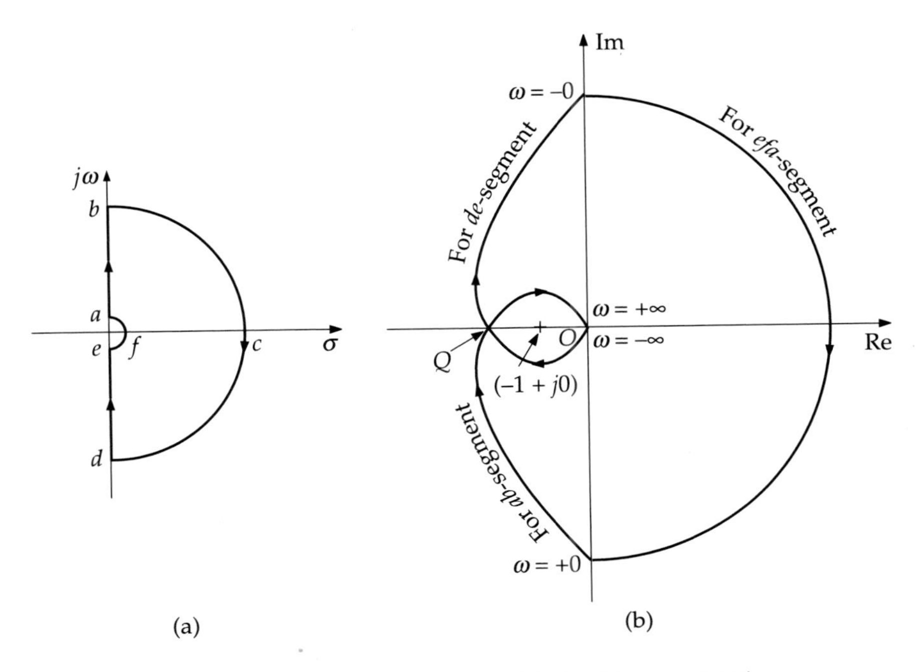

Figure 15.9 (a) Nyquist path; (b) plot (Example 15.5).

(ii) *For bcd-segment:* $s \equiv \lim_{R\to\infty} Re^{j\theta}, \quad +90^\circ < \theta < -90^\circ$

$$G(s)H(s) = \lim_{R\to\infty} \left\{ \frac{12}{Re^{j\theta}(Re^{j\theta}+1)(Re^{j\theta}+2)} \right\} = 0$$

Therefore, this segment maps onto a point at the origin.

(iii) *For de-segment:* The mirror image of the mapping of the *ab*-segment.

(iv) *For efa-segment:* $s \equiv \lim_{\rho\to 0} \rho e^{j\theta} \quad -90^\circ < \theta < +90^\circ$

As this is a type 1 system, the Nyquist plots will close from $\omega = -0$ to $\omega = +0$, sweeping an angle of π, i.e. 180° in the clockwise direction (see Closing Nyquist Plots at page 525). Therefore, the semicircular curve of infinite radius from $\omega = -0$ to $\omega = +0$ maps the segment.

Q-point: Here the curve crosses the Re-axis. Therefore, $\text{Im}\,|G(j\omega)H(j\omega)| = 0$. Now,

$$G(j\omega)H(j\omega) = \frac{12}{j\omega(j\omega+1)(j\omega+2)} = \frac{12}{-3\omega^2 + j(2\omega - \omega^3)}$$

$$= \frac{12\left\{-3\omega^2 - j(2\omega - \omega^3)\right\}}{(-3\omega)^2 + (2\omega - \omega^3)^2}$$

Therefore,

$$\text{Im}\,\{G(\jmath\omega)H(\jmath\omega)\}_{\omega_p} = \left.\frac{-12\,(2\omega-\omega^3)}{(-3\omega)^2+(2\omega-\omega^3)^2}\right|_{\omega_p} = 0$$

or

$$2-\omega_p^2 = 0$$

or

$$\omega_p = \sqrt{2}$$

At this frequency the magnitude is

$$|G(\jmath\omega)H(\jmath\omega)|_{\omega=\sqrt{2}} = \left|\frac{12}{\omega\sqrt{\omega^2+1}\sqrt{\omega^2+4}}\right|_{\omega=\sqrt{2}} = \frac{12}{\sqrt{2}\sqrt{3}\sqrt{6}} = 2$$

Therefore,

$$Q = (-2+\jmath 0)$$

The entire mapping is given in Figure 15.9(b). We find that there is CW encirclement of $(-1+\jmath 0)$ point. Now, since $P=0,\ Z>0$, which indicates the existence of CLTF pole in the RHP. Therefore, the system is unstable.

$$\text{Gm} = -20\log\,(OQ) = -20\log\,(2) = -6.2\text{ dB}$$

EXAMPLE 15.6. For a unity feedback system having an open-loop transfer function

$$G(s) = \frac{K}{s(s+2)(s+6)}$$

determine, using the Nyquist stability criterion, the range of gain K for which the closed-loop system will be stable.

Solution Given $H(s)=1$. Therefore,

$$G(s)H(s) = \frac{K}{s(s+2)(s+6)}$$

So, there is one pole at the origin and none on the $\jmath\omega$-axis. Therefore, the Nyquist path is as given in Figure 15.10(a).

Mapping: (i) *For ab-segment:* $s \equiv \jmath\omega \quad 0<\omega<\infty$

$$G(\jmath\omega)H(\jmath\omega) = \frac{K}{\jmath\omega(\jmath\omega+2)(\jmath\omega+6)} \equiv \frac{K}{\omega\sqrt{\omega^2+4}\sqrt{\omega^2+36}}\angle\left(-90^\circ-\tan^{-1}\frac{\omega}{2}-\tan^{-1}\frac{\omega}{6}\right)$$

The above equation generates the following values of magnitude and angle:

ω	$\lvert G(\jmath\omega)H(\jmath\omega)\rvert$	$\angle G(\jmath\omega)H(\jmath\omega)$
$+0$	∞	-90°
$+\infty$	0	-270°

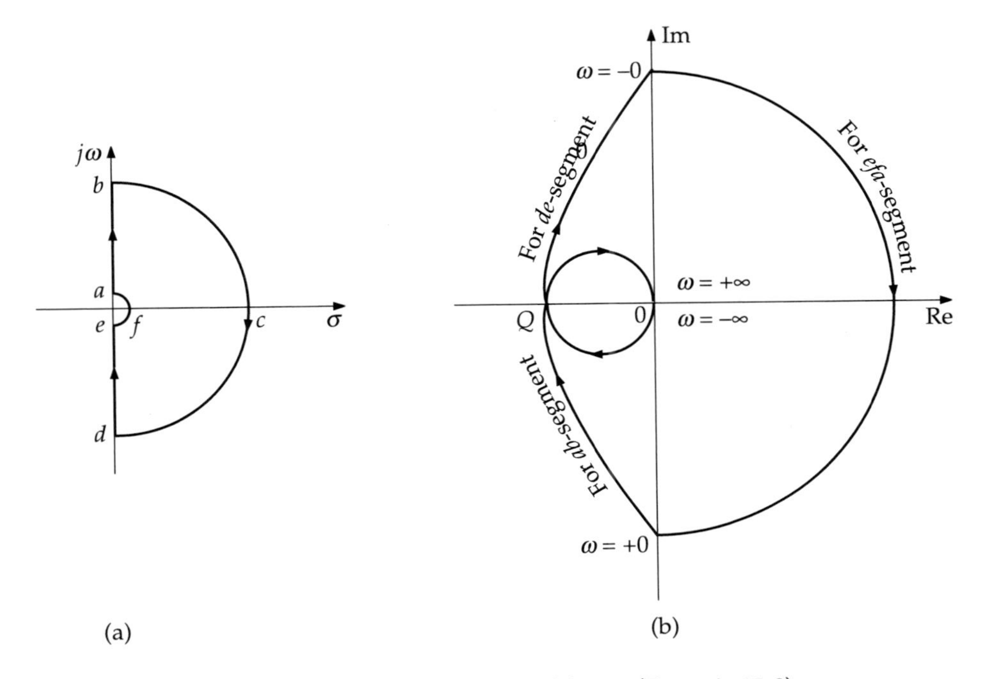

Figure 15.10 (a) Nyquist path; (b) plot (Example 15.6).

The mapping of the segment is shown in Figure 15.10(b) with an appropriate label.

(ii) *For de-segment:* The mirror image of *ab*-segment mapping.

(iii) *For bcd-segment:* $s \equiv \lim_{R\to\infty} Re^{j\theta} \quad +90^\circ < \theta < -90^\circ$

$$G(s)H(s) = \lim_{R\to\infty}\left\{\frac{K}{Re^{j\theta}(Re^{j\theta}+2)(Re^{j\theta}+6)}\right\} = 0$$

Therefore, this segment maps onto a point at the origin.

(iv) *For efa-segment:* $s \equiv \lim_{\rho\to 0} \rho e^{j\theta} \quad -90^\circ < \theta < +90^\circ$

As this is a type 1 system, the Nyquist plots will close from $\omega = -0$ to $\omega = +0$ sweeping an angle of π, i.e. 180° in the clockwise direction (see Closing Nyquist Plots at page 525). Therefore, the semicircular curve of infinite radius from $\omega = -0$ to $\omega = +0$ maps the segment [see Figure 15.10(b)].

Q-point: Here the curve crosses the Re-axis. Therefore, $\text{Im}\,\{G(j\omega)H(j\omega)\}_{\omega_p} = 0$. Now,

$$G(j\omega)H(j\omega) = \frac{K}{j\omega(j\omega+2)(j\omega+6)} = \frac{K}{-8\omega^2 + j(12\omega - \omega^3)} = \frac{K\{-8\omega^2 - j(12\omega-\omega^3)\}}{(-8\omega^2)^2 + (12\omega-\omega^3)^2}$$

Therefore, $$\text{Im}\,\{G(\jmath\omega)H(\jmath\omega)\}_{\omega_p} = \frac{-K\left(12\omega_p - \omega_p^3\right)}{\left(-8\omega_p^2\right)^2 + \left(12\omega_p - \omega_p^3\right)^2} = 0$$

or $$12 - \omega_p^2 = 0$$

or $$\omega_p = 2\sqrt{3}$$

Alternatively

Here the phase of the open-loop transfer function becomes $-180°$. Therefore,

$$-180° = -90° - \tan^{-1}\frac{\omega_p}{2} - \tan^{-1}\frac{\omega_p}{6}$$

or $$90° = \tan^{-1}\frac{\omega_p}{2} + \tan^{-1}\frac{\omega_p}{6} = \tan^{-1}\frac{\dfrac{4\omega_p}{6}}{1 - \dfrac{\omega_p^2}{12}}$$

or $$\frac{\dfrac{4\omega_p}{6}}{1 - \dfrac{\omega_p^2}{12}} = \tan 90° = \infty$$

or $$1 - \frac{\omega_p^2}{12} = 0$$

Therefore, $$\omega_p = 2\sqrt{3}$$

At this frequency the magnitude is

$$|G(\jmath\omega)H(\jmath\omega)|_{\omega_p} = \left.\frac{K}{\omega\sqrt{\omega^2+4}\sqrt{\omega^2+36}}\right|_{\omega_p} = \frac{K}{(2\sqrt{3})(\sqrt{16})(\sqrt{48})} = \frac{K}{96}$$

For stability, this magnitude should be < 1. Therefore, the required condition is $K < 96$.

EXAMPLE 15.7. Determine, through Nyquist's criterion, the stability of the system having open-loop transfer function as

$$G(s)H(s) = \frac{1}{s^2(s+2)}$$

Solution The system has two poles at the origin and none on the $\jmath\omega$-axis. Therefore, the Nyquist path is as given in Figure 15.11(a).

Mapping: (i) *For ab-segment:* $s \equiv \jmath\omega \quad 0 < \omega < \infty$

$$G(\jmath\omega)H(\jmath\omega) = \frac{1}{(\jmath\omega)^2(\jmath\omega+2)} \equiv \frac{1}{\omega^2\sqrt{\omega^2+4}}\angle\left(-180° - \tan^{-1}\frac{\omega}{2}\right)$$

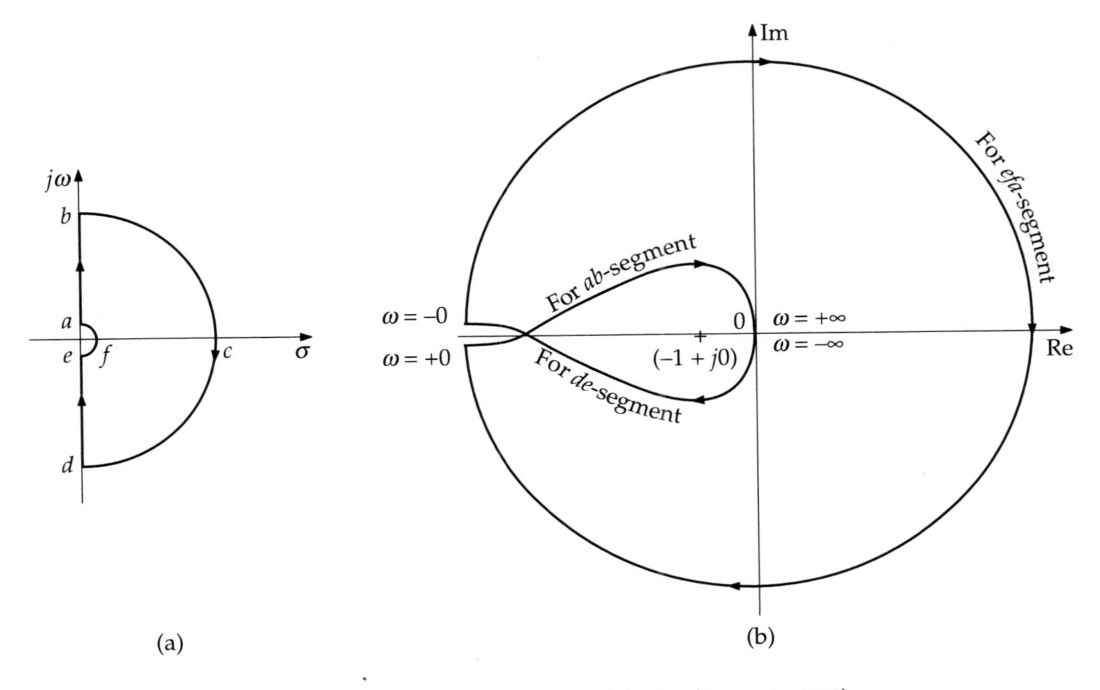

Figure 15.11 (a) Nyquist path; (b) plot (Example 15.7).

The above equation helps generate the following data:

ω	$\lvert G(\jmath\omega)H(\jmath\omega)\rvert$	$\angle G(\jmath\omega)H(\jmath\omega)$
$+0$	∞	-180°
$+\infty$	0	-270°

The corresponding mapping is shown in Figure 15.11(b) with an appropriate label.

(ii) *For de-segment:* The mirror image of ab-segment mapping.

(iii) *For bcd-segment:* $s \equiv \lim\limits_{R\to\infty} Re^{\jmath\theta} \quad +90^\circ < \theta < -90^\circ$

$$G(s)H(s) = \lim_{R\to\infty}\left\{\frac{1}{R^2 e^{\jmath 2\theta}(Re^{\jmath\theta}+2)}\right\} = 0$$

Therefore, this segment maps onto a point at the origin.

(iv) *For efa-segment:* $s \equiv \lim\limits_{\rho\to 0} \rho e^{\jmath\theta} \quad -90^\circ < \theta < +90^\circ$

As this is a type 2 system, the Nyquist plots will close from $\omega = -0$ to $\omega = +0$, sweeping an angle of 2π, i.e. 360° in the clockwise direction (see Closing Nyquist Plots at page 525). Therefore, the circular curve of infinite radius from $\omega = -0$ to $\omega = +0$ maps the segment [see Figure 15.11(b)].

The Nyquist plot shows encirclement of the $(-1 + \jmath 0)$ point twice while the number of RHP poles P in the OLTF is zero. Therefore,

$$2 = Z - 0$$

which indicates that the system, having two poles in the RHP of the CLTF, is unstable.

Note: The points at $\omega = +0$ and $\omega = -0$ are separate in the plot and, therefore, the two encirclements are not apparent. Actually, the two points are one and the same and apart from the inner encirclement, there is another by the outer circle.

EXAMPLE 15.8. Given

$$G(s)H(s) = \frac{K}{s(1 + sT_1)(1 + sT_2)}$$

Find the condition of stability for the system from the Nyquist plot.

Solution The system has one pole at the origin and none on the $\jmath\omega$ -axis. Therefore, the Nyquist path is as given in Figure 15.12(a).

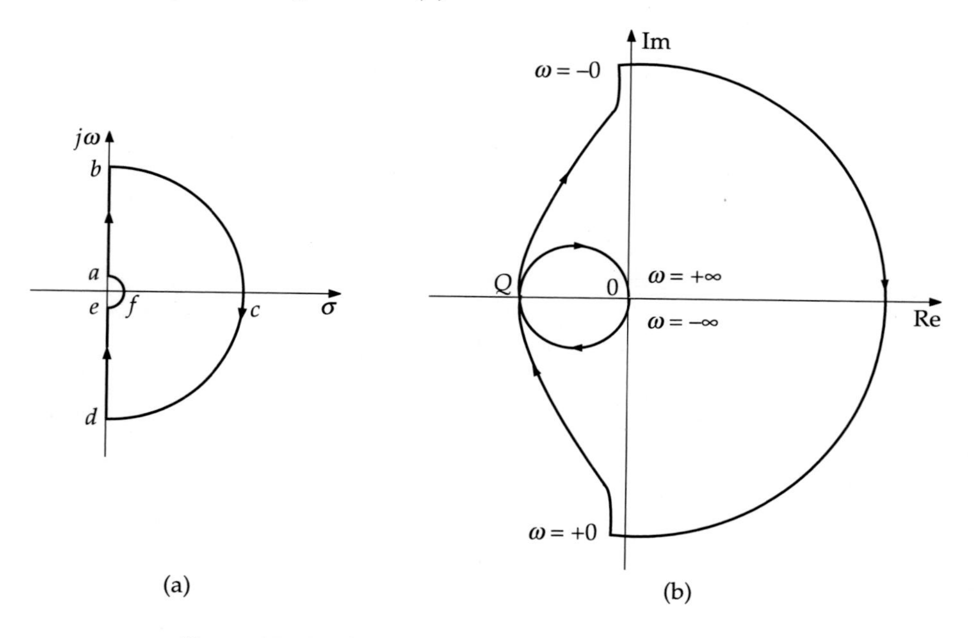

Figure 15.12 (a) Nyquist path; (b) plot (Example 15.8).

Mapping: (i) *For ab-segment:* $s \equiv \jmath\omega \quad 0 < \omega < \infty$

$$G(\jmath\omega)H(\jmath\omega) = \frac{K}{(1 + \jmath\omega T_1)(1 + \jmath\omega T_2)}$$

$$\equiv \frac{K}{\omega\sqrt{1 + \omega^2 T_1^2}\sqrt{1 + \omega^2 T_2^2}} \angle\left(-90^\circ - \tan^1 \omega T_1 - \tan^1 \omega T_2\right)$$

The above equation helps generate the following data:

ω	$\lvert G(\jmath\omega)H(\jmath\omega)\rvert$	$\angle G(\jmath\omega)H(\jmath\omega)$
$+0$	∞	$-90°$
$+\infty$	0	$-270°$

The corresponding mapping is shown in Figure 15.12(b) with an appropriate label.

(ii) *For de-segment:* The mirror image of *ab*-segment mapping.

(iii) *For bcd-segment:* $s \equiv \lim_{R\to\infty} Re^{\jmath\theta} \quad +90° < \theta < -90°$

$$G(s)H(s) = \lim_{R\to\infty} \frac{K}{Re^{\jmath\theta}(1+Re^{\jmath\theta}T_1)(1+Re^{\jmath\theta}T_2)} = 0$$

(iv) *For efa-segment:* $s \equiv \lim_{\rho\to 0} \rho e^{\jmath\theta} \quad -90° < \theta < +90°$

As this is a type 1 system, the Nyquist plots will close from $\omega = -0$ to $\omega = +0$ sweeping an angle of π, i.e. 180° in the clockwise direction (see Closing Nyquist Plots at page 525). Therefore, the semicircular curve of infinite radius from $\omega = -0$ to $\omega = +0$ maps the segment [see Figure 15.12(b)].

Q-point: Here the curve crosses the Re-axis. Therefore, $\text{Im}\,\{G(\jmath\omega)H(\jmath\omega)\}_{\omega_p} = 0$. Now,

$$G(\jmath\omega)H(\jmath\omega) = \frac{K}{(1+\jmath\omega T_1)(1+\jmath\omega T_2)} = \frac{K}{-(T_1+T_2)\omega^2 + \jmath(\omega - T_1T_2\omega^3)}$$

$$= \frac{K\{-(T_1+T_2)\omega^2 - \jmath(\omega - T_1T_2\omega^3)\}}{(T_1+T_2)^2\omega^4 + (\omega - T_1T_2\omega^3)^2}$$

Therefore,

$$\text{Im}\,\{G(\jmath\omega)H(\jmath\omega)\}_{\omega_p} = \left.\frac{(\omega - T_1T_2\omega^3)}{(T_1+T_2)^2\omega^4 + (\omega - T_1T_2\omega^3)^2}\right|_{\omega_p} = 0$$

or

$$1 - T_1T_2\omega_p^2 = 0 \quad \text{or} \quad \omega_p = \frac{1}{\sqrt{T_1T_2}}$$

Alternatively

The phase of the OLTF is $-180°$ when it crosses the Re-axis. The corresponding frequency is the phase crossover frequency ω_p. Therefore,

$$-180° = -90° - \tan^{-1}\omega_pT_1 - \tan^{-1}\omega_pT_2$$

or

$$90° = \tan^{-1}\frac{\omega_p(T_1+T_2)}{1-\omega_p^2T_1T_2}$$

or

$$\frac{\omega_p(T_1+T_2)}{1-\omega_p^2T_1T_2} = \tan 90° = \infty$$

or

$$1 - \omega_p^2T_1T_2 = 0$$

Therefore,

$$\omega_p = \frac{1}{\sqrt{T_1T_2}}$$

At this frequency the magnitude is

$$\left|G(j\omega)H(j\omega)\right|_{\omega=\frac{1}{\sqrt{T_1T_2}}} = \left.\frac{K}{\omega\sqrt{1+\omega^2T_1^2}\sqrt{1+\omega^2T_2^2}}\right|_{\omega=\frac{1}{\sqrt{T_1T_2}}}$$

$$= \frac{K}{\left(\frac{1}{\sqrt{T_1T_2}}\right)\left(\sqrt{1+\frac{T_1}{T_2}}\right)\left(\sqrt{1+\frac{T_2}{T_1}}\right)}$$

$$= \frac{KT_1T_2}{T_1+T_2}$$

For stability, this magnitude should be less than 1. Therefore, the required condition is

$$K < \frac{T_1+T_2}{T_1T_2}$$

EXAMPLE 15.9. A unity feedback system has an open-loop transfer function

$$G(s) = \frac{1}{s(1+2s)(1+s)}$$

Sketch the Nyquist plot for the system and obtain gain and phase margins.

Solution The system has one pole at the origin and none on the $j\omega$-axis. Therefore, the Nyquist path is as given in Figure 15.13(a).

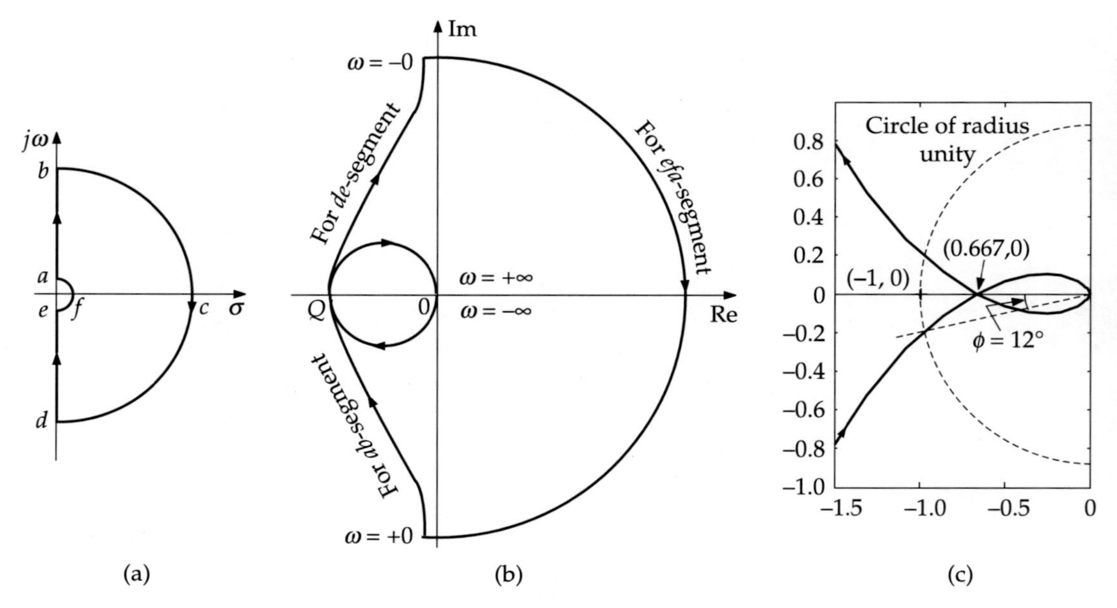

Figure 15.13 (a) Nyquist path; (b) plot; (c) details of phase margin measurement (Example 15.9).

Mapping: (i) *For ab-segment:* $s \equiv \jmath\omega \quad 0 < \omega < \infty$

$$G(\jmath\omega)H(\jmath\omega) = \frac{1}{\jmath\omega(1+\jmath 2\omega)(1+\jmath\omega)}$$

$$\equiv \frac{1}{\omega\sqrt{1+4\omega^2}\sqrt{1+\omega^2}}\angle\left(-90^\circ - \tan^1 2\omega - \tan^1\omega\right)$$

The above equation helps generate the following data:

ω	$\lvert G(\jmath\omega)H(\jmath\omega)\rvert$	$\angle G(\jmath\omega)H(\jmath\omega)$
$+0$	∞	-90°
$+\infty$	0	-270°

The corresponding mapping is shown in Figure 15.13(b) with an appropriate label.

(ii) *For de-segment:* The mirror image of *ab*-segment mapping.

(iii) *For bcd-segment:* $s \equiv \lim_{R\to\infty} Re^{\jmath\theta} \quad +90^\circ < \theta < -90^\circ$

$$G(s)H(s) = \lim_{R\to\infty}\left\{\frac{1}{Re^{\jmath\theta}(1+2Re^{\jmath\theta})(1+Re^{\jmath\theta})}\right\} = 0$$

(iv) *For efa-segment:* $s \equiv \lim_{\rho\to 0}\rho e^{\jmath\theta} \quad -90^\circ < \theta < +90^\circ$

As this is a type 1 system, the Nyquist plots will close from $\omega = -0$ to $\omega = +0$, sweeping an angle of π, i.e. 180° in the clockwise direction. Therefore, the semicircular curve of infinite radius from $\omega = -0$ to $\omega = +0$ maps the segment [see Figure 15.13(b)].

Q-point: Here the curve crosses the Re-axis. Therefore, $\text{Im}\,\{G(\jmath\omega)H(\jmath\omega)\}_{\omega_p} = 0$. Now,

$$G(\jmath\omega)H(\jmath\omega) = \frac{1}{\jmath\omega(1+\jmath 2\omega)(1+\jmath\omega)}$$

$$= \frac{1}{-3\omega^2 + \jmath(\omega - 2\omega^3)}$$

$$= \frac{-3\omega^2 - \jmath(\omega - 2\omega^3)\}}{(-3\omega^2)^2 + (\omega - 2\omega^3)^2}$$

Therefore, $$\text{Im}\,\{G(\jmath\omega)H(\jmath\omega)\}_{\omega_p} = \frac{(\omega_p - 2\omega_p^3)}{(-3\omega_p^2)^2 + (\omega_p - 2\omega_p^3)^2} = 0$$

or $$1 - 2\omega_p^2 = 0$$

or $$\omega_p = \frac{1}{\sqrt{2}}$$

Alternatively

The phase of the OLTF is $-180°$ when it crosses the Re-axis. The corresponding frequency is the phase crossover frequency, ω_p. Therefore,

$$-180° = -90° - \tan^{-1} 2\omega_p - \tan^{-1}\omega_p$$

or
$$90° = \tan^{-1}\frac{3\omega_p}{1-2\omega_p^2}$$

or
$$\frac{3\omega_p}{1-2\omega_p^2} = \tan 90° = \infty$$

or
$$1 - 2\omega_p^2 = 0$$

Thus,
$$\omega_p = \frac{1}{\sqrt{2}}$$

At this frequency the magnitude is

$$\left|G(\jmath\omega)H(\jmath\omega)\right|_{\omega_p} = \left.\frac{1}{\omega\sqrt{1+4\omega^2}\sqrt{1+\omega^2}}\right|_{\omega_p}$$

$$= \frac{1}{\left(\frac{1}{\sqrt{2}}\right)(\sqrt{1+2})\left(\sqrt{1+\frac{1}{2}}\right)} = \frac{2}{3}$$

Therefore,
$$\text{Gm} = -20\log\left(\frac{2}{3}\right)\text{ dB} = 3.521\text{ dB}$$

For measuring the phase margin, a unit circle is drawn with its centre at the origin. It intersects the Nyquist plot at 12° beyond 180°. Hence, the phase margin

$$\text{Pm} = \phi = 12°$$

Note: To measure phase margin from the Nyquist diagram, an accurate plot has to be made with many points on it. It is indeed a time consuming and laborious process. We have shown the relevant area of the plot in Figure 15.13(c).

EXAMPLE 15.10. Determine, through Nyquist's criterion, the stability of the system having open-loop transfer function as

$$G(s)H(s) = \frac{1+4s}{s^2(1+s)(1+2s)}.$$

Solution The system has two poles at the origin and none on the $\jmath\omega$-axis. Therefore, the Nyquist path is as given in Figure 15.14(a).

Mapping: (i) *For ab-segment:* $s \equiv \jmath\omega \quad 0 < \omega < \infty$

$$G(\jmath\omega)H(\jmath\omega) = \frac{1+\jmath 4\omega}{(\jmath\omega)^2(1+\jmath\omega)(1+\jmath 2\omega)}$$

$$\equiv \frac{\sqrt{1+16\omega^2}}{\omega^2\sqrt{1+\omega^2}\sqrt{1+4\omega^2}}\angle(-180° - \tan^{-1}\omega - \tan^{1} 2\omega + \tan^{-1} 4\omega)$$

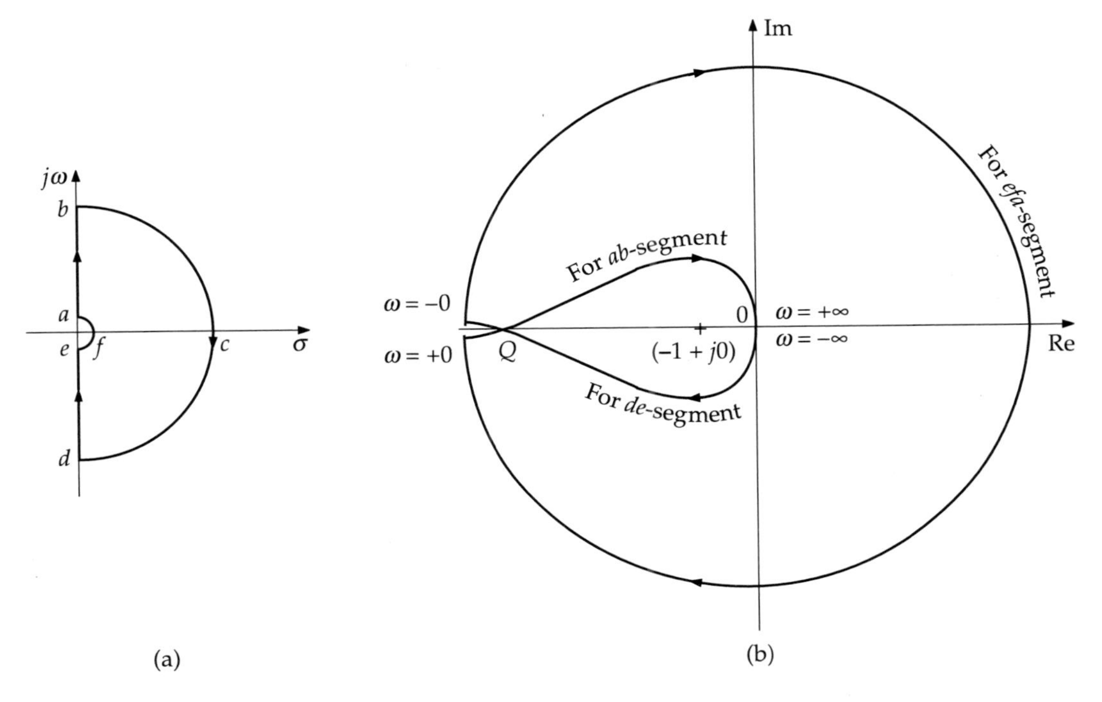

Figure 15.14 (a) Nyquist path; (b) plot (Example 15.10).

The above equation helps generate the following data:

ω	$\lvert G(j\omega)H(j\omega)\rvert$	$\angle G(j\omega)H(j\omega)$
$+0$	∞	-180°
$+\infty$	0	-270°

The corresponding mapping is shown in Figure 15.14(b) with an appropriate label.

(ii) *For de-segment:* The mirror image of *ab*-segment mapping.

(iii) *For bcd-segment:* $s \equiv \lim_{R\to\infty} Re^{j\theta} \quad +90^\circ < \theta < -90^\circ$

$$G(s)H(s) = \lim_{R\to\infty} \frac{1 + 4Re^{j\theta}}{R^2 e^{j2\theta}(1 + Re^{j\theta})(1 + 2Re^{j\theta})}$$

$$= \lim_{R\to\infty} \frac{R\left(\dfrac{1}{R} + 4e^{j\theta}\right)}{R^2 e^{j2\theta} R\left(\dfrac{1}{R} + e^{j\theta}\right) R\left(\dfrac{1}{R} + 2e^{j\theta}\right)}$$

$$= \lim_{R\to\infty} \frac{4e^{j\theta}}{R^3 (e^{j2\theta})(2e^{j\theta})} = 0$$

(iv) *For efa-segment:* $s \equiv \lim_{\rho \to 0} \rho e^{j\theta} \quad -90^\circ < \theta < +90^\circ$

Since this is a type 2 system, the Nyquist plots will close from $\omega = -0$ to $\omega = +0$, sweeping an angle of 2π, i.e. 360° in the clockwise direction (see Closing Nyquist Plots at page 525). Therefore, the circular curve of infinite radius from $\omega = -0$ to $\omega = +0$ maps the segment [see Figure 15.14(b)].

Q point: Since the curve crosses the Re-axis here, $\quad \text{Im}\,\{G(j\omega)H(j\omega)\big|_{\omega_p} = 0$. Now,

$$G(j\omega_p)H(j\omega_p) = \frac{1 + j\,4\omega_p}{(j\omega_p)^2(1 + j\omega_p)(1 + j\,2\omega_p)}$$

$$= \frac{(1 + 4j\omega_p)(1 - j\omega_p)(1 - j\,2\omega_p)}{(j\omega_p)^2(1 + j\omega_p)(1 + j\,2\omega_p)(1 - j\omega_p)(1 - j2\omega_p)}$$

$$= \frac{1 + 10\omega_p^2 + j\omega_p(1 - 8\omega_p^2)}{(j\omega_p)^2(1 + j\omega_p)(1 + j\,2\omega_p)(1 - j\omega_p)(1 - j2\omega_p)} \quad \text{(i)}$$

Equating the imaginary part of the numerator of Eq. (i) to zero, we get

$$1 - 8\omega_p^2 = 0$$

or

$$\omega_p = \sqrt{\frac{1}{8}} = \frac{1}{2\sqrt{2}}$$

Alternatively

The phase of the OLTF is -180° when it crosses the Re-axis. The corresponding frequency is the phase crossover frequency ω_p. Therefore,

$$-180^\circ = -180^\circ - \tan^{-1}\omega_p - \tan^1 2\omega_p + \tan^{-1} 4\omega_p$$

or

$$\tan^{-1} 4\omega_p = \tan^{-1}\omega_p + \tan^1 2\omega_p = \tan^{-1}\frac{3\omega_p}{1 - 2\omega_p^2}$$

or

$$1 - 2\omega_p^2 = \frac{3}{4}$$

Therefore,

$$\omega_p = \frac{1}{2\sqrt{2}}$$

At this frequency, the magnitude is

$$\left.\frac{\sqrt{1 + 16\omega_p^2}}{\omega_p^2\sqrt{1 + \omega_p^2}\sqrt{1 + 4\omega_p^2}}\right|_{\omega_p = \frac{1}{2\sqrt{2}}} = \frac{\sqrt{1 + \dfrac{16}{8}}}{\left(\dfrac{1}{8}\right)^2\sqrt{1 + \dfrac{1}{8}}\sqrt{1 + \dfrac{4}{8}}}$$

$$= 85.3$$

Thus,

$$\text{Gm} = -20\log(85.3)\ \text{dB} = -38.6\ \text{dB}$$

The Nyquist plot shows two encirclements of the $(-1 + \jmath 0)$ point while $P = 0$. Therefore, the system is unstable. The value of the Gm also lends support to it.

EXAMPLE 15.11. Sketch the Nyquist plot and determine the stability of the unity feedback control system given by

$$G(s) = \frac{K}{(1 + sT_1)(1 + sT_2)}$$

Solution The system has no poles at the origin (type 0 system) and none on the $\jmath\omega$-axis. Therefore, the Nyquist path is as given in Figure 15.15(a).

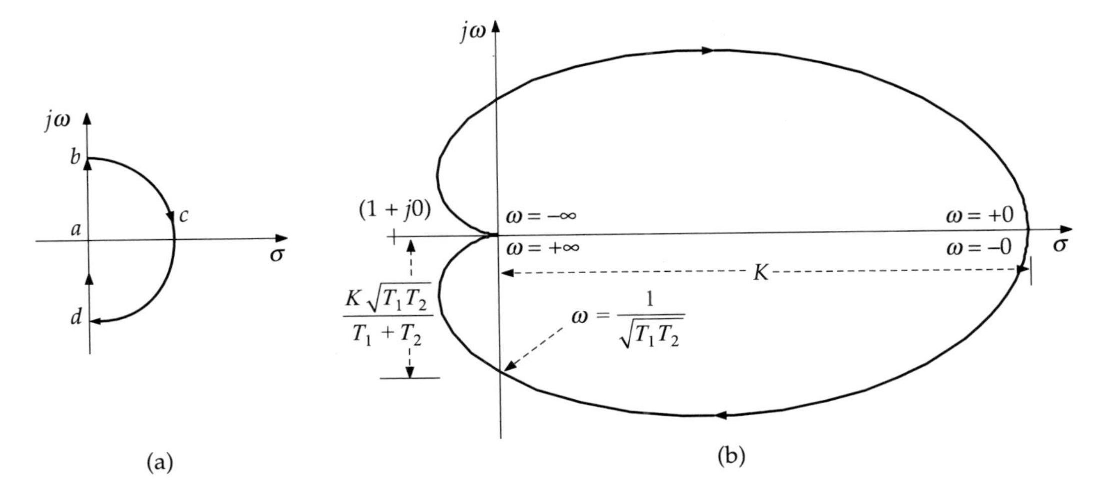

Figure 15.15 (a) Nyquist path; (b) plot (Example 15.11).

Mapping: (i) *For ab-segment:* $s \equiv \jmath\omega \quad 0 < \omega < \infty$

$$G(\jmath\omega)H(\jmath\omega) = \frac{K}{(1 + \jmath\omega T_1)(1 + \jmath\omega T_2)}$$

$$\equiv \frac{K}{\sqrt{1 + \omega^2 T_1^2}\sqrt{1 + \omega^2 T_2^2}} \angle\left(-\tan^{-1}\omega T_1 - \tan^{-1}\omega T_2\right)$$

The above equation helps generate the following data:

ω	$\lvert G(\jmath\omega)H(\jmath\omega)\rvert$	$\angle G(\jmath\omega)H(\jmath\omega)$
$+0$	K	$0°$
$+\infty$	0	$-180°$

The above data indicate that the curve will cross the Im-axis. To find the intercept on the Im-axis, we can isolate the real and imaginary parts of the frequency domain transfer function, equate the real part to 0, and solve the resulting equation to find out the frequency. The magnitude of the frequency domain transfer function at that frequency gives the intercept.

Thus,

$$G(\jmath\omega)H(\jmath\omega) = \frac{K}{(1+\jmath\omega T_1)(1+\jmath\omega T_2)}$$

$$= \frac{K(1-\jmath\omega T_1)(1-\jmath\omega T_2)}{(1+\jmath\omega T_1)(1+\jmath\omega T_2)(1-\jmath\omega T_1)(1-\jmath\omega T_2)}$$

$$= \frac{K(1-\omega^2 T_1 T_2)}{(1+\omega^2 T_1^2)(1+\omega^2 T_2^2)} - \jmath\left\{\frac{K\omega(T_1+T_2)}{(1+\omega^2 T_1^2)(1+\omega^2 T_2^2)}\right\}$$

Equating the real part to zero to obtain the frequency for marginal stability of the system, we get

$$\left.\frac{K(1-\omega^2 T_1 T_2)}{(1+\omega^2 T_1^2)(1+\omega^2 T_2^2)}\right|_{\omega_n} = 0$$

or

$$1-\omega_n^2 T_1 T_2 = 0$$

or

$$\omega_n = \frac{1}{\sqrt{T_1 T_2}}$$

Alternatively

We can find ω_n from the consideration that the phase where the OLTF crosses the Im-axis is $-270°$. Therefore,

$$-270° = -\tan^{-1}\omega_n T_1 - \tan^{-1}\omega_n T_2$$

$$= -\tan^{-1}\frac{\omega_n(T_1+T_2)}{1-\omega_n^2 T_1 T_2}$$

or

$$\frac{\omega_n(T_1+T_2)}{1-\omega_n^2 T_1 T_2} = \tan 270° = \infty$$

or

$$1-\omega_n^2 T_1 T_2 = 0$$

Thus,

$$\omega_n = \frac{1}{\sqrt{T_1 T_2}}$$

The magnitude at this frequency is

$$\left.\frac{K}{\sqrt{1+\omega^2 T_1^2}\sqrt{1+\omega^2 T_2^2}}\right|_{\omega=\frac{1}{\sqrt{T_1 T_2}}} = \frac{K\sqrt{T_1 T_2}}{T_1+T_2}$$

(ii) *For bcd-segment:* $s \equiv \lim_{R\to\infty} Re^{j\theta} \quad +90^\circ < \theta < -90^\circ$

$$G(s)H(s) = \lim_{R\to\infty} \frac{K}{(1+T_1Re^{j\theta})(1+T_2Re^{j\theta})bigl}$$
$$= \lim_{R\to\infty} \frac{K}{R\left(\frac{1}{R}+T_1e^{j\theta}\right)R\left(\frac{1}{R}+T_2e^{j\theta}\right)}$$
$$= \lim_{R\to\infty} \frac{K}{R^2(T_1e^{j\theta})(T_2e^{j\theta})} = 0$$

(iii) *For de-segment:* The mirror image of *ab*-segment mapping.

Since this is a type 0 system, the points $\omega = +0$ and $\omega = -0$ are coincident (see Closing Nyquist Plots at page 525). Taking these factors into consideration, the Nyquist plot, as shown in Figure 15.15(b), is constructed.

The plot shows no encirclement of the $(1 + j0)$ point while $P = 0$. Therefore, the system is stable.

EXAMPLE 15.12. Draw the Nyquist plot for the unity feedback control system with the open-loop transfer function of

$$G(s) = \frac{10(s+2)}{s(s-1)}$$

Using the Nyquist stability criterion, determine the stability of the closed-loop system.

Solution The open-loop transfer function has a pole at the origin and another pole in the RHP. So, $P = 1$ and the Nyquist path is as shown in Figure 15.16(a).

Mapping: (i) *For ab-segment:* $s \equiv j\omega \quad 0 < \omega < \infty$

$$G(j\omega)H(j\omega) = \frac{10(j\omega+3)}{j\omega(j\omega-1)} \tag{i}$$
$$= \frac{10\sqrt{\omega^2+9}}{\omega\sqrt{\omega^2+1}} \angle\left[\tan^{-1}\frac{\omega}{3} - 90^\circ - (\pi - \tan^{-1}\omega)\right] \tag{ii}$$

From Eq. (ii), we can construct the following table:

ω	$\lvert G(j\omega)H(j\omega)\rvert$	$\angle G(j\omega)H(j\omega)$
$+0$	∞	-270°
$+\infty$	0	-90°

The data help us draw the plot for the segment as labelled in Figure 15.16(b). It is obvious from the plot that the curve intersects the Re-axis. To find the intercept on the Re-axis, we rationalize the RHS of Eq. (i), pick out the imaginary part of the numerator, and equate it zero to find the phase cross-over frequency ω_p as follows:

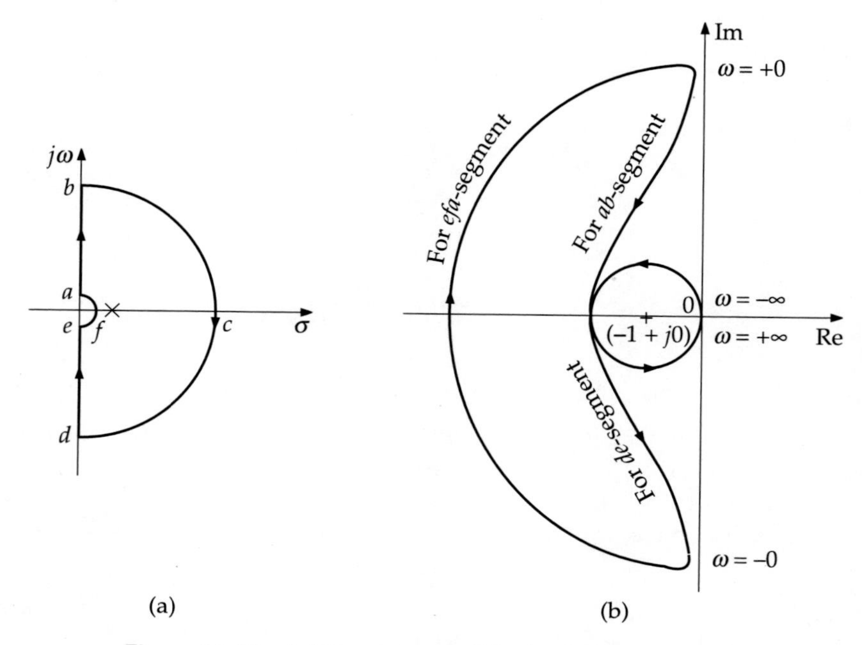

Figure 15.16 (a) Nyquist path; (b) plot (Example 15.12).

$$G(j\omega)H(j\omega) = \frac{10(j\omega+3)}{j\omega(j\omega-1)} = \frac{10(j\omega+3)(-j)(-j\omega-1)}{\omega(j\omega-1)(-j\omega-1)} = \frac{10[-4\omega + j(3-\omega^2)]}{\omega(j\omega-1)(-j\omega-1)}$$

$\Rightarrow$
$$3 - \omega_p^2 = 0$$

or
$$\omega_p = \sqrt{3}$$

Alternatively

The phase of the OLTF is $-180°$ when it crosses the Re-axis. The corresponding frequency is the phase crossover frequency ω_p. Therefore,

$$-180° = -90° - \pi + \tan^{-1}\frac{\omega_p}{3} + \tan^{-1}\omega_p$$

or
$$90° = \tan^{-1}\frac{(4/3)\omega_p}{1-(\omega_p^2/3)}$$

or
$$\frac{(4/3)\omega_p}{1-(\omega_p^2/3)} = \tan 90° = \infty$$

or
$$1 - \frac{\omega_p^2}{3} = 0$$

Therefore,
$$\omega_p = \sqrt{3}$$

The magnitude at ω_p is

$$\left.\frac{10\sqrt{\omega^2+9}}{\omega\sqrt{\omega^2+1}}\right|_{\omega_p} = \frac{10\sqrt{12}}{\sqrt{3}\sqrt{4}} = 10$$

(ii) *For bcd-segment:* $s \equiv \lim_{R\to\infty} Re^{j\theta} \quad +90^\circ < \theta < -90^\circ$

$$G(s)H(s) = \lim_{R\to\infty} \frac{10(Re^{j\theta}+3)}{Re^{j\theta}(Re^{j\theta}-1)} = \lim_{R\to\infty} \frac{10R\left(e^{j\theta}+\dfrac{3}{R}\right)}{R^2 e^{j\theta}\left(e^{j\theta}-\dfrac{1}{R}\right)}$$

$$= 0$$

Therefore, the segment maps onto a point at the origin.

(iii) *For the de-segment:* A mirror image of the curve obtained for the *ab*-segment is drawn.

(iv) *For the efa-segment:* $s \equiv \lim_{\rho\to 0} \rho e^{j\theta} \quad -90^\circ > \theta > +90^\circ$

Since this is a type 1 system, the Nyquist plots will close from $\omega = -0$ to $\omega = +0$, sweeping an angle of π i.e. 180° in the clockwise direction (see Closing Nyquist Plots at page 525). Therefore, the semicircular curve of infinite radius from $\omega = -0$ to $\omega = +0$ maps the segment [see Figure 15.16(b)].

We observe from the figure that there is, *CCW encirclement* of the critical point by the Nyquist plot. So, $N = -1$. From the Nyquist criterion,

$$N = Z - P$$

$$-1 = Z - 1 \quad \text{or} \quad Z = 0$$

Thus, the system is stable.

Note: The example shows that, despite the presence of a pole in the RHP in the open-loop transfer function, the closed-loop system can be stable. We will consider more such examples.

EXAMPLE 15.13. The open-loop transfer function of a unity feedback system is given by

$$G(s) = \frac{8s}{(s-1)(s-2)}$$

Find the closed-loop stability of the system by the Nyquist method.

Solution The open-loop transfer function possesses two poles in the RHP. Therefore, $P = 2$. There is a zero at the origin. So the Nyquist path is as shown in Figure 15.17(a).

Mapping: (i) *For ab-segment:* $s \equiv j\omega \quad 0 < \omega < \infty$

$$G(j\omega)H(j\omega) = \frac{j8\omega}{(j\omega-1)(j\omega-2)} \tag{i}$$

$$= \frac{8\omega}{\sqrt{\omega^2+1}\sqrt{\omega^2+4}} \angle\left[90^\circ - (\pi - \tan^{-1}\omega) - \left(\pi - \tan^{-1}\frac{\omega}{2}\right)\right]$$

$$= \frac{8\omega}{\sqrt{\omega^2+1}\sqrt{\omega^2+4}} \angle\left(-270^\circ + \tan^{-1}\omega + \tan^{-1}\frac{\omega}{2}\right) \tag{ii}$$

From Eq. (ii), we can construct the following table:

ω	$\lvert G(\jmath\omega)H(\jmath\omega)\rvert$	$\angle G(\jmath\omega)H(\jmath\omega)$
$+0$	0	$-270°$
$+\infty$	0	$-90°$

We note that the magnitude is 0 both when $\omega \to 0$ and $\omega \to \infty$, while the angle varies from $-270°$ to $-90°$. Therefore, it is evident that the polar plot has to cross the Re-axis at a point.

Intersection in the Re-axis: To find the intercept, as before, we rationalize the complex open-loop transfer function and equate the imaginary part of the numerator to zero. The phase crossover frequency ω_p found thereby is used to calculate the magnitude of the intercept. Thus,

$$G(\jmath\omega)H(\jmath\omega) = \frac{\jmath 8\omega}{(\jmath\omega - 1)(\jmath\omega - 2)}$$

$$= \frac{8(\jmath\omega)(-\jmath\omega - 1)(-\jmath\omega - 2)}{(\jmath\omega - 1)(\jmath\omega - 2)(-\jmath\omega - 1)(-\jmath\omega - 2)}$$

$$= \frac{8(\jmath\omega)(-\omega^2 + 2 + \jmath 3\omega)}{(\jmath\omega - 1)(\jmath\omega - 2)(-\jmath\omega - 1)(-\jmath\omega - 2)}$$

$$= \frac{8\big[-3\omega^2 + \jmath\omega(-\omega^2 + 2)\big]}{(\jmath\omega - 1)(\jmath\omega - 2)(-\jmath\omega - 1)(-\jmath\omega - 2)}$$

Equating the imaginary part of the numerator to zero, we get

$$\omega_p^2 = 2 \quad \text{or} \quad \omega_p = \sqrt{2}$$

Alternatively

The phase of the OLTF is $-180°$ when it crosses the Re-axis. The corresponding frequency is the phase crossover frequency ω_p. Therefore,

$$-180° = -270° + \tan^{-1}\omega_p + \tan^{-1}\frac{\omega_p}{2}$$

or

$$90° = \tan^{-1}\frac{(3/2)\omega_p}{1 - (\omega_p^2/2)}$$

or

$$\frac{(3/2)\omega_p}{1 - (\omega_p^2/2)} = \tan 90° = \infty$$

or

$$1 - \frac{\omega_p^2}{2} = 0$$

Therefore,

$$\omega_p = \sqrt{2}$$

The magnitude at ω_p is

$$\frac{8\sqrt{2}}{\sqrt{2+1}\sqrt{2+4}} = \frac{8}{3} = 2.67$$

Thus, the ab-segment of the Nyquist path maps onto a complete curve (because it starts from 0 and ends at 0) intersecting the Re-axis at a distance of 2.67, as shown in Figure 15.17. Thereby, it encircles the critical point in a CCW direction.

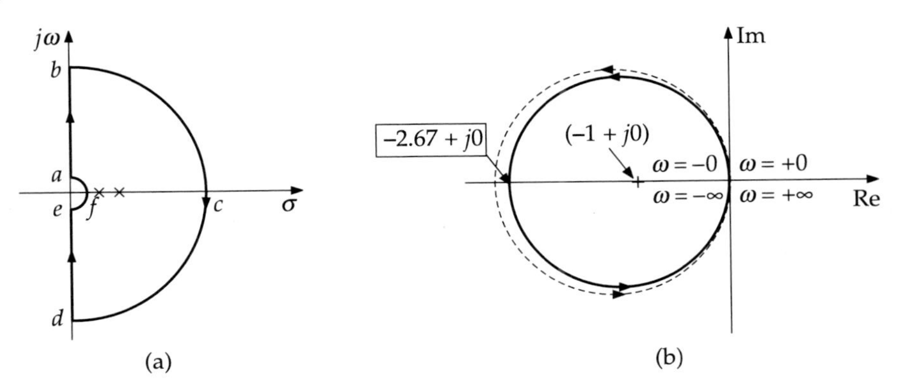

Figure 15.17 (a) Nyquist path; (b) plot (Example 15.13). The dashed curve, which maps to the *de*-segment of the Nyquist path should actually coincide with the solid curve that maps to the *ab*-segment. To show its existence, it has been displaced to some extent.

The de-segment, which maps to form a mirror image of that of the ab-segment, also encircles the critical point in a CCW direction. Since this is a type 0 system, the points $\omega = -0$ and $\omega = +0$ are coincident (see Closing Nyquist Plots at page 525).

Thus, $N = -2$ and, according to the Nyquist criterion,

$$N = Z - P$$
$$-2 = Z - 2 \quad \text{or} \quad Z = 0$$

The result indicates that the closed-loop system is stable.

EXAMPLE 15.14. Examine by the Nyquist method the stability of a unity feedback system having the open-loop transfer function of

$$G(s) = \frac{2(s+3)}{s(s-1)}$$

Solution The open-loop transfer function indicates the presence of a pole in the RHP and another at the origin. Hence, $P = 1$ and the Nyquist path is represented by Figure 15.18(a).

Mapping: (i) *For ab-segment:* $\quad s \equiv \jmath\omega \quad 0 < \omega < \infty$

$$G(\jmath\omega)H(\jmath\omega) = \frac{2(\jmath\omega + 3)}{\jmath\omega(\jmath\omega - 1)} \tag{i}$$

$$= \frac{2\sqrt{\omega^2+9}}{\omega\sqrt{\omega^2+1}}\angle\left[\tan^{-1}\frac{\omega}{3} - 90^\circ - (\pi - \tan^{-1}\omega)\right]$$

$$= \frac{2\sqrt{\omega^2+9}}{\omega\sqrt{\omega^2+1}}\angle\left(\tan^{-1}\frac{\omega}{3} - 270^\circ + \tan^{-1}\omega\right) \tag{ii}$$

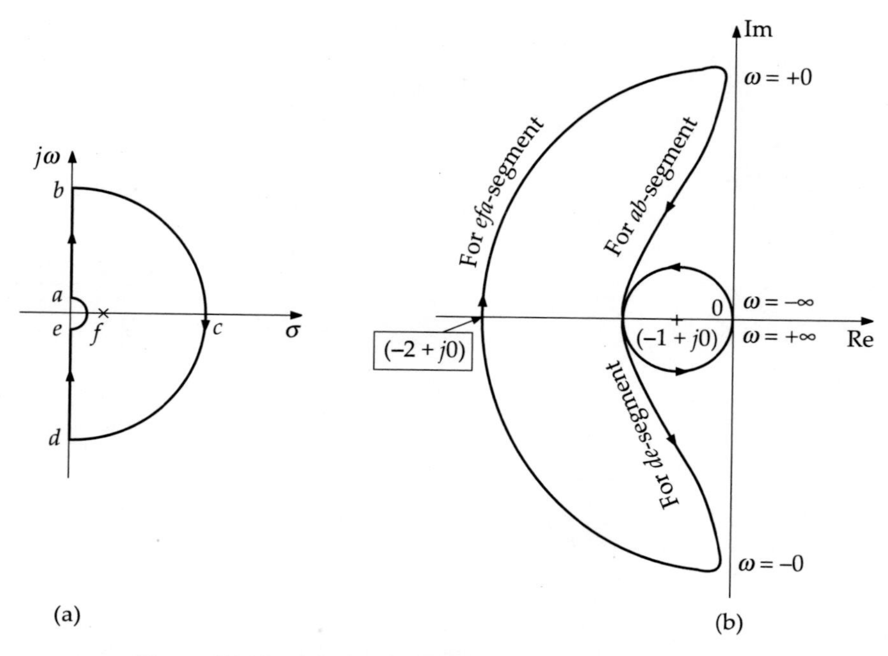

Figure 15.18 (a) Nyquist path; (b) plot (Example 15.14).

From Eq. (ii), we can construct the following table:

ω	$\lvert G(j\omega)H(j\omega)\rvert$	$\angle G(j\omega)H(j\omega)$
$+0$	∞	$-270°$
$+\infty$	0	$-90°$

The data amply indicate that the polar plot intersects the Re-axis at some point. To figure that out, we proceed in the usual way of rationalizing the complex open-loop transfer function and equating the imaginary part of the numerator to zero. This will give us the phase crossover frequency ω_p, for which we need to calculate the magnitude. Thus, from Eq. (i),

$$G(j\omega)H(j\omega) = \frac{2(j\omega + 3)}{j\omega(j\omega - 1)}$$

$$= \frac{2(j\omega + 3)(-j)(-j\omega - 1)}{\omega(j\omega - 1)(-j\omega - 1)}$$

$$= \frac{-4\omega + j(-\omega^2 + 3)}{\omega(j\omega - 1)(-j\omega - 1)}$$

Equating the imaginary part of the numerator to zero, we get

$$\omega_p^2 = 3$$

or

$$\omega_p = \sqrt{3}$$

Alternatively

The phase of the OLTF is -180° when it crosses the Re-axis. The corresponding frequency is the phase crossover frequency ω_p. Therefore,

$$-180^\circ = -270^\circ + \tan^{-1}\frac{\omega_p}{3} + \tan^{-1}\omega_p$$

or

$$90^\circ = \tan^{-1}\frac{(4/3)\omega_p}{1-(\omega_p^2/3)}$$

or

$$\frac{(4/3)\omega_p}{1-(\omega_p^2/3)} = \tan 90^\circ = \infty$$

or

$$1 - \frac{\omega_p^2}{3} = 0$$

Therefore,

$$\omega_p = \sqrt{3}$$

The magnitude at ω_p is

$$\left.\frac{2\sqrt{\omega^2+9}}{\omega\sqrt{\omega^2+1}}\right|_{\omega_p} = \frac{2\sqrt{3+9}}{\sqrt{3}\sqrt{3+1}} = 2$$

With these available data, we make a polar plot that maps the ab-segment of the Nyquist path and is appropriately labelled in Figure 15.18(b).

(ii) *For bcd-segment:* $s \equiv \lim_{R\to\infty} Re^{j\theta} \quad +90^\circ < \theta < -90^\circ$

$$G(s)H(s) = \lim_{R\to\infty}\frac{2(Re^{j\theta}+3)}{Re^{j\theta}(Re^{j\theta}-1)} = \lim_{R\to\infty}\frac{2R\left(e^{j\theta}+\dfrac{3}{R}\right)}{R^2e^{j\theta}\left(e^{j\theta}-\dfrac{1}{R}\right)} = 0$$

Therefore, the segment maps to a point at the origin.

(iii) *For the de-segment:* A mirror image of the curve obtained for the ab-segment is drawn.

(iv) *For the efa-segment:* $s \equiv \lim_{\rho\to 0}\rho e^{j\theta} \quad -90^\circ > \theta > +90^\circ$

Since this is a type 1 system, the Nyquist plots will close from $\omega = -0$ to $\omega = +0$, sweeping an angle of π, i.e. 180° in the clockwise direction (see Closing Nyquist Plots at page 525). Therefore, the semicircular curve of infinite radius from $\omega = -0$ to $\omega = +0$ maps the segment [see Figure 15.18(b)].

We observe from the figure that there is a *CCW encirclement* of the critical point by the Nyquist plot. So, $N = -1$. From the Nyquist criterion,

$$N = Z - P$$
$$-1 = Z - 1 \quad \text{or} \quad Z = 0$$

Thus, the system is stable.

EXAMPLE 15.15. The open-loop transfer function of a unity feedback system is given by

$$G(s) = \frac{s+2}{(s+1)(s-1)}$$

Examine the stability of the closed-loop system by Nyquist criteria.

Solution The open-loop transfer function consists of a pole in the RHP and none at the origin or on the Im-axis. Therefore, $P = 1$, and the Nyquist path is as shown in Figure 15.19(a).

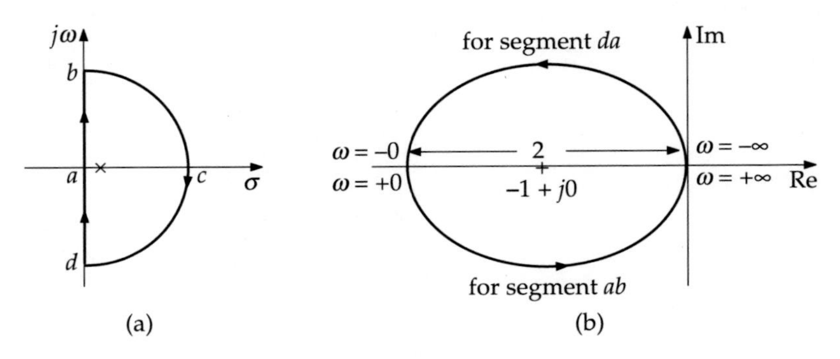

Figure 15.19 (a) Nyquist path; (b) plot (Example 15.15).

Mapping: (i) *For ab-segment:* $s \equiv j\omega \quad 0 < \omega < \infty$. The frequency-domain open-loop transfer function can be written as follows:

$$G(j\omega)H(j\omega) = \frac{\sqrt{\omega^2+4}}{\sqrt{\omega^2+1}\sqrt{\omega^2+1}} \angle\left[\tan^{-1}\frac{\omega}{2} - \tan^{-1}\omega - (\pi - \tan^{-1}\omega)\right]$$

$$= \frac{\sqrt{\omega^2+4}}{\sqrt{\omega^2+1}\sqrt{\omega^2+1}} \angle\left(\tan^{-1}\frac{\omega}{2} - \pi\right) \tag{i}$$

From Eq. (i) we can build the following table:

ω	$\lvert G(j\omega)H(j\omega)\rvert$	$\angle G(j\omega)H(j\omega)$
$+0$	2	$-180°$
$+\infty$	0	$-90°$

These data help us construct the polar plot for the ab-segment [see 'for segment ab'-labelled part of Figure 15.19(b)].

(ii) *For da-segment:* The da-segment of the Nyquist path maps to form a mirror image of that of the ab-segment. Since this is a type 0 system, the points $\omega = -0$ and $\omega = +0$ are coincident (see Closing Nyquist Plots at page 525).

Thus, the Nyquist plot shows a *CCW encirclement* of the critical point so that $N = -1$. From the Nyquist criterion,

$$N = Z - P$$
$$-1 = Z - 1 \quad \text{or} \quad Z = 0$$

Therefore, the system is stable.

EXAMPLE 15.16. A unity feedback system has the open-loop transfer function of

$$G(s) = \frac{2}{s(1-2s)}$$

Draw the Nyquist plot and determine the closed-loop stability.

Solution The open-loop transfer function consists of one pole in the RHP and another at the origin. Therefore, $P = 1$ and the Nyquist path is as shown in Figure 15.20(a).

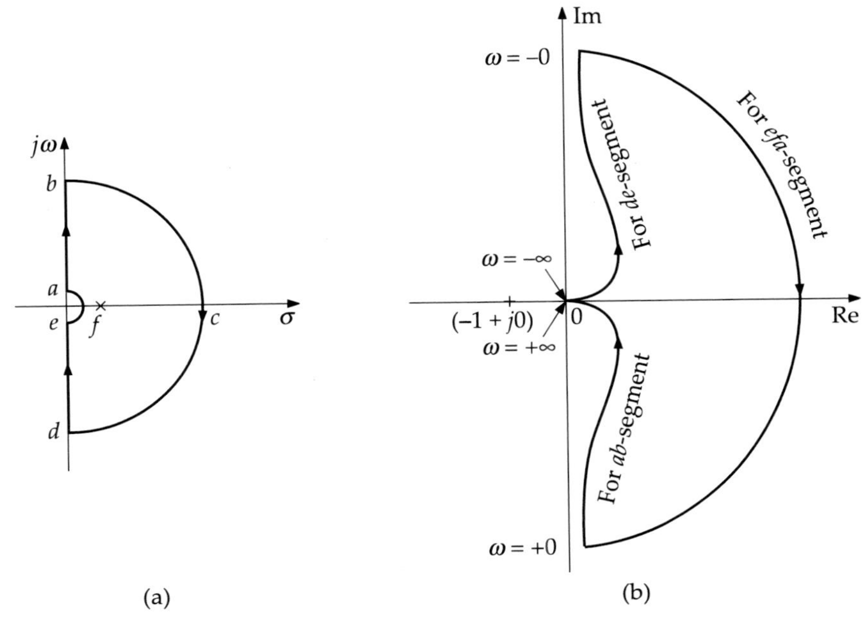

Figure 15.20 (a) Nyquist path; (b) plot (Example 15.16).

Mapping: (i) *For ab-segment:* $s \equiv j\omega \quad 0 < \omega < \infty$.

The open-loop transfer function can then be written as

$$\begin{aligned} G(j\omega)H(j\omega) &= \frac{2}{j\omega(1 - j2\omega)} \\ &= \frac{2}{\omega\sqrt{1 + 4\omega^2}} \angle[-90^\circ + \tan^{-1} 2\omega] \end{aligned} \tag{i}$$

From Eq. (i), we can construct the following table:

ω	$\lvert G(j\omega)H(j\omega)\rvert$	$\angle G(j\omega)H(j\omega)$
$+0$	∞	-90°
$+\infty$	0	0°

The corresponding polar plot is shown with an appropriate label in Figure 15.20(b).

(ii) *For bcd-segment:* $s \equiv \lim_{R\to\infty} Re^{j\theta} \quad +90^\circ > \theta > -90^\circ$

$$G(s)H(s) \equiv \lim_{R\to\infty} \frac{2}{Re^{j\theta}(1 - 2Re^{j\theta})} = 0$$

Thus, this segment of the Nyquist path maps onto a point at the origin in the Nyquist plot.

(iii) *For de-segment:* This segment maps to a mirror image of that of the ab-segment.

(iv) *For efa-segment:* $s \equiv \lim_{\rho \to 0} \rho e^{j\theta}, \quad -90^\circ > \theta > +90^\circ$

Since this is a type 1 system, the Nyquist plots will close from $\omega = -0$ to $\omega = +0$, sweeping an angle of π, i.e. 180° in the clockwise direction (see Closing Nyquist Plots at page 525). Therefore, the semicircular curve of infinite radius from $\omega = -0$ to $\omega = +0$ maps the segment [see Figure 15.20(b)].

From the plot we observe that the curve is CW throughout and that there is no encirclement of the critical point, i.e. $N = 0$. So, from the Nyquist criterion,

$$N = Z - P$$

$$0 = Z - 1 \quad \text{or} \quad Z = 1 \tag{ii}$$

Equation (ii) indicates the presence of a pole in the RHP of the CLTF. Therefore, the closed-loop system is unstable.

Note: This example demonstrates that a system can be unstable even if there is no encirclement of the critical point by the Nyquist plot.

EXAMPLE 15.17. Check the stability of the closed-loop system by the Nyquist criterion when open-loop transfer function of the unity feedback system is given by

$$G(s) = \frac{10}{s^2(1 + 0.2s)(1 + 0.5s)}$$

Solution The open-loop transfer function possesses no poles in the RHP and two poles at the origin. Therefore, $P = 0$, and the Nyquist path is as shown in Figure 15.21.

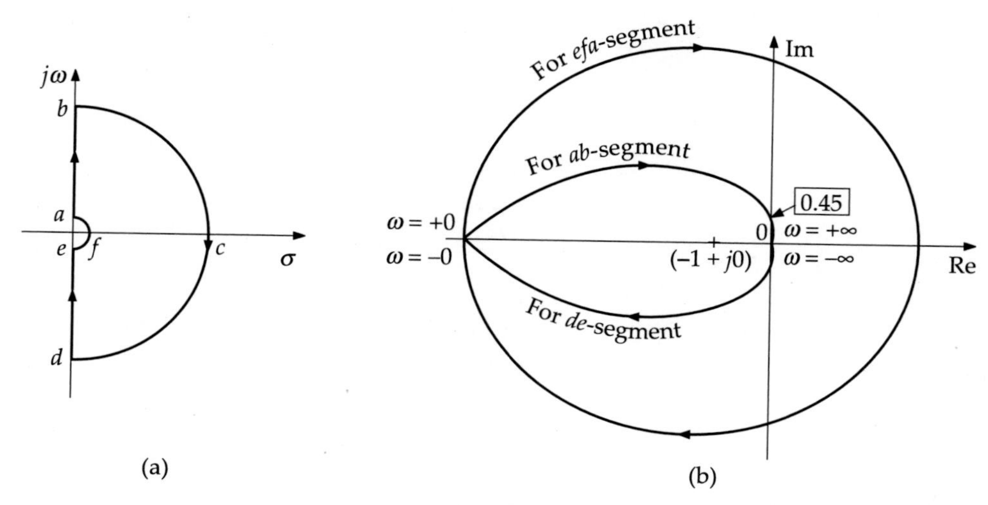

Figure 15.21 (a) Nyquist path; (b) plot (Example 15.17).

Mapping: (i) *For ab-segment:* $s \equiv \jmath\omega \quad 0 < \omega < \infty$. The open-loop transfer function can be then written as

$$G(\jmath\omega)H(\jmath\omega) = \frac{10}{(\jmath\omega)^2(1+\jmath0.2\omega)(1+\jmath0.5\omega)}$$

$$= \frac{10}{\omega^2\sqrt{1+0.04\omega^2}\sqrt{1+0.25\omega^2}}\angle(-180^\circ - \tan^{-1}0.2\omega - \tan^{-1}0.5\omega) \qquad \text{(i)}$$

From Eq. (i), we can construct the following table:

ω	$\lvert G(\jmath\omega)H(\jmath\omega)\rvert$	$\angle G(\jmath\omega)H(\jmath\omega)$
$+0$	∞	-180°
$+\infty$	0	-360°

From the two angles it is apparent that if a polar plot has to traverse from -180° at magnitude ∞ to -360° at magnitude 0, it has to intersect the Im-axis somewhere. So, we calculate the intercept by first rationalizing the frequency-domain open-loop transfer function and then equating the real part of the numerator to zero to find the corresponding frequency. This frequency, for which the system is marginally stable is, in fact, its natural frequency of oscillation ω_n because here the damping ratio $\zeta = 0$. From this frequency, we calculate the magnitude. Thus,

$$G(\jmath\omega)H(\jmath\omega) = \frac{10}{(\jmath\omega)^2(1+\jmath0.2\omega)(1+\jmath0.5\omega)}$$

$$= \frac{10(1-\jmath0.2\omega)(1-\jmath0.5\omega)}{(\jmath\omega)^2(1+\jmath0.2\omega)(1+\jmath0.5\omega)(1-\jmath0.2\omega)(1-\jmath0.5\omega)}$$

$$= \frac{10[(1-0.1\omega^2)-\jmath0.7\omega]}{(\jmath\omega)^2(1+\jmath0.2\omega)(1+\jmath0.5\omega)(1-\jmath0.2\omega)(1-\jmath0.5\omega)}$$

Hence,
$$1 - 0.1\omega_n^2 = 0$$

or
$$\omega_n^2 = \frac{1}{0.1}$$

or
$$\omega_n = \sqrt{10}$$

Alternatively

At the point where the polar plot intersects the Im-axis, its phase is -270°. So, we have

$$-270^\circ = -180^\circ - \tan^{-1}0.2\omega_n - \tan^{-1}0.5\omega_n$$

or
$$90^\circ = \tan^{-1}\frac{0.7\omega_n}{1-0.1\omega_n^2}$$

or
$$\frac{0.7\omega_n}{1-0.1\omega_n^2} = \tan 90^\circ = \infty$$

or
$$1 - 0.1\omega_n^2 = 0$$

Therefore,
$$\omega_n = \sqrt{10}$$

The corresponding magnitude is

$$|G(\jmath\omega)H(\jmath\omega)|_{\omega_n} = \left.\frac{10}{\omega^2\sqrt{1+0.04\omega^2}\sqrt{1+0.25\omega^2}}\right|_{\omega_n}$$

$$= \frac{10}{10\sqrt{1.4}\sqrt{3.5}} = 0.45$$

With these available data, we draw the polar plot corresponding to ab-segment of the Nyquist path. It is shown with an appropriate label in Figure 15.21(b).

(ii) *For de-segment:* The mapping for this segment is the mirror image of that for the ab-segment. These two plots complete the encirclement of the critical point once. We are not calculating the contribution of the bcd-segment which will be a point at the origin.

(iii) *For efa-segment:* $s \equiv \lim_{\rho\to 0} \rho e^{\jmath\theta} \quad -90^\circ > \theta > +90^\circ$

Since this is a type 2 system, the Nyquist plots will close from $\omega = -0$ to $\omega = +0$, sweeping an angle of 2π, i.e. 360° in the clockwise direction (see Closing Nyquist Plots at page 525). Therefore, the circular curve of infinite radius from $\omega = -0$ to $\omega = +0$ maps the segment.

The complete Nyquist plot is shown in Figure 15.21(b). From the plot we observe that the curve is CW throughout and that there are two encirclements of the critical point, i.e. $N = 2$. So, from the Nyquist criterion,

$$N = Z - P$$

$$2 = Z - 0$$

or
$$Z = 2 \tag{ii}$$

Equation (ii) indicates the presence of two poles in the RHP of the CLTF. Therefore, the closed-loop system is unstable.

EXAMPLE 15.18. The open-loop transfer function of a system is given by

$$G(s)H(s) = \frac{K(1+s)}{1-s}$$

Determine through the Nyquist formalism the value of K that will make the closed-loop system stable.

Solution The open-loop transfer function possesses one pole in the RHP and no pole at the origin. Therefore, $P = 1$, and the Nyquist path is as shown in Figure 15.22(a).

Mapping: (i) *For ab-segment:* $s \equiv \jmath\omega \quad 0 < \omega < \infty$

The open-loop transfer function can then be written as

$$G(\jmath\omega)H(\jmath\omega) = \frac{K(1+\jmath\omega)}{1-\jmath\omega}$$

$$= \frac{K\sqrt{1+\omega^2}}{\sqrt{1+\omega^2}}\angle 2\tan^{-1}\omega \tag{i}$$

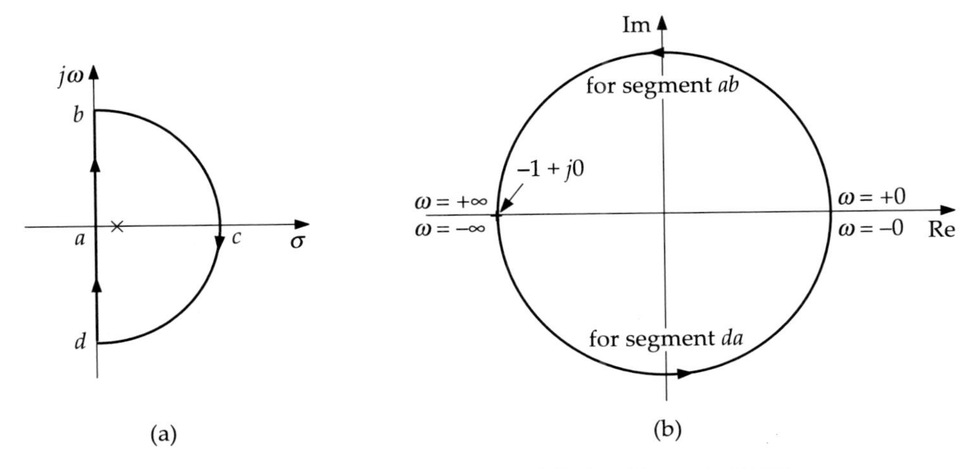

Figure 15.22 (a) Nyquist path; (b) plot (Example 15.18).

We can construct the following table from Eq. (i):

ω	$\|G(j\omega)H(j\omega)\|$	$\angle G(j\omega)H(j\omega)$
$+0$	K	$0°$
1	K	$90°$
$+\infty$	K	$180°$

The polar plot for the above data is a semicircular curve of radius K.

The da-segment will map to form a mirror image of this curve. Since this is a type 0 system, the points $\omega = -0$ and $\omega = +0$ are coincident (see Closing Nyquist Plots at page 525).

So, the complete Nyquist plot is given by Figure 15.22(b). It shows a CCW encircling. For $K = 1$, it passes through the critical point making the system marginally stable. The system is stable if $K > 1$ when the Nyquist plot encircles the critical point and, therefore, $N = -1$. Then,

$$N = Z - P$$
$$-1 = Z - 1 \quad \text{or} \quad Z = 0$$

This means that for $K > 1$ there will be no RHP pole in the CLTF and, therefore, the system will be stable.

EXAMPLE 15.19. The open-loop transfer function of a feedback control system is given by

$$G(s)H(s) = \frac{-1}{2s(1 - 20s)}$$

Determine the stability of the closed-loop system by Nyquist criteria.

Solution The open-loop transfer function consists of a pole in the RHP and another at the origin. So, $P = 1$ and the Nyquist path is as shown in Figure 15.23(a).

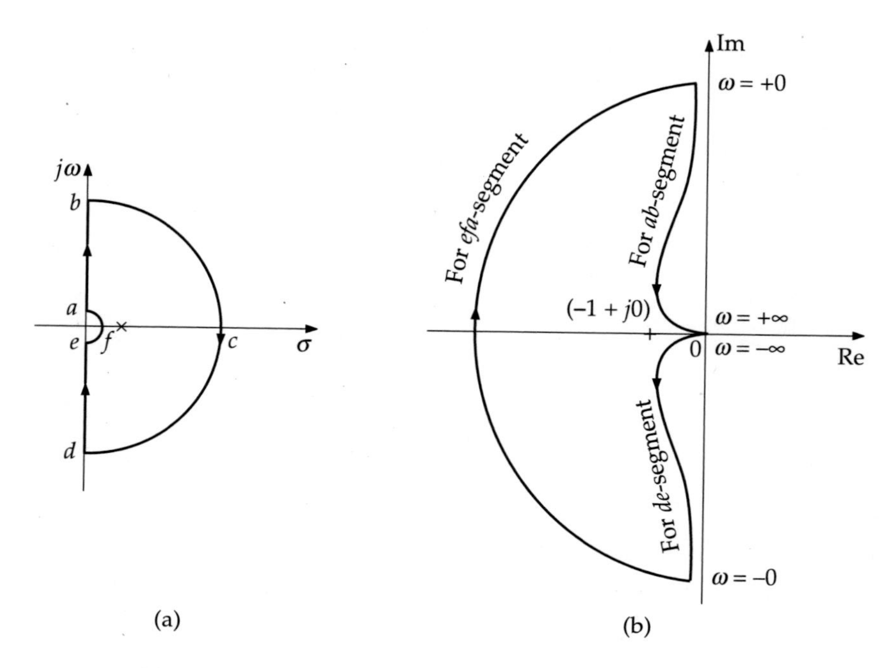

Figure 15.23 (a) Nyquist path; (b) plot (Example 15.19).

Mapping: (i) *For ab-segment:* $s \equiv \jmath\omega \quad 0 < \omega < \infty$

The open-loop transfer function can then be written as

$$\begin{aligned} G(\jmath\omega)H(\jmath\omega) &= \frac{-1}{\jmath 2\omega(1 - \jmath 20\omega)} \\ &= \frac{1}{2\omega\sqrt{1 + 400\omega^2}} \angle \tan^{-1}\frac{0}{-1} - 90^\circ + \tan^{-1} 20\omega \\ &= \frac{1}{2\omega\sqrt{1 + 400\omega^2}} \angle(\pi - 0^\circ) - 90^\circ + \tan^{-1} 20\omega \\ &= \frac{1}{2\omega\sqrt{1 + 400\omega^2}} \angle\left(180^\circ - 90^\circ + \tan^{-1} 20\omega\right) \end{aligned} \tag{i}$$

From Eq. (i), we can construct the following table:

ω	$\lvert G(\jmath\omega)H(\jmath\omega)\rvert$	$\angle G(\jmath\omega)H(\jmath\omega)$
$+0$	∞	90°
$+\infty$	0	180°

With these available data, we draw the polar plot corresponding to *ab*-segment of the Nyquist path. It is shown with an appropriate label in Figure 15.23(b).

(ii) *For bcd-segment:* $s \equiv \lim_{R\to\infty} Re^{j\theta} \quad +90^\circ > \theta > -90^\circ$

$$G(s)H(s) \equiv \lim_{R\to\infty} \frac{-1}{2Re^{j\theta}(1 - 20Re^{j\theta})} = 0$$

Thus, this segment of the Nyquist path maps onto a point at the origin in the Nyquist plot.

(iii) *For de-segment:* This segment maps onto a mirror image of that of the ab-segment.

(iv) *For efa-segment:* $s \equiv \lim_{\rho\to 0} \rho e^{j\theta} \quad -90^\circ > \theta > +90^\circ$

Since this is a type 1 system, the Nyquist plots will close from $\omega = -0$ to $\omega = +0$, sweeping an angle of π, i.e. 180° in the clockwise direction (see Closing Nyquist Plots at page 525). Therefore, the semicircular curve of infinite radius from $\omega = -0$ to $\omega = +0$ maps the segment.

The complete Nyquist plot is shown in Figure 15.20(b). From the plot we observe that the CW curve encircles the critical point once, i.e. $N = 1$. So, from the Nyquist criterion,

$$N = Z - P$$

$$1 = Z - 1$$

or

$$Z = 2 \tag{ii}$$

Equation (ii) indicates the presence of two poles in the RHP of the CLTF. Therefore, the closed-loop system is unstable.

EXAMPLE 15.20. Sketch the Nyquist plot for the open-loop transfer function

$$G(s)H(s) = \frac{s+1}{s^2(s-2)}$$

and determine the closed-loop stability of the system.

Solution This is a type 2 system having a pole in the RHP. Therefore, $P = 1$. The frequency domain transfer function can be written as

$$G(j\omega)H(j\omega) = \frac{j\omega + 1}{-\omega^2(j\omega - 2)}$$

$$\equiv \frac{\omega^2 + 1}{\omega^2\sqrt{\omega^2 + 4}} \angle\left[\tan^{-1}\omega - 180^\circ - \left(\pi - \tan^{-1}\frac{\omega}{2}\right)\right] \tag{i}$$

The Nyquist path for the system is given in Figure 15.24(a).

Mapping: (i) *For ab-segment:*

From Eq. (i) we get the following data:

ω	M	ϕ
$+0$	∞	-360°
$+\infty$	0	-180°

The polar plot of these data is shown between $\omega = +0$ and $\omega = +\infty$ in Figure 15.24(b).

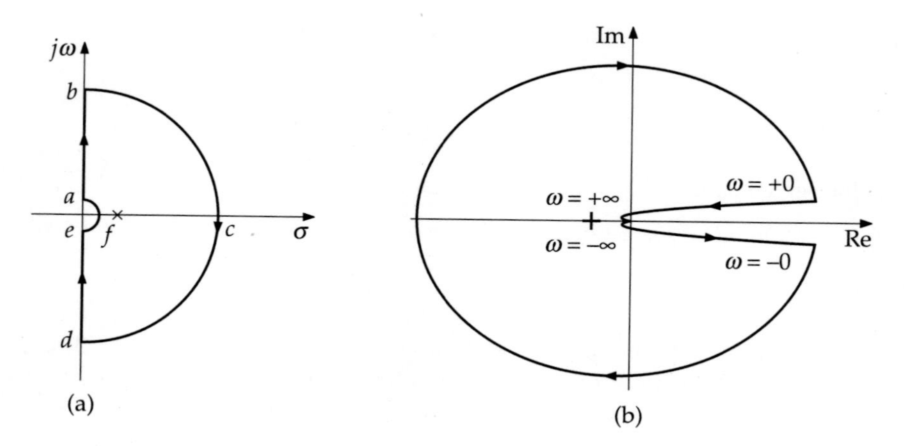

Figure 15.24 (a) Nyquist path; (b) Nyquist plot (Example 15.20). The critical point is shown with a + sign on the negative Re-axis.

(ii) *For bcd-segment:* In the usual way it can be seen that this segment maps onto a point at the origin.

(iii) *For de-segment:* This segment maps to the plot between $\omega = -\infty$ and $\omega = -0$ in Figure 15.24(b), which is a mirror image of the mapping of the ab-segment.

(iv) *For efa-segment:* $s \equiv \lim_{\rho \to 0} \rho e^{j\theta} \quad -90° > \theta > +90°$

Since this is a type 2 system, the Nyquist plots will close from $\omega = -0$ to $\omega = +0$, sweeping an angle of 2π, i.e. 360° in the clockwise direction (see Closing Nyquist Plots at page 525). Therefore, the circular curve of infinite radius from $\omega = -0$ to $\omega = +0$ maps the segment.

The complete Nyquist plot is shown in Figure 15.24(b). From the plot we observe that the curve is CW throughout and that there is one encirclement of the critical point, i.e. $N = 1$. So, from the Nyquist criterion,

$$N = Z - P$$

$$1 = Z - 1$$

or

$$Z = 2 \tag{i}$$

Equation (i) indicates the presence of two poles in the RHP of the CLTF. Therefore, the closed-loop system is unstable.

EXAMPLE 15.21. Construct the complete Nyquist plot for a unity feedback control system whose open-loop transfer function is

$$G(s)H(s) = \frac{K}{s(s^2 + 2s + 2)}$$

Find the maximum value of K for which the system is stable.

Solution This is a type 1 system with no OLTF pole in the RHP. The quadratic pole factorizes to $(s+1+j1)$ and $(s+1-j1)$. Therefore, there exists no pole on the Im-axis. So, the Nyquist path is as shown in Figure 15.25(a).

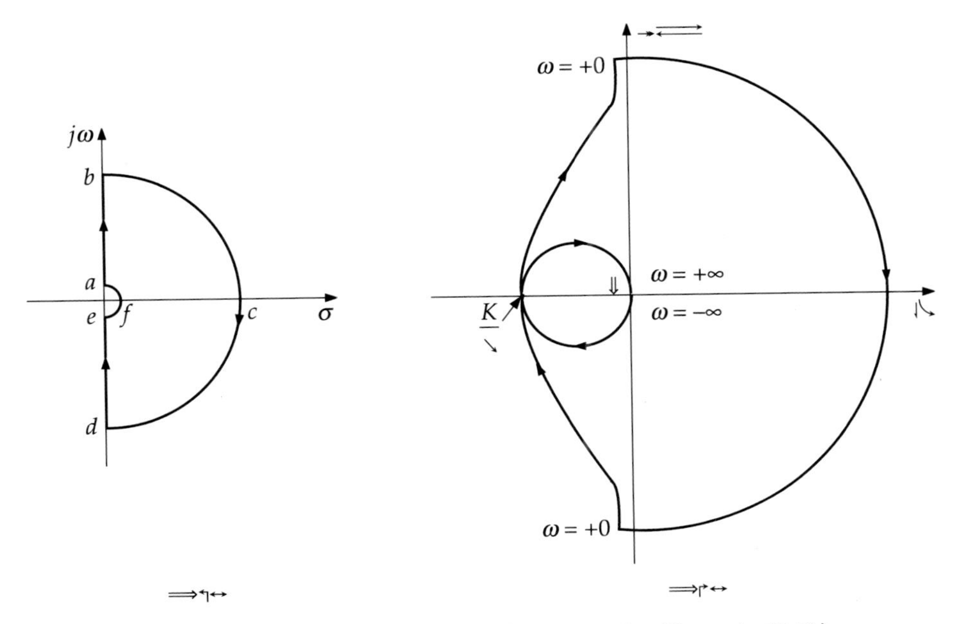

Figure 15.25 (a) Nyquist path; (b) Nyquist plot (Example 15.21).

Mapping: (i) *For ab-segment:*

The frequency domain representation of the open-loop transfer function is

$$G(j\omega)H(j\omega) = \frac{K}{j\omega(-\omega^2 + j2\omega + 2)}$$

$$\equiv \frac{K}{\omega\sqrt{(2-\omega^2)^2 + 4\omega^2}} \angle\left(-90^\circ - \tan^{-1}\frac{2\omega}{2-\omega^2}\right) \tag{i}$$

From Eq. (i), we get the following values:

For $\omega = +0$ $\quad |G(j\omega)H(j\omega)| = \infty$ $\quad \angle G(j\omega)H(j\omega) = -90^\circ$

For $\omega = +\infty$ $\quad |G(j\omega)H(j\omega)| = 0$ $\quad \angle G(j\omega)H(j\omega) = -270^\circ$[7]

The mapping is shown in Figure 15.25(b) with proper labelling.

[7] The calculation is as follows:

$$-90^\circ - \tan^{-1}\left.\frac{2}{\frac{2}{\omega} - \omega}\right|_{\omega=\infty} = -90^\circ - \tan^{-1}\left.\frac{2}{-\omega}\right|_{\omega=\infty}$$

$$= -90^\circ - \pi + \tan^{-1}\left.\frac{2}{\omega}\right|_{\omega=\infty}$$

$$= -90^\circ - \pi = -270^\circ$$

(ii) *For bcd-segment:* $s = \lim_{R\to\infty} Re^{j\theta} \quad +90^\circ > \theta > -90^\circ$

Since this is a strictly proper transfer function, this segment maps onto a point at the origin as usual.

(iii) *For de-segment:* The mapping of this segment is a mirror image of that of the *ab*-segment. In Figure 15.25(b), it is the curve marked with $\omega = -\infty$ to $\omega = -0$.

(iv) *For efa-segment:* $s = \lim_{\rho\to 0} \rho e^{j\theta} \quad -90^\circ > \theta > +90^\circ$

Since this is a type 1 system, the Nyquist plots will close from $\omega = -0$ to $\omega = +0$, sweeping an angle of π, i.e. 180° in the clockwise direction (see Closing Nyquist Plots at page 525). Therefore, the semicircular curve of infinite radius from $\omega = -0$ to $\omega = +0$ maps the segment.

To find where the curve intersects the Re-axis, we find the phase crossover frequency ω_p and calculate the magnitude M at that frequency as follows:

$$-90^\circ - \tan^{-1}\frac{2\omega_p}{2-\omega_p^2} = -180^\circ$$

$$\Rightarrow \quad \frac{2\omega_p}{2-\omega_p^2} = \tan 90^\circ = \infty$$

$$\Rightarrow \quad 2-\omega_p^2 = 0$$

Therefore,

$$\omega_p = \sqrt{2}$$

Thus,

$$M\Big|_{\omega_p=\sqrt{2}} = \frac{K}{\sqrt{2}\sqrt{8}} = \frac{K}{4} \qquad \text{[from Eq. (i)]}$$

The complete Nyquist plot is shown in Figure 15.25(b). For the stability of the system, this intercept magnitude should be less than the magnitude of the critical point, i.e.

$$\frac{K}{4} < 1$$

or

$$K < 4$$

EXAMPLE 15.22. (a) Draw the Nyquist plot for the system

$$G(s) = \frac{s+1}{s(s-1)}$$

by plotting G along the closed contour shown in Figure 15.26(a). Note the indentation of this contour at the origin is to the left.

(b) What is the encirclement criterion for stability of this system with the type of contour chosen?

(c) Using the criterion, determine the range of the gain K of the proportional controller over which the closed-loop system will be stable.

Solution (a) This is a type 1 system with an OLTF pole in the RHP. Therefore, $P = 1$. In the frequency domain, the open-loop transfer function can be written as

$$G(j\omega)H(j\omega) = \frac{j\omega+1}{j\omega(j\omega-1)} = \frac{\sqrt{\omega^2+1}}{\omega\sqrt{\omega^2+1}}\angle\left[\tan^{-1}\omega - 90^\circ - (\pi - \tan^{-1}\omega)\right] \qquad \text{(i)}$$

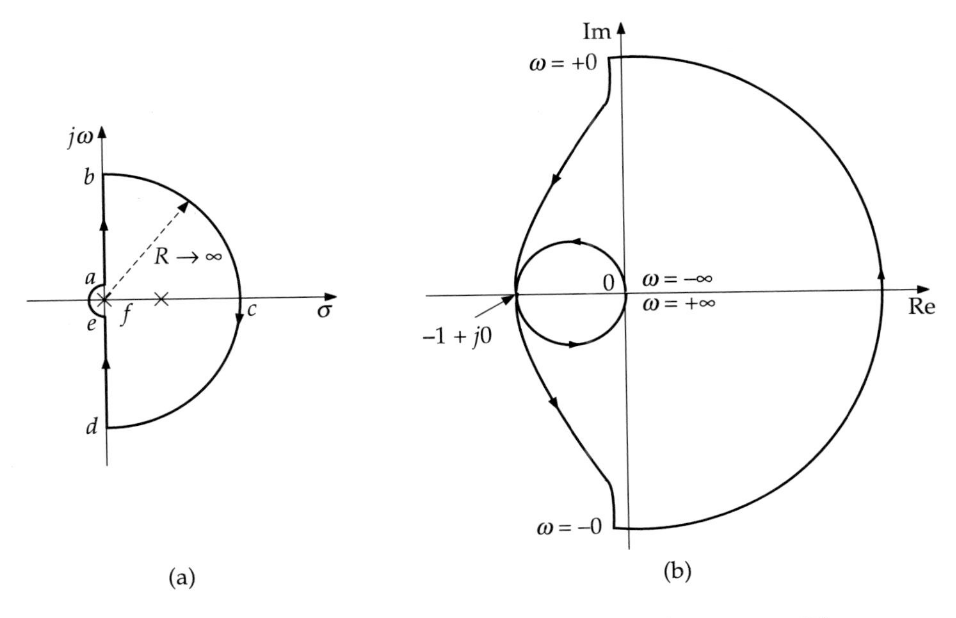

Figure 15.26 (a) Nyquist path; (b) Nyquist plot (Example 15.22).

Mapping: (i) *For ab-segment:* $s \equiv \jmath\omega \quad +0 < \omega < +\infty$

We get the following data from Eq. (i):

ω	M	ϕ
$+0$	∞	$-270°$
$+\infty$	0	$-90°$

The corresponding curve is shown in Figure 15.26(b) with proper labelling.

(ii) *For bcd-segment:* $\quad s = \lim\limits_{R\to\infty} Re^{\jmath\theta} \quad +90° > \theta > -90°$

Since this is a strictly proper transfer function, this segment maps onto a point at the origin as usual.

(iii) *For de-segment:* The mapping of this segment is a mirror image of that of the ab-segment. In Figure 15.26(b), it is the curve marked with $\omega = -\infty$ to $\omega = -0$.

(iv) *For efa-segment:* $\quad s = \lim\limits_{\rho\to 0} \rho e^{\jmath\theta} \quad -90° > \theta > +90°$

Since this is a type 1 system, and that the Nyquist path includes the pole at the origin, the Nyquist plots will close from $\omega = -0$ to $\omega = +0$, sweeping an angle of π, i.e. 180° in the *counterclockwise* direction (see Closing Nyquist Plots at page 525). Therefore, the semicircular curve of infinite radius from $\omega = -0$ to $\omega = +0$ maps the segment.

To find where the curve intersects the Re-axis, we find the phase crossover frequency ω_p and calculate the magnitude M at that frequency as follows:

$$2\tan^{-1}\omega_p - 270° = -180°$$

$$\Rightarrow \qquad \tan^{-1}\omega_p = 45°$$

Therefore,

$$\omega_p = 1$$

So,

$$M\Big|_{\omega_p=\sqrt{2}} = \frac{1}{1} = 1 \qquad \text{[from Eq. (i)]}$$

The complete Nyquist plot is shown in Figure 15.25(b).

(b) The Nyquist criterion for stability of this system with this type of contour chosen is the same as given by Eq. (15.7) stated in page 520.

(c) Since there are two CCW encirclements, $N = -2$ and since one open-loop transfer function pole is included on the Nyquist path, $P = 2$. So, from Eq. (15.7),

$$-2 = Z - 2 \quad \text{or} \quad Z = 0$$

As can be seen from the plot, it passes through the critical point. Therefore, the system as such is marginally stable. If a proportional controller of gain K were included, the condition for stability would be $K < 1$.

Nyquist Plot for System with Transportation Lag

EXAMPLE 15.23. For the unity feedback system shown in the following figure, the phase crossover occurs at a frequency of $\omega = 2.448$ rad/s.

(a) Determine the transport delay D.

(b) Find the limiting value of the proportional controller gain K for the stable operation and sketch the corresponding Nyquist plot.

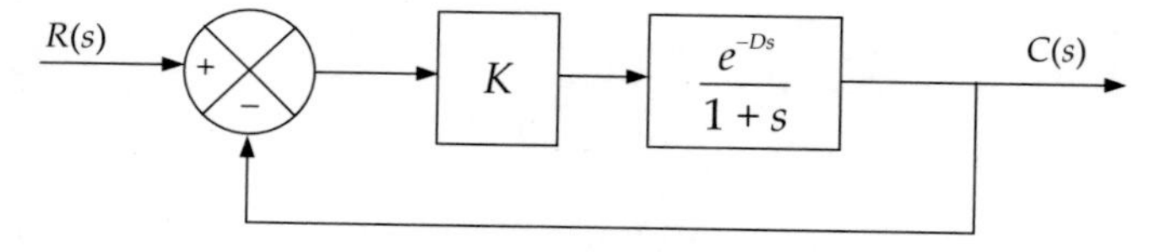

Solution (a) The open-loop transfer function is given by

$$G(s)H(s) = \frac{Ke^{-Ds}}{1+s}$$

$$\Rightarrow \quad G(j\omega)H(j\omega) = \frac{Ke^{-j D\omega}}{1+j\omega} \equiv \frac{K}{\sqrt{1+\omega^2}}\angle(-D\omega - \tan^{-1}\omega) \qquad \text{(i)}$$

$$\left[\because \; e^{-jD\omega} = \cos D\omega - j\sin D\omega \equiv |\sqrt{\sin^2 D\omega + \cos^2 D\omega}|\angle \tan^{-1}\frac{-\sin D\omega}{\cos D\omega} = |1|\angle - D\omega\right]$$

Corresponding to the phase crossover frequency ω_p, we get from Eq. (i) the relation

$$-D\omega_p - \tan^{-1}\omega_p = -180° = -\pi$$

$$\Rightarrow \qquad 2.448D + \tan^{-1} 2.448 = \pi$$

Therefore,
$$D = \frac{\pi - 1.183}{2.448} = 0.8 \text{ s}$$

(b) The open-loop transfer function in the frequency domain is given by

$$G(j\omega)H(j\omega) = \frac{K}{\sqrt{1+\omega^2}}\angle(-0.8\omega - \tan^{-1}\omega) \tag{ii}$$

From Eq. (ii), we obtain the following data:

ω	M	ϕ
$+0$	K	$0°$
$+\infty$	0	$-\infty$

From Eq. (ii), we also observe that as ω varies between 0 and $+\infty$, the phase angle ϕ varies continuously between $0°$ and $-\infty$. This variation indicates that the Nyquist plot will assume the form of a spiral as shown in Figure 15.27.

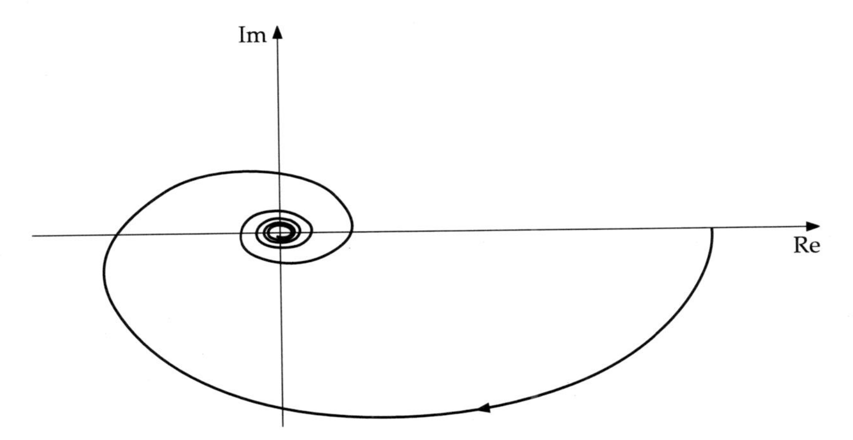

Figure 15.27 Indicated Nyquist plot for $\omega = +0$ to $+\infty$ (Example 15.23).

The points of intersection of the plot at the negative Re-axis are determined by the relation

$$-0.8\omega - \tan^{-1}\omega = -(2n+1)\pi \qquad n = 0,\ 1,\ 2, \cdots$$

The first intercept for $\omega > 0$ will take place when $n = 0$. This condition yields

$$-0.8\omega - \tan^{-1}\omega = -\pi \tag{iii}$$

To solve this equation, we use the approximate relation

$$\tan^{-1}x \approx \frac{\pi}{2} - \frac{1}{x} \quad \text{valid for } x > 1$$

Thus, from Eq. (iii),

$$\frac{\pi}{2} - \frac{1}{\omega} + 0.8\omega = \pi$$

$\Rightarrow$

$$0.8\omega^2 - \frac{\pi}{2}\omega - 1 = 0$$

Therefore,

$$\omega = 2.4696, \ -0.5061$$

The negative frequency being unrealistic, we get, from the magnitude condition for stability, the inequality

$$\left.\frac{K}{\sqrt{1+\omega^2}}\right|_{\omega=2.4696} < 1$$

or

$$\frac{K}{2.6644} < 1$$

or

$$K < 2.6644$$

Therefore, the limiting value of the proportional controller gain K is approximately 2.66. The Nyquist plot for a gain of 2.6 is displayed in Figure 15.28.

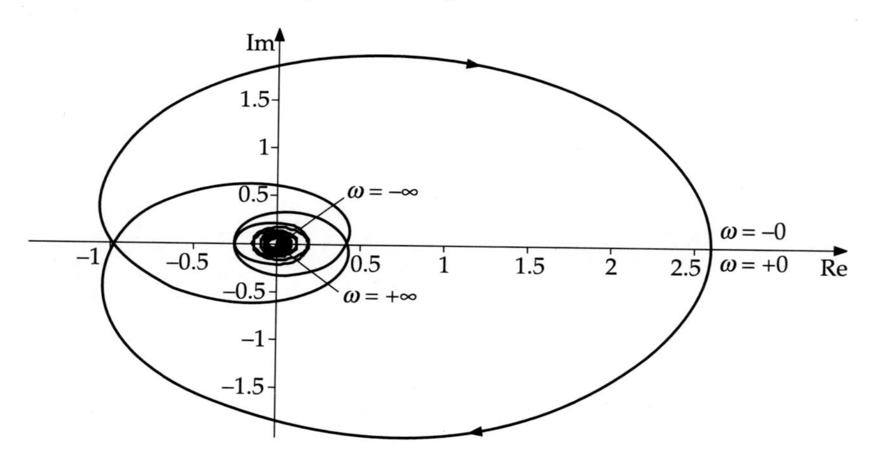

Figure 15.28 Nyquist plot for a gain of 2.6 (Example 15.23).

15.9 CLOSED-LOOP FREQUENCY RESPONSE FROM OPEN-LOOP TRANSFER FUNCTION

Consider a closed-loop transfer function of a unity feedback system in the frequency domain, given by

$$\frac{C(j\omega)}{R(j\omega)} = \frac{G(j\omega)}{1+G(j\omega)} \tag{15.8}$$

$G(\jmath\omega)$ being a complex quantity, it can be written in the form of

$$G(\jmath\omega) \equiv x + \jmath y$$

Then, Eq. (15.8) can be rewritten as

$$\frac{C(\jmath\omega)}{R(\jmath\omega)} \equiv \frac{x + \jmath y}{1 + x + \jmath y} \tag{15.9}$$

M-Circles

The magnitude part of the complex quantity given by Eq. (15.9) is

$$M = \left| \frac{x + \jmath y}{1 + x + \jmath y} \right| = \frac{\sqrt{x^2 + y^2}}{\sqrt{(1 + x)^2 + y^2}}$$

or

$$M\left(\sqrt{(1 + x)^2 + y^2}\right) = \sqrt{x^2 + y^2} \tag{15.10}$$

Squaring both sides of Eq. (15.10) and then regrouping the terms, we get

$$M^2(1 + x^2 + 2x + y^2) = x^2 + y^2$$

or

$$(1 - M^2)x^2 + (1 - m^2)y^2 - 2M^2x = M^2 \tag{15.11}$$

Dividing both sides of Eq. (15.11) by $(1 - M^2)$, we obtain

$$x^2 + y^2 - \frac{2M^2}{1 - M^2}x = \frac{M^2}{1 - M^2} \tag{15.12}$$

Adding $\left(\frac{M^2}{1 - M^2}\right)^2$ to both sides of Eq. (15.12) and then rearranging, we have

$$x^2 + y^2 - \frac{2M^2}{1 - M^2}x + \left(\frac{M^2}{1 - M^2}\right)^2 = \frac{M^2}{1 - M^2} + \left(\frac{M^2}{1 - M^2}\right)^2$$

or

$$\left(x - \frac{M^2}{1 - M^2}\right)^2 + y^2 = \frac{M^2}{(1 - M^2)^2} \tag{15.13}$$

Equation (15.13) represents a circle having its centre at

$$\left(\frac{M^2}{1 - M^2},\ 0\right)$$

and radius

$$r = \left| \frac{M}{1 - M^2} \right|$$

However, Eq. (15.13) is invalid for $M = 1$. But substituting $M = 1$ in Eq. (15.11), we get $x = -\frac{1}{2}$ which represents a straight line parallel to the Im-axis passing through the point $(-\frac{1}{2},\ \jmath 0)$ in the $G(\jmath\omega)$-plane.

For other values of M, we get a family of circles in the $G(\jmath\omega)$-plane. These are called *M-circles* or *M-loci*. To figure out the characteristics of these circles, we have calculated the location of centres and values of radii of a few loci and presented them in Table 15.3 and shown the circles in Figure 15.29.

Table 15.3 Centres and Radii of a Few *M*-circles

M	*Centre* $x = \dfrac{M^2}{1-M^2},\ y = 0$ (Only x values given)	*Radius* $r = \left\lvert \dfrac{M}{1-M^2} \right\rvert$
0.4	0.190	0.476
0.6	0.563	0.938
0.8	1.778	2.222
1.2	−3.273	2.727
1.3	−2.449	1.884
1.4	−2.042	1.458
1.6	−1.641	1.026
2.0	−1.333	0.067
3.0	−1.125	0.375
5.0	−1.042	0.208

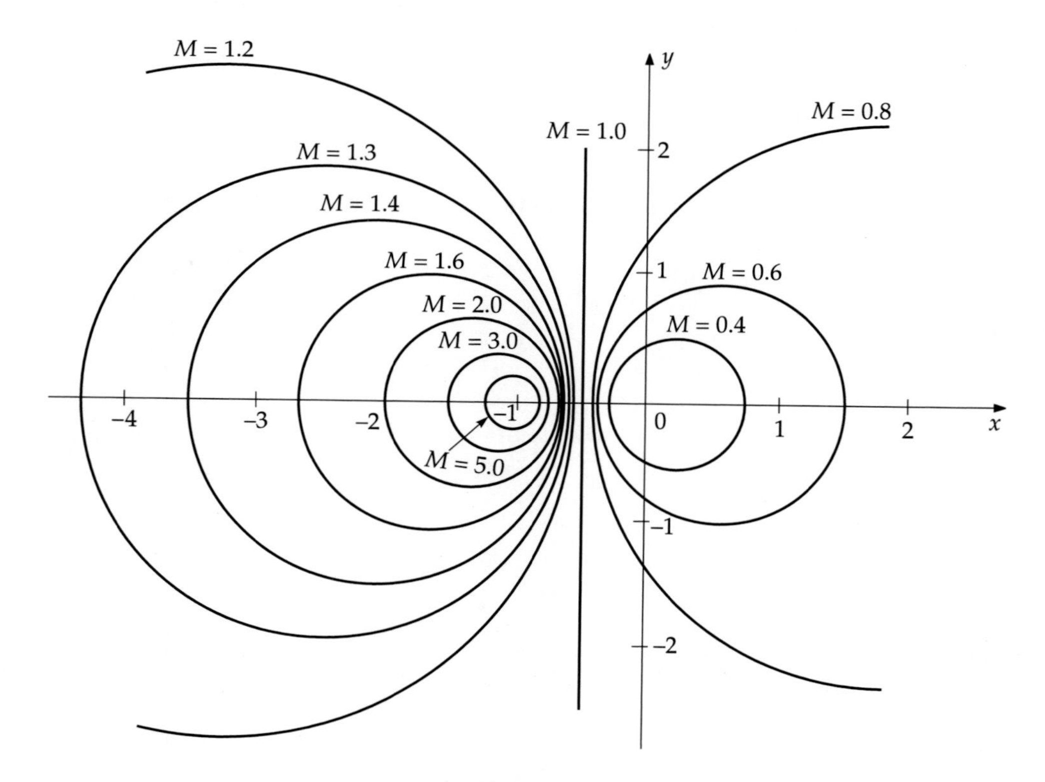

Figure 15.29 *M*-circles.

It can be observed from Table 15.3 as well as Figure 15.29 that

If $M > 1$: The centres of circles lie to the left of $(-1 + \jmath 0)$. The radii of circles become smaller as M grows bigger.

If $M < 1$: The centres of circles lie to the right of $(-1 + \jmath 0)$. The radii of circles become bigger as M grows bigger.

If $M \to \infty$: The circle degenerates to a point at $(-1 + \jmath 0)$. This is in agreement with the fact that when the polar plot passes through $(-1 + \jmath 0)$, the system becomes marginally stable (or marginally unstable) and the resonant peak M_r becomes infinite.

The following closed-loop data may be derived from the polar plot of $G(\jmath\omega)$ along with the M-circle grid superposed on it:

Closed-loop data	*How to obtain*
The value of the magnitude M (i.e. $\lvert G(\jmath\omega)\rvert$) at the frequency denoted by the $\lvert G(\jmath\omega)\rvert$-curve	From the intersections of $G(\jmath\omega)$-plot and constant M-circles
The value of the resonant peak M_r and resonant frequency at the tangent point of the $G(\jmath\omega)$-curve	From the M-circle with the smallest radius which is tangent to the $G(\jmath\omega)$-curve
The value of the bandwidth as determined from the frequency denoted on the $G(\jmath\omega)$-curve	From the intersection of the polar plot with the $M = 0.707$ circle

Figure 15.30 shows the polar plot of $G(\jmath\omega)$ for a control system for different values of K.

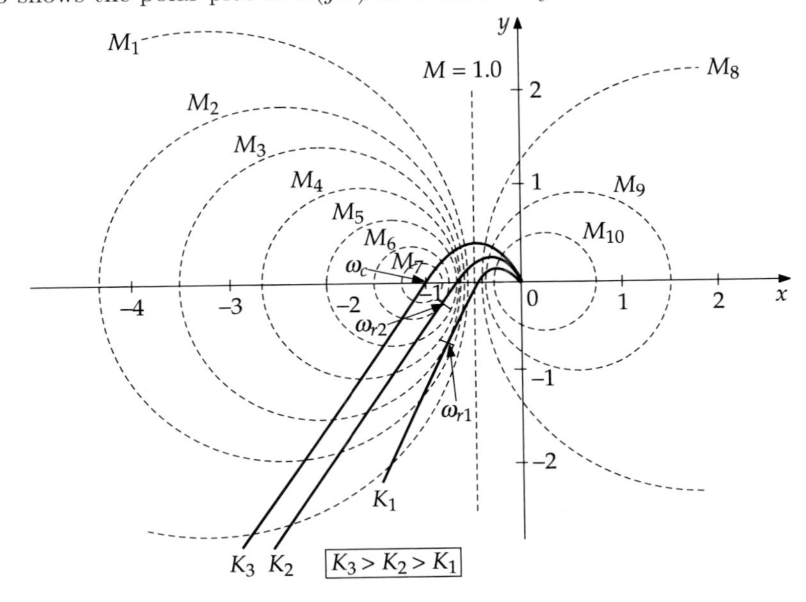

Figure 15.30 Polar plots and M-circles.

We can determine

1. The corresponding M_r and ω_r values by noting the points where each polar plot touches M-circles.
2. The M values at frequencies denoted on the $G(j\omega)$-curves from the intersection points of each polar plot with different M-loci.

In this way we can determine the magnitude part of the closed-loop frequency response for each value of K as shown in Figure 15.31.

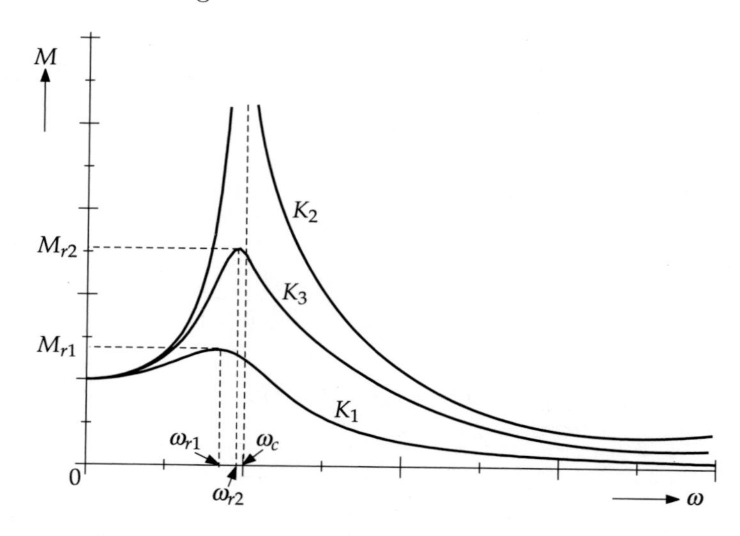

Figure 15.31 Magnitude part of the closed-loop frequency response derived from polar plots and M-circles.

N-Circles

For the phase part of the closed-loop transfer function, we get from Eq. (15.9) the relation

$$\phi = \arg\left|\frac{x + jy}{1 + x + jy}\right| = \tan^{-1}\frac{y}{x} - \tan^{-1}\frac{y}{1+x}$$

$$= \tan^{-1}\frac{\dfrac{y}{x} - \dfrac{y}{1+x}}{1 + \left(\dfrac{y}{x}\right)\left(\dfrac{y}{1+x}\right)}$$

$$= \tan^{-1}\frac{y}{x^2 + x + y^2} \tag{15.14}$$

Taking tangents of both sides of Eq. (15.14), we get

$$\tan\phi = \frac{y}{x^2 + x + y^2} \tag{15.15}$$

Let us put $\tan\phi = N$. Then, from Eq. (15.15),

$$x^2 + x + y^2 - \frac{y}{N} = 0 \tag{15.16}$$

Adding $\frac{1}{4} + \frac{1}{4N^2}$ to both sides of Eq. (15.16), we obtain

$$x^2 + x + y^2 - \frac{y}{N} + \frac{1}{4} + \frac{1}{4N^2} = \frac{1}{4} + \frac{1}{4N^2}$$

or

$$\left(x + \frac{1}{2}\right)^2 + \left(y - \frac{1}{2N}\right)^2 = \sqrt{\frac{N^2+1}{4N^2}} \tag{15.17}$$

Equation (15.17) represents a circle having its centre at

$$\left(-\frac{1}{2}, \frac{1}{2N}\right)$$

and radius

$$r = \left|\sqrt{\frac{N^2+1}{4N^2}}\right|$$

For different values of N, they produce a family of circles which are called *N-circles* or *N-loci*. Values of centres and radii for different values of ϕ and hence different values of N are presented in Table 15.4. The corresponding N-circles are shown in Figure 15.32.

Table 15.4 Data for Constant N-circles

ϕ (degree)	$N = \tan\phi$	*Centre* $x = \frac{1}{2}$, $y = \frac{1}{2N}$ (Only y values given)	*Radius* $r = \left\|\sqrt{\frac{N^2+1}{4N^2}}\right\|$
20	0.364	1.374	1.462
30	0.577	0.867	1.001
40	0.839	0.596	0.778
60	1.732	0.289	0.577
80	5.671	0.087	0.501
−20	−0.364	−1.374	1.462
−30	−0.577	−0.867	1.001
−40	−0.839	−0.596	0.778
−60	−1.732	−0.289	0.577
−80	−5.671	−0.087	0.501

Equation (15.17) is satisfied by ($x = 0$, $y = 0$) and ($x = 1$, $y = 0$) irrespective of the value of N. Hence all circles pass through the origin and the $(-1 + \jmath 0)$ point.

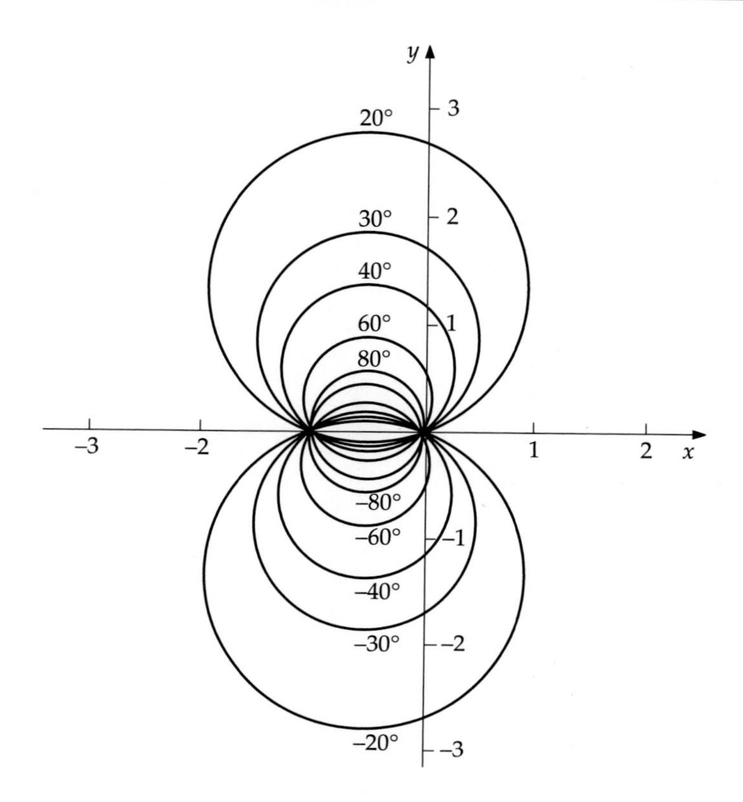

Figure 15.32 Constant *N*-circles.

As discussed in the case of M-circles, we can determine the phase part of the closed-loop frequency response from the polar plot of $G(j\omega)$ with constant N-circles. However, M-circles which provide information of M_r, ω_r and Bw are generally of more use in studies of relative stability of systems.

Nichols Chart

The Nichols[8] Chart is a tool for the designer to read off closed-loop gain and phase directly from a plot of open-loop logarithmic gain and phase, parametrized by frequency. It has proved to be arguably the most useful closed-loop system design tool in the control field.

In polar plots, magnitude and phase are plotted in the G-plane. The M-circle and N-circle grid can be superimposed on it to gather information on closed-loop frequency response of a system. In design problems, it is frequently needed to alter loop gain or add series and feedback compensators[9] to the system. In such cases, the polar plot has to be recalculated. The Bode plot, in which the plot is shifted upwards or downwards depending upon the change in gain,

[8]Nathaniel B Nichols (1914–97) was an American engineer and researcher.
[9]See Chapter 16 at page 575.

has a distinct advantage in this respect. But Bode plots as such do not allow calculation of closed-loop frequency response from plots of open-loop transfer functions. Nichols combined the advantages of the Bode plot and polar plot by devising a plot of magnitude in dB vs. phase in degrees (also known as *gain-phase plot*) and superimposing the circles of M (in dB) and N (in degree) on it.

Suppose we have the open-loop transfer function of a system given by

$$G(s)H(s) = \frac{K}{(s+a)(s+b)(s+c)}$$

Its magnitude vs. phase plot looks like what is shown in Figure 15.33(a).

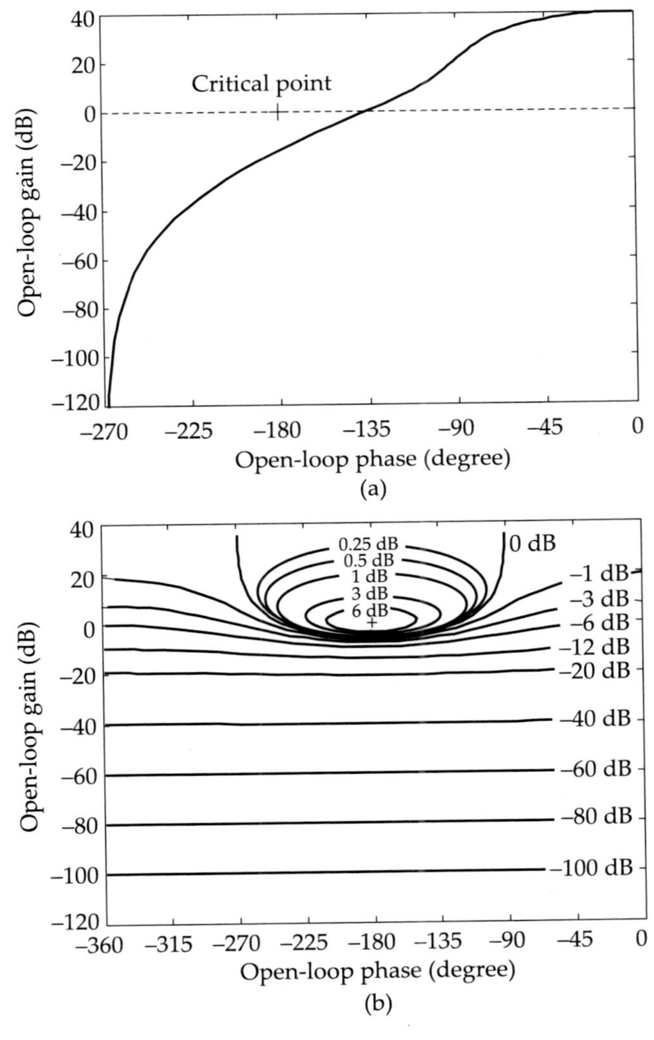

Figure 15.33 (a) Magnitude vs. phase plot; (b) Nichols chart grid for magnitude.

The Nichols chart for magnitude is shown in Figure 15.33(b). These are M-circles plotted in dBs. A similar chart is prepared for the phase in degrees. This grid, when superimposed on the gain-phase plot (see Figure 15.34), generates data for the closed-loop frequency response, much in the same way M- and N- circles do for the polar plot.

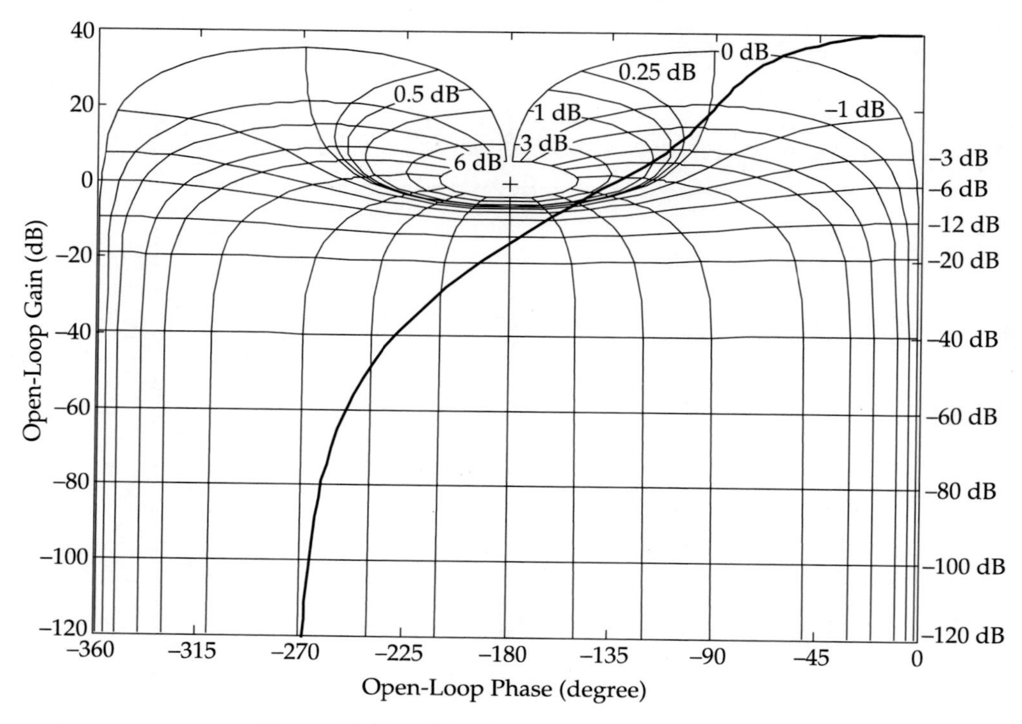

Figure 15.34 Gain-phase plot on Nichols chart.

REVIEW QUESTIONS

15.1 Choose the correct answer:

(i) In the plot shown in the following figure, the numbers of encirclement of A and B are

(a) 2, 2 (b) 2, 1 (c) 1, 2 (d) 1, 1

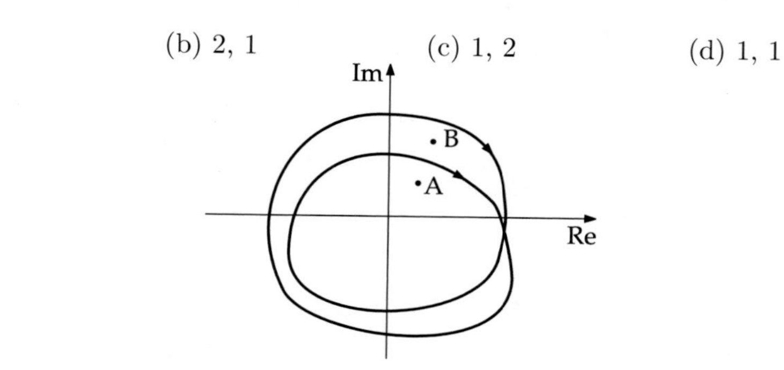

(ii) The Nichols chart helps us determine
   (a) open-loop frequency response
   (b) closed-loop frequency response
   (c) open-loop response in time domain
   (d) closed-loop response in time domain

(iii) For the given Nyquist plot the system is
   (a) stable (b) unstable
   (c) marginally stable (d) none of the above

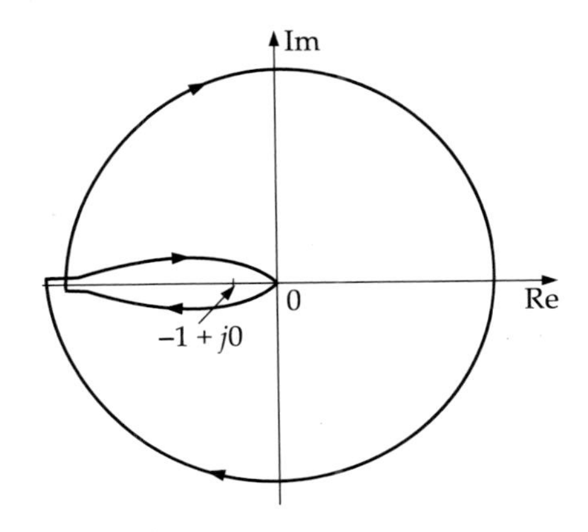

(iv) With the help of the Nyquist criterion, the stability of the closed-loop system
   (a) cannot be determined
   (b) can be determined by studying the behaviour of the plot around $(1 + \jmath 0)$
   (c) can be determined by studying the behaviour of the plot around $(1 + \jmath 0)$
   (d) can be determined by studying whether the plot crosses the Im-axis

15.2 Find the Nyquist plot for

$$G(s)H(s) = \frac{K}{s^2(1 + s\tau)}$$

where $K,\ \tau > 0$.

15.3 For the transfer function

$$G(\jmath\omega) = \frac{K}{(1 + \jmath 2\omega)(1 + \jmath 4\omega)(1 + \jmath 6\omega)}$$

draw the Nyquist plot.

15.4 Examine the closed-loop stability of the system whose open-loop transfer function is

$$G(s)H(s) = \frac{50}{s(s + 1)(s + 2)}$$

using the Nyquist criterion.

15.5 State and explain the principle of argument.

15.6 (a) State the principle of argument and its extension to Nyquist criterion.
(b) Draw the Nyquist plot for the following system

$$G(s)H(s) = \frac{5}{s(1-s)}$$

Discuss the closed-loop stability of the system.

15.7 (a) Define the Nyquist stability criterion. State the principle from which the criterion originated.
(b) The open-loop transfer function of a unity feedback control system is given by

$$G(s) = \frac{K}{(1+s)(1+2s)(1+3s)}$$

(i) Sketch the Nyquist plot.
(ii) From the plot find the value of K above which the system becomes unstable. Also find the critical frequency ω_c.

15.8 (a) State the Nyquist stability criterion.
(b) Consider a system with loop gain

$$G(s)H(s) = \frac{K(s+2)}{s^2(s+1)}$$

Obtain the Nyquist plot and comment on its open-loop stability.

15.9 Consider a system with an open-loop transfer function

$$G(s)H(s) = \frac{2s+1}{s^2(s+1)(s+2)}$$

(a) Draw the Nyquist contour and the corresponding mapping of $G(s)H(s)$.
(b) Comment on the closed-loop stability of the system.

15.10 Write short notes on:
(a) Nyquist stability criterion
(b) Principle of arguments
(c) Constant M-circles
(d) Nyquist plot
(e) Nichols chart

Answers to Selected Questions

15.1. (i) (b); (ii) (a); (iii) (b); (iv) (c)
15.6. (b) Stable
15.7. (b) $K < 10$, $\omega_g = 1$ rad/s
15.8. (b) Unstable
15.9. (b) Unstable

CHAPTER 16

Compensators

16.1 WHAT THEY ARE

The performance of a control system is judged from the following characteristics it possesses:

1. Speed of response
2. Steady-state error
3. Damping factor
4. Peak overshoot
5. Stability.

In case the performance of a control system falls short of expectations, an additional device may be added to the system to make good, or compensate, for the shortfall of performance. Such devices are called *compensators*.

16.2 TYPES AND EFFECTS

Depending upon the location of the compensating network, compensation is divided into the following three categories:

1. Series or cascade compensation
2. Feedback or parallel compensation
3. Load or series-parallel compensation.

Their configurations and functions are elaborated now.

Series or Cascade Compensation

If the compensating network is included in feed-forward path, it is called *series* or *cascade compensation*. The corresponding block diagam is shown in Figure 16.1.

The addition of a compensating network in the forward path basically adjusts the gain of a system. As a result,

1. The speed of response increases.
2. Peak overshoot decreases.
3. The stability of the system may worsen.

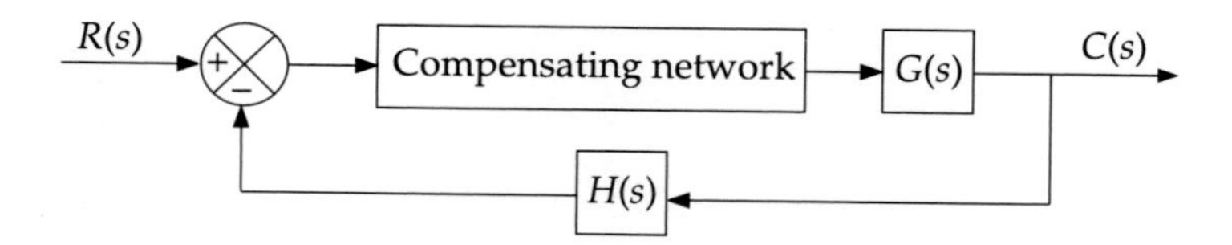

Figure 16.1 Series or cascade compensation. The compensating network is included in the feed-forward path.

Feedback or Parallel Compensation

When the compensating network is included in the feedback path (Figure 16.2), it is called *feedback* or *parallel compensation.*

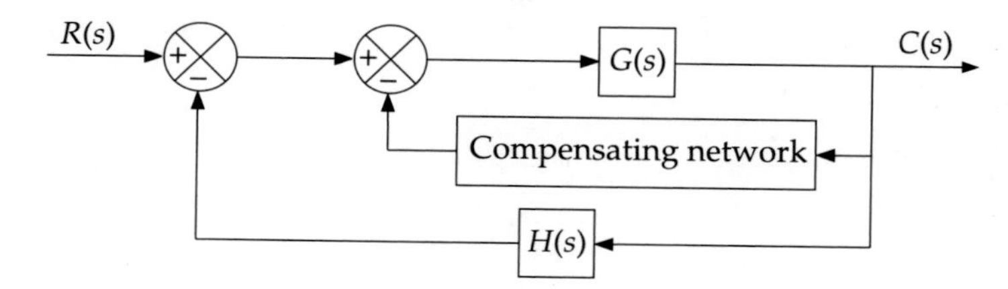

Figure 16.2 Feedback or parallel compensation. The compensating network is included in the feedback path.

A compensating network in the feedback path may

1. lower the steady-state error;
2. lower the speed of response; or
3. improve the stability of the system.

Load or Series-Parallel Compensation

Some situations may call for a combination of series and feedback compensation. Such type of compensation is called *load* or *series-parallel compensation.* The corresponding block diagram is shown in Figure 16.3

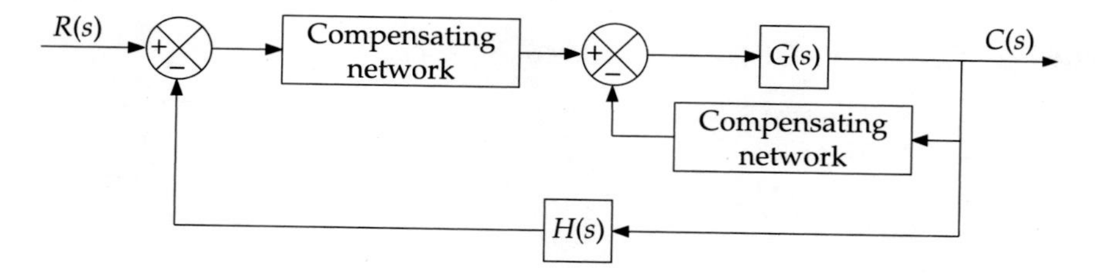

Figure 16.3 Load or series-parallel compensation. Compensating networks are included in the feed-forward as well as feedback path.

Compensating networks introduce additional poles and/or zeros in the transfer function of the system. The effects of such additions are shown in the following table.

Addition of poles	Addition of zeros
1. Pulls the root locus towards the RHP. As a result, the stability is lowered.	1. Pulls the root locus towards LHP. As a result, the stability improves.
2. Increases settling time.	2. Lowers settling time.
3. Lowers steady-state error, i.e., improves accuracy.	3. Increases steady-state error, i.e., lowers accuracy.

16.3 TYPES OF COMPENSATION NETWORKS

With this background on compensation and its effects, let us study the three types of compensating networks (or compensators) that are used in electrical systems:

1. Lead compensation network
2. Lag compensation network
3. Lead-lag compensation network

16.4 LEAD COMPENSATION NETWORK

A simple circuit is shown in Figure 16.4.

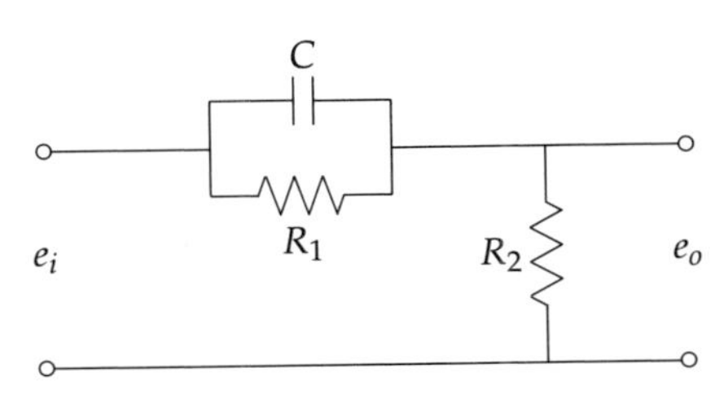

Figure 16.4 Lead compensator.

Considering its Laplace transform equivalent, the transfer function of the circuit can be worked out as follows:

$$I(s) = \frac{E_i(s)}{\left(R_1 \| \frac{1}{sC}\right) + R_2} = \frac{E_i(s)}{\frac{R_1}{sR_1C+1} + R_2} = \frac{sR_1C+1}{sR_1R_2C + R_1 + R_2} \tag{16.1}$$

Also,

$$I(s) = \frac{E_o(s)}{R_2} \tag{16.2}$$

Combining Eqs. (16.1) and (16.2), we get

$$G(s) \equiv \frac{E_o(s)}{E_i(s)} = \frac{R_2(sR_1C+1)}{sR_1R_2C+R_1+R_2}$$

$$= \frac{R_1R_2C\left(s+\dfrac{1}{R_1C}\right)}{R_1R_2C\left(s+\dfrac{R_1+R_2}{R_1R_2C}\right)}$$

$$\equiv \frac{s+(1/T)}{s+(1/\alpha T)} \tag{16.3}$$

where $$T = R_1C, \quad \alpha = \frac{R_2}{R_1+R_2}$$

We may note here that T is the time-constant and α, which is < 1, is an attenuation constant. In the frequency domain, the transfer function can be written as

$$G(\jmath\omega) = \frac{\jmath\omega+(1/T)}{\jmath\omega+(1/\alpha T)} = \frac{\alpha(1+\jmath\omega T)}{1+\jmath\alpha\omega T} \quad \alpha < 1 \tag{16.4}$$

From Eqs. (16.3) and (16.4), we observe the following:

1. The lead network has a zero at $s = -1/T$ and a pole at $s = -1/\alpha T$. Since $\alpha < 1$, the zero is dominant, i.e. it is located nearer to the imaginary axis [see Figure 16.5(a)].
2. The numerator introduces a phase angle of $\tan^{-1}\omega T$ and the denominator, $-\tan^{-1}(\alpha\omega T)$. Since $\alpha < 1$, the net phase angle is positive and hence *the output leads the input.*
3. The polar plot of the network transfer function can be constructed [see Figure 16.5(b)] from

$$G(\jmath\omega) = \left|\frac{\alpha\sqrt{1+\omega^2T^2}}{\sqrt{1+\alpha^2\omega^2T^2}}\right| \angle\left[\tan^{-1}\omega T - \tan^{-1}(\alpha\omega T)\right] \tag{16.5}$$

   Equation (16.5) helps us generate the following data:

ω	$\lvert G(\jmath\omega)\rvert$	$\angle G(\jmath\omega)$
0	α	0°
$1/T$ ($\alpha = 0.9$)	0.95	3°
∞	1	0°

4. The phase angle ϕ is given by

$$\phi = \tan^{-1}\omega T - \tan^{-1}(\alpha\omega T) \tag{16.6}$$

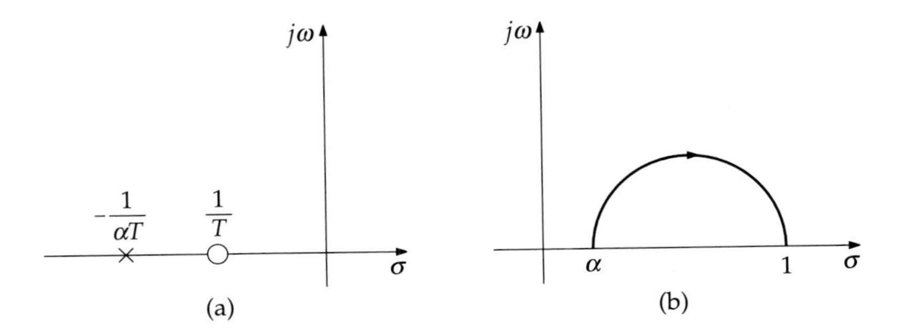

Figure 16.5 (a) Pole-zero plot; (b) polar plot of the lead network transfer function.

If ω_m is the frequency for which $\phi = \phi_{\max}$, then

$$\left.\frac{d\phi}{d\omega}\right|_{\omega=\omega_m} = 0$$

$$\Rightarrow \quad \frac{T}{1+\omega_m^2 T^2} - \frac{\alpha T}{1+\alpha^2\omega_m^2 T^2} = 0$$

$$\Rightarrow \quad \frac{T}{1+\omega_m^2 T^2} = \frac{\alpha T}{1+\alpha^2\omega_m^2 T^2}$$

$$\Rightarrow \quad \omega_m^2(\alpha T^2 - \alpha^2 T^2) = 1-\alpha$$

$$\Rightarrow \quad \omega_m = \frac{1}{T\sqrt{\alpha}} = \left(\frac{1}{\sqrt{T}}\right)\left(\frac{1}{\sqrt{\alpha T}}\right) \tag{16.7}$$

Again, from Eq. (16.6),

$$\phi = \tan^{-1}\frac{\omega T(1-\alpha)}{1+\alpha\omega^2 T^2}$$

or

$$\tan\phi = \frac{\omega T(1-\alpha)}{1+\alpha\omega^2 T^2} \tag{16.8}$$

Plugging in the value of ω_m in Eq. (16.8), we get

$$\tan\phi_{\max} = \frac{\frac{1}{\sqrt{\alpha}}(1-\alpha)}{1+1} = \frac{1-\alpha}{2\sqrt{\alpha}}$$

Therefore,

$$\sin\phi_{\max} = \frac{1-\alpha}{1+\alpha}$$

$$\Rightarrow \quad \alpha = \frac{1-\sin\phi_{\max}}{1+\sin\phi_{\max}} \tag{16.9}$$

From Eq. (16.7) we note that the frequency ω_m corresponding to the maximum phase angle is the geometric mean of the two corner frequencies. Also, from Eqs. (16.5) and (16.7), at ω_m

$$|G(j\omega)|_{\omega=\omega_m} = \frac{\alpha\sqrt{1+\omega_m^2 T^2}}{\sqrt{1+\alpha^2\omega_m^2 T^2}} = \frac{\alpha\sqrt{1+(1/\alpha)}}{\sqrt{1+\alpha}} = \sqrt{\alpha}$$

Therefore,

$$20\log|G(j\omega)|_{\omega=\omega_m} = 20\log(\sqrt{\alpha}) = -10\log\left(\frac{1}{\alpha}\right)\text{ dB} \tag{16.10}$$

5. From the above discussions, we can draw the Bode plot for the network transfer function as in Figure 16.6.

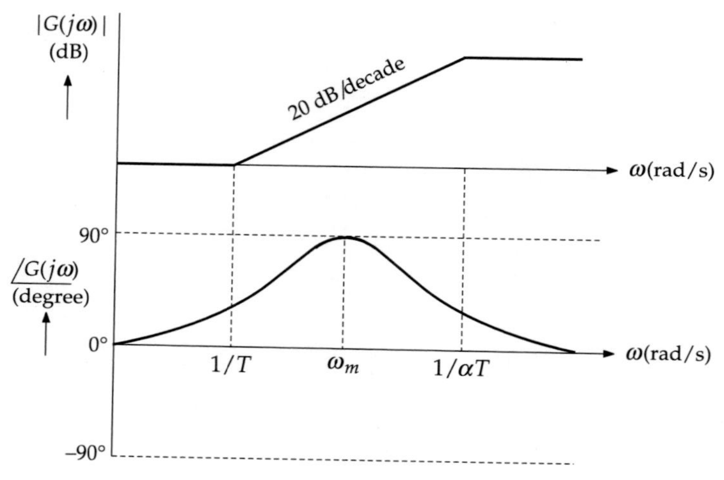

Figure 16.6 Bode plot for lead network transfer function.

6. At $\omega = 0$ (dc), the gain is α, which is < 1. So, the circuit acts like an attenuator.

Effects of Incorporating Lead Network

1. Bandwidth (hence speed) improves because the network moves the system to a higher gain crossover frequency. This also improves the relative stability of the system because the phase margin increases as a result of increasing ω_g.
2. Damping increases because of addition of a dominant zero and a pole. As a result, peak overshoot and settling time decrease.

Designing Lead Network

Step 1. Draw the Bode plot of the system open-loop transfer function with the gain constant K adjusted according to the prescribed steady-state error, if necessary.

Step 2. Determine the gain and phase margins of the uncompensated system from the Bode plot. If ϕ is the Pm of the uncompensated system, ϕ_r is the required Pm, and τ is the tolerance introduced, then

$$\phi_m = \phi_r - \phi + \tau$$

Since each lead network normally provides a maximum phase shift of 60°, if $\phi_m >$ 60°, two identical networks, each contributing $\phi_m/2$, are required.

Step 3. The value of the α-parameter is calculated using Eq. (16.9).

Step 4. The corresponding required gain is calculated using Eq. (16.10).

Step 5. Once the required gain is figured out, the corresponding ω_m is found from the Bode plot.

Step 6. Now that α and ω_m are known, the corner frequencies for pole and zero namely, $1/\alpha T$ and $1/T$, are found using Eq. (16.7).

Step 7. The transfer function of the lead network is written using corner frequencies for the pole and zero.

The following examples will make the procedure clear.

EXAMPLE 16.1. Design a phase lead compensator for a unity feedback control system having open-loop transfer function

$$G(s) = \frac{50}{s(1+0.4s)}$$

so as to have a phase margin of 30°.

Solution The frequency domain transfer function is

$$G(j\omega) = \frac{50}{j\omega\left(1+\dfrac{j\omega}{2.5}\right)}$$

The corner frequencies are 0 and 2.5 for the pole factor.

The phase angle is

$$\phi = \angle G(j\omega) = -90° - \tan^{-1}(0.4\omega)$$

The above relation gives the following data:

ω	0.1	0.5	1.0	5.0	10.0	11.0	15.0	50.0	100.0
ϕ	−92.3°	−101.3°	−111.8°	−153.4°	−166.0°	−167.0°	−170.5°	−177.1°	−178.5

The Bode plots (Figure 16.7) give $\omega_g = 11$ rad/s and $\omega_p = \infty$. Phase margin = 180° − 167° = 13°.

To increase Pm to 30°, ω_g has to be moved to a higher value. The required phase lead is $\phi_m = (30° - 13°) + 3°$ (tolerance) = 20°. The corresponding value of the α-parameter of the lead network is given by [see Eq. (16.9)]

$$\alpha = \frac{1-\sin\phi_{\max}}{1+\sin\phi_{\max}} = \frac{1-\sin 20°}{1+\sin 20°} = 0.49$$

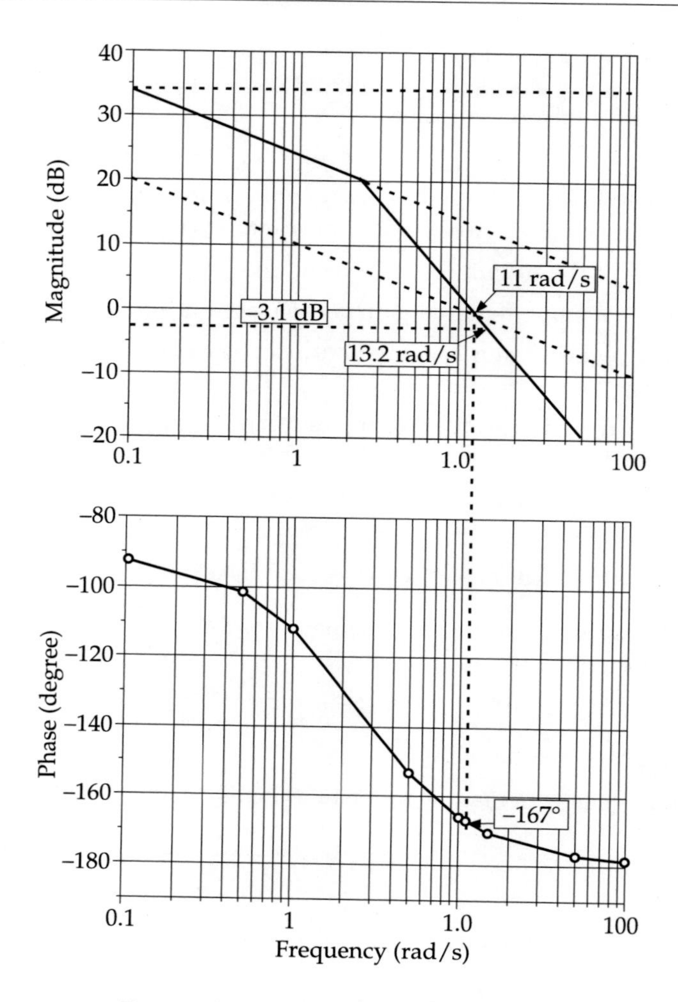

Figure 16.7 Bode plots (Example 16.1).

Now, from Eq. (16.10),

$$20 \log |G(\jmath\omega)|_{\omega=\omega_m} = -10 \log \left(\frac{1}{\alpha}\right) \text{ dB} = -3.1 \text{ dB}$$

From the Bode plot, −3.1 dB gain occurs at $\omega_m = 13.2$ rad/s. Therefore, for the lead network, upper and lower corner frequencies become [see Eq. (16.7)]

$$\frac{1}{\alpha T} = \frac{\omega_m \sqrt{\alpha}}{\alpha} = \frac{\omega_m}{\sqrt{\alpha}} = \frac{13.2}{\sqrt{0.49}} = 18.86 \text{ rad/s}$$

$$\frac{1}{T} = \omega_m \sqrt{\alpha} = 13.2\sqrt{0.49} = 9.24 \text{ rad/s}$$

Therefore, the required open-loop transfer function for the lead network is

$$G'(s) = \frac{s + \dfrac{1}{T}}{s + \dfrac{1}{\alpha T}} = \frac{s + 9.24}{s + 18.86} = 0.49 \times \frac{1 + \dfrac{s}{9.24}}{1 + \dfrac{s}{18.86}}$$

EXAMPLE 16.2. Design a cascade compensation for a unity feedback system whose transfer function is

$$G(s) = \frac{K}{s(1 + 0.1s)(1 + 0.001s)}$$

It should fulfil the following specifications:

(i) Phase margin $\geq 45°$

(ii) Velocity error constant $K_v = 1000 \text{ s}^{-1}$

Solution We know that

$$K_v = \lim_{s \to 0} \{s \cdot G(s)H(s)\} \qquad \text{[see Eq. (7.56) at page 261]}$$

$$= \lim_{s \to 0} \left\{ s \cdot \frac{K}{s(1 + 0.1s)(1 + 0.001s)} \right\}$$

Therefore, $\qquad K_v = K$

Since $K_v = 1000$ (given), we have

$$G(s) = \frac{1000}{s(1 + 0.1s)(1 + 0.001s)}$$

The frequency domain transfer function is

$$G(j\omega) = \frac{1000}{j\omega(1 + 0.1j\omega)(1 + 0.001j\omega)}$$

Corner frequencies are 0, 10 and 1000 for the pole factors. The phase angle is

$$\phi = \angle G(j\omega) = -90° - \tan^{-1}(0.1\omega) - \tan^{-1}(0.001\omega)$$

The above relation generates the following data:

ω	1.0	5.0	10.0	50.0	100.0	200.0	300.0	500.0	1000.0
ϕ	−95.77	−116.85	−135.57	−171.55	−180	−188.45	−194.79	−205.42	−224.43

The Bode plots (Figure 16.8) give $\omega_g = 100$ rad/s and $\omega_p = 100$ rad/s. Phase margin $= 180° - 180° = 0°$. To increase Pm to 45°, ω_g has to be moved to a higher value. The required phase lead is

$$\phi_{\max} = 45° + 5° \text{ (tolerance)} = 50°$$

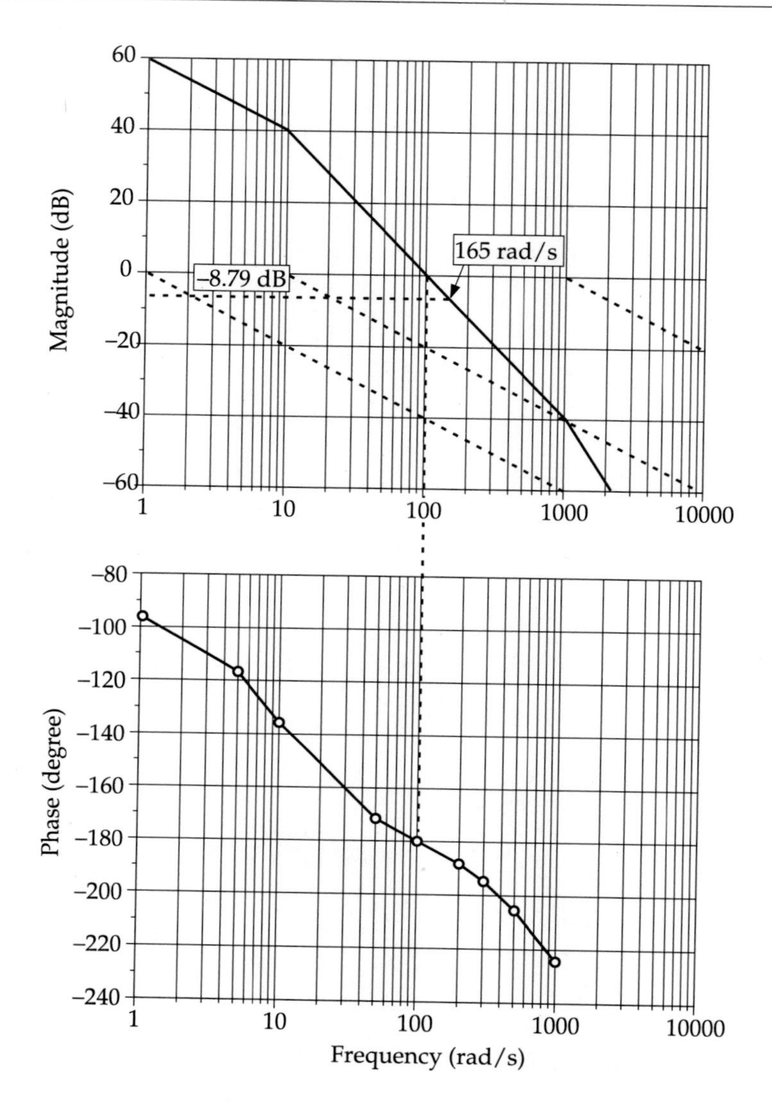

Figure 16.8 Bode plot (Example 16.2).

The corresponding value of the α-parameter of the lead network is given by [see Eq. (16.9)]

$$\alpha = \frac{1 - \sin \phi_{\max}}{1 + \sin \phi_{\max}} = \frac{1 - \sin 50°}{1 + \sin 50°} = 0.132$$

Now, from Eq. (16.10),

$$20 \log |G(\jmath\omega)|_{\omega=\omega_m} = -10 \log \left(\frac{1}{\alpha}\right) \text{ dB} = -8.79 \text{ dB}$$

From the Bode plot, -8.79 dB gain occurs at $\omega_m = 165$ rad/s. Therefore, for the lead network, the upper and lower corner frequencies become [see Eq. (16.7)]

$$\frac{1}{\alpha T} = \frac{\omega_m\sqrt{\alpha}}{\alpha} = \frac{\omega_m}{\sqrt{\alpha}} = \frac{165}{\sqrt{0.132}} = 454.147 \text{ rad/s}$$

$$\frac{1}{T} = \omega_m\sqrt{\alpha} = 165(\sqrt{0.132}) = 59.947 \text{ rad/s}$$

Therefore, the required open-loop transfer function for the lead network is

$$G'(s) = \frac{s+\dfrac{1}{T}}{s+\dfrac{1}{\alpha T}} = \frac{s+59.947}{s+454.147} = 0.13 \times \frac{1+\dfrac{s}{59.9}}{1+\dfrac{s}{454.1}}$$

16.5 LAG COMPENSATION NETWORK

A simple lag compensation circuit is shown in Figure 16.9.

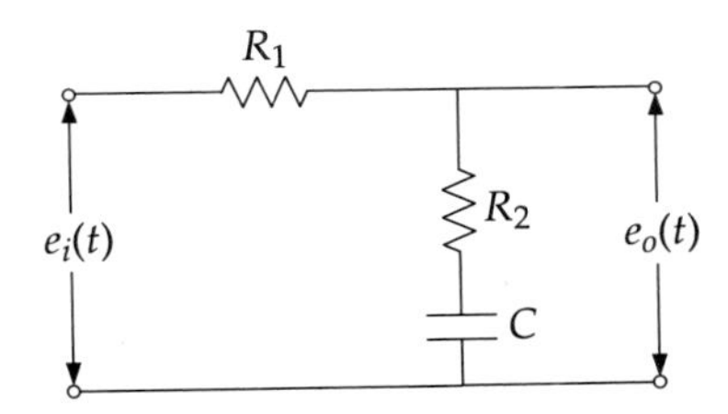

Figure 16.9 Lag compensator circuit.

Applying KVL to the Laplace transformed circuit

$$E_i(s) = \left(R_1 + R_2 + \frac{1}{sC}\right) I(s) \tag{16.11}$$

$$E_o(s) = \left(R_2 + \frac{1}{sC}\right) I(s) \tag{16.12}$$

From Eqs. (16.11) and (16.12), we get, for the transfer function of the lag compensation network, the equation

$$G(s) = \frac{E_o(s)}{E_i(s)} = \frac{R_2 + \dfrac{1}{sC}}{R_1 + R_2 + \dfrac{1}{sC}} = \frac{R_1}{R_1+R_2} \cdot \frac{s + \dfrac{1}{R_2C}}{s + \dfrac{1}{(R_1+R_2)C}}$$

$$\Rightarrow \qquad G(s) \equiv \frac{1}{\beta} \cdot \frac{s+\dfrac{1}{T}}{s+\dfrac{1}{\beta T}} \tag{16.13}$$

where

$$\beta = \frac{R_1 + R_2}{R_2} \quad \text{which is always } > 1$$

$$T = R_2 C = \text{time constant} \tag{16.14}$$

The frequency domain transfer function is, therefore,

$$G(\jmath\omega) = \frac{1}{\beta} \cdot \frac{\jmath\omega + \dfrac{1}{T}}{\jmath\omega + \dfrac{1}{\beta T}} = \frac{\jmath\omega T + 1}{\jmath\omega\beta T + 1} \tag{16.15}$$

From Eqs. (16.12) and (16.14), we observe the following:

1. The lag network transfer function has a zero at $s = -1/T$ and a pole at $-1/\beta T$. Since $\beta > 1$, the pole is dominant because it is located nearer to the Im-axis [see Figure 16.10(a)].

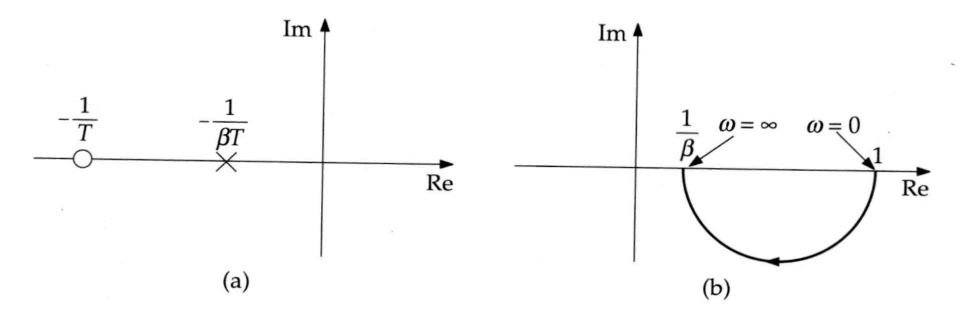

Figure 16.10 (a) Pole-zero plot; (b) polar plot of the lag network transfer function.

2. The phase angles introduced by the numerator and denominator are $\tan^{-1}\omega T$ and $\tan^{-1}(\beta\omega T)$, respectively. Since $\beta > 1$, the latter angle is bigger and, therefore, the net phase angle is negative. Hence the name *lag network.*
3. The polar plot of the lag network transfer function can be constructed from

$$G(\jmath\omega) = \frac{\jmath\omega T + 1}{\jmath\omega\beta T + 1} \equiv \frac{\sqrt{\omega^2 T^2 + 1}}{\sqrt{\omega^2\beta^2 T^2 + 1}} \angle\left[\tan^{-1}\omega T - \tan^{-1}(\omega\beta T)\right] \tag{16.16}$$

Equation (16.16) yields the following data:

ω	$\lvert G(\jmath\omega)\rvert$	$\angle G(\jmath\omega)$
0	1	0°
$1/T$ ($\beta = 1.1$)	−0.64	−3°
∞	$1/\beta$	0°

The consequent polar plot is shown in Figure 16.10(b).

4. The phase angle ϕ is given by

$$\phi = \tan^{-1}\omega T - \tan^{-1}(\omega\beta T) = \tan^{-1}\frac{\omega T(1-\beta)}{1+\beta\omega^2T^2} \tag{16.17}$$

If ω_m is the frequency for which $\phi = \phi_{\max}$, then

$$\left.\frac{d\phi}{d\omega}\right|_{\omega=\omega_m} = 0$$

$$\Rightarrow \quad \frac{T}{1+\omega_m^2T^2} - \frac{\beta T}{1+\omega_m^2\beta^2T^2} = 0$$

$$\Rightarrow \quad \frac{T}{1+\omega_m^2T^2} = \frac{\beta T}{1+\omega_m^2\beta^2T^2}$$

$$\Rightarrow \quad \omega_m^2(\beta T^2 - \beta^2T^2) = 1-\beta$$

$$\Rightarrow \quad \omega_m = \frac{1}{T\sqrt{\beta}} = \sqrt{\left(\frac{1}{T}\right)\left(\frac{1}{\beta T}\right)} \tag{16.18}$$

We note that Eqs. (16.18) and (16.7) are similar. Therefore, we can write the following equations as well:

$$\tan\phi_{\max} = \frac{1-\beta}{2\sqrt{\beta}}$$

$$\Rightarrow \quad \sin\phi_{\max} = \frac{1-\beta}{1+\beta}$$

$$\Rightarrow \quad \beta = \frac{1-\sin\phi_{\max}}{1+\sin\phi_{\max}}$$

From Eq. (16.18) we note that the frequency ω_m, corresponding to the maximum phase angle, is the geometric mean of the two corner frequencies. Also, from Eq. (16.16) at ω_m,

$$|G(j\omega)|_{\omega=\omega_m} = \frac{\sqrt{\omega^2T^2+1}}{\sqrt{\omega^2\beta^2T^2+1}} = \frac{\sqrt{\frac{1}{\beta}+1}}{\beta+1} = \frac{1}{\sqrt{\beta}}$$

Therefore,

$$20\log|G(j\omega)|_{\omega=\omega_m} = 20\log\left(\frac{1}{\sqrt{\beta}}\right) = 10\log\left(\frac{1}{\beta}\right) \text{ dB} \tag{16.19}$$

5. From the above discussion, Bode plots for the lag network can be drawn as those shown in Figure 16.11.
6. For $\omega = 0$ (dc), the circuit gain is unity, whereas at high frequency, the gain is $1/\beta$. We know that $\beta > 1$. Its value is generally 10 or higher.

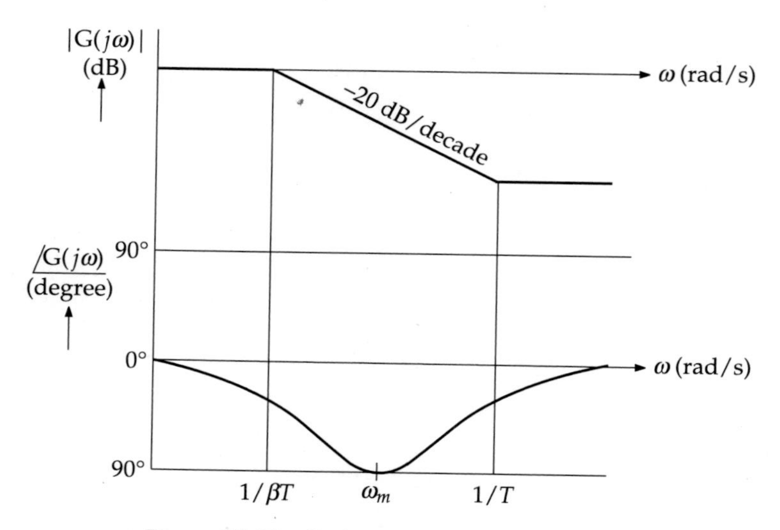

Figure 16.11 Bode plots for the lag network.

Effects of Incorporating Lag Compensator

The effects of incorporating lag compensators are:

1. The network allows low frequencies but attenuates high frequencies. As a result, it improves the steady-state performance.
2. The gain crossover frequency is lowered. Which is why the bandwidth (i.e. the speed of response) decreases, increasing the peak overshoot and settling time.

Designing Lag Network

To design a lag network, the following steps may be followed:

Step 1. Draw the Bode plot of the system open-loop transfer function with gain constant K adjusted according to the prescribed steady-state error, if necessary.

Step 2. Determine the phase margin ϕ of the uncompensated system from the Bode plot. If ϕ_r is the required Pm and τ, the tolerance to be introduced, then

$$\phi = \phi_r + \tau$$

Step 3. Determine the frequency from the phase curve where ϕ is the phase margin. This is the new gain crossover frequency ω_g'.

Step 4. Check the gain at ω_g' from the gain curve of the Bode plot. The gain curve has to be adjusted to make it 0 dB at ω_g'. The necessary attenuation has to be provided by the factor $1/\beta$ in the transfer function [see Eq. (16.13)] to keep the shape of the Bode plot unchanged. Therefore,

$$\text{Change necessary in gain} = 20 \log (1/\beta) = -20 \log \beta$$

Thus, β is obtained from this relation.

Step 5. Usually, the upper corner frequency $1/T$ is placed about a decade below the new gain crossover frequency ω_g'. So, calculate T from the relation

$$\frac{1}{T} = \frac{\omega_g'}{10}$$

Step 6. Once β and T are known, the corner frequencies for pole and zero, viz. $1/\beta T$ and $1/T$, are known and, hence, the transfer function for the network can be written.

EXAMPLE 16.3. Design a phase-lag compensating network for the unity feedback system given by

$$G(s) = \frac{K}{s(1+0.1s)(1+0.2s)}$$

so that

$$K_v = 30 \text{ s}^{-1} \qquad \text{Pm} \geq 40°$$

Solution We know from Eq. (7.56) that $K_v = \lim_{s\to 0} sG(s)H(s)$. Here, since $H(s) = 1$,

$$K_v = \lim_{s\to 0} sG(s) = \lim_{s\to 0} \left\{ s \cdot \frac{K}{s(1+0.1s)(1+0.2s)} \right\} = K$$

Thus, the transfer function becomes

$$G(s) = \frac{30}{s\left(1+\dfrac{s}{10}\right)\left(1+\dfrac{s}{5}\right)}$$

The Bode plots of the function are given in Figure 16.12.

From the Bode plot we find that the uncompensated Pm is $-18°$. But our required Pm is $40°$, and the tolerance to be introduced is $5°$. So, the Pm to be made is

$$40° + 5° = 45°$$

This Pm is available at $\omega = 2.3$ rad/s (see the Bode phase plot). The corresponding Gain is 22 dB. Thus, we need to move the gain plot by -22 dB so that it crosses the 0 dB line there to make 2.3 rad/s the gain crossover frequency and, consequently, Pm $= 45°$. Therefore,

$$-20 \log \beta = -22$$

or

$$\beta = 12.59 \tag{i}$$

Also,

$$T = \frac{10}{2.3} = 4.35 \tag{ii}$$

With the help of available data from Eqs. (i) and (ii), we can write the transfer function for the lag compensating network as [see Eq. (16.13)]

$$G(s) = \frac{1}{\beta} \cdot \frac{s+\dfrac{1}{T}}{s+\dfrac{1}{\beta T}} = \frac{1}{12.59} \cdot \frac{s+\dfrac{1}{4.35}}{s+\dfrac{1}{(4.35)(12.59)}} = \frac{4.35s+1}{54.77s+1}$$

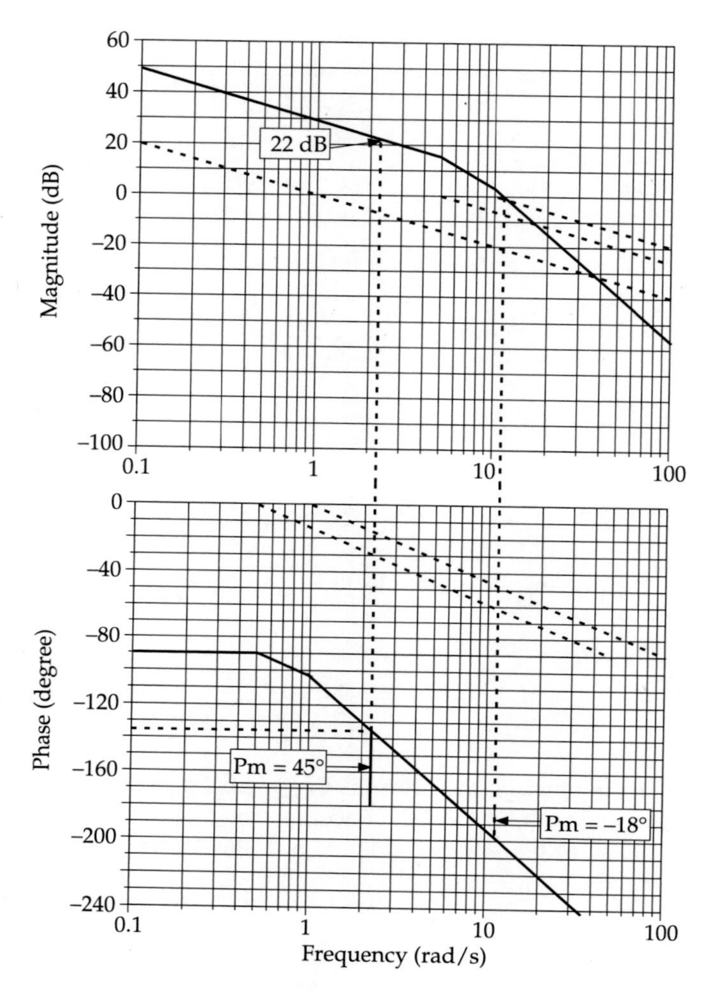

Figure 16.12 Bode plot (Example 16.3).

16.6 LEAD-LAG COMPENSATION NETWORK

A simple electrical circuit for the lead-lag compensation network is shown in Figure 16.13.

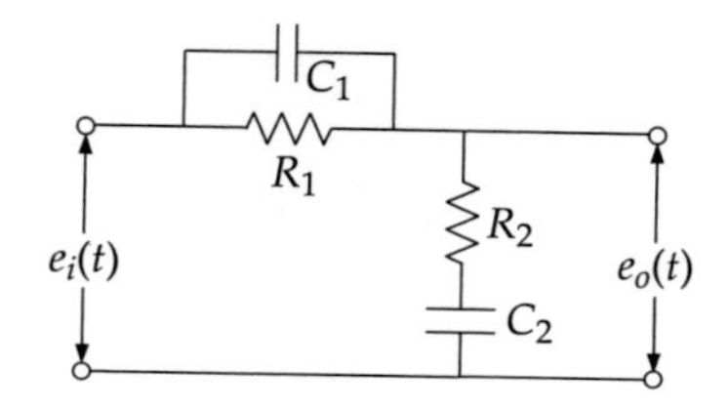

Figure 16.13 Simple lead-lag compensation circuit.

Applying KVL to the Laplace transform equivalent of the circuit, we get

$$E_i(s) = \left[\left(R_1 \| \frac{1}{sC_1}\right) + R_2 + \frac{1}{sC_2}\right] I(s)$$

$$= \left(\frac{R_1}{sR_1C_1 + 1} + R_2 + \frac{1}{sC_2}\right) I(s)$$

$$= \frac{sR_1C_2 + sR_2C_2(sR_1C_1 + 1) + (sR_1C_1 + 1)}{sC_2(sR_1C_1 + 1)} I(s) \tag{16.20}$$

$$E_o(s) = \left(R_2 + \frac{1}{sC_2}\right) I(s) = \frac{sR_2C_2 + 1}{sC_2} I(s) \tag{16.21}$$

With the help of Eqs. (16.20) and (16.21), we get the transfer function of the circuit as

$$G(s) \equiv \frac{E_o(s)}{E_i(s)} = \frac{sR_2C_2 + 1}{sC_2} I(s) \cdot \frac{sC_2(sR_1C_1 + 1)}{sR_1C_2 + sR_2C_2(sR_1C_1 + 1) + (sR_1C_1 + 1)}$$

$$= \frac{(sR_2C_2 + 1)(sR_1C_1 + 1)}{sR_1C_2 + (sR_2C_2 + 1)(sR_1C_1 + 1)}$$

$$= \frac{(R_1C_1)(R_2C_2)}{(R_1C_1)(R_2C_2)} \cdot \frac{\left(s + \frac{1}{R_1C_1}\right)\left(s + \frac{1}{R_2C_2}\right)}{\left[\frac{s}{R_1C_1} + \left(s + \frac{1}{R_1C_1}\right)\left(s + \frac{1}{R_2C_2}\right)\right]}$$

$$= \frac{\left(s + \frac{1}{R_1C_1}\right)\left(s + \frac{1}{R_2C_2}\right)}{s^2 + s\left(\frac{1}{R_1C_1} + \frac{1}{R_2C_2} + \frac{1}{R_2C_1} + \frac{1}{R_1R_2C_1C_2}\right)} \tag{16.22}$$

We can write Eq. (16.22) as

$$G(s) \equiv \frac{\left(s + \frac{1}{T_1}\right)\left(s + \frac{1}{T_2}\right)}{\left(s + \frac{1}{\alpha T_1}\right)\left(s + \frac{1}{\beta T_2}\right)} \tag{16.23}$$

where

$$T_1 = R_1C_1 \tag{16.24}$$

$$T_2 = R_2C_2 \tag{16.25}$$

$$\frac{1}{\alpha T_1} + \frac{1}{\beta T_2} = \frac{1}{R_1C_1} + \frac{1}{R_2C_2} + \frac{1}{R_2C_1} \tag{16.26}$$

Therefore, comparing the denominators of Eq. (16.26), we get

$$\alpha\beta T_1T_2 = R_1R_2C_1C_2$$

which, when compared with Eqs. (16.24) and (16.25), gives

$$\alpha\beta = 1 \tag{16.27}$$

Equation (16.27) helps us rewrite Eq. (16.23) as

$$G(s) \equiv \frac{\left(s+\dfrac{1}{T_1}\right)\left(s+\dfrac{1}{T_2}\right)}{\left(s+\dfrac{\beta}{T_1}\right)\left(s+\dfrac{1}{\beta T_2}\right)} = \beta \cdot \frac{(sT_1+1)(sT_2+1)}{(sT_1+\beta)(s\beta T_2+1)} \tag{16.28}$$

This is a lead and lag compensator combination, which is displayed by the pole-zero and polar plots (Figure 16.14).

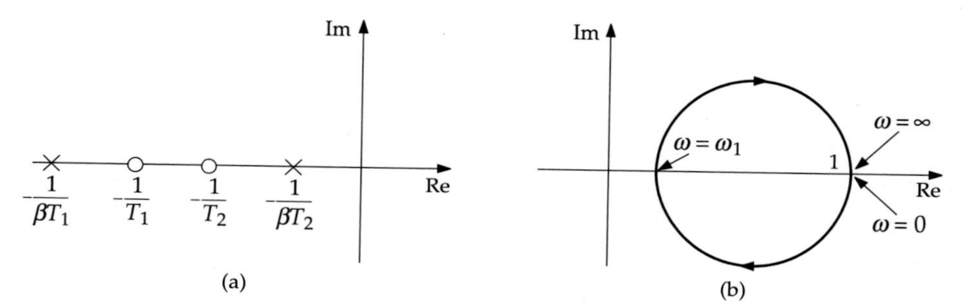

Figure 16.14 (a) Pole-zero; (b) polar plot of lead-lag compensator.

The same combination is also discernible in the Bode plot (Figure 16.15).

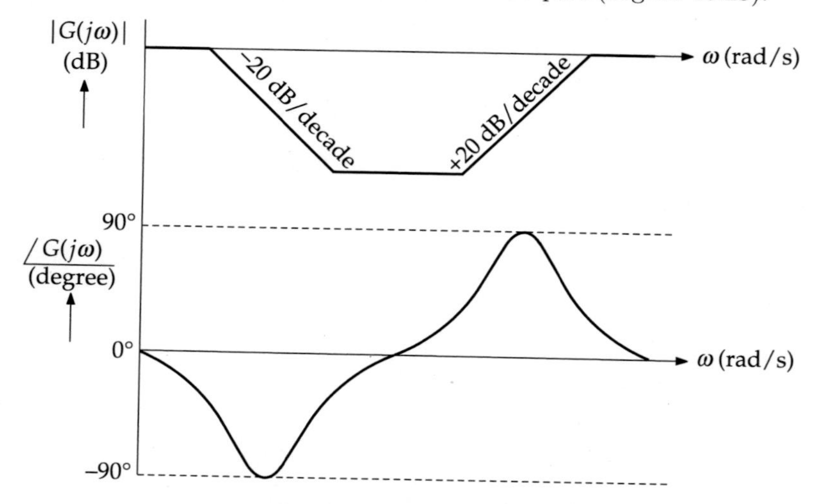

Figure 16.15 Bode plot of lead-lag compensator.

Effects of Lead-Lag Compensation

1. Bandwidth (i.e. speed) improves owing to the existence of the lead network.
2. Steady-state performance (i.e. accuracy) improves owing to the lag network.

EXAMPLE 16.4. A lead-lag compensating network transfer function is given by

$$G(s) = K\beta \cdot \frac{(1+sT_1)(1+sT_2)}{(\beta+sT_2)(1+s\beta T_2)}$$

Show that the phase angle of $G(\jmath\omega)$ becomes zero at the frequency $\omega = \dfrac{1}{\sqrt{T_1T_2}}$.

Solution The frequency domain transfer function is

$$G(\jmath\omega) = K\beta \cdot \frac{(1+\jmath\omega T_1)(1+\jmath\omega T_2)}{(\beta+\jmath\omega T_1)(1+\jmath\omega\beta T_2)}$$

$$\equiv K\beta \frac{\sqrt{1+\omega^2T_1^2}\sqrt{1+\omega^2T_2^2}}{\sqrt{\beta^2+\omega^2T_1^2}\sqrt{1+\omega^2\beta^2T_2^2}}$$

$$\angle\left[\tan^{-1}(\omega T_1) + \tan^{-1}(\omega T_2) - \tan^{-1}\left(\frac{\omega T_1}{\beta}\right) - \tan^{-1}(\omega\beta T_2)\right]$$

Thus, the phase angle ϕ is given by

$$\phi = \tan^{-1}(\omega T_1) + \tan^{-1}(\omega T_2) - \tan^{-1}\left(\frac{\omega T_1}{\beta}\right) - \tan^{-1}(\omega\beta T_2)$$

$$= \tan^{-1}\frac{\omega(T_1+T_2)}{1-\omega^2T_1T_2} - \tan^{-1}\frac{\omega\left(\dfrac{T_1}{\beta}+\beta T_2\right)}{1-\omega^2T_1T_2}$$

We observe that the denominators of two inverse functions are the same. At $\omega = \dfrac{1}{\sqrt{T_1T_2}}$, it turns out to be

$$1-\omega^2T_1T_2 = 1-\left(\frac{1}{\sqrt{T_1T_2}}\right)^2 T_1T_2 = 0$$

Therefore,

$$\phi = \tan^{-1}\infty - \tan^{-1}\infty = 90^\circ - 90^\circ = 0^\circ$$

REVIEW QUESTIONS

16.1 Choose the correct answer

(i) The transfer function of a compensator is given by $\dfrac{1+s}{1+0.5s}$. The compensator is a

(a) lag-lead network
(b) lead network
(c) lag network
(d) none of the above

(ii) The transfer function of a compensator is given by $G(s) = \dfrac{s+z}{s+p}$. It will act as a phase lag compensator if

(a) $z > p$ (b) $z < p$
(c) $z = p$ (d) none of the above

(iii) A lead compensator
(a) increases the bandwidth
(b) decreases the bandwidth
(c) has nothing to do with bandwidth
(d) makes the bandwidth negative

(iv) $G(s) = \dfrac{s+5}{s+15}$ is the transfer function of a lead compensator. The maximum phase shift that can be obtained from it is

(a) 15° (b) 30° (c) 45° (d) 60°

(v) A lag compensator
(a) allows low frequencies but attenuates high frequencies
(b) allows high frequencies but attenuates low frequencies
(c) allows low and high frequencies equally
(d) attenuates low and high frequencies equally

16.2 A unity feedback system is characterized by an open-loop transfer function of

$$G(s) = \frac{10}{s(1+0.5s)(1+0.1s)}$$

(a) Draw the asymptotic Bode plot of the system and determine the phase and gain margins.
(b) A lead compensator having a transfer function of $(1 + 0.25s)/(1 + 0.025s)$ is introduced in series with the plant. Determine the new phase and gain margins.
(c) Comment on the improvement in system response.

16.3 The open-loop transfer function of a unity feedback system is

$$G(s) = \frac{K}{s(1+0.2s)}$$

It is required that the velocity error constant should at least be 20 and the phase margin, 44°. Does the system meet the required specifications? If not, design the compensating network to satisfy the required specifications.

16.4 Write short notes on the following:

(a) Lead compensator
(b) Lag compensator

Answers to Selected Questions

16.1. (i) (c); (ii) (b); (iii) (a); (iv) (b); (v) (a)

16.2. (a) Gm = 1.58 dB, $\omega_p = 4.47$ rad/s, Pm = 3.94°, $\omega_g = 4.08$ rad/s
(b) Gm = 17.5 dB, $\omega_p = 17.4$ rad/s, Pm = 38.6°, $\omega_g = 5.19$ rad/s

16.3. $G'(s) = \dfrac{s + 5.36}{s + 10.93}$

CHAPTER 17

State-space Analysis

17.1 INTRODUCTION

So far, we have considered SISO linear time-invariant systems which could be dealt with by the conventional transfer function approach. However, this method is not suitable for application to multiple input multiple output (MIMO) systems as well as time-varying and nonlinear systems. With n inputs and m outputs, we would otherwise have to write down mn Laplace transforms to encode all the information about a MIMO system. The state-space representation provides a convenient and compact way to model and analyze those systems. In contrast to the conventional theory which is a predominantly frequency domain approach, the present theory is a time domain approach. And, unlike the frequency domain approach, the use of the state-space representation is not limited to systems with linear components and zero initial conditions. "State-space" refers to the space whose axes are the state variables. The state of the system can be represented as a vector within that space.

Before we discuss the approach further, we shall familiarize ourselves with a few definitions.

17.2 STATE-SPACE DEFINITIONS

State. If there exists a smallest set of variables—called the *state variables*—whose values and interrelationships at $t = t_0$ along with the input for $t \geq t_0$ completely determine the dynamic behaviour of a system at $t \geq 0$, then this set is known as the *state* of the system.

Suppose we have a set of variables — $x_1(t)$, $x_2(t)$, $\ldots$, $x_n(t)$ — which completely describe the behaviour of a dynamic system such that with a knowledge of the initial values at $t = t_0$ and the input at $t \geq t_0$, the future behaviour of the system is completely determinable. Then these variables are called state variables, and the set, the state.

State vector. The n state variables that completely describe the behaviour of a particular system constitute n components of a vector called the *state vector*.

State-space. One can postulate an n-dimensional space having n-axes representing n state variables and indicate points there to represent a state of the system at a certain time. This space is called the *state-space*.

17.3 STATE-SPACE REPRESENTATION

Let us consider a MIMO system as given in Figure 17.1. Here, $u_1, u_2, \ldots, u_m$ inputs produce $y_1, y_2, \ldots, y_n$ outputs and $x_1, x_2, \ldots, x_p$ state variables describe the system behaviour.

Input, output and state variables are all time-dependent. We define them in vector forms as

$$\mathbf{u}(t) = \begin{bmatrix} u_1(t) \\ u_2(t) \\ \vdots \\ u_m(t) \end{bmatrix} \qquad \mathbf{y}(t) = \begin{bmatrix} y_1(t) \\ y_2(t) \\ \vdots \\ y_n(t) \end{bmatrix} \qquad \mathbf{x}(t) = \begin{bmatrix} cx_1(t) \\ x_2(t) \\ \vdots \\ x_p(t) \end{bmatrix} \tag{17.1}$$

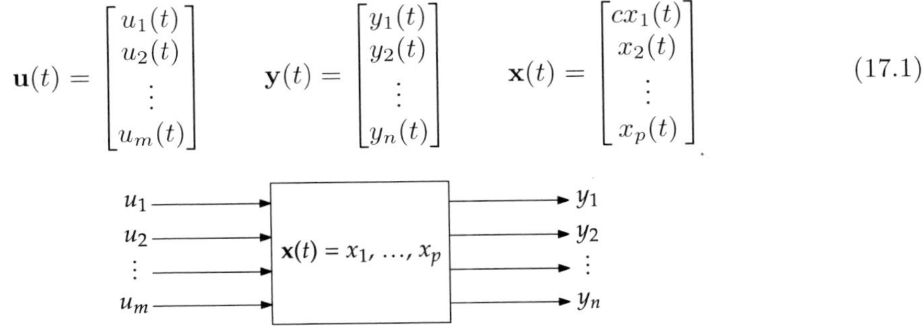

Figure 17.1 MIMO system.

Let these vectors be interrelated as

$$\dot{\mathbf{x}}(t) = \mathbf{A}(t)\mathbf{x}(t) + \mathbf{B}(t)\mathbf{u}(t) \quad \text{[state equation]} \tag{17.2}$$

$$\mathbf{y}(t) = \mathbf{C}(t)\mathbf{x}(t) + \mathbf{D}(t)\mathbf{u}(t) \quad \text{[output equation]} \tag{17.3}$$

Equations (17.2) and (17.3) together are called state-space representation (Figure 17.2 gives a block diagrammatic repesentation).

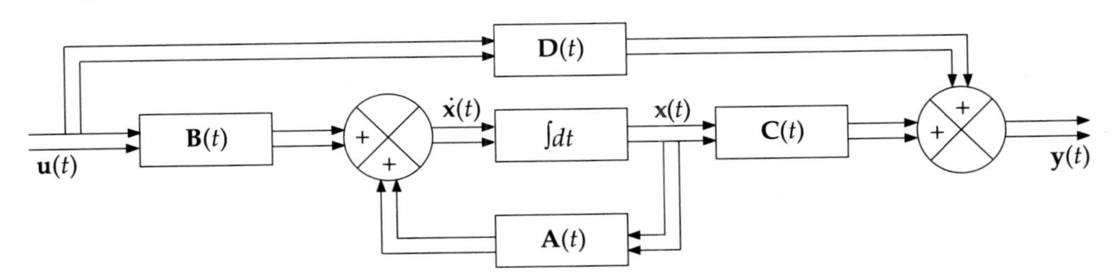

Figure 17.2 Block diagram of state-space representation. Double lines indicate multiple signals.

In these equations,

$\mathbf{x}$ = state vector
$\mathbf{y}$ = output vector
$\mathbf{u}$ = input (or control) vector
$\mathbf{A}$ = state (or system) matrix
$\mathbf{B}$ = input (or control) matrix
$\mathbf{C}$ = output matrix, and
$\mathbf{D}$ = direct transmission matrix[1]

[1] Also called *feedthrough* (or *feedforward*) matrix.

D is usually chosen as the zero matrix in control systems where power amplification is desired. This implies that the system input is chosen not to have direct feedthrough or transmission to the output.

Salient Features of State-Space Representation

The salient features of the state-space representation are:

1. The system is described in terms of n first order differential equations.
2. The equations may form a vector-matrix differential equation.
3. The use of vector-matrix notation simplifies the mathematical representation of the system. Multiple inputs, outputs and state variables can be dealt with without difficulty or complication.
4. The method is particularly suitable for solution using digital computers.

Apart from the SISO continuous time-invariant systems, which we will use to gain familiarity with the approach, there are other forms of the state-space model (see Table 17.1).

Table 17.1 Different State-space Models

System type	*State-space model*
Continuous time-invariant	$\dot{\mathbf{x}}(t) = \mathbf{A}\mathbf{x}(t) + \mathbf{B}\mathbf{u}(t)$ $\mathbf{y}(t) = \mathbf{C}\mathbf{x}(t) + \mathbf{D}\mathbf{u}(t)$
Continuous time-variant	$\dot{\mathbf{x}}(t) = \mathbf{A}(t)\mathbf{x}(t) + \mathbf{B}(t)\mathbf{u}(t)$ $\mathbf{y}(t) = \mathbf{C}(t)\mathbf{x}(t) + \mathbf{D}(t)\mathbf{u}(t)$
Discrete time-invariant	$\dot{\mathbf{x}}(k+1) = \mathbf{A}\mathbf{x}(k) + \mathbf{B}\mathbf{u}(k)$ $\mathbf{y}(k) = \mathbf{C}\mathbf{x}(k) + \mathbf{D}\mathbf{u}(k)$
Discrete time-variant	$\dot{\mathbf{x}}(k+1) = \mathbf{A}(k)\mathbf{x}(k) + \mathbf{B}(k)\mathbf{u}(k)$ $\mathbf{y}(k) = \mathbf{C}(k)\mathbf{x}(k) + \mathbf{D}(k)\mathbf{u}(k)$
Laplace domain of continuous time-invariant	$s\mathbf{X}(s) = \mathbf{A}\mathbf{X}(s) + \mathbf{B}\mathbf{U}(s)$ $\mathbf{Y}(s) = \mathbf{C}\mathbf{X}(s) + \mathbf{D}\mathbf{U}(s)$
z-domain of discrete time-invariant	$z\mathbf{X}(z) = \mathbf{A}\mathbf{X}(z) + \mathbf{B}\mathbf{U}(z)$ $\mathbf{Y}(z) = \mathbf{C}\mathbf{X}(z) + \mathbf{D}\mathbf{U}(z)$

How the equations are framed can be gauged from a consideration of LTI systems. Here, derivatives of state variables at any instant t are linear combinations of system states and inputs. So, they can be written as

$$\begin{aligned}
\dot{x}_1(t) &= [a_{11}x_1(t) + a_{12}x_2(t) + \cdots + a_{1n}x_n(t)] + [b_{11}u_1(t) + b_{12}u_2(t) + \cdots + b_{1m}u_m(t)] \\
\dot{x}_2(t) &= [a_{21}x_1(t) + a_{22}x_2(t) + \cdots + a_{2n}x_n(t)] + [b_{21}u_1(t) + b_{22}u_2(t) + \cdots + b_{2m}u_m(t)] \\
&\vdots \\
\dot{x}_n(t) &= [a_{n1}x_1(t) + a_{n2}x_2(t) + \cdots + a_{nn}x_n(t)] + [b_{n1}u_1(t) + b_{n2}u_2(t) + \cdots + b_{nm}u_m(t)]
\end{aligned} \tag{17.4}$$

We can write Eq. (17.4) in matrix form as

$$\dot{\mathbf{x}}(t) = \mathbf{A}\mathbf{x}(t) + \mathbf{B}\mathbf{u}(t) \tag{17.5}$$

where

$$\mathbf{A} = \begin{bmatrix} a_{11} & a_{12} & \cdots & a_{1n} \\ a_{21} & a_{22} & \cdots & a_{2n} \\ \vdots & \vdots & & \vdots \\ a_{n1} & a_{n2} & \cdots & a_{nn} \end{bmatrix} \text{ is an } n \times n \text{ matrix}$$

$$\mathbf{B} = \begin{bmatrix} b_{11} & b_{12} & \cdots & b_{1m} \\ b_{21} & b_{22} & \cdots & b_{2m} \\ \vdots & \vdots & & \vdots \\ b_{n1} & b_{n2} & \cdots & b_{nm} \end{bmatrix} \text{ is an } n \times m \text{ matrix}$$

and $\dot{\mathbf{x}}$, $\mathbf{x}$ and $\mathbf{u}$ are column vectors of dimensions $n \times 1$, $n \times 1$ and $m \times 1$, respectively, as given in Eq. (17.1). In a similar way, we can write the output variables as linear combinations of system states and inputs at any instant t as

$$\begin{aligned} y_1(t) &= [c_{11}x_1(t) + c_{12}x_2(t) + \cdots + c_{1n}x_n(t)] + [d_{11}u_1(t) + d_{12}u_2(t) + \cdots + d_{1m}u_m(t)] \\ y_2(t) &= [c_{21}x_1(t) + c_{22}x_2(t) + \cdots + c_{2n}x_n(t)] + [d_{21}u_1(t) + d_{22}u_2(t) + \cdots + d_{2m}u_m(t)] \\ &\vdots \\ y_p(t) &= [c_{n1}x_1(t) + c_{n2}x_2(t) + \cdots + c_{pn}x_n(t)] + [d_{n1}u_1(t) + d_{n2}u_2(t) + \cdots + d_{pm}u_m(t)] \end{aligned} \tag{17.6}$$

Equation (17.6) can be written in matrix form as

$$\mathbf{y}(t) = \mathbf{C}\mathbf{x}(t) + \mathbf{D}\mathbf{u}(t) \tag{17.7}$$

where

$$\mathbf{C} = \begin{bmatrix} c_{11} & c_{12} & \cdots & c_{1n} \\ c_{21} & c_{22} & \cdots & c_{2n} \\ \vdots & \vdots & & \vdots \\ c_{n1} & c_{n2} & \cdots & c_{pn} \end{bmatrix} \text{ is a } p \times n \text{ matrix}$$

$$\mathbf{D} = \begin{bmatrix} d_{11} & d_{12} & \cdots & d_{1m} \\ d_{21} & d_{22} & \cdots & d_{2m} \\ \vdots & \vdots & & \vdots \\ d_{p1} & d_{p2} & \cdots & d_{pm} \end{bmatrix} \text{ is a } p \times m \text{ matrix}$$

and $\mathbf{y}(t)$ is a column vector of dimension $p \times 1$. Thus, Eqs. (17.5) and (17.7) together constitute state equations for an LTI MIMO system. More often than not, $\mathbf{D}$ is a null matrix and $\mathbf{C}$ is a $1 \times n$ row vector.

We work out a few examples to see how to set up state-space models for systems.

EXAMPLE 17.1. Obtain the state-space model for the system defined by

$$\frac{d^3c(t)}{dt^3} + 5\frac{d^2c(t)}{dt^2} + 8\frac{dc(t)}{dt} + 12c(t) = 5r(t)$$

where $c(t)$ = output and $r(t)$ = input.

Solution Let us choose the state variables as

$$x_1 = c(t) \tag{i}$$

$$x_2 = \frac{dc(t)}{dt}$$

$$x_3 = \frac{d^2c(t)}{dt^2}$$

Thus, we have the following set of equations

$$\dot{x}_1 = x_2$$

$$\dot{x}_2 = x_3$$

$$\dot{x}_3 = -5x_3 - 8x_2 - 12x_1 + 5r(t)$$

This set of equations can be written in the following matrix form:

$$\begin{bmatrix} \dot{x}_1 \\ \dot{x}_2 \\ \dot{x}_3 \end{bmatrix} = \begin{bmatrix} 0 & 1 & 0 \\ 0 & 0 & 1 \\ -12 & -8 & -5 \end{bmatrix} \begin{bmatrix} x_1 \\ x_2 \\ x_3 \end{bmatrix} \begin{bmatrix} 0 \\ 0 \\ 5 \end{bmatrix} [r(t)]$$

$$c(t) = \begin{bmatrix} 1 & 0 & 0 \end{bmatrix} \begin{bmatrix} x_1 \\ x_2 \\ x_3 \end{bmatrix} \qquad \text{[from Eq. (i)]}$$

These matrices are in the standard form of the state-space, i.e.

$$\dot{\mathbf{x}}(t) = \mathbf{A}\mathbf{x}(t) + \mathbf{B}\mathbf{u}(t)$$

$$\mathbf{y}(t) = \mathbf{C}\mathbf{x}(t) + \mathbf{D}\mathbf{u}(t)$$

where

$$\mathbf{A} = \begin{bmatrix} 0 & 1 & 0 \\ 0 & 0 & 1 \\ -12 & -8 & -5 \end{bmatrix} \qquad \mathbf{B} = \begin{bmatrix} 0 \\ 0 \\ 5 \end{bmatrix}$$

$$\mathbf{C} = \begin{bmatrix} 1 & 0 & 0 \end{bmatrix} \qquad \mathbf{D} = [0] = \text{null matrix}$$

The signal flow graph of Figure 17.3 also represents the set of equations.

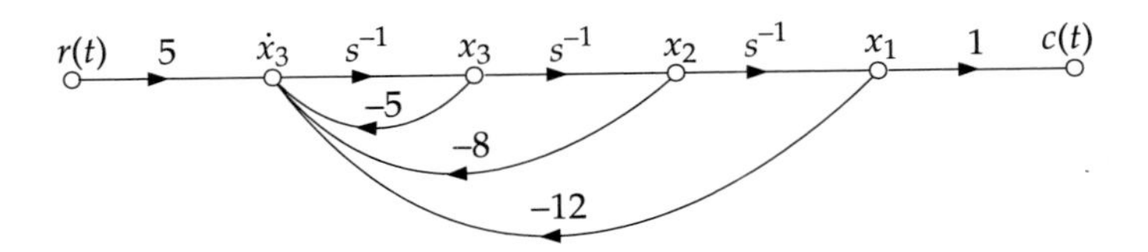

Figure 17.3 Signal flow graph (Example 17.1)

EXAMPLE 17.2. Obtain the state model for the network given in Figure 17.4.

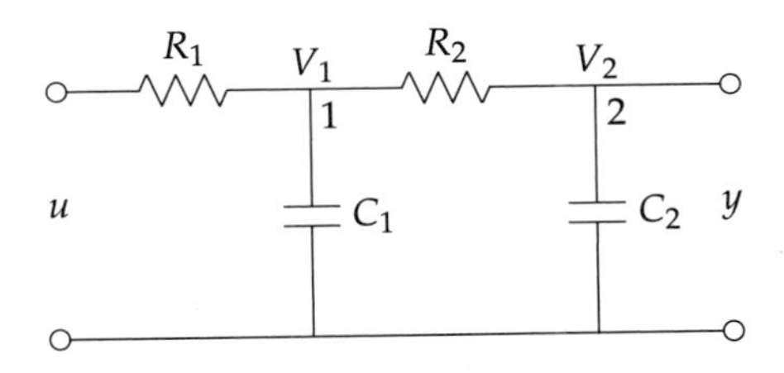

Figure 17.4 Network (Example 17.2).

Solution Considering the circuit, we can choose V_1 and V_2, the voltages across two capacitors (energy storage devices) as state vectors. Applying KCL at node 1, we get

$$\frac{V_1 - u}{R_1} + C_1\frac{dV_1}{dt} + \frac{V_1 - V_2}{R_2} = 0$$

$$\Rightarrow \qquad \frac{dV_1}{dt} = -\left(\frac{1}{R_1C_1} + \frac{1}{R_2C_1}\right)V_1 + \frac{1}{R_2C_1}V_2 + \frac{1}{R_1C_1}u \qquad \text{(i)}$$

Applying KCL at node 2, we get

$$\frac{V_2 - V_1}{R_2} + C_2\frac{dV_2}{dt} = 0$$

$$\Rightarrow \qquad \frac{dV_2}{dt} = \frac{1}{R_2C_2}V_1 - \frac{1}{R_2C_2}V_2 \qquad \text{(ii)}$$

and,

$$y = V_2 \qquad \text{(iii)}$$

To obtain the state-space representation, we write Eqs. (i) to (iii) in matrix form as follows:

$$\begin{bmatrix} \dot{V}_1 \\ \dot{V}_2 \end{bmatrix} = \begin{bmatrix} -\left(\frac{1}{R_1C_1} + \frac{1}{R_2C_1}\right) & \frac{1}{R_2C_1} \\ \frac{1}{R_2C_2} & -\frac{1}{R_2C_2} \end{bmatrix} \begin{bmatrix} V_1 \\ V_2 \end{bmatrix} + \begin{bmatrix} \frac{1}{R_1C_1} \\ 0 \end{bmatrix} [u]$$

$$[y] = \begin{bmatrix} 0 & 1 \end{bmatrix} \begin{bmatrix} V_1 \\ V_2 \end{bmatrix} + [0][u]$$

EXAMPLE 17.3. Write state variables in matrix form for the circuit given in the following figure.

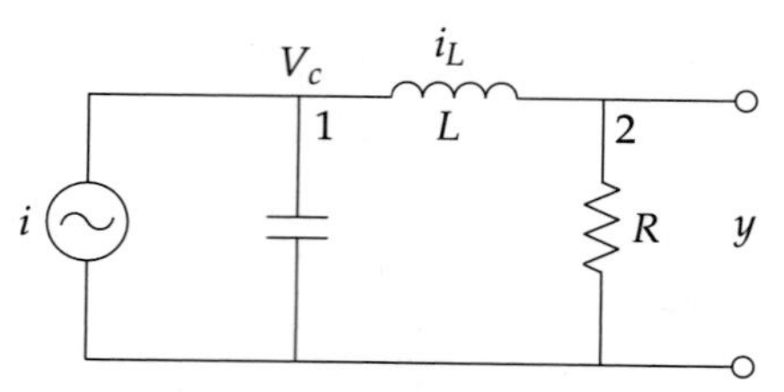

Solution In the given circuit, there are two energy storage devices[2]—the capacitor C and the inductor L. Therefore, we can choose the voltage across the capacitor V_c and the current through the inductor i_L as state vectors. Now, at node 1, from KCL,

$$C\frac{dV_C}{dt} + i_L + i = 0$$

$$\Rightarrow \qquad \frac{dV_C}{dt} = -\frac{1}{C}i_L - \frac{1}{C}i \tag{i}$$

At node 2, from KVL,

$$L\frac{di_L}{dt} + i_L R - V_c = 0$$

$$\Rightarrow \qquad \frac{di_L}{dt} = -\frac{R}{L}i_L + \frac{1}{L}V_C \tag{ii}$$

and,

$$y = Ri_L \tag{iii}$$

To obtain the state-space representation, we express Eqs. (i) to (iii) in matrix form as:

$$\begin{bmatrix} \dot{V}_c \\ \dot{i}_L \end{bmatrix} = \begin{bmatrix} 0 & -\dfrac{1}{C} \\ \dfrac{1}{L} & -\dfrac{R}{L} \end{bmatrix} + \begin{bmatrix} -\dfrac{1}{C} \\ 0 \end{bmatrix}$$

$$[y] = \begin{bmatrix} 0 & R \end{bmatrix} \begin{bmatrix} V_c \\ i_L \end{bmatrix} + [0][u]$$

17.4 NON-UNIQUENESS OF STATE-SPACE MODEL

The representation of any system in the state-space model is not unique. Alternative representations are possible for a given system. Consider a system given by

$$\dot{\mathbf{x}} = \mathbf{Ax} + \mathbf{Bu} \tag{17.8}$$

$$\mathbf{y} = \mathbf{Cx} + \mathbf{Du} \tag{17.9}$$

where $\dot{\mathbf{x}}$, $\mathbf{x}$, $\mathbf{u}$, $\mathbf{y}$ are all functions of time t.

[2]Actually, we are looking for derivative factors which are necessary to construct the first equation of the state model. The current passing through a capacitor, $i_C = C(dV_C/dt)$ and the voltage across an inductor, $V_L = L(di_L/dt)$ are suitable in this respect.

We will see that matrices $\mathbf{A}$, $\mathbf{B}$, $\mathbf{C}$, $\mathbf{D}$ and the state variables $\mathbf{x}$ are not unique for the interrelationship between $\mathbf{u}$ and $\mathbf{y}$.

Consider a transformation

$$\mathbf{x} = \mathbf{Mz}$$

where $\mathbf{M}$ is a non-singular constant matrix. Using this transformation, we can obtain, for Eq. (17.8), the relation

$$\mathbf{M\dot{z}} = \mathbf{AMz} + \mathbf{Bu}$$

$$\Rightarrow \qquad \mathbf{\dot{z}} = \mathbf{M}^{-1}\mathbf{AMz} + \mathbf{M}^{-1}\mathbf{Bu}$$

$$\equiv \mathbf{Pz} + \mathbf{Qu} \tag{17.10}$$

where

$$\mathbf{P} \equiv \mathbf{M}^{-1}\mathbf{AM} \qquad \text{and} \quad \mathbf{Q} \equiv \mathbf{M}^{-1}\mathbf{B}$$

Similarly, Eq. (17.9) can be written as

$$\mathbf{y} = \mathbf{CMz} + \mathbf{Du} \equiv \mathbf{Rz} + \mathbf{Du} \tag{17.11}$$

where

$$\mathbf{R} \equiv \mathbf{CM}$$

Thus, from Eqs. (17.10) and (17.11), we see that the state and output equations can be written as

$$\mathbf{\dot{z}} = \mathbf{Pz} + \mathbf{Qu}$$

$$\mathbf{y} = \mathbf{Rz} + \mathbf{Du}$$

where the state variable vector as well as state, input and output matrices are all different.

17.5 TRANSFER FUNCTION AND STATE MODEL CANONICAL REALIZATIONS

The state model can be readily derived from the transfer function of the system through its signal flow graph. However, for the same transfer function, there is more than one set of state variables and, therefore, there is more than one possible form of the signal flow graphs. In general, a strictly proper[3] transfer function can be written as

$$G(s) \equiv \frac{Y(s)}{U(s)} = \frac{s^m + b_{m-1}s^{m-1} + \cdots + b_1 s + b_0}{s^n + a_{n-1}s^{n-1} + \cdots + a_1 s + a_0} \tag{17.12}$$

where $n \geq m$ and all the coefficients are real positive numbers. If we multiply the numerator and denominator of Eq. (17.12) by s^{-n}, we get

$$G(s) = \frac{s^{-(n-m)} + b_{m-1}s^{-(n-m+1)} + \cdots + b_1 s^{-(n-1)} + b_0 s^{-n}}{1 + a_{n-1}s^{-1} + \cdots + a_1 s^{-(n-1)} + a_0 s^{-n}} \tag{17.13}$$

[3]See Section 3.2 at page 106 for definition.

Equation (17.13) is in the form of Mason's gain formula[4] which reduces to the following form when all the forward paths touch the feedback loops:

$$G(s) = \frac{\sum_k p_k \Delta_k}{1 - \sum_q L_q} = \frac{\text{Sum of forward (gain} \times \text{path factors)}}{1 - \text{Sum of feedback loop gains}}$$

A number of signal flow graphs can represent a transfer function. Out of them we will consider only two which are of interest to us.

Let us consider a transfer function given by

$$G(s) = \frac{b_3 s^3 + b_2 s^2 + b_1 s + b_0}{s^4 + a_3 s^3 + a_2 s^2 + a_1 s + a_0} = \frac{b_3 s^{-1} + b_2 s^{-2} + b_1 s^{-3} + b_0 s^{-4}}{1 + a_3 s^{-1} + a_2 s^{-2} + a_1 s^{-3} + a_0 s^{-4}} \tag{17.14}$$

From Eq. (17.14), we note that

1. The system is of 4th order. Therefore, we need four state variables—x_1, x_2, x_3, x_4.
2. The signal flow graph must have four integrators because the power of s, which indicates integration, has a maximum value of 4.

Now, we will see how $G(s)$ of Eq. (17.14) can be represented by two signal flow graph forms, namely,

1. Controllable canonical form, and
2. Observable canonical form

Let us consider these two forms in detail and work out a few problems thereafter.

Controllable Canonical Form (CCF)

The signal flow graph, shown in Figure 17.5, may represent the transfer function. This is readily verified by noting that the product of forward gains and path factors are $b_3 s^{-1}$, $b_2 s^{-2}$, $b_1 s^{-3}$, and $b_0 s^{-4}$, as required to provide the numerator of the transfer function, and the denominator is equal to 1 minus the sum of the loop gains.

The state variables are identified in Figure 17.5 as the output of each energy storage element, i.e. the output of each integrator. In order to obtain the set of first order differential equations representing the state model of Figure 17.5, we have written $x_4 = \dot{x}_3$, $x_3 = \dot{x}_2$, $x_2 = \dot{x}_1$ because they automatically follow from the signal flow graph. Thus, we can obtain the following set of first order differential equations describing the state model

$$\begin{aligned} \dot{x}_1 &= x_2 \\ \dot{x}_2 &= x_3 \\ \dot{x}_3 &= x_4 \\ \dot{x}_4 &= -a_0 x_1 - a_1 x_2 - a_2 x_3 - a_3 x_4 + u \end{aligned} \tag{17.15}$$

[4] See Eq. (4.7) at page 139.

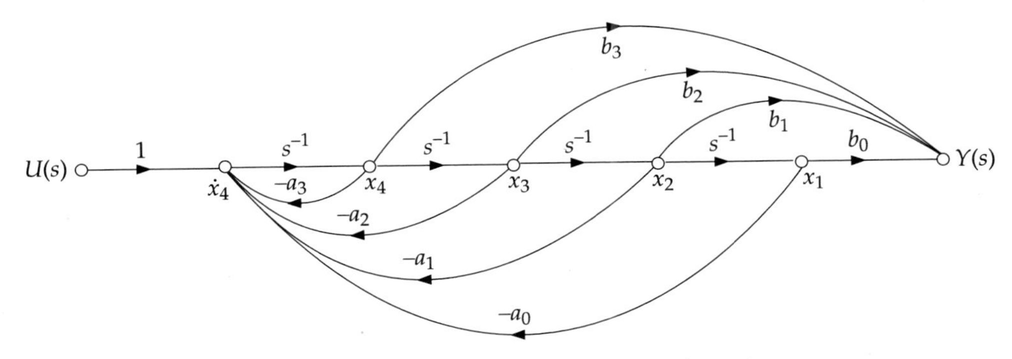

Figure 17.5 Signal flow graph for Eq. (17.14).

Furthermore, the output is simply

$$y(t) = b_0 x_1 + b_1 x_2 + b_2 x_3 + b_3 x_4$$

In matrix form, we have from Eq. (17.15)

$$\dot{\mathbf{x}} = \begin{bmatrix} 0 & 1 & 0 & 0 \\ 0 & 0 & 1 & 0 \\ 0 & 0 & 0 & 1 \\ -a_0 & -a_1 & -a_2 & -a_3 \end{bmatrix} \mathbf{x} + \begin{bmatrix} 0 \\ 0 \\ 0 \\ 1 \end{bmatrix} \mathbf{u}$$

$$\mathbf{y} = \begin{bmatrix} b_0 & b_1 & b_2 & b_3 \end{bmatrix} \mathbf{x} + [0]\mathbf{u}$$

This state-space realization, which incorporates *phase variable* feedback and zeros in the *output* path, is called *controllable canonical form* (CCF) because the resulting model is guaranteed to be controllable[5]. The realization is also known as *phase-variable canonical form.*

Observable Canonical Form (OCF)

The signal flow graph structure of Figure 17.5 is not a unique representation of Eq. (17.14). Another signal flow graph which represents Eq. (17.14) equally well is shown in Figure 17.6. In this case, the forward path factors are obtained by feeding forward the signal $U(s)$. Then the output signal $y(t)$ is equal to the first state variable $x_1(t)$. The flow graph has product of forward gains and path factors as $b_0 s^{-4}$, $b_1 s^{-3}$, $b_2 s^{-2}$, $b_3 s^{-1}$, and all forward paths touch feedback loops. Therefore, the resulting transfer function is indeed equal to Eq. (17.14). Using the signal flow graph of Figure 17.6, we obtain the following set of equations:

$$\begin{aligned} \dot{x}_1 &= -a_3 x_1 + x_2 + b_3 u \\ \dot{x}_2 &= -a_2 x_1 + x_3 + b_2 u \\ \dot{x}_3 &= -a_1 x_1 + x_4 + b_1 u \\ \dot{x}_4 &= -a_0 x_1 + b_0 u \end{aligned} \qquad (17.16)$$

[5]See Section 17.9 at page 642 for the meaning of Controllability and Observability.

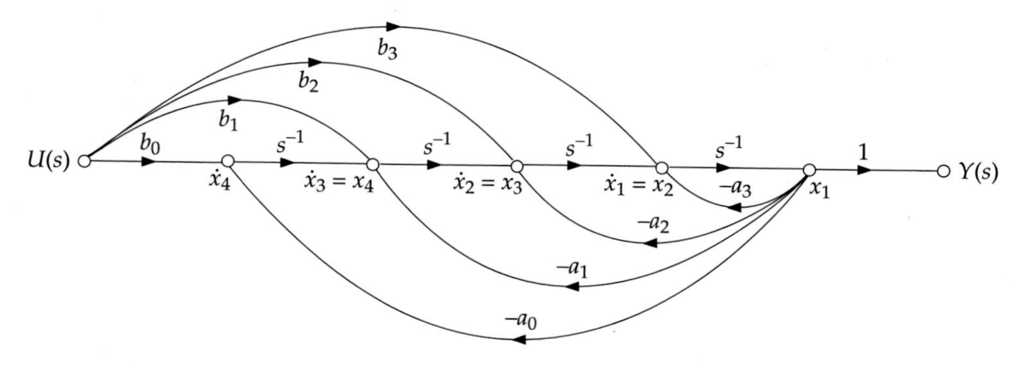

Figure 17.6 Alternative signal flow graph for Eq. (17.14).

$$y(t) = x_1 \tag{17.17}$$

Equations (17.16) and (17.17) can be expressed in matrix forms as

$$\dot{\mathbf{x}} = \begin{bmatrix} -a_3 & 1 & 0 & 0 \\ -a_2 & 0 & 1 & 0 \\ -a_1 & 0 & 0 & 1 \\ -a_0 & 0 & 0 & 0 \end{bmatrix} \mathbf{x} + \begin{bmatrix} b_3 \\ b_2 \\ b_1 \\ b_0 \end{bmatrix} \mathbf{u}$$

$$\mathbf{y} = \begin{bmatrix} 1 & 0 & 0 & 0 \end{bmatrix} \mathbf{x} + [0]\mathbf{u}$$

This state-space realization, which incorporates *output* feedback and zeros in the *input* path, is called *observable canonical form* (OCF) because the resulting model is guaranteed to be observable.

Note: The signal flow graphs presented in Figures 17.5 and 17.6 represent the same transfer function given by Eq. (17.14). But the state variables of two graphs are not identical because the structures of the two graphs are different.

Let us consider a few examples to gain familiarity with the two forms of state models.

EXAMPLE 17.4. A unity feedback control system is shown in Figure 17.7. Obtain state matrices for its transfer function through CCF and OCF.

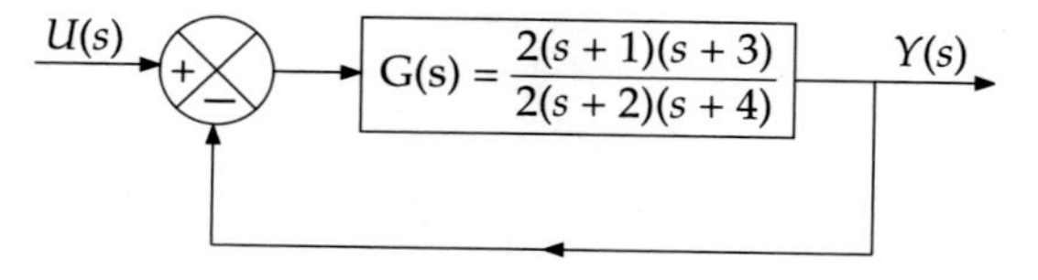

Figure 17.7 Control system (Example 17.4).

Solution The closed-loop transfer function of the system is

$$T(s) \equiv \frac{U(s)}{Y(s)} = \frac{2s^2 + 8s + 6}{s^3 + 8s^2 + 16s + 6} = \frac{2s^{-1} + 8s^{-2} + 6s^{-3}}{1 + 8s^{-1} + 16s^{-2} + 6s^{-3}}$$

The CCF signal flow graph is given in Figure 17.8. From this graph, we get the following equations and their matrix representations:

$$\dot{x}_1 = x_2$$
$$\dot{x}_2 = x_3$$
$$\dot{x}_3 = -6x_1 - 16x_2 - 8x_3 + u$$
$$y = 6x_1 + 8x_2 + 2x_3$$

$$\begin{bmatrix} \dot{x}_1 \\ \dot{x}_2 \\ \dot{x}_3 \end{bmatrix} = \begin{bmatrix} 0 & 1 & 0 \\ 0 & 0 & 1 \\ -6 & -16 & -8 \end{bmatrix} \begin{bmatrix} x_1 \\ x_2 \\ x_3 \end{bmatrix} + \begin{bmatrix} 0 \\ 0 \\ 1 \end{bmatrix} [u]$$

$$[y] = \begin{bmatrix} 6 & 8 & 2 \end{bmatrix} \begin{bmatrix} x_1 \\ x_2 \\ x_3 \end{bmatrix} + [0][u]$$

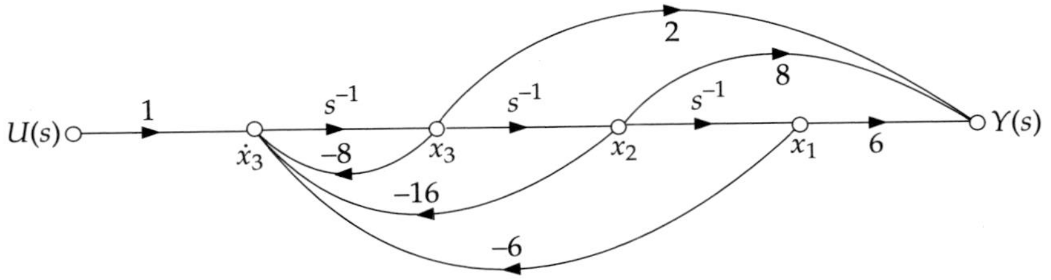

Figure 17.8 CCF signal flow graph (Example 17.4).

The OCF signal flow graph is given in Figure 17.9. From this graph, we get the following equations and their matrix representations:

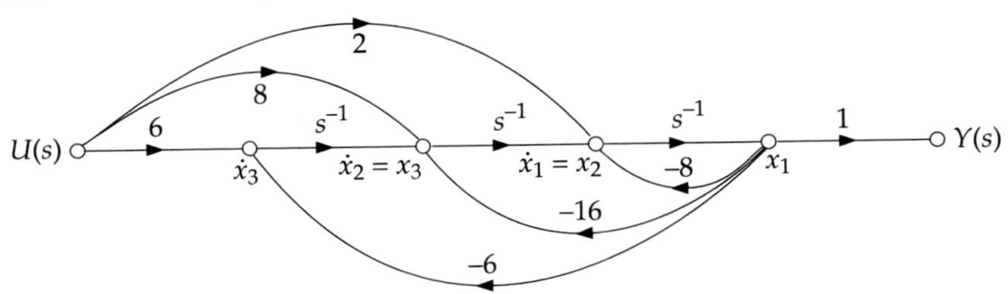

Figure 17.9 OCF signal flow graph (Example 17.4).

$$\dot{x}_1 = -8x_1 + x_2 + 2u$$
$$\dot{x}_2 = -16x_1 + x_3 + 8u$$
$$\dot{x}_3 = -6x_1 + 6u$$
$$y = x_1$$

$$\begin{bmatrix} \dot{x}_1 \\ \dot{x}_2 \\ \dot{x}_3 \end{bmatrix} = \begin{bmatrix} -8 & 1 & 0 \\ -16 & 0 & 1 \\ -6 & 0 & 0 \end{bmatrix} \begin{bmatrix} x_1 \\ x_2 \\ x_3 \end{bmatrix} \begin{bmatrix} 2 \\ 8 \\ 6 \end{bmatrix} [u]$$

$$[y] = \begin{bmatrix} 1 & 0 & 0 \end{bmatrix} \begin{bmatrix} x_1 \\ x_2 \\ x_3 \end{bmatrix} = [0][u]$$

EXAMPLE 17.5. Express the following closed-loop transfer function in CCF and OCF and obtain the corresponding state matrices.

$$\frac{Y(s)}{U(s)} = \frac{5s^2 + 2s + 6}{s^3 + 7s^2 + 11s + 8}$$

Solution The transfer function can be written in the following form:

$$\frac{Y(s)}{U(s)} = \frac{5s^{-1} + 2s^{-2} + 6s^{-3}}{1 + 7s^{-1} + 11s^{-2} + 8s^{-3}}$$

The CCF signal flow graph is given in Figure 17.10.

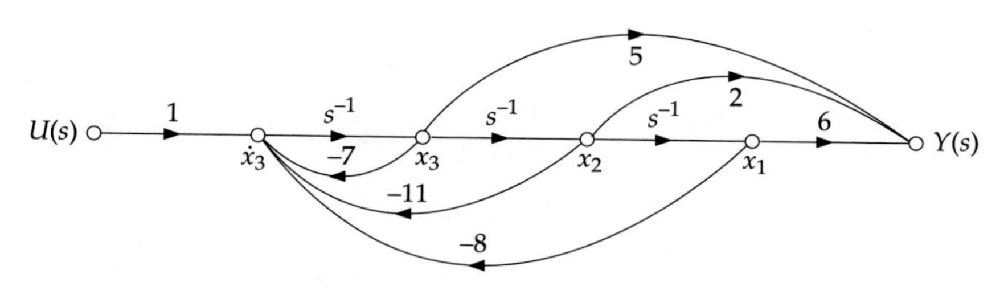

Figure 17.10 CCF signal flow graph (Example 17.5).

From the above graph, we get the following equations and their matrix representations:

$$\dot{x}_1 = x_2$$

$$\dot{x}_2 = x_3$$

$$\dot{x}_3 = -8x_1 - 11x_2 - 7x_3 + u$$

$$y = 6x_1 + 2x_2 + 5x_3$$

$$\begin{bmatrix} \dot{x}_1 \\ \dot{x}_2 \\ \dot{x}_3 \end{bmatrix} = \begin{bmatrix} 0 & 1 & 0 \\ 0 & 0 & 1 \\ -8 & -11 & -7 \end{bmatrix} \begin{bmatrix} x_1 \\ x_2 \\ x_3 \end{bmatrix} + \begin{bmatrix} 0 \\ 0 \\ 1 \end{bmatrix} [u]$$

$$[y] = \begin{bmatrix} 6 & 2 & 5 \end{bmatrix} \begin{bmatrix} x_1 \\ x_2 \\ x_3 \end{bmatrix} + [0][u]$$

The OCF signal flow graph is given in Figure 17.11.

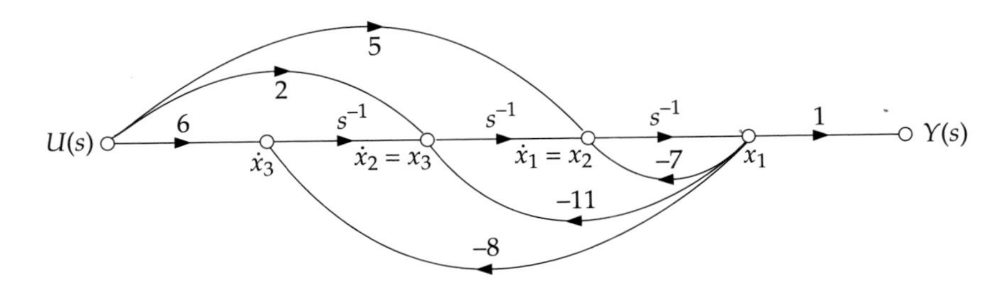

Figure 17.11 OCF signal flow graph (Example 17.5).

From the above graph, we get the following equations and their matrix representations:

$$\dot{x}_1 = -7x_1 + x_2 + 5u$$
$$\dot{x}_2 = -11x_1 + x_3 + 2u$$
$$\dot{x}_3 = -8x_1 + 6u$$
$$y = x_1$$

$$\begin{bmatrix} \dot{x}_1 \\ \dot{x}_2 \\ \dot{x}_3 \end{bmatrix} = \begin{bmatrix} -7 & 1 & 0 \\ -11 & 0 & 1 \\ -8 & 01 & 0 \end{bmatrix} \begin{bmatrix} x_1 \\ x_2 \\ x_3 \end{bmatrix} + \begin{bmatrix} 5 \\ 2 \\ 6 \end{bmatrix} [u]$$

$$[y] = \begin{bmatrix} 1 & 0 & 0 \end{bmatrix} \begin{bmatrix} x_1 \\ x_2 \\ x_3 \end{bmatrix} + [0][u]$$

EXAMPLE 17.6. Express the following closed-loop transfer function in CCF and OCF and obtain the corresponding state matrices:

$$\frac{Y(s)}{U(s)} = \frac{1}{s^2 + 2s + 3}$$

Solution The transfer function can be written in the following form:

$$\frac{Y(s)}{U(s)} = \frac{s^{-2}}{1 + 2s^{-1} + 3s^{-2}}$$

The CCF signal flow graph is given in Figure 17.12.

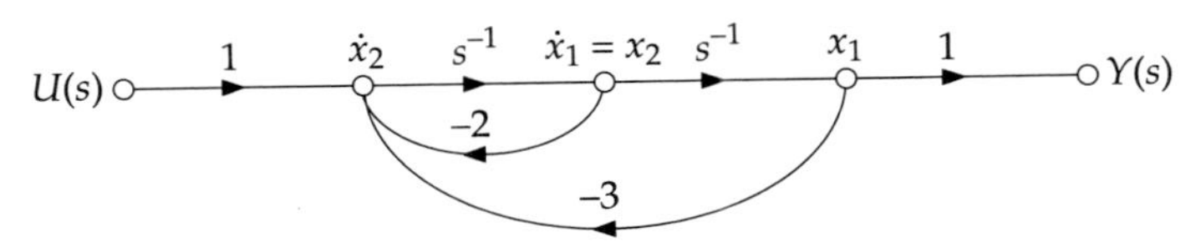

Figure 17.12 CCF signal flow graph (Example 17.6).

From the graph shown in Figure 17.12, we get the following equations and their matrix representations:

$$\dot{x}_1 = x_2$$
$$\dot{x}_2 = -3x_1 - 2x_2 + u$$
$$y = x_1$$
$$\begin{bmatrix}\dot{x}_1\\ \dot{x}_2\end{bmatrix} = \begin{bmatrix}0 & 1\\ -3 & -2\end{bmatrix}\begin{bmatrix}x_1\\ x_2\end{bmatrix} + \begin{bmatrix}0\\ 1\end{bmatrix}[u]$$
$$[y] = \begin{bmatrix}1 & 0\end{bmatrix}\begin{bmatrix}x_1\\ x_2\end{bmatrix} + [0][u]$$

The OCF signal flow graph is given in Figure 17.13.

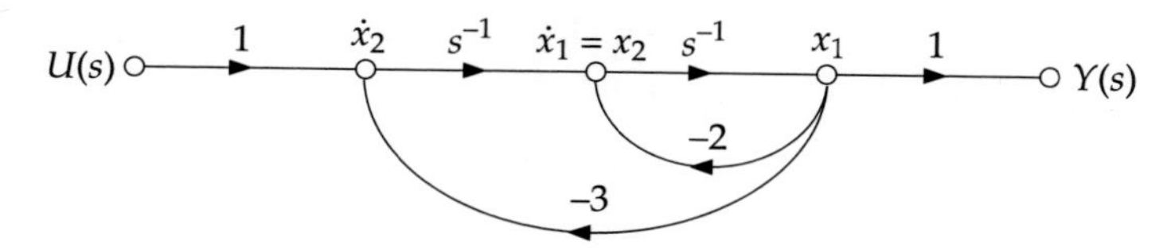

Figure 17.13 OCF signal flow graph (Example 17.6).

From the above graph, we get the following equations and their matrix representations:

$$\dot{x}_1 = -2x_1 + x_2$$
$$\dot{x}_2 = -3x_1 + u$$
$$y = x_1$$
$$\begin{bmatrix}\dot{x}_1\\ \dot{x}_2\end{bmatrix} = \begin{bmatrix}-2 & 1\\ -3 & 0\end{bmatrix}\begin{bmatrix}x_1\\ x_2\end{bmatrix} + \begin{bmatrix}0\\ 1\end{bmatrix}[u]$$
$$[y] = \begin{bmatrix}1 & 0\end{bmatrix}\begin{bmatrix}x_1\\ x_2\end{bmatrix} + [0][u]$$

17.6 DECOMPOSITION OF TRANSFER FUNCTION TO FIND STATE REPRESENTATION

We have seen how we can find state representations from transfer functions through their canonical realizations. Now we will see how we can find them from the decomposition of transfer functions. The decomposition is done in two ways as follows:

1. Cascade decomposition
2. Parallel decomposition

Let us consider the methods in detail.

Cascade Decomposition

Suppose the transfer function is given as the product of first-order terms as follows:

$$\frac{Y(s)}{U(s)} = K\frac{(s+b_1)(s+b_2)}{(s+a_1)(s+a_2)} \tag{17.18}$$

Equation (17.18) can be written as

$$\frac{Y(s)}{U(s)} = K\frac{1+b_1s^{-1}}{1+a_1s^{-1}} \cdot \frac{1+b_2s^{-1}}{1+a_2s^{-1}} \tag{17.19}$$

The signal flow graph for Eq. (17.19) can be drawn as shown in Figure 17.14.

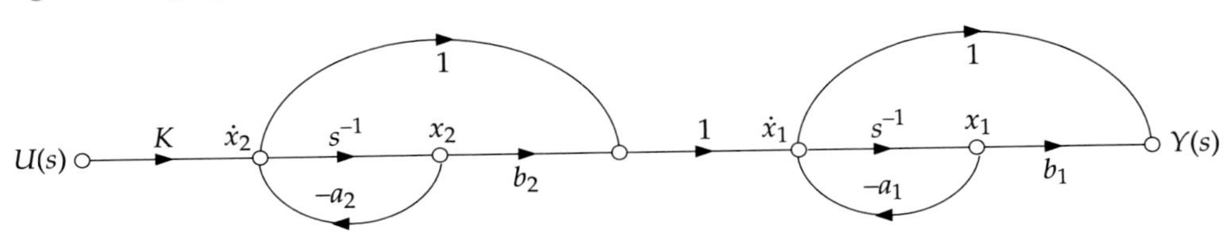

Figure 17.14 Signal flow graph for Eq. (17.19).

From the signal flow graph, $\dot{x}_1$, $\dot{x}_2$ and y can be found.

Let us consider a few examples.

EXAMPLE 17.7. Realize a state model representation by cascade decomposition of the following transfer function:

$$\frac{Y(s)}{U(s)} = \frac{1}{(s+2)(s+3)(s+4)}$$

Solution The transfer function can be written as

$$\frac{Y(s)}{U(s)} = \frac{s^{-1}}{1+2s^{-1}} \cdot \frac{s^{-1}}{1+3s^{-1}} \cdot \frac{s^{-1}}{1+4s^{-1}}$$

The corresponding signal flow graph can be drawn as shown in Figure 17.15.

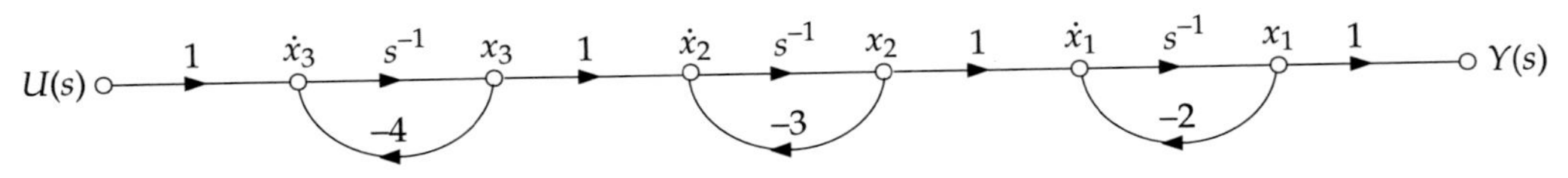

Figure 17.15 Signal flow graph (Example 17.7).

From the above graph, we can write the following equations:

$$\dot{x}_3 = -4x_3 + u$$

$$\dot{x}_2 = -3x_2 + x_3$$

$$\dot{x}_1 = -2x_1 + x_2$$

$$y = x_1$$

Their matrix representations are as follows:

$$\begin{bmatrix} \dot{x}_3 \\ \dot{x}_2 \\ \dot{x}_1 \end{bmatrix} = \begin{bmatrix} -2 & 1 & 0 \\ 0 & -3 & 1 \\ 0 & 0 & 4 \end{bmatrix} \begin{bmatrix} x_1 \\ x_2 \\ x_3 \end{bmatrix} + \begin{bmatrix} 0 \\ 0 \\ 1 \end{bmatrix} [u]$$

$$[y] = \begin{bmatrix} 1 & 0 & 0 \end{bmatrix} \begin{bmatrix} x_1 \\ x_2 \\ x_3 \end{bmatrix}$$

EXAMPLE 17.8. Realize a state model representation by cascade decomposition of the transfer function

$$\frac{Y(s)}{U(s)} = \frac{5(s+1)(s+2)}{(s+4)(s+5)}$$

Solution The transfer function can be written as

$$\frac{Y(s)}{U(s)} = 5 \cdot \frac{1+s^{-1}}{1+4s^{-1}} \cdot \frac{1+2s^{-1}}{1+5s^{-1}}$$

The corresponding signal flow graph can be drawn as shown in Figure 17.16.

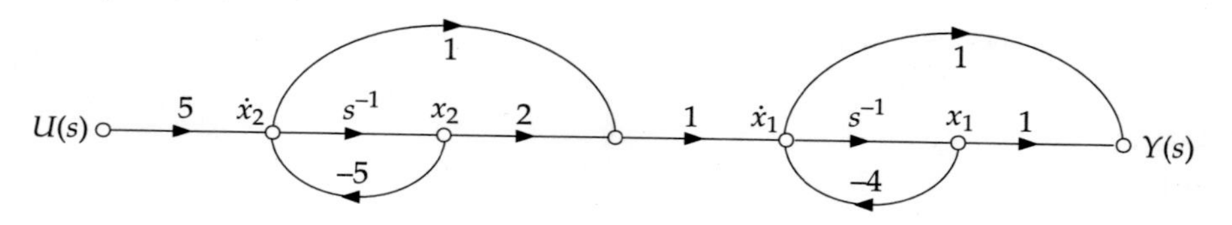

Figure 17.16 Signal flow graph (Example 17.8).

From the graph we can write the following equations:

$$\dot{x}_2 = -5x_2 + u$$
$$\dot{x}_1 = -4x_1 + 2x_2$$
$$y = x_1$$

Therefore, the matrix or state model representation of the transfer function is obtained as

$$\begin{bmatrix} \dot{x}_2 \\ \dot{x}_1 \end{bmatrix} = \begin{bmatrix} -4 & 2 \\ 0 & -5 \end{bmatrix} \begin{bmatrix} x_1 \\ x_2 \end{bmatrix} + \begin{bmatrix} 0 \\ 1 \end{bmatrix} [u]$$

$$[y] = \begin{bmatrix} 1 & 0 \end{bmatrix} \begin{bmatrix} x_1 \\ x_2 \end{bmatrix}$$

Parallel Decomposition

The parallel decomposition method is based on decomposition of the transfer function in partial fractions. We present two examples to explain the procedure.

EXAMPLE 17.9. Realize a state model representation for the following transfer function by its parallel decomposition:

$$\frac{Y(s)}{U(s)} = \frac{1}{(s+2)(s+3)(s+4)}$$

Solution The transfer function can be written as

$$\frac{Y(s)}{U(s)} \equiv \frac{A}{s+2} + \frac{B}{s+3} + \frac{C}{s+4}$$

The constants A, B and C can be found out as follows:

$$A = (s+2)\left.\frac{Y(s)}{U(s)}\right|_{s=-2} = \left.\frac{1}{(s+3)(s+4)}\right|_{s=-2} = \frac{1}{2}$$

$$B = (s+3)\left.\frac{Y(s)}{U(s)}\right|_{s=-3} = \left.\frac{1}{(s+2)(s+4)}\right|_{s=-3} = -1$$

$$C = (s+4)\left.\frac{Y(s)}{U(s)}\right|_{s=-4} = \left.\frac{1}{(s+2)(s+3)}\right|_{s=-4} = \frac{1}{2}$$

So, finally, the transfer function can be written as

$$\frac{Y(s)}{U(s)} = \frac{1}{2}\cdot\frac{s^{-1}}{1+2s^{-1}} - \frac{s^{-1}}{1+3s^{-1}} + \frac{1}{2}\cdot\frac{s^{-1}}{1+4s^{-1}}$$

The corresponding signal flow graph can be drawn as shown in Figure 17.17.

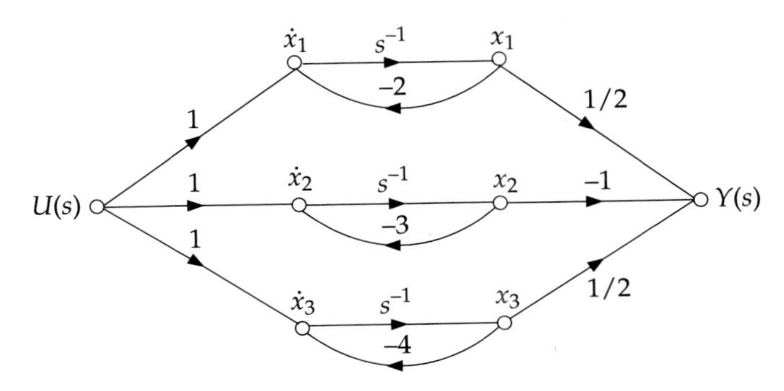

Figure 17.17 Signal flow graph (Example 17.9).

From the graph we can write the following equations:

$$\dot{x}_1 = -2x_1 + u$$

$$\dot{x}_2 = -3x_2 + u$$

$$\dot{x}_3 = -4x_3 + u$$

$$y = \frac{1}{2}x_1 - x_2 + \frac{1}{2}x_3$$

Therefore, the matrix or state model representation of the transfer function is as follows:

$$\begin{bmatrix} \dot{x}_1 \\ \dot{x}_2 \\ \dot{x}_3 \end{bmatrix} = \begin{bmatrix} -2 & 0 & 0 \\ 0 & -3 & 0 \\ 0 & 0 & -4 \end{bmatrix} \begin{bmatrix} x_1 \\ x_2 \\ x_3 \end{bmatrix} \begin{bmatrix} 1 \\ 1 \\ 1 \end{bmatrix} + [u]$$

$$[y] = \begin{bmatrix} \frac{1}{2} & -1 & \frac{1}{2} \end{bmatrix} \begin{bmatrix} x_1 \\ x_2 \\ x_3 \end{bmatrix}$$

EXAMPLE 17.10. Realize a state model representation for the following transfer function by parallel decomposition:

$$\frac{Y(s)}{U(s)} = \frac{(s+1)}{(s+2)(s+3)(s+5)}$$

Solution The transfer function can be written as

$$\frac{Y(s)}{U(s)} \equiv \frac{A}{s+2} + \frac{B}{s+3} + \frac{C}{s+5}$$

The constants A, B and C can be figured out as follows:

$$A = (s+2)\left.\frac{Y(s)}{U(s)}\right|_{s=-2} = \left.\frac{(s+1)}{(s+3)(s+5)}\right|_{s=-2} = -\frac{1}{3}$$

$$B = (s+3)\left.\frac{Y(s)}{U(s)}\right|_{s=-3} = \left.\frac{(s+1)}{(s+2)(s+5)}\right|_{s=-3} = 1$$

$$C = (s+5)\left.\frac{Y(s)}{U(s)}\right|_{s=-5} = \left.\frac{(s+1)}{(s+2)(s+3)}\right|_{s=-5} = -\frac{2}{3}$$

So, finally the transfer function can be written as

$$\frac{Y(s)}{U(s)} = -\frac{1}{3} \cdot \frac{s^{-1}}{1+2s^{-1}} + \frac{s^{-1}}{1+3s^{-1}} - \frac{2}{3} \cdot \frac{s^{-1}}{1+5s^{-1}}$$

The corresponding signal flow graph can be drawn as shown in Figure 17.18. From the graph we can write the following equations:

$$\dot{x}_1 = -2x_1 + u$$

$$\dot{x}_2 = -3x_2 + u$$

$$\dot{x}_3 = -5x_3 + u$$

$$y = -\frac{1}{3}x_1 + x_2 - \frac{2}{3}x_3$$

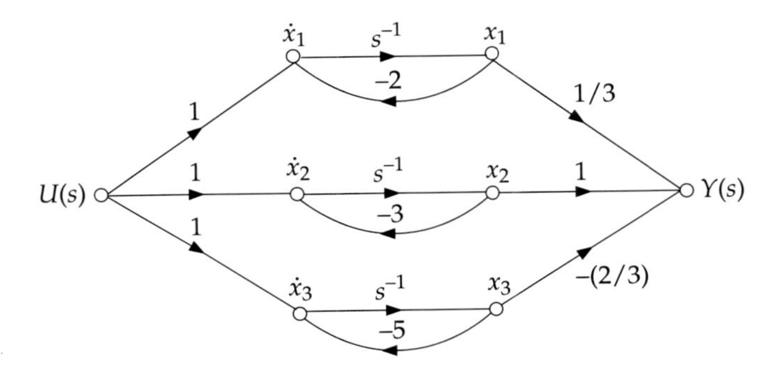

Figure 17.18 Signal flow graph (Example 17.10).

Therefore, the state matrices are

$$\begin{bmatrix} \dot{x}_1 \\ \dot{x}_2 \\ \dot{x}_3 \end{bmatrix} = \begin{bmatrix} -2 & 0 & 0 \\ 0 & -3 & 0 \\ 0 & 0 & -5 \end{bmatrix} \begin{bmatrix} x_1 \\ x_2 \\ x_3 \end{bmatrix} + \begin{bmatrix} 1 \\ 1 \\ 1 \end{bmatrix} [u]$$

$$[y] = \begin{bmatrix} -\frac{1}{3} & 1 & -\frac{2}{3} \end{bmatrix} \begin{bmatrix} x_1 \\ x_2 \\ x_3 \end{bmatrix}$$

17.7 TRANSFER FUNCTION FROM STATE MODEL

We know that, by definition,

$$\text{Transfer function} = \left.\frac{\text{Laplace transform of output}}{\text{Laplace transform of input}}\right|_{\text{initial conditions}=0}$$

Let the state equations be

$$\dot{\mathbf{x}} = \mathbf{Ax} + \mathbf{Bu}$$

$$\mathbf{y} = \mathbf{Cx} + \mathbf{Du}$$

The Laplace transform of the equations are

$$s\mathbf{X}(s) = \mathbf{AX}(s) + \mathbf{BU}(s) \tag{17.20}$$

$$\mathbf{Y}(s) = \mathbf{CX}(s) + \mathbf{DU}(s) \tag{17.21}$$

where we have assumed that $\mathbf{X}(0) = 0$. Now, from Eq. (17.20), we get

$$[s\mathbf{I} - \mathbf{A}]\mathbf{X}(s) = \mathbf{BU}(s)$$

or

$$\mathbf{X}(s) = [s\mathbf{I} - \mathbf{A}]^{-1}\mathbf{BU}(s) \tag{17.22}$$

where **I** is the identity matrix. Substituting Eq. (17.22) in Eq. (17.21), we get

$$\mathbf{Y}(s) = \mathbf{C}[s\mathbf{I} - \mathbf{A}]^{-1}\mathbf{B}\mathbf{U}(s) + \mathbf{D}\mathbf{U}(s)$$

Therefore, the transfer function

$$\mathbf{G}(s) = \frac{\mathbf{Y}(s)}{\mathbf{U}(s)} = \mathbf{C}[s\mathbf{I} - \mathbf{A}]^{-1}\mathbf{B} + \mathbf{D}$$

Clearly, $\mathbf{G}(s)$ must have $m \times n$ dimensionality, and thus has mn elements. So, for every input there are m transfer functions with one for each output. Which is why the state-space representation can easily be the preferred choice for multiple-input, multiple-output (MIMO) systems.

If **D** is a null matrix, which is true when **Y** and **U** are not directly connected, then

$$\mathbf{G}(s) = \frac{\mathbf{Y}(s)}{\mathbf{U}(s)} = \mathbf{C}[s\mathbf{I} - \mathbf{A}]^{-1}\mathbf{B} \tag{17.23}$$

From the definition of an inverse matrix[6], we get from Eq. (17.23)

$$\mathbf{G}(s) = \frac{\mathbf{Y}(s)}{\mathbf{U}(s)} = \mathbf{C}\frac{\text{Adj }[s\mathbf{I} - \mathbf{A}]}{|s\mathbf{I} - \mathbf{A}|}\mathbf{B}$$

Thus, the characteristic equation of the system is given by

$$|s\mathbf{I} - \mathbf{A}| = 0$$

Obviously, the roots of the characteristic equation are the eigenvalues of the matrix **A**.

We work out a few examples to familiarize ourselves with the procedure.

EXAMPLE 17.11. The state representation of a system is given by

$$\dot{\mathbf{x}} = \begin{bmatrix} 0 & 1 \\ 0 & -3 \end{bmatrix}\mathbf{x} + \begin{bmatrix} 1 \\ 0 \end{bmatrix}\mathbf{u}$$

$$\mathbf{y} = \begin{bmatrix} 1 & 0 \end{bmatrix}\mathbf{x}$$

The transfer function $Y(s)/U(s)$ of the system is

(a) $\dfrac{1}{s}$ (b) $\dfrac{1}{s(s+3)}$ (c) $\dfrac{1}{s+3}$ (d) $\dfrac{1}{s^2}$

[6]See Section 1.3 at page 26.

Solution

$$[s\mathbf{I} - \mathbf{A}] = \begin{bmatrix} s & -1 \\ 0 & s+3 \end{bmatrix}$$

$$\det\,[s\mathbf{I} - \mathbf{A}] = s^3 + 3s$$

$$\text{Adj}\,[s\mathbf{I} - \mathbf{A}] = \begin{bmatrix} s+3 & 0 \\ 1 & s \end{bmatrix}^T = \begin{bmatrix} s+3 & 1 \\ 0 & s \end{bmatrix}$$

$$[s\mathbf{I} - \mathbf{A}]^{-1} = \frac{1}{s(s+3)} \begin{bmatrix} s+3 & 1 \\ 0 & s \end{bmatrix}$$

$$G(s) = [1 \quad 0]\,\frac{1}{s(s+3)} \begin{bmatrix} s+3 & 1 \\ 0 & s \end{bmatrix} \begin{bmatrix} 1 \\ 0 \end{bmatrix} = \frac{1}{s(s+3)}\,[1 \quad 0] \begin{bmatrix} s+3 \\ 0 \end{bmatrix}$$

$$= \frac{1}{s(s+3)} \times (s+3) = \frac{1}{s}$$

Ans: (a).

EXAMPLE 17.12. Find the transfer function for the system given by

$$\begin{bmatrix} \dot{x}_1 \\ \dot{x}_2 \end{bmatrix} = \begin{bmatrix} 0 & 1 \\ -3 & -2 \end{bmatrix} \begin{bmatrix} x_1 \\ x_2 \end{bmatrix} + \begin{bmatrix} 0 \\ 1 \end{bmatrix} [u]$$

$$[y] = [1 \quad 0] \begin{bmatrix} x_1 \\ x_2 \end{bmatrix}$$

Solution Here,

$$\mathbf{A} = \begin{bmatrix} 0 & 1 \\ -3 & -2 \end{bmatrix} \qquad \mathbf{B} = \begin{bmatrix} 0 \\ 1 \end{bmatrix} \qquad \mathbf{C} = [1 \quad 0]$$

Therefore, from Eq. (17.23),

$$\mathbf{G}(s) = [1 \quad 0]\,[s\mathbf{I} - \mathbf{A}]^{-1} \begin{bmatrix} 0 \\ 1 \end{bmatrix}$$

Now,

$$[s\mathbf{I} - \mathbf{A}] = \begin{bmatrix} s & 0 \\ 0 & s \end{bmatrix} - \begin{bmatrix} 0 & 1 \\ -3 & -2 \end{bmatrix} = \begin{bmatrix} s & -1 \\ 3 & s+2 \end{bmatrix}$$

$$\text{Adj}[s\mathbf{I} - \mathbf{A}] = \begin{bmatrix} s+2 & -3 \\ 1 & s \end{bmatrix}^{\text{T}} = \begin{bmatrix} s+2 & 1 \\ -3 & s \end{bmatrix}$$

$$|s\mathbf{I} - \mathbf{A}| = s(s+2) + 3$$

So,

$$[s\mathbf{I} - \mathbf{A}]^{-1} = \frac{\begin{bmatrix} s+2 & 1 \\ -3 & s \end{bmatrix}}{s(s+2)+3} = \frac{1}{s^2 + 2s + 3} \begin{bmatrix} s+2 & 1 \\ -3 & s \end{bmatrix}$$

Therefore, the transfer function is

$$\mathbf{G}(s) = \begin{bmatrix} 1 & 0 \end{bmatrix} \frac{1}{s^2 + 2s + 3} \begin{bmatrix} s+2 & 1 \\ -3 & s \end{bmatrix} \begin{bmatrix} 0 \\ 1 \end{bmatrix}$$

$$= \begin{bmatrix} 1 & 0 \end{bmatrix} \frac{1}{s^2 + 2s + 3} \begin{bmatrix} 1 \\ s \end{bmatrix}$$

or

$$G(s) = \frac{1}{s^2 + 2s + 3}$$

EXAMPLE 17.13. Find the transfer function for the system given by

$$\begin{bmatrix} \dot{x}_1 \\ \dot{x}_2 \end{bmatrix} = \begin{bmatrix} 0 & 1 \\ -2 & -3 \end{bmatrix} \begin{bmatrix} x_1 \\ x_2 \end{bmatrix} + \begin{bmatrix} 0 \\ 1 \end{bmatrix} [u]$$

$$[y] = \begin{bmatrix} 1 & 0 \end{bmatrix} \begin{bmatrix} x_1 \\ x_2 \end{bmatrix}$$

Solution Here,

$$\mathbf{A} = \begin{bmatrix} 0 & 1 \\ -2 & -3 \end{bmatrix} \qquad \mathbf{B} = \begin{bmatrix} 0 \\ 1 \end{bmatrix} \qquad \mathbf{C} = \begin{bmatrix} 1 & 0 \end{bmatrix}$$

Therefore, from Eq. (17.23)

$$\mathbf{G}(s) = \begin{bmatrix} 1 & 0 \end{bmatrix} [s\mathbf{I} - \mathbf{A}]^{-1} \begin{bmatrix} 0 \\ 1 \end{bmatrix}$$

Now,

$$[s\mathbf{I} - \mathbf{A}] = \begin{bmatrix} s & 0 \\ 0 & s \end{bmatrix} - \begin{bmatrix} 0 & 1 \\ -2 & -3 \end{bmatrix} = \begin{bmatrix} s & -1 \\ 2 & s+3 \end{bmatrix}$$

$$\text{Adj } [s\mathbf{I} - \mathbf{A}] = \begin{bmatrix} s+3 & -2 \\ 1 & s \end{bmatrix}^{\mathrm{T}} = \begin{bmatrix} s+3 & 1 \\ -2 & s \end{bmatrix}$$

$$|s\mathbf{I} - \mathbf{A}| = s(s+3) + 2$$

Hence,

$$[s\mathbf{I} - \mathbf{A}]^{-1} = \frac{\begin{bmatrix} s+3 & 1 \\ -2 & s \end{bmatrix}}{s(s+3)+2} = \frac{1}{s^2 + 3s + 2} \begin{bmatrix} s+3 & 1 \\ -2 & s \end{bmatrix}$$

Therefore, the transfer function is

$$\mathbf{G}(s) = \begin{bmatrix} 1 & 0 \end{bmatrix} \frac{1}{s^2 + 3s + 2} \begin{bmatrix} s+3 & 1 \\ -2 & s \end{bmatrix} \begin{bmatrix} 0 \\ 1 \end{bmatrix}$$

$$= \begin{bmatrix} 1 & 0 \end{bmatrix} \frac{1}{s^2 + 3s + 2} \begin{bmatrix} 1 \\ s \end{bmatrix}$$

or

$$G(s) = \frac{1}{s^2 + 3s + 2}$$

EXAMPLE 17.14. A control system is described by the state equations

$$\begin{bmatrix} \dot{x}_1 \\ \dot{x}_2 \end{bmatrix} = \begin{bmatrix} -5 & -1 \\ 3 & -1 \end{bmatrix} \begin{bmatrix} x_1 \\ x_2 \end{bmatrix} + \begin{bmatrix} 2 \\ 5 \end{bmatrix} [u]$$

$$[y] = \begin{bmatrix} 1 & 2 \end{bmatrix} \begin{bmatrix} x_1 \\ x_2 \end{bmatrix}$$

Obtain the transfer function of the system.

Solution Here,

$$\mathbf{A} = \begin{bmatrix} -5 & -1 \\ 3 & -1 \end{bmatrix} \qquad \mathbf{B} = \begin{bmatrix} 2 \\ 5 \end{bmatrix} \qquad \mathbf{C} = \begin{bmatrix} 1 & 2 \end{bmatrix}$$

Therefore, from Eq. (17.23)

$$\mathbf{G}(s) = \begin{bmatrix} 1 & 2 \end{bmatrix} [s\mathbf{I} - \mathbf{A}]^{-1} \begin{bmatrix} 2 \\ 5 \end{bmatrix}$$

Now,

$$[s\mathbf{I} - \mathbf{A}] = \begin{bmatrix} s & 0 \\ 0 & s \end{bmatrix} - \begin{bmatrix} -5 & -1 \\ 3 & -1 \end{bmatrix} = \begin{bmatrix} s+5 & 1 \\ -3 & s+1 \end{bmatrix}$$

$$\text{Adj}[s\mathbf{I} - \mathbf{A}] = \begin{bmatrix} s+1 & 3 \\ -1 & s+5 \end{bmatrix}^{\mathrm{T}} = \begin{bmatrix} s+1 & -1 \\ 3 & s+5 \end{bmatrix}$$

$$|s\mathbf{I} - \mathbf{A}| = (s+1)(s+5) + 3$$

So,

$$[s\mathbf{I} - \mathbf{A}]^{-1} = \frac{\begin{bmatrix} s+1 & -1 \\ 3 & s+5 \end{bmatrix}}{(s+1)(s+5)+3} = \frac{1}{s^2+6s+8} \begin{bmatrix} s+1 & -1 \\ 3 & s+5 \end{bmatrix}$$

Therefore, the transfer function is

$$\mathbf{G}(s) = \begin{bmatrix} 1 & 2 \end{bmatrix} \frac{1}{s^2+6s+8} \begin{bmatrix} s+1 & -1 \\ 3 & s+5 \end{bmatrix} \begin{bmatrix} 2 \\ 5 \end{bmatrix}$$

$$= \begin{bmatrix} 1 & 2 \end{bmatrix} \frac{1}{s^2+6s+8} \begin{bmatrix} 2(s+1)-5 \\ 6+5(s+5) \end{bmatrix}$$

or

$$G(s) = \frac{2(s+1) - 5 + 12 + 10(s+5)}{s^2+6s+8} = \frac{12s+59}{s^2+6s+8}$$

EXAMPLE 17.15. For a dynamic system, the matrices for a state space model are given by

$$\mathbf{A} = \begin{bmatrix} 0 & 1 \\ -2 & -3 \end{bmatrix} \qquad \mathbf{B} = \begin{bmatrix} 0 \\ 1 \end{bmatrix} \qquad \mathbf{C} = \begin{bmatrix} 1 & 0 \end{bmatrix}$$

The system poles are located at

(a) $-2,\ -1$ (b) $-2,\ -3$ (c) $-3,\ -1$ (d) $-1,\ 0$

Solution We may calculate the transfer function to locate the poles as follows:

$$[s\mathbf{I} - \mathbf{A}] = \begin{bmatrix} s & 0 \\ 0 & s \end{bmatrix} - \begin{bmatrix} 0 & 1 \\ -2 & -3 \end{bmatrix} = \begin{bmatrix} s & -1 \\ 2 & s+3 \end{bmatrix}$$

$$\det\,[s\mathbf{I} - \mathbf{A}] = s^2 + 3s + 2$$

$$\text{Adj}\,[s\mathbf{I} - \mathbf{A}] = \begin{bmatrix} s+3 & -2 \\ 1 & s \end{bmatrix}^T = \begin{bmatrix} s+3 & 1 \\ -2 & s \end{bmatrix}$$

$$[s\mathbf{I} - \mathbf{A}]^{-1} = \frac{1}{s^2+3s+2}\begin{bmatrix} s+3 & 1 \\ -2 & s \end{bmatrix}$$

$$G(s) = \frac{1}{s^2+3s+2}\begin{bmatrix} 1 & 0 \end{bmatrix}\begin{bmatrix} s+3 & 1 \\ -2 & s \end{bmatrix}\begin{bmatrix} 0 \\ 1 \end{bmatrix} = \frac{1}{s^2+3s+2}\begin{bmatrix} 1 & 0 \end{bmatrix}\begin{bmatrix} 1 \\ s \end{bmatrix}$$

$$= \frac{1}{(s+1)(s+2)}$$

Therefore, the poles are at $-1,\ -2$.
Ans: (a).

EXAMPLE 17.16. Consider the single-input single-output systems with state and output equations:

System 1:

$$\begin{bmatrix} \dot{x}_1 \\ \dot{x}_2 \end{bmatrix} = \begin{bmatrix} 0 & 1 \\ -2 & -3 \end{bmatrix}\begin{bmatrix} x_1 \\ x_2 \end{bmatrix} + \begin{bmatrix} 0 \\ 2 \end{bmatrix} u_1(t)$$

$$y_1 = \begin{bmatrix} 1 & 2 \end{bmatrix}\begin{bmatrix} x_1 \\ x_2 \end{bmatrix}$$

System 2:

$$\begin{bmatrix} \dot{x}_3 \\ \dot{x}_4 \end{bmatrix} = \begin{bmatrix} -3 & 1 \\ 0 & -3 \end{bmatrix}\begin{bmatrix} x_3 \\ x_4 \end{bmatrix} + \begin{bmatrix} 0 \\ 1 \end{bmatrix} u_2(t)$$

$$y_2 = \begin{bmatrix} 1 & 1 \end{bmatrix}\begin{bmatrix} x_3 \\ x_4 \end{bmatrix}$$

The two systems are connected in cascade by applying the output of system 1 to system 2, that is, by making $u_2(t) = y_1(t)$. Determine the transfer function of the combined system.

Solution *System 1:*

$$\mathbf{A} = \begin{bmatrix} 0 & 1 \\ -2 & -3 \end{bmatrix}, \qquad \mathbf{B} = \begin{bmatrix} 0 \\ 2 \end{bmatrix}, \qquad \mathbf{C} = \begin{bmatrix} 1 & 2 \end{bmatrix}$$

$$[s\mathbf{I} - \mathbf{A}] = \begin{bmatrix} s & 0 \\ 0 & s \end{bmatrix} - \begin{bmatrix} 0 & 1 \\ -2 & -3 \end{bmatrix} = \begin{bmatrix} s & -1 \\ 2 & s+3 \end{bmatrix}$$

$$\text{Adj}\,[s\mathbf{I} - \mathbf{A}] = \begin{bmatrix} s+3 & -2 \\ 1 & s \end{bmatrix}^T = \begin{bmatrix} s+3 & 1 \\ -2 & s \end{bmatrix}$$

$$\det\,[s\mathbf{I} - \mathbf{A}] = s^2 + 3s + 2$$

$$[s\mathbf{I} - \mathbf{A}]^{-1} = \frac{1}{s^2 + 3s + 2}\begin{bmatrix} s+3 & 1 \\ -2 & s \end{bmatrix}$$

Therefore,

$$G_1(s) = \begin{bmatrix} 1 & 2 \end{bmatrix} \frac{1}{s^2 + 3s + 2}\begin{bmatrix} s+3 & 1 \\ -2 & s \end{bmatrix}\begin{bmatrix} 0 \\ 2 \end{bmatrix}$$

$$= \frac{1}{s^2 + 3s + 2}\begin{bmatrix} 1 & 2 \end{bmatrix}\begin{bmatrix} 2 \\ 2s \end{bmatrix} = \frac{2(1+2s)}{s^2 + 3s + 2}$$

System 2:

$$\mathbf{A} = \begin{bmatrix} -3 & 1 \\ 0 & -3 \end{bmatrix}, \qquad \mathbf{B} = \begin{bmatrix} 0 \\ 1 \end{bmatrix}, \qquad \mathbf{C} = \begin{bmatrix} 1 & 1 \end{bmatrix}$$

$$[s\mathbf{I} - \mathbf{A}] = \begin{bmatrix} s & 0 \\ 0 & s \end{bmatrix} - \begin{bmatrix} -3 & 1 \\ 0 & -3 \end{bmatrix} = \begin{bmatrix} s+3 & -1 \\ 0 & s+3 \end{bmatrix}$$

$$\text{Adj } [s\mathbf{I} - \mathbf{A}] = \begin{bmatrix} s+3 & 0 \\ 1 & s+3 \end{bmatrix}^T = \begin{bmatrix} s+3 & 1 \\ 0 & s+3 \end{bmatrix}$$

$$\det\ [s\mathbf{I} - \mathbf{A}] = s^2 + 6s + 10$$

$$[s\mathbf{I} - \mathbf{A}]^{-1} = \frac{1}{s^2 + 6s + 10}\begin{bmatrix} s+3 & 1 \\ 0 & s+3 \end{bmatrix}$$

Therefore,

$$G_2(s) = \begin{bmatrix} 1 & 1 \end{bmatrix} \frac{1}{s^2 + 6s + 10}\begin{bmatrix} s+3 & 1 \\ 0 & s+3 \end{bmatrix}\begin{bmatrix} 0 \\ 1 \end{bmatrix}$$

$$= \frac{1}{s^2 + 6s + 10}\begin{bmatrix} 1 & 1 \end{bmatrix}\begin{bmatrix} 1 \\ s+3 \end{bmatrix} = \frac{s+4}{s^2 + 6s + 10}$$

The transfer function of the cascaded systems is, therefore

$$G(s) = G_1(s)G_2(s) = \frac{2(1+2s)(s+4)}{(s^2 + 3s + 2)(s^2 + 6s + 10)}$$

EXAMPLE 17.17. A state-space representation for the transfer function

$$\frac{Y(s)}{U(s)} = \frac{s+6}{s^2 + 5s + 6}$$

is

$$\mathbf{x} = \mathbf{Ax} + \mathbf{Bu}$$

$$\mathbf{y} = \mathbf{Cx}$$

where

$$\mathbf{A} = \begin{bmatrix} 0 & 1 \\ -6 & -5 \end{bmatrix} \qquad \mathbf{B} = \begin{bmatrix} 0 \\ 1 \end{bmatrix}$$

The value of $\mathbf{C}$ is

(a) $\begin{bmatrix} 1 \\ 2 \end{bmatrix}$ (b) $\begin{bmatrix} 6 & 1 \end{bmatrix}$ (c) $\begin{bmatrix} 1 & 2 \\ -1 & 0 \end{bmatrix}$ (d) $\begin{bmatrix} 1 & -5 \end{bmatrix}$

Solution The given transfer function can be written as

$$\frac{Y(s)}{U(s)} = \frac{s^{-1} + 6s^{-2}}{1 + 5s^{-1} + 6s^{-2}}$$

The CCF representation of this transfer function is given in the following figure:

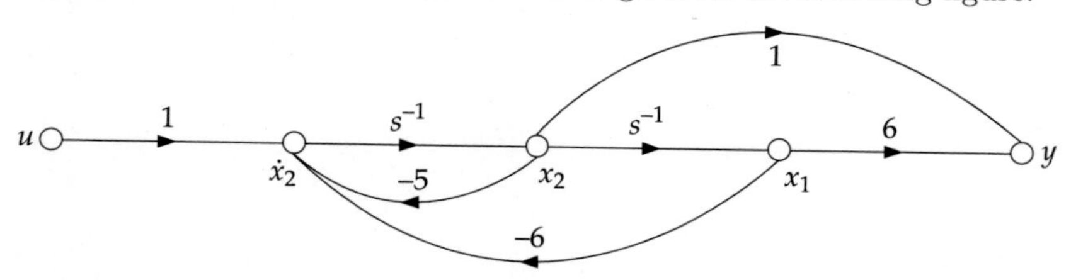

From the signal flow graph, we get the following equations:

$$\begin{aligned} \dot{x}_1 &= x_2 \\ \dot{x}_2 &= -6x_1 - 5x_2 + u \\ y &= 6x_1 + x_2 \end{aligned} \tag{i}$$

The matrix representation of Eq. (i) is

$$\begin{bmatrix} \dot{x}_1 \\ \dot{x}_2 \end{bmatrix} = \begin{bmatrix} 0 & 1 \\ -6 & -5 \end{bmatrix} \begin{bmatrix} x_1 \\ x_2 \end{bmatrix} + \begin{bmatrix} 0 \\ 1 \end{bmatrix} u$$

$$y = \begin{bmatrix} 6 & 1 \end{bmatrix} \begin{bmatrix} x_1 \\ x_2 \end{bmatrix} \tag{ii}$$

From Eq. (ii), we get $\mathbf{C} = \begin{bmatrix} 6 & 1 \end{bmatrix}$.
Ans: (b).

EXAMPLE 17.18. A linear time-invariant single-input single-output system has a state space model given by

$$\frac{d\mathbf{x}}{dt} = \mathbf{F}\mathbf{x} + \mathbf{G}u$$

$$y = \mathbf{H}\mathbf{x}$$

where $\mathbf{F} = \begin{bmatrix} 0 & 1 \\ -4 & -2 \end{bmatrix} \quad \mathbf{G} = \begin{bmatrix} 0 \\ 1 \end{bmatrix} \quad \mathbf{H} = \begin{bmatrix} 1 & 0 \end{bmatrix}$

Here, $\mathbf{x}$ is the state vector, u is the input and y is the output. The damping ratio of the system is

(a) 0.25 (b) 0.5 (c) 1 (d) 2

Solution The transfer function is calculated as follows:

$$[s\mathbf{I} - \mathbf{F}] = \begin{bmatrix} s & -1 \\ 4 & s+2 \end{bmatrix}$$

$$\det\,[s\mathbf{I} - \mathbf{F}] = s^2 + 2s + 4$$

$$\text{Adj}\,[s\mathbf{I} - \mathbf{F}] = \begin{bmatrix} s+2 & -4 \\ 1 & s \end{bmatrix}^T = \begin{bmatrix} s+2 & 1 \\ -4 & s \end{bmatrix}$$

$$[s\mathbf{I} - \mathbf{F}]^{-1} = \frac{1}{s^2 + 2s + 4} \begin{bmatrix} s+2 & 1 \\ -4 & s \end{bmatrix}$$

$$G(s) = \begin{bmatrix} 1 & 0 \end{bmatrix} \frac{1}{s^2 + 2s + 4} \begin{bmatrix} s+2 & 1 \\ -4 & s \end{bmatrix} \begin{bmatrix} 0 \\ 1 \end{bmatrix}$$

$$= \frac{1}{s^2 + 2s + 4} \begin{bmatrix} 1 & 0 \end{bmatrix} \begin{bmatrix} 1 \\ s \end{bmatrix} = \frac{1}{s^2 + 2s + 4}$$

From the denominator of the transfer function, we get

$$\omega_n^2 = 4 \quad \Rightarrow \quad \omega_n = 2$$

$$2\zeta\omega_n = 2 \quad \Rightarrow \quad \zeta = \frac{2}{2\omega_n} = 0.5$$

Ans: (b).

17.8 SOLVING STATE EQUATIONS

If the system is time-invariant, the state equation is given by

$$\dot{\mathbf{x}}(t) = \mathbf{A}\mathbf{x}(t) + \mathbf{B}\mathbf{u}(t) \tag{17.24}$$

In case the system is free-running with no application of control forces, $\mathbf{Bu}(\mathrm{t}) = 0$. Hence,

$$\dot{\mathbf{x}}(t) = \mathbf{A}\mathbf{x}(t) \tag{17.25}$$

where $\mathbf{A}$ is a constant matrix. Equations (17.24) and (17.25) are called *non-homogeneous* and *homogeneous* state equations, respectively. We will discuss methods of solving them in the reverse order. But before doing that, let us consider what is called the *state transition matrix* (STM) and how to calculate it because STM is necessary to solve both homogeneous and non-homogeneous equations.

State Transition Matrix (STM)

Let us consider Eq. (17.25). Its Laplace transform is

$$s\mathbf{X}(s) - \mathbf{X}(0) = \mathbf{A}\mathbf{X}(s)$$

$$\Rightarrow \qquad (s\mathbf{I} - \mathbf{A})\mathbf{X}(s) = \mathbf{X}(0)$$

Therefore,

$$\mathbf{X}(s) = (s\mathbf{I} - \mathbf{A})^{-1}\mathbf{X}(0) \equiv \mathbf{\Phi}(s)\mathbf{X}(0) \tag{17.26}$$

where

$$\mathbf{\Phi}(s) = (s\mathbf{I} - \mathbf{A})^{-1} \tag{17.27}$$

is called the *resolvent matrix*. Now,

$$\mathbf{\Phi}(s) = (s\mathbf{I} - \mathbf{A})^{-1} = \frac{1}{s}\left(\mathbf{I} - \frac{\mathbf{A}}{s}\right)^{-1} = \frac{1}{s}\left(\mathbf{I} + \frac{\mathbf{A}}{s} + \frac{\mathbf{A^2}}{s^2} + \cdots\right)$$

Therefore, $$\mathbf{\Phi}(s) = \frac{\mathbf{I}}{s} + \frac{\mathbf{A}}{s^2} + \frac{\mathbf{A}^2}{s^3} + \cdots \tag{17.28}$$

The inverse Laplace transform of Eq. (17.28) yields

$$\mathscr{L}^{-1}\{\mathbf{\Phi}(s)\} = \boldsymbol{\phi}(t) = \mathbf{I} + \mathbf{A}t + \frac{\mathbf{A}^2t^2}{2!} + \frac{\mathbf{A}^3t^3}{3!} + \cdots = e^{\mathbf{A}t} \tag{17.29}$$

Taking the inverse Laplace transform of Eq. (17.26) and substituting Eq. (17.29) therein, we get

$$\mathbf{x}(t) = \boldsymbol{\phi}(t)\mathbf{x}(0) = e^{\mathbf{A}t}\mathbf{x}(0) \tag{17.30}$$

The matrix $\boldsymbol{\phi}$, which helps us to know the system behaviour when it transits from $t = 0$ to $t = t$, is called the *state transition matrix* (STM). Six important properties of the STM are listed in Table 17.2.

Table 17.2 Important Properties of the STM

I	$\boldsymbol{\phi}(0) = e^{\mathbf{A}0} = 1$
II	$\boldsymbol{\phi}(t) = e^{\mathbf{A}t} = e^{-(-\mathbf{A}t)} = [\boldsymbol{\phi}(-t)]^{-1}$
III	$\boldsymbol{\phi}^{-1}(t) = (e^{\mathbf{A}t})^{-1} = \boldsymbol{\phi}(-t)$
IV	$\boldsymbol{\phi}(t_1 + t_2) = e^{\mathbf{A}t_1} \cdot e^{\mathbf{A}t_2} = \boldsymbol{\phi}(t_1) \cdot \boldsymbol{\phi}(t_2) = \boldsymbol{\phi}(t_2) \cdot \boldsymbol{\phi}(t_1)$
V	$[\boldsymbol{\phi}(t)]^n = (e^{\mathbf{A}t})^n = e^{\mathbf{A}(nt)} = \boldsymbol{\phi}(nt)$
VI	$\boldsymbol{\phi}(t_2 - t_1)\boldsymbol{\phi}(t_1 - t_0) = \boldsymbol{\phi}(t_2 - t_0) = \boldsymbol{\phi}(t_1 - t_0)\boldsymbol{\phi}(t_2 - t_1)$

The STM can be obtained by three methods:

1. Inverse Laplace transform method
2. Series summation method
3. $\mathbf{\Lambda}$-matrix method

Let us consider these methods now.

1. Inverse Laplace transform method. We know that the resolvent matrix $\mathbf{\Phi}(s)$ is given by [see Eq. (17.27)]

$$\mathbf{\Phi}(s) = (s\mathbf{I} - \mathbf{A})^{-1}$$

Its inverse Laplace transform yields $\boldsymbol{\phi}(t)$, i.e.

$$\boldsymbol{\phi}(t) = \mathscr{L}^{-1}\{\mathbf{\Phi}(s)\}$$

Therefore, the STM can be obtained by an inverse Laplace transform of the resolvent matrix.

The following examples will make the procedure clear.

EXAMPLE 17.19. Find the STM for a system described by $\dot{\mathbf{x}}(t) = \mathbf{A}\mathbf{x}(t) + \mathbf{B}\mathbf{u}(t)$, where

$$\mathbf{A} = \begin{bmatrix} 1 & 0 \\ 1 & 1 \end{bmatrix} \qquad \mathbf{B} = \begin{bmatrix} 1 \\ 1 \end{bmatrix}$$

Solution Here,

$$[s\mathbf{I} - \mathbf{A}] = \begin{bmatrix} s & 0 \\ 0 & s \end{bmatrix} - \begin{bmatrix} 1 & 0 \\ 1 & 1 \end{bmatrix} = \begin{bmatrix} s-1 & 0 \\ -1 & s-1 \end{bmatrix}$$

$$\text{Adj } [s\mathbf{I} - \mathbf{A}] = \begin{bmatrix} s-1 & 1 \\ 0 & s-1 \end{bmatrix}^{\mathrm{T}} = \begin{bmatrix} s-1 & 0 \\ 1 & s-1 \end{bmatrix}$$

$$|s\mathbf{I} - \mathbf{A}| = (s-1)^2$$

Therefore,

$$\mathbf{\Phi}(s) = [s\mathbf{I} - \mathbf{A}]^{-1} = \frac{\begin{bmatrix} s-1 & 0 \\ 1 & s-1 \end{bmatrix}}{(s-1)^2} = \begin{bmatrix} \dfrac{1}{s-1} & 0 \\ \dfrac{1}{(s-1)^2} & \dfrac{1}{s-1} \end{bmatrix}$$

We know that

$$\mathscr{L}^{-1}\left\{\frac{1}{s-1}\right\} = e^t$$

$$\mathscr{L}^{-1}\left\{\frac{1}{(s-1)^2}\right\} = te^t$$

Thus,

$$\boldsymbol{\phi}(t) = \mathscr{L}^{-1}\{\mathbf{\Phi}(s)\} = \begin{bmatrix} e^t & 0 \\ te^t & e^t \end{bmatrix}$$

EXAMPLE 17.20. The state model of a system is given by

$$\dot{\mathbf{x}}(t) = \begin{bmatrix} 0 & 1 \\ -2 & 3 \end{bmatrix} \mathbf{x}(t)$$

Find the STM of the system.

Solution Here,

$$[s\mathbf{I} - \mathbf{A}] = \begin{bmatrix} s & 0 \\ 0 & s \end{bmatrix} - \begin{bmatrix} 0 & 1 \\ -2 & 3 \end{bmatrix} = \begin{bmatrix} s & -1 \\ 2 & s-3 \end{bmatrix}$$

$$\text{Adj}[s\mathbf{I} - \mathbf{A}] = \begin{bmatrix} s-3 & -2 \\ 1 & s \end{bmatrix}^{\mathrm{T}} = \begin{bmatrix} s-3 & 1 \\ -2 & s \end{bmatrix}$$

$$|s\mathbf{I} - \mathbf{A}| = s^2 - 3s + 2 = (s-1)(s-2)$$

Therefore,

$$\boldsymbol{\Phi}(s) = [s\mathbf{I} - \mathbf{A}]^{-1} = \frac{\begin{bmatrix} s-3 & 1 \\ -2 & s \end{bmatrix}}{(s-1)(s-2)} = \begin{bmatrix} \dfrac{s-3}{(s-1)(s-2)} & \dfrac{1}{(s-1)(s-2)} \\ \dfrac{-2}{(s-1)(s-2)} & \dfrac{s}{(s-1)(s-2)} \end{bmatrix}$$

$$= \begin{bmatrix} \dfrac{2}{s-1} - \dfrac{1}{s-2} & -\dfrac{1}{s-1} + \dfrac{1}{s-2} \\ \dfrac{2}{s-1} - \dfrac{2}{s-2} & -\dfrac{1}{s-1} + \dfrac{2}{s-2} \end{bmatrix}$$

The inverse Laplace transform of the resolvent matrix $\boldsymbol{\Phi}(s)$ yields

$$\boldsymbol{\phi}(t) = \begin{bmatrix} 2e^{t} - e^{2t} & -e^{t} + e^{2t} \\ 2e^{t} - 2e^{2t} & -e^{t} + 2e^{2t} \end{bmatrix}$$

EXAMPLE 17.21. A system is described by the following state and output equations:

$$\frac{dx_1(t)}{dt} = -3x_1(t) + 2u(t)$$

$$\frac{dx_2(t)}{dt} = -2x_2(t) + u(t)$$

$$y(t) = x_1(t)$$

where $u(t)$ is the input and $y(t)$ is the output.

(a) The system transfer function is

(i) $\dfrac{s+2}{s^2+5s-6}$ (ii) $\dfrac{s+3}{s^2+5s+6}$ (iii) $\dfrac{2s+5}{s^2+5s+6}$ (iv) $\dfrac{2s-5}{s^2+5s-6}$

(b) The state-transition matrix of the above system is

(i) $\begin{bmatrix} e^{-3t} & 0 \\ e^{-2t} + e^{-3t} & e^{-2t} \end{bmatrix}$ (ii) $\begin{bmatrix} e^{-3t} & e^{-2t} - e^{-3t} \\ 0 & e^{-2t} \end{bmatrix}$

(iii) $\begin{bmatrix} e^{-3t} & e^{-2t} + e^{-3t} \\ 0 & e^{-2t} \end{bmatrix}$ (iv) $\begin{bmatrix} e^{3t} & e^{-2t} - e^{-3t} \\ 0 & e^{-2t} \end{bmatrix}$

Solution The given equations can be written in matrix form as follows:

$$\begin{bmatrix} \dot{x}_1 \\ \dot{x}_2 \end{bmatrix} = \begin{bmatrix} -3 & 1 \\ 0 & -2 \end{bmatrix} \begin{bmatrix} x_1 \\ x_2 \end{bmatrix} + \begin{bmatrix} 2 \\ 1 \end{bmatrix} u(t)$$

$$y(t) = \begin{bmatrix} 1 & 0 \end{bmatrix} \begin{bmatrix} x_1 \\ x_2 \end{bmatrix}$$

Let us write the matrices as

$$\mathbf{A} = \begin{bmatrix} -3 & 1 \\ 0 & -2 \end{bmatrix} \qquad \mathbf{B} = \begin{bmatrix} 2 \\ 1 \end{bmatrix} \qquad \mathbf{C} = \begin{bmatrix} 1 & 0 \end{bmatrix}$$

(a) Then,

$$[s\mathbf{I} - \mathbf{A}] = \begin{bmatrix} s+3 & -1 \\ 0 & s+2 \end{bmatrix}$$

$$\det\,[s\mathbf{I} - \mathbf{A}] = (s+2)(s+3)$$

$$\text{Adj}\,[s\mathbf{I} - \mathbf{A}] = \begin{bmatrix} s+2 & 0 \\ 1 & s+3 \end{bmatrix}^T = \begin{bmatrix} s+2 & 1 \\ 0 & s+3 \end{bmatrix}$$

$$[s\mathbf{I} - \mathbf{A}]^{-1} = \frac{1}{(s+2)(s+3)} \begin{bmatrix} s+2 & 1 \\ 0 & s+3 \end{bmatrix} \tag{i}$$

$$\Rightarrow \qquad G(s) = \begin{bmatrix} 1 & 0 \end{bmatrix} \frac{1}{(s+2)(s+3)} \begin{bmatrix} s+2 & 1 \\ 0 & s+3 \end{bmatrix} \begin{bmatrix} 2 \\ 1 \end{bmatrix}$$

$$= \frac{1}{(s+2)(s+3)} \begin{bmatrix} 1 & 0 \end{bmatrix} \begin{bmatrix} 2(s+2)+1 \\ s+3 \end{bmatrix}$$

$$= \frac{2(s+2)+1}{(s+2)(s+3)} = \frac{2s+5}{s^2+5s+6}$$

Ans: (iii).

(b) From Eq. (i),

$$\mathbf{\Phi}(s) = [s\mathbf{I} - \mathbf{A}]^{-1} = \frac{1}{(s+2)(s+3)} \begin{bmatrix} s+2 & 1 \\ 0 & s+3 \end{bmatrix}$$

$$= \begin{bmatrix} \dfrac{1}{s+3} & \dfrac{1}{s+2} - \dfrac{1}{s+3} \\ 0 & \dfrac{1}{s+2} \end{bmatrix}$$

$$\Rightarrow \qquad \boldsymbol{\phi}(t) = \begin{bmatrix} e^{-3t} & e^{-2t} - e^{-3t} \\ 0 & e^{-2t} \end{bmatrix}$$

Ans: (ii).

2. Series summation method. We have seen from Eq. (17.29) that

$$\boldsymbol{\phi}(t) = \mathbf{I} + \mathbf{A}t + \frac{\mathbf{A}^2t^2}{2!} + \frac{\mathbf{A}^3t^3}{3!} + \cdots$$

This series can be worked out to find $\boldsymbol{\phi}(t)$. The method is particularly suitable for digital computation, as will be clear from the following discussion.

Let us put $\mathbf{P} = \mathbf{A}t$. Then,

$$\begin{aligned}\boldsymbol{\phi}(t) &= \mathbf{I} + \mathbf{P} + \frac{\mathbf{P}^2}{2!} + \frac{\mathbf{P}^3}{3!} + \cdots \\ &= \mathbf{I} + \mathbf{P} + \frac{\mathbf{P}(\mathbf{P})}{2} + \frac{\mathbf{P}}{3}\left[\frac{\mathbf{P}(\mathbf{P})}{2}\right] + \cdots \end{aligned} \tag{17.31}$$

Equation (17.31) shows that each term in the series contains the preceding term and a multiplier in a regular order. If we denote the terms in the series as T_1, T_2, ..., they are

$$\begin{aligned} T_1 &= \mathbf{I} \\ T_2 &= \mathbf{P}(\mathbf{I}) = \frac{\mathbf{P}}{2-1}(T_1) \\ T_3 &= \frac{\mathbf{P}}{2}(\mathbf{P}) = \frac{\mathbf{P}}{3-1}(T_2) \\ &\vdots \\ T_n &= \frac{\mathbf{P}}{n-1}(T_{n-1}) \end{aligned}$$

The series may converge soon if the terms contain exponential factors. However, for manual calculation of STM, it is not so suitable because we have to calculate it through Eq. (17.29). We consider a simple example.

EXAMPLE 17.22. Find $\boldsymbol{\phi}(t)$ for the system given by the state equation

$$\dot{\mathbf{x}}(t) = \begin{bmatrix} -1 & 1 \\ 0 & -2 \end{bmatrix} \mathbf{x}(t) + \begin{bmatrix} 0 \\ 1 \end{bmatrix} \mathbf{u}(t)$$

through the series summation method.

Solution Given

$$\mathbf{A} = \begin{bmatrix} -1 & 1 \\ 0 & -2 \end{bmatrix}$$

we have

$$\mathbf{A}^2 = \begin{bmatrix} -1 & 1 \\ 0 & -2 \end{bmatrix} \begin{bmatrix} -1 & 1 \\ 0 & -2 \end{bmatrix} = \begin{bmatrix} 1 & -3 \\ 0 & 4 \end{bmatrix}$$

$$\mathbf{A}^3 = \begin{bmatrix} -1 & 1 \\ 0 & -2 \end{bmatrix} \begin{bmatrix} 1 & -3 \\ 0 & 4 \end{bmatrix} = \begin{bmatrix} -1 & 7 \\ 0 & -8 \end{bmatrix}$$

Therefore,

$$\boldsymbol{\phi}(t) = \begin{bmatrix} 1 & 0 \\ 0 & 1 \end{bmatrix} + t\begin{bmatrix} -1 & 1 \\ 0 & -2 \end{bmatrix} + \frac{t^2}{2}\begin{bmatrix} 1 & -3 \\ 0 & 4 \end{bmatrix} + \frac{t^3}{6}\begin{bmatrix} -1 & 7 \\ 0 & -8 \end{bmatrix} + \cdots$$

$$= \begin{bmatrix} 1 & 0 \\ 0 & 1 \end{bmatrix} + \begin{bmatrix} -t & t \\ 0 & -2t \end{bmatrix} + \begin{bmatrix} \frac{t^2}{2} & -\frac{3t^2}{2} \\ 0 & 2t^2 \end{bmatrix} + \begin{bmatrix} -\frac{t^3}{6} & \frac{7t^3}{6} \\ 0 & -\frac{4t^3}{3} \end{bmatrix} + \cdots$$

$$= \begin{bmatrix} 1 - t + \frac{t^2}{2} - \frac{t^3}{6} + \cdots & t - \frac{3}{2}t^2 + \frac{7}{6}t^3 - \cdots \\ 0 & 1 - 2t + 2t^2 - \frac{4}{3}t^3 + \cdots \end{bmatrix}$$

$$= \begin{bmatrix} 1 - t + \frac{t^2}{2} - \frac{t^3}{6} + \cdots & \left(1 - t + \frac{t^2}{2} - \frac{t^3}{6} + \cdots\right) - \left(1 - 2t + \frac{(2t)^2}{2} - \frac{(2t)^3}{3!} + \cdots\right) \\ 0 & 1 - 2t + \frac{(2t)^2}{2} - \frac{(2t)^3}{3!} + \cdots \end{bmatrix}$$

$$= \begin{bmatrix} e^t & e^t - e^{2t} \\ 0 & e^{2t} \end{bmatrix}$$

3. $\boldsymbol{\Lambda}$-matrix method. The $\boldsymbol{\Lambda}$-matrix method is based on finding a nonsingular matrix $\mathbf{M}$, called the *modal matrix*, such that the transformation

$$\mathbf{x}(t) = \mathbf{Mz}(t) \tag{17.32}$$

transforms the linear system given by

$$\dot{\mathbf{x}}(t) = \mathbf{Ax}(t) + \mathbf{Bu}(t) \tag{17.33}$$

to

$$\dot{\mathbf{z}}(t) = \boldsymbol{\Lambda}\mathbf{z}(t) + \boldsymbol{\Gamma}\mathbf{u}(t)$$

We can verify this from the following manipulations. Let us substitute Eq. (17.32) into Eq. (17.33). Then we get

$$\mathbf{M}\dot{\mathbf{z}}(t) = \mathbf{AMz}(t) + \mathbf{Bu}(t) \tag{17.34}$$

Now, we left multiply Eq. (17.34) by $\mathbf{M}^{-1}$ to obtain

$$\dot{\mathbf{z}}(t) = \mathbf{M}^{-1}\mathbf{AMz}(t) + \mathbf{M}^{-1}\mathbf{Bu}(t) \equiv \boldsymbol{\Lambda}\mathbf{z}(t) + \boldsymbol{\Gamma}\mathbf{u}(t) \tag{17.35}$$

where

$$\boldsymbol{\Lambda} \equiv \mathbf{M}^{-1}\mathbf{AM} \tag{17.36}$$

$$\boldsymbol{\Gamma} \equiv \mathbf{M}^{-1}\mathbf{B}$$

The transformation, given by the RHS of Eq. (17.36) is called *similarity transformation*, and the state equation given by Eq. (17.35) is known as the *canonical form*. We will presently see that the similarity transformation diagonalizes **A** having its eigenvalues[7] as its diagonal elements, i.e.

$$\mathbf{\Lambda} = \begin{bmatrix} \lambda_1 & 0 & 0 & 0 \\ 0 & \lambda_2 & 0 & 0 \\ \vdots & \vdots & \vdots & \vdots \\ 0 & 0 & 0 & \lambda_n \end{bmatrix}$$

Now, the question is how to obtain the modal matrix **M**.

Modal matrix. It can be shown that **M** can be formed from the eigenvectors[8] $\mathbf{p}_i$ of **A** as

$$\mathbf{M} = \begin{bmatrix} \mathbf{p}_1 & \mathbf{p}_2 & \cdots & \mathbf{p}_n \end{bmatrix}$$

where $\mathbf{p}_i$ is the eigenvector associated with the eigenvalue λ_i. Because, then for $\mathbf{p}_1$,

$$\mathbf{A}\mathbf{p}_1 = \lambda_1 \mathbf{p}_1$$

That is,

$$\begin{bmatrix} a_{11} & a_{12} & \cdots & a_{1n} \\ a_{21} & a_{22} & \cdots & a_{2n} \\ \vdots & \vdots & & \vdots \\ a_{n1} & a_{n2} & \cdots & a_{nn} \end{bmatrix} \begin{bmatrix} p_{11} \\ p_{12} \\ \vdots \\ p_{n1} \end{bmatrix} = \lambda_1 \begin{bmatrix} p_{11} \\ p_{12} \\ \vdots \\ p_{n1} \end{bmatrix} \tag{17.37}$$

Similarly,

$$\mathbf{A}\mathbf{p}_2 = \lambda_2 \mathbf{p}_2$$

and so on.

Diagonalization of A by modal matrix. Interchanging the sides of Eq. (17.37) and forming the $n \times n$ matrices, we get

$$\begin{bmatrix} \lambda_1 p_{11} & \lambda_2 p_{12} & \cdots & \lambda_n p_{1n} \\ \lambda_1 p_{21} & \lambda_2 p_{22} & \cdots & \lambda_n p_{2n} \\ \vdots & \vdots & & \vdots \\ \lambda_1 p_{n1} & \lambda_2 p_{n2} & \cdots & \lambda_n p_{nn} \end{bmatrix} \begin{bmatrix} p_{11} \\ p_{12} \\ \vdots \\ p_{n1} \end{bmatrix} = \begin{bmatrix} a_{11} & a_{12} & \cdots & a_{1n} \\ a_{21} & a_{22} & \cdots & a_{2n} \\ \vdots & \vdots & & \vdots \\ a_{n1} & a_{n2} & \cdots & a_{nn} \end{bmatrix} \begin{bmatrix} p_{11} & p_{12} & \cdots & p_{1n} \\ p_{21} & p_{22} & \cdots & p_{2n} \\ \vdots & \vdots & & \vdots \\ p_{n1} & p_{n2} & \cdots & p_{nn} \end{bmatrix}$$

or

$$\begin{bmatrix} p_{11} & p_{12} & \cdots & p_{1n} \\ p_{21} & p_{22} & \cdots & p_{2n} \\ \vdots & \vdots & & \vdots \\ p_{n1} & p_{n2} & \cdots & p_{nn} \end{bmatrix} \begin{bmatrix} \lambda_1 & 0 & \cdots & 0 \\ 0 & \lambda_2 & \cdots & 0 \\ \vdots & \vdots & & \vdots \\ 0 & 0 & \cdots & \lambda_n \end{bmatrix} = \begin{bmatrix} a_{11} & a_{12} & \cdots & a_{1n} \\ a_{21} & a_{22} & \cdots & a_{2n} \\ \vdots & \vdots & & \vdots \\ a_{n1} & a_{n2} & \cdots & a_{nn} \end{bmatrix} \begin{bmatrix} p_{11} & p_{12} & \cdots & p_{1n} \\ p_{21} & p_{22} & \cdots & p_{2n} \\ \vdots & \vdots & & \vdots \\ p_{n1} & p_{n2} & \cdots & p_{nn} \end{bmatrix} \tag{17.38}$$

[7]See Section 1.3 on page 30.
[8]See Section 1.3 on page 30.

Equation (17.38) can be written as

$$\mathbf{M\Lambda} = \mathbf{AM} \tag{17.39}$$

Left multiplying both sides of Eq. (17.39) by $\mathbf{M}^{-1}$, we get

$$\mathbf{\Lambda} = \mathbf{M}^{-1}\mathbf{AM} \tag{17.40}$$

So, we see that the modal matrix $\mathbf{M}$, formed by the eigenvectors of $\mathbf{A}$, diagonalizes it by the similarity transformation.

Modal matrix for CCF-type A. However, if the matrix $\mathbf{A}$ is of the CCF-type (or phase-variable canonical form), it is known that the modal matrix may be the Vandermonde matrix given by

$$\begin{bmatrix} 1 & 1 & \cdots & 1 \\ \lambda_1 & \lambda_2 & \cdots & \lambda_n \\ \lambda_1^2 & \lambda_2^2 & \cdots & \lambda_n^2 \\ \vdots & \vdots & & \vdots \\ \lambda_1^{n-1} & \lambda_2^{n-1} & \cdots & \lambda_n^{n-1} \end{bmatrix} \tag{17.41}$$

where $\lambda_1,\ \lambda_2,\ \cdots \lambda_n$ are the eigenvalues of $\mathbf{A}$.

Obtaining the STM. The STM is obtained from the similarity transformation of $e^{\mathbf{\Lambda}t}$ by the modal matrix, i.e.

$$\boldsymbol{\phi}(t) = e^{\mathbf{A}t} = \mathbf{M}e^{\mathbf{\Lambda}t}\mathbf{M}^{-1}$$

Proof. We know from Eq. (17.40) that

$$\mathbf{\Lambda} = \mathbf{M}^{-1}\mathbf{AM}$$

Therefore,

$$\begin{aligned} \mathbf{A} &= \mathbf{M\Lambda M}^{-1} \\ \mathbf{A}^2 &= (\mathbf{M\Lambda M}^{-1})(\mathbf{M\Lambda M}^{-1}) \\ &= \mathbf{M\Lambda}(\mathbf{M}^{-1}\mathbf{M})\mathbf{\Lambda M}^{-1} \\ &= \mathbf{M\Lambda}^2\mathbf{M}^{-1} \\ &\ \ \vdots \\ \mathbf{A}^n &= \mathbf{M\Lambda}^n\mathbf{M}^{-1} \end{aligned} \tag{17.42}$$

Thus,

$$\begin{aligned} \boldsymbol{\phi}(t) = e^{\mathbf{\Lambda}t} &= \mathbf{I} + \mathbf{A}t + \frac{\mathbf{A}^2t^2}{2!} + \cdots \\ &= \mathbf{I} + \mathbf{M\Lambda M}^{-1}t + \frac{\mathbf{M\Lambda}^2\mathbf{M}^{-1}t^2}{2!} + \cdots \qquad \text{[from Eq. (17.42)]} \\ &= \mathbf{I} + \mathbf{M}(\mathbf{\Lambda}t)\mathbf{M}^{-1} + \mathbf{M}\left[\frac{(\mathbf{\Lambda}t)^2}{2!}\right]\mathbf{M}^{-1} + \cdots \\ &= \mathbf{M}e^{\mathbf{\Lambda}t}\mathbf{M}^{-1} \end{aligned}$$

Let us work out a problem to make ourselves familiar with the procedure.

EXAMPLE 17.23. The state model of a system is given by

$$\dot{\mathbf{x}}(t) = \begin{bmatrix} 0 & 1 \\ -2 & 3 \end{bmatrix} \mathbf{x}(t)$$

Find the eigenvalues, the eigenvectors and the STM.

Solution Let

$$\mathbf{A} = \begin{bmatrix} 0 & 1 \\ -2 & 3 \end{bmatrix}$$

Then,

$$|\lambda\mathbf{I} - \mathbf{A}| = \left| \lambda \begin{bmatrix} 1 & 0 \\ 0 & 1 \end{bmatrix} - \begin{bmatrix} 0 & 1 \\ -2 & 3 \end{bmatrix} \right|$$

$$= \begin{vmatrix} \lambda & -1 \\ 2 & \lambda - 3 \end{vmatrix}$$

$$= \lambda^2 - 3\lambda + 2$$

Solving the equation

$$\lambda^2 - 3\lambda + 2 = 0$$

we get the eigenvalues of $\mathbf{A}$:

$$\lambda_1 = 1 \quad \text{and} \quad \lambda_2 = 2$$

If $\mathbf{p}_1$ is the eigenvector associated with the eigenvalue $\lambda_1 = 1$, it must satisfy the equation

$$[\lambda_1\mathbf{I} - \mathbf{A}]\mathbf{p}_1 = 0$$

That is,

$$\left[1 \begin{bmatrix} 1 & 0 \\ 0 & 1 \end{bmatrix} - \begin{bmatrix} 0 & 1 \\ -2 & 3 \end{bmatrix}\right] \begin{bmatrix} p_{11} \\ p_{21} \end{bmatrix} = 0$$

$$\Rightarrow \qquad \begin{bmatrix} 1 & -1 \\ 2 & -2 \end{bmatrix} \begin{bmatrix} p_{11} \\ p_{21} \end{bmatrix} = 0$$

This gives us the equations

$$p_{11} - p_{21} = 0$$

$$2p_{11} - 2p_{21} = 0$$

both of which yield

$$p_{11} = p_{21}$$

Let us put both the equations equal to 1 so that the eigenvector associated with the eigenvalue $\lambda_1 = 1$ is

$$\mathbf{p}_1 = \begin{bmatrix} 1 \\ 1 \end{bmatrix}$$

In a similar way, the matrix equation for the eigenvalue $\lambda_2 = 2$ is obtained as

$$\begin{bmatrix} 2 & -1 \\ 2 & -1 \end{bmatrix} \begin{bmatrix} p_{12} \\ p_{22} \end{bmatrix} = 0$$

which yields identical equations involving p_{12} and p_{22}. They give the relation

$$p_{22} = 2p_{12}$$

Let us put $p_{12} = 1$. Then, $p_{22} = 2$. So, the eigenvector associated with the eigenvalue $\lambda_2 = 2$ is

$$\mathbf{p}_2 = \begin{bmatrix} 1 \\ 2 \end{bmatrix}$$

Therefore, the modal matrix is

$$\mathbf{M} = \begin{bmatrix} \mathbf{p}_1 & \mathbf{p}_2 \end{bmatrix} = \begin{bmatrix} 1 & 1 \\ 1 & 2 \end{bmatrix}$$

The inverse of the modal matrix is

$$\mathbf{M}^{-1} = \frac{\text{Adj}[\mathbf{M}]}{|\mathbf{M}|} = \frac{\begin{bmatrix} 2 & -1 \\ -1 & 1 \end{bmatrix}^{\text{T}}}{1}$$

$$= \begin{bmatrix} 2 & -1 \\ -1 & 1 \end{bmatrix}$$

Note: 1. Since the matrix $\mathbf{A}$ is of the CCF-type, we could straightaway construct the modal matrix as the Vandermonde matrix

$$\mathbf{M} = \begin{bmatrix} 1 & 1 \\ \lambda_1 & \lambda_2 \end{bmatrix} = \begin{bmatrix} 1 & 1 \\ 1 & 2 \end{bmatrix}$$

without finding out the eigenvectors. However, we need to find out the eigenvectors because the problem specifically asks for them.

2. If $\mathbf{M}$ is correctly formed, it should diagonalize $\mathbf{A}$ through its similarity transformation with the eigenvalues as its diagonal elements. We observe that

$$\mathbf{M}^{-1}\mathbf{A}\mathbf{M} = \begin{bmatrix} 2 & -1 \\ -1 & 1 \end{bmatrix} \begin{bmatrix} 0 & 1 \\ -2 & 3 \end{bmatrix} \begin{bmatrix} 1 & 1 \\ 1 & 2 \end{bmatrix}$$

$$= \begin{bmatrix} 2 & -1 \\ -1 & 1 \end{bmatrix} \begin{bmatrix} 1 & 2 \\ 1 & 4 \end{bmatrix}$$

$$= \begin{bmatrix} 1 & 0 \\ 0 & 2 \end{bmatrix}$$

The result shows that the modal matrix is correct. Now,

$$\mathbf{\Lambda} = \begin{bmatrix} 1 & 0 \\ 0 & 2 \end{bmatrix}$$

$$e^{\mathbf{\Lambda}t} = \begin{bmatrix} e^t & 0 \\ 0 & e^{2t} \end{bmatrix}$$

Therefore, the STM is given by

$$\begin{aligned} \boldsymbol{\phi}(t) = \mathbf{M}e^{\mathbf{\Lambda}t}\mathbf{M}^{-1} &= \begin{bmatrix} 1 & 1 \\ 1 & 2 \end{bmatrix} \begin{bmatrix} e^t & 0 \\ 0 & e^{2t} \end{bmatrix} \begin{bmatrix} 2 & -1 \\ -1 & 1 \end{bmatrix} \\ &= \begin{bmatrix} 1 & 1 \\ 1 & 2 \end{bmatrix} \begin{bmatrix} 2e^t & -e^t \\ e^{-2t} & e^{2t} \end{bmatrix} \\ &= \begin{bmatrix} 2e^t - e^{2t} & -e^t + e^{2t} \\ 2e^t - 2e^{2t} & -e^t + 2e^{2t} \end{bmatrix} \end{aligned}$$

Once $\boldsymbol{\phi}(t)$ is found out, homogeneous and nonhomogeneous equations can be easily solved.

Homogeneous Equation

For a free-running system, $\mathbf{Bu}(t) = 0$. Hence the state equation becomes

$$\dot{\mathbf{x}}(t) = \mathbf{Ax}(t)$$

We know[9] that its solution is given by

$$\mathbf{x}(t) = \boldsymbol{\phi}(t)\mathbf{x}(0)$$

where $\mathbf{x}(0)$ is the initial condition. So, to solve homogeneous equations, all we need to do is to find out the STM and multiply it with the given initial condition. Let us work out a couple of problems.

EXAMPLE 17.24. Find $\mathbf{\Phi}(s)$, $\boldsymbol{\phi}(t)$, $\boldsymbol{\phi}^{-1}(t)$, and the solution to the system given by the state equation

$$\begin{bmatrix} \dot{x}_1 \\ \dot{x}_2 \end{bmatrix} = \begin{bmatrix} -1 & 1 \\ 0 & -2 \end{bmatrix} \begin{bmatrix} x_1 \\ x_2 \end{bmatrix} + \begin{bmatrix} 0 \\ 1 \end{bmatrix} u$$

with the initial state

$$\begin{bmatrix} x_1 \\ x_2 \end{bmatrix} = \begin{bmatrix} 1 \\ 0 \end{bmatrix}$$

Solution We note that although the state equation contains the term involving u, its value is not given. So, we need to solve it as a homogeneous equation. Here,

$$\mathbf{A} = \begin{bmatrix} -1 & 1 \\ 0 & -2 \end{bmatrix}$$

[9] See Eq. (17.30) on page 624.

$$[s\mathbf{I}-\mathbf{A}] = \begin{bmatrix} s & 0 \\ 0 & s \end{bmatrix} - \begin{bmatrix} -1 & 1 \\ 0 & -2 \end{bmatrix} = \begin{bmatrix} s+1 & -1 \\ 0 & s+2 \end{bmatrix}$$

$$\text{Adj}\,[s\mathbf{I}-\mathbf{A}] = \begin{bmatrix} s+2 & 0 \\ 1 & s+1 \end{bmatrix}^{\mathrm{T}} = \begin{bmatrix} s+2 & 1 \\ 0 & s+1 \end{bmatrix}$$

$$|s\mathbf{I}-\mathbf{A}| = (s+1)(s+2)$$

Therefore,

$$\mathbf{\Phi}(s) = [s\mathbf{I}-\mathbf{A}]^{-1} = \frac{\begin{bmatrix} s+2 & 1 \\ 0 & s+1 \end{bmatrix}}{(s+1)(s+2)}$$

$$= \begin{bmatrix} \dfrac{s+2}{(s+1)(s+2)} & \dfrac{1}{(s+1)(s+2)} \\ 0 & \dfrac{s+1}{(s+1)(s+2)} \end{bmatrix}$$

$$= \begin{bmatrix} \dfrac{1}{s+1} & \dfrac{1}{(s+1)(s+2)} \\ 0 & \dfrac{1}{s+2} \end{bmatrix}$$

Now,

$$\mathscr{L}^{-1}\{\mathbf{\Phi}_{11}(s)\} = \mathscr{L}^{-1}\left\{\frac{1}{s+1}\right\} = e^{-t}$$

$$\mathscr{L}^{-1}\{\mathbf{\Phi}_{12}(s)\} = \mathscr{L}^{-1}\left\{\frac{1}{(s+1)(s+2)}\right\} = \mathscr{L}^{-1}\left\{\frac{1}{s+1} - \frac{1}{s+2}\right\} = e^{-t} - e^{-2t}$$

$$\mathscr{L}^{-1}\{\mathbf{\Phi}_{21}(s)\} = \mathscr{L}^{-1}\{0\} = 0$$

$$\mathscr{L}^{-1}\{\mathbf{\Phi}_{22}(s)\} = \mathscr{L}^{-1}\left\{\frac{1}{s+2}\right\} = e^{-2t}$$

Therefore,

$$\boldsymbol{\phi}(t) = \begin{bmatrix} e^{-t} & e^{-t} - e^{-2t} \\ 0 & e^{-2t} \end{bmatrix}$$

$$\boldsymbol{\phi}^{-1}(t) = \boldsymbol{\phi}(-t) = \begin{bmatrix} e^{t} & e^{t} - e^{2t} \\ 0 & e^{2t} \end{bmatrix}$$

$$\mathbf{x}(t) = \boldsymbol{\phi}(t)\mathbf{x}(0) = \begin{bmatrix} e^{-t} & e^{-t} - e^{-2t} \\ 0 & e^{-2t} \end{bmatrix} \begin{bmatrix} 1 \\ 0 \end{bmatrix} = \begin{bmatrix} e^{-t} \\ 0 \end{bmatrix}$$

EXAMPLE 17.25. The state model of a system is given by

$$\dot{\mathbf{x}}(t) = \begin{bmatrix} 0 & 1 \\ -2 & 3 \end{bmatrix} \mathbf{x}(t)$$

Find the response of the system when

(a) $\mathbf{x}(0) = \begin{bmatrix} 1 \\ 1 \end{bmatrix}$

(b) $\mathbf{x}(0) = \begin{bmatrix} 1 \\ 2 \end{bmatrix}$

Solution We have worked out the STM for the system by the inverse Laplace transform method in Example 17.20 as well as by the $\mathbf{\Lambda}$-matrix method in Example 17.23. It is given by

$$\boldsymbol{\phi}(t) = \begin{bmatrix} 2e^t - e^{2t} & -e^t + e^{2t} \\ 2e^t - 2e^{2t} & -e^t + 2e^{2t} \end{bmatrix}$$

Hence, the responses are

(a)

$$\mathbf{x}(t) = \boldsymbol{\phi}(t)\mathbf{x}(0) = \begin{bmatrix} 2e^t - e^{2t} & -e^t + e^{2t} \\ 2e^t - 2e^{2t} & -e^t + 2e^{2t} \end{bmatrix} \begin{bmatrix} 1 \\ 1 \end{bmatrix}$$

$$= \begin{bmatrix} 2e^t - e^{2t} - e^t + e^{2t} \\ 2e^t - 2e^{2t} - e^t + 2e^{2t} \end{bmatrix} = \begin{bmatrix} e^t \\ e^t \end{bmatrix}$$

This means that $x_1(t) = e^t, \qquad x_2(t) = e^t$

(b)

$$\mathbf{x}(t) = \boldsymbol{\phi}(t)\mathbf{x}(0) = \begin{bmatrix} 2e^t - e^{2t} & -e^t + e^{2t} \\ 2e^t - 2e^{2t} & -e^t + 2e^{2t} \end{bmatrix} \begin{bmatrix} 1 \\ 2 \end{bmatrix}$$

$$= \begin{bmatrix} 2e^t - e^{2t} - 2e^t + 2e^{2t} \\ 2e^t - 2e^{2t} - 2e^t + 4e^{2t} \end{bmatrix} = \begin{bmatrix} e^{2t} \\ 2e^{2t} \end{bmatrix}$$

This means that $x_1(t) = e^{2t}, \qquad x_2(t) = 2e^{2t}$

Non-Homogeneous Equation

The Laplace transform of the non-homogeneous state equation [Eq. (17.24)] is

$$s\mathbf{X}(s) - \mathbf{X}(\mathbf{0}) = \mathbf{AX}(s) + \mathbf{BU}(s)$$

$$\Rightarrow \qquad (s\mathbf{I} - \mathbf{A})\mathbf{X}(s) = \mathbf{X}(0) + \mathbf{BU}(s)$$

$$\Rightarrow \qquad \mathbf{X}(s) = (s\mathbf{I} - \mathbf{A})^{-1}\mathbf{X}(0) + (s\mathbf{I} - \mathbf{A})^{-1}\mathbf{BU}(s) \tag{17.43}$$

The inverse Laplace transform of Eq. (17.43) yields

$$\mathbf{x}(t) = \mathscr{L}^{-1}\{(s\mathbf{I} - \mathbf{A})^{-1}\mathbf{X}(0)\} + \mathscr{L}^{-1}\{(s\mathbf{I} - \mathbf{A})^{-1}\mathbf{BU}(s)\}$$

Therefore, the state vector can be calculated from the following equation:

$$\mathbf{x}(t) = \underbrace{\boldsymbol{\phi}(t)\mathbf{x}(0)}_{\text{Free response}} + \underbrace{\int_0^t \boldsymbol{\phi}(t - \tau)\mathbf{Bu}(\tau)\, d\tau}_{\text{Forced response}}$$

The output vector can be obtained from Eq. (17.3) as follows:

$$\mathbf{y}(t) = \mathbf{C}\,\boldsymbol{\Phi}(t)\mathbf{x}(0) + \int_0^t \mathbf{C}\,\boldsymbol{\Phi}(t-\tau)\mathbf{B}\mathbf{u}(\tau)\,d\tau \tag{17.44}$$

The method will be clear from the following examples.

EXAMPLE 17.26. The state variable representation of a plant is given by

$$\dot{\mathbf{x}} = \mathbf{A}\mathbf{x} + \mathbf{B}\mathbf{u}, \;\; \mathbf{y} = \mathbf{C}\mathbf{x}$$

where $\mathbf{x}$ is the state, $\mathbf{u}$ is the input, and $\mathbf{y}$ is the output. Assuming zero initial conditions, the impulse response of a plant is given by

(a) $\exp\,(\mathbf{A}t)$

(b) $\int \exp\,[\mathbf{A}(t-\tau)]\mathbf{B}\mathbf{u}(\tau)\,d\tau$

(c) $\mathbf{C}\,\exp\,(\mathbf{A}t)\,\mathbf{B}$

(d) $\mathbf{C}\int \exp\,[\mathbf{A}(t-\tau)]\mathbf{B}\mathbf{u}(\tau)\,d\tau$

Solution The given data are:

(i) The initial conditions are zero.

(ii) The input is an impulse function.

The zero initial conditions make the first part of the output [see Eq. (17.44)] zero. The impulse input, which exists only at $t = 0$ with an area of 1, converts the integral in the second part of Eq. (17.44) to

$$\mathbf{C}\exp\,(\mathbf{A}t)\mathbf{B}$$

because

$$\boldsymbol{\Phi}(t) = \exp\,(\mathbf{A}t)$$

Ans: (c).

EXAMPLE 17.27. A system dynamics is given by

$$\dot{\mathbf{x}}(t) = \begin{bmatrix} -1 & 1 \\ 0 & -2 \end{bmatrix}\mathbf{x}(t) + \begin{bmatrix} 0 \\ 1 \end{bmatrix}\mathbf{u}(t)$$

Given the initial vector

$$\mathbf{x}(0) = \begin{bmatrix} -1 \\ 0 \end{bmatrix}$$

and $\mathbf{u}(t)$ is a unit step function. Find the response of the system.

Solution We have already found (see Example 17.24 at page 634) that

$$\boldsymbol{\Phi}(t) = \begin{bmatrix} e^{-t} & e^{-t} - e^{-2t} \\ 0 & e^{-2t} \end{bmatrix}$$

Therefore,

$$\mathbf{x}(t) = \begin{bmatrix} e^{-t} & e^{-t} - e^{-2t} \\ 0 & e^{-2t} \end{bmatrix} \begin{bmatrix} -1 \\ 0 \end{bmatrix} + \int_0^t \begin{bmatrix} e^{-(t-\tau)} & e^{-(t-\tau)} - e^{-2(t-\tau)} \\ 0 & e^{-2(t-\tau)} \end{bmatrix} \begin{bmatrix} 0 \\ 1 \end{bmatrix} d\tau \quad [\because\ \mathbf{u}(t) = 1]$$

$$= \begin{bmatrix} -e^{-t} \\ 0 \end{bmatrix} + \int_0^t \begin{bmatrix} e^{-(t-\tau)} - e^{-2(t-\tau)} \\ e^{-2(t-\tau)} \end{bmatrix} d\tau$$

$$= \begin{bmatrix} -e^{-t} \\ 0 \end{bmatrix} + \begin{bmatrix} e^{-(t-\tau)} - \dfrac{e^{-2(t-\tau)}}{2} \\ \dfrac{e^{-2(t-\tau)}}{2} \end{bmatrix}_0^t$$

$$= \begin{bmatrix} -e^{-t} \\ 0 \end{bmatrix} + \begin{bmatrix} \dfrac{1}{2} - e^{-t} - \dfrac{e^{-2t}}{2} \\ \dfrac{1}{2} - \dfrac{e^{-2t}}{2} \end{bmatrix}$$

$$= \begin{bmatrix} \dfrac{1}{2} - 2e^{-t} - \dfrac{e^{-2t}}{2} \\ \dfrac{1}{2} - \dfrac{e^{-2t}}{2} \end{bmatrix}$$

which means that

$$x_1(t) = \frac{1}{2} - 2e^{-t} - \frac{e^{-2t}}{2}, \quad x_2(t) = \frac{1}{2} - \frac{e^{-2t}}{2}$$

EXAMPLE 17.28. A first order matrix differential equation of a system is given as

$$\begin{bmatrix} \dot{x}_1 \\ \dot{x}_2 \end{bmatrix} = \begin{bmatrix} 0 & 1 \\ -2 & -3 \end{bmatrix} \begin{bmatrix} x_1 \\ x_2 \end{bmatrix} + \begin{bmatrix} 0 \\ 2 \end{bmatrix} u$$

$$y = [1 \quad 1] \begin{bmatrix} x_1 \\ x_2 \end{bmatrix}$$

(a) Find the transfer function of the system.
(b) Find the solution of the states and the output when the input is a unit step input and the initial condition of the states is $\begin{bmatrix} 0 \\ 1 \end{bmatrix}$.

Solution Given

$$\mathbf{A} = \begin{bmatrix} 0 & 1 \\ -2 & -3 \end{bmatrix} \quad \mathbf{B} = \begin{bmatrix} 0 \\ 2 \end{bmatrix} \quad \mathbf{C} = [1 \quad 1]$$

(a)
$$G(s) = \begin{bmatrix}1 & 1\end{bmatrix}[s\mathbf{I}-\mathbf{A}]^{-1}\begin{bmatrix}0\\2\end{bmatrix}$$

$$[s\mathbf{I}-\mathbf{A}] = \begin{bmatrix}s & 0\\0 & s\end{bmatrix} - \begin{bmatrix}0 & 1\\-2 & -3\end{bmatrix} = \begin{bmatrix}s & -1\\2 & s+3\end{bmatrix}$$

$$\det\,[s\mathbf{I}-\mathbf{A}] = s^2+3s+2$$

$$\text{Adj}\,[s\mathbf{I}-\mathbf{A}] = \begin{bmatrix}s+3 & -2\\1 & s\end{bmatrix}^T = \begin{bmatrix}s+3 & 1\\-2 & s\end{bmatrix}$$

$\Rightarrow$
$$[s\mathbf{I}-\mathbf{A}]^{-1} = \frac{1}{s^2+3s+2}\begin{bmatrix}s+3 & 1\\-2 & s\end{bmatrix}$$

Therefore,
$$G(s) = \frac{1}{s^2+3s+2}\begin{bmatrix}1 & 1\end{bmatrix}\begin{bmatrix}s+3 & 1\\-2 & s\end{bmatrix}\begin{bmatrix}0\\2\end{bmatrix}$$
$$= \frac{1}{s^2+3s+2}\begin{bmatrix}1 & 1\end{bmatrix}\begin{bmatrix}2\\2s\end{bmatrix} = \frac{2(s+1)}{s^2+3s+2}$$

(b)
$$\mathbf{\Phi}(s) = [s\mathbf{I}-\mathbf{A}]^{-1} = \frac{\begin{bmatrix}s+3 & 1\\-2 & s\end{bmatrix}}{(s+1)(s+2)}$$
$$= \begin{bmatrix}\dfrac{s+3}{(s+1)(s+2)} & \dfrac{1}{(s+1)(s+2)}\\ \dfrac{-2}{(s+1)(s+2)} & \dfrac{s}{(s+1)(s+2)}\end{bmatrix} = \begin{bmatrix}\dfrac{2}{s+1}-\dfrac{1}{s+2} & \dfrac{1}{s+1}-\dfrac{1}{s+2}\\ \dfrac{2}{s+2}-\dfrac{2}{s+1} & \dfrac{2}{s+2}-\dfrac{1}{s+1}\end{bmatrix}$$

$\Rightarrow$
$$\boldsymbol{\phi}(t) = \begin{bmatrix}2e^{-t}-e^{-2t} & e^{-t}-e^{-2t}\\2e^{-2t}-2e^{-t} & 2e^{-2t}-e^{-t}\end{bmatrix}$$

Therefore,
$$\mathbf{x}(t) = \begin{bmatrix}2e^{-t}-e^{-2t} & e^{-t}-e^{-2t}\\2e^{-2t}-2e^{-t} & 2e^{-2t}-e^{-t}\end{bmatrix}\begin{bmatrix}0\\1\end{bmatrix} + \int_0^t \begin{bmatrix}2e^{-(t-\tau)}-e^{-2(t-\tau)} & e^{-(t-\tau)}-e^{-2(t-\tau)}\\2e^{-2(t-\tau)}-2e^{-(t-\tau)} & 2e^{-2(t-\tau)}-e^{-(t-\tau)}\end{bmatrix}\begin{bmatrix}0\\2\end{bmatrix}d\tau$$
$$= \begin{bmatrix}e^{-t}-e^{-2t}\\2e^{-2t}-e^{-t}\end{bmatrix} + \begin{bmatrix}2\left\{e^{-(t-\tau)}-\dfrac{e^{-2(t-\tau)}}{2}\right\}\\2\left\{\dfrac{2e^{-2(t-\tau)}}{2}-e^{-(t-\tau)}\right\}\end{bmatrix}_0^t$$
$$= \begin{bmatrix}e^{-t}-e^{-2t}\\2e^{-2t}-e^{-t}\end{bmatrix} + \begin{bmatrix}1-2e^{-t}+e^{-2t}\\2e^{-t}-2e^{-2t}\end{bmatrix} = \begin{bmatrix}1-e^{-t}\\e^{-t}\end{bmatrix}$$

EXAMPLE 17.29. (a) Find the state representation corresponding to the following transfer function:

$$\frac{\mathbf{Y}(s)}{\mathbf{U}(s)} = \frac{s+3}{s^2+2s-3}$$

(b) Find the unit step response of the system using state transition matrix if the initial vector is given by

$$\mathbf{x}(0) = \begin{bmatrix} 1 \\ 0 \end{bmatrix}$$

Solution (a) The state representation, as we know, can be found by any of the following method: (i) CCF, (ii) OCF, (iii) cascade decomposition, and (iv) parallel decomposition. We do it by method (i). For this purpose, dividing the numerator and denominator by s^2, the transfer function becomes

$$\frac{\mathbf{Y}(s)}{\mathbf{U}(s)} = \frac{s^{-1} + 3s^{-2}}{1 + 2s^{-1} - 3s^{-2}}$$

The corresponding CCF signal flow graph is given in Figure 17.19.

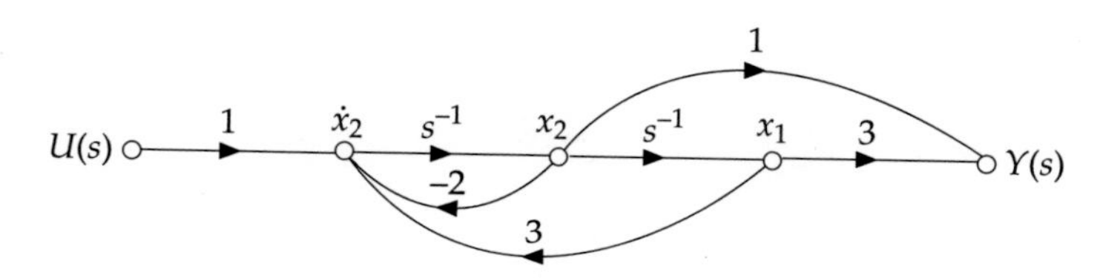

Figure 17.19 Signal flow graph (Example 17.29).

From the signal flow graph,

$$\dot{x}_1 = x_2$$

$$\dot{x}_2 = 3x_1 2x_2 + u$$

$$y = 3x_1 + x_2$$

Therefore, the state representation is

$$\begin{bmatrix} \dot{x}_1 \\ \dot{x}_2 \end{bmatrix} = \begin{bmatrix} 0 & 1 \\ 3 & -2 \end{bmatrix} + \begin{bmatrix} 0 \\ 1 \end{bmatrix} u$$

$$y = \begin{bmatrix} 3 & 1 \end{bmatrix} \begin{bmatrix} x_1 \\ x_2 \end{bmatrix}$$

(b)

$$[s\mathbf{I} - \mathbf{A}] = \begin{bmatrix} s & 0 \\ 0 & s \end{bmatrix} - \begin{bmatrix} 0 & 1 \\ 3 & -2 \end{bmatrix} = \begin{bmatrix} s & -1 \\ -3 & s+2 \end{bmatrix}$$

$$\text{Adj } [s\mathbf{I} - \mathbf{A}] = \begin{bmatrix} s+2 & 3 \\ 1 & s \end{bmatrix}^{\mathrm{T}} = \begin{bmatrix} s+2 & 1 \\ 3 & s \end{bmatrix}$$

$$|s\mathbf{I} - \mathbf{A}| = s^2 + 2s - 3 = (s-1)(s+3)$$

Therefore,

$$\boldsymbol{\Phi}(s) = [s\mathbf{I} - \mathbf{A}]^{-1} = \frac{\begin{bmatrix} s+2 & 1 \\ 3 & s \end{bmatrix}}{(s-1)(s+3)} = \begin{bmatrix} \dfrac{s+2}{(s-1)(s+3)} & \dfrac{1}{(s-1)(s+3)} \\ \dfrac{3}{(s-1)(s+3)} & \dfrac{s}{(s-1)(s+3)} \end{bmatrix}$$

$$= \begin{bmatrix} \dfrac{3}{4}\cdot\dfrac{1}{s-1} + \dfrac{1}{4}\cdot\dfrac{1}{s+3} & \dfrac{1}{4}\cdot\dfrac{1}{s-1} - \dfrac{1}{4}\cdot\dfrac{1}{s+3} \\ \dfrac{3}{4}\cdot\dfrac{1}{s-1} + \dfrac{3}{4}\cdot\dfrac{1}{s+3} & \dfrac{1}{4}\cdot\dfrac{1}{s-1} + \dfrac{3}{4}\cdot\dfrac{1}{s+3} \end{bmatrix}$$

$$\boldsymbol{\phi}(t) = \begin{bmatrix} \frac{3}{4}e^{t} + \frac{1}{4}e^{-3t} & \frac{1}{4}e^{t} - \frac{1}{4}e^{-3t} \\ \frac{3}{4}e^{t} - \frac{3}{4}e^{-3t} & \frac{1}{4}e^{t} + \frac{3}{4}e^{-3t} \end{bmatrix}$$

$$= \frac{1}{4}\begin{bmatrix} 3e^{t} + e^{-3t} & e^{t} - e^{-3t} \\ 3e^{t} - 3e^{-3t} & e^{t} + 3e^{-3t} \end{bmatrix}$$

Alternatively

The STM can be worked out by the $\boldsymbol{\Lambda}$-matrix method as follows. Here,

$$\boldsymbol{\Lambda} = \begin{bmatrix} 0 & 1 \\ 3 & -2 \end{bmatrix}$$

The eigenvalues of the matrix can be found from the equation

$$|\lambda\mathbf{I} - \mathbf{A}| = 0$$

$$\Rightarrow \qquad \begin{bmatrix} \lambda & -1 \\ -3 & \lambda+2 \end{bmatrix} = 0$$

$$\Rightarrow \qquad \lambda^2 + 2\lambda - 3 = 0$$

Therefore, the eigenvalues are 1 and -3. Since the matrix $\mathbf{A}$ is of the CCF-type, the modal matrix is the Vandermonde matrix given by

$$\mathbf{M} = \begin{bmatrix} 1 & 1 \\ \lambda_1 & \lambda_2 \end{bmatrix} = \begin{bmatrix} 1 & 1 \\ 1 & -3 \end{bmatrix}$$

Its inverse is

$$\mathbf{M}^{-1} = \frac{\text{Adj}\,[\mathbf{M}]}{|\mathbf{M}|} = \frac{\begin{bmatrix} -3 & -1 \\ -1 & 1 \end{bmatrix}^{\mathrm{T}}}{-3-1} = \frac{1}{4}\begin{bmatrix} 3 & 1 \\ 1 & -1 \end{bmatrix}$$

The diagonal $\boldsymbol{\Lambda}$-matrix is

$$\boldsymbol{\Lambda} = \begin{bmatrix} \lambda_1 & 0 \\ 0 & \lambda_2 \end{bmatrix} = \begin{bmatrix} 1 & 0 \\ 0 & -3 \end{bmatrix}$$

Hence,

$$e^{\mathbf{\Lambda}t} = \begin{bmatrix} e^t & 0 \\ 0 & e^{-3t} \end{bmatrix}$$

Therefore, the STM is

$$\begin{aligned}\boldsymbol{\phi}(t) = \mathbf{M}e^{\mathbf{\Lambda}t}\mathbf{M}^{-1} &= \begin{bmatrix} 1 & 1 \\ 1 & -3 \end{bmatrix}\begin{bmatrix} e^t & 0 \\ 0 & e^{-3t} \end{bmatrix}\frac{1}{4}\begin{bmatrix} 3 & 1 \\ 1 & -1 \end{bmatrix} \\ &= \frac{1}{4}\begin{bmatrix} 1 & 1 \\ 1 & -3 \end{bmatrix}\begin{bmatrix} 3e^t & e^t \\ e^{-3t} & -e^{-3t} \end{bmatrix} \\ &= \frac{1}{4}\begin{bmatrix} 3e^t + e^{-3t} & e^t - e^{-3t} \\ 3e^t - 3e^{-3t} & e^t + 3e^{-3t} \end{bmatrix}\end{aligned}$$

Thus, the unit step response is obtained as follows:

$$\begin{aligned}\mathbf{x}(t) &= \frac{1}{4}\begin{bmatrix} 3e^t + e^{-3t} & e^t - e^{-3t} \\ 3e^t - 3e^{-3t} & e^t + 3e^{-3t} \end{bmatrix}\begin{bmatrix} 1 \\ 0 \end{bmatrix} + \int_0^t \frac{1}{4}\begin{bmatrix} 3e^{(t-\tau)} + e^{-3(t-\tau)} & e^{(t-\tau)} - e^{-3(t-\tau)} \\ 3e^{(t-\tau)} - 3e^{-3(t-\tau)} & e^{(t-\tau)} + 3e^{-3(t-\tau)} \end{bmatrix}\begin{bmatrix} 0 \\ 1 \end{bmatrix} d\tau \\ &= \frac{1}{4}\begin{bmatrix} 3e^t + e^{-3t} \\ 3e^t - 3e^{-3t} \end{bmatrix} + \frac{1}{4}\int_0^t \begin{bmatrix} e^{(t-\tau)} - e^{-3(t-\tau)} \\ e^{(t-\tau)} + 3e^{-3(t-\tau)} \end{bmatrix} \\ &= \frac{1}{4}\begin{bmatrix} 3e^t + e^{-3t} \\ 3e^t - 3e^{-3t} \end{bmatrix} + \frac{1}{4}\begin{bmatrix} e^{(t-\tau)} - \dfrac{e^{-3(t-\tau)}}{3} \\ e^{(t-\tau)} + e^{-3(t-\tau)} \end{bmatrix}_0^t \\ &= \frac{1}{4}\begin{bmatrix} 3e^t + e^{-3t} \\ 3e^t - 3e^{-3t} \end{bmatrix} + \frac{1}{4}\begin{bmatrix} \frac{2}{3} - e^t + \frac{1}{3}e^{-3t} \\ 2 - e^t - e^{-3t} \end{bmatrix} \\ &= \frac{1}{4}\begin{bmatrix} \frac{2}{3} + 2e^t + \frac{4}{3}e^{-3t} \\ 2 + 2e^t - 4e^{-3t} \end{bmatrix} \\ &= \begin{bmatrix} \frac{1}{6} - \frac{1}{2}e^t + \frac{1}{3}e^{-3t} \\ \frac{1}{2} + \frac{1}{2}e^t - e^{-3t} \end{bmatrix}\end{aligned}$$

17.9 CONTROLLABILITY AND OBSERVABILITY

A system is said to be controllable if there exists some finite control vector $\mathbf{u}$ that will bring the system to any specified state $\mathbf{x}$ in the state space within a finite length of time $t = t_f$ from any state $\mathbf{x}(0)$ at $t = 0$.

Observability is the dual of controllability. Thus, the system is said to be observable at any time $t_f > 0$, a knowledge of the output $\mathbf{y}$, and the system matrices enables us to determine the initial state $\mathbf{x}(0)$.

According to these concepts, any system can be divided into the following four parts (see Figure 17.20):

A : Completely controllable and observable
B : Controllable but unobservable
C : Uncontrollable but observable
D : Uncontrollable and unobservable

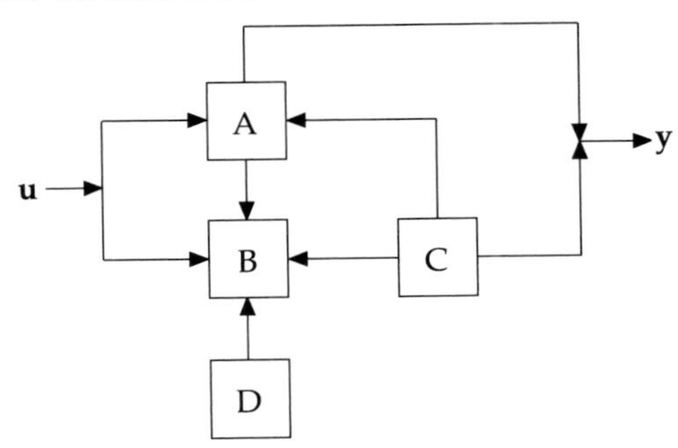

Figure 17.20 Four possible parts of a system.

Controllability Criterion

If the system is characterized by

$$\dot{\mathbf{x}} = \mathbf{A}\mathbf{x} + \mathbf{B}\mathbf{u}$$

then it is controllable *iff* the $n \times n$ matrix

$$\mathbf{G} = \left[\mathbf{B} : \mathbf{A}\mathbf{B} : \mathbf{A}^2\mathbf{B} : \cdots : \mathbf{A}^{n-1}\mathbf{B}\right] \tag{17.45}$$

is of rank[10] n.

Observability Criterion

If the system is characterized by dynamical equations such as

$$\dot{\mathbf{x}} = \mathbf{A}\mathbf{x} + \mathbf{B}\mathbf{u}$$

$$\mathbf{y} = \mathbf{C}\mathbf{x} + \mathbf{D}\mathbf{u}$$

then it is observable *iff* the $n \times n$ matrix

$$\mathbf{H} = \left[\mathbf{C}^{\mathrm{T}} : \mathbf{A}^{\mathrm{T}}\mathbf{C}^{\mathrm{T}} : \left(\mathbf{A}^{\mathrm{T}}\right)^2 \mathbf{C}^{\mathrm{T}} : \cdots : \left(\mathbf{A}^{\mathrm{T}}\right)^{n-1} \mathbf{C}^{\mathrm{T}}\right] \tag{17.46}$$

is of rank n.

EXAMPLE 17.30. Find the controllability and observability of the system

$$\begin{bmatrix} \dot{x}_1 \\ \dot{x}_2 \end{bmatrix} = \begin{bmatrix} -2 & 1 \\ 0 & -3 \end{bmatrix} \begin{bmatrix} x_1 \\ x_2 \end{bmatrix} + \begin{bmatrix} 4 \\ 1 \end{bmatrix} u$$

$$y = \begin{bmatrix} 1 & 0 \end{bmatrix} \begin{bmatrix} x_1 \\ x_2 \end{bmatrix}$$

[10]For definition, see Section 1.3 on page 24.

Solution Here,

$$\mathbf{A} = \begin{bmatrix} -2 & 1 \\ 0 & -3 \end{bmatrix} \qquad \mathbf{B} = \begin{bmatrix} 4 \\ 1 \end{bmatrix} \qquad \mathbf{C} = \begin{bmatrix} 1 & 0 \end{bmatrix}$$

Therefore,

$$\mathbf{A}^{\mathrm{T}} = \begin{bmatrix} -2 & 0 \\ 1 & -3 \end{bmatrix} \qquad \mathbf{C}^{\mathrm{T}} = \begin{bmatrix} 1 \\ 0 \end{bmatrix}$$

To examine the controllability of the system, we find that

$$\mathbf{AB} = \begin{bmatrix} -2 & 1 \\ 0 & -3 \end{bmatrix} \begin{bmatrix} 4 \\ 1 \end{bmatrix} = \begin{bmatrix} -8+1 \\ 0+3 \end{bmatrix} = \begin{bmatrix} -7 \\ 3 \end{bmatrix}$$

We need not calculate other terms of $\mathbf{G}$ of Eq. (17.45) which will be a 2×2 matrix. So,

$$\mathbf{G} = [\mathbf{B} : \mathbf{AB}] \begin{bmatrix} 4 & -7 \\ 1 & 3 \end{bmatrix}$$

$$\det \mathbf{G} = \begin{vmatrix} 4 & -7 \\ 1 & 3 \end{vmatrix} = 19$$

Since $\mathbf{G}$ is non-singular and therefore of rank 2, the system is controllable.

To examine the observability of the system, we observe that

$$\mathbf{A}^{\mathrm{T}}\mathbf{C}^{\mathrm{T}} = \begin{bmatrix} -2 & 0 \\ 1 & -3 \end{bmatrix} \begin{bmatrix} 1 \\ 0 \end{bmatrix} = \begin{bmatrix} -2 \\ 1 \end{bmatrix}$$

We need not calculate further terms of $\mathbf{H}$ of Eq. (17.46) which will be a 2×2 matrix. Thus,

$$\mathbf{H} = \left[\mathbf{C}^{\mathrm{T}} : \mathbf{A}^{\mathrm{T}}\mathbf{C}^{\mathrm{T}}\right] = \begin{bmatrix} 1 & -2 \\ 0 & 1 \end{bmatrix}$$

$$\det \mathbf{H} = \begin{vmatrix} 1 & -2 \\ 0 & 1 \end{vmatrix} = 3$$

Since $\mathbf{H}$ is non-singular, it is of rank 2. Therefore, the system is observable.

REVIEW QUESTIONS

17.1 Choose the correct answer:

(i) A state variable approach converts an nth order system into
- (a) n second order differential equations
- (b) two differential equations
- (c) n first order differential equations
- (d) a lower order system

(ii) The matrix $\mathbf{A}$ in $\dot{\mathbf{x}} = \mathbf{A}\mathbf{x}$ for the system described by the differential equation $\ddot{y} + 2\dot{y} + 3y = 0$ is

(a) $\begin{bmatrix} 1 & 0 \\ -1 & -2 \end{bmatrix}$ (b) $\begin{bmatrix} 1 & 0 \\ -2 & -1 \end{bmatrix}$

(c) $\begin{bmatrix} 0 & 1 \\ -3 & -2 \end{bmatrix}$ (d) $\begin{bmatrix} 0 & 1 \\ -2 & 1 \end{bmatrix}$

(iii) If $\mathbf{\Lambda} = \begin{bmatrix} 1 & 0 \\ 0 & 2 \end{bmatrix}$, then $e^{\mathbf{\Lambda}t}$ will be

(a) $\begin{bmatrix} e^t & 0 \\ 0 & e^{2t} \end{bmatrix}$ (b) $\begin{bmatrix} e^t & 0 \\ 0 & 2e^t \end{bmatrix}$

(c) $\begin{bmatrix} 0 & e^t \\ e^{2t} & 0 \end{bmatrix}$ (d) $\begin{bmatrix} e^{2t} & 0 \\ 0 & e^t \end{bmatrix}$

(iv) If $\mathbf{A} = \begin{bmatrix} 0 & 1 \\ -2 & 3 \end{bmatrix}$, then the eigenvalues of $\mathbf{A}$ are

(a) 0 and 3 (b) 1 and -2

(c) 1 and 2 (d) 1 and 4

(v) $\mathbf{A}$ is an $n \times n$ matrix. For the system to be controllable, the rank of the controllability matrix should be

(a) n (b) $> n$ (c) $\geq n$ (d) $\leq n$

17.2 A system is described by the state-space equation

$$\begin{bmatrix} \dot{x}_1 \\ \dot{x}_2 \end{bmatrix} = \begin{bmatrix} 0 & 1 \\ 0 & 0 \end{bmatrix} \begin{bmatrix} x_1 & x_2 \end{bmatrix} + \begin{bmatrix} 0 \\ 1 \end{bmatrix} \mathbf{u}$$

$$\mathbf{y} = \begin{bmatrix} 1 & 0 \end{bmatrix} \begin{bmatrix} x_1 & x_2 \end{bmatrix}$$

Find out the transfer function of the system.

17.3 (a) Derive the expression for the transfer function from the state model

$$\dot{\mathbf{x}} = \mathbf{A}\mathbf{x} + \mathbf{B}\mathbf{u}$$

$$\mathbf{y} = \mathbf{C}\mathbf{x} + \mathbf{D}\mathbf{u}$$

(b) For the electrical network shown in the following figure, determine the state model. Consider i_1, i_2 and V_c as state variables. The output variables are i_1 and i_2.

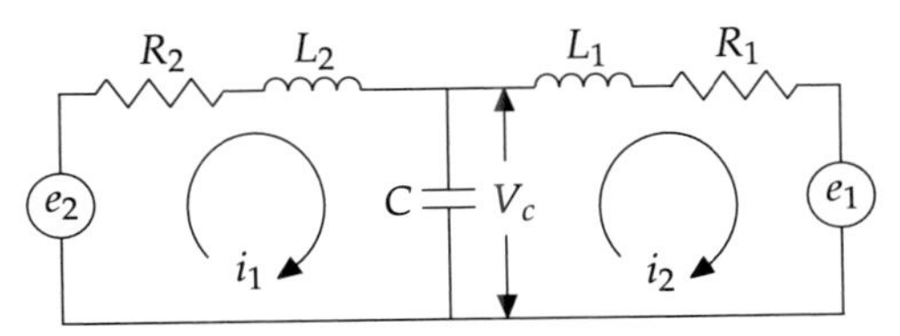

17.4 (a) The state equation of a control system is given by

$$\begin{bmatrix} \dot{x}_1 \\ \dot{x}_2 \end{bmatrix} = \begin{bmatrix} 0 & 1 \\ -2 & -3 \end{bmatrix} \begin{bmatrix} x_1 \\ x_2 \end{bmatrix}$$

Obtain the state transition matrix.

(b) Obtain the eigenvalues and eigenvectors for a system described by

$$\dot{\mathbf{x}} = \begin{bmatrix} 0 & 1 & 0 \\ 3 & 0 & 2 \\ -12 & -7 & -6 \end{bmatrix} \begin{bmatrix} x_1 \\ x_2 \\ x_3 \end{bmatrix} + \begin{bmatrix} 1 \\ 0 \\ 2 \end{bmatrix}$$

$$\mathbf{y} = \begin{bmatrix} 1 & 0 & 0 \end{bmatrix} \mathbf{x}$$

17.5 Obtain the state equation in matrix form for the network shown in the following figure:

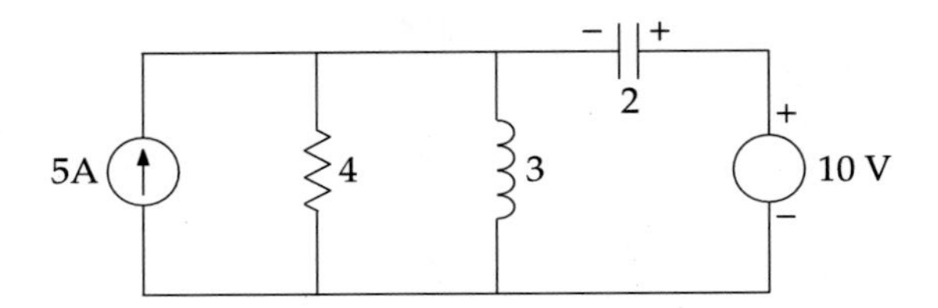

17.6 (a) Define state variable and explain its importance and use in the mathematical modelling of a system.

(b) Define state transition matrix and state its properties.

(c) Solve the following state equation:

$$\begin{bmatrix} \dfrac{d}{dt}x_1(t) \\ \dfrac{d}{dt}x_2(t) \end{bmatrix} = \begin{bmatrix} 0 & 1 \\ -2 & -3 \end{bmatrix} \begin{bmatrix} x_1(t) \\ x_2(t) \end{bmatrix} + \begin{bmatrix} 0 \\ 1 \end{bmatrix} r(t)$$

Assume $r(t) = 1$ for $t > 0$ and $r(t) = 0$ for $t = 0$.

17.7 (a) Find out a state-space model for the electrical network shown in the following figure. The system input is the voltage source $v_i(t)$ and the output is $v_o(t)$, the voltage across the capacitor C.

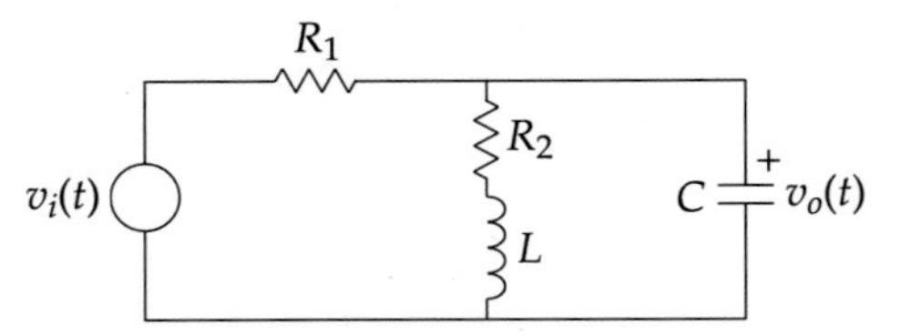

(b) Assuming zero initial condition, derive the transfer function $V_o(s)/V_i(s)$ from the state-space model.

17.8 A linear time-invariant system is described by the following state model:

$$\frac{dx_2}{dt} = -x_1 - x_2 + u$$

$$\frac{dx_1}{dt} = x_2$$

and $$y = x_1$$

Here, $u(t)$ is the system input, $y(t)$ is the system output, and $[x_1(t)x_2(t)]^T$ is the state vector.

(a) If $u(t)$ is a unit step, calculate the rise time, peak time, per cent peak overshoot, and settling time (2% band) for the system.

(b) Write down the sinusoidal transfer function of the system and calculate the resonant frequency, resonant peak, and −3 dB bandwidth.

17.9 The following figure shows the signal flow graph of a system.

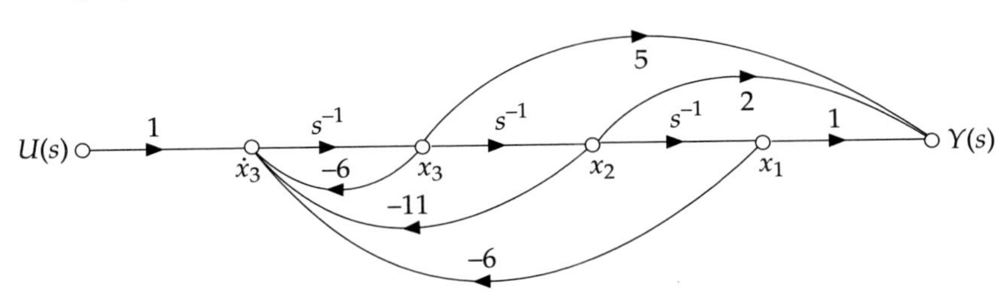

(a) Identify the forward paths and loops, and calculate the corresponding path gains and loop gains.

(b) Using Mason's gain formula, obtain the transfer function $X_6(s)/X_1(s)$.

(c) If one of the poles of the system is at $s = -1$, obtain a state-space model of the system using canonical variables.

Answers to Selected Questions

17.1. (i) (b); (ii) (c); (iii) (a); (iv) (c); (v) (a)

17.2 $G(s) = \dfrac{1}{s^2}$

17.3. (b) $$\begin{bmatrix} \dot{i}_1 \\ \dot{i}_2 \\ \dot{V}_c \end{bmatrix} = \begin{bmatrix} -\dfrac{R_2}{L_2} & 0 & -\dfrac{1}{L_2} \\ 0 & -\dfrac{R_1}{L_1} & -\dfrac{1}{L_1} \\ -\dfrac{1}{C} & -\dfrac{1}{C} & 0 \end{bmatrix} \begin{bmatrix} i_1 \\ i_2 \\ V_c \end{bmatrix} + \begin{bmatrix} 0 & -\dfrac{1}{L_2} \\ -\dfrac{1}{L_1} & 0 \\ 0 & 0 \end{bmatrix}, \quad y = \begin{bmatrix} 1 & 0 & 0 \\ 0 & 1 & 0 \\ 0 & 0 & 0 \end{bmatrix} \begin{bmatrix} i_1 \\ i_2 \\ V_c \end{bmatrix}$$

17.4. (a) $\boldsymbol{\Phi}(t) = \begin{bmatrix} 2e^{-t} - e^{-2t} & e^{-t} - e^{-2t} \\ 2e^{-2t} - 2e^{-t} & 2e^{-2t} - e^{-t} \end{bmatrix}$

(b) For $\lambda_1 = -1$, $\mathbf{p}_1 = \begin{bmatrix} 1 \\ -1 \\ -1 \end{bmatrix}$; $\lambda_2 = -2$, $\mathbf{p}_2 = \begin{bmatrix} 2 \\ -4 \\ 1 \end{bmatrix}$; $\lambda_3 = -3$, $\mathbf{p}_3 = \begin{bmatrix} 1 \\ -3 \\ 3 \end{bmatrix}$

17.5. $\begin{bmatrix} \dot{i}_L \\ \dot{V}_c \end{bmatrix} = \begin{bmatrix} 0 & -\frac{1}{3} \\ -\frac{1}{2} & 0 \end{bmatrix} \begin{bmatrix} i_L \\ V_c \end{bmatrix} + \begin{bmatrix} -\frac{10}{3} \\ \frac{5}{2} \end{bmatrix} u(t)$

17.6. (c) $x(t) = \begin{bmatrix} -\frac{1}{2} + e^t - \frac{e^{2t}}{2} \\ e^t - e^{2t} \end{bmatrix}$ assuming $\mathbf{x}(0) = \begin{bmatrix} 0 \\ 0 \end{bmatrix}$

17.7. (a) $\begin{bmatrix} \dot{i}_L \\ \dot{v}_o \end{bmatrix} = \begin{bmatrix} -\frac{R_2}{L} & \frac{1}{L} \\ -\frac{1}{C} & -\frac{1}{R_1 C} \end{bmatrix} \begin{bmatrix} i_L \\ v_a \end{bmatrix} + \begin{bmatrix} -\frac{2}{L} \\ \frac{1}{R_2 C} \end{bmatrix} v_i \qquad y = \begin{bmatrix} 0 & 1 \end{bmatrix} \begin{bmatrix} i_L \\ v_a \end{bmatrix}$

(b) $G(s) = \dfrac{sL + 2R_1 + R_2}{s^2 R - 1LC + s\big(R_1 R_2 C + L\big) + R_1 + R_2}$

17.8. (a) $t_r = 2.42$ s, $t_p = 3.63$ s, $M_p = 12.31\%$, $t_s = 8$ s

(b) $G(\jmath\omega) = \dfrac{1}{-\omega^2 + \jmath\omega + 1}$, $\omega_r = 0.707$ rad/s, $M_r = 1.15$, Bw $= 1.27$ rad/s

17.9. (a) Forward paths and gains: $p_1 = s^{-3}$, $p_2 = 2s^{-2}$, $p_3 = 5s^{-1}$

Loops and gains: $L_1 = -6s^{-1}$, $L_2 = -11s^{-2}$, $L_3 = 6s^{-3}$

(b) $\dfrac{Y(s)}{U(s)} = \dfrac{5s^2 + 2s + 1}{s^3 + 6s^2 + 11s + 6}$

(c) $\begin{bmatrix} \dot{x}_1 \\ \dot{x}_2 \\ \dot{x}_3 \end{bmatrix} = \begin{bmatrix} 0 & 1 & 0 \\ 0 & 0 & 1 \\ -6 & -11 & -6 \end{bmatrix} \begin{bmatrix} x_1 \\ x_2 \\ x_3 \end{bmatrix} + \begin{bmatrix} 0 \\ 0 \\ 1 \end{bmatrix} u$

$$y = \begin{bmatrix} 1 & 2 & 5 \end{bmatrix} \begin{bmatrix} x_1 \\ x_2 \\ x_3 \end{bmatrix}$$

CHAPTER 18

Discrete Time Systems

18.1 INTRODUCTION

So far we have considered analogue systems with continuous control signals. With the advent of digital computers and a very powerful digital electronics, control signals are sent or the feedback is received in the form of discrete pulses. Systems incorporating digital monitoring and control are collectively called *discrete time systems.*

The numbers that enter a digital control system do so at some fixed time interval, T, called the *sampling period.* Consider the typical digital control system presented in Figure 18.1.

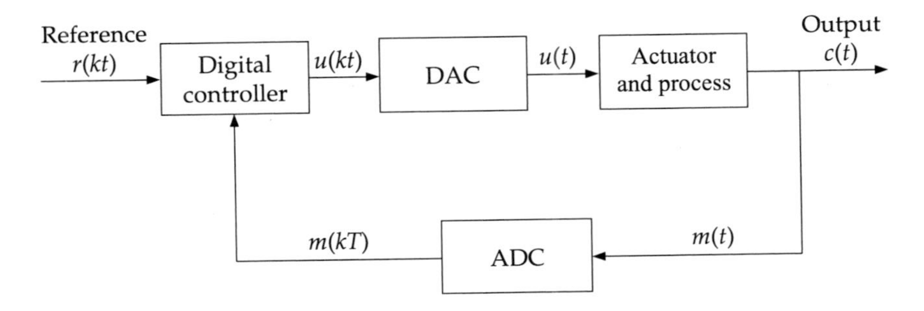

Figure 18.1 Typical digital control system.

Here, the reference input $r(kT)$ is a train of sampled values. The variables $r(kT)$, $u(kT)$ and $m(kT)$ are discrete signals in contrast to continuous signals $u(t)$, $c(t)$ and $m(t)$.

The sampling is done by a sampler which is basically a switch that becomes ON for an instant every T seconds (see Figure 18.2). So it can be thought of as producing a train of impulse functions $\delta(t)$.

When we sample a signal $r(t)$ to produce a sampled signal $r^*(t)$, we actually produce a train of impulses of amplitude $r(kT)$, starting at $t = 0$ and spaced T seconds apart. In Section 1.5 at page 50 we have discussed a mathematical tool, called the *z-transform*, to deal with such signals.

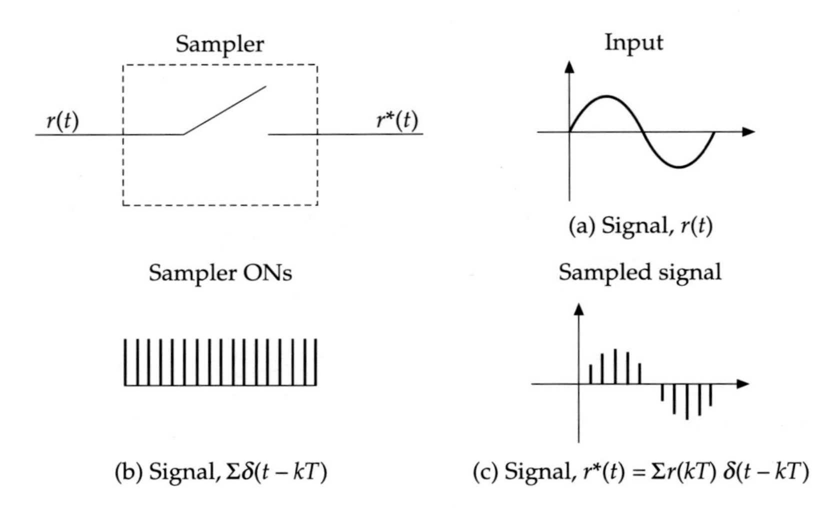

Figure 18.2 The sampling process.

18.2 TRANSFER FUNCTIONS OF DISCRETE DATA SYSTEMS

The discrete data systems can be divided into the following five categories:

1. Simple digital system
2. Cascaded elements separated by a sampler
3. Cascaded elements not separated by a sampler
4. Zero order hold circuit
5. Closed-loop discrete data

We consider them in that order.

Simple Digital System

The input and output of a simple digital system consists of a train of signals as shown in Figure 18.3.

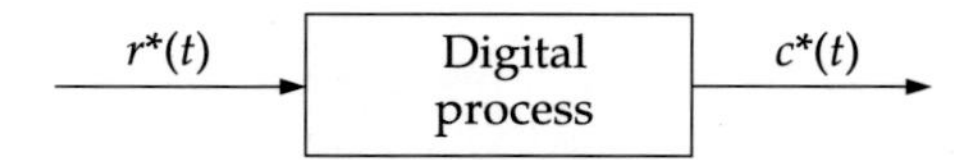

Figure 18.3 Simple digital system.

The Laplace transformed equation of such a system can be written as

$$C^*(s) = G^*(s)\, R^*(s)$$

Its z-transformed form is

$$C(z) = G(z)\, R(z)$$

Cascaded Elements Separated by a Sampler

A digital system having cascaded linear elements separated by a sampler is shown in Figure 18.4.

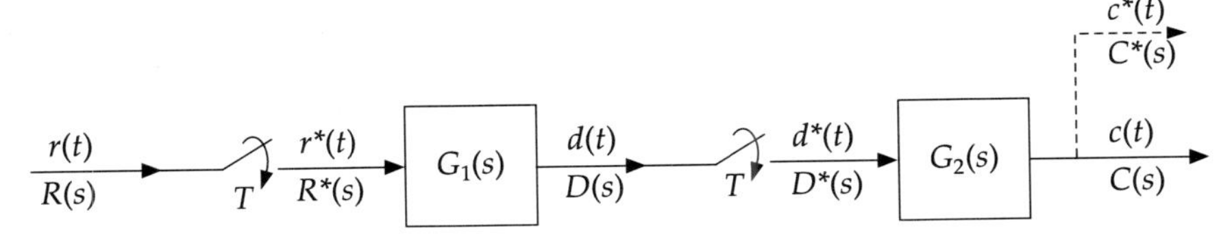

Figure 18.4 Cascaded systems separated by a sampler.

For this system we have the following equations:

$$D(s) = G_1(s)R^*(s)$$
$$D^*(s) = G_1^*(s)R^*(s)$$
$$C(s) = G_2(s)D^*(s) = G_2(s)G_1^*(s)R^*(s)$$

Therefore,

$$C^*(s) = G_2^*(s)G_1^*(s)R^*(s) \tag{18.1}$$

Equation (18.1) leads to

$$C(z) = G_2(z)G_1(z)R(z) \tag{18.2}$$

The import of Eq. (18.2) is that the z-transform of two linear elements separated by a sampler equals the product of z-transforms of two individual transfer functions.

Cascaded Elements Not Separated by a Sampler

The situation is as shown in Figure 18.5.

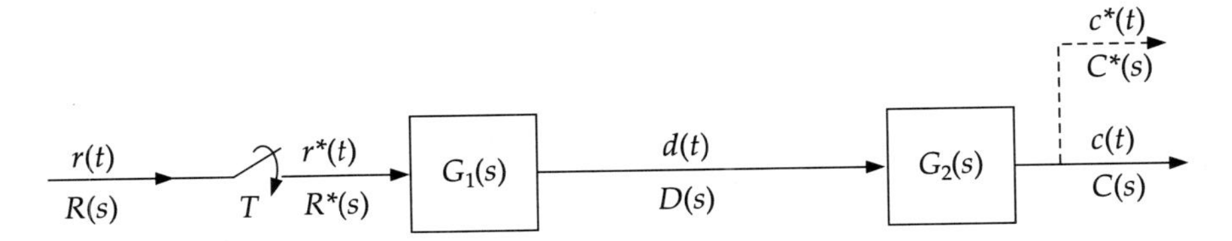

Figure 18.5 Cascaded systems not separated by a sampler.

In this case the relevant equations are

$$C(s) = G_1(s)G_2(s)R^*(s)$$
$$C^*(s) = [G_1(s)G_2(s)]^*R^*(s) \tag{18.3}$$

Let us rewrite Eq. (18.3) in a slightly different way to emphasize that the functions within the square brackets have to be multiplied first before they are sampled. Then,

$$C^*(s) = [G_1G_2]^*(s)R^*(s)$$

which leads to

$$C(z) = [G_1G_2](z)R(z)$$

where $[G_1G_2](z)$ is the z-transform of the product of $G_1(s)$ and $G_2(s)$. We should keep in mind that

$$\left[G_1G_2\right]^*(s) \neq G_1^*(s)\, G_2^*(s)$$

$$\left[G_1G_2\right](z) \neq G_1(z)\, G_2(z)$$

Zero Order Hold Circuit

A zero order hold circuit (often called ZOH) holds the input signal from a switch for a duration T, thus giving the output signal the shape of a unit step function of duration T (see Figure 18.6).

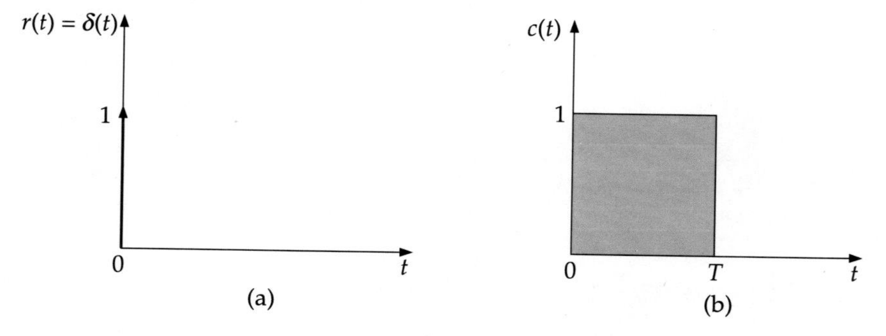

Figure 18.6 (a) Input; (b) output of a ZOH.

In fact, it acts as an interface or filter which converts impulses into an analogue signal. Its action can be visualized from Figure 18.7.

The transfer function of such a circuit can be obtained as follows. We know that

$$u(t) = \begin{cases} 1 & t \geq 0 \\ 0 & t < 0 \end{cases}$$

$$u(t-T) = \begin{cases} 1 & (t-T) \geq 0 \\ 0 & (t-T) < 0 \end{cases}$$

Since the output lasts only for T seconds, it can be represented as

$$c(t) = u(t) - u(t-T) \tag{18.4}$$

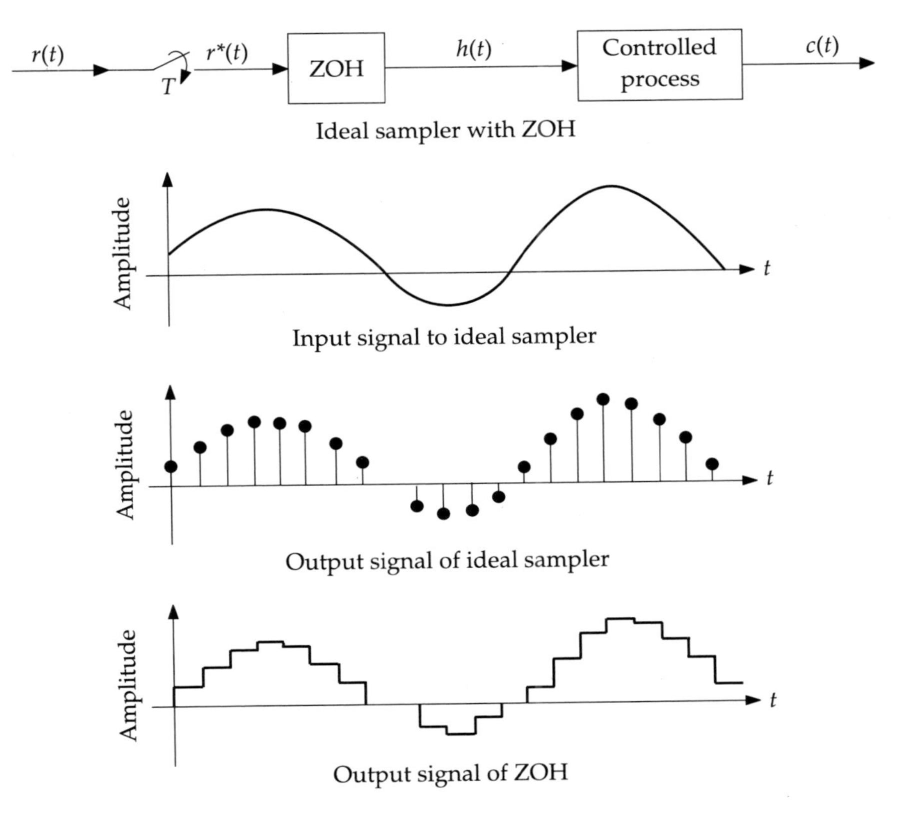

Figure 18.7 Action of a ZOH.

Therefore, the Laplace transform of Eq. (18.4) is

$$\begin{aligned}\mathscr{L}\{c(t)\} \equiv C(s) &= \mathscr{L}\{u(t)\} - \mathscr{L}\{u(t-T)\} \\ &= \frac{1}{s} - \frac{e^{-sT}}{s} \qquad \text{[by time shifting[1] property]}\end{aligned} \tag{18.5}$$

The input $r(t)$ is an impulse function and its Laplace transform is 1. Therefore,

$$\text{Transfer function } G(s) = \frac{C(s)}{R(s)} = \frac{1 - e^{-sT}}{s}$$

Closed-Loop Discrete Data

A closed-loop digital control system contains a switching system over and above the other components of an analogue control system. Such a system is shown in Figure 18.8.

[1]See Section 1.2 at page 5.

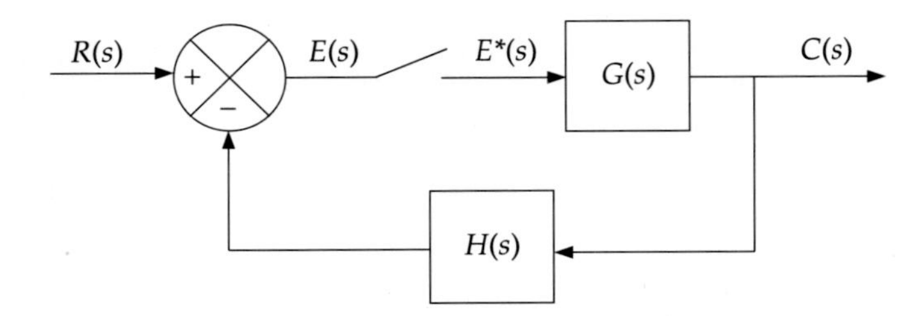

Figure 18.8 A closed-loop digital control system.

In a non-sampled system,

$$E(s) = R(s) - C(s)H(s)$$

$$C(s) = E(s)G(s)$$

Therefore,

$$E(s) = R(s) - E(s)G(s)H(s) \tag{18.6}$$

But, in a discrete time system, $E(s)$ is being sampled which means that all signals passing through the switch are being sampled. Thus, for a discrete time system, we can write Eq. (18.6) as

$$E^*(s) = R^*(s) - E^*(s)G^*(s)H^*(s)$$

or

$$E^*(s)\left[1 + G^*(s)H^*(s)\right] = R^*(s) \tag{18.7}$$

Also,

$$C^*(s) = E^*(s)G^*(s) \tag{18.8}$$

Combining the results of Eqs. (18.7) and (18.8), we get

$$\frac{C^*(s)}{R^*(s)} = \frac{G^*(s)}{1 + G^*(s)H^*(s)}$$

In terms of z-transform,

$$\frac{C(z)}{R(z)} = \frac{G(z)}{1 + G(z)H(z)} \tag{18.9}$$

18.3 STABILITY OF DIGITAL CONTROL SYSTEMS

From Eq. (18.9) we observe that the characteristic equation of a digital control system is

$$1 + G(z)H(z) = 0 \tag{18.10}$$

The roots of Eq. (18.10) are the poles of the discrete time system.

We know that in an analogue system, the characteristic equation is given by

$$1 + G(s)H(s) = 0$$

and we have seen that the poles of such a system must lie in the LHP of the s-plane to make the system stable. The discrete time system maps with the analogue system through the relation

$$z = e^{sT}$$

For a complex analysis, as usual, we put $s = \jmath\omega$ and see

$$\begin{aligned} z = e^{\jmath\omega T} &= \cos\omega T + \jmath\sin\omega T \\ &\equiv \sqrt{\cos^2\omega T + \sin^2\omega T}\angle\tan^{-1}\frac{\sin\omega t}{\cos\omega T} \\ &= \sqrt{\cos^2\omega T + \sin^2\omega T}\angle\tan^{-1}\tan\omega T \end{aligned}$$

$\Rightarrow$

$$z = |1|\angle\omega T$$

Thus, the LHP of the s-plane corresponds to a unit circle in the z-plane (see Figure 18.9).

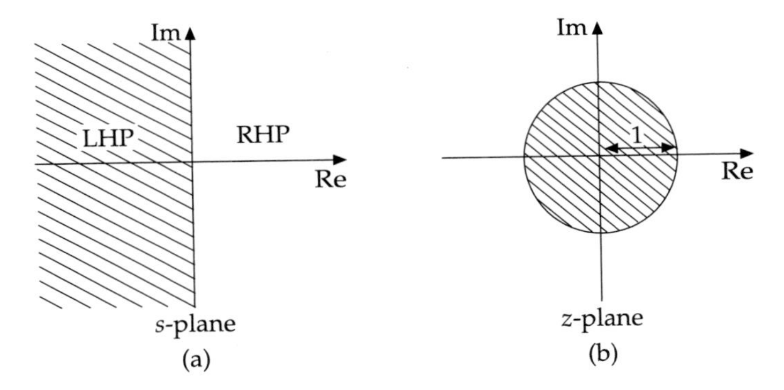

Figure 18.9 Correspondence of (a) LHP of s-plane; (b) unit circle of z-plane.

The foregoing discussion can be summarized as follows:

1. The $\jmath\omega$-axis in the s-plane maps to the unit circle in the z-plane.
2. The right-half of the s-plane maps to outside of the unit circle in the z-plane.
3. The left-half of the s-plane maps to inside of the unit circle in the z-plane.

Therefore, a discrete time system is stable if the poles of the z-plane closed-loop transfer function lie inside a unit circle.

With this background of the z-transform, let us work out a few problems.

EXAMPLE 18.1. The transfer function of a system is given by

$$G(s) = \frac{10}{s(s+1)(s+3)}$$

If the system is sampled, find

(a) the z-transfer function of the system, and
(b) check its stability (T = sampling time).

Solution (a) Let

$$G(s) \equiv \frac{A}{s} + \frac{B}{s+1} + \frac{C}{s+3}$$

Then,

$$A = sG(s)|_{s=0} = \left.\frac{10}{(s+1)(s+3)}\right|_{s=0} = 3.33$$

$$B = (s+1)G(s)|_{s=-1} = \left.\frac{10}{s(s+3)}\right|_{s=-1} = -5$$

$$C = (s+3)G(s)|_{s=-3} = \left.\frac{10}{s(s+1)}\right|_{s=-3} = 1.67$$

Thus,

$$G(s) = \frac{3.33}{s} - \frac{5}{s+1} + \frac{1.67}{s+3}$$

Therefore,

$$G(z) = \frac{3.33z}{z-1} - \frac{5z}{z-e^{-T}} + \frac{1.67z}{z-e^{-3T}} \tag{i}$$

(b) It is apparent from Eq. (i) that the poles of $G(z)$ are at 1, e^{-T} and e^{-3T}. e^{-T} and $e^{-3T} < 1$ for $T > 0$. Therefore, they lie within the $|z| = 1$ circle. But the pole at 1 touches the unity circle. Thus, the system is marginally stable.

EXAMPLE 18.2. Find the transfer function for the arrangement given in Figure 18.10.

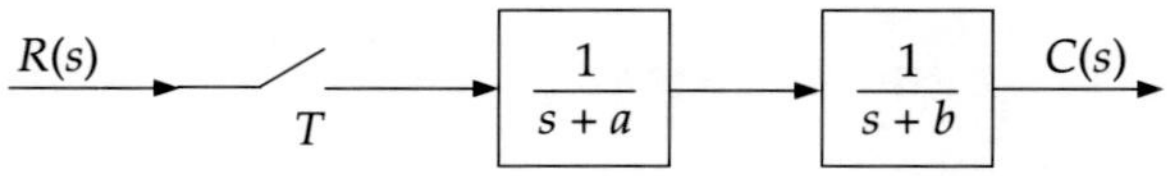

Figure 18.10 Discrete time system (Example 18.2).

Solution Here,

$$G(s) = \frac{1}{(s+a)(s+b)} = \frac{1}{b-a}\left[\frac{1}{s+a} - \frac{1}{s+b}\right]$$

Hence,

$$g(t) \equiv \mathscr{L}^{-1}\{G(s)\} = \frac{1}{b-a}\left[e^{-at} - e^{-bt}\right]$$

For a discrete time system, $t = kT$, where $k = 0,\ 1,\ 2,\ \ldots, \infty$. Therefore,

$$g(kT) = \frac{1}{b-a}\left[e^{-akT} - e^{-bkT}\right]$$

Thus,

$$G(z) \equiv \mathcal{Z}\{g(kT)\} = \frac{1}{b-a}\left[\mathcal{Z}\{e^{-akT}\} - \mathcal{Z}\{-e^{-bkT}\}\right]$$

$$= \frac{1}{b-a}\left[\frac{z}{z-e^{-aT}} - \frac{z}{z-e^{-bT}}\right]$$

$$= \frac{1}{b-a} \cdot \frac{z\left(e^{-aT} - e^{-bT}\right)}{\left(z-e^{-aT}\right)\left(z-e^{-bT}\right)}$$

EXAMPLE 18.3. Find the transfer function for the arrangement given in Figure 18.11.

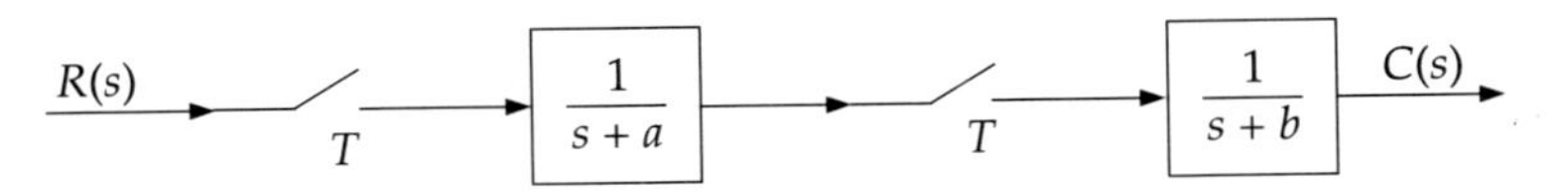

Figure 18.11 Discrete time system (Example 18.3).

Solution The difference between the previous example and the present example is that here both the processes are made discrete separately. Hence,

$$G_1(s) = \frac{1}{s+a}$$

$$\Rightarrow \qquad g_1(t) = e^{-at}$$

$$\Rightarrow \qquad g_1(kT) = e^{-akT}$$

Therefore,

$$G_1(z) = \frac{z}{z - e^{-aT}} \tag{i}$$

Similarly,

$$G_2(z) = \frac{z}{z - e^{-bT}} \tag{ii}$$

Combining Eqs. (i) and (ii), we get the z-transform of the composite process as

$$G_1(z)G_2(z) = \frac{z^2}{\left(z - e^{-aT}\right)\left(z - e^{-bT}\right)}$$

EXAMPLE 18.4. Consider the open-loop sampled data system given in Figure 18.12.

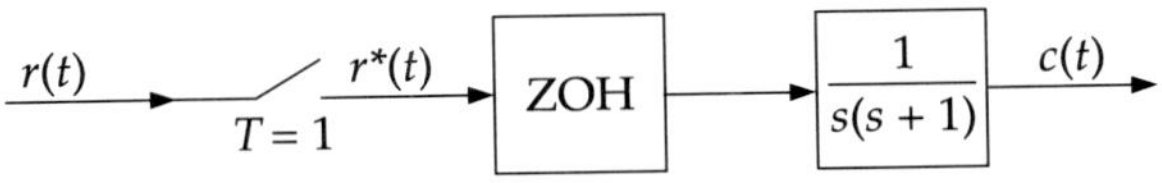

Figure 18.12 Open-loop sampled data system (Example 18.4).

Find the following:

(a) The z-transfer function of the system

(b) The response of the system to a unit impulse.

Solution

(a) Incorporating the transfer function of the ZOH [see Eq. (18.5)], the open loop transfer function of the system is

$$G(s) = \frac{1 - e^{-sT}}{s^2(s+1)}$$

Expanding it into partial fractions, we get

$$G(s) = \left(1 - e^{-sT}\right)\left(\frac{1}{s^2} - \frac{1}{s} + \frac{1}{s+1}\right)$$

Therefore,

$$\begin{aligned}
G(z) \equiv \mathcal{Z}\{G(s)\} &= \left(1 - z^{-1}\right)\mathcal{Z}\left\{\frac{1}{s^2} - \frac{1}{s} + \frac{1}{s+1}\right\} \\
&= \left(1 - z^{-1}\right)\left[\frac{Tz}{(z-1)^2} - \frac{z}{z-1} + \frac{z}{z-e^{-T}}\right] \\
&= \frac{z-1}{z}\left[\frac{Tz\left(z-e^{-T}\right) - z(z-1)\left(z-e^{-T}\right) + z(z-1)^2}{(z-1)^2\left(z-e^{-T}\right)}\right] \\
&= \frac{\left(ze^{-T} - z + Tz\right) + \left(1 - e^{-T} - Te^{-T}\right)}{(z-1)\left(z-e^{-T}\right)}
\end{aligned}$$

(b) When $T = 1$, we have

$$\begin{aligned}
G(z) &= \frac{ze^{-1} + 1 - 2e^{-1}}{(z-1)\left(z-e^{-1}\right)} \\
&= \frac{0.3678z + 0.2644}{(z-1)(z-0.3678)} \\
&= \frac{0.3678z + 0.2644}{z^2 - 1.3678z + 0.3678} \qquad \text{(i)}
\end{aligned}$$

By long division of Eq. (i), we find

$$G(z) = 0.3678z^{-1} + 0.7675z^{-2} + 0.9145z^{-3} + \cdots \qquad \text{(ii)}$$

Since $R(z) = 1$ for an impulse input, $C(z) = G(z)$. So, we can write

$$\begin{aligned}
C(z) \equiv C(kT) &= C(0) + C(T)z^{-1} + C(2T)z^{-2} + C(3T)z^{-3} + \cdots \qquad \text{(iii)} \\
&= G(z)
\end{aligned}$$

Comparing the coefficients Eqs. (ii) and (iii), we obtain $C(kT)$ as follows:

$$\begin{aligned}
C(0) &= 0 \\
C(T) &= 0.3678 \\
C(2T) &= 0.7675 \\
C(3T) &= 0.9145
\end{aligned}$$

REVIEW QUESTIONS

18.1 (a) In brief, discuss z- and s- domain relationship.
(b) From this, describe stability criteria of a system in the z-domain.
(c) In brief, discuss zero-order hold circuit.
(d) $x[n] = -a^n u[-n-1]$. Then calculate $x(z)$ and also its ROC.

18.2 (a) Find the z-transform of unit step.
(b) Discuss in brief s-domain to z-domain mapping of a system.
(c) If the z-transfer function of a system is given, explain how you will comment about the stability of the system.

18.3 Explain the meaning of sampling and hold. Obtain the unit-step response of the system shown in the following figure.

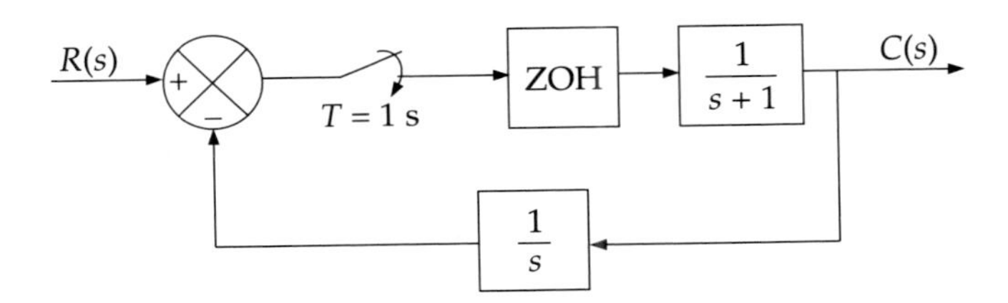

18.4 (a) What is a sampled data system?
(b) Express the output $c(t)$ in the form of zero-order hold (ZOH) sampled data system given in the following figure.

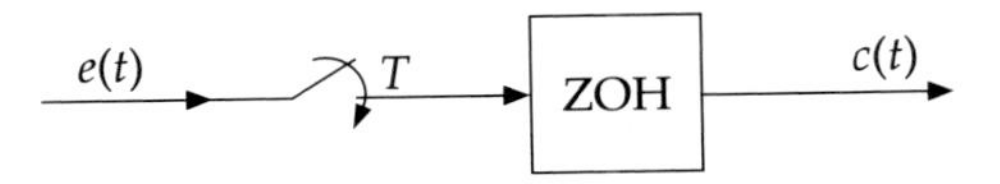

18.5 (a) Find the z-transforms of
(i) $f(k) = ke^{-3k}$
(ii) $f(k) = k^2$
(b) Find $C(z)/R(z)$ for the sampled data system shown in the following figure.

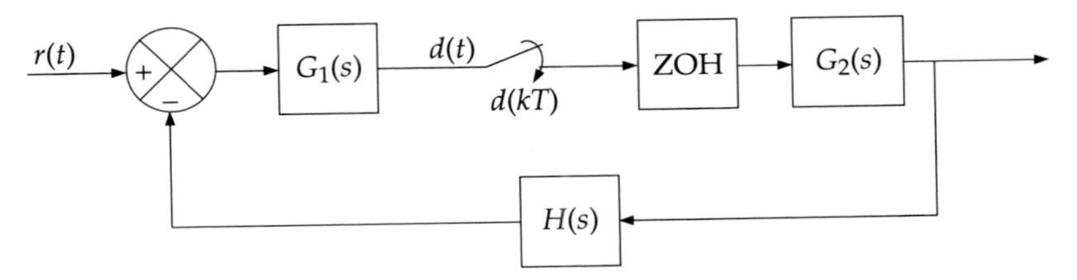

Answers to Selected Questions

18.1. (d) $\dfrac{z}{z-a}$ ROC: $|z| < a$

18.3. $C(1) = 0,\ C(2) = 0.5335,\ C(3) = 0.4193,\ C(4) = 0.1332,\ C(5) = -0.0495,$
$C(6) = -0.0879,\ C(7) = -0.0508,\ C(8) = -0.0076,\ C(9) = 0.0127,\ C(10) = 0.0128,$
$C(11) = 0.0054$

18.5. (a) $\dfrac{Tze^{-3T}}{\left(z-e^{-3T}\right)^2}$; (b) $\dfrac{T^2 z(z+1)}{(z-1)^3}$

Appendix A

Laplace Transform Table

Serial No.	*Time function, $f(t)$*	*Laplace transform, $F(s)$*
1	Unit impulse at $t = 0$, $\delta(t)$	1
2	Double impulse at $t = 0$, $\frac{d}{dt}\delta(t)$	s
3	$\delta(t-a)$	$e^{-as} \quad a \geq 0$
4	Unit step, $u(t)$	$\frac{1}{s}$
5	$u(t-a)$	$\frac{e^{-as}}{s}$
6	$f(t) = \begin{cases} 0 & \text{for } t < a \\ 1 & \text{for } a < t < b \\ 0 & \text{for } t > b \end{cases}$	$\frac{1}{s}\left(e^{-as} - e^{-bs}\right)$
7	t	$\frac{1}{s^2}$
8	$t^n \quad n = 1, 2, 3 \ldots$	$\frac{n!}{s^{n+1}}$
9	e^{-at}	$\frac{1}{s+a}$
10	te^{-at}	$\frac{1}{(s+a)^2}$
11	$t^n e^{-at} \quad n = 1, 2, 3 \ldots$	$\frac{n!}{(s+a)^{n+1}}$
12	$\frac{t^{n-1}}{(n-1)!} e^{-at} \quad n = 1, 2, 3 \ldots$	$\frac{1}{(s+a)^n}$
13	$1 - e^{-at}$	$\frac{a}{s(s+a)}$

Serial No.	*Time function,* $f(t)$	*Laplace transform,* $F(s)$
14	$\dfrac{e^{-at}-e^{-bt}}{b-a}$	$\dfrac{1}{(s+a)(s+b)}$
15	$\dfrac{be^{-bt}-ae^{-at}}{b-a}$	$\dfrac{s}{(s+a)(s+b)}$
16	$\sin\omega t$	$\dfrac{\omega}{s^2+\omega^2}$
17	$\sin(\omega t+\phi)$	$\dfrac{s\sin\phi+\omega\cos\phi}{s^2+\omega^2}$
18	$\cos(\omega t+\phi)$	$\dfrac{s\cos\phi-\omega\sin\phi}{s^2+\omega^2}$
19	$t\sin\omega t$	$\dfrac{2\omega s}{(s^2+a^2)^2}$
20	$\dfrac{1}{t}\sin\omega t$	$\tan^{-1}\left(\dfrac{\omega}{t}\right)$
21	$\cos\omega t$	$\dfrac{s}{s^2+\omega^2}$
22	$t\cos\omega t$	$\dfrac{s^2-\omega^2}{(s^2+\omega^2)^2}$
23	$\dfrac{1-\cos\omega t}{\omega^2}$	$\dfrac{1}{s(s^2+\omega^2)}$
24	$\dfrac{1}{\omega^3}(\sin\omega t-\omega t\cos\omega t)$	$\dfrac{1}{(s^2+a^2)^2}$
25	$\dfrac{1}{2\omega}(\sin\omega t+\omega t\cos\omega t)$	$\dfrac{s^2}{(s^2+a^2)^2}$
26	$\sinh\omega t$	$\dfrac{\omega}{s^2-\omega^2}$
27	$\cosh\omega t$	$\dfrac{s}{s^2-\omega^2}$
28	$\dfrac{\cos\omega_1 t-\cos\omega_2 t}{\omega_2^2-\omega_1^2}, \quad \omega_1^2\neq\omega_2^2$	$\dfrac{s}{(s^2+\omega_1^2)(s^2+\omega_2^2)}$
29	$e^{-at}\sin\omega t$	$\dfrac{\omega}{(s+a)^2+\omega^2}$
30	$e^{-at}\cos\omega t$	$\dfrac{s+a}{(s+a)^2+\omega^2}$
31	$\dfrac{1}{\sqrt{\pi t}}$	$\dfrac{1}{\sqrt{s}}$
32	$\dfrac{1}{\sqrt{\pi t}}e^{-\omega t}$	$\dfrac{1}{\sqrt{s+\omega}}$
33	$2\sqrt{\dfrac{t}{\pi}}$	$\dfrac{1}{\sqrt{s^3}}$
34	$\dfrac{f(t)}{t}$	$\int_s^\infty f(u)\,du$

Appendix B

z-Transform Table

Serial No.	*Time function* $f(t),\ t > 0$	*z-transform* $F(z)$
1	$\delta(t)$	1
2	$\delta(t - kT)$	z^{-k}
3	$u(t)$	$\dfrac{z}{z-1}$
4	$u(t - T)$	$\dfrac{1}{z-1}$
5	$u(t - nT) \quad n = 1,\ 2,\ 3, \ldots$	$\dfrac{z}{z^n(z-1)}$
6	$u(t) - u(t - T)$	1
7	$u(t) - u(t - 2T)$	$1 + \dfrac{1}{z}$
8	$u(t - nT) - u(t - \overline{n+1}T) \quad n = 1,\ 2,\ 3, \ldots$	$\dfrac{1}{z}\ n$
9	$\dfrac{T}{t}\ u(t - T)$	$\ln\left(\dfrac{z}{z-1}\right)$
10	t	$\dfrac{Tz}{(z-1)^2}$
11	t^2	$\dfrac{T^2 z(z+1)}{(z-1)^3}$
12	t^3	$\dfrac{T^3 z(z^2 + 4z + 1)}{(z-1)^4}$

Serial No.	*Time function* $f(t),\ t > 0$	*z-transform* $F(z)$
13	$t^n \quad n = 1, 2, 3, \ldots$	$\lim_{a\to 0}(-1)^n \frac{\partial^n}{\partial a^n}\left(\frac{z}{z-e^{-aT}}\right)$
14	e^{-at}	$\frac{z}{z-e^{-aT}}$
15	te^{-at}	$\frac{Tze^{-aT}}{(z-e^{-aT})^2}$
16	$t^n e^{-at} \quad n = 1, 2, 3, \ldots$	$(-1)^n \frac{\partial^n}{\partial a^n}\left(\frac{z}{z-e^{-aT}}\right)$
17	$1 - e^{-at}$	$\frac{z(1-e^{-aT})}{(z-1)(z-e^{-aT})}$
18	$-\frac{1}{a}\left[\delta(t) - a^{\frac{t}{T}}\right]$	$\frac{1}{z-a}$
19	$\frac{1}{(a-b)}\left[a^{\left(\frac{t}{T}-1\right)} - b^{\left(\frac{t}{T}-1\right)}\right]$	$\frac{1}{(z-a)(z-b)}$
20	$\frac{1}{(a-b)}\left[a^{\frac{t}{T}} - b^{\frac{t}{T}}\right]$	$\frac{z}{(z-a)(z-b)}$
21	$\frac{1}{(a-b)}\left[a^{\left(\frac{t}{T}+1\right)} - b^{\left(\frac{t}{T}+1\right)}\right]$	$\frac{z^2}{(z-a)(z-b)}$
22	$\left(\frac{T}{t}\right)!$	$e^{\frac{1}{z}}$
23	$\sin\omega t$	$\frac{z\sin\omega T}{z^2 - 2z\cos\omega T + 1}$
24	$\cos\omega t$	$\frac{z^2 - z\cos\omega T}{z^2 - 2z\cos\omega T + 1}$
25	$\sinh\omega t$	$\frac{z\sinh\omega T}{z^2 - 2z\cosh\omega T + 1}$
26	$\cosh\omega t$	$\frac{z(z-\cosh\omega T)}{z^2 - 2z\cosh\omega T + 1}$
27	$e^{-at}\sin\omega t$	$\frac{ze^{-aT}\sin\omega T}{z^2 - 2ze^{-aT}\cos\omega T + e^{-2aT}}$
28	$e^{-at}\cos\omega t$	$\frac{z^2 - ze^{-aT}\cos\omega T}{z^2 - 2ze^{-aT}\cos\omega T + e^{-2aT}}$
29	$e^{-at}\sinh\omega t$	$\frac{ze^{-aT}\sinh\omega T}{z^2 - 2ze^{-aT}\cosh\omega T + e^{-2aT}}$
30	$e^{-at}\cosh\omega t$	$\frac{z^2 - ze^{-aT}\cosh\omega T}{z^2 - 2ze^{-aT}\cosh\omega T + e^{-2aT}}$

Appendix C

Convolution

C.1 DEFINITION

A convolution is an integral that expresses the amount of overlap of one function g as it is shifted over another function f. It therefore *blends* one function with another. Convolution of two functions f and g over a finite range $[0, t]$ is given by

$$f \star g = \int_0^t f(\tau)g(t-\tau)\, d\tau$$

where the symbol $\star$ (sometimes also written as $f \otimes g$) denotes convolution of f and g. Convolution is often taken over an infinite range, i.e.

$$f \star g = \int_{-\infty}^{\infty} f(\tau)g(t-\tau)\, d\tau = \int_{-\infty}^{\infty} g(\tau)f(t-\tau)\, d\tau$$

C.2 UTILITY OF CONVOLUTION IN CONTROL SYSTEMS

Convolution is a powerful tool for determining the output of a system to *any* input. If the input to a system is $r(t)$, and the impulse response of that system is $g(t)$, then we can determine the output of the system, $c(t)$, from the convolution

$$c(t) = \int_0^t r(t-\tau)g(\tau)\, d\tau$$

Proof. Let us consider a linear time-invariant (LTI) system (Figure C.1) with an input $r(t)$ and an output $c(t)$.

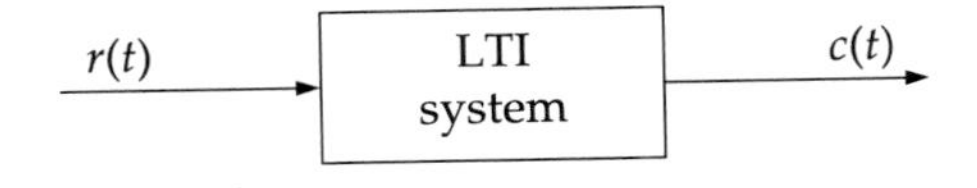

Figure C.1 LTI system with input and output.

Now we consider the same system with an impulse input $\delta(t)$. The output is then, by definition, the impulse response $g(t)$. If we use a delayed impulse into the system, the output is just the impulse response with the same delay.

Then,

$$\text{Input} = \delta(t - \tau)$$

$$\text{Output} = g(t - \tau)$$

Because of linearity, if we scale the input by any factor, the output will be scaled by the same factor. Then, if we integrate the input, the output is also integrated. So, let us first scale the input (and hence the output) by the factor $r(\tau)\, d\tau$. Then, we integrate the input and consequently the output. In this case, we will take the lower and upper limits of the integration to be 0 and t; but they could just as well have been $+\infty$ and $-\infty$. The situation is, therefore,

$$\text{Input} = \int_0^t \delta(t - \tau)\, d\tau$$

$$\text{Output} = \int_0^t g(t - \tau)\, d\tau$$

The δ-function is zero everywhere except at $t = \tau$. And, since it is used as a unit impulse function, its value is 1 at that instant. Hence,

$$\text{Input} = \int_0^t \delta(t - \tau) r(\tau) d\tau = r(t)$$

$$\text{Output} = \int_0^t g(t - \tau) r(\tau)\, d\tau$$

From our initial proposition (see Figure C.1), the output equals $c(t)$ when the input is $r(t)$. Thus, we are led to the convolution theorem

$$c(t) = \int_0^t r(t - \tau) g(\tau)\, d\tau = \int_0^t r(\tau) g(t - \tau)\, d\tau$$

We note that in this method we do not need to have recourse to Laplace transform and then, its inverse to find the time-domain response of a system.

Now, we will see from the following example how the method really works.

EXAMPLE C.1. Consider a system described by the equation

$$\frac{dc(t)}{dt} + 2c(t) = r(t)$$

Its Laplace transform, with all initial conditions set to zero, is

$$(s + 2)C(s) = R(s)$$

So, the transfer function is given by

$$G(s) = \frac{C(s)}{R(s)} = \frac{1}{s + 2} \tag{i}$$

From the inverse Laplace transform of Eq. (i), we get the impulse response of the system as

$$g(t) = e^{-2t} \tag{ii}$$

The plot of the response is given in Figure C.2.

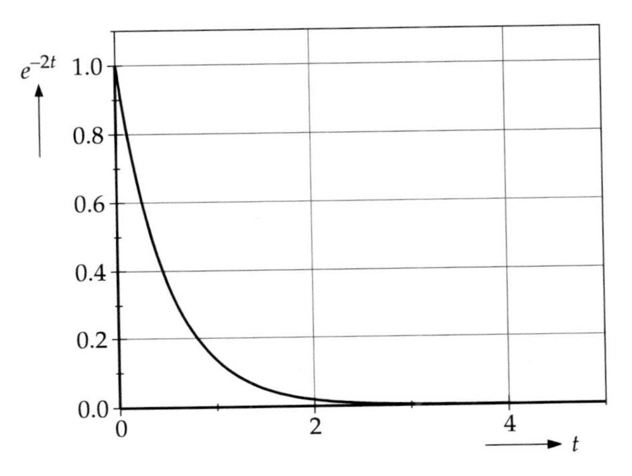

Figure C.2 Plot of Eq. (ii).

We will use convolution to find the response of this system to the input given by

$$r(t) = \begin{cases} 1 & 0 < t < 1 \\ 0 & \text{otherwise} \end{cases} \tag{iii}$$

The plot of the input function of Eq. (iii) is given in Figure C.3.

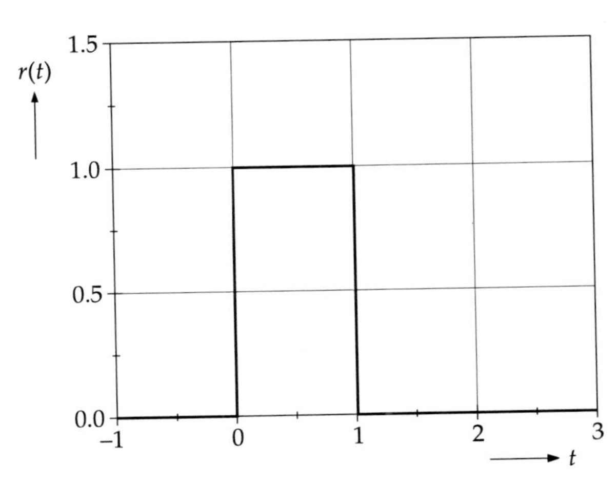

Figure C.3 Plot of Eq. (iii).

We will use the convolution to see how the system of Eq. (i) responds to the input given by Eq. (iii). The integral is given by

$$c(t) = \int_{-\infty}^{\infty} r(\tau)g(t-\tau)\, d\tau$$

Since $r(t)$ has three distinct regions, namely, $t < 0$, $0 < t < 1$ and $t > 1$, we need to split the integral into three parts.

Part I: $t < 0$. For a particular value of t, $g(t-\tau)$ exists up to $\tau = t$. So, for negative value of t, $g(t-\tau)$ exists for a negative value of τ, where $r(\tau) = 0$. Therefore, there is no overlap between $r(\tau)$ and $g(t-\tau)$. Consequently, the value of the integral is zero. The situation for $t = -0.5$ is shown in Figure C.4.

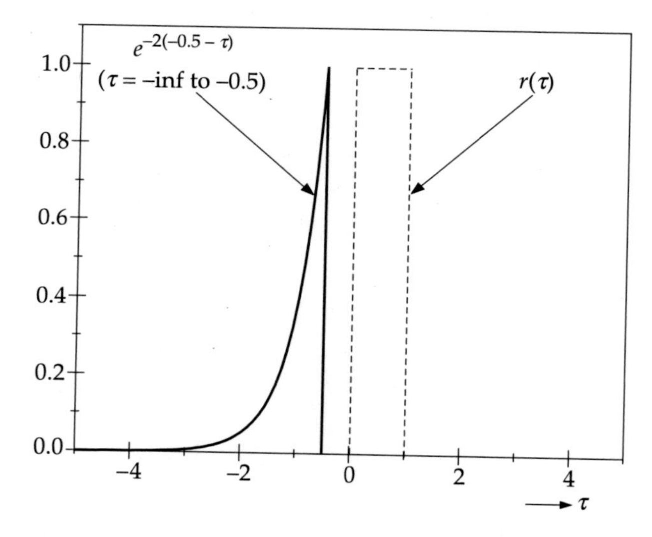

Figure C.4 The input and the impulse function $g(t-\tau)$ at $\tau = -0.5$.

Part II: $0 < t < 1$. In this region, both $r(\tau)$ and $g(t-\tau)$ exist for all values of τ from 0 to t. Therefore, we need to evaluate the integral for $\tau = 0$ to t. Graphically, the situation at $t = 0.5$ looks like that of Figure C.5 where the area of integration is shown by shading it.

The result of the integration is as follows:

$$c(t) = \int_0^t r(\tau)g(t-\tau)\, d\tau = \int_0^t (1)e^{-2(t-\tau)}\, d\tau = e^{-2t}\int_0^t e^{2\tau}\, d\tau$$

$$= e^{-2t}\left[\frac{1}{2}e^{2\tau}\right]_0^t = e^{-2t}\cdot\frac{1}{2}(e^{2t}-1)$$

$$= \frac{1}{2}(1-e^{-2t})$$

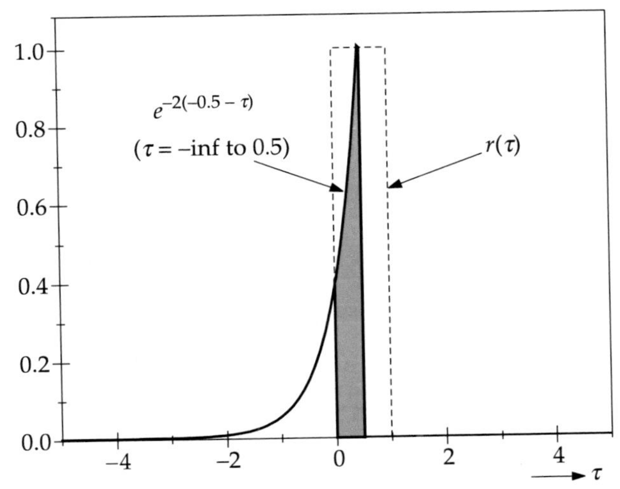

Figure C.5 The input and the impulse function $g(t-\tau)$ at $\tau = 0.5$. The area of integration is shown by the shaded portion.

Part III: $t > 1$. We know that $r(\tau) = 0$ for $\tau > 1$. So, here we need to evaluate the integral for $\tau = 0$ to 1. The situation can be visualized from the graph shown for $\tau = 1.2$ in Figure C.6. The area of integration has been shown by shading it.

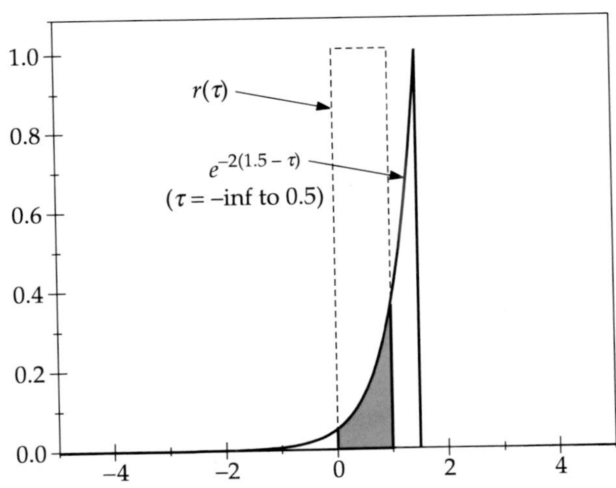

Figure C.6 The input and the impulse function $g(t-\tau)$ at $\tau = 1.5$. The shaded portion indicates the area of integration.

We can now evaluate the integral as

$$c(t) = \int_0^1 r(\tau)g(t-\tau)\,d\tau = \int_0^1 (1)e^{-2(t-\tau)}d\tau$$

$$= e^{-2t}\int_0^1 e^{2\tau}\,d\tau = e^{-2t}\left[\frac{1}{2}e^{2\tau}\right]_0^1 = e^{-2t}\cdot\frac{1}{2}(e^2-1)$$

The output for the input given by Eq. (iii) to the system having transfer function given by Eq. (i) is

$$c(t) = \begin{cases} 0 & \text{for } t < 0 \\ \dfrac{1}{2}(1 - e^{-2t}) & \text{for } 0 < t < 1 \\ e^{-2t} \cdot \dfrac{1}{2}(e^2 - 1) & \text{for } t > 1 \end{cases} \tag{iv}$$

This result is shown in Figure C.7.

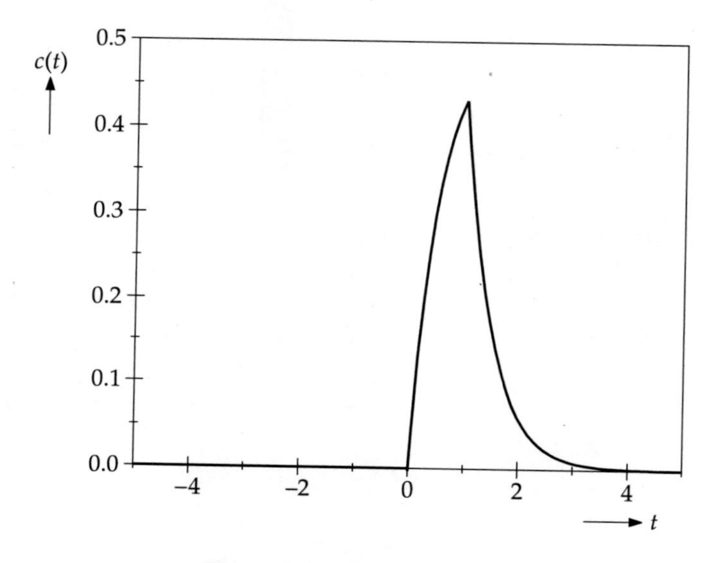

Figure C.7 Plot of Eq. (iv).

Appendix D

MatLab Primer

Although our aim is to make the reader capable of handling Control System studies with the help of the MatLab software marketed by Mathworks Inc., Natick, Massachusetts, USA, we present the primer with preliminaries of the relevant mathematics and graphics interspersed in order that the reader may gain control over the material fruitfully.

D.1 PRELIMINARIES

Like any other software, MatLab opens when its icon on the desktop is double-clicked. It opens three windows, namely,

1. Workspace
2. Command History
3. Command Window

It is the Command Window where the cursor blinks after the command prompt `>>`. We need to type commands after the command prompt. To quit MatLab, we type `exit` or `quit` at the command prompt, or click the ⊠ button at the top right-hand corner of the window.

MatLab has many toolboxes which extend the basic functions of the software to different application areas. We are concerned here with the Control Systems Toolbox. But as already stated, before exploring that, we need to know a few basic operations of MatLab.

Creating Vectors

To create a vector, which means a set of numbers with a name 'a', such as

```
a = 1 2 3 4 5 6 7 8
```

we type in the command window at the command prompt (>>)

```
>>  a=[1 b̸ 2 b̸ 3 b̸ 4 b̸ 5 b̸ 6 b̸ 7 b̸ 8]<--
```

where b̸ indicates a blank space and <-- indicates 'press Enter'.

Note: Blank spaces are essential between numbers to allow MatLab to separate them.

MatLab will return

```
a =
    1   2   3   4   5   6   7   8
```

Alternatively, we could have typed

```
>> a = [1,2,3,4,5,6,7,8]
```

to get the same result. Suppose we want to create a vector having elements between 0 and 3, spaced evenly with an increment of 0.5. We should type

```
>> x = 0:0.5:3<--
```

MatLab will return

```
x =
    0   0.5000   1.0000   1.5000   2.0000   2.5000   3.0000
```

Now, suppose we want to add 3 to each element of `a` and store it with a name `b`. We write

```
>> b = a+3<--
```

MatLab returns

```
b =
    4   5   6   7   8   9   10   11
```

If we want to add vectors `a` and `b` to produce another vector `c`, we simply write

```
>> c = a+b<--
```

and MatLab returns

```
c =
    5   7   9   11   13   15   17   19
```

Note: 1. MatLab is case sensitive. Which means `a` and `A` are different variable names to MatLab.
2. Vectors can be added only if they are of equal length.

If we type

```
>> d = a+x<--
```

MatLab will write

```
??? Error using ==> +
Matrix dimensions must agree.
```

MatLab writes 'Matrix' because it treats all numbers as matrices. Hence the name *MatLab*.

Henceforward, we will omit the Enter symbol (`<--`) while writing commands.

Dealing with Polynomials

Writing polynomials is similar to writing a vector. Suppose we want to deal with the polynomial

$$s^5 + 2s^4 - 3s^3 + 5s^2 - s + 10$$

we simply type the coefficients of the polynomial in the descending order of s as:

```
>>s = [1 2 -3 5 -1 10]
```

and we get a return of

```
s =
   1   2   -3   5   -1   10
```

Note: If there is one or more *missing terms* in the polynomial, we have to write 0's for them.

So, for the polynomial

$$s^5 - 3s^3 + 5s^2 + 10$$

we need to type

```
>>snew = [1 0 -3 5 0 10]
```

Value of a polynomial. We can use the `polyval` command to know the value of the polynomial for a particular value of the variable s. Our polynomial is named s and we want its value for $s = 5$, say. Then we write

```
>>polyval(s,5)
ans =
      4130
```

Instead of creating the polynomial first with a name and then finding out the value, we could have done it in one command as follows:

```
>>polyval([1 2 -3 5 -1 10],5)
ans =
      4130
```

The '`ans`' part is the return from MatLab. Therefore, the syntax for the command is

```
polyval(P,X)
```

where `P` is the polynomial and `X` is the value of the variable. There are other ways in which the command can be used. Those can be found by typing

```
>>help ployval
```

when MatLab will respond with a big write-up as follows:

```
POLYVAL Evaluate polynomial.
  Y = POLYVAL(P,X), when P is a vector of length N+1 whose elements are
  the coefficients of a polynomial, is the value of the polynomial
  evaluated at X.

  Y = P(1)*X^N + P(2)*X^(N-1) + ... + P(N)*X + P(N+1)

  If X is a matrix or vector, the polynomial is evaluated at all points
  in X. See also POLYVALM for evaluation in a matrix sense.

  Y = POLYVAL(P,X,[ ],MU) uses XHAT = (X-MU(1))/MU(2) in place of X.
  The centring and scaling parameters MU are optional outputs computed
  by POLYFIT.

  [Y,DELTA] = POLYVAL(P,X,S) or [Y,DELTA] = POLYVAL(P,X,S,MU) uses the
  optional output structure S provided by POLYFIT to generate error
  estimates, Y +/- delta. If the errors in the data input to POLYFIT are
  independent normal with constant variance, Y +/- DELTA contains at least
  50% of the predictions.

  See also POLYFIT, POLYVALM.

  Overloaded methods
  help gf/polyval.m
```

For every command, we can get help from MatLab in this way. The other way is to use the drop down menu item 'Help' and click 'MatLab Help'.

Roots of a polynomial. The command and the MatLab response are

```
>> roots(s)
ans =
   -3.4093
    1.0994 + 1.0053i
    1.0994 - 1.0053i
   -0.3948 + 1.0797i
   -0.3948 - 1.0797i
```

Multiplication and division of polynomials.

```
>> a=[1 2 3];
>> b=[3 4 5];
>> c=conv(a,b)
c =
     3    10    22    22    15
```

Here a and b are two polynomials and c is the result of their multiplication.

Note: We have used semicolons after the first two commands. This we did to suppress immediate printing of the result of the command. Had we not used semicolons, the session would have looked like

```
>> a=[1 2 3]
a =
    1   2   3
>> b=[3 4 5]
b =
    3   4   5
>> c=conv(a,b)
c =
    3   10   22   22   15
```

which we wanted to avoid.

For division, we write

```
>> [quo,rem]=deconv(c,a)
quo =
    3   4   5
rem =
    0   0   0   0   0
```

We assigned two variables `quo` and `rem` to get the quotient and remainder of the division of polynomial `c` by polynomial `a`. The result is obvious.

Addition of polynomials. If the polynomials are of the same order, this can be done through the built-in command of vector addition of MatLab. However, for addition of polynomials of different orders, we have to write our own programme in MatLab.

Creation of polynomials. Suppose we want to create a polynomial having roots -2, -3, and 1. We can create it by the MatLab command

```
>> [x]=poly([-2 -3 1])
x =
    1   4   1   -6
```

That means the polynomial is $x^3 + 4x^2 + x - 6$. To check if the roots are so, we type

```
>> roots(x)
ans =
    -3.0000
    -2.0000
     1.0000
```

Yes, they are verified. Obviously, the command `roots` and `poly` are inverse to each other. So the commands

```
>> x=1:20;
>> roots(poly(x))
```

should generate numbers 1 to 20. Let us see

```
ans =
    20.0003
    18.9970
    18.0118
    16.9695
    16.0509
    14.9319
    14.0684
    12.9472
    12.0345
    10.9836
    10.0063
     8.9983
     8.0003
     7.0000
     6.0000
     5.0000
     4.0000
     3.0000
     2.0000
     1.0000
```

They are indeed so though from 8 upwards the values are approximate because of the *round off* or *truncation* error.

Writing Functions and Plotting Them

MatLab has a host of built-in scientific functions like sin, cos, tan, asin ($\sin^{-1}$), acos, atan, log, log10, exp, sqrt etc. The value of π (3.1416) is available under the name `pi`. Complex numbers can be written with `i` or `j`.

Suppose we wish to plot a graph of $y = \sin 3\pi x$ for $0 \leq x \leq 1$. We do this by sampling the function at a sufficiently large number of points and then joining up the points (x, y) by straight lines. Suppose we take $N + 1$ points equally spaced a distance h apart. Then,

```
>> N = 10; h = 1/N; x = 0:h:1;
```

defines the set of points $x = 0, h, 2h, \ldots, 1 - h, 1$. Alternatively, we may use the command `linspace`. The general form of the command is `linspace (a,b,n)` which generates $n + 1$ equispaced points between a and b, inclusive. So, in this case, we would use the command

```
>> x = linspace (0,1,11);
```

The corresponding y values are computed by

```
>> y = sin(3*pi*x);
```

and finally, we can plot the points with

```
>> plot(x,y)
```

The result is shown in Figure D.1(a), where it is clear that the value of N is too small.

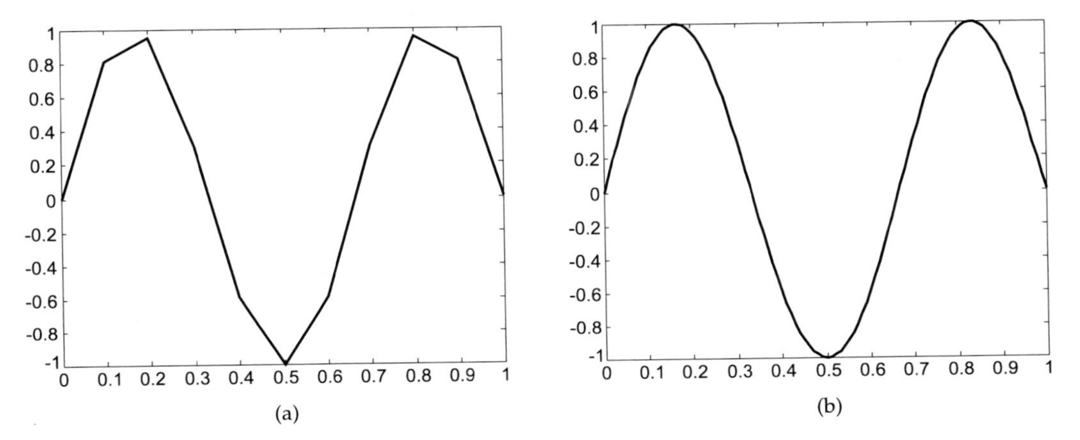

Figure D.1 Two plots with different h values: (a) with $h = 0.1$; (b) with $h = 0.01$.

Now we change the value of N to 100 and see the result [Figure D.1(b)].

```
>> N = 100; h = 1/N; x = 0:h:1;
>> y = sin(3*pi*x); plot(x,y)
```

Plotting titles and labels. To put a title and label the axes, we use

```
>> title('Graph of y = sin(3pi x)')
>> xlabel('x axis')
>> ylabel('y-axis')
```

The *strings* enclosed in single quotes can be anything we choose.

Grid. A dotted grid may be added by

```
>> grid
```

This can be removed by using either `grid` again, or `grid off`.

Multi-plots. Several graphs may be drawn on the same figure. A descriptive legend may be included. Let us try the following commands:

```
>> plot(x,y,'-',x,cos(3*pi*x),'--')
>> legend('Sin curve','Cos curve')
>> title('Multi-plot')
>> xlabel('x axis'), ylabel('y axis')
>> grid
```

These will produce Figure D.2.

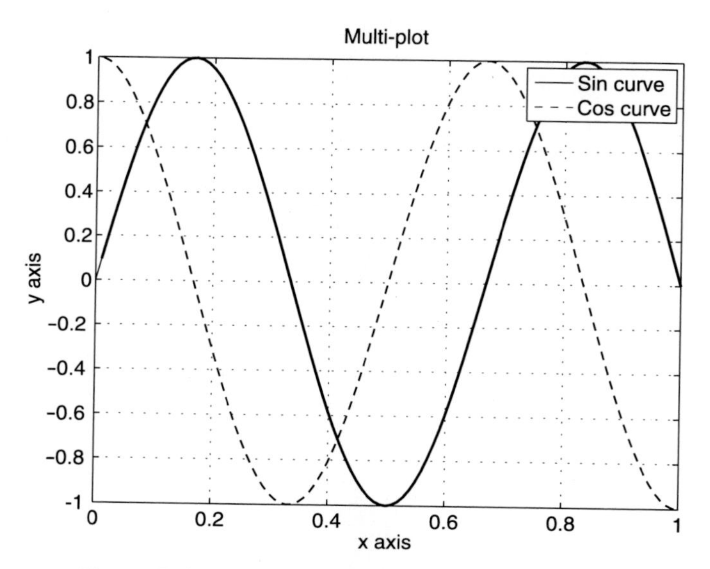

Figure D.2 Multi-plot with legend and axes labels.

Hold. A second `plot` command clears the graphics window before plotting the current graph. This is not convenient if we wish to add further graphs to the figure at some later stage. To stop the window being cleared, we can issue commands:

```
>> N=100; h=1/N;x=0:h:1;
>> y=x;
>> plot(x,y,'-'), hold on
>> plot(x,sin(2*y),'kx'), hold off
```

Note:

1. `hold on` holds the current picture; `hold off` releases it (but does not clear the window, which can be done with `clf`). Like `grid`, the command `hold` toggles the hold state.
2. k of `'kx'` indicates that the colour of points should be black and x indicates that the points will have x marks. Table D.1 gives relevant data.

Table D.1 Options for colours and styles of plots. The upper rows indicate commands and the lower rows indicate their implications.

Colour	y	m	c	r	g	b	w	k	
	Yellow	Magenta	Cyan	Red	Green	Blue	White	Black	
Style	.	o	x	+	-	*	:	-.	--
	Point	Circle	x-mark	Plus	Solid	Star	Dotted	Dashdot	Dashed

Zooming. We often need to *zoom in* on some portion of a plot in order to see it in more detail. Clicking on the *Zoom in* or *Zoom out* button on the figure window is simplest, but we can also use the command

```
>> zoom
```

We can click the zoom in button ⊕ on the top bar and point it to the relevant position on the plot. It will zoom in by a factor of 2. We may repeat it to any desired level.

Clicking the ⊖ button will zoom out by a factor of 2.

`zoom off` turns off the zoom capability.

Subplot. The graphics window may be split into an $m \times n$ array of smaller windows into which we may plot one or more graphs. The windows are counted 1 to mn row-wise, starting from the top left. Both `hold` and `grid` work on the current subplot. If we try the following commands:

```
>> subplot(221), plot(x,y)
>> xlabel('x'),ylabel('sin 3 pi x')
>> subplot(222), plot(x,cos(3*pi*x))
>> xlabel('x'),ylabel('cos 3 pi x')
>> subplot(223), plot(x,sin(6*pi*x))
>> xlabel('x'),ylabel('sin 6 pi x')
>> subplot(224), plot(x,cos(6*pi*x))
>> xlabel('x'),ylabel('cos 6 pi x')
```

then we should get Figure D.3.

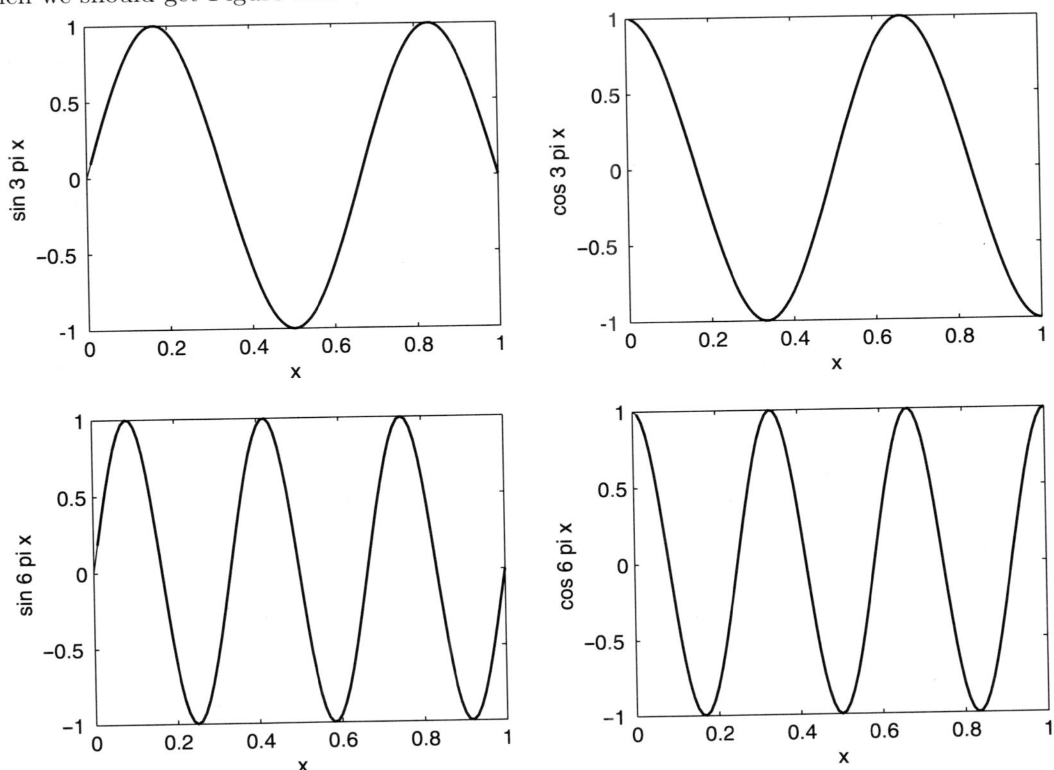

Figure D.3 Having four graphs in the same plot.

Note: `subplot(221)` [or `subplot(2,2,1)`] specifies that the window should be split into a 2×2 array.

D.2 MATRICES

A matrix can be entered by any of the following three ways:

1. Separating each row by a semicolon

```
>> a=[1 2 3;4 5 6;7 8 9]
a =
1 2 3
4 5 6
7 8 9
```

2. By pressing Enter (`<--`) after each row. The `<--` sign is shown only to indicate the action. MatLab obviously will not display it.

```
>> a=[1 2 3<--
4 5 6<--
7 8 9]<--
a =
1 2 3
4 5 6
7 8 9
```

3. By clicking on the New Variable icon (leftmost icon) in Workspace browser. Once we enter a variable name, we may double click on the icon associated with the variable name (in Workspace browser) to bring up the Array Editor. From the Array Editor, we may enter the values as we would in a typical spreadsheet programme like Excel. Suppose our variable name is `a`. We click on the Command Window, type `a` and press Enter. Our matrix is there.

We can also specify a range of numbers in a defined vector or matrix using the colon operator. The colon causes MatLab to step in sequence through the numbers specified. For example,

```
>> a=[1:3;4:6;7:9]
a =
     1     2     3
     4     5     6
     7     8     9
```

Ones matrix. The `ones` function creates a matrix whose elements are all ones. Typing `ones(m,n)` creates an `m` row by `n` column matrix of ones. Thus,

```
>> ones(3,3)
ans =
     1     1     1
     1     1     1
     1     1     1
```

To create a ones matrix that is of the same size as an existing matrix `X`, we can use `ones(size(X))`. For example,

```
>> ones(size(a))
ans =
     1     1     1
     1     1     1
     1     1     1
```

Zeros matrix. The `zeros` function is similar to the `ones` function. Typing `zeros(m, n)` creates an `m`-by-`n` matrix of zeros, and `zeros(size(X))` will create a matrix of zeros, which is of the same size as `X` defined earlier.

Addition and subtraction. The steps are as follows:

```
>> a=[1 2 3;4 5 6;7 8 9];
>> b=[0 1 0;1 1 0;0 0 1];
>> c=a+b
c =
  1 3 3
  5 6 6
  7 8 10
>> d=a-b
d =
  1 1 3
  3 4 6
  7 8 8
```

Multiplication. Two matrices can be multiplied by the $*$ operator. We have to remember the rule of matrix multiplication while doing it.

Examples:

(a)

```
>> e=a*b
e =
  2 3 3
  5 9 6
  8 15 9
```

(b)

```
>> a=[1 2 3]
a =
  1 2 3
>> b=[1 3;2 4;4 5]
b =
  1 3
  2 4
  4 5
>> c=a*b
c =
  17 26
```

Element-by-element multiplication. We may note that this is different from multiplication of two matrices. The corresponding elements of two matrices can be multiplied by using the `.*` operator as follows:

```
>> a=[1 2;3 4]
a =
   1 2
   3 4
>> b=[4 5;6 7]
b =
   4 5
   6 7
>> c=a.*b
c =
   4 10
  18 28
```

Division: *Right matrix division.* Dividing matrix `b` into matrix `a` is done by the slash (/) operator.

Example:

```
>> a=[1:3;4:6;7,8,0]
a =
     1     2     3
     4     5     6
     7     8     0
>> b=[2,4,6;0,3,7;9,8,1]
b =
     2     4     6
     0     3     7
     9     8     1
>> c=a/b
c =
    0.5000         0         0
    3.6875   -2.2500   -0.3750
   -8.3125    6.7500    2.6250
```

Left matrix division. Dividing `a` into `b` is done by the backslash (\) operator. This is equivalent to `inv(a)*b`. Note that `x = c` is the solution to `a*x=b`.

```
>> c=a\b
c =
   -4.5556   -5.3333   -4.5556
    5.1111    5.6667    4.1111
   -1.2222   -0.6667    0.7778
>> x=inv(a)*b
```

```
x =
 -4.5556   -5.3333   -4.5556
  5.1111    5.6667    4.1111
 -1.2222   -0.6667    0.7778
```

Element-by-element division. The right division is done by the ./ operator.

```
>> c=a./b
c =
  0.5000    0.5000    0.5000
     Inf    1.6667    0.8571
  0.7778    1.0000         0
```

Note: c_{21} is ∞ because of a division by 0. Similarly, the left division may be done by the .\ operator.

Transpose of a matrix. A matrix can be transposed by using the ' operator.

```
>> a=[1 2 3;4 5 6;7 8 9]
a =
  1 2 3
  4 5 6
  7 8 9
>> b=a'
b =
  1 4 7
  2 5 8
  3 6 9
```

Matrix inversion. A matrix can be inverted by using the inv(x) command, where x is the matrix.

```
>> a=[1 0 1;1 1 0;0 1 1]
a =
  1 0 1
  1 1 0
  0 1 1
>> inv(a)
ans =
  0.5000 0.5000 -0.5000
 -0.5000 0.5000 0.5000
  0.5000 -0.5000 0.5000
```

Eigenvalues. The eigenvalues of a can be found as follows:

```
>> eig(a)
ans =
  0.5000 + 0.8660i
  0.5000 - 0.8660i
  2.0000
```

Characteristic polynomial of a matrix.

```
>> p=poly (a)
p =
  1.0000 -3.0000 3.0000 -2.0000
```

We know that the roots of a characteristic polynomial are the eigenvalues of the matrix. We verify

```
>> roots(p)
ans =
2.0000
0.5000 + 0.8660i
0.5000 - 0.8660i
```

These results tally with those found earlier.

Determinant of a matrix.

```
>> det(a)
ans =
2
```

D.3 LAPLACE TRANSFORM

The Laplace transform of a function $f(t)$ can be obtained with the MatLab function `laplace`. MatLab performs Laplace transform symbolically. So, first we need to define the variable t as a *symbol*. The following examples will make the procedure clear.

Example 1: Find the Laplace transform of $f(t) = 3e^{-5t}$.

```
>> syms t
>> f=3*exp(-5*t)
f =
   3/exp(5*t)
>> l=laplace(f)
l =
   3/(s + 5)
```

Example 2: Find the Laplace transform of $5\dfrac{d^2x(t)}{dt^2}$.

```
>> laplace(5*diff(sym('x(t)'), 2))
```

Note: The function $x(t)$ is defined as a symbol with the embedded command `sym`. The number 2 at the end indicates that the second derivative of the function $x(t)$ is required. The MatLab returns

```
ans =
5*s^2*laplace(x(t), t, s) - 5*s*x(0) - 5*D(x)(0)
```

Here, $x(0)$'s are the initial conditions.

Example 3: Find the inverse Laplace transform of

$$C(s) = \frac{1}{s} - \frac{2}{s+3} + \frac{1}{s+4}$$

```
>> syms s
>> ilaplace(1/s-2/(s+3)+1/(s+4))
```

The MatLab returns

```
ans =
1/exp(4*t) - 2/exp(3*t) + 1
```

With this background on MatLab, let us see how we can use it to study control systems.

D.4 CONTROL SYSTEM STUDIES

This is a mere introduction to the wealth of information one can avail oneself of through control system toolbox of MatLab. Remember, we will consider the LTI systems.

Writing Transfer Functions

Suppose the transfer function is

$$G(s) = \frac{s+2}{s^2 + 9s + 20} \tag{D.1}$$

We can write the transfer function by any of the following four ways:

1.
```
>> num=[1 2];
>> den=[1 9 20];
>> sys=tf(num,den)
```

   when MatLab returns

```
Transfer function:
    s + 2
--------------
s^2 + 9 s + 20
```

2. Alternatively, we can have the same result by typing

```
>> sys=tf([1 2],[1 9 20])
Transfer function:
    s + 2
--------------
s^2 + 9 s + 20
```

3.
```
>> s=tf('s');
>> sys=(s+2)/(s^2+9*s+20)
Transfer function:
   s + 2
--------------
s^2 + 9 s + 20
```

4. We may also print the transfer function with the help of the **printsys** command as follows:

```
>> num=[1 2]; den=[1 9 20]; printsys(num,den)
num/den =
   s + 2
--------------
s^2 + 9 s + 20
```

But it does not store any value for the **num/den** ratio.

We may extract the zero, pole, constant (**zpk**) data from the transfer function as follows:

```
>> [z,p,k]=zpkdata(sys,'v')
z =
   -2
p =
   -5.0000
   -4.0000
k =
   1
```

So, we know that the transfer function has a zero at $s = -2$ (which is obvious from the transfer function) and two poles at $s = -4$ and $s = -5$. This means that Eq. (D.1) can be rewritten as

$$G(s) = \frac{s+2}{(s+4)(s+5)}$$

Quite often, transfer functions are given in this form. Then it can be entered in MatLab in two ways:

1.
```
>> sys=tf([1 2],conv([1 4],[1 5]))
Transfer function:
   s + 2
--------------
s^2 + 9s + 20
```

2.
```
>> sys=zpk(-2,[-4,-5],1)
Zero/pole/gain:
   (s+2)
-----------
(s+4) (s+5)
```

Cascade, parallel and feedback connections of transfer functions. The transfer functions can be cascaded (series connection) or connected in parallel by using the commands

```
>> sys=series(sys1, sys2) (or sys=sys1*sys2)
>> sys=parallel(sys1,sys2) (or sys=sys1 + sys2)
```

The transfer function of a feedback system having forward path transfer function g and feedback transfer function h can be found by writing

```
>> sys=feedback(g,h,-1)
```

Note: The feedback sign is to be mentioned. If positive (or regenerative) feedback is required, the last quantity should be mentioned as `+1`. If it is unity feedback, a shorter command as

```
>> sys=feedback(g,1)
```

works. But there, a negative feedback is assumed for 1 or +1, and positive feedback for −1. To avoid this ambiguity, it is better to write

```
>> sys=feedback(g,1,-1)
```

for negative unity feedback, and

```
>> sys=feedback(g,1,1)
```

for positive unity feedback, rather than using the shorter version for unity feedback.

Apart from the `zpkdata` command cited before, the dc gain, poles and zeros of a transfer function can be found individually by the commands

```
>> k=dcgain(sys)
>> p=pole(sys)
>> z=zero(sys)
```

Pole-zero Plotting

Pole-zero plotting of transfer function can be made by using the command

```
>> pzmap(sys)
```

when the plot will be shown on a separate window.

Responses for Step, Impulse and Other Inputs

To find how a transfer function responds to a step input, we can write

```
>> sys=zpk(-2,[-4, -5],20)
Zero/pole/gain:
  20 (s+2)
-----------
(s+4) (s+5)
>> step(sys)
```

We get the plot of Figure D.4 in a new window. The dc gain of the transfer function is 2. So, the response shows an underdamped response around that value.

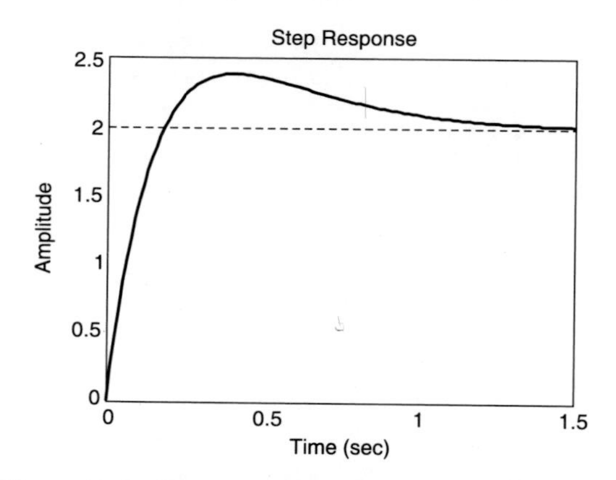

Figure D.4 Step response of underdamped system.

To obtain the impulse response, we replace *step* with *impulse* in the code. For doing this, we do not have to retype the whole thing. We can find old commands just by pressing the ↑ key on the keyboard. Once we get the required command, we just replace the required word and Enter. To find response for other inputs, we may use the `lsim` command. Its syntax is

```
lsim(sys, u, t)
```

where `sys` is the output, `u`, the input function and `t`, the time vector. We can also specify a colour, line style, and marker for each system, as in

```
lsim(sys,'.-',u,t)
```

We watch its usage for a sine function input to a system having an input delay of 1.5 s.

```
>> sys=tf(1,[6 1], 'InputDelay',1.5)
Transfer function:
                    1
exp(-1.5*s) * -------
              6 s + 1
>> t=0:0.1:10;
>> y=sin(t);
>> lsim(sys, 'k .', y, t)
```

Note: An input delay (transportation lag) of, say, 1.5 s in the transfer function can be introduced by using the `InputDelay` command.

This should generate the graph of Figure D.5, where the input (`y`) and output of the `sys` are shown in the same plot. The input is the big sinusoidal oscillation between −1 and 1, and the response is the dotted curve in the middle. We observe the poor fidelity of the system with an input delay!

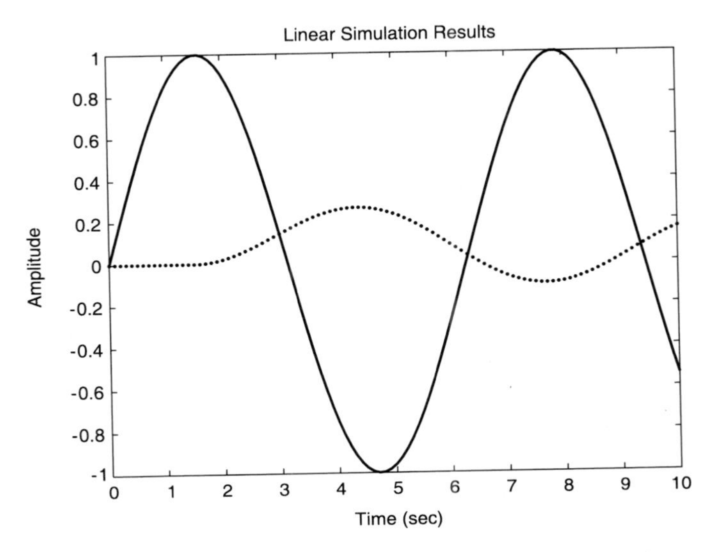

Figure D.5 Response to sine input with linear simulation (`lsim`) function.

Note: Our given `sys` in `lsim` is in the s-space while `y` is in t-space. Obviously, MatLab performs an inverse Laplace transform of the `sys` output to compare it with the input.

Impulse and Step Responses of Systems of Different Types

Type 1. A type 1 system has one pole at the origin. Let the transfer function of the system be

$$G(s) = \frac{10}{s}$$

To study the impulse and step responses of this transfer function and its unity feedback ones, we type the following commands:

```
>> g=tf(10,[1 0]);
>> sys=feedback(g,1,-1)
>> subplot(421);
>> impulse(g)
>> subplot(422);
>> impulse(sys)
>> subplot(423);
>> step(g)
>> subplot(424);
>> step(sys)
```

We should get the following figure:

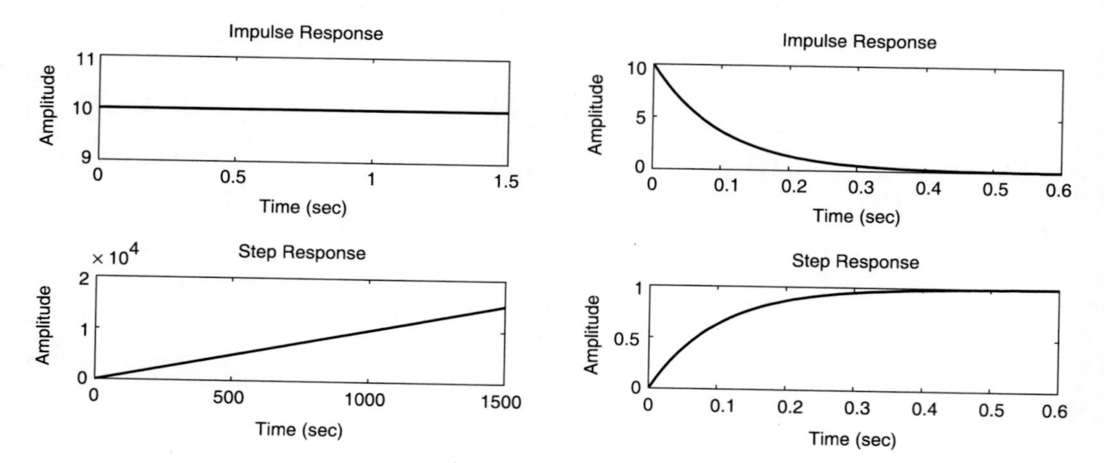

Note:

1. The impulse response should be a constant 10 because its inverse Laplace transform is so. This is shown in the upper left figure where MatLab draws a straight line at $y = 10$.
2. With unity feedback, it becomes a Type 0 system and it is stable. MatLab shows it in the upper right figure.
3. For the step input, theoretically, it should make the system unstable because the input generates another pole at the origin. The lower left figure drawn by MatLab shows a diverging output with time, which indicates an unstable system.
4. But with unity feedback, the step response becomes stable (see the lower right figure) because then the transfer function becomes

$$\frac{10}{s+10}$$

Type 2. Let the system be

$$G(s) = \frac{10}{s^2}$$

Obviously, both the responses of the system without feedback should indicate an unstable system. But with unity feedback, both the responses become marginally stable (i.e. oscillatory). We can check these by typing the following commands:

```
>> sys=tf(10,[1 0 0]);
>> sysf=feedback(sys,1,-1);
>> t=0:.01:10;
>> subplot(421);
>> impulse(sys,t);
>> title('Impulse response without feedback');
>> subplot(422);
>> impulse(sysf,t);
>> title('Impulse response with feedback');
>> subplot(423);
```

```
>> step(sys,t);
>> title('Step response without feedback');
>> subplot(424);
>> step(sysf,t);
>> title('Step response with feedback');
```

The results are shown in Figure D.6.

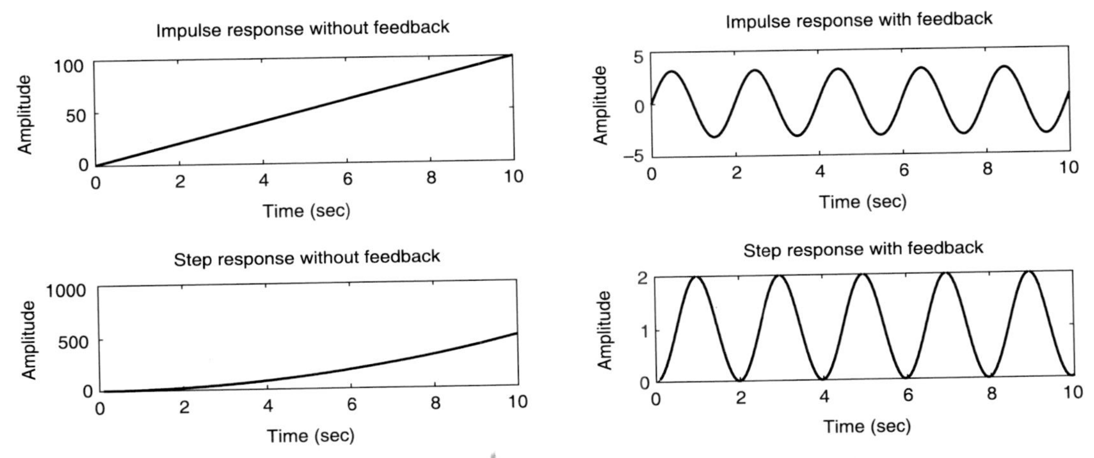

Figure D.6 Impulse and step responses of Type 2 system without and with feedback.

Note: In the previous example, we did not restrict `impulse` and `step` commands with a time vector from 0 to 10. Here we have to do it to increase the resolution of the graph. Observe the result if it is not done.

Type 3. Here, let the transfer function be

$$G(s) = \frac{10}{s^3}$$

From the following commands, we observe how the system without feedback is unstable for both impulse and step inputs, but both stabilize when feedback is used.

```
>> sys=tf(10,[1 0 0 0]);
>> sysf=feedback(sys,1,-1);
>> t=0:0.001:2.5;
>> [y,t]=impulse(sys,t);
>> [yf,t]=impulse(sysf,t);
>> plot(t,y,'k-',t,yf,'k--')
>> legend('Without feedback','With feedback')
>> xlabel('Time (s)'),ylabel('Amplitude')
>> title('Impulse Response of Type 3 System')
```

The results are displayed in Figure D.7. Similarly, for the step input, graphs are plotted and the results are shown in Figure D.8.

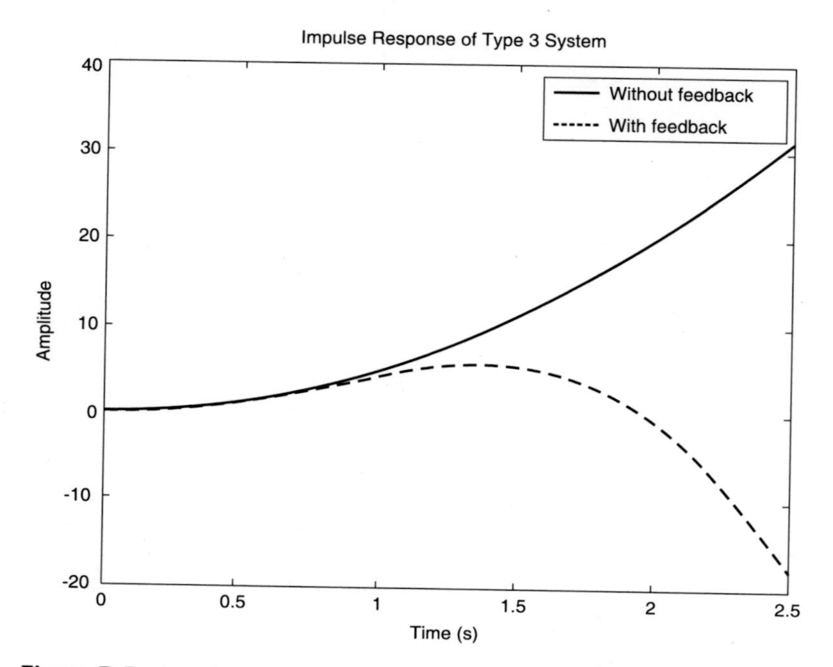

Figure D.7 Impulse response of Type 3 system without and with feedback.

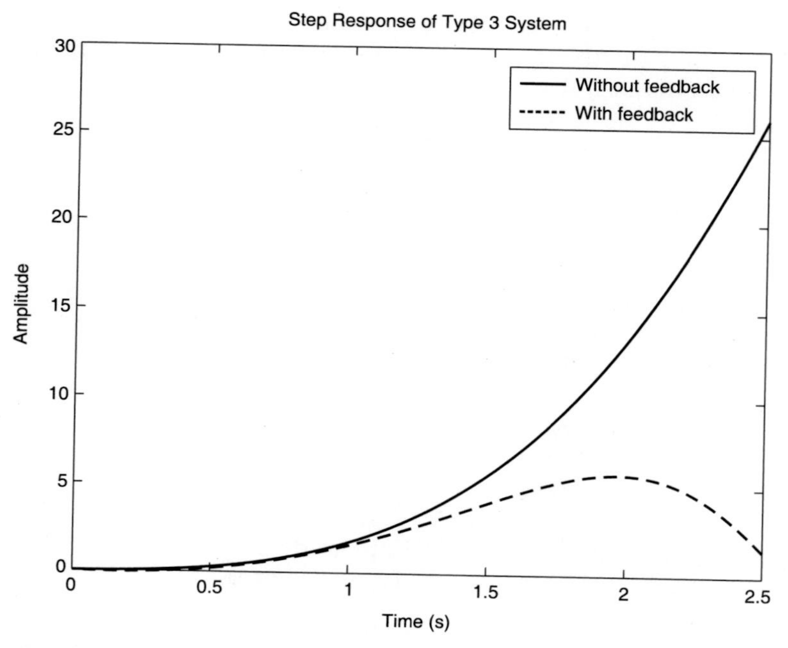

Figure D.8 Step response of Type 3 system without and with feedback.

Note: Here we have made multi-plots on the same graph by preserving `y, yf, t` values and plotting them by the `plot` command. We have also added labels for x and y axes, inserted legends for curves, and added titles to the graphs.

D.5 ROOT LOCUS

The MatLab command `rlocus` generates root loci for the given transfer functions. It has a few syntaxes such as the following:

`rlocus(sys)`	computes and plots the root locus of the SISO LTI model `sys`.
`rlocus(sys,k)`	uses a user-specified vector `k` of gain values.
`rlocus(sys1,sys2,...)`	draws the root loci of multiple LTI models `sys1, sys2,...` on a single plot. Colour, line style and marker for each model can be specified as `rlocus(sys1,'r',sys2,'y:',sys3,'gx')`.
`[r,k] = rlocus(sys)` or `r = rlocus(sys,k)`	returns the matrix **r** of complex root locations for the gains k. The matrix **r** has length k columns and its j-th column lists the closed-loop roots for the gain $k(j)$.

Let us consider the open-loop transfer function

$$G(s)H(s) = \frac{s+7}{s(s+5)(1+15)(s+20)} \tag{D.2}$$

The following commands produce the plot shown in Figure D.9.

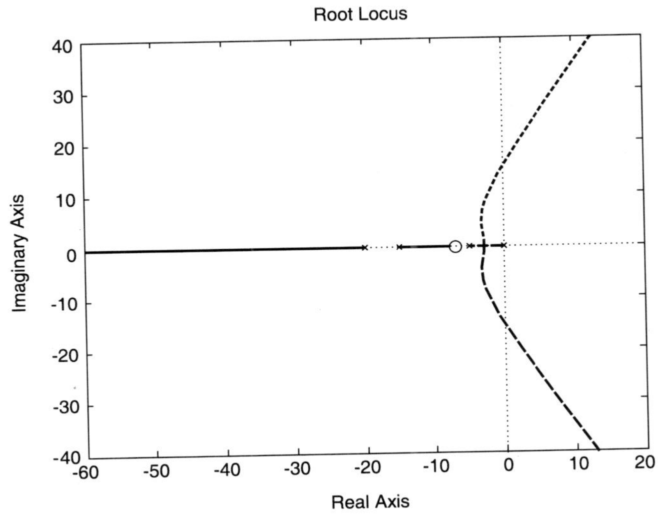

Figure D.9 Root locus diagram of Eq. (D.2).

```
>> sys=zpk([-7],[0,-5,-15,-20],[1])
Zero/pole/gain:
         (s+7)
---------------------
s (s+5) (s+15) (s+20)
>> rlocus(sys)
```

Note: We have ignored K while defining the transfer function and put 1 as the gain because MatLab automatically increases the gain while drawing the RL. The plot shows different loci for different roots, each locus being denoted by a particular shape, e.g. dotted line, etc.

Sizing Axes

The automatically generated diagram of Figure D.9 produces a lot of space on the left and along the imaginary axis, which is not desirable. On observing this figure we feel it is better if the horizontal axis remains confined between -25 and 5, and the vertical axis, between -20 and 20. So we add one command to restrict the axes. Also, we note that the `rlocus` command accepts the numerator and the denominator in the argument though it is not mentioned in the MatLab help prompt. We type the commands

```
>> num=[1 7];
>> den=conv(conv([1 0],[1 5]),conv([1 15],[1 20]));
>> rlocus(num,den);axis([-25 5 -20 20])
```

They generate Figure D.10 which looks better.

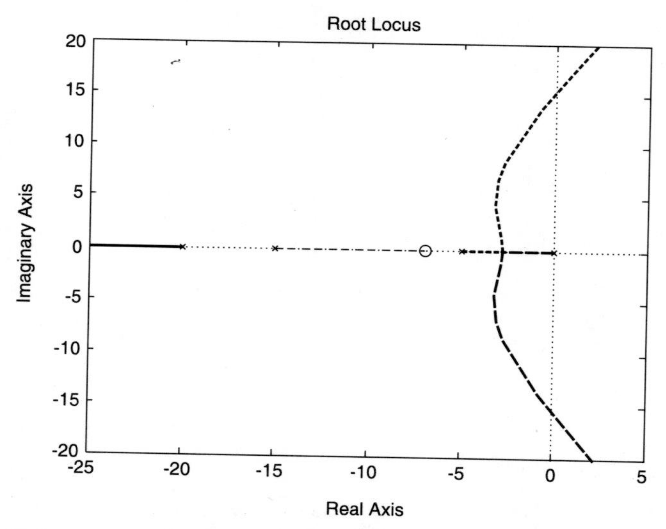

Figure D.10 Properly sized root locus diagram of Eq. (D.2).

Note: Instead of sizing the diagram to our desired limits by **axis** command, the diagram can be resized by right-clicking at a white space in the diagram when a menu pops up. We may click Properties→Limits in subsequent menus and put the desired limits for X-Limits and Y-Limits.

Drawing Grids

Suppose we want to find the gain at which the damping ratio ζ equals 0.6 and the part of the root locus where natural frequency ω_n is 6 rad/s. That can be done by giving the following additional commands:

```
>>z=0.6;
>>w=6;
>>rlocus(num,den);axis([-25 5 -20 20]); sgrid(z,w)
```

This will draw a grid showing two lines, where $\zeta = 0.6$ and a semicircle within which $\omega_n = 6$ rad/s. To know gain, pole location and % overshoot at $\zeta = 0.6$, we right-click on the locus, where $\zeta = 0.6$ line intercepts it. At once a data marker pops up to show these values (see Figure D.11).

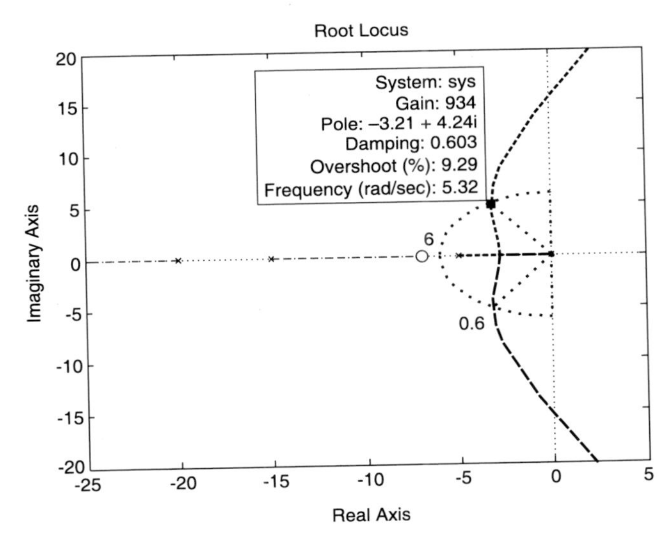

Figure D.11 Incorporating grid in root locus diagram of Eq. (D.2).

Note: The general grid can also be put on the diagram by checking the Grid item in the Property Editor: Root Locus menu that pops up on right-clicking.

D.6 BODE PLOT

We can use **bode** command for generating Bode plots of a given system. The command has many variations, of which the following are important for us:

`bode(sys)`	draws the Bode plot of the LTI model `sys`.
`bode(sys,{wmin,wmax})`	draws the Bode plot for frequencies between `wmin` and `wmax` (in rad/s).
`[mag,phase]= bode(sys,w)` and `[mag,phase,w]= bode(sys)`	return the response magnitudes and phases in degrees (along with the frequency vector `w` if unspecified). No plot is drawn on the screen.
`bode(sys1,sys2,...,w)`	graphs the Bode response of multiple LTI models `sys1`, `sys2`,... on a single plot. The frequency vector is optional. We may specify a colour, line style, and marker for each model, as `bode(sys1,'r',sys2,'y--',sys3,'gx')`.

Now, let us draw Bode plots for the system having an open-loop transfer function of

$$G(s)H(s) = \frac{1}{(0.2s+1)(s+1)(s+0.1)} \tag{D.3}$$

Our commands are

```
>> g1=tf(1,[0.2 1])

Transfer function:
      1
  ---------
  0.2 s + 1
>> g2=zpk([],[-1, -0.1],[1])

Zero/pole/gain:
       1
  -------------
  (s+1) (s+0.1)
>> g=g1*g2

Zero/pole/gain:
         5
 -------------------
 (s+5) (s+1) (s+0.1)
>> %Note how we converted the tf to Bode form
>> %though for MatLab it was not necessary
>> bode(g);grid
```

Note:
1. The lines beginning with a % sign are not executable commands—they are mere comments.
2. The `grid` command generates the vertical and horizontal lines in the graphs.

The generated Bode plot is depicted in Figure D.12. We need to keep in mind that MatLab does not generate asymptotic plots; it generates actual plots.

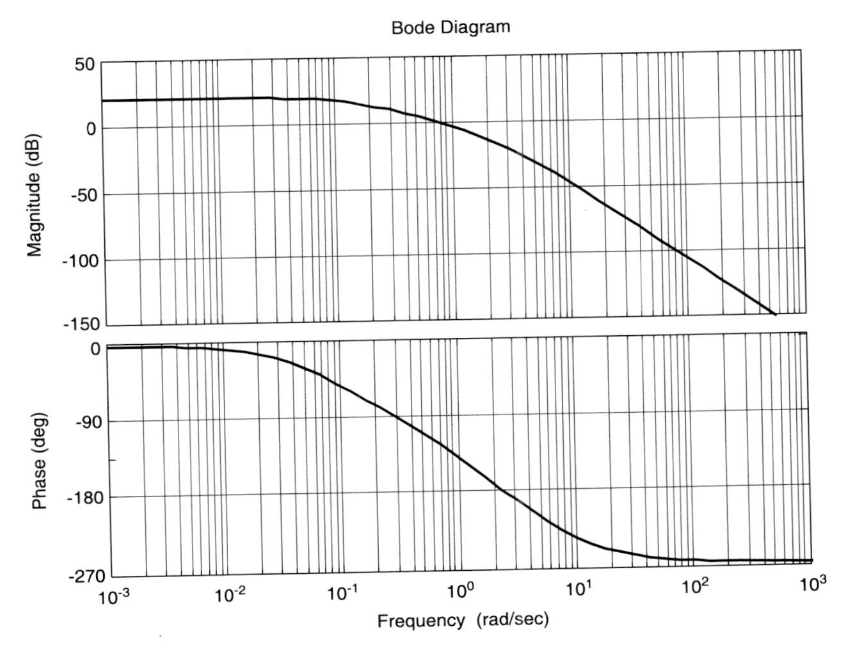

Figure D.12 Bode diagram of the OLTF given by Eq. (D.3).

If only the magnitude plot is what we require, we should write

```
>> bodemag(sys)
```

Stability Margins

To know the values of gain and phase margins from Bode plots, the following command may be used:

```
>> margin(g)
```

Figure D.13 showing Gm, Pm, ω_p and ω_g is generated by MatLab for the OLTF given by Eq. (D.3). We find that it is a stable system with positive margins and $\omega_p > \omega_g$.

Now suppose we increase the gain 10 times and see what the Bode plot looks like. We write

```
>> gnew=g*10
Zero/pole/gain:
          50
-------------------
(s+5) (s+1) (s+0.1)
>> margin(gnew)
```

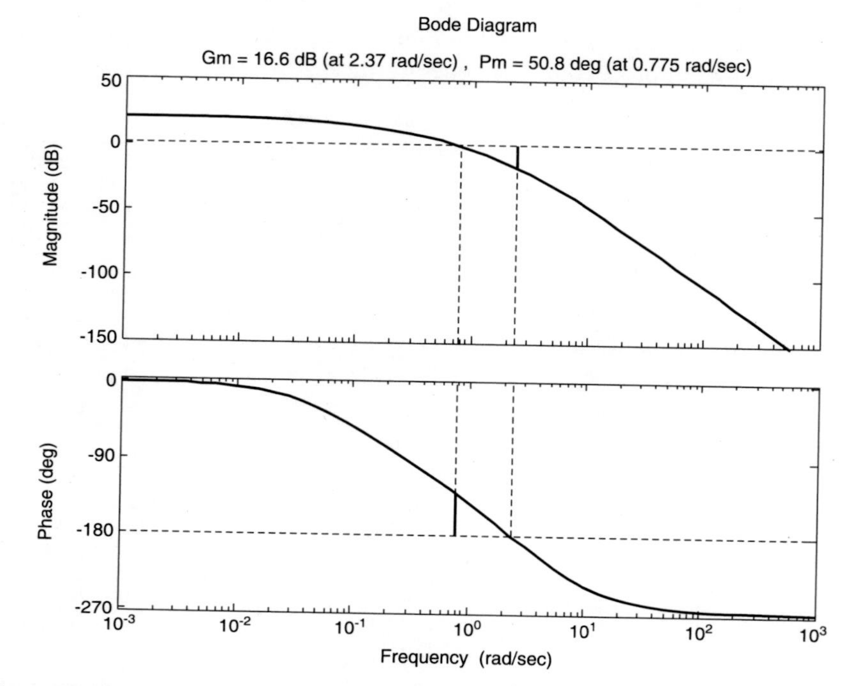

Figure D.13 Gain and phase margin determination for the OLTF given by Eq. (D.3).

The resulting plot is given in Figure D.14.

We observe that the phase plot has remained the same (which it should because gain does not contribute anything to the phase), but the system has become unstable with negative margins.

When we say that the system has become unstable, we mean that its closed-loop response is unstable. Let us check the unity feedback closed-loop responses for the stable system **g** and unstable system **gnew**. First, we test the unity feedback of **g**. Then we test the step response of **gnew**.

```
>> gf=feedback(g,1,-1)

Zero/pole/gain:
                5
----------------------------------
(s+5.23) (s^2 + 0.8696s + 1.052)
>> gnewf=feedback(gnew,1,-1)

Zero/pole/gain:
                50
-----------------------------------
(s+6.446) (s^2 - 0.3465s + 7.834)
>> t=0:0.01:15;
```

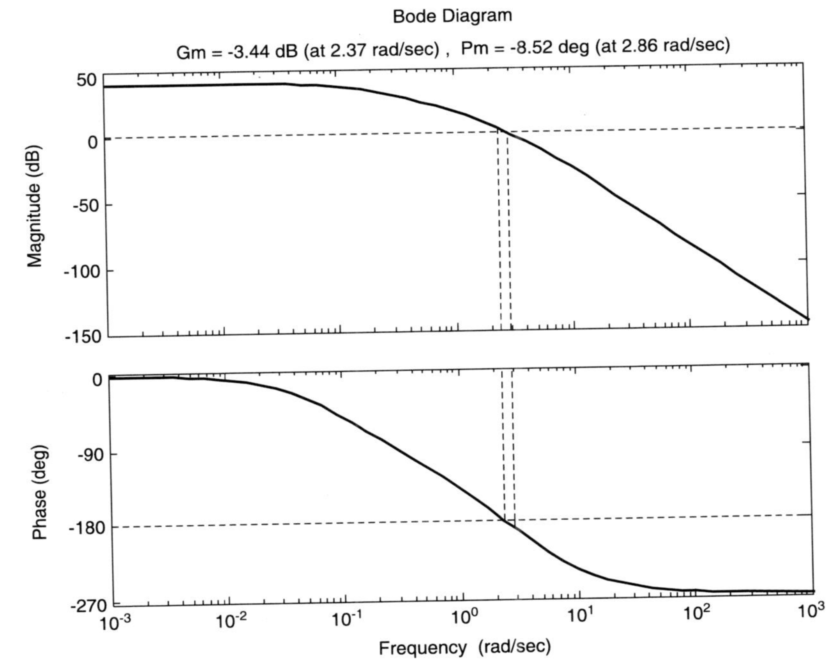

Figure D.14 Effect of increasing gain.

```
>> subplot(2,1,1);
>> step(gf,t)
>> title('Step response of gf')
>> subplot(2,1,2);
>> step(gnewf,t)
>> title('Step response of gnewf')
```

The plots are given in Figure D.15. Indeed, we find that while the step response of `gf` is stable, that of `gnewf` is unstable.

Static Error Constants

In the previous example, we found `g` to be a stable system. In this system, the initial part of the magnitude plot is horizontal which indicates that it is a type 0 system. Its transfer function, defined by Eq. (D.3), also corroborates it. For a type 0 system, we know that the initial gain equals K_p. We right-click on the plot to see that the gain is 20 dB. Hence,

$$20 \log K_p = 20$$

or

$$K_p = 10$$

Therefore, the steady-state error is

$$e_{ss} = \frac{1}{1 + K_p} = \frac{1}{11} = 0.1$$

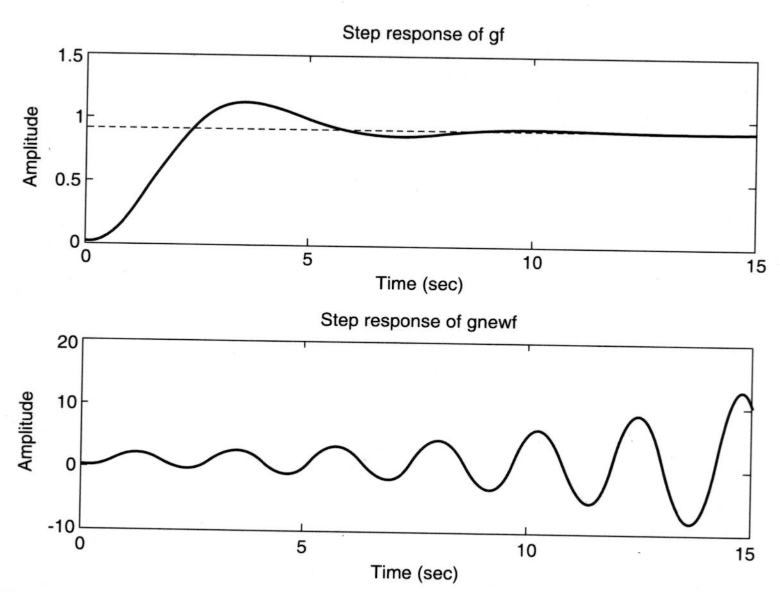

Figure D.15 Effect of increasing gain. The upper curve indicates the step response when the gain was not increased and the lower curve indicates the same when the gain is increased 10 times.

We observe from the unit step response of g that the steady value is nearly 0.9.

To determine K_v, let us consider a type 1 system given by

$$G(s)H(s) = \frac{10(s+2)}{s(s+1)(s+4)}$$

We will find from the Bode magnitude plot where the initial -20 dB/decade part intercepts the 0 dB line. The corresponding frequency will give the K_v value.

```
>> sys=zpk([-2],[0,-1,-4],[10])

Zero/pole/gain:
      10 (s+2)
   -------------
    s (s+1) (s+4)
>> bodemag(sys);grid
```

The plot is given in Figure D.16. By extending the initial -20 dB/decade line to 0 dB intersection, we find the frequency is 5 rad/s. So, $K_v = 5$. This means that the steady-state error for a unit ramp input is

$$e_{ss} = \frac{1}{K_v} = \frac{1}{5} = 0.2$$

To verify, we plot the ramp response of the closed-loop transfer function as follows:

```
>> sysf=feedback(sys,1,-1);
>> t=0:.1:5; ramp=t; lsim(sysf,ramp,t);grid
```

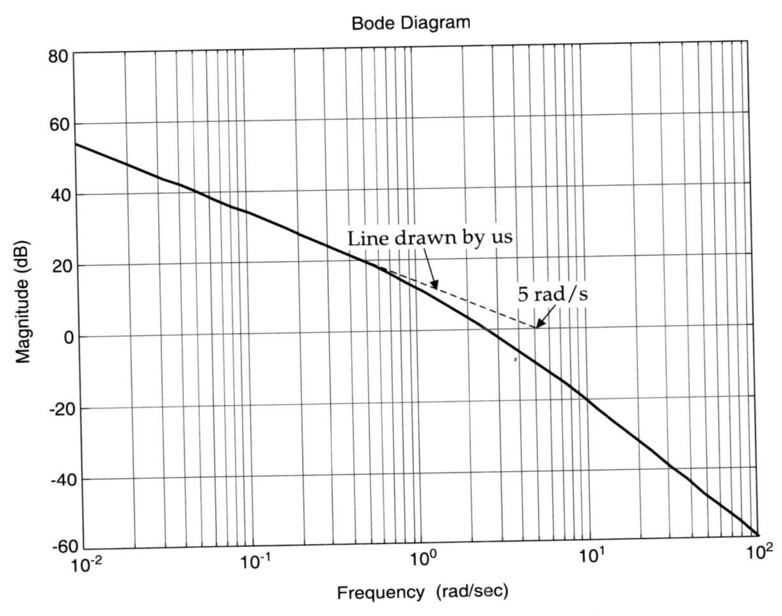

Figure D.16 K_v determination from Bode diagram.

The plot of Figure D.17 is produced by MatLab. It really shows that the steady-state error is around 0.2. In a similar way, K_a can be evaluated from the Bode magnitude plot of type 2 systems.

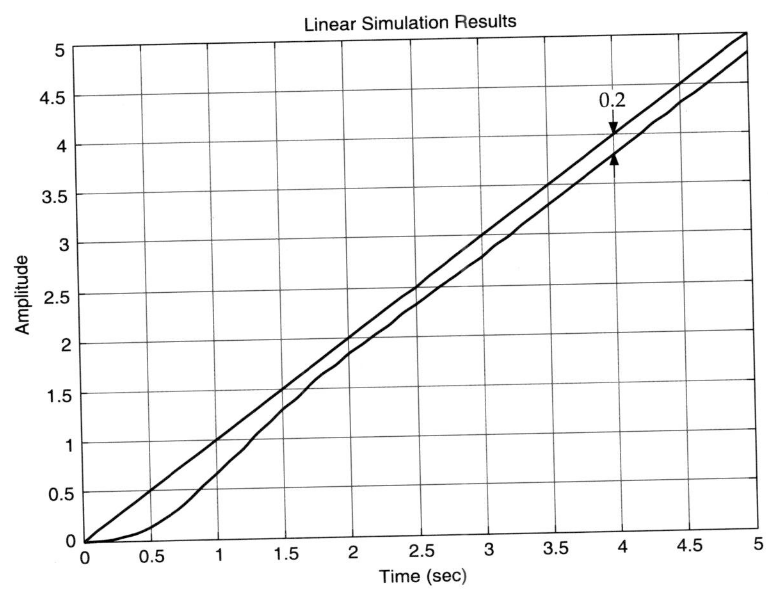

Figure D.17 Ramp response plot of closed-loop function. Arrows are drawn by us to show e_{ss}.

D.7 NYQUIST PLOT

As in the case of Bode plots, Nyquist plots can be obtained from `nyquist` command. The variations that MatLab offers are as follows:

`nyquist(sys)`	draws the Nyquist plot of the LTI model `sys`.
`nyquist(sys,wmin,wmax)`	draws the Nyquist plot for frequencies between `wmin` and `wmax` (in rad/s).
`[re,im]=nyquist(sys,w)` and `[re,im,w]=nyquist(sys)`	return the real and imaginary parts of the frequency response (along with the frequency vector `w` if unspecified). No plot is drawn on the screen.
`nyquist(sys1,sys2,...,w)`	graphs the Nyquist response of multiple LTI models `sys1, sys2,...` on a single plot. The frequency vector `w` is optional. We may specify a colour, line style, and marker for each model, as `nyquist(sys1,'r',sys2,'y--',sys3,'gx')`.

Although not mentioned in the MatLab help menu in its list of syntax, the `nyquist` command can be invoked as

```
>> nyquist([1],[1 0.5 1])
```

to generate the Nyquist plot for the transfer function

$$G(s) = \frac{1}{s^2 + 0.5s + 1}$$

as shown in Figure D.18.

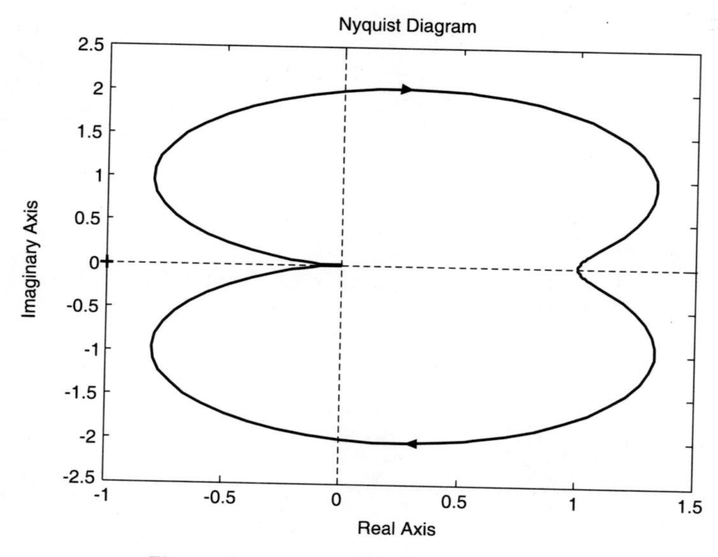

Figure D.18 Nyquist plot without grid.

Note: 1. We have changed the colour of the Nyquist plot from blue to black and the axes, from continuous lines to broken lines.
2. MatLab indicates the critical point with a + sign in red. We have also changed its colour.

M-Circles on Nyquist plot and stability study. M-circles can be superposed on the plot with the help of the `grid` command.

```
>> nyquist([1 10 24],[1 -8 15])
>> grid
```

The result is shown in Figure D.19. We observe that there are two counterclockwise encirclements of the critical point. This means that $N = 2$. Now, let us find out the number of poles of the open-loop transfer function in the RHP.

```
>> roots([1 -8 15])
ans =
  5
  3
```

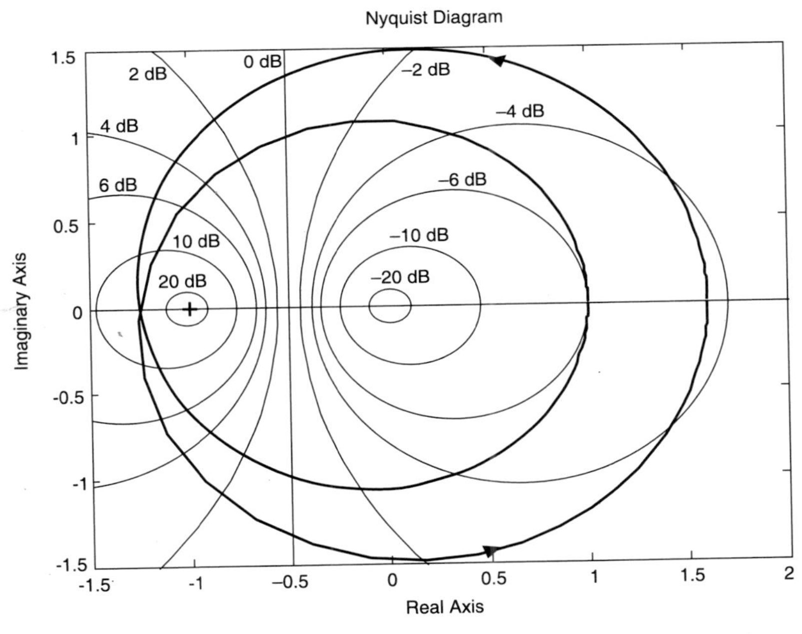

Figure D.19 Nyquist plot showing CCW encirclement of critical point.

The answer says that there are five roots, three of which lie in the LHP. So, there are two poles in the RHP indicating $P = 2$. Since these are CCW encirclements, we find from the formula

$$-N = Z - P$$

that $Z = 0$. So, it indicates a stable closed-loop system[1].

[1]See Section 15.6 at page 520.

We now check the closed-loop step response of the system to see if it is stable.

```
>> sys=tf([1 10 24],[1 -8 15])
Transfer function:
     s^2 + 10 s + 24
     ---------------
     s^2 - 8 s + 15
>> clsys=feedback(sys,1,-1);
>> step(clsys)
```

The MatLab response is shown in Figure D.20. We may see that the system becomes more and more stable if we increase the gain. However, if we decrease the gain to, say, 0.5, what happens? Let us see

```
>> sys1=0.5*sys; nyquist(sys1)
```

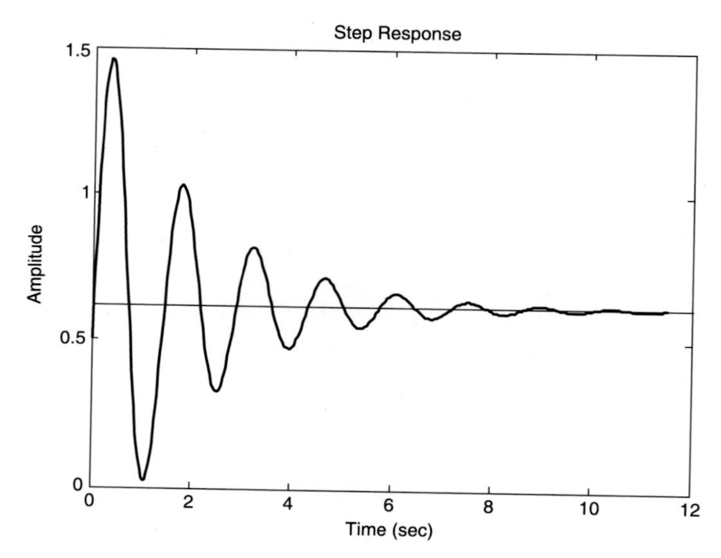

Figure D.20 Step response of the closed-loop transfer function.

The Nyquist diagram (Figure D.21) now shows no CCW encirclement. Which means that because of two RHP open-loop poles, the system has become unstable.

Let us check the step response.

```
>> clsys1=feedback(sys1,1,-1); step(clsys1)
```

The result, shown in Figure D.22, indicates that the closed-loop system is unstable.

Gain margin from Nyquist plot. Let us find out Gm for the open-loop transfer function

$$G(s) = \frac{50}{s^3 + 9s^2 + 30s + 40}$$

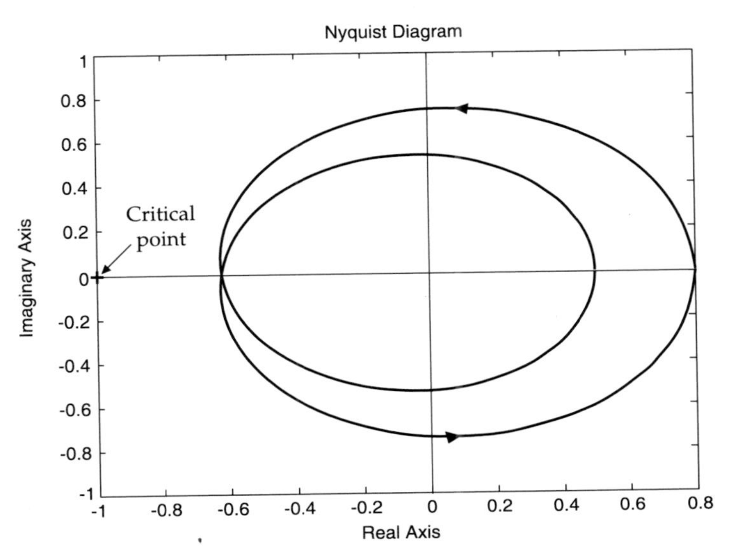

Figure D.21 Nyquist diagram of sys for gain = 0.5.

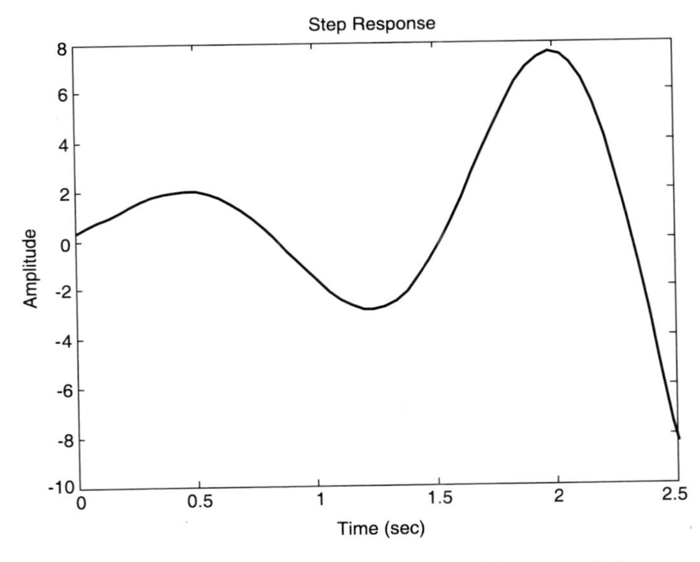

Figure D.22 Step response of sys with gain = 0.5.

The Nyquist plot is shown in Figure D.23. By right-clicking at the point where the curve cuts the real axis (i.e. at the phase crossover frequency), we find that the value of the real part of the coordinate is –0.216. Therefore,

$$\text{Gm} = 20\log\left(\frac{1}{0.216}\right) = 13.31 \text{ dB}$$

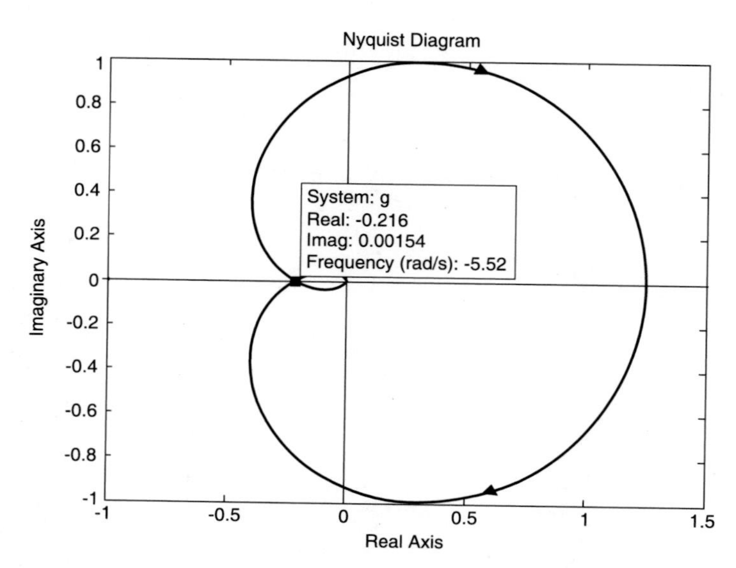

Figure D.23 Nyquist diagram: Gain margin determination.

Note: The `nyquist` command does not always produce the diagrams we need. For example, for the open-loop transfer function given by

$$G(s)H(s) = \frac{10}{s^2(1 + 0.2s)(1 + 0.5s)} \tag{D.4}$$

the diagram that MatLab produces with `nyquist` command is shown in Figure D.24.

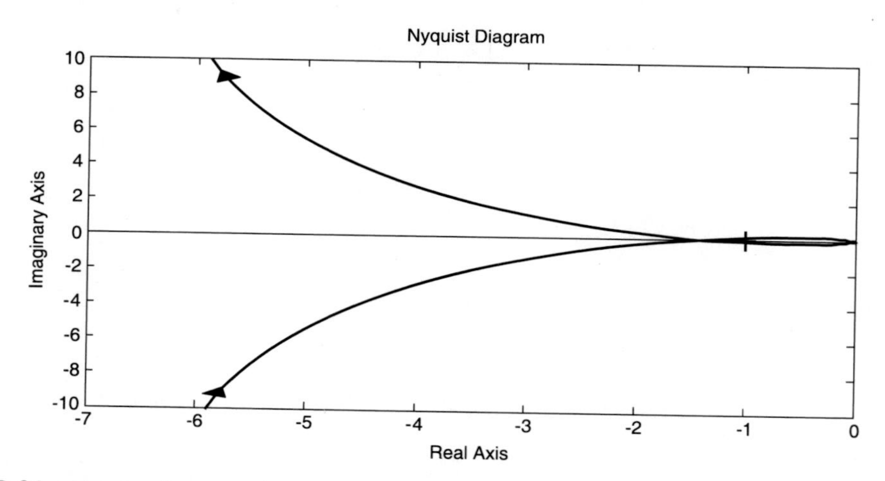

Figure D.24 Nyquist diagram of the system given by Eq. (D.4) produced by the `nyquist` command.

It is obvious that the diagram shows one CW encirclement of the critical point. But, actually, the number of encirclements is 2, as can be seen from Example 15.17 (page 552). In fact, MatLab fails to draw the arc which sweeps with an infinite radius from $+180°$ to $-180°$.

Two interesting commands for studying LTI systems are `ltiview` and `sisotool`. Let us consider `ltiview` first.

D.8 LTIVIEW

We will consider the following syntaxes of `ltiview`:

`ltiview(sys1, sys2, ..., sysn)`	opens an LTI Viewer containing the step response of the LTI models `sys1, sys2, ..., sysn`. We may specify a distinctive colour, line style, and marker for each system.
`ltiview(plottype,sys1, sys2, ..., sysn)`	further specifies which responses to plot in the LTI Viewer. `plottype` may be any of the following strings (or a combination thereof): 1. `step` Step response 2. `impulse` Impulse response 3. `bode` Bode diagram 4. `bodemag` Bode Magnitude diagram 5. `nyquist` Nyquist plot 6. `nichols` Nichols plot

To see how the first syntax works, we type the following commands:

```
>> sys=tf([1,10,24],[1,-8,15]);
>> sys1=zpk(-2,[-4,-5],20);
>> sys2=feedback(sys,1,-1);
>> ltiview(sys1,'k-.',sys2,'k--')
```

MatLab produces Figure D.25. Here, `k` indicates black colour of lines and `-.`, `--`, line shapes.

If we right-click on the diagram, we will find a pop-up menu showing seven items, namely, Plot types, Systems, Characteristics, Grid, Normalize, Full view (ticked) and Properties. Under Plot types, we will see nine items in which Step is ticked. This indicates that the plots are step responses of systems, which is the default option. We may choose other options by clicking on any other option.

Under Characteristics, there are four options—Peak response, Settling time, Rise time, and Steady state. If we click on Peak response, Figure D.26 is produced. The markers in the figure are generated by placing the cursor at the appropriate areas.

To check how the second syntax works, we type

```
>> ltiview({'step';'bode'},sys1,'k-.',sys2,'k--')
```

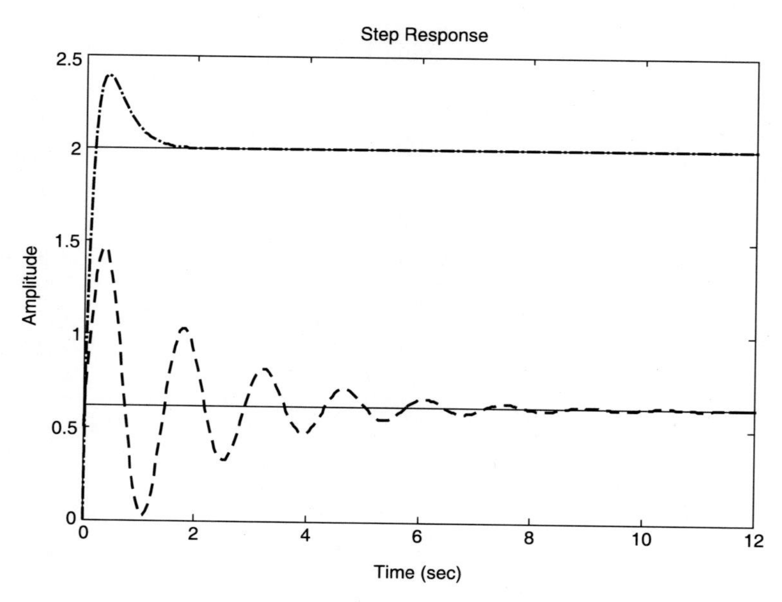

Figure D.25 Step response of systems produced by `ltiview`.

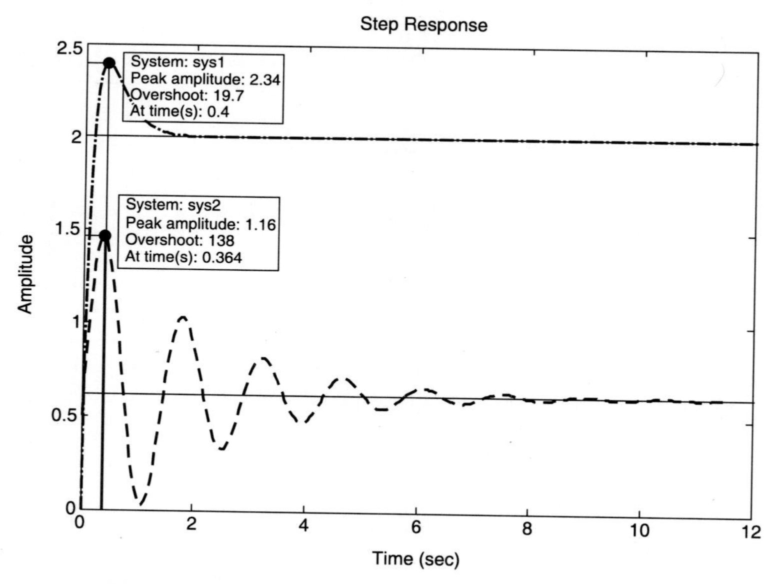

Figure D.26 Peak response of systems produced by `ltiview`.

MatLab produces Figure D.27.

Instead, if the following command is given:

```
>> ltiview({'nyquist'},sys1,'k-.',sys2,'k--')
```

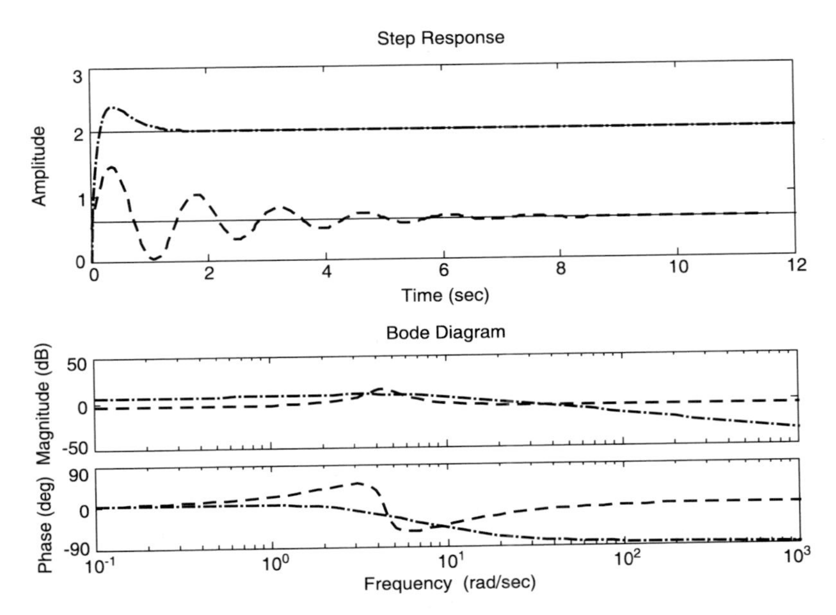

Figure D.27 Step responses and Bode diagrams of systems produced by `ltiview`.

MatLab will produce Figure D.28.

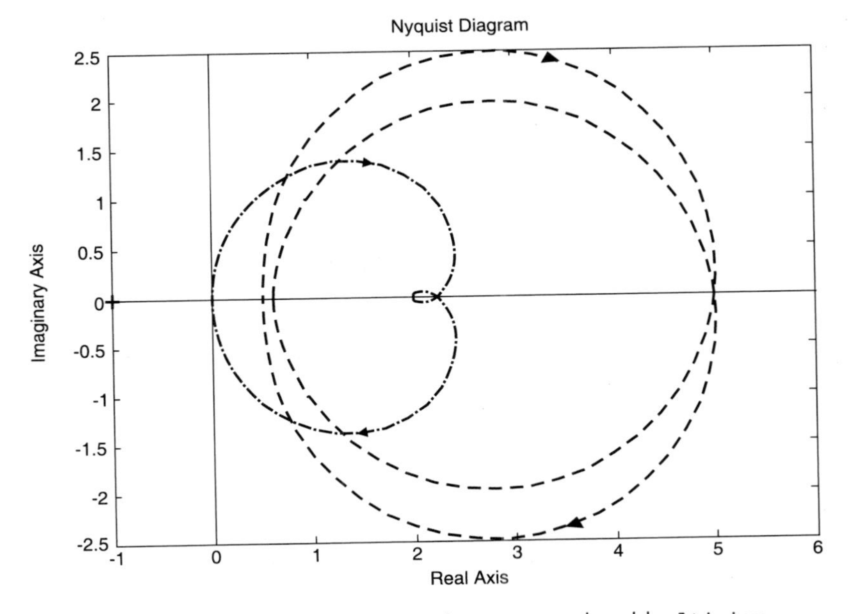

Figure D.28 Nyquist diagrams of systems produced by `ltiview`.

D.9 SISOTOOL

The `sisotool` is a more versatile command. It offers graphics user interface (GUI), not only for most of the plots offered by `ltiview`, but also for other facilities.

1. To add/remove real/complex poles/zeros.
2. To add integrator/derivative at the desired locations on the plot.
3. To study open-loop/closed-loop behaviour of the system by placing a desired compensator at the desired position in the loop.
4. To design SISO compensators with the help of root locus/Bode/Nichols plots of the open-loop system.

The default loop configuration is given in Figure D.29.

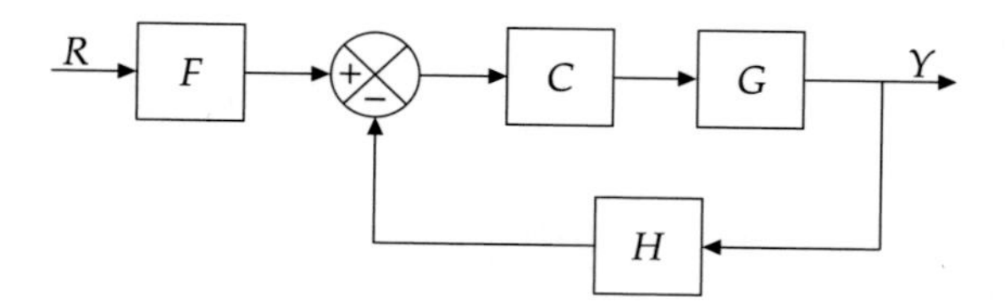

Figure D.29 Default loop configuration of `sisotool`.

MatLab offers the following variations of the `sisotool` command:

`sisotool(g,c,h,f)`	where `g, c, h, f` are defined in Figure D.29. If only `g` is defined, the command should be `sisotool(g)` and so on.
`sisotool(views,g,...)`	`views` can be any of the following strings (or combination) of: `'rlocus'`, `'bode'`, `'nichols'`, `'filter'`

To import plant data into `sisotool`, the *Import* item from the *File* menu is to be selected.
Let us consider the following examples regarding the usage of `sisotool`:

```
>> sys=zpk(-2,[-4,-5],20);
>> sisotool(sys)
```

These produce Figure D.30. Again,

```
>> sisotool({'nichols','bode'},sys)
```

produces Figure D.31. A host of interactive operations, such as addition of poles/zeros, conversion to discrete form, response to step command, etc. can be carried out with the generated diagram. For example, poles/zeros at the desired locations can be added in three ways:

1. By left-clicking outside the diagram when a dialogue box pops up.
2. From the drop down Edit menu: Edit→ Open-loop Bode/Nichols→Add Pole/Zero.
3. By right-clicking on the diagram when the same drop down menu pops up.

Analysis and Tools menus offer step response study and conversion facilities, among others. The versatility of `sisotool` can be gauged only by exploring it.

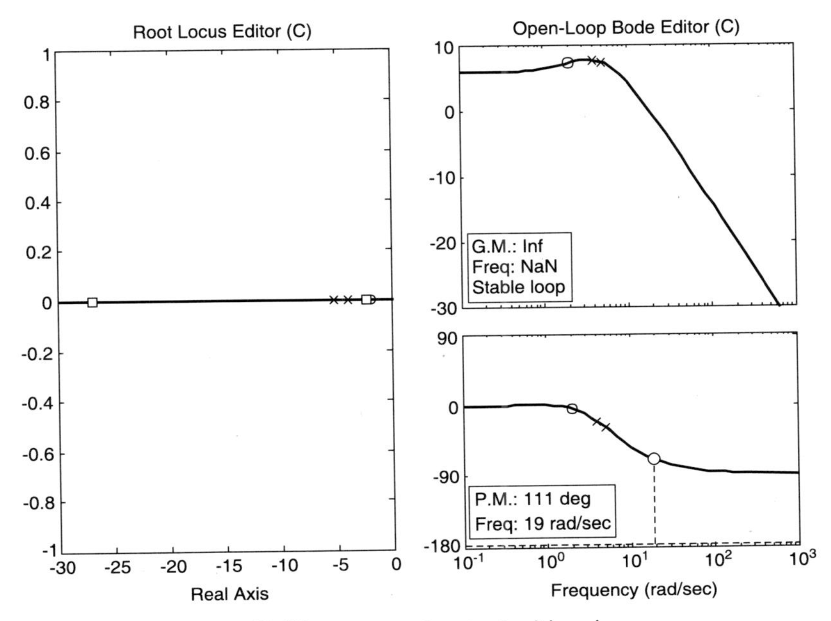

Figure D.30 `sisotool` output with only `sys`.

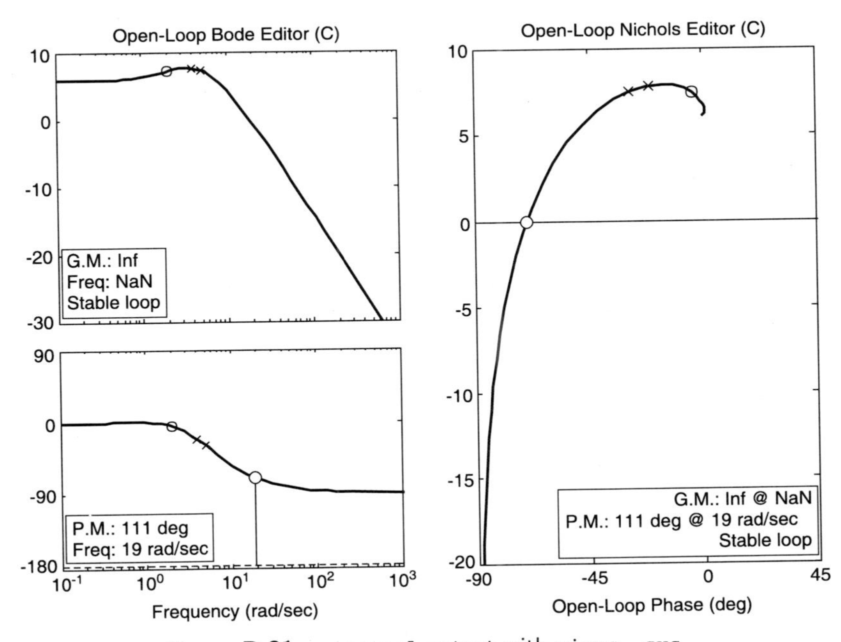

Figure D.31 `sisotool` output with `views, sys`.

D.10 STATE-SPACE

A state-space system can be built by using the ss command. Suppose, our state model is given by

$$\mathbf{A} = \begin{bmatrix} 0 & 1 & 0 \\ 0 & 0 & 1 \\ -12 & -8 & 5 \end{bmatrix} \qquad \mathbf{B} = \begin{bmatrix} 0 \\ 0 \\ 5 \end{bmatrix} \qquad \mathbf{C} = \begin{bmatrix} 1 & 0 & 0 \end{bmatrix} \qquad \mathbf{D} = [0] = \text{null matrix}$$

We want to find the step response of this system. So, we proceed as follows:

```
>> a=[0 1 0;0 0 1;-12 -8 -5];
>> b=[0;0;5];
>> c=[1 0 0];
>> sys=ss(a,b,c,0)
a =
        x1   x2   x3
  x1     0    1    0
  x2     0    0    1
  x3   -12   -8   -5
b =
        u1
  x1     0
  x2     0
  x3     5
c =
        x1   x2   x3
  y1     1    0    0
d =
        u1
  y1     0
Continuous-time model.
>> step(sys)
```

The step response is shown in Figure D.32. This being a SISO system, only one output is generated.

Next we consider a MIMO system, having two inputs, given by

$$\mathbf{A} = \begin{bmatrix} -2 & 1 \\ 0 & -3 \end{bmatrix} \qquad \mathbf{B} = \begin{bmatrix} 4 & 1 \\ 3 & 2 \end{bmatrix} \qquad \mathbf{C} = \begin{bmatrix} 1 & 0 \end{bmatrix} \qquad \mathbf{D} = [0]$$

To observe its step response, we write the following code:

```
>> a=[-2 1;0 -3];b=[4 1;3 2];c=[1 0];
>> sys=ss(a,b,c,0);step(sys)
```

The MatLab-generated response is shown in Figure D.33. Two responses are produced by two inputs.

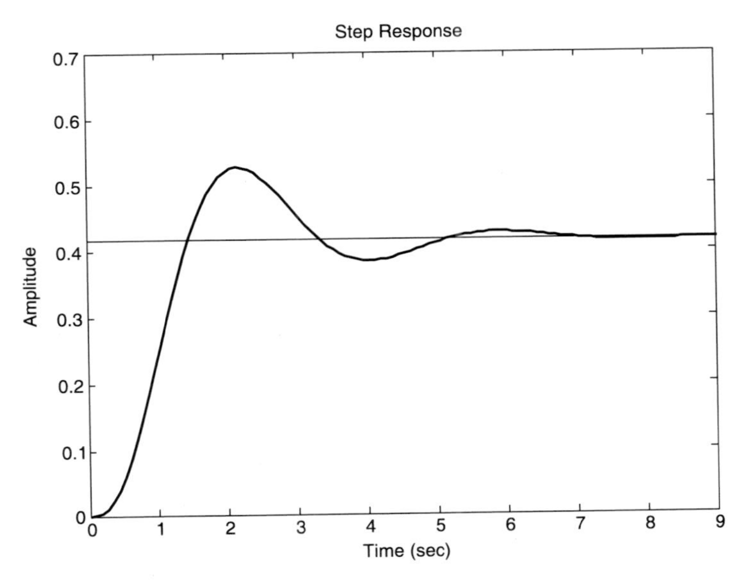

Figure D.32 Step response for state-space SISO system.

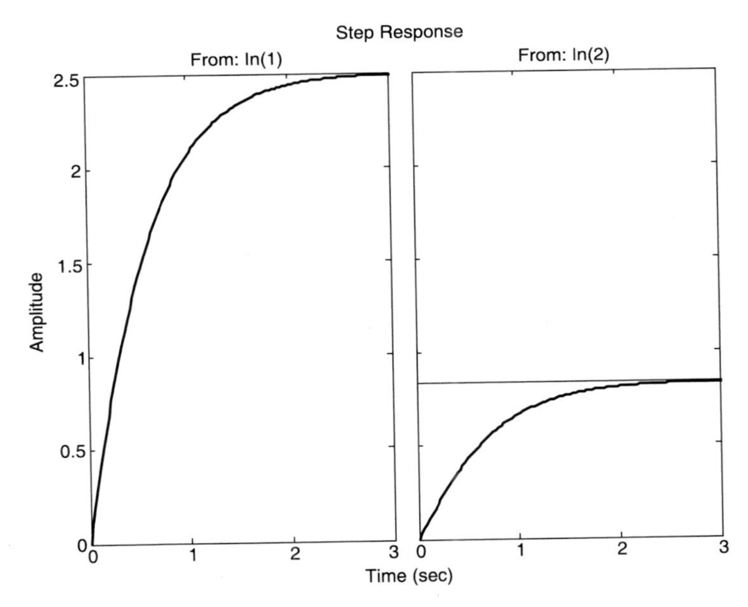

Figure D.33 Step response for state-space MIMO system.

Generating controllability and observability matrices. MatLab generates controllability and observability matrices[2] through the following commands:

---

[2]See Section 17.9 at page 642.

`co = ctrb(a,b)`	returns the controllability matrix $\begin{bmatrix}\mathbf{b} & \mathbf{ab} & \mathbf{a^2b}\ldots\end{bmatrix}$
`co = ctrb(sys)`	returns the controllability matrix of the state-space model `sys` with realisation `(a, b, c, d)`. This is equivalent to `ctrb(sys.a, sys.b)`
`ob = obsv(a,c)`	returns the observability matrix $\begin{bmatrix}\mathbf{c} & \mathbf{ca} & \mathbf{ca^2}\ldots\end{bmatrix}$
`ob = obsv(sys)`	returns the observability matrix of the state-space model `sys` with realization `(a, b, c, d)`. This is equivalent to `obsv(sys.a, sys.c).`

Suppose we want to check the controllability and observability of the following state-space system:

$$\mathbf{A} = \begin{bmatrix} -2 & 1 \\ 0 & 3 \end{bmatrix} \qquad \mathbf{B} = \begin{bmatrix} 4 \\ 1 \end{bmatrix}, \qquad \mathbf{C} = \begin{bmatrix} 1 & 0 \end{bmatrix}$$

We can generate controllability and observability matrices as follows:

```
>> a=[-2 1;0 -3];
>> b=[4;1];
>> c=[1 0];
>> con=ctrb(a,b)
con =
     4    -7
     1    -3
>> %check rank of the matrix
>> rank(con)
ans =
     2
>> obs=obsv(a,c)
obs =
     1     0
    -2     1
>> rank(obs)
ans =
     2
```

We draw our inference from the ranks of the matrices that the system is both controllable and observable.

Transfer function from state-space model. The transfer function of a SISO system can be derived from its state model by using the `tf` command as follows:

```
>> sys=ss(a,b,c,0)
a =
              x1       x2
      x1      -2        1
      x2       0       -3
```

```
b =
            u1
   x1        4
   x2        1
c =
            x1      x2
   y1        1       0
d =
            u1
   y1        0
Continuous-time model.
>> tfsys=tf(sys)
Transfer function:
        4 s + 13
     -------------
     s^2 + 5 s + 6
```

Canonical realizations. The following commands are useful in realizing the canonical forms:

`csys = canon(sys,type)`	computes a canonical state-space realization `csys` of the LTI model `sys`. The string type selects the type of canonical form: `'modal'`: Modal canonical form where the system eigenvalues appear on the diagonal. The state matrix $\mathbf{A}$ must be diagonalizable. `'companion'`: Companion canonical form where the characteristic polynomial appears in the right column.
`[csys,t] = canon(sys,type)`	also returns the state transformation matrix $\mathbf{t}$ relating the new state vector $\mathbf{z}$ to the old state vector $\mathbf{x}$ by $\mathbf{z} = \mathbf{tx}$. This syntax is only meaningful when `sys` is a state-space model.
`sys = ss2ss(sys,t)`	performs the similarity transformation $\mathbf{z} = \mathbf{tx}$ on the state vector $\mathbf{x}$ of the state-space model `sys`. The resulting state-space model is described by: $\dot{\mathbf{z}} = \left[\mathbf{t}^{-1}\mathbf{at}\right]\mathbf{z} + [\mathbf{tb}]\mathbf{u}$ $\mathbf{y} = \left[\mathbf{c}^{-1}\mathbf{t}\right]\mathbf{z} + \mathbf{du}$ `ss2ss` is applicable to both continuous- and discrete-time models. For LTI arrays `sys`, the transformation $\mathbf{t}$ is performed on each individual model in the array.

Let our transfer function be

$$\frac{Y(s)}{U(s)} = \frac{3s^2 + 2s + 7}{s^3 + 5s^2 + 12s + 5}$$

Its `companion` and `modal` realizations are as follows:

```
>> yu=tf([3 2 7],[1 5 12 5])
Transfer function:
      3s^2 + 2 s + 7
----------------------
s^3 + 5 s^2 + 12 s + 5
>> cyuc=canon(yu,'companion')
a =
            x1     x2     x3
    x1       0      0     -5
    x2       1      0    -12
    x3       0      1     -5

b =
            u1
    x1       1
    x2       0
    x3       0
c =
            x1     x2     x3
    y1       3    -13     36
d =
            u1
    y1       0
Continuous-time model.
>> cyum=canon(yu,'modal')
a =
            x1         x2          x3
    x1  -2.242      2.159           0
    x2  -2.159     -2.242           0
    x3       0          0     -0.5162
b =
            u1
    x1  -5.884
    x2  -4.704
    x3   4.225
c =
             x1         x2         x3
    y1  0.08048    -0.5501     0.2097
d =
            u1
    y1       0
Continuous-time model.
```

For the modal form, the eigenvalues should appear on the diagonal of the **A** matrix. We check.

```
>> eig(yu)
ans =
   -2.2419 + 2.1585i
   -2.2419 - 2.1585i
   -0.5162
```

Note: Two of the eigenvalues are complex conjugates with diagonals of the **A** matrix comprising the real parts and the off-diagonal terms constituting the imaginary parts.

Let us now find STMs for `cyuc` and `cyum`.

```
>> [cyucr,t]=canon(cyuc,'companion')
a =
               x1       x2       x3
   x1           0        0       -5
   x2           1        0      -12
   x3           0        1       -5
b =
               u1
   x1           1
   x2           0
   x3           0
c =
               x1       x2       x3
   y1           3      -13       36
d =
               u1
   y1           0
Continuous-time model.
t =
    1    0    0
    0    1    0
    0    0    1
>> [cyumr,t]=canon(cyum,'modal')
a =

               x1        x2        x3
   x1      -2.242     2.159         0
   x2      -2.159    -2.242         0
   x3           0         0   -0.5162
b =
               u1
   x1      -8.321
   x2      -6.652
   x3       4.225
```

```
c =
                x1        x2        x3
   y1    0.05691    -0.389    0.2097
d =
                u1
   y1            0
Continuous-time model.
t =
       1.4142         0         0
            0    1.4142         0
            0         0    1.0000
```

Response to initial conditions. The response of a state model can be plotted by using the `initial` command when the initial condition matrix is given. Let our system be given by

$$\frac{Y(s)}{U(s)} = \frac{s+3}{s^2+2s-3}$$

and the initial vector by

$$\mathbf{x}(0) = \begin{bmatrix} 1 \\ 0 \end{bmatrix}$$

We can plot the response through the following code:

```
>> sys=tf([1,3],[1,2,-3])

Transfer function:
    s + 3
-------------
s^2 + 2 s - 3
>> x0=[1;0]

x0 =

    1
    0
>> sys_ss=canon(sys,'companion')
a =
           x1     x2
   x1       0      3
   x2       1     -2
b =
           u1
   x1       1
   x2       0
```

```
c =
         x1    x2
  y1      1     1
d =
         u1
  y1      0
Continuous-time model.
>> initial(sys_ss,x0)
```

The plot produced is shown in Figure D.34.

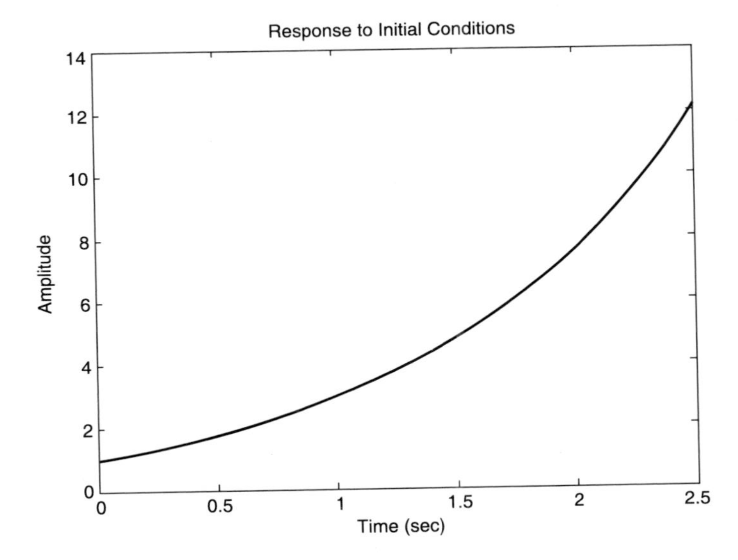

Figure D.34 Response to initial condition.

D.11 DISCRETE SYSTEMS

Let us find the ZOH sampling of the transfer function

$$G(s) = \frac{e^{-1.5s}}{6s + 1} \tag{D.5}$$

with a sampling time of two seconds and find the step response of the discrete model. Our code, where we have included comments with a % sign at the beginning, is as follows.

Note: The transport lag (i.e. input delay) is included in the transfer function:

```
>> g=tf(1,[6 1],'InputDelay',1.5)
```

```
Transfer function:
                        1
    exp(-1.5*s) * -------
                   6 s + 1

>> % continuous to discrete conversion
>> h=c2d(g,2)

Transfer function:
    0.07996 z + 0.2035
    ------------------
      z^2 - 0.7165 z

Sampling time: 2
>> step(h)
```

The result is shown in Figure D.35.

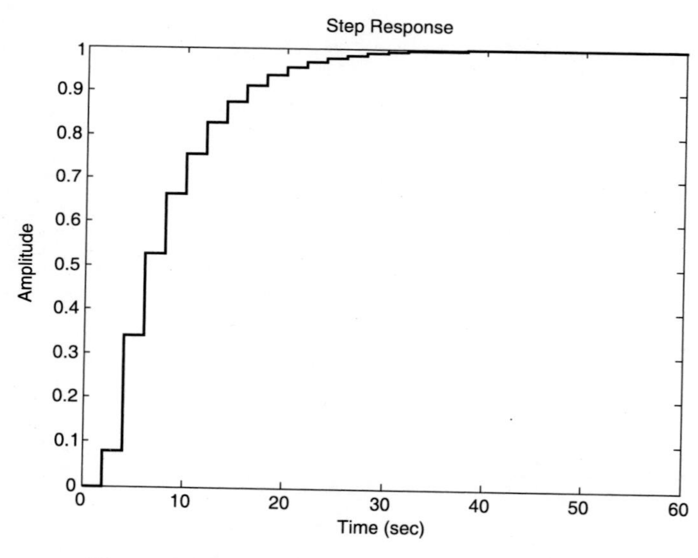

Figure D.35 Discrete step response to Eq. (D.5).

Now we undertake a little more complicated task. We check the step response, both for continuous and discrete models by plotting them on the same graph.

```
>> % step response for g for 40 sec stored
>> % under yc,tc
>> [yc,tc]=step(g,40);
>> % step response for h for 40 sec stored
>> % under yd,td
>> [yd,td]=step(h,40);
```

```
>> % plot yc,tc values with '-' (dash) and
>> % yd,td values with 'o'
>> plot(tc,yc,'-',td,yd,'o')
```

After doing this exercise, the graph obtained is as depicted in Figure D.36.

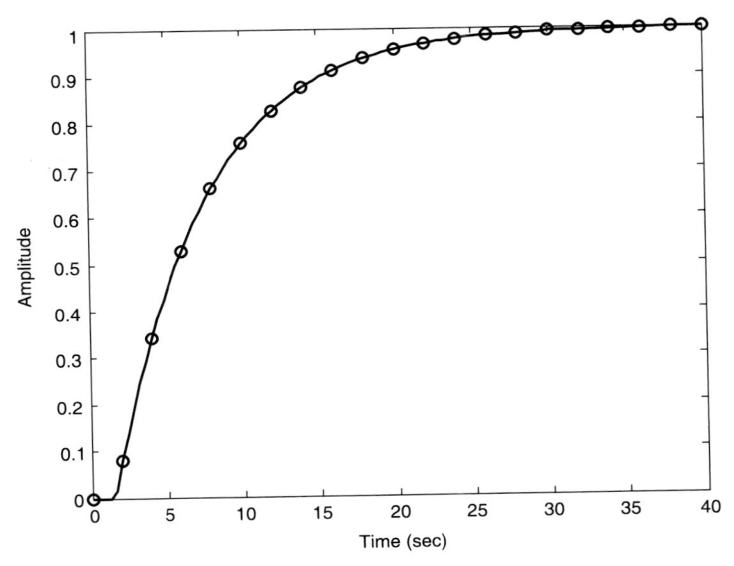

Figure D.36 Agreement between continuous and discrete models.

Note: The continuous and discrete models yield compatible results. For the impulse response, however, the continuous model output and ZOH sampled data may not agree well with such sampling rate. Hence, less sampling time should help resolve the issue.

D.12 CONTROLLER ACTIONS

Suppose the open-loop transfer function of our system is

$$G(s) = \frac{1}{s^2 + 10s + 20} \tag{D.6}$$

Its response to a unit step input is obtained as follows and shown in Figure D.37.

```
>> sys=tf(1,[1 10 20])

Transfer function:

        1
---------------
s^2 + 10 s + 20

>> step(sys)
```

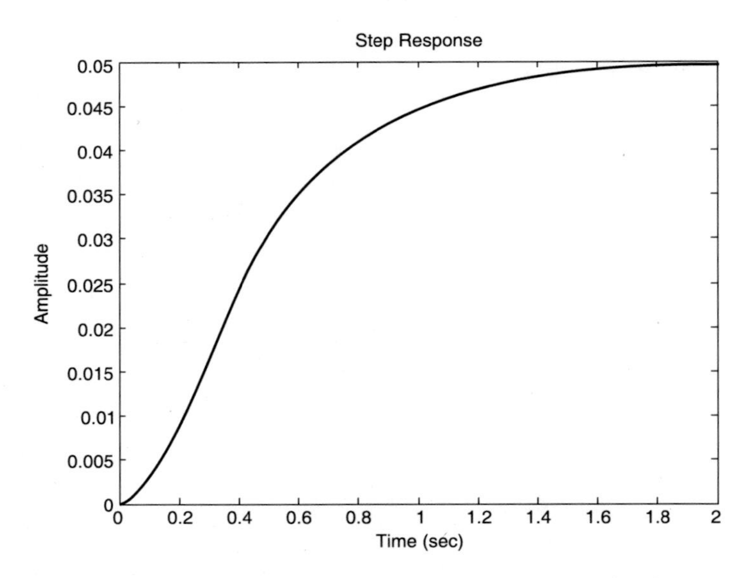

Figure D.37 Step response of the system of Eq. (D.6).

We observe that the plant produces an output of 0.05 when the input is 1. So, the steady-state error is 0.95, which is high.

P Control

We introduce a proportional controller to reduce the rise time as well as steady-state error. With $K_p = 300$, our closed-loop transfer function becomes

$$G(s) = \frac{K_p}{s^2 + 10s + (20 + K_p)} = \frac{300}{s^2 + 10s + 320} \tag{D.7}$$

The step response of Eq. (D.7) obtained through the following code is shown in Figure D.38.

```
>> sys=tf(300,[1 10,320])

Transfer function:
       300
----------------
s^2 + 10 s + 320

>> step(sys)
```

Figure D.38 shows that both rise time and stead-state error situations have improved with the introduction of a proportional controller.

PI Control

Now we introduce an integral controller with $K_i = 50$, say. We know that this introduces a factor $\left[K_p + \left(K_i/s\right)\right]$ in the numerator of the open-loop transfer function.

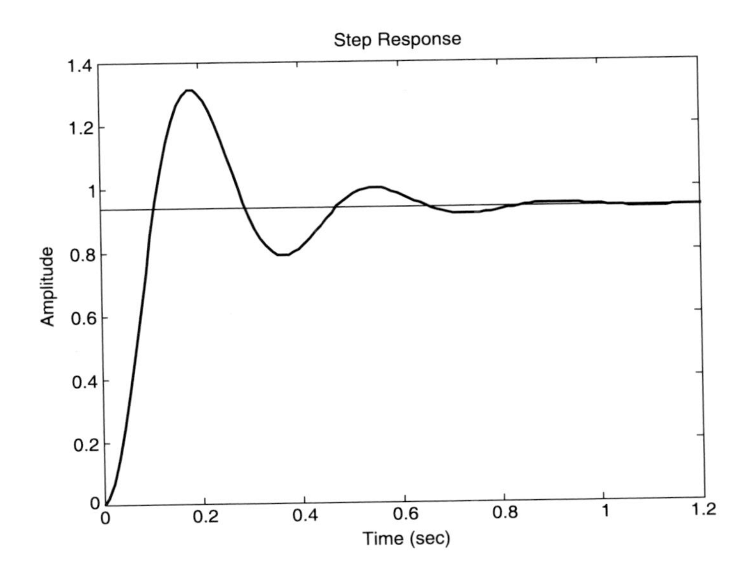

Figure D.38 Step response with proportional controller.

So, the closed-loop transfer function becomes

$$\frac{C(s)}{R(s)} = \frac{K_p s + K_i}{s^3 + 10s^2 + \left(K_p + 20\right)s + K_i}$$

$$= \frac{300s + 50}{s^3 + 10s^2 + 320s + 50}$$

The corresponding step response, obtained through the following MatLab commands, is shown in Figure D.39.

```
>> sys=tf([300 50],[1 10 320 50])

Transfer function:
        300 s + 50
--------------------------
s^3 + 10 s^2 + 320 s + 50

>> step(sys)
```

Obviously, then;

- The rise time and steady-state error are reduced
- The overshoot and settling time have increased

with the addition of integral control.

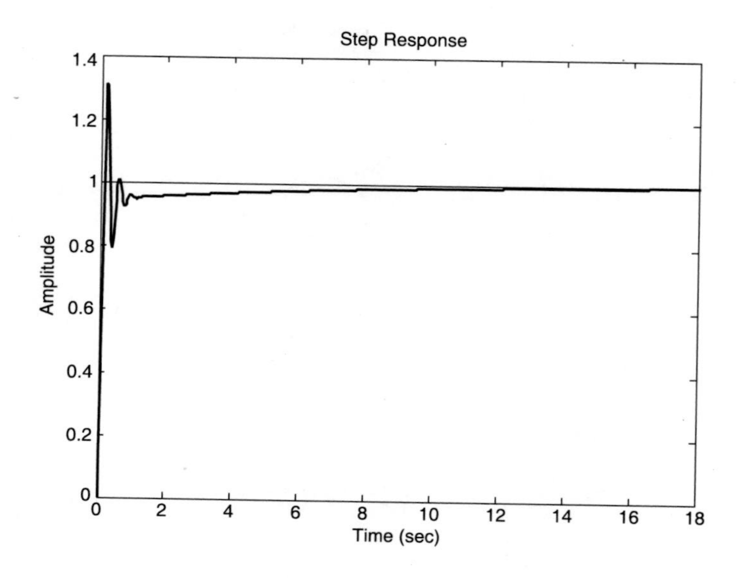

Figure D.39 Step response with PI control.

PD Control

Next, we remove the integral controller and introduce a derivative controller with $K_d = 15$. So the closed-loop transfer function is now

$$G(s) = \frac{K_d s + K_p}{s^2 + (10 + K_d)s + (20 + K_p)}$$

$$= \frac{15s + 300}{s^2 + 25s + 320} \tag{D.8}$$

The step response of Eq. (D.8), obtained through the following commands, is illustrated in Figure D.40.

```
>> sys=tf([15 300],[1 25 320])

Transfer function:
   15 s + 300
----------------
s^2 + 25 s + 320

>> step(sys)
```

Figure D.40 shows that, with the introduction of derivative action

- Both overshoot and settling time have improved
- The steady-state error is considerable

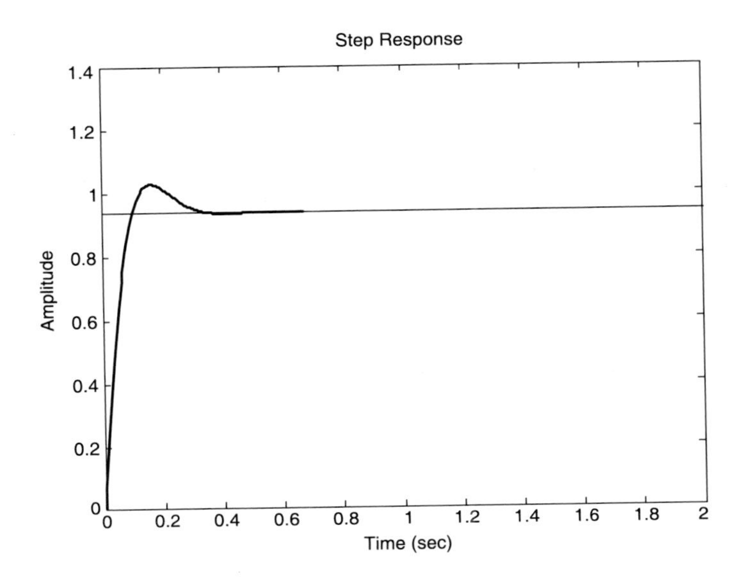

Figure D.40 Step response with PD control.

PID Control

With PID control, the closed-loop transfer function of the system assumes the form

$$\frac{C(s)}{R(s)} = \frac{K_d s^2 + K_p s + K_i}{s^3 + \left(10 + K_d\right)s^2 + \left(20 + K_p s\right) + K_i}$$

Now, let us check the step response:

```
>> kp=300;
>> ki=300;
>> kd=100;
>> num=[kd kp ki];
>> den=[1 10+kd 20+kp ki];
>> t=0:0.05:2;
>> step(num,den,t)
```

The result is shown in Figure D.41. The combined action of PID control has produced:

- A low system response time
- A low settling time
- No overshoot
- No steady-state error

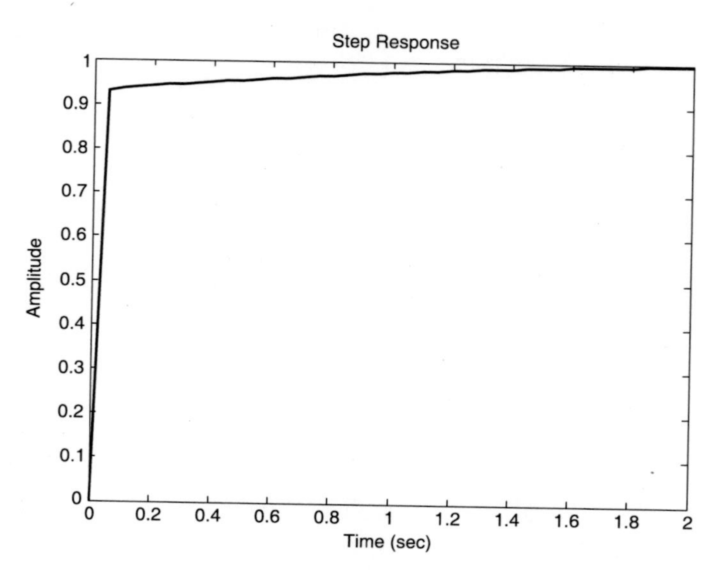

Figure D.41 PID controlled step response.

D.13 NICHOLS CHART

The Nichols chart is considered the best to assess the closed-loop behaviour of systems from their open-loop transfer functions. We can draw it for any system through MatLab pretty easily by the command `nichols(sys)`.

Let our system be

$$G(s) = \frac{10}{s(s+1)}$$

As we type the commands

```
>> gn=zpk([],[0, -1],[10]);
>> nichols(gn)
```

the diagram (Figure D.42) pops up in another window. The chart produced is without grid and, therefore, it is not of much help. To superpose the grid on it, we type

```
>> ngrid
```

when the resized diagram is superposed with the grid as shown in Figure D.43.

This is a brief introduction to the control system toolbox of MatLab. MatLab is also of help in myriad other applications.

D.14 WRITING APPLICATION PROGRAMMES

We can also write application specific programmes in MatLab and store them as M-files. To create an M-file, we need to go to the 'File' menu and select New→Blank M-file. An editor window will pop up. We use this editor to write programmes. We give a specific example here.

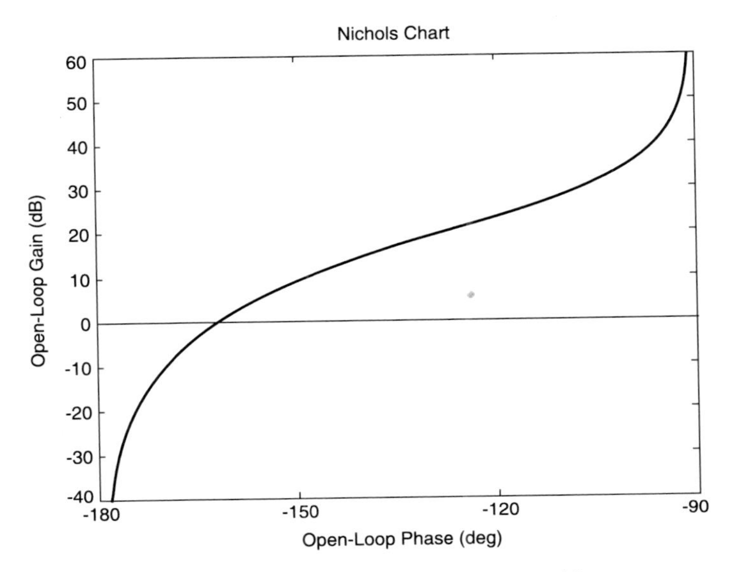

Figure D.42 Nichols chart without grid.

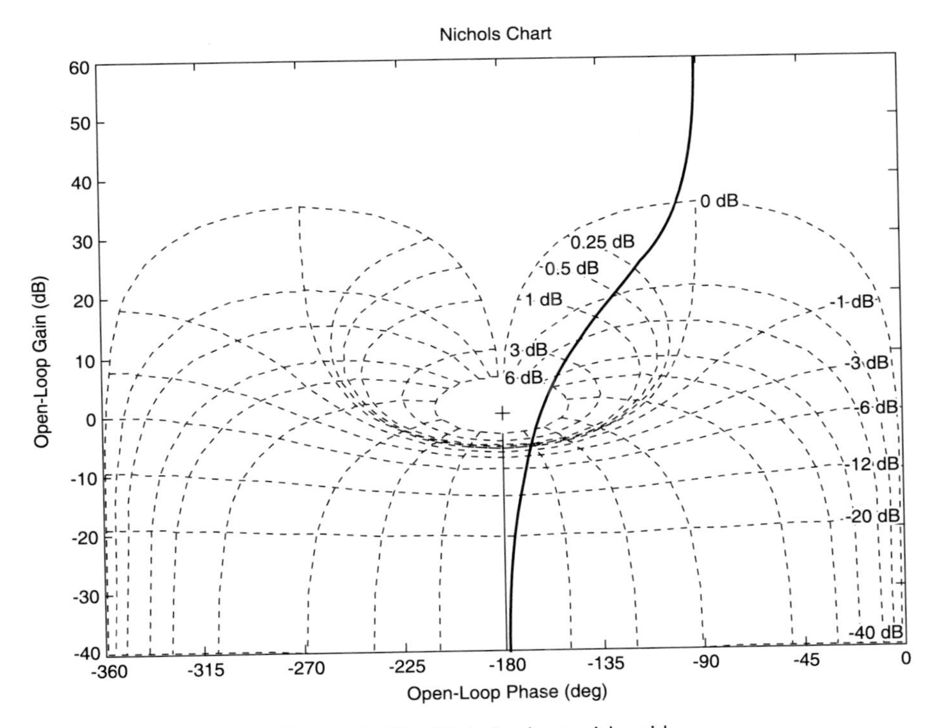

Figure D.43 Nichols chart with grid.

Consider the LCR circuit of Example 3.8 at page 97. We reproduce the figure here.

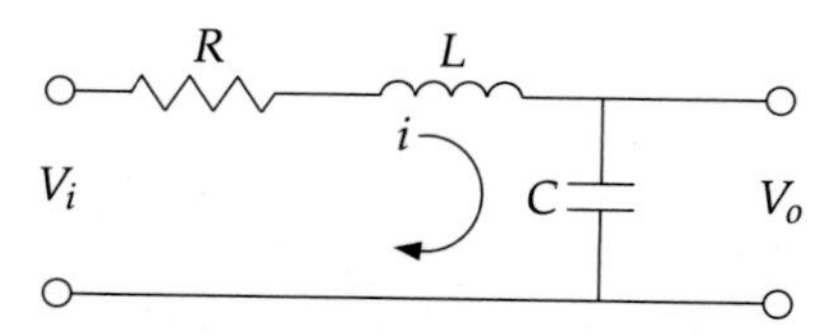

The state model of the system can be formulated from the following equations:

$$v_o = \frac{Q}{C}$$

or

$$\dot{v_o} = \frac{1}{C} i \tag{i}$$

$$v_i = L\frac{di}{dt} + Ri + \frac{Q}{C}$$

or

$$\dot{i} = -\frac{R}{L} i - \frac{1}{L} v_o + \frac{1}{L} v_i \tag{ii}$$

Equations (i) and (ii) can be written in matrix forms as follows:

$$\begin{bmatrix} \dot{v_o} \\ \dot{i} \end{bmatrix} = \begin{bmatrix} 0 & \dfrac{1}{C} \\ -\dfrac{1}{L} & -\dfrac{R}{L} \end{bmatrix} \begin{bmatrix} v_o \\ i \end{bmatrix} + \begin{bmatrix} 0 \\ \dfrac{1}{L} \end{bmatrix} [v_o]$$

$$y = \begin{bmatrix} 1 & 0 \end{bmatrix} \begin{bmatrix} v_o \\ i \end{bmatrix} + [0][v_o]$$

Thus, we have

$$\mathbf{A} = \begin{bmatrix} 0 & \dfrac{1}{C} \\ -\dfrac{1}{L} & -\dfrac{R}{L} \end{bmatrix}, \qquad \mathbf{B} = \begin{bmatrix} 0 \\ \dfrac{1}{L} \end{bmatrix}$$

$$\mathbf{C} = \begin{bmatrix} 1 & 0 \end{bmatrix}, \qquad \mathbf{D} = [0]$$

So, we write the following programme in the editor:

```
%LCR circuit state-space model
%
L=input('Enter inductor value:');
C=input('Enter capacitor value:');
R=input('Enter resistor value:');
A=[0 1/C; -1/L -R/L];
```

```
B=[0;1/L];
C=[1 0];
D=[0];
step(A,B,C,D);grid
```

and save it with the name `lcr`. MatLab automatically puts the extension `.m` to it. Next in the command window we type the command `lcr` when MatLab asks for values of L, C, R as follows:

```
>> lcr
Enter inductor value:2<--
Enter capacitor value:3<--
Enter resistor value:1<--
```

Our typed-in values are 2, 3 and 1, respectively. The step response produced by MatLab is shown in the Figure D.44.

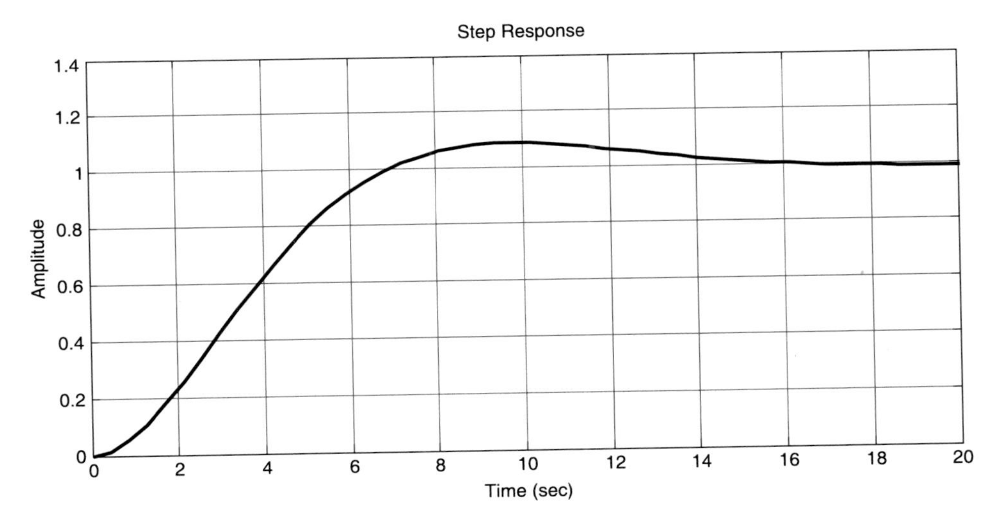

Figure D.44 Step response produced by MatLab.

Index